D0408359

SURVEYING FOR CIVIL ENGINEERS

Closing the horizon. (*From a photograph by J. Wayman Williams, Jr., in The Princton Engineer.*)

Surveying for Civil Engineers

PHILIP KISSAM, C.E.
Professor of Civil Engineering
Princeton University

New York Toronto London
McGRAW-HILL BOOK COMPANY, INC.
1956

SURVEYING FOR CIVIL ENGINEERS

Library of Congress Catalog Card Number 55-6158

THE MAPLE PRESS COMPANY, YORK, PA.

To my students,
who have taught me more
than I like to admit

PREFACE

Once the need for a construction project has been established without question, the degree of its success or failure depends, to a very large measure, on the fundamental economics of construction and operation provided in its original plan. The larger the project, the more these economies depend on the adaptation of the plan to the topographic possibilities available and on the perfection of the execution of the plan. It follows that surveys that adequately determine the topographic details, and location surveys that accurately control the desired construction, are essential elements in a successful enterprise.

Surveys of this nature are expensive and, since they cannot be readily mechanized, their costs will continue to rise with respect to costs not so dependent on man-hours. It seems evident that now, more than ever, the engineer must become familiar with the newest types of surveying equipment and techniques, as well as with the theory and application of the advanced types of surveying procedures that are most likely to produce the greatest topographic coverage and the best accuracies at a minimum cost.

This book has been written to cover this field. The subject of each chapter has been selected with great care to fit this scheme. Each topic introduced is thoroughly described, the theory is rigorously developed, and numerical examples are always included to illustrate its applications.

It is hoped that practicing engineers will find, in this book, the information necessary to select and to carry out the best available procedures for whatever surveying requirements they may encounter.

The book is arranged so that, if it is to be used as a text for courses in surveying, once Part I has been covered a course may be terminated wherever desired. A minimum course, however, should include Part II and at least some of the sections in the chapter on aerial mapping.

In addition to the emphasis placed on new or improved instruments and methods, the theory of errors and the adjustments of survey nets have been stressed so that the accuracies of survey methods can be quickly appraised and the best use can be made of survey data. A very practical method of utilizing the theory of probability has been introduced in Chap. 1 so that the reader will have available at once a means of selecting, by simple tests, the instruments and methods that will give the desired accuracies. The complete development of the theory has been postponed to Chap. 21. Simple methods of net adjustment are introduced in Chap. 6, more rigorous methods in Chaps. 17 to 19, and the theory and application of the method of least squares is thoroughly covered in Chap. 22. The first sections in Chap. 22 give examples of the application of least squares to surveying problems often encountered that are simple to understand and that clearly demonstrate the theory involved without the need for extensive computation.

Subjects not listed in the chapter heads but which are often required in surveying practice are covered in the chapters where they are most applicable. For example, the adjustment of a sextant is found in the chapter on hydrographic surveys; geodetic concepts appear in the chapter on state coordinate systems; and the effect of steep slopes on the sag in tapes is demonstrated in the chapter on mine surveys.

It is hoped that the aim of this book has been achieved and that the material presented will further advance the economies already attained in engineering construction.

ACKNOWLEDGMENTS

The author's heartfelt thanks go to Robert R. Singleton, consulting engineer, who has gone over this book with a fine-tooth comb, suggested many improvements, and caught innumerable errors. The author is very grateful to Hugh C. Mitchell, of the U.S. Coast and Geodetic Survey (retired) who helped edit some of the original manuscript, particularly that of Chap. 12; to Lewis V. Judson of the United States National Bureau of Standards whose words are quoted in Chap. 3; to Edwin L. Kimble, a friend and former student, who reviewed much of the material for Chap. 9; to Andrew H. Holt of Worcester Polytechnic Institute who carefully reviewed, and made many suggestions for, Chap. 10; to Lester S. Holmes, who reviewed Chap. 10 from a legal standpoint; to L. G. Simmons of the U.S. Coast and Geodetic Survey who helped with Chap. 12; to Norman F. Braaten of the same bureau whose new and important method of level net adjustment is included in Chap. 17; to Walter C. Johnson of Princeton University who carefully reviewed Chap. 21; to my dear wife Dorothy, who named the book; and last but not least to Virginia Nonziato and Euthie Anthony, who turned the author's scrawl into beautifully typed pages.

PHILIP KISSAM

CONTENTS

PART I. INSTRUMENTS AND METHODS FOR LARGE SURVEYS

PART II. OPERATIONS

PART III. PROCEDURES FOR PRECISE CONTROL

PART IV

PART V. APPENDIX

LIST OF REFERENCE TABLES AND CHARTS

Part I. Instruments and Methods for Large Surveys

CHAPTER 1

INTRODUCTION, ERRORS, AND PLANNING A SURVEY

INTRODUCTION

1-1. The Purpose of This Textbook. This book is designed to cover surveying operations which, because of their size, permanence, importance, or the need for high accuracy, have special requirements for precise measurements, efficient procedures, or special operations.

All the material is applicable to plane surveys. However, certain geodetic operations are included because of their application to plane surveying. Geodetic concepts, procedures, and instruments are covered. These include the principle of the spheroid, the derivation, and several applications of the theory of least squares, direction theodolites, and geodetic levels. The geodetic reductions included are only those applicable to the state systems of plane coordinates. The methods of reduction of large triangulation nets like those of the U.S. Coast and Geodetic Survey have been omitted as they are seldom handled by practicing engineers.

It is assumed that the reader of this text is familiar with the instruments and methods ordinarily used in surveys of limited extent. In fact, this text really constitutes a continuation of the author's book entitled "Surveying,"[1] which covers the subject; accordingly, material in that text is seldom repeated here.

1-2. Accuracy and Planning. It is nearly always possible to increase the accuracy or to reduce the costs of a survey by proper planning. This is especially true of surveys of the types covered in this book, which are usually supported by sufficient funds to cover the initial expenditures for the special equipment and procedures necessary to reduce the cost of attaining the desired accuracy.

The key to proper planning is a thorough understanding of the means of producing accuracy and the behavior of errors. Before taking up planning, these subjects will be discussed.

[1] Philip Kissam, "Surveying—Instruments and Methods for Surveys of Limited Extent," McGraw-Hill Book Company, Inc., New York, 1947.

ACCURACY AND ERRORS

1-3. The Elements in Accuracy. Accuracy depends on three elements: precise instruments, precise methods, and good planning. Precise instruments are not absolutely necessary but they save time and therefore provide economies. Precise methods *must* be used. They eliminate or reduce the effect of all types of errors. Good planning is the greatest element in economy and a very important element in obtaining accuracy. It includes the proper choice and arrangement of survey control and the proper choice of instruments and methods for each operation.

All three of these elements—instruments, methods, and planning—can be evaluated only by the economies they provide in producing the necessary results at the desired accuracy. No choice among these elements can be made without an estimate of the errors to be expected. It follows that successful survey operations are impossible without a thorough understanding of the nature of errors.

1-4. Definitions. The words **precision** and **accuracy** must be defined before errors and accuracy can be readily discussed. Precision is the degree of perfection *used* in the instruments, the methods, and the observations. When the precision of several operations is known, it will be shown that the accuracy of the results can be estimated.

Accuracy is the degree of perfection *obtained.* Actual results, therefore, must be used to compute accuracy. When the accuracy of the results compares unfavorably with its estimated value it can usually be assumed that faults exist which should be corrected.

1-5. Errors. Errors are of three general types: blunders, systematic errors, and accidental errors.

A **blunder** is a mistake in the determination of a value. In the technical sense it does not apply to incorrect methods. The art of eliminating blunders is one of the most important elements in surveying. It gives rise to the three following basic rules which govern surveying. If one of these rules is broken, the value of the survey is usually destroyed.

1. Every value to be recorded in the field must be checked by some independent field observation.

2. Once this check indicates that there is no blunder, the field record must never be changed or destroyed. Notes of necessary changes made later must be written with a colored pencil.

3. An over-all check must be applied to every control survey. As many over-all checks as possible must be arranged for in planning the work and every over-all check made possible by the arrangement of the control must be computed and applied.

A **systematic error** is an error that, under the same conditions, will always be of the same size and sign. Systematic errors can be detected *only by recognizing conditions that will create them.* This makes them insidious and therefore dangerous. Every effort must be made to recognize the existence of conditions that create them. When these conditions are recognized steps must be taken to neutralize them. Two basic rules of surveying hinge on systematic errors.

1. All surveying equipment must be designed and used so that whenever possible systematic errors will be *automatically* eliminated.

2. All systematic errors that cannot be surely eliminated by this means must be evaluated and their relationship to the conditions that cause them must be determined. When such a study indicates these errors are not negligible, field data must be recorded that measure the conditions causing these errors so that proper corrections can be computed and applied.

For example, in leveling, the instrument must first be adjusted so that the line of sight is as nearly horizontal as possible when the bubble is centered. Despite this precaution, the horizontal lengths of the plus and minus sights from each instrument position should be kept as nearly equal as possible. These operations satisfy Rule 1, above.

In ordinary leveling the errors that remain are known to be negligible and thus they are evaluated by experience. In precise leveling, Rule 2 is complied with more definitely. Each day the actual error of the instrument is determined by a careful peg test. The length of each sight is measured by stadia and a correction to the results is applied that is proportional to the size of the error and to the difference in the sum of the lengths of the backsights and the sum of the lengths of the foresights. This operation eliminates the error and satisfies Rule 2.

The **accidental error** of a single determination is the difference between (1) the true value of the quantity and (2) a determination that is free from blunders or systematic errors. Accidental errors represent the limit of precision in the determination of a value. They obey the laws of chance and therefore must be handled according to the mathematical laws of probability. The proper treatment of accidental errors is not complex but the fundamental theory depends on lengthy mathematical derivations. Chapter 21 is devoted to this subject. Near the end of that chapter, on page 623, are the formulas developed by these derivations. These formulas are not difficult to use and will be referred to repeatedly.

From the theory of probability emerges the theory of least squares. This subject is covered in Chap. 22. It provides the means of treating accidental errors by mathematically correct procedures.

Despite every effort to eliminate blunders and systematic errors, the errors in a survey are never absolutely pure accidental errors. Very small blunders are impossible to detect and it is impossible to recognize all the conditions that may create small systematic errors. Nevertheless, experience proves that when surveys are properly conducted, the application of the laws of chance to the errors of the survey will give the nearest approach to perfection. In other words, the errors of a properly conducted survey can be treated as accidental errors.

1-6. Measures of Precision and Estimates of Accuracy. To avoid repetition in this discussion, accidental errors will be referred to merely as errors.

TABLE 1-1. COMPUTATIONS OF ERRORS FROM MEASUREMENTS OF THE SAME QUANTITY

Values	Errors	(100 Er)²	Values	Errors	(100 Er)²
2372.21	−0.25	625	2372.62	+0.16	256
2372.59	+0.13	169	2372.12	−0.34	1156
2372.22	−0.24	576	2371.95	−0.51	2601
2372.78	+0.32	1024	2372.63	+0.17	289
2372.40	−0.06	36	2372.24	−0.22	484
2371.95	−0.51	2601	2372.09	−0.37	1369
2372.65	+0.19	361	2372.45	−0.01	1
2372.89	+0.43	1849	2372.67	+0.21	441
2373.00	+0.54	2916	2372.93	+0.47	2209
2372.45	−0.01	1	2372.34	−0.12	144
2372.24	−0.22	484	2371.89	−0.57	3249
2372.18	−0.28	784	2372.52	+0.06	36
2372.86	+0.40	1600	2372.78	+0.32	1024
2372.38	−0.08	64	2372.45	−0.01	1
2371.93	−0.53	2809	2372.70	+0.24	576
2372.38	−0.08	64	2372.61	+0.15	225
2373.25	+0.79	6241	2372.72	+0.26	676
2372.51	+0.05	25	2372.52	+0.06	36
2372.15	−0.31	961	2372.67	+0.21	441
2372.05	−0.41	1681	2372.43	−0.03	9
		Sum	94,898.40	Sum	40,094
		Av.	2,372.46	÷ 39 =	1,028

$$\frac{\Sigma v^2}{n-1} = 0.1028$$

$$\sqrt{\frac{\Sigma v^2}{n-1}} = 0.321$$

Because of the laws of chance, when a quantity is measured a number of times with the same precision, the errors will tend to differ according to a certain pattern. Table 1-1 shows the results of 40 measurements of the same distance. The error of each measurement has been computed by subtracting the average of all the measurements from each individual

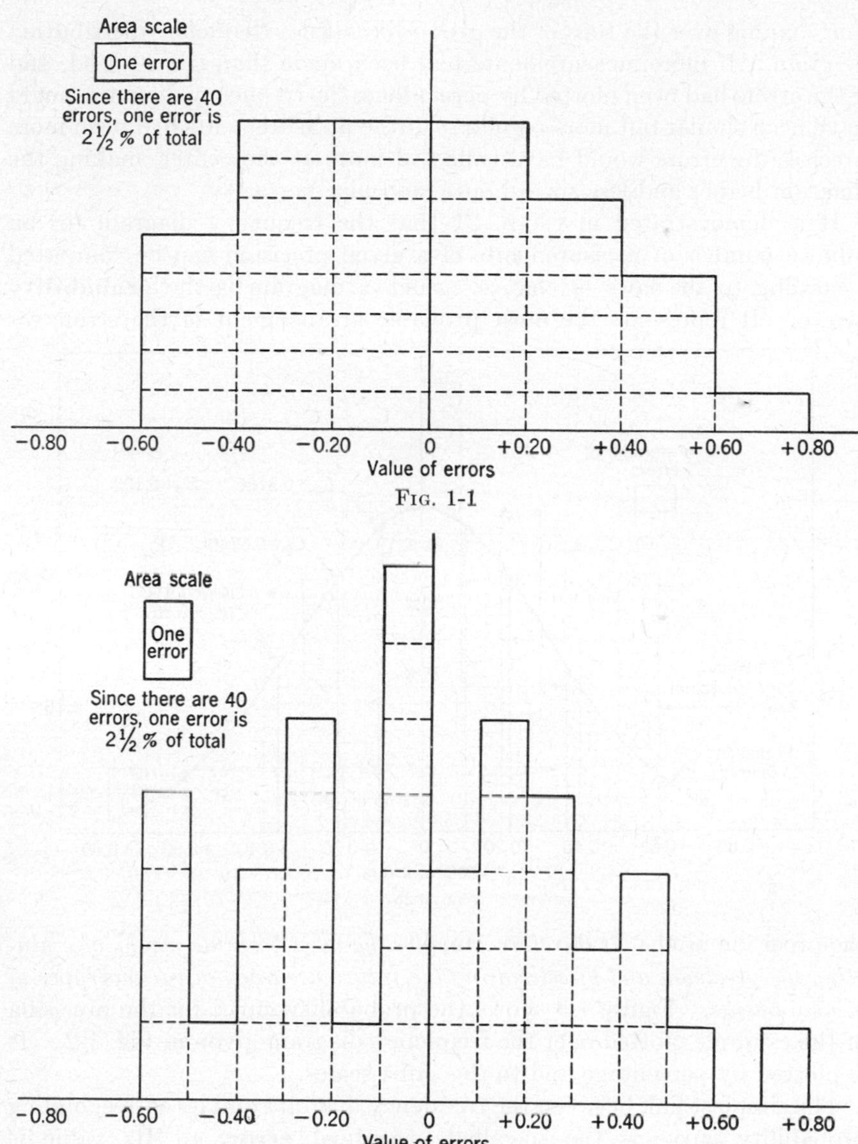

FIG. 1-1

FIG. 1-2

measurement. The pattern of these errors can be determined by collecting them into groups according to their values. Figures 1-1 and 1-2 show two such arrangements. In Fig. 1-1 the errors have been collected by groups ranging over intervals of 0.20 ft and in Fig. 1-2 by ranges of 0.10 ft.

In both figures the same scales are used. The sizes of the errors are represented along the X axis and a given area is assigned to an error. The line running over the tops of the groups is called a frequency-distribution diagram. If more measurements had been made than the 40 used, and if the errors had been plotted by percentage, the frequency diagram would have been similar but more regular. If the measurements had been more precise, the errors would have collected more at the center, making the diagram higher and less spread out—and vice versa.

It is demonstrated in Chap. 21 that the frequency diagram for an infinite number of measurements of a given precision can be computed according to the laws of chance. Such a diagram is the **probability curve.** It represents the most probable arrangement of the errors for

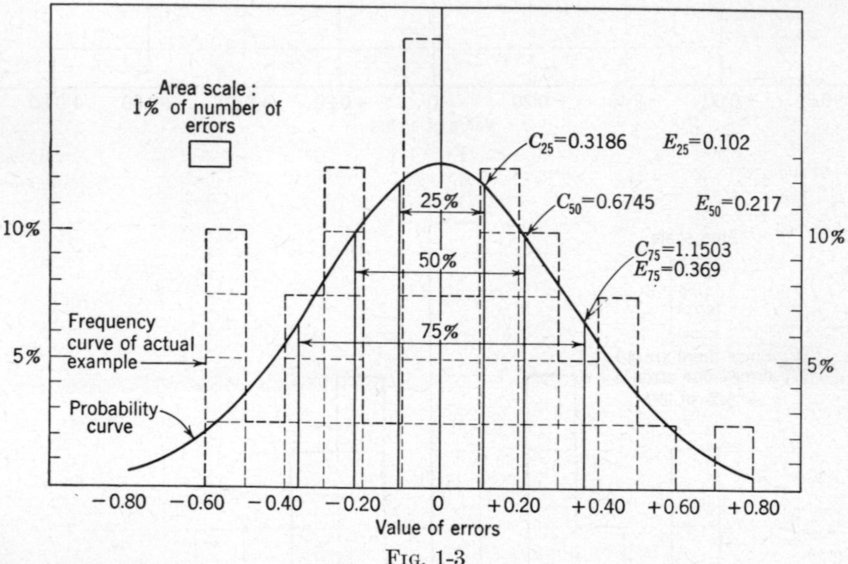

FIG. 1-3

the precision used. *It therefore provides the most accurate means of evaluating the precision and of estimating the future accuracy of various types of measurements.* Figure 1-3 shows the probability curve for the precision of the example plotted over the frequency diagram given in Fig. 1-2. It is plotted by percentage and to the same scales.

The simplest link between the frequency diagram and its corresponding probability curve is the so-called **standard error,** σ. Its value is expressed as follows:

$$\sigma = \sqrt{\frac{\Sigma v^2}{n - 1}}$$

where Σv^2 = sum of the squares of the errors; errors are computed from the average of the measurements

n = number of measurements and therefore number of errors

Table 1-1 shows the computation of σ (= 0.321 ft) for the example.

Once σ is known, the entire probability curve can be found. The coordinates for the probability curve can be computed in a number of ways. The most useful concept is the relationship between any selected size of error and the area under the curve between the limits established

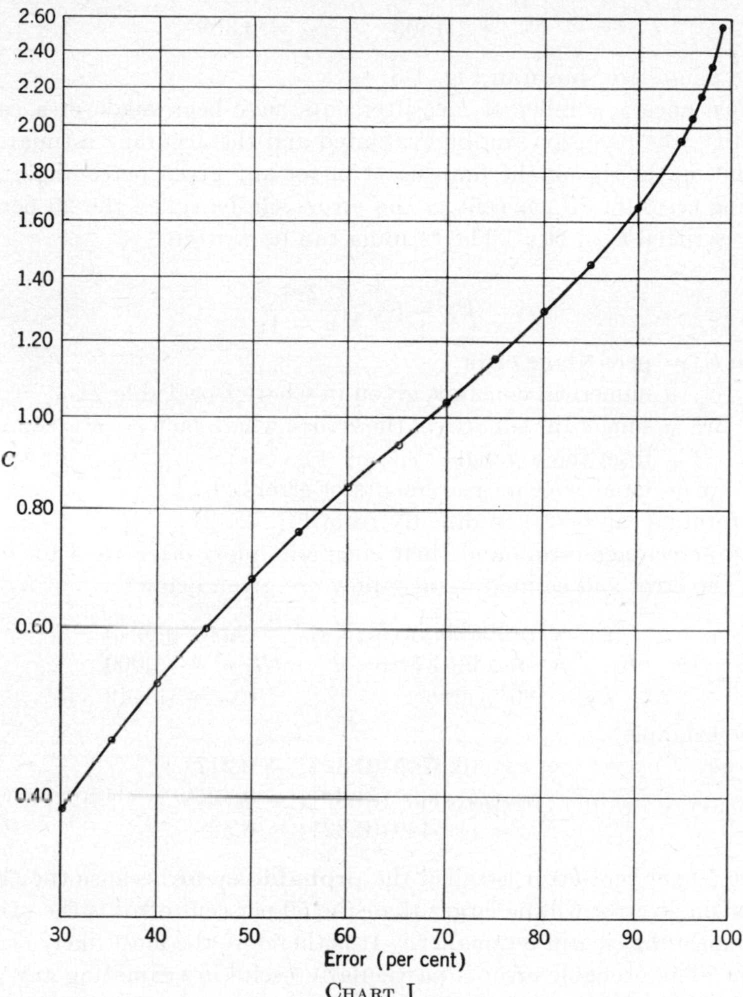

CHART I

by the plus and minus values of the selected error. It will be remembered that such an area represents a certain percentage of all the errors.

It is shown in Chap. 21 that the size of the error that limits a certain percentage of the errors can be found by multiplying σ by a given constant C appropriate to that percentage. Table 21-3 and Chart I give these constants.

For example, from Table 21-3 the constant for 25 per cent of the errors is 0.3186, and the value of σ (0.321 ft) multiplied by this value is 0.102 ft. Thus 25 per cent of the errors will probably fall between ± 0.102 ft.

Likewise, using subscripts to show the percentages involved,

$$C_{50} = 0.6745 \qquad E_{50} = 0.218$$
$$C_{75} = 1.1503 \qquad E_{75} = 0.369$$

These points are shown in Fig. 1-3.

Thus, once a number of measurements have been made of a certain quantity, the precision can be evaluated and the accuracy estimated by computing the size of the limiting error for any given percentage. The limiting error for 25 per cent of the errors can be called the 25 per cent error (written E_{25}), etc. The formula can be written

$$E_p = C_p \sqrt{\frac{\Sigma v^2}{n - 1}} \tag{1-1}$$

where E_p = percentage error
C_p = numerical constant given in Chart I or Table 21-3
Σv^2 = sum of the squares of the errors, where each error is computed from the average
n = number of measurements or errors

This formula can be taken directly from Eq. (21-27).

The percentage errors and their constants most often used for evaluating the error and estimating accuracy are given below:

$$E_{50} = \text{probable error}, \epsilon \qquad C_{50} = 0.6745$$
$$E_{68.27} = \text{standard error}, \sigma \qquad C_{68.27} = 1.0000$$
$$E_{90} = 90\% \text{ error} \qquad C_{90} = 1.6449$$

In the example

$$\epsilon = (0.6745)(0.321) = 0.217'$$
$$\sigma = (1.0000)(0.321) = 0.321'$$
$$E_{90} = (1.6449)(0.321) = 0.528'$$

The 50 per cent error is called the **probable error** because the chance that a single error will be larger than the 50 per cent error is the same as the chance that it will be smaller. It is therefore the most likely error to occur. The probable error is particularly useful in evaluating any given set of measurements as it states the error *expected* in the results. It is frequently used to specify desired precisions or accuracies. It is designated in this text by small epsilon, ϵ, as stated previously.

The 68.27 per cent error is called the **standard error.** It is slightly easier to compute as the constant involved is unity. It is being used more and more, particularly to *compare* precisions or accuracies. It is designated in this text by small sigma, σ, as stated previously.

The 90 per cent error is used in this text to designate the precision of a measuring instrument, or of a method. When an instrument or a method is chosen for a certain purpose, it is important to know the maximum error to be expected. Theoretically there is no limit to the size of the error that *might* occur. Accordingly the maximum error (the 100 per cent error) cannot be used. Also if too precise an instrument or method is chosen, the costs will usually be too high. It is usually cheaper to repeat a small portion of the measurements that do not reach the desired accuracy than to attempt to maintain a level of accuracy that will always be accepted. Usually the precision of measurements should be held at such a level that 10 per cent of the work must be repeated. Accordingly, when a certain value is established as the maximum allowable error, instruments and methods should be chosen so that their 90 per cent error is this value.

1-7. The Accumulation of Errors. When a measured quantity is made up of the results of several observations, each of a known precision, the precision of the quantity can be estimated by computation.

The equations for making these computations are given in Chap. 21 on page 623.

Example 1. When the value of a quantity is made up of the sum or difference of two or more measurements of the same precision, Eq. (21-23) may be used. Equation (21-23) may be written

$$E_{pq} = E_{pa}\sqrt{n}$$

where E_{pq} = percentage error of the value of the quantity
E_{pa} = known percentage error of each of a set of similar observations
n = number of similar observations that together give the value

Assume that a length of 900 ft was measured by a 100-ft tape and it is known that the 90 per cent error of measuring 100 ft E_{90a} is ± 0.01 ft. The 90 per cent error of measuring the 900 ft, E_{90q}, would be found by the following substitution:

$$E_{90q} = \pm 0.01\sqrt{9}$$
$$E_{90q} = \pm 0.03$$

The 50 per cent error for measuring 100 ft would be

$$E_{50a} = \frac{C_{50}}{C_{90}} \times 0.01 = \frac{0.6745}{1.6449} \times 0.01 = 0.0041$$

The 50 per cent error for measuring 900 ft could be computed now by two methods

$$E_{50q} = E_{50a}\sqrt{9} \quad \text{or} \quad E_{50q} = \frac{0.6745}{1.6449}E_{90q}$$
$$= 0.0041\sqrt{9} \qquad\qquad = \frac{0.6745}{1.6449}0.03$$
$$= 0.0123 \qquad\qquad = 0.0123$$

Example 2. When the value of a quantity is determined by the sum or difference of two or more measurements of different precision, Eq. (21-22) may be used to find the precision of the entire measurement. Equation (21-22) may be written

$$E_{90s} = \sqrt{E_{90a}^2 + E_{90b}^2}$$

Find the 90 per cent error in a measurement of 1100 ft measured in two sections, one 900 ft long measured with the tape in the previous example and one 200-ft section measured with a 50-ft tape used so that the 90 per cent error of one tape length is 0.0075 ft.

$$E_{90a} \text{ (for 900 ft)} = \pm 0.03 \text{ (as found before)}$$
$$E_{90b} \text{ (for 200 ft)} = \pm 0.0075\sqrt{4} = \pm 0.015$$
$$E_{90s} = \sqrt{(\pm 0.03)^2 + (\pm 0.015)^2} = \pm 0.034$$

Example 3. When the value of a quantity is the average of several measurements of the same precision, Eq. (21-24) may be used to find the precision of this average. Equation (21-24) may be written

$$\epsilon_0 = \frac{\epsilon_a}{\sqrt{n}} = 0.6745\sqrt{\frac{\Sigma v^2}{n(n-1)}}$$

Required to find the probable error of the average of the list of 40 measurements in Table 1-1.

$$\epsilon_0 = 0.6745\sqrt{\frac{4.0094}{40(40-1)}} = 0.0342$$

Thus, if the estimated 90 per cent error, for example, is known for the various measurements in a survey, it is possible to compute the estimated 90 per cent error of the completed work by the application of the equations on page 8. The 90 per cent error of any means of measurement can be found by making repeated measurements of the same quantity.

Example 4. Required to determine the number of repetitions necessary with a 1-minute transit in order to close a triangle with an error of 3 seconds or less. Arrange a test as shown in Fig. 1-4.

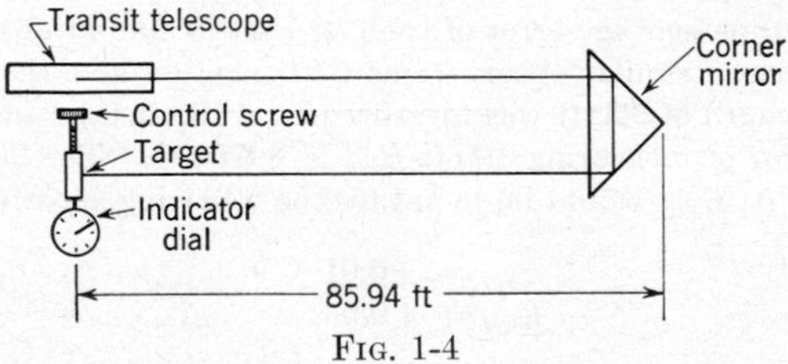

FIG. 1-4

1. Move the target off line and reset it on the cross hair about thirty times. The resulting angles can be computed from the target positions and hence the value of σ can be computed for pointing errors.

2. Set the vernier repeatedly at one vernier setting. Set the target on line each time. Determine the angles from the target movements. This will give the σ for the combined error of setting the vernier and pointing.

3. Move the target to different positions and read each angle. This will give the σ for the combined error of reading and pointing. Assume that the resulting probable errors ($0.6745\,\sigma$) are as follows:

Pointing	0.2″
Setting the vernier and pointing	3.5″
Reading the vernier and pointing	10.0″

First compute the required probable error of one angle from the required 90 per cent error of the triangle. Let E_{90} and ϵ refer to the error per angle; then

$$E_{90} = \sqrt{3} \times 3'' \qquad E_{90} = 1.73''$$

$$\epsilon = \frac{0.6745}{1.6449} E_{90} = 0.709''$$

When an angle is measured by R repetitions, the setting-pointing and reading-pointing errors come in once each, the pointing error alone $(2R - 2)$ times. The total error for measuring an angle is then divided by R. The error in measuring an angle by R repetitions is therefore

$$\epsilon = 0.709 = \frac{1}{R} \sqrt{(3.5)^2 + (10)^2 + (2R - 2)(0.2)^2}$$

Solving, $R = 15$ or 8 D.R.

If some triangulation has been completed and the results show that the probable error of closure is 3.07 seconds for the various triangles, when the angles were measured 3 D.R., the combined errors of setting, pointing, and reading, V, can be computed as follows:

$$\frac{3.07}{\sqrt{3}} = \tfrac{1}{6} \sqrt{V^2 + 12(0.2)^2}$$

Solving for V, $V = 10.6''$

Example 5. Required to find the proper length of sights for a level to produce third-order accuracy. Assume that repeated readings of the rod from the level set up at 100 ft gave a probable error of 0.002 ft.

$$E_{90} = \frac{1.6449}{0.6745} \epsilon = (2.44)(0.002) = 0.00488 \text{ ft per 100 ft}$$

$$E_{90} = 0.0000488 \text{ per ft}$$

Let L = length of sight. Then

$$0.05 = 0.0000488L \sqrt{\frac{5280}{L}}$$

Solving, $L = 199'$

1-8. Measures of Accuracy. The accuracy of the final results of a survey can be expressed in terms of the various per cent errors when sufficient measurements are made of the same quantity to compute σ, as in the case of base-line measurements. Usually, however, the errors in the various field checks (called the errors of closure) are the only criteria available to indicate accuracy. These are used to express the degree of accuracy and they are undoubtedly familiar to the reader. Table 1-2 gives the specifications for orders of accuracy generally accepted throughout the United States.

1-9. Rejection of Observations. The rejection of data is based on the principle that all blunders should be eliminated and all accidental errors should be retained. If an accidental error, no matter how large, is rejected, the accuracy of the result will be reduced.

Large blunders can be easily recognized and eliminated. Small blunders can never be identified but they have little effect. It is the large errors that may or may not be blunders that give the difficulty.

Sometimes errors beyond a certain *established* size are rejected automatically. This is a reasonably safe procedure as it will not remove as many accidental errors as blunders in the long run. The process to be avoided is the arbitrary selection of errors to be rejected. This is a great temptation and always leads to poor results.

TABLE 1-2. ORDER OF ACCURACY DEFINED BY BOARD OF SURVEYS AND MAPS OF THE FEDERAL GOVERNMENT, MAY 9, 1933

Type of survey	Type of error	Limits of error Order of accuracy			
		First	Second	Third	Fourth
Triangulation	Maximum angular closure per triangle	3″	5″	10″	No appreciable error in resulting map
	Average angular closure per triangle	1″	3″	5″	
	Error in length of base as computed through triangles after angles are adjusted	1/25,000	1/10,000	1/5,000	
	Probable error* of final base length	1/1,000,000	1/500,000	1/200,000	
Traverse	Position closure after angles are adjusted	1/25,000	1/10,000	1/5,000	
Leveling	Error of closure, ft, divided by square root of distance leveled in miles	0.012	0.025	0.050	

* Probable error of the mean $= 0.6745 \sqrt{\frac{\Sigma v^2}{n(n-1)}}$

where n = number of measurements

Σv^2 = sum of the squares of the quantities by which each measurement differs from the mean of the measurements

The mean of the measurements is used for the final base length.

Many criteria of rejection have been suggested; the criterion proposed by Chauvenet[1] is given here. This criterion determines whether or not the largest error of a set should be rejected. If such an error is then eliminated, the remaining observations are treated alone. The criterion is then applied to the largest error in the new set. The process can be continued until the largest remaining error is retained according to the rule.

[1] William Chauvenet, "A Manual of Spherical and Practical Astronomy," vol. II, p. 564, J. B. Lippincott Company, Philadelphia, 1885.

1-10. Chauvenet's Criterion.[1] Table 1-3 gives a list of the errors in the original example (see Table 1-1) arranged according to absolute size. In this table the position of the first error really covers a position range from 0 to 1. Its exact position should be expressed by ½ (the average of 0 and 1) in the 40 positions of the set. In like manner error No. 40 covers

TABLE 1-3

No.	Error	No.	Error	No.	Error	No.	Error	No.	Error
1	−0.01	9	−0.08	17	+0.21	25	−0.28	33	+0.43
2	−0.01	10	−0.08	18	+0.21	26	−0.31	34	+0.47
3	−0.01	11	−0.12	19	−0.22	27	+0.32	35	−0.51
4	−0.03	12	+0.13	20	−0.22	28	+0.32	36	−0.51
5	+0.05	13	+0.15	21	−0.24	29	−0.34	37	−0.53
6	+0.06	14	+0.16	22	+0.24	30	−0.37	38	+0.54
7	−0.06	15	+0.17	23	−0.25	31	+0.40	39	−0.57
8	+0.06	16	+0.19	24	+0.26	32	−0.41	40	+0.79

the range from 39 to 40 with its exact position at 39½. Its percentage position therefore would be $\frac{39\frac{1}{2}}{40}$ or 98.75 per cent. The most likely size of this per cent error would be computed by Eq. (1-1), thus:

$$\begin{aligned} E_{98.75} &= C_{98.75}\sigma \\ &= (2.50)(0.321) \\ &= 0.8025 \end{aligned}$$

Chauvenet points out that any error larger than 0.8025 in this case has less than a 50 per cent chance to occur and therefore can be safely rejected since 0.8025 is the most likely size. But the largest error, No. 40, is 0.79. This would be retained because it is less than the limit 0.8025. The general formula can be written

$$\frac{p}{100} = \frac{\frac{1}{2}[n + (n - 1)]}{n}$$

or

$$p = 100\,\frac{2n - 1}{2n}$$

where p is the per cent of the center of the range of the largest error. Then

$$E_p = C_p \sqrt{\frac{\Sigma v^2}{n - 1}} \tag{1-1}$$

where E_p in this case represents the maximum error admitted.

[1] The theoretical accuracy of this criterion has been questioned, but it has had a long history of useful application to surveying measurements. A more rigorous method, which requires considerably more computation, gives nearly the same results.

SIGNALS AND OBSERVATIONS

1-11. Human Errors. No two observers have exactly the same skill or speed, nor do they make accidental errors of the same size. Observers can even introduce systematic errors in operations like bisecting a large

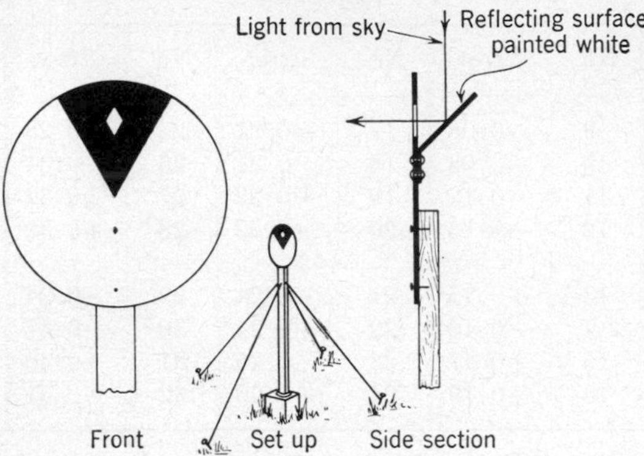

FIG. 1-5. A traverse target. It is galvanized iron, painted black and white. A diamond-shaped hole gives a view of sky-illuminated reflector. It is easily identified and can be accurately pointed at any distance.

signal or indicating the moment that a star makes contact with a cross hair. These systematic errors can be measured, and the corrections found for them are called **human equations.** Certain surveying techniques have been developed that reduce the size of the accidental and systematic errors of all observers. They apply to signals and to observations.

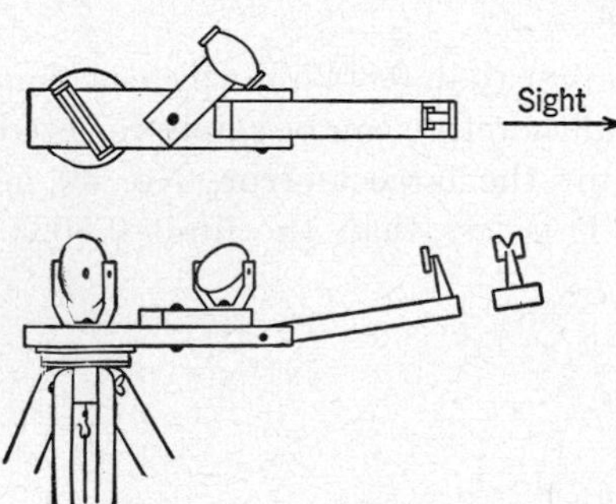

FIG. 1-6. Homemade heliotrope. From the hole in the mirror silver, the front sight is aimed at the transit position. The mirrors are arranged to reflect the sun on the front sight. Usually the main mirror must be covered with cardboard except for a ¼-in. hole at center to reduce the light.

1-12. Signals. A signal or **target** marks the position of a vertical or a horizontal line, sometimes both. *It should have certain definite characteristics* for the greatest accuracy. These are listed in the following paragraphs:

1. The signal must be easily identifiable so that it can be quickly picked up. Sometimes quite a large signal must be used for identification when the part of the signal pointed (the target) is quite small (see Fig. 1-5).

2. All markings on the signal and the target and preferably the entire outline of the signal itself should be symmetrical with respect to the line it marks.

3. The greatest length of the target should be parallel to the line it marks.

4. The target should be free from **phase.** Phase is caused by unequal illumination. The observer tends to point on the brightest or the darkest part rather than the center. It follows that all surfaces presented to the instrument must be flat.

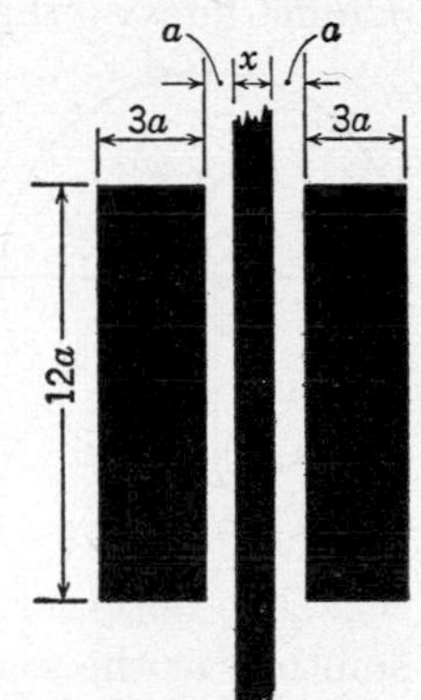

x = width of cross hair, usually about 2½ seconds

a = 2½ 7 seconds

FIG. 1-7. Ideal target.

5. A target should present the greatest possible contrast. This can be accomplished in a wide variety of ways. Lights and **heliotropes** give the greatest contrast. Silhouettes formed against the sky are excellent. Targets for long distances should be black and white. Yellow is excellent for short range and for identification. When a signal or target has different backgrounds both black and white or black and yellow should be used.

A heliotrope (see Fig. 1-6) is a device having two mirrors and an arrangement for sighting. It is used to reflect the sun toward the instrument. It is excellent in haze or smoke for short ranges and in clear air over great distances. It must be continually manned so that one of the mirrors can be turned to follow the sun.

6. The bisection principle must be used. If the target is smaller than the cross hair, the pointing may vary throughout the difference in the size of the cross hair and the size of the target. Fortunately, if the target is larger than the cross hair, the target can be bisected with great accuracy. Apparently, an observer is capable of extremely accurate bisection, so that the target can safely be considerably larger than the cross hair. By using two or three sizes, the target can be used at any range (see Fig. 1-7).

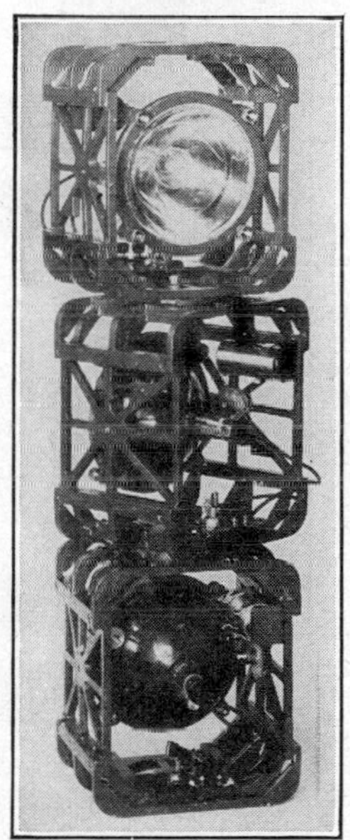

FIG. 1-8. Signal lamps arranged for three observing stations. (*U.S. Coast and Geodetic Survey.*)

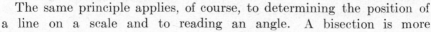

When the target must be smaller than the cross hair, as is the case when observing a light (Fig. 1-8), a heliotrope, or a star, bisection can be provided by using two parallel cross hairs in the reticule. Transits used entirely at night are so equipped. When a glass reticule is used, half of the cross hair is usually double (Fig. 1-9).

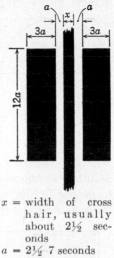

The same principle applies, of course, to determining the position of a line on a scale and to reading an angle. A bisection is more accurate than a coincidence. Accordingly, it is better to place a pair of parallel lines so that a single line bisects the interval between them than to bring two lines to coincidence.

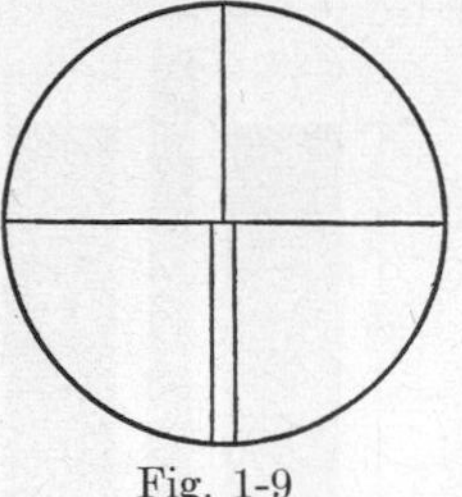

Fig. 1-9

Tests at Princeton University show that the 97.5 per cent error of pointing a target designed as indicated in Fig. 1-7 is ½ second arc (note Figs. 1-10 and 1-11).

1-13. Observations. An observer performs two types of operations. He *points on a target* or he *reads a scale* (usually estimating the last figure). Two rules should be followed in these operations:

1. The observer must be able to control the movement that sets the pointing, or the scale reading. He can never direct someone else to perform this for him with the same accuracy. For example, he can bring the cross hair on the target more accurately than he can direct an assistant to move a target into line with the cross hair. Also, he can read a level rod more accurately than he can direct the rodman to set the target.

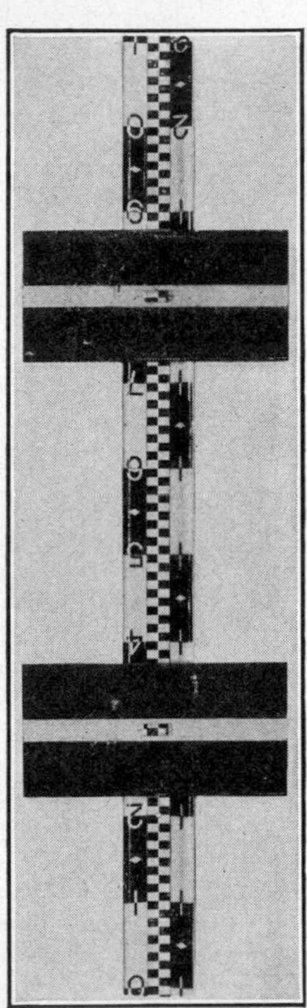

FIG 1-10. Geodetic leveling rod with targets for river crossings—face view. (*U.S. Coast and Geodetic Survey.*)

FIG. 1-11. Precise triangulation target mounted on top of automatic signal lamp. (*U.S. Coast and Geodetic Survey.*)

Of course, in staking out operations, it is always necessary to direct someone to move a target. If accuracy is required, the position of the target should be subsequently checked by an observation on it.

2. When the observer points a target or reads a scale he should perform the operation smoothly, quickly, but without hurry. If he tries to perfect the pointing or the reading for more than a minimum length of time, he will begin to fuss and tire and he will actually reduce the accuracy of his work.

PLANNING THE SURVEY

1-14. Planning Can Only Be Outlined. Without a comprehensive understanding of most surveying practices a survey cannot be successfully planned. Planning is outlined here so that the importance of the subsequent material may be gaged.

1-15. Steps in Planning a Survey. Once the project is established and the desired results of the survey are known, planning the survey follows the steps listed.

1. Choice of accuracies required
2. Study of existing control
3. Reconnaissance for triangulation, traverse, and leveling, including the selection of probable locations for stations and benchmarks
4. Choice of instruments and methods
5. Choice of computation and drafting procedure

At this point the costs can be estimated, the personnel organized, and the work commenced.

1-16. Accuracy of Basic Control. The choice of accuracy for the basic control depends on an estimate of the use to be made of the work. As the basic control comprises only a small part of the cost of the survey, it is well to err on the side of too high accuracy. It is dangerous to lower the accuracy even though there is no apparent reason for high accuracy at this time. Experience shows that once the survey data are available they will be used for every conceivable purpose often far beyond the original requirements.

In general, second-order accuracy should be used for both horizontal and vertical basic control. First-order would be desirable only in rare instances and third-order is usually a little too inaccurate for safety.

Under normal conditions a location survey will be made for construction throughout some of the area covered by a map. In the vicinity of such construction second-order control must be established so that the construction surveys can also be based on it.

In localities where it is *certain* that no construction will occur, the desired map accuracy ordinarily can be used as a basis of choice. Maps can seldom be drawn with the smaller measurements laid out to better than 1 part in 300 so that horizontal closures of 1:1000 would be sufficient, and basic control should not be better than third-order. Third-order leveling is so easy to attain that it can be safely used without undue costs.

1-17. Other Criteria for Accuracy. In certain cases the accuracy ratio of the map is not so important as the accuracy of position. For example, it is often specified that the maximum error in position on the map shall be $\frac{1}{50}$ in. If the scale were 1 in. = 100 ft the maximum absolute error allowable would be 2 ft. If the extent of the area mapped were 10 miles or 50,000 ft, the accuracy ratio would be 1:25,000, requiring first-order for basic control.

When structures are to be designed to carry water or other liquids in open channels or by gravity flow without pressure, leveling must be particularly accurate. Basic control for such work should often be first-order if the distances involved are large.

Sometimes the limit of error of position in the construction survey is the criterion. Surveys for bridges, tunnels, city property, complicated steel erection, and complicated real estate developments are examples. Any of these may require first-order measurements.

1-18. Accuracy in Supplementary Control. Horizontal secondary control accuracies are usually from 1:2000 up to third-order. Leveling is usually held to third-order. Error of position is more often the criterion than an accuracy ratio. When shorter lines are run between stations of the basic control, larger closure errors will still give positions within the required limits. For example, if a control traverse 500 ft long were used to locate a railroad track to an absolute position of 0.20 ft, 1:2500 would be sufficient.

1-19. Accuracy in Topographic Measurements. Mapping accuracy is the usual criteria for ties for topography. However, the requirements for certain measurements are sometimes far beyond mapping accuracy and must be shown on the map by actual dimensions. Certain clearances, distances between buildings, and the like are examples.

The choice of the contour interval is a very important decision. The usefulness of the survey and its cost are in a large measure dependent on it.

The largest interval that will give the necessary information should always be used. When the contours are determined by ground surveys, the interval may be varied almost at will to correspond with the particular requirements of each part of the mapped area. In general, a 5-ft interval is necessary where earthwork is planned. When the cuts or fills are extensive but not deep, 2-ft contours are required. Maps of areas for water storage or gravity flow must sometimes have 2-ft and even 1-ft contours, depending on the particular information required.

1-20. Specified Accuracies for Photogrammetry. Aerial mapping requires special equipment and special skills. The work nearly always must be performed by organizations that are specialists in the field. The necessary ground control is established either by these organizations or by the engineer who will be in charge of the construction surveys. The second plan is often the more satisfactory, as then the necessary accuracies can be established at the start.

The major specifications for an air map are the *contour interval* and the *map scale*. Unless otherwise stated it is assumed that the elevations of a profile placed anywhere on the map shall agree with the elevations of a corresponding profile measured on the ground within one-half the contour interval at 90 per cent of the points checked and that the horizontal positions of points on the map shall be correct within $\frac{1}{50}$ in.

The cost of air mapping increases almost inversely as the square of the contour interval except in the unusual case when the horizontal accuracy controls. Frequently the scale for the map decided on is larger than necessary for purposes of accuracy. For example, it might be found to be convenient in planning construction to have plenty of room to work on the map so that a scale of 1 in. = 200 ft is selected, despite the fact that an error of 10 ft would be negligible. In this case an error of position on the map of $\frac{1}{20}$ in. would do no harm. If the scale is specified only, the map will be constructed so that the positions shown will be correct to $\frac{1}{50}$ in., $2\frac{1}{2}$ times as accurately as required.

Accordingly, when contracting for air mapping, the contour interval and the *actual horizontal accuracy really needed* must be carefully considered and carefully stated in addition to the map scale, if the cost is to be kept to a minimum. If certain parts of the map require greater accuracies, special treatment should be given them rather than to make them the criterion for the whole survey. If the special areas are large, they can be mapped by air using more accurate methods, otherwise ground methods can be used to obtain the higher accuracy.

1-21. Existing Control. A careful search of available records should be made to discover any horizontal or vertical control that may exist in the neighborhood of the survey and to determine its importance and its accuracy. Connections for position, azimuth, and elevation should *always* be made no matter how inaccurate or unimportant the control may be. When these connections are made, all the data from the original survey will be available if desired. If the existing control is accurate, it may be used to check or actually to control the work. Such connections make it possible to utilize the same coordinate system and the same level datum that is generally used in the neighborhood. This still further facilitates utilization of the data.

Whenever practical, connections should be made with control that is connected with the U.S. Coast and Geodetic Survey triangulation stations and benchmarks. It will be possible then to use the state systems of plane coordinates and mean sea level datum (see Chap. 12).

1-22. Reconnaissance. The first operation in reconnaissance, after the area has been viewed, is to search for and, if possible, to recover existing control survey marks. This is difficult and discouraging as the marks are seldom easy to find. Experience indicates that most survey marks remain undisturbed so that time spent on their recovery is seldom wasted. Reconnaissance for the basic control triangulation is next. It is made in accordance with the methods described in Chaps. 7 and 17. Reconnaissances for traverse and leveling are described in Chap. 7. It should be emphasized that, during reconnaissance, stakes or other marks should be set for most of the control stations and benchmarks and short descriptions should be written so that they may be found. As soon as possible, and *before the survey is started* permanent monuments should be set in these positions.

1-23. Choice of Instruments and Methods. The choice of the proper instruments and methods depends on the accuracies that must be obtained. As is demonstrated in later chapters, precise instruments save time but are not always economical. For example, if a second-order triangulation system of a few stations constituted the basic control, an ordinary engineer's 1-minute transit could be used for measuring the angles. Many repetitions of the angles would be required but as there are only a few stations, the cost would not be high. The various available instruments and the methods of using them are described in later chapters. A proper selection is made in accordance with the capabilities of each.

1-24. Choice of Computation and Drafting Procedure. It is important at the start to set up an effective method of handling the incoming field data. A standard processing program should be established and standard computation forms selected. Several are recommended in later chapters. The order of procedure should be based on these forms. The system of filing and keeping records suggested in "Surveying," Chap. 14, is recommended.

PROBLEMS

1-1. The following measurements were made of a certain length: 3625.6 ft; 3627.0 ft; 3627.4 ft; 3626.3 ft; 3627.7 ft; 3626.2 ft.

a. Find the 90% error.

b. Would the method be appropriate for an accuracy of 1:3000?

c. Approximately what percentage of the distances of about 3600 ft would have to be remeasured?

d. What lengths would this method be appropriate for, for an accuracy of 1:3000?

1-2. In a level run of ½ mile the following differences in elevation were determined: 23.487; 23.481; 23.460; 23.479; 23.467; 23.458.

a. Find the order of accuracy that this method will produce.

b. What percentage of the runs will have to be repeated for approximately second-order accuracy?

1-3. Four angles that closed the horizon were measured repeatedly with a transit. The probable error of a single set of four angles making a horizon closure was found to be 0.2 second. What would be the estimated 90% error of a triangle closure using the same procedure to measure the angles with this instrument?

CHAPTER 2

PRECISE LEVELS AND DIFFERENTIAL LEVELING

PRECISE LEVELS

2-1. Extent of This Chapter. This chapter covers precise levels, precise rods, the process of adjusting the elevations of a line of single-wire precise differential levels, and reciprocal leveling. Other methods of leveling and the adjustment of level nets are reserved for later chapters. The discussions of level vials, telescopes, and three-screw leveling heads are applicable to these parts on other instruments.

2-2. The Level Vial. One of the most sensitive devices employed in surveying is the level vial (see Figs. 2-1 to 2-3). It is used to determine

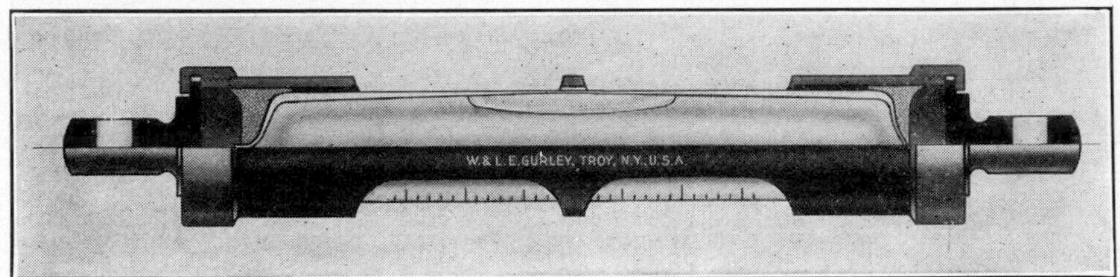

FIG. 2-1. A level vial mounted so that it may be rotated on its horizontal axis. (*W. & L. E. Gurley.*)

the direction of gravity for even the most precise astronomical instruments. The level vial is a sealed glass tube nearly filled with a spirit having a low freezing point and a low viscosity. The interior surface of the glass is ground to a barrel shape so that its longitudinal cross section is composed of two circles of large radius (R in Fig. 2-2), and the lateral sections are small circles of decreasing diameter toward the ends of the tube. The entrapped air that collects at the highest point in the tube to form the bubble or **blister** is forced together into its familiar shape by the weight and the surface tension of the liquid.

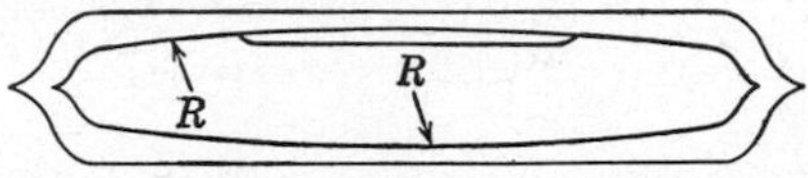

FIG. 2-2. Longitudinal section of a level vial showing curvature exaggerated.

The interior surface of the vial can be more mathematically described as the surface generated by part of a segment of a circle rotated about its chord. The chord forms the **axis of the level vial** and in this text the term axis of the vial is so defined, contrary to the usage in many works on surveying.

2-3. Sensitivity. The sensitivity of the vial depends on the length of the radius R. It is defined as the vertical angle through which the axis of the vial must be tilted to cause the bubble to move one division.

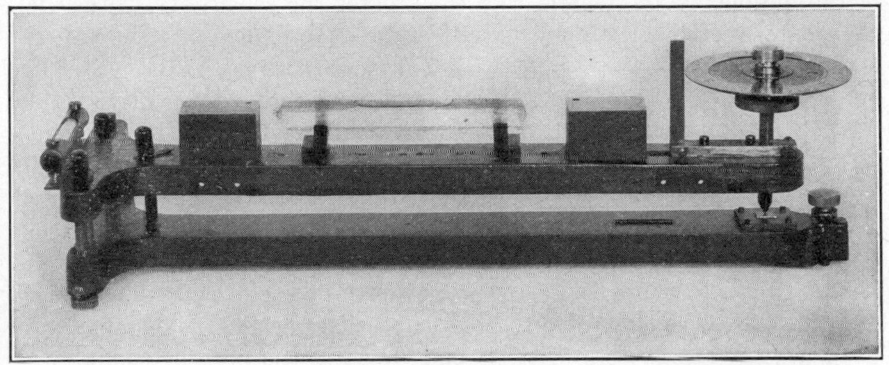

FIG. 2-3. A level tryer with a vial in place. (*W. & L. E. Gurley.*)

Usually the graduations are spaced at 2-mm intervals so that the sensitivity is usually expressed in terms of seconds per 2-mm units.

2-4. To Measure the Sensitivity. The sensitivity can be measured before the vial is mounted by a **level tryer** (Fig. 2-3), but it can be more accurately determined when the vial is in place on the instrument. Hold a level rod at a known distance from the instrument (see Fig. 2-4). Level the instrument so that the bubble is three or four divisions off-center and note the rod reading. Move the bubble to a point three or four divisions off-center in the opposite direction and note the reading. Then

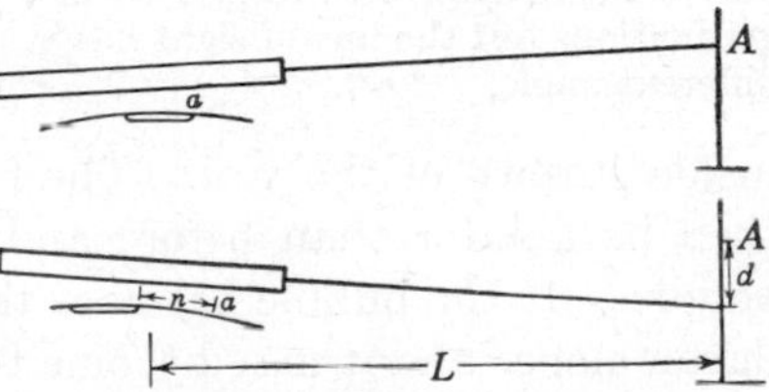

FIG. 2-4. Determining the sensitivity of a level.

$$s = \frac{1}{0.00000485} \frac{d}{L} \frac{1}{n}$$

where s = sensitivity in seconds per division
d = difference between rod readings
L = distance from center of instrument to rod
n = number of divisions of total bubble movement

2-5. Spirits. The spirits ordinarily used are ether, chloroform, alcohol, or a hydrocarbon like pentane. They must be of such low viscosity that they will move quickly in response to a very slight change in angle. In vials of great sensitivity, the spirits should have the lowest practical viscosity. In less sensitive vials higher viscosities are often used to bring the bubble quickly to rest. Often mixtures are used to obtain the desired qualities.

Spirits of this nature usually have high coefficients of thermal expansion. At warm temperatures they expand and thus shorten the length of the bubble. As a general rule the greater the sensitivity of the vial the greater are the variations in length for the same changes in temperature.

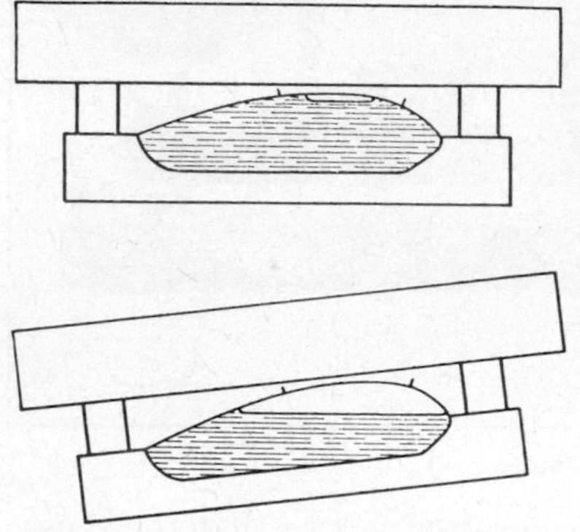

FIG. 2-5. Effect of variation in the radius of a vial when bubble lengthens. Note that in both views the bubble is centered with respect to the graduations but the line of sight has a different angle.

2-6. Chambered Vials. The radius of curvature of a sensitive vial is so great that it can seldom be made absolutely constant throughout the length of the vial. (The radius of a 10-second-per-2-mm vial is about 130 ft.) If the radius is greater toward one end of the vial than the other, when the bubble expands, the end toward the longer radius will lengthen more than the opposite end. Under these circumstances, when the bubble is centered, the line of sight will not be leveled as before (see Fig. 2-5). Also, when the bubble lengthens, the ends of the bubble will be in contact with steeper slopes on the inside surface of the vial. The forces tending to center the bubble will then be greater than before and the bubble may become too lively to handle. If the bubble becomes too short its ends will be in contact with flatter slopes and it may become too sluggish.

To remedy these conditions, a vial for a precise instrument usually is made with a **chamber** at one end (see Fig. 2-6) connected to the main vial by an orifice near the bottom of the tube. When the chamber is held

FIG. 2-6. Section of a chambered level vial. The length of the bubble may be regulated by allowing air to escape from the chamber to the vial and vice versa.

upward, air from the bubble escapes slowly into the chamber. When it is held downward, the air is returned. The length of the bubble thus can be regulated as desired.

2-7. Other Factors Involved. The graduations on the vial should be placed so that when the bubble is centered, its surface is parallel to the axis of the barrel-shaped interior of the bubble tube. Since the tube is smaller at the ends, if the bubble is not symmetrically placed, the ends will have different shapes so that when the bubble changes size, one end will move farther than the other. If it is then centered, the line of sight will take a different direction.

Dirt, surface roughness, or crystals from impurities in the spirit will change the shape of the meniscus and make the bubble movements irregular. Sometimes impurities in the spirit etch the glass after a period of time.

The slightest temperature gradient along the length of the level vial changes the position of the bubble. It vaporizes the spirit more rapidly at the warmer end, causing the bubble to move toward it as the spirit is forced out by the increased vapor pressure. It also expands the spirit at a different rate from the glass. The extreme sensitivity of the bubble to differential heating can be tested by holding a finger on the glass for a few seconds. The bubble will move a considerable distance. It follows that the bubble must be carefully protected from differential heating. Since radiant heat is the chief source of differential heating, the vial in a precise instrument should be completely enclosed.

The vial should be mounted so that no strain, particularly bending strain, can reach it. To be absolutely safe, the metal tube that holds the vial also should be free from bending strain even during adjustment.

2-8. Optics. Good optics are of greater importance in a level than in a transit. In leveling, the exact position of the cross hair on the rod graduations must be deciphered; in using a transit it is necessary only to establish coincidence (or a bisection) between the cross hair and the target, a much less difficult process.

The best test for optics is to compare the view through one telescope with that through another. Good optical design requires such a complicated balance between so many possibilities that no one feature is significant by itself. For example, high magnification would seem to be always an advantage. But a telescope of lower magnification with good optics might give a far clearer view of the rod than one of poorer design but of greater magnification.

Stray light is unfocused light that reaches the observer. It tends to dim the image as the house lights will dim the picture on a movie screen. Some stray light is caused by rays that enter the objective at too great an angle to be focused. Some is caused by light reflected by the lens surfaces. The black interiors and disks with holes in the center, called "stops," absorb some stray light. Coated lens surfaces eliminate much of the reflected light.

The optics should be of such quality that the minimum change in level noticeable by movement of the cross hairs on the rod is approximately equal to the minimum change in level that can be detected by movement of the bubble.

In general, the best optics can be obtained with an internal-focusing telescope, an inverting eyepiece, a fairly large objective lens, and a fairly long telescope with coated optics. Below is a table illustrating normal combinations for different types of levels:

Type of level	Diameter of objective, in.	Angle of inclination per 2 mm of bubble run, seconds of arc	Power, diameters
Construction	1–1½	15–40	15–20
Engineer's	1⅜–1¾	8–20	20–35
Engineer's precise	1½–1⅞	3–4	30–40
Precise (geodetic)	1¾–2	2	38–43

2-9. Stadia Hairs. Most precise levels are equipped with stadia hairs. They serve five functions.

1. The rod is read at each hair and the average used for the value of the rod reading. This gives increased accuracy.

2. The values of the two half intercepts are compared to prevent blunders.

3. The full intercepts are used to check the unbalance between each pair of sights at any instrument position.

4. The sum of the forward intercepts is compared with the sum of the back intercepts to determine the total unbalance in the line. Correction can be applied in accordance with this value.

5. The total length of the line between benchmarks is used to distribute the errors between **junction points**, and the total length of the **link** between junction points is used to weight the line in a simultaneous adjustment of a level net (see Chap. 17). These lengths can be obtained from the stadia intercepts.

Since the determination of actual distance does not need to be very accurate, the stadia multiplier constant can be quite small. This has the advantages of bringing the cross hairs nearer to the center of the field where the optics are better, and of reducing the spread between the lines of sight so that greater differences in elevation can be observed. In some instruments the stadia ratio of 1 ft per 100 ft is used, but 0.6 and 0.3 ft per 100 ft are much more desirable.

2-10. Glass Reticules. Both spider-web and glass reticules are used in modern levels. There seem to be advantages in each type. Spider web, properly stretched when installed, remains in tension. It is straight, smooth, and opaque. However, the patterns available are limited to systems of lines running entirely across the field.

Glass reticules can be marked in any desired pattern. Stadia lines are often made short so that they cannot be confused with the central line (see Fig. 2-7). Even a graduated scale can be used. But under bright light the marks on a glass reticule sometimes produce a "ghost" from the reflection of the two surfaces. This appears as bright white lines superimposed beside the black ones. Also any speck of dust on the same side of the glass as the lines is exactly in focus and therefore seen clearly. Accordingly, glass reticules should be used only in internal-focusing telescopes that can be made dustproof and installed under dust-controlled conditions. Finally, a glass reticule adds two reflecting surfaces to the optics, each of which increase the stray light that tends to dim the image.

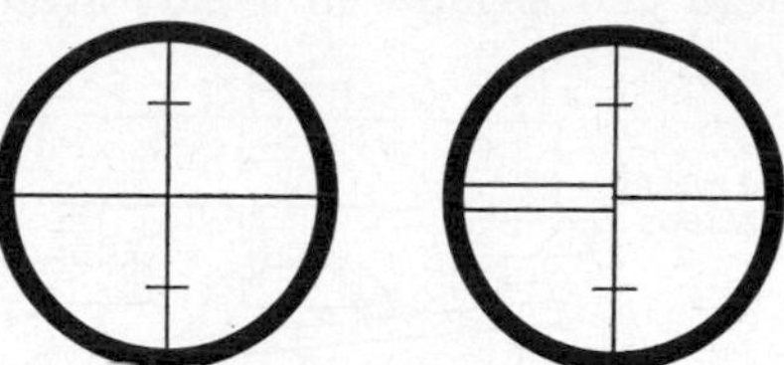

FIG. 2-7. Designs for glass level reticules.

2-11. Three-screw Leveling. Many of the more precise instruments have three leveling screws instead of four (Fig. 2-8). With this arrangement, each screw can be turned independently, which simplifies leveling and makes the process quicker and smoother. It is accordingly very popular with many who are familiar with it. The height of the instru-

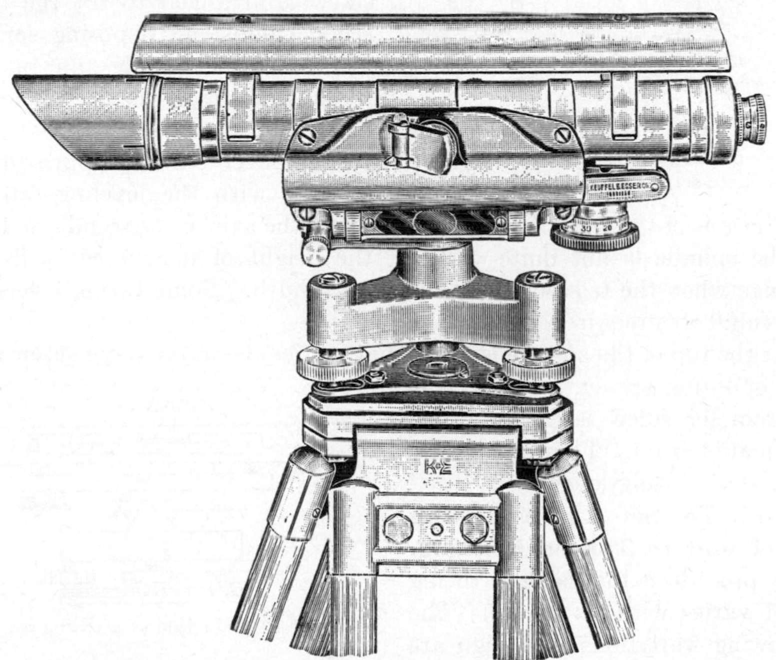

FIG. 2-8. A precise level with three leveling screws. (*Keuffel & Esser Co.*)

ment changes when any one screw is turned (Fig. 2-9) so that three leveling screws can only be used with **tilting levels.** These are described in the next paragraph. A three-screw leveling instrument is usually attached to the tripod by a large screw that comes up through the tripod head and engages in a part attached to a spring that either holds the instrument down or holds down the feet of the leveling screws.

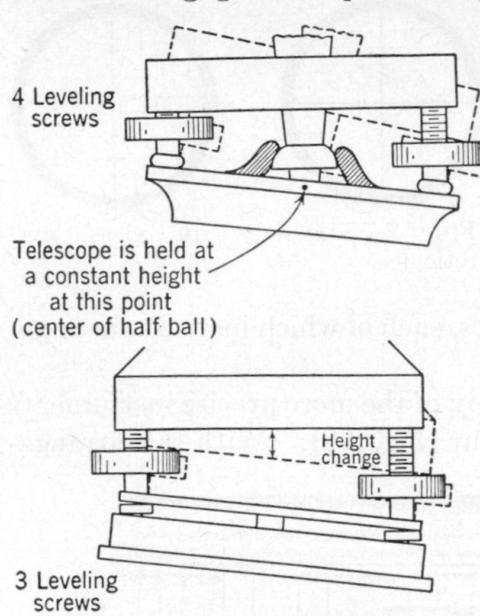

FIG. 2-9. With four leveling screws the height of instrument remains constant. With three screws it changes.

2-12. The Tilting Level. The telescope of a tilting level (Fig. 2-10) is mounted on horizontal trunnions usually attached to the telescope tube. After the instrument has been set up and leveled approximately, it can be precisely leveled when pointing in any given direction by slightly tilting the telescope. Usually a micrometer screw is arranged to tilt the telescope against an opposing spring. The vertical spindle fits in the leveling head, as usual, and ordinarily a circular level is attached to the level bar for approximate leveling with the leveling screws.

The axis of the trunnions should intersect the axis of the spindle so that if the spindle is not quite vertical, the height of instrument will not change when the telescope is turned in azimuth. Some tilting levels do not fulfill this requirement.

At the top of the spindle is a bar that carries the micrometer screw and the opposing spring. Usually the micrometer screw is fitted with a graduated drum. Figures 2-11 and 2-12 show tilting levels.

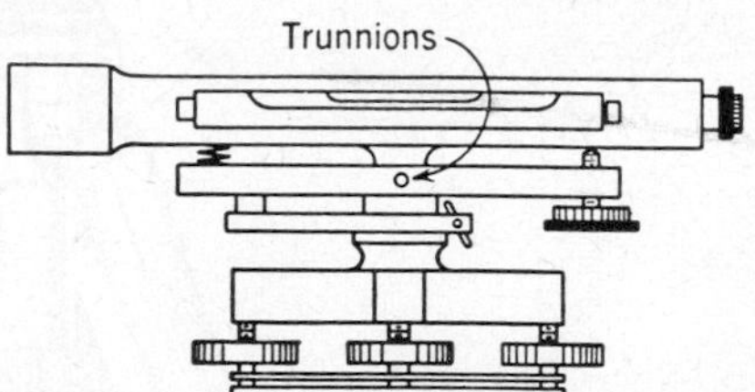

FIG. 2-10. Principle of a tilting level.

2-13. To Set Up a Tilting Level and to Take a Reading. The procedure for using a tilting level varies with its design. The following variations in design are usual: (1) three or four leveling screws, (2) circular level or none, and (3) trunnion axis intersecting spindle axis or not.

2-14. Three-screw Leveling Details. In every case the purpose of the main leveling screws is to set the spindle vertical. The use of four leveling screws is familiar to the reader. Three leveling screws are

FIG. 2-11. A Wild N3 Precise Level. (*Henry Wild Surveying Instruments Supply Co. of America.*)

Fig. 2-12. A typical tilting level. (*Keuffel & Esser Co.*)

extremely easy to use and several techniques are employed. Usually the instrument is first leveled along one pair of leveling screws in the ordinary way by turning both screws simultaneously in opposite directions. It is then leveled along a line at 90° to the previous line by using the third screw alone.

When the trunnion axis intersects the spindle axis, the spindle need be set only vertical enough to prevent excess leveling with the micrometer screw. The indications of the circular level are sufficiently accurate to obtain this result.

When the trunnion axis is in front of or behind the axis of the spindle, the instrument must be leveled accurately enough to prevent the trunnion axis from changing height by more than a negligible amount when the telescope is turned in azimuth.

In this case the spindle must be made vertical according to the indications of the main bubble. Whenever it is necessary to set up with the main bubble, as in this case, or when there is no circular bubble, it is very helpful to know the setting for the micrometer screw that will center the main bubble when the spindle is *exactly* vertical. This setting is called the **reversing point.** Usually a pencil mark is placed on the micrometer drum to show this position.

2-15. To Find the Reversing Point. Turn the micrometer screw to approximately the center of its run. Record the reading of the drum or, if there are no graduations, mark its position with a pencil. Center the main bubble with the leveling screws. Turn the telescope 180° in azimuth. Center the bubble with the micrometer screw and note the second drum reading or mark its position. The reversing point is the drum position that is halfway between these two positions. Mark the reversing point or record the reading.

The process should be repeated as a check. When the drum is set at the reversing point it should be possible to level the instrument so that the main bubble remains centered at all azimuths.

2-16. To Take a Rod Reading. Once the instrument is set up each rod reading is taken as follows:

1. Focus on the rod.
2. Bring the vertical cross hair exactly on the rod.
3. Center the bubble.
4. Read the rod.
5. Check the bubble.

It is important to remember that with a three-screw instrument, the leveling screws must not be touched once the first sight has been taken. The movement of any one of them will change the height of the instrument.

With a level vial of high sensitivity, the last movement of the bubble is so slow that it apparently creeps. The observer should wait a few seconds before taking a rod reading to assure himself that the bubble will remain centered.

2-17. Optical Devices. Since the leveling is perfected for each sight with the micrometer screw, there is no need for the levelman to move to the side of the instrument to use the leveling screws. It is of considerable advantage, therefore, to provide a means of viewing the bubble directly from the eye end of the telescope. This can be accomplished with a simple mirror, but better devices are usually provided. The U.S. Coast and Geodetic Survey geodetic level has a viewer that makes it possible to see the bubble with the left eye while the right eye observes the rod through the telescope. The device apparently brings the two ends of the bubble together by cutting out the center. The device is adjustable, so that if the bubble changes length the ends of the bubble can always be centered in the field of view (see Fig. 2-13).

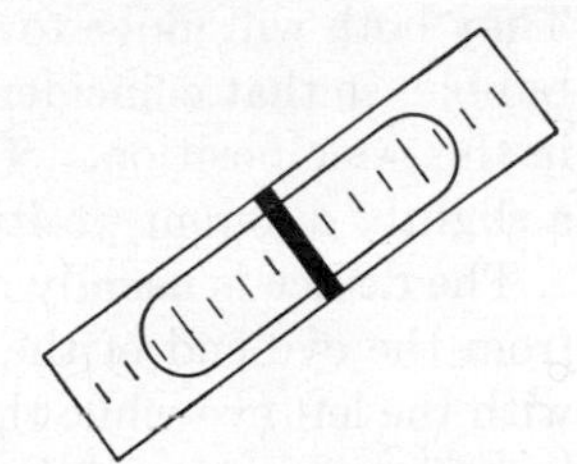

FIG. 2-13. Appearance of bubble in U.S. Coast and Geodetic Survey geodetic level as seen by the left eye.

2-18. Coincidence Bubble. An optical device can be arranged to split the bubble longitudinally and turn one end around so that it appears

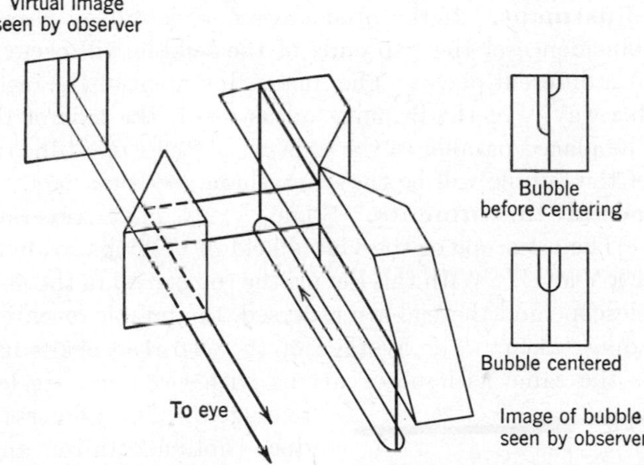

FIG. 2-14. Arrangement of prisms in a coincidence reading level. The arrow on the bubble indicates the direction that the bubble must be moved to be centered and therefore to cause coincidence. If the *prism assembly* is moved by the adjusting screw *in the opposite direction,* coincidence will also be obtained.

adjacent to the other end (note Fig. 2-14). When the bubble moves in leveling, one end, of course, actually moves toward the rest of the bubble and the other end moves away from it. In the optical device shown, the two ends accordingly appear to move in opposite directions. There is only one position for the bubble that will cause the two ends to appear to coincide. The instrument is leveled by causing this coincidence. If the bubble should change size the two ends of the bubble will actually move equal amounts in opposite directions (provided the vial is properly made). They both will move toward or both will move away from the rest of the bubble, so that coincidence will still occur when the instrument is leveled in the same position. The point of coincidence will, of course, appear in a slightly different position in the field of view.

The device is usually arranged so that the coincidence can be observed from the eye end of the telescope and frequently so that it can be seen with the left eye while the right eye can look through the telescope. The instruments shown in Figs. 2-8, 2-11, and 2-12 have coincidence bubbles. In some instruments the coincidence is observed *inside the telescope.*

The coincidence bubble presents several advantages:

1. No graduations are required.
2. One point is observed instead of two.
3. The bubble is set not by a *comparison* of the ends, but by *coincidence.*
4. Any movement of the bubble is apparently doubled.

Obviously the bubble can be centered very precisely with this device. Certainly the accuracy of setting is over twice that of conventional methods and perhaps considerably more.

2-19. Adjustment. If the prism assembly is moved parallel to the vial, the coincidence of the two ends of the bubble will occur when the bubble is at a different place. The final adjustment of the instrument is made in this way. In the primary adjustment, the axis of the bubble tube must be placed parallel to the axis of the telescope tube so that the two ends of the bubble will be the same shape (see Sec. 2-7).

2-20. Special Instruments. Some levels have **reversion vials** (Fig. 2-1). The telescope or the vial is held in bearings so that it can be rotated like a Y level. With this device the rod is read in the normal way; then the telescope and the vial are reversed, the bubble recentered, and a second reading taken. The average of the two readings is used. The accuracy is the same as if observations with two separate levels were averaged. The reversal process does not neutralize any instrumental errors.

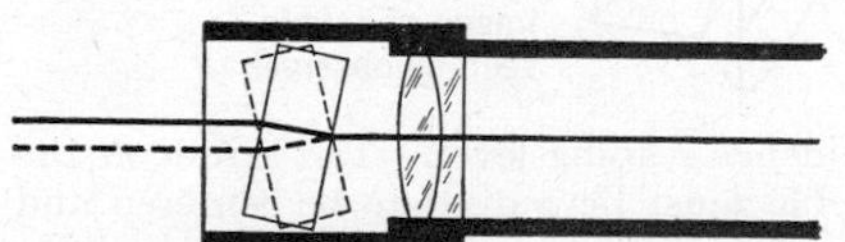

FIG. 2-15. Plano-parallel mounted in front of objective lens. It moves the line of sight up and down parallel to itself.

2-21. Optical Micrometer. A device called an **optical micrometer** or a **telemeter** (Fig. 2-15) can be either attached to the instrument (Fig. 2-16) or built into the instrument itself. A typical example of the device consists of a thick piece of optical glass with its two surfaces ground exactly flat and parallel (called a plano-parallel). It is mounted on a horizontal axis that is perpendicular to the line of sight. A graduated drum is used to rotate the glass on its axis.

When the plano-parallel is rotated, the line of sight moves up and down parallel to itself. The accuracy of this translation (that is, its lack of rotation) is dependent only on the accuracy of the grinding of the glass. It is independent of the mounting.

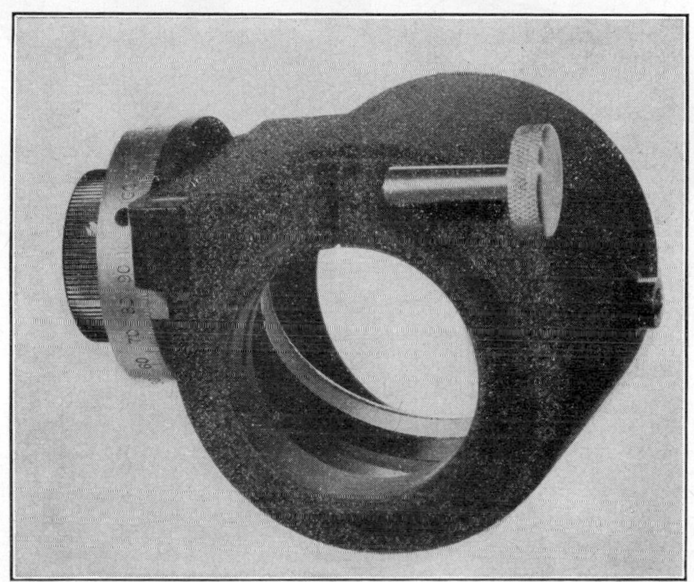

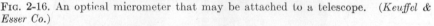

FIG. 2-16. An optical micrometer that may be attached to a telescope. (*Keuffel & Esser Co.*)

The drum is calibrated so that it gives the vertical movement of the line of sight.

When the level is pointed on the rod and leveled, the line of sight is moved to the lowest division that can be reached. The value of this division plus the reading of the drum gives the actual rod reading. No estimation is required.

2-22. The Pendulum Level. A fundamental innovation in design has been brought out by a division of the Zeiss Company. The instrument is called the Zeiss Opton Ni 2. It is a level that utilizes an inverted pendulum instead of a spirit level to give the direction of gravity. The pendulum is supported by four wires and operates a mirror prism that effectively keeps the images of points at the height of the instrument[1]

[1] Actually at the height of the anallactic point (see Sec. 7-37).

on the cross hair when the telescope is tilted. Thus the instrument is *automatic*. It is only necessary to level the instrument with a circular level to the point where the pendulum is free to operate. Thereafter, the line of sight is always kept horizontal by the pendulum.

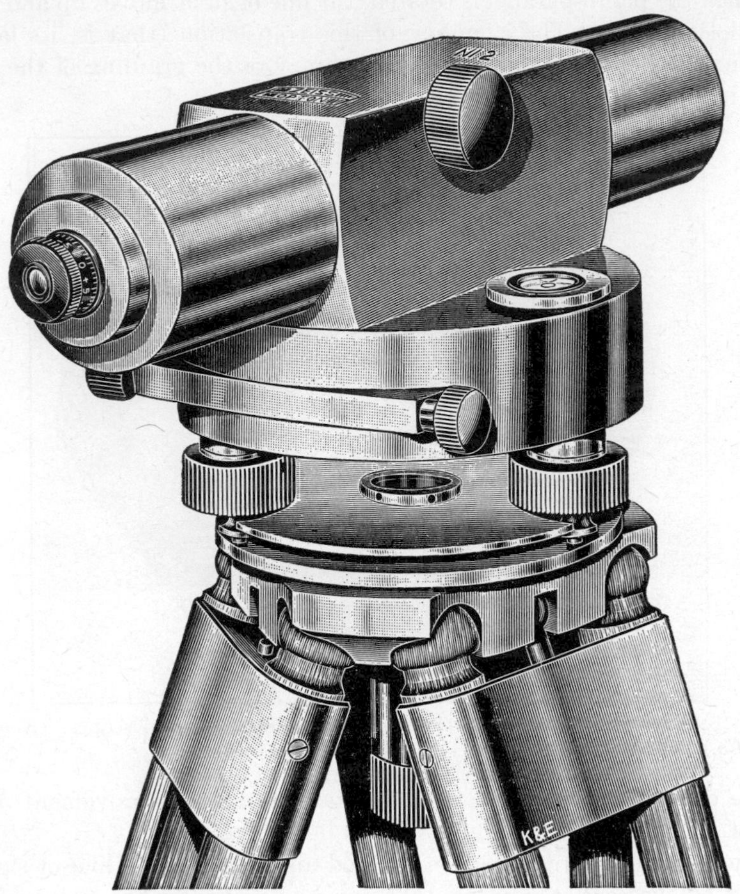

FIG. 2-17. A self-leveling level. The Zeiss Ni 2. (*Keuffel & Esser Co.*)

Figure 2-17 shows an engraving of the instrument. Figure 2-18 shows its operation schematically. The upper drawing shows the telescope horizontal. When the telescope is slightly tilted downward as shown in the lower view, the pendulum moves until its center of gravity is over the intersection of the lines of the wire supports. This moves the pendulum mirror so that the level line of sight is brought to the cross hair. The instrument is, of course, very fast and easy to use. Tests indicate that it is capable of high accuracy.

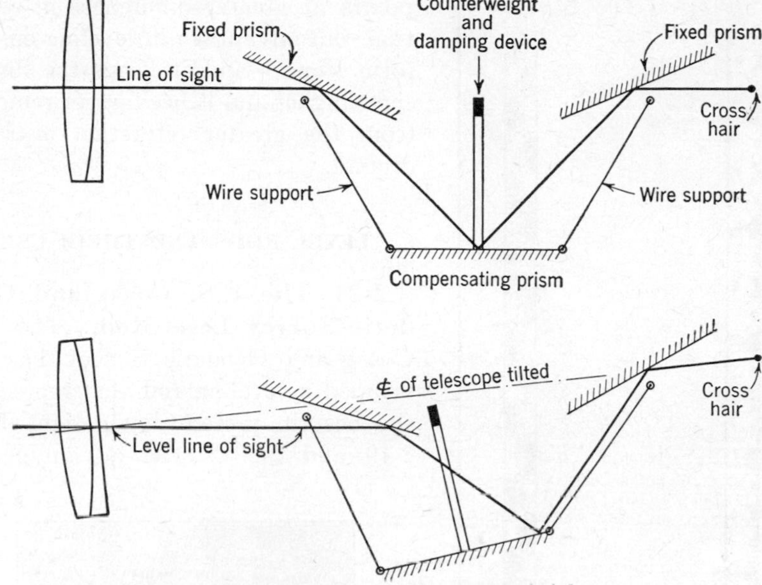

FIG. 2-18. When the telescope is tilted the line of sight remains level.

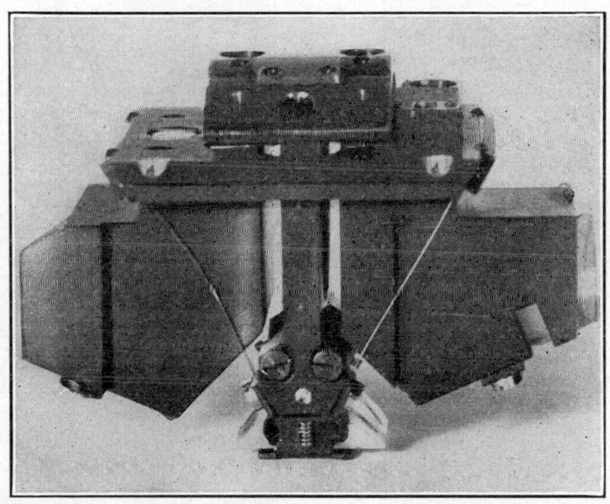

FIG. 2-18*a*. The compensator in the Zeiss Ni 2 level. (*U.S. Geological Survey and Keuffel & Esser Co.*)

2-23. Tripods. When an optical device to observe the bubble is provided, there is no need to look down on the bubble to read it. This affords an opportunity to use a high tripod so that the levelman can stand erect when observing. A high tripod makes it possible to observe turning

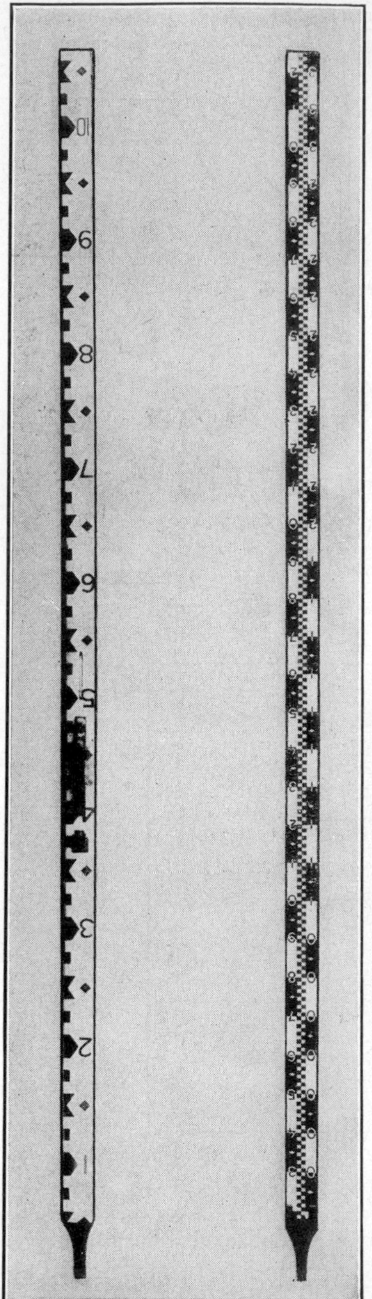

FIG. 2-19. U.S. Coast & Geodetic Survey precise level rod. (*U.S. Coast and Geodetic Survey.*)

points at greater differences in elevation—an advantage in leveling on uniform slopes—and it keeps the line of sight higher and hence farther removed from the greater refraction near the ground.

LEVEL RODS AND THEIR USE

2-24. The U.S. Coast and Geodetic Survey Level Rod. The U.S. Coast and Geodetic Survey has developed a leveling rod that represents the best in present-day design (Figs. 2-19 and 2-20). The advantages of

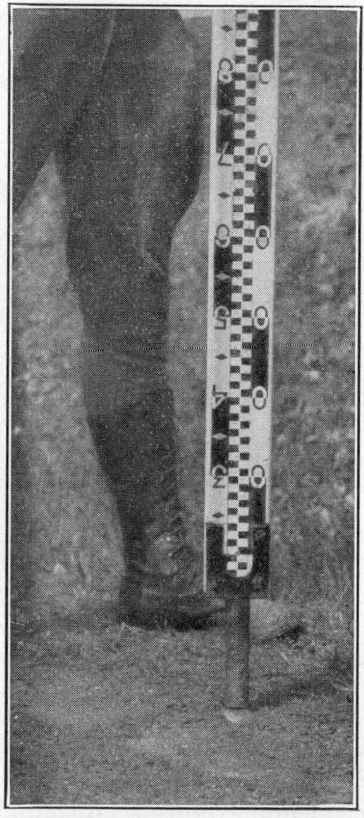

FIG. 2-20. Precise rod in use. (*U.S. Coast and Geodetic Survey.*)

this rod are obvious. A description of the rod is given here to serve as an illustration of the features that should be included in a rod if the maximum in leveling accuracy is desired.

The measuring element in the rod is a 25- by 2-mm strip of invar (coefficient of expansion about 0.000001 per degree Fahrenheit) graduated in centimeters and screwed and doweled to a solid-metal foot piece. A pine staff about 3½ in. wide and 10 ft 6 in. long, treated to prevent moisture changes, is screwed to the foot plate and supports the invar strip at its upper end through a stiff spring. The invar strip is loosely held in a shallow channel routed out throughout the length of the staff. It is graduated in the metric system. The centimeters are marked by a checkerboard pattern so that the cross hair always falls on a white space where its exact position can be most accurately estimated. The graduations extend to 3.25 m (10.7 ft) above the base of the foot piece. The numbers are placed on the staff. A thermometer is mounted in a slot so that it is in contact with the back of the invar strip and may be read from the back of the rod. A circular level is used to keep the rod vertical. The invar strip is divided under tension by a special device which refers the marks to the base of the foot plate.

The index correction, the temperature coefficient of expansion, and the length per meter at a standard temperature are determined by the U.S. Bureau of Standards for each individual rod.

The back of the staff rod is graduated in feet and is read only for a check.

The length of the rod has been carefully chosen. In benchmark leveling the length of sights is reduced below the desired values more often by slopes than by any other cause. The height of the instrument above the ground is the limiting factor. It cannot be conveniently greater than the height of the observer's eye, which averages about 5.4 ft. In order to utilize this height fully, the rod should be twice this value, or about 10.8 ft (3.3 m). Then, on a uniform slope, where the backsights and foresights are equal, the whole rod will be used. There are many advantages in longer rods but they are not nearly so important in benchmark leveling as in ordinary leveling; since each foot added to the length of the rod makes its transportation that much more difficult, the criterion described should control.

FIG. 2-21. A precise, engine-divided rod with graduations on a metal strip. It is divided into hundredths of a yard. (*Keuffel & Esser Co.*)

2-25. Commercial Rods. Precise rods with the necessary features may be purchased from instrument makers. The Keuffel & Esser Co. makes the precise rod shown in Fig. 2-21. It is engine-divided in hundredths of a yard on an invar strip which is firmly attached to the metal shoe by a special connection and supported at the top by a strong spring. A rod level can be attached and a thermometer can be placed so that its bulb is in contact with the back of the invar strip. The rod is used for three-wire leveling. Frequently, however, there is not enough control leveling for any project to warrant purchasing a special rod for the purpose. Fairly good results can be obtained by carefully calibrating ordinary rods at frequent intervals.

2-26. One-piece Wood Rods. One-piece wood rods with the graduations painted on the wood give very good results. They should be frequently calibrated against a steel tape (corrected for temperature), and they should be stored in unheated shelters to reduce changes in moisture content.

2-27. Metagrad[1] Rod. The Keuffel & Esser Co. manufactures a two-piece rod with the markings painted on a steel ribbon (see Fig. 2-22). It is arranged so that when properly operated it gives the same readings that would be obtained from a solid-steel rod. When temperatures are recorded, the actual values can be accurately computed.

Figure 2-23 shows how the rod operates. The principle is based on the assumption that, since the two parts of the rod are made of the same kind of wood and have been exposed to the same conditions, they expand and contract by equal amounts.

There are three positions in which a two-piece rod is used as shown in the views *a*, *b*, and *c*, respectively. In each case two schematic drawings have been made showing exaggerated changes in the size of the wood. One drawing shows the wood shrunk; the companion drawing shows it expanded. It can be seen that because of the way the steel strips and the back index mark are connected, the change in length has no effect on the steel so long as the back index is set for 7 ft for low rod and 13 ft for high rod.

In view *a* note that the 7-ft mark remains at the same elevation when the wood is expanded or shrunk, and that the back of the rod always reads 7 ft.

In view *b* note that the reading is continuous up the front of the rod no matter what may be the size of the wood, so long as the rod is set at 13 ft.

In view *c* note that at any particular setting of the back of the rod (in this case 10 ft), the target will be that distance above the foot of the rod no matter what the size of the wood may be.

2-28. Two-piece Wood Rods. The accuracy of leveling with an ordinary two-piece wood rod can be considerably improved by the following precautions:

1. Store the rod under shelter but where the open air can reach it. This will preserve a more uniform moisture content.

2. Calibrate the rod with a steel tape corrected for temperature. The values required are the average length of a foot and the setting for high rod that most nearly eliminates an index correction for the upper position.

[1] Registered U.S. Patent Office.

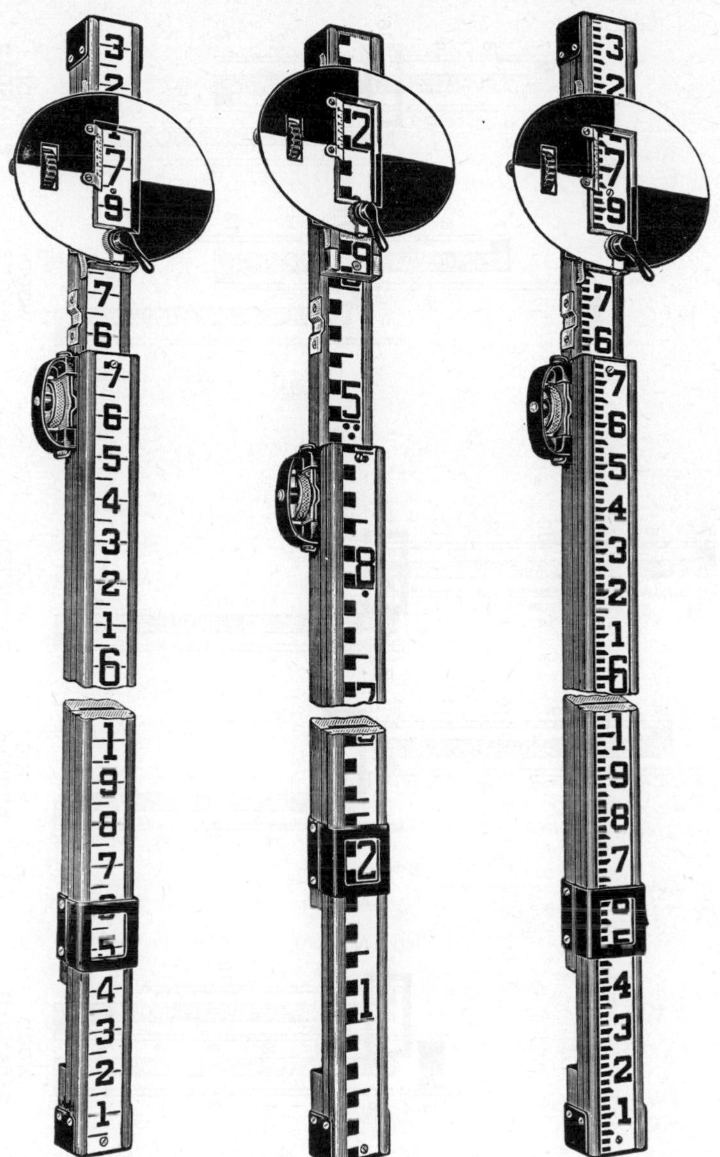

FIG. 2-22. Metagrad Rods. The readings are the same as though the rod were solid steel. (*Keuffel & Esser Co.*)

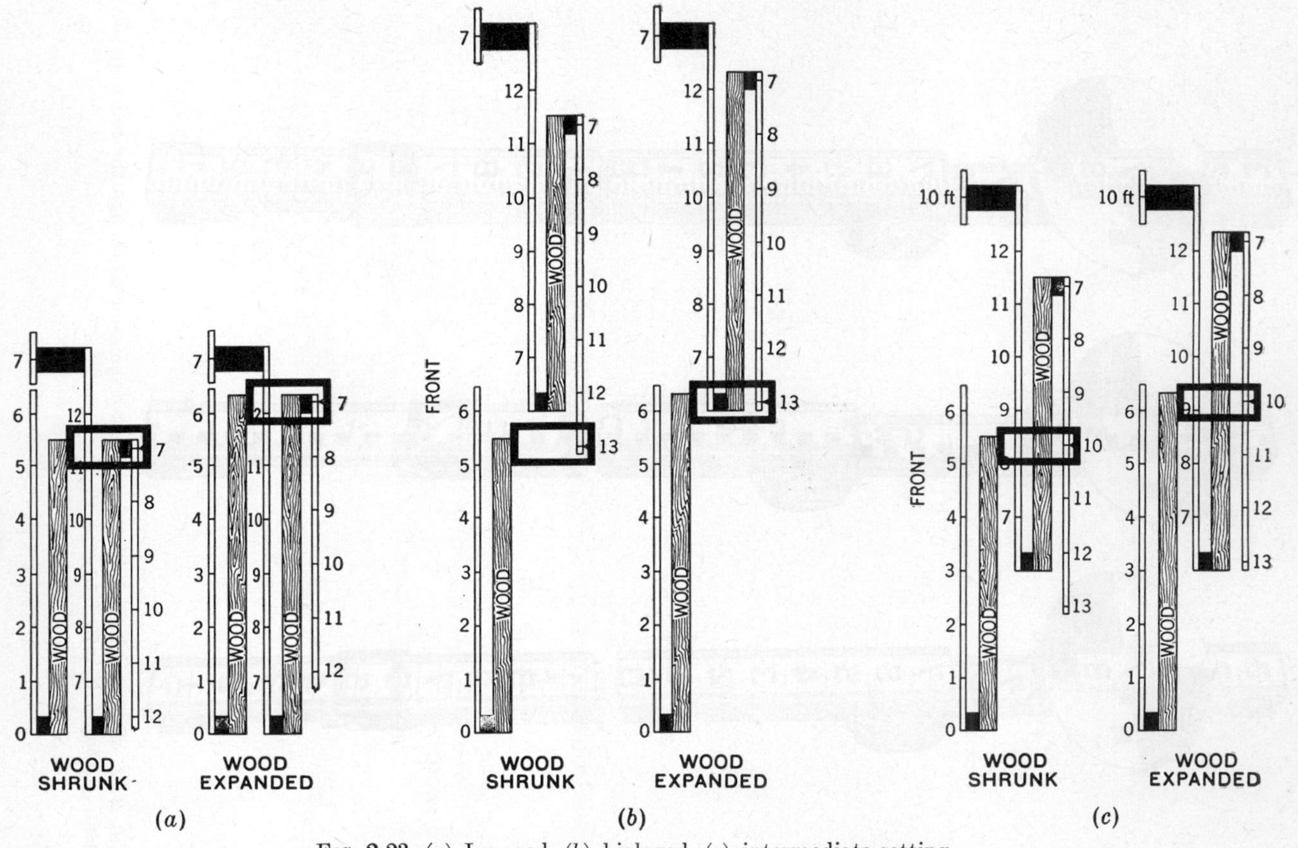

FIG. 2-23. (a) Low rod, (b) high rod, (c) intermediate setting.

For example, assume the following values, measured on the face of the rod with the rod fully extended:

	Readings	Difference	Actual length
Lower section.........	0.100– 6.700	6.600	6.607
Upper section.........	6.700–12.700	6.000	6.009
Sums................	....	12.600	12.616

$$\text{Average length per foot of rod} = \frac{12.616}{12.000} = 1.00127$$

The rod should then be set so that the distance between the 0.100-ft mark and the 12.700-ft mark is exactly 12.616 ft. When this setting is made, the rod can be treated exactly like a one-piece rod. This assumes that the difference between the values of the length per foot in the two sections of the rod are due to irregularities in marking the rod rather than to constant errors. The index reading at the back of the rod at this setting is recorded and used to set high rod until the rod is calibrated again.

2-29. Method of Applying Rod Corrections. Before the actual procedure for applying rod corrections is described, the fundamental principle should be explained. If the rod is uniformly in error throughout its length so that the error in any reading is proportional to the value of the reading, then the *difference in elevation* between two benchmarks determined with this rod will be affected by the error in exactly this same proportion. This is true no matter how much the intervening turning points may have carried the level line up or down in elevation.

For example, assume that the length per foot of the rod is 1.002 ft and the difference in elevation between two benchmarks 10 miles apart was found to be +50 ft when this rod was used. The corrected difference would be computed as follows:

$$50 \text{ ft} \times 1.002 = 50.100 \text{ ft}$$

2-30. Application of Rod Corrections. There are three rod corrections that can be applied to the determination of a difference in elevation:

c_i = correction for rod-index error
c_r = correction for rod length
c_t = correction for rod thermal expansion

c_i is computed by adding the rod-index correction once for each time a rod is used for a plus sight and subtracting it once for each time the rod is used for a minus sight. Obviously, when only one rod is used, the value of c_i is zero. When two rods are used and properly alternated, it is also zero.

c_r is computed by first finding the correction per nominal foot from the standardization data and multiplying the correction per foot by the difference in elevation.

c_t is computed from the coefficient of expansion of the rod, as shown in the following section.

Example. Length of rod from end to 0.5-ft mark, 0.502; index correction = +0.002 ft.

Length of rod from 0.5-ft mark to 9.5-ft mark, 9.007 at 68°F. Rod length = 1.000778 ft per nominal foot at 68°F.

Coefficient of expansion = 0.00000645 per degree Fahrenheit.

Field data:

Mean difference in elevation, 50 ft

Temperature average, 88°F

$$c_i = 0 \qquad\qquad 0$$
$$c_r = 50 \times 0.000778 = +0.0389$$
$$c_t = 50 \times 0.00000645(88 - 68) = +0.0064$$

Total correction = +0.0453

Corrected difference in elevation 50.045

2-31. Use of Two Rods. Considerable time can be saved when two rods are used. One can be held on the back turning point and one on the forward turning point for each instrument set up. The delay occasioned by moving a single rod forward after the plus sight is taken is thus eliminated. If, between benchmarks, each rod is used as many times for the plus sight as for the minus sight, the index corrections are eliminated and the remaining rod corrections can in each case be the average of like corrections for the rods.

It is considered best practice to alternate the rods at each setup. For example, rod No. 1 might be used for the plus sight at the first setup and then for the minus sight for the next setup, etc. A still more rapid method is to keep one rod forward for half of the distance and the other rod forward for the remainder of the distance between benchmarks. When there is an odd number of setups, the same rod is used for both sights on the last setup.

2-32. Turning Points. When good turning points cannot be found, stakes, foot pins, or foot plates are used. A hammer must be carried to drive stakes or foot pins. Footplates (Fig. 2-24) are suitable where masonry is available to support them, as along a concrete highway pavement. They always present the danger that they may be unwittingly disturbed. Carefully chosen high points in the masonry are not so precise, but they are safer.

FIG. 2-24. Foot plate.

2-33. Benchmarks. Typical benchmarks are chisel cuts or bronze disks with shanks that may be grouted in place. These are placed on the top of natural rocks, at the base of masonry or steel structures or on concrete monuments set for this purpose. They are often spikes in trees, poles or wooden structures. Usually the top of the mark is the height for which the elevation is recorded, but these surfaces are subject to wear. The most permanent benchmarks are placed on vertical surfaces, with a line to mark the exact height, but these are difficult to use.

If a benchmark is to be permanent, it must be supported by a structure that reaches below frost and that has ceased to settle. The foundations of old buildings that have permanently settled are excellent. But retaining walls, including street curbs, are dangerous as they move for years under lateral earth pressures. Any structure that resists lateral pressure is questionable. These include bases for individual railroad signals or poles that support wire.

Important benchmarks should, if possible, be established in groups of three, all visible from a single instrument setup. Movement of any one can then be recognized.

The markers used to indicate that a point is a benchmark should display numbers or letters for identification and record. When the final elevation is determined, it should be stamped on the mark.

2-34. Descriptions of Benchmarks. As soon as a benchmark is established a description of it should be recorded. The description should sometimes state how to reach the benchmark from some well-known landmark such as a city hall or an important street or highway intersection. The following information should be stated, in this order:

1. Name
2. General locality

3a. Position with respect to surrounding, easily recognized objects, preferably with measured distances

3b. Position by plus and offset from a traverse line

4. The object upon which it is placed
5. Exactly what it is
6. Identifying marks
7. Elevation (when determined)

Example 1. W2 Princeton, Mercer Co. N.J. On the campus of Princeton University, at the main entrance on the north side of Nassau Hall, 7 ft. east of the center of the entrance, 8 inches out from the wall, and 2.33 ft. from the base of the bronze tiger on the east side of the entrance, set in the capstone of the landing. A standard U.S. C. & G.S. B.M. disk stamped W-2 1924. Elevation 218.622 ft.*

Example 2. Mon 177 Princeton, N.J. At the intersection of Nassau Street and Washington Road, 17 ft. south of the south curb line of Nassau Street, 9 ft. west of the west curb line of Washington Road, 3 ft. northeast from iron fence, set in a

* Adapted from a description by the U.S. Coast and Geodetic Survey.

concrete monument flush with the ground. A standard U.S. C. & G.S. & S.S. disk stamped 177. Elevation 193.953 ft.†

ERRORS

2-35. Systematic Errors. The chief systematic errors in leveling and the methods of correcting for them have already been described. There are two minor types that should be eliminated as much as possible.

Within a few feet above the ground where level lines are run, there often exist considerable variations in the refraction coefficient of the air. When levels are run on a slope, the uphill sights will be consistently nearer the ground than the downhill sights and thus pass through air having different refraction coefficients. Refraction variations occur where evaporation is rapid, when the air temperature is changing, when the ground is at a different temperature from the air, and when "air boiling" is evident. It is well to avoid leveling on continuous slopes under these conditions.

H. S. Rappleye mentions[1] that while it is difficult to see how large errors could accumulate, systematic errors may be caused by differences in illumination on the rods. When two rods are used, it is well to read the forward rod first at every second setup to compensate for possible settlement of the instrument.

2-36. Balanced Sights. The horizontal lengths of the plus and minus sights from each instrument position must, of course, be equal. This eliminates the effect of any residual error in the instrument adjustment, the effect of earth curvature, and the effect of constant refraction. When the stadia intercept is determined for each sight, the exact value of any unbalance is known and corrections for these effects can be applied.

2-37. Accidental Errors. There are three major sources of accidental errors in leveling: (1) centering the bubble, (2) reading the rod, and (3) variations in the refraction of the air. The effect of the first two of these types of errors is obviously proportional to the length of the sight. The effect of the third certainly increases with the length of the sight. The exact relationship cannot be determined as it depends on how near the instrument these variations in refraction occur. As the effect usually is small, it is assumed to vary with the length of sight.

In analyzing leveling errors, since they all behave in the same way, it is of little use to consider them separately. It is better to treat the result of all three of them as a single value.

In analyzing the errors in a line of levels it is necessary to assume that the sights are very nearly the same length throughout the line.

† Adapted from a description by the N.J. Geodetic Control Survey.

[1] U.S. Coast and Geodetic Survey Special Publication 239, "Manual of Geodetic Leveling," p. 37., 1948.

Let E_r = combination of all three accidental errors in one sight per unit length of sight

E_s = error of a complete instrument setup

l = length of the sights

Then by Eq. (21-23),

$$E_s = lE_r\sqrt{2} \tag{2-1}$$

Let E = error in a line of levels

n = number of setups

L = length of the line of levels

Then by Eq. (21-23),

$$E = E_s\sqrt{n} \tag{2-2}$$

But

$$n = \frac{L}{2l}$$

Substituting in Eq. (2-2),

$$E = lE_r\sqrt{2}\sqrt{\frac{L}{2l}}$$

$$E = E_r\sqrt{Ll} \tag{2-3}$$

Equation (2-3) expresses the behavior of accidental errors in leveling. It shows that the errors increase with the square root of the length of the line and also with the square root of the length of the sights, despite the fact that with longer sights fewer setups are required.

This equation explains why the accuracy in leveling is measured in terms of the square root of the length of the line; it gives a basis for estimating the length of sight that should be used; and it provides a means of estimating the accuracy that any level instrument will produce in an actual line of levels.

Example 1. Assume that it were necessary to decide what should be the maximum length of sight for a given level to obtain second-order accuracy.

There are a number of ways to test the level. A simple, though somewhat inconclusive method, is used for illustration. Thirty or forty readings on one rod position were taken from a single instrument setup. The errors were

$$\sqrt{\frac{\Sigma v^2}{n-1}} = 0.00001743 \text{ ft}$$

where the v values are the *unit errors*, that is, each error in the rod readings was divided by the length of the sight. From Eq. (1-1),

$$E_{90r} = 0.00002867 \text{ ft}$$

From Table 1-2, for second-order leveling

$$E = 0.025\sqrt{M} \qquad \text{ft}$$

where M is the number of miles leveled. From Eq. (2-3),

$$0.025\sqrt{M} = 0.00002867\sqrt{M \times 5280 \times l}$$

$$l = 144 \text{ ft}$$

Example 2. Assume that actual level runs showed consistent closures of $0.022\sqrt{M}$ ft. What length sights can now be safely used? From Eq. (2-3),

$$\sqrt{\frac{l}{144}} = \frac{0.025}{0.022}$$
$$l = 186 \text{ ft}$$

Example 3. What would be the percentage of reruns if the level in Example 1 were used for second-order leveling with 200-ft sights? From Eq. (1-1),

$$E_p = 0.00001743C$$

From Table 1-2,

$$E = 0.025\sqrt{M} \qquad \text{ft}$$

From Eq. (2-1),

$$0.025\sqrt{M} = 0.00001743C\sqrt{M \times 5280 \times 200}$$
$$C = 1.40$$

From Table 21-3 or Chart I, the desired accuracy would be reached in 84 ± % of the runs. Therefore, there would be 16 ± % reruns.

SINGLE-WIRE PRECISE DIFFERENTIAL LEVELING

2-38. Refinements for Precision. The method of differential leveling with an engineer's level and the form of notes used are given in "Surveying," Chap. VIII. Three modifications of this method are required to obtain precise results.

1. The rod must be calibrated (see Secs. 2-25 ff.) and preferably equipped with a rod level.

2. The level instrument should, preferably, have a vial sensitivity of better than 10 seconds per 2 mm for direct reading, or 20 seconds per 2 mm for coincidence reading.

Whatever instrument is used, the length of the sights must be limited according to its capabilities (see Sec. 2-37).

3. Certain specifications must control the results accepted throughout the work.

2-39. Specifications for Single-wire Precise Leveling

1. A **section** is the leveling between two permanent benchmarks. It should not be greater than 2 miles in length.

2. Each section must be leveled in two directions.

3. Each section must close within the order of accuracy required.

4. The accumulated errors of closure at any intermediate point should never exceed the limit for the line.

5. When the limit of error is exceeded in a section, further runs should be made until the closure between the average of the forward runs and the average of the backward runs shall be less than the required closure.

6. A run must be rejected when it differs from the average of all the runs by more than 1.5 times the required closure. No run may be rejected that differs by less.

Example. Figure 2-25 shows the results of a line of levels held to second-order accuracy. The field notes for the forward run for the first section (Mon 3100 to

BM LEVELING		Sect. Mon 3100-BM 30				⊼ Smith Rod Jones	Date Overcast Rod 89° F
Sta.	+	H.I.	−	Rod	Elev.		
Mon 3100	1.023	269.347			268.324		Level K+E 13892
TP 1	2.103	260.090	11.360		257.987	$C_t = (89-68)(36.5)(.00000645)$	
TP 2	0.742	250.365	10.467		249.623	= +.005	
TP 3	1.034	239.635	11.764		238.601		
BMA			7.784		231.851	Diff −36.473	
	4.902		41.375			Diff −36.473	

LEVEL LINE ADJUSTMENT (2nd ORDER) Mon 3100 − BM 30						Date	
Section	Length in Miles Max. Er.	Observed Differences	Divergence of B Runs Part.	Acc.	Mean Difference	Cor. for Rod	Elevations Corrected Differences
Mon 3100	0.2	F −36.473					268.324
BM A	.016	B +36.468	+5	+5	−36.470	+.005	−36.475
BM A	0.3	F −10.647					231.849
BM B	.019	B +10.668	(−21)				
		F −10.645	(−22)				
		B +10.641	−8	−3	−10.650	+.001	−10.651
BM B	0.5	F +18.239					221.198
BM C	.025	B −18.212	(−27)				
		F +18.216	−16	−19	+18.220	+.003	+18.223
BM C	0.3	F +16.035					239.421
BM D	.019	(B −16.098)	(+63)				
		F +16.040					
		B −16.038	0	−19	+16.038	+.002	+16.040
BM D	0.4	F −57.029					255.461
BM 30	.022	B +57.019	+10	−9	−57.024	+.008	−57.032
Total	1.7						198.429
	.046						

FIG. 2-25

B.M. A) are given at the top. The difference in elevation in the field notes is computed both by the algebraic sum of the plus and minus sights and by the difference in the elevations. The results, in each case, are −36.473.

It is assumed that a steel-faced rod was used that is correct in length at 68°F. $c_t = +0.005$. The plus sign indicates that the absolute value of the difference should be increased. If the temperature is quite different for the various runs for the section, c_t should be applied to each run separately. Otherwise it can be applied, as shown, after the mean difference is determined.

The main part of Fig. 2-25 shows the form for the adjustment of a level line. The entries for each column are described in the following paragraphs under the title for the column.

Section. The names of the benchmarks at the two ends of each section are given in the order of the forward direction.

Length in Miles. The length in miles of each section is recorded.

Maximum Error. The limit for second-order leveling for twice the length of the section is computed from Table 1-1 or taken from Table I. Twice the length of the section is used as the leveling is run twice (forward and backward).

Observed Difference. The observed difference for each run is taken from the field notes. F and B are used to designate forward and backward runs.

Divergence of B Runs. Partial. The closures are computed by determining the divergence of the B run from the F run. For example, in section Mon 3100 to B.M. *A*, the substitution of the B run (with its sign changed) for the F run would cause B.M. *A* to have an elevation 0.005 ft greater. Hence the divergence of the B run is +0.005 ft. It is written in units of one-thousandth of a foot.

The divergence takes the sign of the smaller difference except when both differences have the same sign. This can occur only when the differences are nearly zero. In every case the divergence takes the sign of the correction that is applied to the F run to obtain the average of the two runs.

The computations of the divergences for three of the sections are given in Computation 2-1.

COMPUTATION 2-1

Mon 3100	F −36.473		
B.M. *A*	B +36.468		
	Diff. +0.005		
B.M. *A*	F −10.647	F −10.647	F −10.647
B.M. *B*	B −10.668	F −10.645	F −10.645
	Diff. −0.021	Av. F −10.646	Av. F −10.646
		F −10.646	B +10.668
		B +10.668	B +10.641
		Diff. −0.022	Av. B +10.654
			F −10.646
			B +10.654
			Diff. −0.008
B.M. *C*	F +16.035	F +16.035	
B.M. *D*	B −16.098	F +16.040	
	Diff. +0.063	Av. F +16.038	
		B −16.098	F +16.038
		B −16.038	B −16.038
		Av. B −16.068	Diff. 0
		Av. of all four 16.053	
		Reject −16.098	

The differences that are rejected and the divergences that are not used are circled in Fig. 2-25. Note that the first B run between B.M. *C* and B.M. *D* (−16.098) was rejected as it differed from the average of all four differences by 0.045 ft. This is greater than the rejection limit of 1.5(0.019 ft) = 0.029 ft.

Divergence of B Runs Accumulated. The accumulated divergence is the running total of the partial divergence. When it shows a continuous trend, some systematic error is present. When it becomes larger than the limit of 0.046 for the whole line, a careful analysis of the conditions existing when the line was run should be made. Those sections should be rejected where such systematic errors are suspected.

Mean Difference. To compute the mean difference, find the average of all the accepted forward differences and also find the average of all the accepted backward differences with their sign changed. The mean of these two averages is the mean difference. For example,

B.M. *A* to B.M. *B*:

Av. of F differences..........................	−10.646
Av. of B differences with sign changed........	−10.654
Mean..	−10.650

B.M. *B* to B.M. *C*:

Av. of F differences..........................	+18.228
Av. of B differences with sign changed........	+18.212
Mean..	+18.220

Corrections for Rod. These are taken from the field notes if they have not already been applied.

Elevations and Corrected Differences. These are computed as shown.

UNBALANCED SIGHTS

2-40. Corrections for Curvature and Refraction. When a perfectly adjusted instrument is leveled, the line of sight is tangent to a level

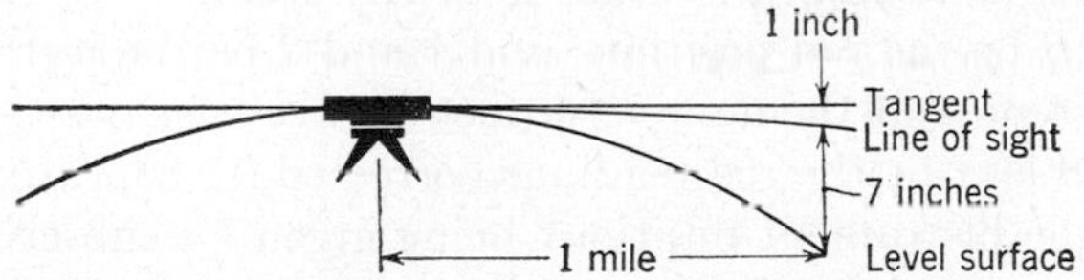

FIG. 2-26. Effect of curvature and refraction.

surface at the instrument. Because of the curvature of the earth, this level surface falls away from the tangent approximately 8 in. in 1 mile. The refraction of the curved atmosphere bends the line of sight downward an average of 1 in. in 1 mile. The combined effect of the two causes the level surface to fall away from the line of sight 7 in. at 1 mile. The distance varies nearly as the square of the distance (see Fig. 2-26). This may be written

$$C_c = -0.00000002092d^2$$

where C_c = correction to rod reading for combined curvature and refraction, ft

d = distance from instrument to rod, ft

Values for C_c based on this formula are given in Table II. When a plus sight and a minus sight are very different in length, each must be corrected by C_c.

2-41. Instrumental Error (Fig. 2-27). No matter how carefully a level may be adjusted, a residual error of adjustment will remain. Also, its value may change from time to time. When a plus sight and a minus sight are very different in length, a correction for this error should be applied. The effect of this error on a rod reading varies with the distance from the instrument.

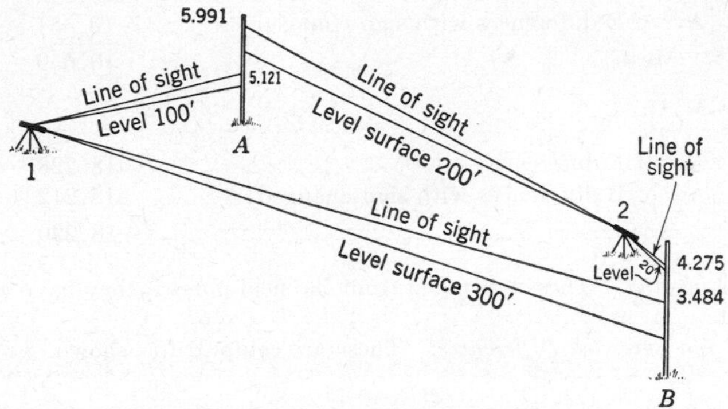

FIG. 2-27. Determining the instrument correction.

To determine the correction, choose two firm objects 200 to 300 ft apart. Take a rod reading on both from two instrument positions, one near one of the marks and the other near the other.

Let A and B be the rod positions and 1 and 2 be the instrument positions. Let d represent the number of feet between the positions given by subscripts and let r be the rod readings corrected for curvature, the mark sighted and the instrument positions being given by subscripts. Let C_i represent the correction per foot of distance to be applied to each rod reading. The true difference between the rod readings can be expressed by the readings from each instrument position:

$$\begin{aligned}\text{True diff.} &= r_{1A} + d_{1A}C_i - (r_{1B} + d_{1B}C_i) \\ &= r_{2A} + d_{2A}C_i - (r_{2B} + d_{2B}C_i)\end{aligned}$$

Equating and solving for C_i,

$$C_i = -\frac{(r_{1A} - r_{1B}) - (r_{2A} - r_{2B})}{(d_{1A} - d_{1B}) - (d_{2A} - d_{2B})} \tag{2-3}$$

Example 1. Refer to Fig. 2-27 and assume the rod readings are as given below:

Inst. pos.	Rod A				Rod B			
	Dist. d	Actual read.	Curve cor.	r	Dist. d	Actual read.	Curve cor.	r
1	100	5.121	0	5.121	300	3.484	−0.002	3.482
2	200	5.991	−0.001	5.990	20	4.275	0	4.275

$$C_i = \frac{(5.121 - 3.482) - (5.990 - 4.275)}{(100 - 300) - (200 - 20)}$$

$$C_i = -0.0002$$

Thus a rod reading of 5.628 on a rod 200 ft from the instrument would be corrected as follows:

$$5.628 - (0.0002)(200) = 5.588$$

When, in a line of levels, the lengths of the backsights and the foresights are measured, the final elevation can be corrected for the instrumental error by applying the instrument correction. The correction is multiplied by the sum of the lengths of the plus sights minus the sum of the lengths of the minus sights and applied to the observed elevation.

Example 2

Sum of lengths:

Plus sights	2300 ft
Minus sights	2000
Difference	300 ft

Elevations:

Observed elevation	203.741 ft
300 × (−0.0002)	−0.060
Corrected elevation	203.681 ft

2-42. Reciprocal Leveling. When it is necessary to carry leveling over a river, ravine, or any obstacle requiring a long sight, special methods must be used to obtain accuracy and to eliminate the effect of the following: (1) error in instrument adjustment, (2) combined effect of the earth's curvature and the refraction of the atmosphere, and (3) variations in the average refraction.

Accuracy can be obtained by repeated observations. Instrumental error and curvature and refraction can be eliminated by applying the corrections described in the last two sections. But more accurate results can be had by arranging the work so that these effects are eliminated automatically. The levels can be carried from a mark on one side of the obstacle to a mark on the other side by finding the difference in elevation from two instrument positions, one near each mark. The average is the true difference. This method can be used for distances up to 500 ft.

If possible, two levels should be used and the observations made simultaneously so that the conditions of refraction will be the same for observations in both directions. A second set of observations should be made with the instruments interchanged and the results averaged to eliminate the errors of instrument adjustment (see Fig. 2-28).

When the distance is greater than 500 ft, the following method[1] is recommended. Two rods and two levels should be used and each rod should be equipped with two targets. Because of methods of manufacture or wear on the shoe of the rods, the zero of the graduations may not coincide exactly with the bottom of the shoes. Readings should be taken

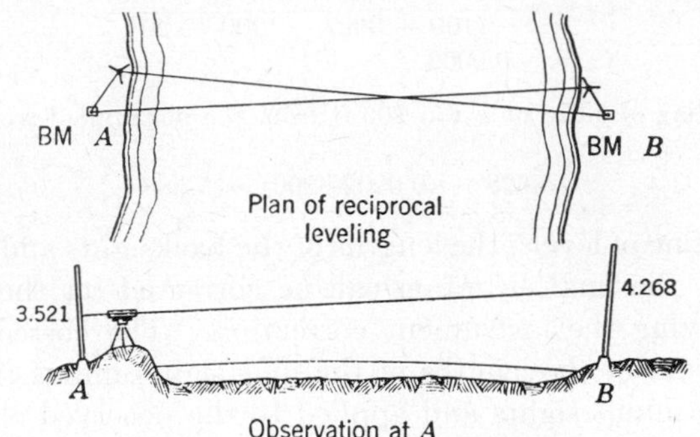

Fig. 2-28. Reciprocal leveling.

on both rods held on the same point from a level set up as near as possible. If there should be a difference, a correction must be applied to the readings of one of the rods.

2-43. Procedure for Precise Reciprocal Leveling. Choose a location for the work where the lines of sight will be high enough above the ground to avoid any large and irregular refraction effects that may exist near the ground. Establish a benchmark at each side of the obstacle to be crossed. Guy the rods permanently over both benchmarks. Set up near benchmark A, and read and record a plus sight on A. Pointing at B across the river with the instrument level, direct the rodman by signals to bring one target approximately on the line of sight. When this is accomplished the rodman at B sets the two targets as close together as convenient so that the line of sight will strike somewhere between them.

Method for an Engineer's Level

With the leveling screws, point on the upper target and read the position of the two ends of the bubble. To take each reading, count the number of graduations from the zero mark on the bubble vial to the end of the bubble, estimating the final tenths (see Fig. 2-29). Take the direction toward the objective lens as plus. The average of the readings of the two ends of the bubble will give the number of graduations the bubble has shifted from its center position. Repeat, pointing on the lower target. The observations require very little time. At least five pointings should be made

[1] This method is a modified form of the procedure described by Henry G. Avers in U.S. Coast and Geodetic Survey Special Publication 140, "Manual of First-order Leveling," pp. 33–35, 1935.

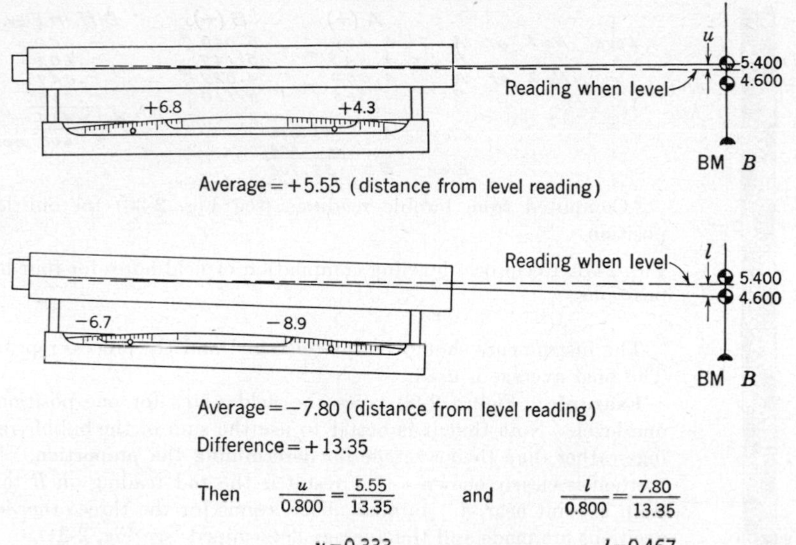

FIG. 2-29. Reciprocal leveling. Method of reading bubble and computing proportion.

Reciprocal Leveling		BM−A to BM−B			
Rod on	A	4.420		Level	No 1
		Ends of	Bubble		
Pointing	Upper	Target	Lower	Target	on B
	+6.8	+4.3	−6.7	−8.9	
	+6.9	+4.8	−7.0	−9.1	
	+6.9	+4.9	−6.8	−8.9	
	+7.0	+5.0	−7.0	−9.0	
	+7.0	+5.0	−6.9	−8.8	
Sums	+58.6		−79.1		
Difference		137.7			
Targets on B		5.400	4.600	Diff	.800
Proportion		+58.6 / 137.7	× .800	= +.340	
		−79.1 / 137.7	x .800	= −.460	
				.800	ck.
Targets on B		5.400		4.600	
Subtract		+ .340		− .460	
Rod on B		5.060		5.060	ck.

FIG. 2-30. Reciprocal leveling. Example of field notes for one level position.

alternately on each target. When these are complete, check the reading on *A*. The reading on rod *B* when the bubble is centered can be now computed by proportion, using the shifts of the bubble from center and the positions of the targets on the rods.

While the observations are being made at *A* the same observations should be carried on at *B*, arranging the work so that both observers begin and end at nearly the same time in order to obtain the same conditions of refraction.

		A (+)	B (−)	Diff. in Elev.
Level No. 1 at	A	4.420	5.060 #	−.640
	B	4.545 #	5.147	−.602
" No. 2 at	A	4.382	4.829 #	−.447
	B	4.365 #	5.110	−.745
				4) 2.434
				− .608 aver.

Elev	A	40.316
		− .608
Elev	B	39.708

[#] Computed from bubble readings (see Fig. 2-30) for one level position.

Fig. 2-31. Reciprocal leveling compilation of field notes for four level positions.

The instruments should be interchanged and the process repeated. The final average is used.

Example. Figure 2-30 gives the field notes for one position of one level. Note that it is better to use the sum of the bubble readings rather than their average for determining the proportion. The method is clearly shown. The result is the rod reading on B taken from a point near A. Similar observations for the three other level positions are made and the average determined (see Fig. 2-31).

The accuracy of this method is limited by the uniformity of the vials. If they are not ground to a perfect arc, the method has a tendency to cancel the error. Second-order accuracy should be obtained.

Method for Tilting Level

2-44. Use of Drum Readings. When a tilting level, equipped with a graduated drum on the micrometer screw, is available, the drum readings are taken instead of the bubble readings. A single observation consists of taking the drum readings as follows:

1. When the center cross hair is on the top target
2. When the bubble is centered
3. When the center cross hair is on the lower target

About 25 or 30 observations are taken.

Assume the following average drum readings on targets set at 5.400 and 6.200, respectively:

On top target.............. 12.5
Bubble centered............ 23.6
On lower target............ 47.3

Then $$\frac{47.3 - 23.6}{47.3 - 12.5} \times (6.200 - 5.400) = 0.545$$

and $$5.400 + 0.545 = 5.945 \text{ rod reading}$$

Figure 2-32 shows a U.S. Coast and Geodetic Survey rod with targets for a reciprocal leveling.

Fig. 2-32. Geodetic leveling rod with river-crossing targets in position. (*U.S. Coast and Geodetic Survey.*)

PROBLEMS

2-1. Assume the following calibration data for two steel-faced rods; use 0.00000645 for the coefficient of expansion per degree Fahrenheit

	No. 1	No. 2
Length at 70°F from 0 to 0.500 ft.	0.494	0.498
Length at 70°F from 0.500 to 10.500 ft.	10.003	9.995

Ten setups were made with No. 1 rod used for all plus sights and No. 2 rod for all minus sights. Average temperature 40°F given, elevation B.M. *A*, 67.362. Observed elevation B.M. *B* 92.487. Compute elevation B.M. *B*.

2-2. Write a description of a hypothetical benchmark.

2-3. In order to obtain a quick indication of the precision of a certain level, it was set up 150 ft from a rod and eight rod readings were taken. The instrument was releveled between each reading. The readings obtained are as follows: 4.552, 4.542, 4.545, 4.544, 4.537, 4.536, 4.530, 4.548. (*Note:* $\sqrt{5280} = 72.66361$.)

a. What is the value of σ?

b. What is the value of σ_r, the standard unit error?

c. What is the value of E_{90r}, the 90 per cent unit error?

d. What is E_{90}, the 90 per cent error, at 1 mile with 100-ft sights?

e. What order of accuracy is this?

f. What percentage of reruns would be required with 100-ft sights for second-order accuracy?

g. What percentage of reruns would be required with 200-ft sights for third-order accuracy?

h. What length sights should be used for third-order accuracy?

2-4. Find the value of C_i, the slope of the line of sight of an instrument when it is leveled, from the following observations. From one instrument position the reading on a rod at *A* was 3.729 and on a rod at *B*, 4.286. The distance to *A* was 20 ft and to *B* 220 ft. From another instrument position the reading on the rod at *A* was 4.368 and the rod at *B* was 4.908. The distance to *A* was 200 ft and to rod *B*, 15 ft.

2-5. In reciprocal leveling with an engineer's level, assume the data given.

Data for Prob. 2-5

Level No. 1 — Rod on *A* 5.692 — Rod on *B*

Upper target at 6.1		Lower target at 5.2	
+1.8	+4.6	−6.2	−3.4
+2.1	+4.8	−6.5	−3.7
+2.0	+4.8	−6.3	−3.5
+2.1	+4.7	−6.4	−3.6
+1.9	+4.4	−6.3	−3.4

Level No. 1 — Rod on *B* 6.629 — Rod on *A*

Upper target at 7.2		Lower target at 6.3	
+3.7	+6.5	−6.4	−1.5
+3.5	+6.2	−6.2	−1.4
+3.5	+6.3	−6.3	−1.4
+3.4	+6.3	−6.5	−1.2
+3.6	+6.4	−6.3	−1.3

If *A* has the elevation 102.386, what is the elevation of *B*?

CHAPTER 3

HORIZONTAL MEASUREMENT

STANDARDS

3-1. Introduction. This chapter describes methods of making horizontal measurements of second-order accuracy suitable for control traverse and for base lines of small triangulation nets. More precise methods are described in Chap. 7, Triangulation.

3-2. Units of Length. The physical standard of length in the United States is the International prototype meter, which is a bar of platinum-iridium alloy composed of nine parts platinum and one part iridium. It is maintained at Sevres, France, by the International Bureau of Weights and Measures. Thirty identical bars were made from the same ingot, all were intercompared, and one was chosen as the international meter. This one and two others as witnesses were retained at the Bureau; the remainder—27 in number—were designated as national prototype meters and were distributed by lot to the various countries that subscribe to the maintenance of the Bureau. The United States received two of these national prototypes in 1890 (Nos. 21 and 27).

Previously, in 1866, Congress had enacted a law that legalized the use of the metric system in the United States and defined the meter as 39.37 in. Prompted by the obvious advantages of using the international meter which was such a carefully made and widely distributed standard, T. C. Mendenhall, Superintendent of the U.S. Coast and Geodetic Survey, in his capacity as Superintendent of Standard Weights and Measures, ordered that the international meter should be the basis of official length determination by the United States government and that the inch should be taken as such a length that exactly 39.37 in. should be equivalent to 1 m. This order, which had the approval of the Secretary of the Treasury, was dated Apr. 5, 1893. The standardization of surveying tapes by the U.S. National Bureau of Standards has since then been based on the international meter, and this standard has been widely accepted throughout the country.

It will be noted that British measures are based on the imperial yard, which is approximately 1 part in 250,000 shorter than the yard accepted in the United States based on the international meter.

The ratios accepted in the United States are as follows (all are exact values):

$^{3937}/_{1200}$ ft = 1 international meter
1 Gunter's chain = 66 ft
1 link = 0.01 Gunter's chain
1 rod = 0.25 Gunter's chain
1 acre = 10 square Gunter's chains
1 mile = 5280 ft

Certain ratios for less common units of measure are legally established in various localities. Some of these are

1 vara = 33⅓ in. (in Texas)
1 vara = 33 in. (in California)
1 vara = 33.372 in. (in Florida)
1 vara = 837 mm or 32.99312 in. (in Mexico); these values are not equivalent
1 arpent = 0.8507 acres or 192.500 ft (in Missouri)
1 arpent = 0.84625 acres or 191.994 ft (in Louisiana, Mississippi, Alabama, and northwest Florida)

Note: One square arpent in linear measure very nearly equals one arpent in square measure.

The inch is little used in surveying. Dimensions for construction, however, are usually given in feet and inches. The inch had considerable vogue in surveys of city lots in times gone by and now survives in the descriptions and survey records of such lots. Today, except for construction purposes, the foot is entirely decimalized. Tenths, hundredths, and thousandths of a foot are used, and 100 ft is often called a "station."

3-3. Standardization of Tapes. The lengths of measuring tapes and other measuring devices can be determined by submitting them to the United States National Bureau of Standards for comparison with standards maintained by that Bureau. Certain regulations govern this service.

3-4. National Bureau of Standards Certificate. The National Bureau of Standards, upon the payment of a nominal fee, will determine lengths between end marks of measuring tapes and lengths between various other marks that are appropriate to the equipment available.

Dr. Lewis V. Judson, Chief, Length Section, National Bureau of Standards, writes under date of Apr. 24, 1950:

Our present bench is graduated for comparisons of a tape supported on a horizontal surface at each foot point from 0 to 10, at each 5-foot point from 10 to 100, at each 10-foot point from 100 to 150, and at the 33, 66, and 99-foot points. There are also graduations at each 1 meter point from 0 to 50 meters. For calibrating a tape when supported at points, there are graduations at each 10-foot point from 0 to 100, and at the 25, 75, and 150-foot points; also at each 5-meter point from 0 to 50 meters. Our new tape bench will have graduations over a 200-foot interval.

When a tape conforms to the specification—shown in Fig. 3-1—of the Bureau, the Bureau will also issue a certificate of the type shown in Fig. 3-2.

SPECIFICATIONS, STANDARDIZATION, AND USE OF STEEL TAPES

1. SPECIFICATIONS FOR STANDARD STEEL TAPES

A steel tape is considered as standard when it has been tested by the National Bureau of Standards and found to conform to the following specifications: It shall be made of a single piece of metal ribbon, and none of the graduations shall be on pieces of solder or on sleeve attached to the tape or wire loops, spring balances, tension handles, or other attachments liable to be detached or changed in shape. The error in the total length of the tape, when supported horizontally throughout its length at the standard temperature of 68° F (20° C) and at standard tension, shall not be more than 0.1 inch per 100 feet (2 millimeters per 25 meters). The standard tension is 10 pounds (4.5 kilograms) for tapes 25 to 100 feet or from 10 to 30 meters in length and 20 pounds (9 kilograms) for tapes longer than 100 feet or 30 meters.

2. CERTIFICATION OF TAPES

Tapes conforming to the above specifications will be certified by the National Bureau of Standards and a precision seal showing year of test will be placed on each tape. For tapes not conforming to the specifications, a report will be issued. The Bureau's serial number on a tape simply signifies that it has been tested by the Bureau and either a certificate or a report issued.

3. COMPARISON OF TAPES WITH THE STANDARDS

(*a*) Unless otherwise stated, the comparisons of this tape with the Bench Standard have been made at the center of the lines on the edge to which the shortest graduations are ruled. If all the graduations extend entirely across the tape, the ends farthest from the observer when the zero of the tape is at his left hand are used.

(*b*) On tapes which have been cut off at the zero mark, the extreme end of the steel ribbon has been taken as the zero point and not the center of any line that may be at that point.

(*c*) On tapes which have the zero point on a loop attached to the steel ribbon at the end, the zero has been taken at the outside of this loop unless noted to the contrary.

4. PRECAUTIONS IN THE USE OF TAPES

In the case of tapes for precision work, attention should be paid to the temperature and to the tension and corrections made for variation from the conditions given in the certificate or report. The accuracy of the balance used with the tape should be checked by comparison with a calibrated balance.

FIG. 3-1. The specifications issued by the National Bureau of Standards.

The graduations with which the tape is compared are based on the length of the International prototype meter. Lengths stated in feet are converted from this meter by the ratio given in previous sections. The tape will be tested in a horizontal position and supported as requested with the requested tension applied. Other details of the test are given in Fig. 3-1.

If requested, the Bureau will determine the weight per linear unit of the tape and also its thermal coefficient of expansion. These data are not usually requested for steel tapes because the unit weight can usually be determined with sufficient accuracy by any careful test, and the coeffi-

Form 579
141211

U. S. DEPARTMENT OF COMMERCE
WASHINGTON

National Bureau of Standards

Certificate

FOR

100-Foot Steel Tape

NBS No. 6200

Maker's Identification Mark: Lufkin #850

SUBMITTED BY

The Lufkin Rule Co.,
Saginaw, Michigan
for
Princeton University, School of Engineering,
Princeton, New Jersey

This tape has been compared with the standards of the United States. It complies with the specifications for a standard tape, and the intervals indicated have the following lengths at 68° Fahrenheit (20° centigrade under the conditions given below:

Supported on a Horizontal Flat Surface:

Tension	Interval	Length
10 pounds	(0 to 100 feet)	100.000 feet

Supported at the 0 and 100-foot Points:

Tension	Interval	Length
20 pounds	(0 to 100 feet)	100.001 feet

Lyman J. Briggs, Director

Test No. II-1/Tw 84837
Date of completion of test: December 23, 1938

The comparisons of this tape with the United States Bench Standard were made at a temperature of 81° Fahrenheit and in reducing to 68° Fahrenheit (20° centigrade), the coefficient of expansion of the tape is assumed to be 0.00000645 per degree Fahrenheit (0.0000116 per degree centigrade).

FIG. 3-2. A certificate issued by the National Bureau of Standards.

cient of expansion may be assumed to be 0.00000645 per degree Fahrenheit. The Bureau will also determine Young's modulus E for the tape. This should be determined for invar tapes, whose coefficients vary slightly. For steel tapes, except for very precise measurements, it is sufficiently accurate to assume that this value is 28,000,000 psi.

3-5. A Standard Tape. The length of every tape used in the field for precise work should be controlled by comparison with a tape certified by the National Bureau of Standards. The certified tape thus serves as a standard tape for the work.

It should be specially purchased for the purpose, kept in a protected place, and never used for field measurements. The manufacturer should be advised that it is to be used as a standard. In addition to the specifications of the National Bureau of Standards, the following specifications should be supplied to the manufacturer. They are written for a 100-ft steel tape. Appropriate changes should be made for other lengths.

1. The ribbon shall be old, seasoned stock about $\frac{1}{4}$ to $\frac{1}{16}$ in. wide and about 0.014 in. thick.

2. Marks shall be placed at the 0 and 100-ft points only. They shall be on the metal ribbon and not at the ends. They shall extend from the center of the tape to the edge farthest from the observer when the zero of the tape is at his left hand.

3. The marks shall be about 0.004 in. wide.

4. The tape is to be standardized by the National Bureau of Standards (1) fully supported and at a tension of 10 lb, and (2) supported at the 0 and 100-ft points and at a tension of 20 lb.

3-6. Standard Temperature. It is generally accepted throughout the United States that the value given for the length of a tape is its length at 68°F (20°C). This value should always be used. It is generally assumed that the thermal coefficient of expansion of the steel of which tapes are made is 0.00000645 per degree Fahrenheit.

3-7. Standard Tension and Support. Since the tension applied and the type of support used can be varied at will, many combinations are customary among surveyors. It is important to select and then to standardize on the best of these combinations so that the chances of blunders are reduced.

It is recommended that two standards be adopted for steel tapes. The tape shall be supported throughout with 10 lb tension or at 100-ft intervals with 20 lb tension. These values should be used for field measurements, for standardizing field tapes, and for tapes certified by the Bureau of Standards.

A tape of average weight will have very nearly the same correction for both standards. When these standards are chosen, there is little chance that incorrect values will be used in the field.

3-8. Correction Tension. Some surveyors prefer to eliminate tape corrections by using a special tension for each tape that makes its length exact. This can be called a **correction tension.** It is a difficult system to operate for several reasons: (1) if more than one field tape is available, each tape must be identified before the proper tension can be applied; (2)

often a correction tension for a tape fully supported is so little that the tape is not fully straightened out; a very slight change in tension will then cause a serious error in the length; and (3) frequently the correction tension for a tape supported at the ends is so high that it is impractical.

FIELD TAPES

3-9. Field Tapes. All precise field measurements of horizontal lengths are made today with steel or invar tapes. Steel tapes are semi-tempered bands of tough, flexible steel which has a thermal coefficient of expansion of very nearly 0.00000645 per degree Fahrenheit. Accurate results can be obtained with steel tapes if the measurements are made at

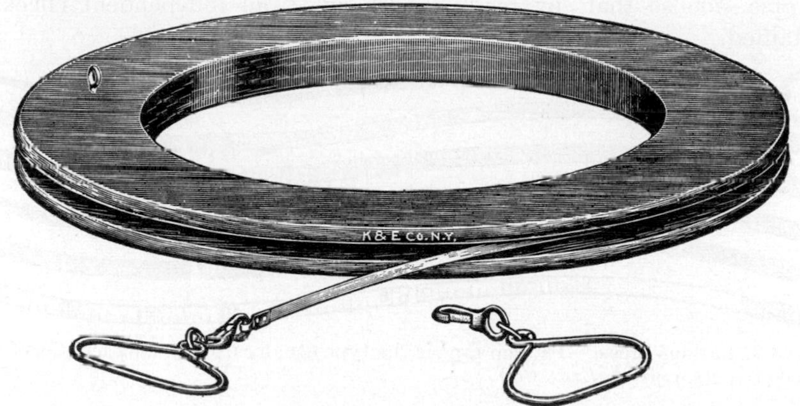

FIG. 3-3. An invar tape on a reel. (*Keuffel & Esser Co.*)

night or on cloudy or even hazy days when there is little radiant heat. At these times the tape and air temperatures are nearly the same so that the temperature of the tape can be accurately determined.

Invar tapes have one advantage only. Their coefficient of expansion is about 0.00000055 per degree Fahrenheit, or less than one-tenth that of steel. The coefficient may vary from tape to tape and it can change in a given tape, particularly if the tape is handled roughly. Invar tapes are easily bent and damaged. They must be kept on reels of large diameter to avoid kinking (Fig. 3-3) and they must be handled with great care. Very accurate results are possible with invar tapes even when the measurements are made in bright sunlight if proper procedures are followed and if the coefficients of expansion of the tapes are checked frequently. The difficulty of handling them prevents their use for other than long, precise measurements and they are, accordingly, usually graduated only at infrequent intervals.

3-10. Steel Tapes. The 100-ft steel tapes used for control measurements are usually $\frac{1}{4}$ to $\frac{5}{16}$ in. wide and 0.012 to 0.025 in. thick. When supported at the 0 and 100-ft points, lightweight tapes give greater accuracy even in the wind, as they are less dependent on exact tension. Heavier tapes are much less likely to kink and to break. Usually 100-ft tapes should be graduated to 0.01 ft throughout their length. The 0 and 100-ft marks should be on the body of the tape and not at the ends. Graduations extending beyond the 100-ft length are not recommended except for special purposes. Such graduations are used to correct for slope or for temperature or, with tapes graduated only at each foot, to give the decimal parts of the foot when less than one tape length is measured. The end of these extensions can be mistaken, all too easily, for the zero mark. Sometimes tapes are graduated in the metric system on the reverse side so that, by reading both sides, an independent check is obtained.

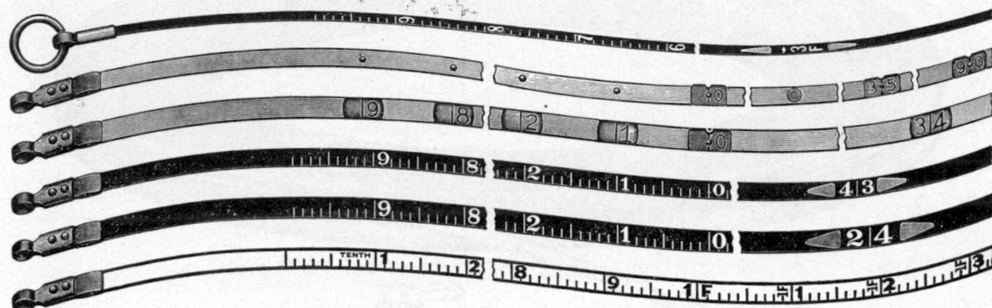

Fig. 3-4. Various tapes. The top tape is the type usually used in long lengths of 200 to 500 ft. (*Keuffel & Esser Co.*)

Steel tapes longer than 100 ft are usually flat wire bands $\frac{1}{8}$ in. wide and 0.025 in. thick (see Fig. 3-4). The most popular lengths are 300 and 500 ft, although others are common. As they are not easy to handle for short measurements, long tapes are frequently graduated only at the 5-ft points to reduce their cost. They are more useful, however, when they are graduated to 0.01 ft throughout.

Long tapes offer a means of very rapid, accurate measurement. The time necessary to mark the tape lengths is reduced and the accidental errors are almost negligible. Systematic errors, however, are more difficult to avoid. Long tapes are difficult to standardize, difficult to support at measurable slopes, and likely to have different temperatures throughout their lengths. Since they always require a number of intermediate supports, there is always the danger that friction may introduce different tensions at different spans. When these tapes are supported throughout, friction may create a variable tension; however, since there is no sag, variations in tension change the length only by the elongation in the actual steel, and the results are usually more accurate.

3-11. Standardizing Field Tapes. It is nearly impossible to manufacture tapes that are exactly correct in length. Also, for a short time after manufacture, molecular changes may occur that affect tape lengths. When tapes are used in the field, wear reduces the cross section so that their weight and therefore the sag is reduced, and the metal itself stretches farther under a given tension. For these reasons, the lengths of field tapes should be tested from time to time.

While a number of methods of testing tapes may occur to the reader, it will be found that the only practical method is to keep available at least one special 100-ft tape that has been certified by the Bureau of Standards and that is never used in the field. As mentioned before, this tape should be graduated only at the 0 and 100-ft points so that it cannot be mistaken for a field tape. A base must be constructed to facilitate comparisons between this tape and the field tape.

3-12. Bases for Standardizing Field Tapes. It can be stated categorically that it is impossible to establish two points that remain exactly 100 ft apart. Even the ground changes size from time to time. It follows that the field tapes must be compared with the standard tape itself and not with a length laid out by it.

A base for making these comparisons can be constructed quite easily, but certain requirements must be rigidly followed.

1. *The air temperature must be uniform.* It is nearly impossible to find a suitable location indoors. Radiant heat and temperature gradients prevent accurate determination of the tape temperature during the test. A site must be chosen outdoors, preferably on the north side of an unheated building where shade is assured and no radiant heat from the building can reach the tape.

2. *The tape must be supported as it will be supported in use.* Usually two sets of equipment must be constructed, one for supporting the tape throughout its length and one for supporting it at the 0 and 100-ft points.

3. *A scale must be provided.* The tapes cannot easily be compared directly. The length of the standard tape must be read on a scale and the readings of each of the field tapes compared with it. The scale should be about 0.1 ft long. It should be made of metal, with graduations cut at intervals of 0.001 or 0.002 ft. Decimal parts of an inch should not be used. Several manufacturers can cut such scales to order.

4. *A weight must be used for tension.* A weight connected to the tape by a piano wire running over a ball-bearing pulley will assure the exact tension.

Many different types of bases can be constructed. Figure 3-5 shows an arrangement for supporting the tape at the end marks. Figure 3-6 shows an arrangement for full support. The mark at the turnbuckle end should be a cross, cut in metal, with one line of the cross perpendicular to the tape. After the tape is in place, the turnbuckle is turned until the mark on the tape coincides with the intersection of the cross lines.

3-13. Procedure for Standardizing Field Tapes. Usually several field tapes are standardized at one time. If they are to be supported throughout, they are unreeled and laid out parallel to the base to pick up ground temperature. If they

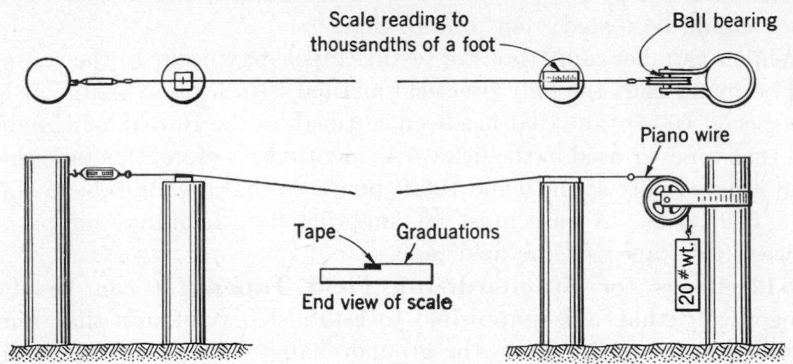

FIG. 3-5. Base for comparing tapes supported at the ends.

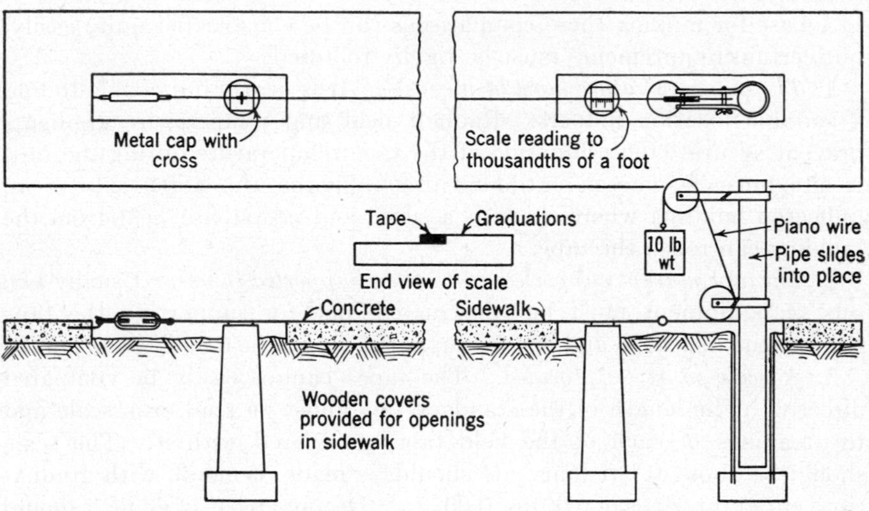

FIG. 3-6. Base for comparing tapes when fully supported.

are to be supported at the end marks only, they are hung clear of the ground. They are then thoroughly cleaned. The standard tape is put in place, one end is brought to the cross mark with the turnbuckle, the scale reading of the 100-ft point is recorded, and the appropriate air or ground temperature is noted. The procedure is repeated for all the field tapes except that, when testing a tape with a two-point support, a thermometer is attached at the point where it is attached in the field to obtain the working catenary. When all the field tapes have been tested, a second reading is obtained for the standard tape.

The tape lengths are computed as shown in Table 3-1. The values in Column 4 of Table 3-1 are taken from Table III in the Appendix. The sign is changed and they are multiplied by 100. The value 99.9493 in Column 5 is derived from the entries in Columns 3 and 4 for the standard tape, as follows:

$$100.007 - \frac{0.052 + 0.0052 + 0.055 + 0.0032}{2} = 99.9493$$

The length of each field tape is the sum of the entries in Columns 3, 4, and 5. For Tape 2 this would be 0.036 + 0.0052 + 99.9493 = 99.990.

Long tapes must be standardized in 100-ft sections. Grips must be arranged to hold the tape. They should be faced with wood or brass to prevent deforming the steel.

TABLE 3-1. COMPUTATIONS FOR STANDARDIZING TAPES*

Tape no. (1)	Temp., °F (2)	Reading (3)	Temp. cor. (4)	Add (5)	Tape length at 68°F (6)
S1	60	0.052	+0.0052		100.007†
2	60	0.036	+0.0052	99.9493	99.990
15	61	0.068	+0.0045	99.9493	100.022
28	61	0.033	+0.0045	99.9493	99.987
46	62	0.065	+0.0039	99.9493	100.018
S1	63	0.055	+0.0032		100.007†

* Taken chiefly from the table supplied by the author for the ASCE Manual of Engineering Practice No. 10.

† Length of Standard Tape S1 supplied by the U.S. National Bureau of Standards.

3-14. To Determine the Correction Tension. When the system of using a correction tension is adopted, the Bureau of Standards is usually requested to determine the correction tension of the standard tape. This is not necessary, because the only purpose of the standard tape is to determine the scale reading that is exactly 100 ft from the reference mark at the time the tapes are to be tested.

The standard tape is placed on the base to obtain the scale reading for 100 ft at 68°F.

When each field tape is placed on the base, the temperature is read and the proper scale reading for that temperature is computed from the data just determined. The weights are varied until the tape gives this corrected scale reading. The record is shown in Table 3-2.

The entries in Column 3 of Table 3-2 are the values found in Table III in the Appendix multiplied by 100. The entry 0.067 in Column 4 is derived as follows:

$$0.063 - (-0.004) = 0.067$$

Long tapes must be standardized in 100-ft lengths and the average tension used in the field.

TABLE 3-2. COMPUTATIONS FOR STANDARDIZING TAPES BY CORRECTING TENSION

Tape no. (1)	Temp., °F (2)	Temp. cor. (3)	Reading for 100 ft at 68°F (4)	Scale reading (5)	Tension, lb (6)
S1	62	−0.004	.....	0.063	18*
2	60	−0.005	0.067	0.062	22
15	61	−0.005	0.067	0.062	15
28	61	−0.005	0.067	0.062	23
46	62	−0.004	0.067	0.063	16
S1	63	−0.003	.....	0.063	18*

* This value is the tension required to give the standard tape S1 a length of 100 ft. It is supplied by the U.S. National Bureau of Standards.

3-15. Spring-balance Handles. Spring-balance handles should read to at least 20 lb. They should be tested occasionally. The test should be made in a horizontal position to prevent the weight of the moving parts from affecting the reading. The best method is to hook them to the wire that lifts the weight on the tape-standardizing equipment and record the reading.

3-16. Thermometers. The tape thermometers used should be designed so that their bulbs come in contact with the tape (Fig. 3-7). Even if they are equipped with clips, they should be attached to the tape

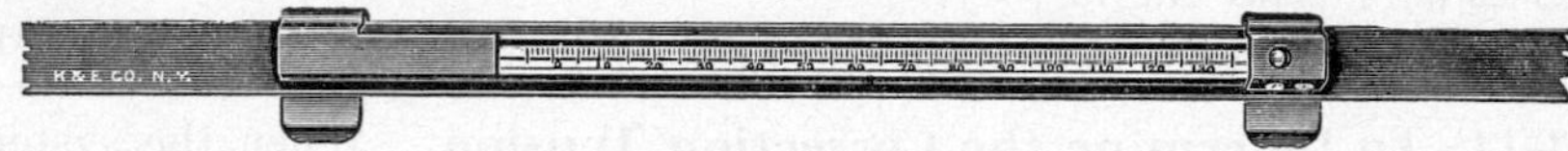

FIG. 3-7. A tape thermometer. The bracket has a light spring that forces the bulb against the tape. (*Keuffel & Esser Co.*)

with white, medical adhesive tape. They should be tested once for accuracy against a standard thermometer. If it is found that the liquid column separates in use, they should be rejected.

TAPING

3-17. Taping with Plumb Bobs. Horizontal measurements are usually made with a steel tape and plumb bobs. With spring-balance handles and corrections for temperature, these methods will give third-order accuracy.

3-18. Methods for Second-order Accuracy. When accuracies of 1:10,000 or better are required, plumb bobs must be eliminated, the slopes of the tape must be measured, the actual tape temperatures must be accurately determined, and the tension must be carefully controlled.

If possible, sunny days should be avoided. It is very difficult to determine tape temperatures accurately when the tapes are subjected to radiant heat. However, a slight haze will eliminate most of this difficulty. The following methods of precise taping will be covered in this chapter: the method based on full support, a semiapproximate method, and the method using taping tripods.

TAPES SUPPORTED THROUGHOUT

3-19. Advantages of Full Support. Taping with the tape supported throughout affords the least expensive and the most rapid method of obtaining horizontal measurements of second-order accuracy. Practically no equipment need be carried, wind does not affect the work, and since there is no sag, errors in tension have little effect on accuracy. Of course, the tape must rest on the ground, as the construction of an artificial support would be prohibitively expensive. Fortunately, proper surfaces are usually available. Railroad rails, street and highway pavements, and sidewalks are best. Graded roadways, even when unpaved, graded fields without too much vegetation, and even graded paths can be used successfully. The tape need not touch the ground continuously. If the tape is supported at only 20 ft intervals, the error for an average ¼-in. tape would be only 1 part in 25,000. Ten-foot spans would introduce an error of 1 part in 100,000. Flat wire tapes of smaller cross section are even less affected.

Whenever possible, stations should be placed so that proper support is available to permit this method to be used. If the stations cannot be placed on line with a proper surface, they can often be placed nearby and connected by short ties.

3-20. Short Ties. Figure 3-8 shows some examples of short ties. Enough angles should be measured to obtain an angle closure and to make it possible to compute the projections of the measured lengths upon the required length. Frequently, the angle between the major measured length and the required length cannot be determined accurately from the angles alone but must be computed from the linear measurements.

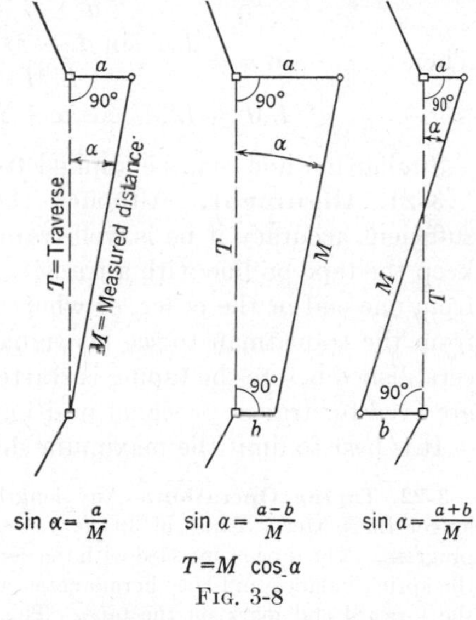

FIG. 3-8

Figure 3-9 is a typical case in which the angles cannot be relied on. L and M are traverse stations. Angles L and M are traverse angles.

LM is the length of the traverse course to be determined, and $L'M'$ is the measured distance. The tie measurements consist of the lengths LL' and MM' and the angles a, b, c, and d. α is the angle between the measured distance and the required length.

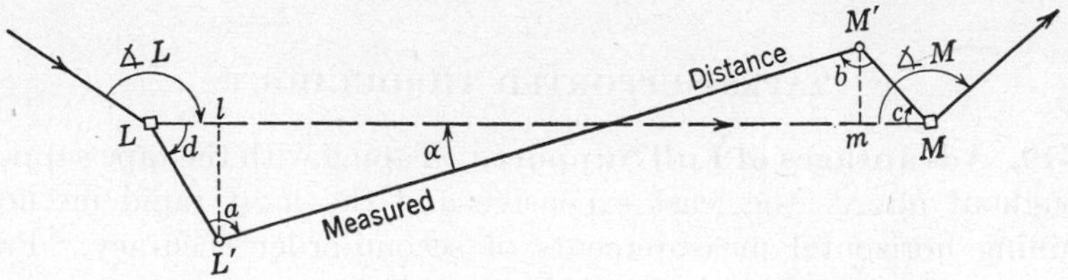

FIG. 3-9

Angle α can be computed from either of the following:

$$\alpha = 180° - (a + d)$$
$$\alpha = 180° - (b + c)$$

but none of these measured angles can be relied on as each has one short side. Instead, the following is used:

First adjust a, b, c, and d by equal quantities so that

$$a + d = b + c$$

Then
$$\sin \alpha = \frac{LL' \sin d + M'M \sin c}{L'M'}$$
$$LM = L'M' \cos \alpha + LL' \cos d + M'M \cos c$$

Similar methods can be applied to other types of ties.

3-21. Alignment. Although the rear tapeman can give line with sufficient accuracy if he is well-trained and is very careful, it is safer to keep the tape on line with a transit. When part of the line is not visible from one end or the other, or when the tapemen are likely to get too far from the transitman to see his signals, intermediate line points must be established before the taping is started. These, as well as the end points, are used for transit or signal positions.

It is best to limit the maximum slope of the tape to 15 per cent.

3-22. Taping Operation. Any length of tape may be used. As a rule, the longer the better. Tapes of 300 ft are easy to handle and long enough for rapid progress. The tape is unreeled with the zero at the rear. The head tapeman attaches the spring balance and the thermometer, which should be placed about 1 ft back of the forward end mark on the tape. The head tapeman stretches the tape for the approximate distance and places a keel mark or a scratch on the ground *on line* according to signals from the transitman. The mark should be just short of the final distance and within ½ in. of precise line. With the help of the rear tapeman, the head tapeman clears the tape of any obstacle, straightens it, and gets it on line throughout its length. Tapes up to 300 ft can be held clear of the ground momentarily if both men stretch it high above their heads. When a longer tape is used, a middle tapeman must assist in this operation.

When the head tapeman is satisfied that the tape lies straight and over his line mark, he applies the correct tension. He can steady his hand against the ground to regulate the spring balance. When the rear tapeman feels the tension come on, he brings the zero graduation to the mark and presses the end of the tape beyond the zero mark hard against the ground to anchor it. When ready, he calls "mark."

The head tapeman adjusts the tension and marks the point. He checks the mark twice. First he relieves the tension and then brings it back to the correct value. Then he increases the tension and again brings it back to the correct value. If friction affects the measurement, this operation will discover it.

When he is satisfied with the check, the head tapeman determines and records the thermometer reading, draws a heavy circle around the mark or sets a guard stake to identify it, and writes its station number on the ground or stake. He then sights along the tape to discover any serious breaks in grade. If any exist, he directs the rear tapeman to them. The rear tapeman marks each with keel or with a stake and writes the tape reading beside it. The tape reading should be estimated to the nearest foot. This will be the distance forward from the previous taping point. When the surface is curved in profile, elevations taken at every hundred feet will be found sufficient on any well-paved highway.

When the breaks have been marked, the tapemen carry the tape forward. They keep it off the ground, if possible. The head tapeman attempts to keep on line as he walks, to reduce the time necessary to straighten the tape.

3-23. Marks. If the surface is soft enough, a surveyor's tack should be used for each mark. This applies to bituminous roads and earth. On harder surfaces, a strip of white adhesive tape should be placed in position and a pencil mark used to show the point. Except under unusual circumstances, stakes should not be used to support the tacks, as it takes too long to drive stakes at the correct point.

3-24. Final Partial Tape Length. The last measurement at the end of the line should be made with three men. The transitman can serve, as alignment is not required. The head tapeman walks past the end mark until the zero of the tape is at the last taping mark. He then applies tension. The transitman marks the *tape* with a pencil where it is adjacent to the end mark. The position of the pencil mark is read and checked. If the tape is graduated only at 5-ft points, a short steel tape graduated to hundredths should be used to measure the distance from the last tape mark to the pencil line.

3-25. Crossing Traffic. When the taping must be carried across traffic, it is well to establish two arbitrary marks on line, one on each side of the road and less than 100 ft apart. This distance should be measured during a lull in the traffic with a 100-ft tape graduated throughout.

3-26. Setups and Setbacks. The term **setup** applied to taping is an extra section whose length must be *added* to the total distance measured. The section across a road described in the previous paragraph is a true **setup.** Sometimes the partial tape length from the last taping mark to the end of the line is called a setup.

A **setback** is a section whose length must be *subtracted* from the total distance measured.

Setups and setbacks occur chiefly in precise base-line measurements when the taping is carried over posts—already in position—that have strips of copper for marking the tape ends. When the tape mark begins to run off the copper strips, a setup or setback of a few millimeters is introduced.

3-27. Leveling. Either during the taping or afterward, a profile is run to determine the elevation of each taping mark and each marked break in grade. Elevations are usually taken to hundredths of a foot. The usual methods of leveling and checking the work are used.

3-28. Field Notes. The form of field notes is shown in Fig. 3-10.

3-29. Field Party. The work can be performed with a three-man party. If a fourth man is available, he should assist the head tapeman. He is particularly useful when the taping is carried across traffic.

Traverse Sta. B-C
Tape #8 300 012 0.011 lb/ft 10 lb T

H.C. Smith Date
Rec Dodd Cloudy
R.C. Thomas

Sta.	Length	Temp	h	h^2	2L	Ch		
B 0+0								
							Tape 10.44(40.012)÷3	+ 0.042
							Temp 1044(0.00000645)(-15)	- .101
3+0	300	52	2.82	7.95	600	.0132	Slope	- .209
4+30	130		5.19	26.94	260	.1036	Measured Length	1043.81
6+0	170	53	2.80	7.84	340	.0231	Final Length B-C	1043.54
9+0	300	53	2.25	5.06	600	.0084		
C 10+43.81	143.81	53	4.19	17.56	287.62	.0611		
	1043.81	53				0.2094		

Traverse Sta. B-C
Leveling

⊼ Dodd Rec Smith Date
Rod Thomas Cloudy

Sta.	+	H.I.	−	Rod	Elev.	Diff
BM	4.31	56.59			52.28	
0+0				6.49	50.10	
T.P.3+0	2.63	49.91	9.31		47.28	2.82
4+30				7.82	42.09	5.19
T.P.6+0	4.02	43.31	10.62		39.29	2.80
9+0				6.27	37.04	2.25
T.P.10+43.81			10.46		32.85	4.19

FIG. 3-10. Field notes for taping with the tape supported throughout.

When considerable taping is required, two men should be added to form a level party, so that the leveling can proceed simultaneously.

APPROXIMATE METHOD

3-30. Method. The approximate method will give second-order accuracy if due care and judgment are used continually throughout the work. It is exactly like the method of supporting the tape throughout, as described in the previous sections, except that it can be used over rough ground or where there is high grass or other vegetation. A three-man party is required, consisting of a head, middle, and rear tapeman. A 300-ft tape should be used. A longer tape cannot be handled and a shorter tape is too slow. The rear tapeman gives alignment.

3-31. Smooth Surfaces. Over smooth surfaces the procedure is the same as before except that the rear tapeman gives alignment.

3-32. Long Grass. To place the tape in position over long grass or other vegetation, it must be raised high in the air and then dropped on line. Sometimes it must be shaken sidewise to work it down. Often it is not possible to bring the ends of the tape down to the ground to reach the taping marks without introducing a sharp break in grade near the end of the tape. The plumb bobs should be used if the height is not much more than 1 ft. As the vegetation prevents the wind from reaching the bob, this operation does not seriously interfere with the accuracy. Whenever a plumb bob is used, the middle tapeman must place a mark at the height of the tape on the guard stake (see Fig. 3-11).

While the tape is under tension, the middle tapeman drives a guard stake beside the tape at each break. Usually the tape is supported on the grass above the ground.

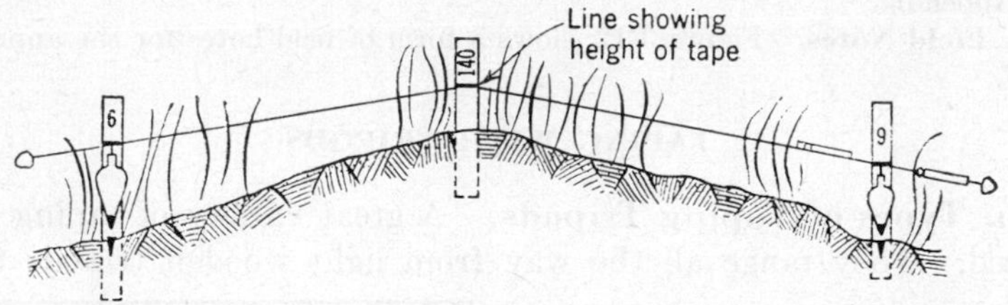

FIG. 3-11. Taping by approximate method. Vegetation often supports tape several inches above the ground.

In this case he places a mark at the height of the tape on the stakes. Guard stakes can be pieces of building lathe or any inexpensive material.

Levels are run over the work. The rod is usually read to tenths of a foot. The rodman must hold his rod at each break. He either holds the bottom of the rod at the line on the stake showing the height of the tape or he holds the rod on the ground beside the stake and notes the position of the mark on the rod. This height is recorded in the level notes. At any measuring point there are usually two heights,

Traverse Sta E-F — Tape #8 300.012 0.011 lb/ft 10 lb Thruout — H.C. Smith, Rec. Dodd, R.C. Thomas — Date, Cloudy

Sta	Length	Temp	Tp Ht	h	$h^2/2L$	C_h		
E 0+0			0.6					
1+20	120		0.2	0.8	0.64/240	0.0027	Tape 9.81 (+0.012) ÷ 3	+ 0.039
							Temp 981 (0.00000645) 18	+ 0.114
3+00	180	86	0.5	1.9	3.61/360	0.0100	Slope	− 0.056
			0.8				Sag 90' (from Chart)	− 0.037
							Measured Length	981.36
6+00	300	85	0.0	4.6	21.16/600	0.0353		
			1.0				Final Length E-F	981.42
8+10	210		0.5	0.8	0.64/420	0.0015		
9+00	SPAN 90	87	0.3	1.1	1.21/180	0.0067		
			0.2					
F 9+81.36	81.36		0.4	0.0	0			
	981.36	86				0.0562		

Traverse Sta E-F Leveling — ⊼ Dodd, Rec. Smith, Rod Thomas — Date, Cloudy

Sta.	+	HI	−	Rod	Elev.	Tp.Ht.	Elev.	Diff
BM	6.27	108.57			102.30			
0+0				3.79	104.78	0.6 ✓	105.4	
1+20				4.21	104.36	0.2 ✓	104.6	0.8
T.P. 3+00	3.41	105.66	6.32		102.25	0.5 ✓	102.7	1.9
3+00						0.8 ✓	103.0	
T.P. 6+00	5.61	103.96		7.31	98.35	0.0 ✓	98.4	4.6
6+00						1.0 ✓	99.4	
8+10				4.30	99.66	0.5 ✓	100.2	0.8
9+00				5.18	98.78	0.3 ✓	99.1	1.1
9+00						0.2 ✓	99.0	
T.P. 9+81.36			5.32		98.64	0.4 ✓	99.0	0.0

FIG. 3-12. Field notes for approximate taping. Note 90-ft span from 8 + 10 to 9 + 00.

one made when the point was set and the other made when the measurement was carried forward. It is frequently wise for the middle tapeman to carry a short rule graduated in tenths of a foot to measure these heights. He records them in the taping record to serve as a check against the level notes.

3-33. Rocky Ground. Rocky ground is usually difficult only because long sections of the tape are unsupported. When this occurs, the middleman should record the tape readings to the nearest foot at each end of any long span. Sag corrections are then applied in the office. It is safe to obtain them by interpolation on Chart II of the Appendix.

3-34. Field Notes. Figure 3-12 shows a form of field notes for the approximate method.

TAPING WITH TRIPODS

3-36. Types of Taping Tripods. A great variety of taping tripods are used. They range all the way from light wooden tripods to very

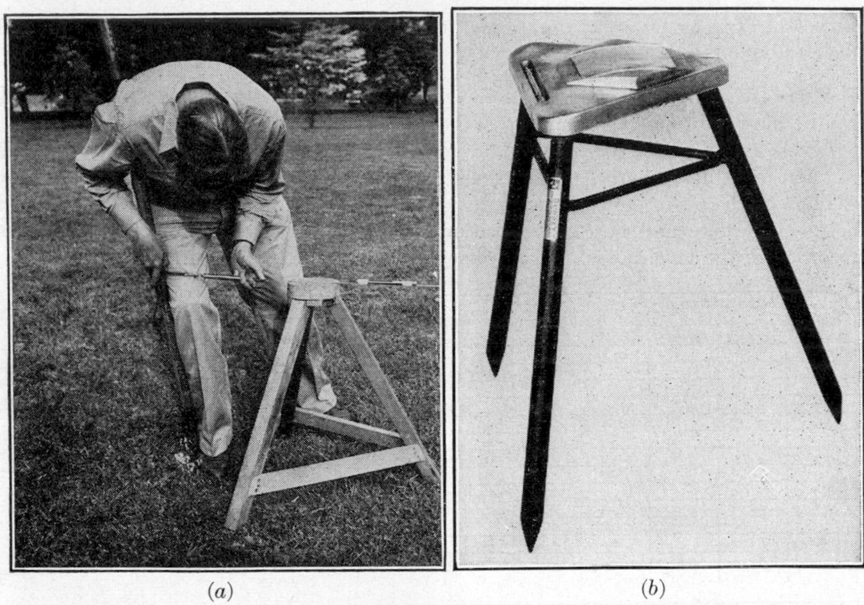

(*a*) (*b*)

FIG. 3-13. (*a*) Wooden taping stool. Head tapeman handling tension, head contact, and temperature reading. (*b*) A metal taping stool with thermometer. (*W. & L. E. Gurley.*)

carefully designed metal devices (see Fig. 3-13). They should be heavy enough to take a firm position, particularly in heavy vegetation. On the other hand they should be light enough to be easily carried. The top should present a surface that can be easily marked for distance and easily set parallel with the tape.

Obviously a compromise is necessary. The U.S. Coast and Geodetic Survey uses the type shown in Fig. 3-14 for base-line or first-order traverse measurement. Many prefer a steel tripod with a curved top having a copper surface, but for second-order measurement light wooden tripods are best. They should be designed to nest together. The top surface should be white pine or some other soft wood so that pins can be easily stuck in it to mark points. Banker's pins are best, as they have T-like tops (Fig. 3-15).

Light tripods can be used in sets of a dozen or so. Since they can be left in position after they have been used for measurement, the level

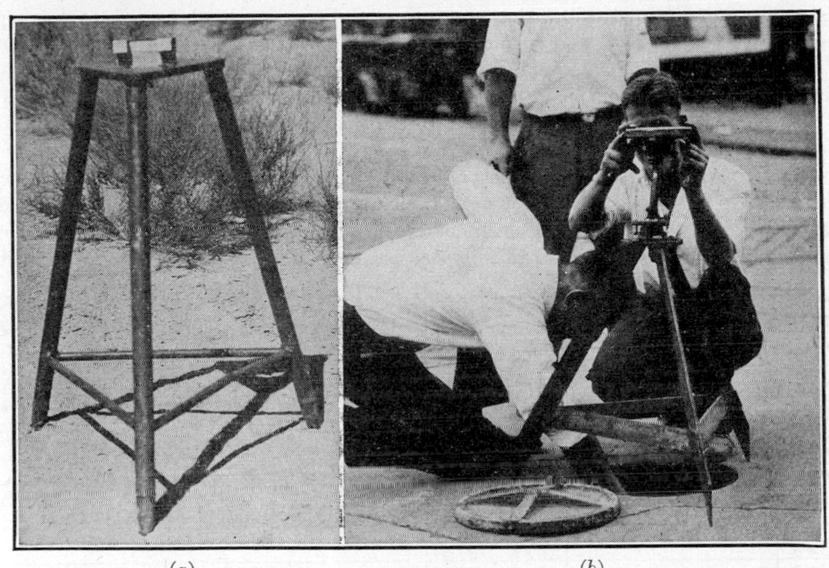

(a) (b)

FIG. 3-14. (a) Taping tripod. (b) Taping tripod with adjustable head in operation. (*U.S. Coast and Geodetic Survey.*)

party does not have to keep up with the taping and therefore sometimes delay it. Also, if a tripod is moved inadvertently, the remeasurement can begin at the last undisturbed tripod.

3-36. Stretchers. Tape stretchers must be used to handle the tape. They are usually wooden rods about 5 ft long shod with metal points. One is used at each end. They should be attached to the tape with a loop of rawhide or string (Fig. 3-13).

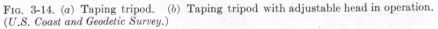

3-37. Alignment. Alignment should be by transit. Intermediate line points should be set if necessary (see previous method). The maximum slope of the tape should be 10 per cent.

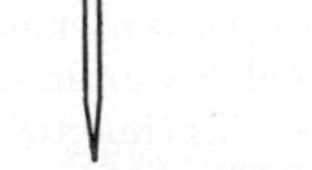

3-38. Taping Operation with 100-ft Tape. Windy days should be avoided. Wind makes the tape vibrate and blows it off line enough to damage the accuracy.

FIG. 3-15

The tripods are distributed at approximately 100-ft intervals; the transit and signals are set in place. The tape is unreeled with the 100-ft mark forward, the spring balance is attached, and the thermometer is taped in position at the 2-ft mark. Thongs are attached at each end.

Sometimes the zero of the tape can be held directly on the starting station marker if the 10 per cent slope limit is not exceeded and if the tape is clear of all contact (except at the supports) when tension is applied. If this is impossible, a taping tripod should be set over the station. The position of the station can be transferred to the tripod most easily as follows. First set up the transit over the point, then place the taping tripod in position and adjust the transit plumb bob so that it will just swing clear of the tripod surface. A mark is made under the plumb-bob point. When the transit is not available, the transfer can be made by plumbing to the top edge of the tripod. A pin should be stuck into the edge point so that it inclines at 45°.

The following is taken chiefly from a description of taping supplied by the author for the ASCE "Manual of Engineering Practice No. 20": The tape end is laid on the starting tripod. The rear tapeman passes his stretcher through the loop (unless the zero mark is to be held in contact with the station marker), places the lower end of the stretcher on the ground against the outside of his right foot, and places the upper end behind his right shoulder. In this position, he leans over the tape to see that the

Traverse Sta. M-N
Tape #3 99.982 0.018 20 lb. 2

Dist.	Temp.	h	h^2	2L	Ch
100	72	2.02	4.08	200	0.0204
100	75	2.53	6.40	200	0.0320
100	80	0.85	0.72	200	0.0036
52.71	80	0	0		0
352.71	76				0.0560

Ch-HC Smith Tension Jones Date
Rec. Dodd Haze
RC Thomas

Tape 3.53(-0.018)	- 0.064
Temp 353 (0.00000645) 8	+ 0.018
Slope	- 0.056
Measured Length	352.71
Final Length M-N	352.61

Traverse Sta. M-W
Leveling

Sta	+	HT		Rod	Elev.
BM	4.56	124.88			120.32
0+0				3.87	121.01
1+0				5.89	118.99
2+0				3.36	121.52
3+0				4.21	120.67
3+52.71				4.21	120.67

Level Black Date
Rod Greene Haze

Diff
2.02
2.53
0.85
0

FIG. 3-16. Field notes for method with taping tripods.

zero graduation is held exactly at the mark. This is readily controlled by adjusting his stance. However, he may find it helpful to grasp the tape near its end and behind the zero graduation and apply a kinking force, just sufficient to control the position of the zero graduation.

The **tension** man passes his stretcher through a 6-in. loop attached to the spring balance, takes the same position employed by the rear tapeman, and applies the correct tension.

The chief of party, who acts as **marker,** places a tripod so that its center is within ½ in. of line (as directed by the transitman) and under the 100-ft graduation. The tension man slides his thong up or down the stretcher until the tape just clears the top of the tripod, and he moves his stretcher slightly left or right to bring the tape to the center of the tripod top. The marker must see that the tape is dry, clean, and free from all obstructions, and he may run a rag lightly along its entire length at this time to remove any moisture or dirt.

The marker gently depresses the tape to touch the marking surface and calls "ready." The rear tapeman assures himself that the zero mark is at the pin and calls "mark." The tension man checks the tension and, if it is correct, also calls "mark." Upon these calls the marker places a pin at the 100-ft graduation. If the tape is vibrating in the wind, the recorder touches the tape lightly to stop it momentarily.

Tension is released slowly, then reapplied for a check. The recorder obtains the temperature from the rear tapeman. If there is no rodman he holds the rod on the tops of the chaining tripods for the levelman and records the rod readings.

A record is made for each individual tape length or partial tape length. It includes the length measured, the temperature, and the data for the slope.

The marker moves back to support the center of the tape as it is carried forward. The tape must be held clear of all contacts by the marker, the tension man, and the rear tapeman. After the second tape length is measured and the levels completed,

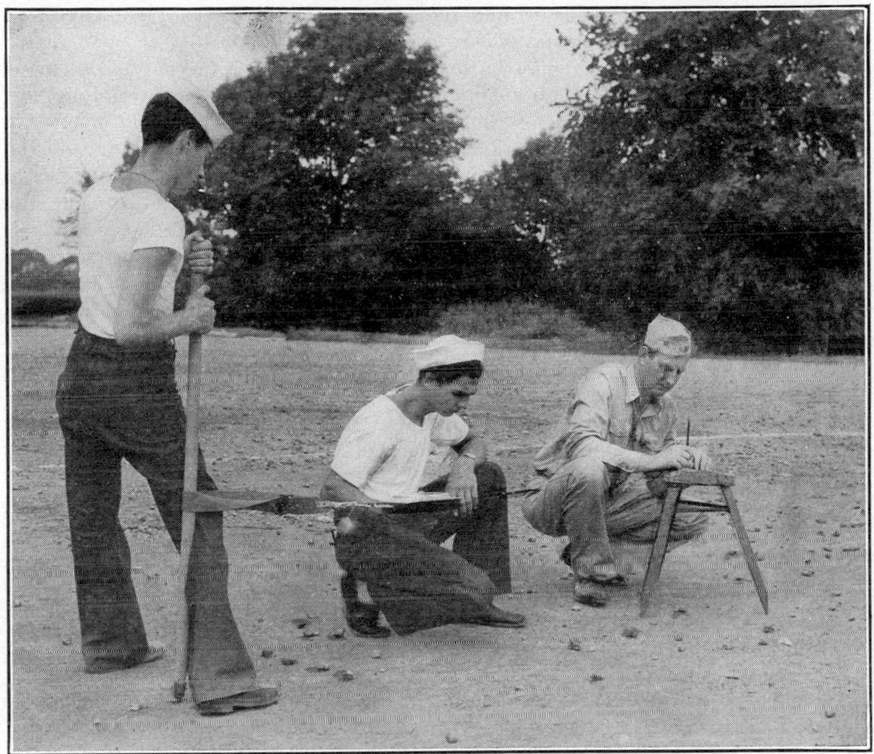

FIG. 3-17. Precise taping. (*U.S. Coast and Geodetic Survey.*)

the recorder may begin picking up tripods. He can carry about five at a time. They are usually distributed at their proper positions by the entire party.

The transit should be brought up to the next line point when the taping has progressed about half the distance from the transit toward it so that the transitman's signals can easily be followed.

For distances of less than a tape length, the tape is read independently by the chief of party and the recorder. If the reverse side of the tape is graduated in meters, the metric reading should be recorded as well.

Upon reaching the end station, a taping tripod is placed over the station and the station position transferred, as described for the beginning station. The head of the tape is carried beyond the end point, the zero being at the back tripod as usual. If the partial tape length (setup) is more than 50 ft, a 20-lb tension is applied. Otherwise a pull of 10 lb is used. This affords a close approximation to the proper proportional application of the tape correction.

When placing the tripods in grass, considerable care must be taken to make sure that they are steady. Sometimes some tufts of grass must be removed. When the tripods cannot be made steady, 2- by 2- by 30-in. stakes should be used instead of the tripods. They can be handled by the marker and recorder, but, if possible, an extra man should be added for this work.

3-39. Backward Measurement. It is best to measure each section in two directions. Although this may not be demanded by the accuracy required, it provides the only proper check against blunders and bad practices. The results, reduced for temperature and slope, should agree within 1 part in 30,000.

3-40. Leveling. Leveling may be done with an engineer's level, the telescope level on a transit, a hand level, or a clinometer. All have been used successfully, but the first two increase both speed and accuracy. When an engineer's level or a transit is employed, rod readings are taken to hundredths of feet on the tops of the tripods. When the instrument must be moved, one of the tripods is usually chosen as a turning point. If the telescope level on the transit is in good adjustment, the levels can sometimes be run while the transit is in position for giving line.

A hand level should be supported by a forked stick. The levelman holds it near the 50-ft mark and takes a reading to tenths of a foot on both tripods for each tape length. Only the difference is recorded. Collapsible foot rules, graduated to tenths of a foot, should be carried by the tension man and the rear tapeman or the recorder, to serve as rods.

A clinometer is most successfully employed when 4-ft taping tripods are used. It is placed on one tripod and sighted at a small target on the next tripod. The angle of slope or the per cent grade is recorded.

3-41. Field Party. The usual field party consists of

1. Chief of party, who acts as marker (head contactman).
2. Recorder, who keeps the tape and level notes, acts as middle tapeman when the tape is being moved, moves the tripods, and sometimes holds the rod.
3. Rear tapeman (rear contactman), who also reads the thermometer.
4. Tension man.
5. Levelman.
6. Rodman (sometimes omitted).
7. Assistant (usually omitted). In heavy traffic he is essential. He directs traffic and keeps the tripods moving rapidly. In swampy ground he assists in staking.

3-42. Taping with a 300-ft Tape. When a 300-ft tape is used, the procedure is exactly the same, except that two intermediate tripods are placed at the 100-ft and the 200-ft marks. These tripods should have a pair of nails about 1 in. apart driven into their tops to keep the tape from falling off. They are lined in with the transit. Since the spans are the same, 20-lb tension is used.

3-43. Form of Field Notes. The form of field notes is shown in Fig. 3-16.

CORRECTIONS TO MEASUREMENT

3-44. Usual Procedure. Under usual circumstances, field measurements are recorded and corrected in accordance with the procedures described in "Surveying." To cover all possibilities, certain other information is given in the following paragraphs.

3-45. Slope Correction. The formula for the slope correction is based on a power series. The derivation is the following:

Let C_h = slope correction
L = length recorded on slope
h = difference in height of two ends
b = horizontal length required

Then
$$L^2 = b^2 + h^2$$
$$b = (L^2 - h^2)^{1/2}$$

Expanding by the binomial theorem,

$$C_h = b - L = -\frac{h^2}{2L} - \frac{h^4}{8L^3} - \frac{h^6}{16L^5} - \frac{5h^8}{128L^7} - \cdots$$

Table IV of the Appendix gives the values of this correction for 100 ft based on the first two terms of this series.

3-46. Tension Correction. The formula for tension correction is derived as follows: E is Young's modulus of elasticity. It is a measure of how much tension is required to stretch a material a certain fraction of its length. More precisely, it is the unit tension in pounds per square inch of the cross-sectional area divided by the unit elongation in feet per foot of length.

Let $t - t_0$ = increase in tension in lb
S = cross-sectional area in sq in.
L = length of tape in ft
C_p = increase in length in ft

Then, by definition,

$$E = \frac{t - t_0}{S} \frac{L}{C_p}$$

or

$$C_p = \frac{L(t - t_0)}{ES}$$

Young's modulus for steel is about 28,000,000 in these units; hence

$$C_p = \frac{L(t - t_0)}{28{,}000{,}000S}$$

3-47. Correction for Sag. The correction for sag C_s is the geometric difference between the actual length L of the tape hanging between two supports and the length of the chord C between the same two points (Fig. 3-18).

The first two terms of the power series expressing the length L of a catenary curve in terms of h and C are the following:

$$L = C\left[1 + \frac{8}{3}\left(\frac{h}{C}\right)^2\right]$$

Hence
$$C_S = C - L = -\frac{8}{3}\frac{h^2}{C}$$

h can be found from statics. If the tape were cut at the center, the exterior force at that point would be equal to the tension t. The moment about A is th. This must balance the moment caused by half the weight

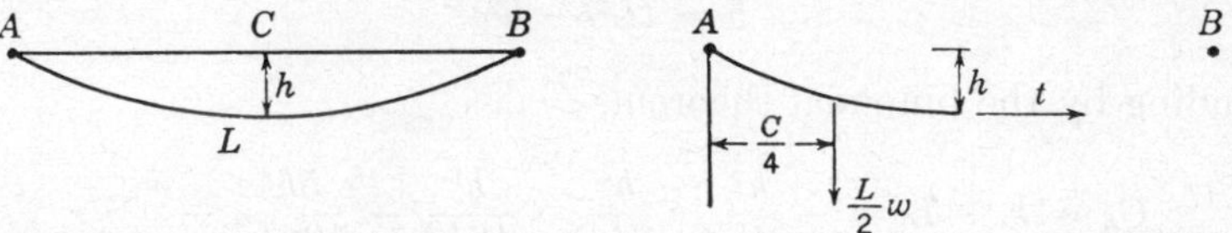

FIG. 3-18. Theory of sag correction.

of the tape acting with a lever arm of $C/4$. When C and L are expressed in feet and w is the weight per foot,

$$th = \frac{L}{2} w \frac{C}{4}$$

$$h = \frac{wLC}{8t}$$

Substituting this in the equation above,

$$C_S = -\frac{w^2L^2C}{24t^2}$$

or
$$C_S = -\frac{w^2L^3}{24t^2} \quad \text{very nearly}$$

Values of C_S computed by this formula are given for various spans and various weights of tape in Chart II of the Appendix.

3-48. Measuring the Weight per Foot and the Cross-sectional Area. The process of determining the weight per foot w and the cross-sectional area S will be demonstrated by an actual example.

Weight of tape, reel, and accessories	1720	g
Less weight of reel and accessories	256	g
Weight of tape	1464	g
1464 × 0.00220462	3.228	lb
Nominal length of tape	300	ft
Length beyond marks and allowance for rings and loops	1.4	ft
Total length	301.4	ft

$$w = \frac{3.228}{301.4} = 0.0107 \text{ lb per ft}$$

A steel bar, 1 in. square, weighs 3.4 lb per ft.

$$S = \frac{0.0107}{3.4} = 0.00315 \text{ sq in.}$$

Check, micrometer-caliper measurement:

Width = 0.125 in.
Thickness = 0.026 in.
Area = 0.00325 sq in.

The first determination, 0.00315 sq in., should be used.

OTHER COMPUTATIONS

3-49. Correction Tension. Sometimes surveyors prefer to eliminate the tape correction by applying the tension required to give the tape the proper length. This can be called the **correction tension.** The method is inapplicable when the correction tension is too large or too small.

After the tape has been standardized by a determination of the tape correction C_t when supported throughout at a tension t_0, the correction tension for the tape supported at the zero and the end mark can be computed from the properties of the tape described in the previous paragraphs.

The correction C_t means that the tape is C_t too long. When the tape is allowed to sag, the error becomes

$$C_t - \frac{w^2L^3}{24t^2}$$

But when the tension is increased to compensate, the error becomes

$$C_t - \frac{w^2L^3}{24t^2} + \frac{L(t - t_0)}{ES}$$

To make the tape the correct length, this expression should be equal to zero

$$C_t - \frac{w^2L^3}{24t^2} + \frac{L(t - t_0)}{ES} = 0$$

Solving for t,

$$t^3 + \left(\frac{ESC_t}{L} - t_0\right)t^2 - \frac{ESw^2L^2}{24} = 0 \qquad (3\text{-}1)$$

In this equation, all the constants are known. It may be solved for t by any of the methods for solving cubic equations. t is the correction tension.

3-50. Normal Tension. Frequently, surveyors prefer to use a tension—when the tape is supported at the end marks—that will make it the same length that it has when supported throughout at its standard tension. This is called **normal tension.** It does not eliminate the tape correction C_t. The expression can be derived from Eq. (3-1) by giving C_t the value zero, since it is not considered.

$$t^3 - t_0t^2 = \frac{ESw^2L^2}{24}$$

$$t\sqrt{t - t_0} = wL\sqrt{\frac{ES}{24}}$$

This equation may be solved for t by any method for solving a cubic equation. It can be solved rather easily on a slide rule as shown in the following example.

Assume $t_0 = 10$, $w = 0.017$, $L = 100$, $E = 28{,}000{,}000$, $S = 0.005$.

$$t\sqrt{t - 10} = 130$$

Set the hairline at 130 on the D scale. Then find the point on the D scale such that when the C index is placed on it, the B scale under the hair line reads 10 less (10 is the value of t_0). This is 29.45. The B scale then reads 19.45. t, the normal tension, is then 29.45 lb.

PROBLEMS

3-1. Complete the field notes given below and compute the final length.

DATA FOR PROB. 3-1
Traverse Sta. G to H

Tape No. 8 299.983 0.012 lb per ft 10 lb tension throughout

Sta.	Length	Temp.	Tape ht.
G 0 + 0			0.2
1 + 13	113		0.0
3 + 00	187	42	{0.5 0.7
3 + 50	50		0.4
4 + 70	Span 120		0.0
6 + 0	130	45	{0.0 1.1
7 + 25	125		0.6
9 + 0	175	47	{0.2 0.0
H 9 + 72.34	72.34		0.0

DATA FOR PROB. 3-1 (*Continued*)
Traverse Sta. *G* to *H*
Leveling

Sta.	+	H.I.	−	Rod	Ground elev.
0 + 0	6.32	58.58			52.26
1 + 13				4.84	53.74
T.P. 3 + 0	7.88	63.78	2.68		55.90
3 + 0					
3 + 50				6.21	57.57
4 + 70				4.32	59.46
T.P. 6 + 0	8.29	69.52	2.55		61.23
6 + 0					
7 + 25				6.26	63.26
9 + 0				4.21	65.31
9 + 0					
T.P. 9 + 72.34			1.01		68.51

3-2. Complete the field notes given below and compute the final length.

DATA FOR PROB. 3-2
Traverse Sta. *D* to *E*
Tape No. 4 300.161 0.010 lb per ft 10 lb tension

Sta.	Length	Temp.	Tape ht.
D 0 + 0			0.4
0 + 92 }	Span		0.6
2 + 04 }			0.1
3 + 0		52	0.3
3 + 0			0.0
4 + 40			0.0
6 + 0		52	1.0
E 6 + 52.37			1.2

Traverse Sta. *G* to *H*
Leveling

Sta.	+	H.I.	−	Rod	Ground elev.
B.M.	1.28	116.62			115.34
0 + 0				2.0	114.6
0 + 92				9.1	107.5
T.P.	2.38	106.58	12.42		104.20
2 + 04				5.7	100.9
3 + 0				10.9	95.7
3 + 0				11.2	95.4
4 + 40				0.5	106.1
T.P.	2.29	99.42	9.45		97.13
6 + 0				3.8	95.6
6 + 52.37				12.1	87.3

3-3. A 300-ft tape weighing 0.013 lb per ft and having a cross-sectional area of 0.0038 sq in. is standardized at 300.003 ft when supported throughout with a tension of 10 lb. What is its length when supported at 0, 100, 200, and 300 ft with a tension of 20 lb?

3-4. A 300-ft tape having a cross-sectional area of 0.0035 sq in. is standardized at 299.980 ft when supported throughout with a tension of 20 lb. What is its length when supported at 0, 100, 200, and 300 ft with a tension of 20 lb?

CHAPTER 4
ANGULAR MEASUREMENT

4-1. Introduction. This chapter covers the accurate measurement of angles.

Units of Angular Measure. There are four systems of angular measure that are used in surveying, as follows:

Sexagesimal System

1 circumference = 360° (degrees of arc)
1 degree = 60′ (minutes of arc)
1 minute = 60″ (seconds of arc)

Centesimal System

1 circumference = 400^g (grads)
1 grad = 100^c (centigrads)
1 centigrad = 100^{cc} (centi-centigrads)

System Based on Hours

1 circumference = 24^h (hours)
1 hour = 60^m (minutes of time)
1 minute = 60^s (seconds of time)

The Coast Artillery System

1 circumference = 6400 mils

The sexagesimal system is used extensively in Europe and almost exclusively in the United States. More complete tables of trigonometric functions are available in this system than in any other. Most angle-measuring instruments are graduated according to this system. Because of the difficulty of computing values in terms of angles that are not expressed decimally, some tables have been written in degrees and decimal parts of a degree, and other tables in degrees and minutes and decimal parts of a minute.

The centesimal system is growing in favor in Europe. Good tables are available and many surveying instruments are graduated in this way. The system facilitates computation and interpolation.

The grad is sometimes spelled *grade*. The centigrad is usually called the centesimal minute and the centi-centigrad the centesimal second. Sometimes the centesimal minutes and seconds are abbreviated by reversed primes and double primes thus: $97.3248^g = 97^g32$`48``.

The system based on hours is used in astronomy and navigation to express angular values measured in a plane perpendicular to the earth's axis. Tables are available. The system is going out of use in navigation and is of interest to the surveyor only in astronomical computations.

The Coast Artillery system is being used more and more for military operations. It facilitates computation and interpolation. Some tables are available. One mil subtends very nearly 1 yd at 1000 yd. The exact value to eight significant digits is

$$1 \text{ mil} = 0.00098174770 \text{ radian}$$

4-2. Definitions. Certain names have been given to surveying instruments according to specific but limited attributes. The names are defined in this section. In common usage these names usually indicate certain specific types of instruments having several attributes. The common meanings are defined in the next section. Except where confusion would result, the meanings in common use are used in this text.

A **theodolite** is the general term for a precise instrument designed to measure horizontal or vertical angles about a point.

A **transit** is a theodolite with a telescope that can be reversed in direction with respect to the rest of the alidade by rotating it about its horizontal axis without removing it from its bearings. Since this process causes it to pass through the vertical, it makes a **transit** through the zenith and hence from one point of reckoning angles to another.

A **repeating theodolite** is one having a graduated horizontal circle that will turn with the alidade or remain stationary with the leveling head and can be precisely set with respect to each one. With this arrangement, when a horizontal angle is turned more than once (repeated), the value of each turn can be successively accumulated on the circle.

A **direction theodolite** is one in which the horizontal circle cannot be made to turn with the alidade. Usually the horizontal circle can be rotated with respect to the leveling head.

A **tacheometer** (or tachymeter) is the general term for an instrument that will measure distance by a single observation. An instrument equipped with stadia hairs is the most common example. Range finders or height finders are also tacheometers.

4-3. Common Usage. Since nearly all theodolites will transit, the word **transit** as used in this country has come to mean the repeating theodolite with a double center of the type generally manufactured and used in the United States. A transit need not have a telescope level, a vertical circle, stadia hairs, or a compass, but today these are usually included. The best type has a full vertical circle, an erecting eyepiece, and a horizontal circle numbered in both directions.

The instrument can be used to measure horizontal and vertical angles, to run levels, to make stadia measurements, to determine the astronomical observations necessary to establish astronomical north, and to stake out straight lines, curves, angles, and grades. The erecting eyepiece makes it especially easy to use for staking-out work. It is capable of measuring horizontal angles to almost any desired degree of accuracy by increasing the number of times the angle is repeated.

When more than one angle is measured at a station to third-order accuracy or better, the transit is slower to operate than a direction instrument. The difference between the two increases with the accuracy desired and with the number of angles measured at the station.

A transit is, therefore, a general purpose instrument designed to perform easily and quickly any surveying operation except accurate length measurement and extensive triangulation.

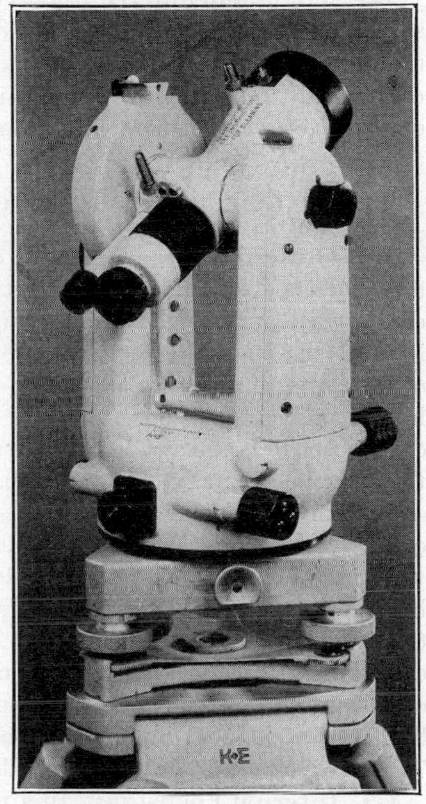

FIG. 4-1. An optical-reading theodolite. Both circles are read through an eyepiece to the left of the telescope. (*Keuffel & Esser Co.*)

The term **theodolite** as used in the United States means a very precise **direction theodolite** designed especially for triangulation and often used for measuring angles in first- or second-order traverse. It has an accurate circle, numbered clockwise, that is read by various types of optical micrometers, and it has other precision features. It is equipped for night observations and usually has a vertical circle for astronomical observations and for the trigonometric leveling usually required in extensive triangulation (Fig. 4-1).

In measuring horizontal angles, the circle remains fixed during a set of observations. The line of sight is pointed successively along each line that forms the side of an angle. The circle is read precisely at each pointing with the precise-reading device and the angles are computed by finding the difference between the directions. To eliminate the errors of the circle and to reduce the effect of accidental errors of observation, after one set of directions has been obtained the circle is rotated to a new position and the process is repeated. A sufficient number of circle positions is used to obtain the desired accuracy.

The term **tacheometer** usually means a theodolite equipped with stadia hairs or a **distance wedge** which accomplishes the same purpose, usually with greater accuracy. Like a transit, it can be used for stadia operations. Since most transits can perform the functions of a tacheometer, the term tacheometer is seldom used in the United States.

PRINCIPAL DESIGN FEATURES OF TRANSITS AND THEODOLITES

4-4. Optics.[1] It can be stated categorically that the optical system of a telescope can never be perfect even in theory. Every design is a compromise among many conflicting desirable objectives. Good optics is the result of skillful compromise, sound design, careful manufacture of the parts, and precise assembly. For example, a telescope may have high magnification, good resolving power, coated optics, be completely color corrected, and yet have a curved field, bad coma, poor focus, and therefore be generally objectionable.

The only satisfactory means of testing the optics of an instrument is to compare the view through the telescope with that of a telescope with which the observer is familiar. The instruments should be set up side by side and various objects observed through both. Bright and dull objects should be observed in bright light and in dim light and at different distances. The appearance of the entire field of view and clarity of the cross hairs and stadia hairs should be noted. Finally, a star or some very bright, concentrated light on a black background should be observed. The relative merits of the optics will be obvious.

4-5. Internal Focusing. In a modern telescope the objective lens is fixed and the telescope is usually focused by moving a small negative lens entirely within the telescope (see Fig. 4-2). Such a telescope is said to be **internal-focusing.**

There have been many misconceptions of this system. It has been pointed out that the extra lens will absorb more light than is necessary. Actually, the light absorbed by such a thin piece of optical glass is negligible. The reflections from the two extra surfaces are more important. They cause a loss of light and introduce stray light in the telescope which tends to dim the image, just as house lights dim a motion picture screen. Sometimes the stray light forms unwanted images of bright parts of the

[1] Cf. Sec. 2-8.

object at various places on the image plane. These are called **flare** or **ghosts.** Fortunately, modern *lens coatings* practically eliminate reflection in the visual range so that the loss of image clarity caused by the extra lens is unimportant.

On the other hand, there are four important advantages of internal focusing.

1. Better optical design is possible. This results in a better image.
2. It is possible to design the optics so that the addition constant $(f + c)$ in the stadia formula is negligible.
3. Since the objective lens is stationary, it can be sealed in position. In an *external*-focusing telescope, when the objective lens is moved forward, a partial vacuum is created in the telescope which draws in dust and moisture. These cloud the optics and increase wear in the focusing draw.

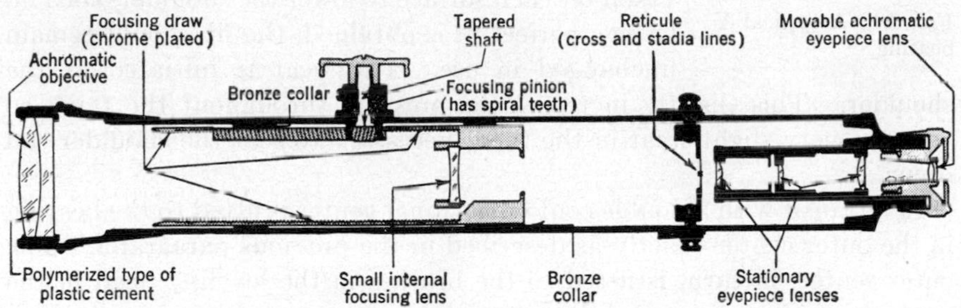

FIG. 4-2. An internal-focusing telescope. *(Keuffel & Esser Co.)*

4. An internal-focusing lens is lighter than an objective lens and therefore its movement does not unbalance the telescope so much.

4-6. Bearings. The accuracy of a theodolite is in a large measure dependent on the telescope axle bearings and the center bearings. They must provide rotation about pure geometric axes. It follows that they must be made without play and designed to prevent play as they wear.

The journal of the best type of telescope axle bearing is usually made of some hard metal turned to a perfect cylinder. It rests on a Y-shaped bearing of softer material. It is usually held in the Y by a small block pressed downward by a spring or screw in the bearing cap. As wear occurs, the journal settles or is forced down deeper in the Y so that no play can develop. The essential element in the design of this bearing is the elimination of any support for the journal at the center, as in a cylindrical bearing. If the journal is supported at the center it cannot drop down to take up wear in the sides (Fig. 4-3).

The journal of the center bearing is usually a hard, tapered shaft recessed for about one-third of its length in the central section. At the top is a flat, horizontal shoulder. In a direction theodolite the journal rests on a tapered bearing in the leveling head, and the shoulder is in contact with the top of the leveling-head bearing. The weight of the alidade is carried on the shoulder while perfect contact is maintained with the tapered bearing (Fig. 4-4).

The two tapers are turned in a precise lathe with the same taper setting so that they are identical. In the best practice they are turned to a size that allows the shoulder to come within two or three thousandths of an inch from its seat at the top of the bearing. The two parts are then lapped together, usually by hand. Lapping consists of working the parts together with a very mild abrasive until they fit each other perfectly. This process is continued until the shoulder makes the proper contact with its seat. Because of the taper, only 0.00005 in. need be taken off each surface to lower the shoulder 0.003 in. Once a perfect fit is obtained, the fit should remain unchanged in use. The wear is initiated at the shoulder. This slightly increases the pressure throughout the taper so that the very slight wear in the taper necessary to seat the shoulder can occur.

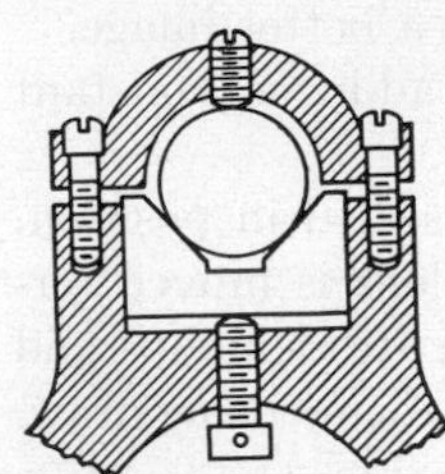

FIG. 4-3. A typical Y bearing.

In a transit with a double center, the inner center is fitted to the bearing in the outer center exactly as described in the previous paragraph. The outer center, in turn, is fitted to the bearing in the leveling head in the same way. The tapers of the outer center must be concentric throughout their length. Usually both tapers are turned without removing the piece from the lathe. Recesses are provided at the center of the bearings to give them stability.

Many other types of bearings are used. Those described have proved their worth over many years in both astronomical and surveying instruments.

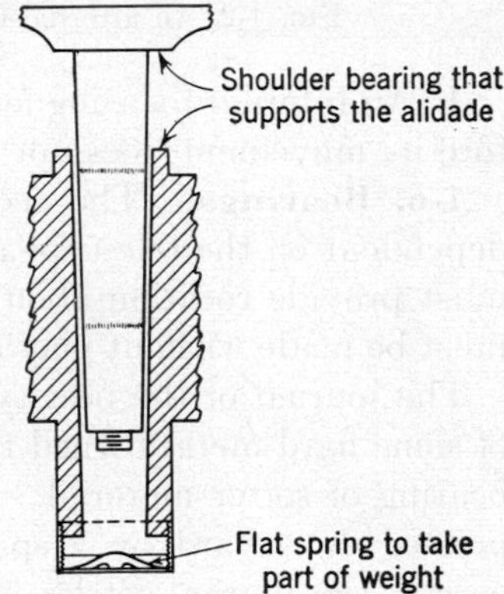

FIG. 4-4. A spindle partly withdrawn from its bearing.

4-7. Tests for Fit and Stability. Play that may cause errors can be discovered quite easily.

Set up the instrument and point at some well-defined object. Twist the tripod plate (footplate) in a horizontal plane just enough to move the cross hairs off the point, and then release it. If the cross hairs fail to return to the point, there is play in the tripod. Loose tripod-head nuts or tripod shoes may be the cause.

When the play has been eliminated, repeat the test by twisting the leveling head. This shows play in the leveling screws. This type of play is more often encountered in three-screw leveling instruments. Usually it can be eliminated by tightening the adjusting screws that regulate the lateral pressure on the leveling screws.

Play in the center bearing or in the clamps is tested by twisting the standards both vertically and horizontally. A very slight twist is sufficient. Finally, the telescope axle bearings can be tested by lightly pushing the telescope off line left and right and up and down.

The outer center of a repeating instrument should be tested for alignment of the tapers as follows. Aim the line of sight at some well-defined point. Loosen both horizontal clamps. Turn the circle about 45°, set both clamps, and bring the vertical cross hair on the point. The horizontal hair should return to the point. Repeat at 45° intervals around the circle.

4-8. Ball Bearings. Ball bearings are used extensively in instruments of European manufacture and in some instruments made in the United States. They will operate with the same lubricant over a very wide temperature range and they give the instrument a smooth, free feel.

The balls used must be specially selected for roundness, and those in each bearing must be matched for size to an exceedingly small tolerance. To prevent play, the bearings must be carefully designed so that the balls are automatically kept in contact with their races. The points of contact between each ball and its races are so small that when a ball bearing is subject to vibration during shipment, pits are sometimes formed under the balls.

4-9. Circles. Circles and verniers are usually made of aluminum or of solid silver mounted on bronze. Glass circles are used extensively in European instruments and are coming into use in the United States.

The accuracy of a circle or vernier depends on the dividing engine used to make it. A circle-dividing engine consists of a wheel about 3 ft in diameter mounted horizontally on a vertical center. As in a theodolite, the center must provide rotation about a pure geometric axis, and it is usually designed on the same principles as a theodolite center. Tapered bearings or ball bearings are used.

The circle to be graduated is mounted at the center of the wheel. The wheel is driven by a worm and gear. The gear is about the same size as the wheel itself, and it practically forms the circumference. The worm that drives the gear is turned by a ratchet wheel and pawl. The movement of the pawl can be regulated to turn the main wheel the desired angle between graduations. When the wheel has been rotated, an automatic cutter cuts a graduation to the proper predetermined length.

When the engine is assembled, and frequently thereafter, careful tests are made to determine the accuracy of the rotation of the wheel throughout its entire circumference. The errors discovered are eliminated by a cam that extends around the wheel, usually below the gear, and is shaped in accordance with the errors measured. The cam follower, through a reducing lever, moves the worm axially just enough to advance or retard the wheel by the amount of correction required. In spite of these precautions slight residual errors always remain. Probably the chief source is play in the center bearing.

4-10. Eccentricity. When the circle is cut, the pattern of the graduations will be centered at the axis of the engine. When the circle is mounted on the theodolite, the center of the pattern should be exactly at the vertical axis of the alidade. The mounting is adjustable so that this result can be attained to very close tolerances. The residual error is called the eccentricity. When an index of the alidade (a vernier or micrometer microscope) travels over a portion of the circle that is too near the axis of the alidade, it covers fewer divisions of the circle than called for by the angle turned. The measured angle is then too small. An index moving on the opposite side will give too large a measured angle. The error varies in a sine curve.

It is shown in the next paragraphs that the effect of eccentricity can be eliminated by reading the circle simultaneously at two or more indices uniformly distributed around the alidade, and then using the average.

4-11. Proof of the Elimination of Eccentricity. Figure 4-5*a* shows a circle turned to any position. e represents the direction and size of the eccentricity (the distance between the center of graduations and the axis of the alidade).

P = point read on circle
T = true angle, the rotation of the alidade from a position at which the index was on line with the eccentricity to the position P
M = measured angle, that is, the angle traversed on the circle
C = correction ($T = M + C$)
r = radius of circle

From the figure

$$\frac{\sin C}{e} = \frac{\sin T}{r}$$

$$C = \frac{e}{r} \sin T \qquad (C \text{ is a small angle}) \tag{4-1}$$

If the circle is read at one other position 180° away at P_1, $T_1 = T + 180°$. By Eq. (4-1),

$$C_1 = \frac{e}{r} \sin T_1$$

Adding, $$C + C_1 = \frac{e}{r} (\sin T + \sin T_1)$$

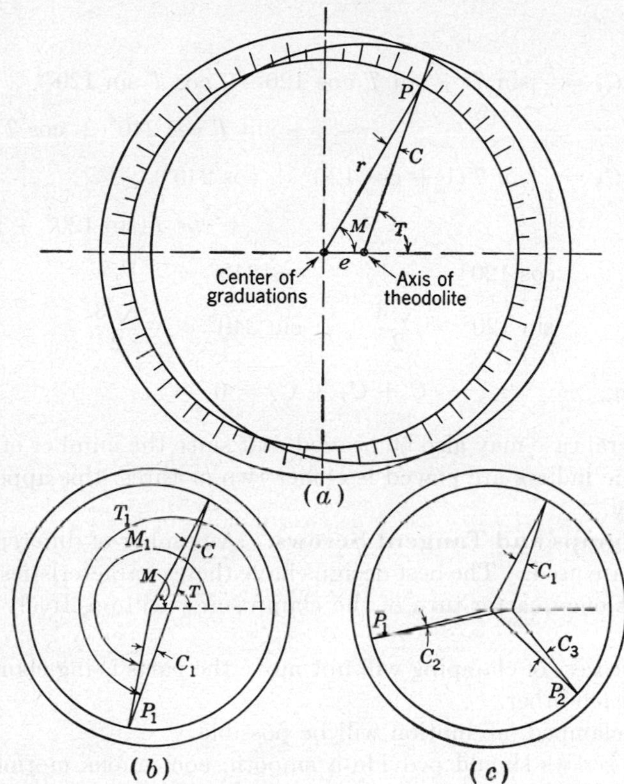

Circle is read at points 180° apart.
$T = M + C$
$T_1 = M_1 + C_1$ (C_1 is a minus value)
But $C_1 = -C$ (base angles of an isosceles triangle).

Circle is read at points 120° apart.
$C + C_1 + C_2 = 0$

FIG. 4-5. The effect of eccentricity.

but $$\sin T_1 = \sin (T + 180°) = -\sin T$$

Substituting, $$C + C_1 = \frac{e}{r} (\sin T - \sin T)$$
$$C + C_1 = 0$$

If the reading is taken at two other positions P_1 and P_2, $T_1 = T + 120°$, and $T_2 = T + 240°$. By Eq. (4-1),

$$C = \frac{e}{r} \sin T$$

$$C_1 = \frac{e}{r} \sin (T + 120°)$$

$$C_2 = \frac{e}{r} \sin (T + 240°)$$

Adding

$$C + C_1 + C_2 = \frac{e}{r}[\sin T + \sin T \cos 120° + \cos T \sin 120° + \sin T \cos 240° + \cos T \sin 240°]$$

$$C + C_1 + C_2 = \frac{e}{r}[\sin T(1 + \cos 120° + \cos 240°) + \cos T(\sin 120° + \sin 240°)]$$

but

$$\cos 120° = -\tfrac{1}{2} \qquad \cos 240° = -\tfrac{1}{2}$$

$$\sin 120° = \frac{\sqrt{3}}{2} \qquad \sin 240° = -\frac{\sqrt{3}}{2}$$

Substituting,

$$C + C_1 + C_2 = 0$$

The general case may also be proved, but since the number of positions at which the indices are placed is either two or three, this appears to be unnecessary.

4-12. Clamps and Tangent Screws. A number of different designs for clamps are used. The best designs have these characteristics:

1. About one-quarter turn of the clamp screw will positively clamp or free it.
2. The process of clamping will not move the parts being clamped with respect to each other.
3. Once clamped, no motion will be possible.

Tangent screws should provide a smooth, continuous motion in each direction.

4-13. Three- and Four-screw Leveling Heads. A four-screw leveling head holds the instrument firmly to its base. This prevents any possible movement of the instrument in operation. On the other hand, when the leveling screws are tightened, the end of each forearm is forced upward while the center of the leveling head is held down by the half-ball. This tends to deform the center bearing into ellipses at the top and bottom. Both play and friction may be introduced (Fig. 4-6).

A three-screw leveling head is sometimes held to its base by the weight of the instrument alone. Often a flat spring acting on a shoulder at the foot of each leveling screw is provided to increase the pressure, or a central spiral spring is arranged to pull the leveling head downward and thus to increase effectively the weight of the instrument (see Fig. 4-1). Each leveling screw can be turned independently and none can deform the bearing. The position of the instrument in azimuth depends on the fit of the threads of the leveling screws. If any play exists in these threads, the instrument is free to rotate slightly. Usually, adjusting screws are provided to tighten the threads in the leveling head so that their play can be eliminated (see Fig. 4-17).

4-14. Reticules. Spider-web cross hairs or lines etched on glass are both successfully used for reticules. Spider web, properly mounted under tension, seldom sags. It is opaque and does not interfere with the remainder of the image. No other filament has proved to be so easily adapted to the purpose. Spider-web reticule patterns are limited to lines extending entirely across the field of view. The spider web can be replaced in the field.

To remove the reticule, remove the eyepiece and two opposite, cross-hair adjusting screws. With the other two screws, rotate the reticule until it can be seen, edge on, in the telescope. Thread a stick of soft wood into the adjusting-screw hole. Remove the other adjusting screws and take out the reticule.

To replace a cross hair, remove the old hair from its groove with alcohol. Cement a length of spider web on dividers with shellac. Stretch the web slightly. Cement in place in the grooves with shellac. It should be placed exactly in the grooves with the aid of a reading glass or a microscope. The eyepiece itself can be used for this purpose.

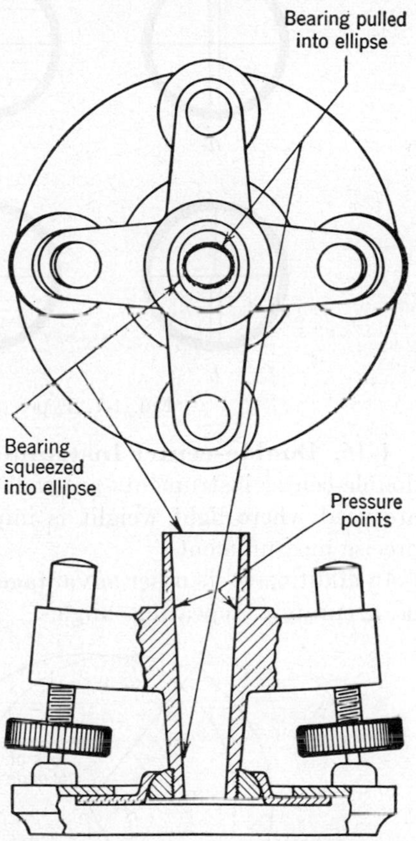

FIG. 4-6. The effect of tightening the leveling screws of a four-screw instrument.

Glass reticules may have any reticule pattern desired. They are permanent and trouble-free. However, the surfaces reflect light, and unless they are coated they increase stray light. In strong light they sometimes show a bright reflection of the pattern beside the pattern itself. Any speck of dust on the etched side of the glass will be in the image plane and therefore appear as a large spot on the object. Accordingly, telescopes with glass reticules should not be opened except where vacuum apparatus is available to eliminate dust.

4-15. Reticule Patterns. Figure 4-7 shows a number of typical reticule patterns. Pattern *a* shows the double vertical wire used chiefly for observing signal lights. The light usually appears too small to be bisected with a single wire but may easily be set between the two wires.

Pattern *b* is the counterpart on a glass reticule. The single wire may be used for large signals and the double wire for small signals. *c* and *d* show stadia wires in spider web and on glass, respectively. The short wires can never be confused with the central horizontal wire. *e* and *f* are types used to observe the sun. The sun's image closely fits the square.

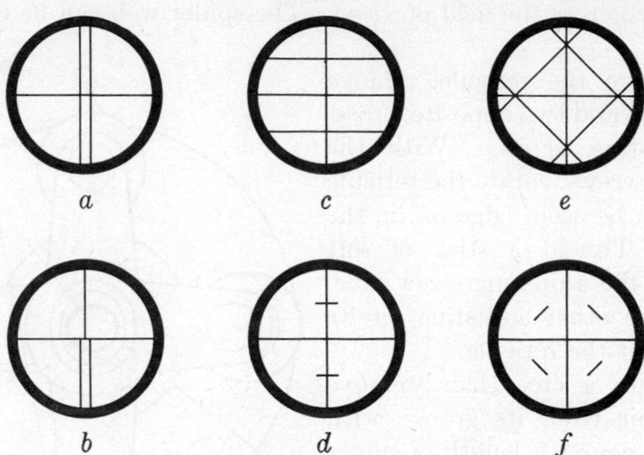

Fig. 4-7. Types of reticule patterns.

4-16. Double-center Instruments. It is assumed that the typical, double-center instruments are well-known to the reader. Small models are used where light weight is important. Large models are used for precise measurements.

In addition to its other advantages, the double center provides a means of accurately repeating angles. Many direction theodolites can be

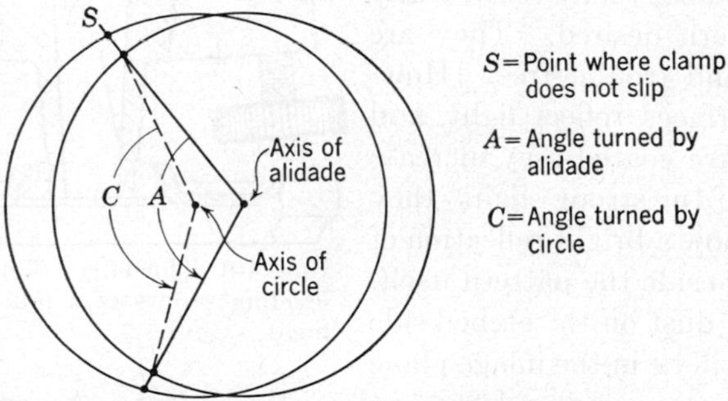

Fig. 4-8. Note the difference between the actual angle *A* and the measured value *C*.

equipped with a proper arrangement of clamps that make it possible to repeat angles, but the results contain a certain systematic error. This error does not occur when a double center is used.

With a double center, only one bearing operates at a time. With other types, when the alidade is being returned to the first side of the angle, the alidade rotates on the center-spindle bearing and the circle rotates on its own bearing. If these bearings are not concentric, an error is introduced as explained under the heading Other Theodolites (see Fig. 4-8).

With a double center, an eccentricity between the inner and outer bearings will have no effect, as each moves independently in true rotation. Theoretically, of course, a slight parallax might be introduced by moving the position of the inner center, but this is too slight in comparison with the lengths of sights to be effective.

THE DIRECTION THEODOLITE

4-17. Distinctive Features. The distinctive features of a direction theodolite that affect its operation are the following:

1. Three-screw leveling.
2. Inverting eyepiece.
3. A single horizontal clamp-and-tangent screw which controls the rotation of the alidade with respect to the leveling head.
4. A finely divided circle that can be rotated around a friction bearing in the leveling head but has no mechanical connection with the alidade.
5. Various types of micrometer microscopes for reading the circle. Usually two, but sometimes three, are provided, uniformly distributed around the alidade.

4-18. The Parkhurst Theodolite. Figure 4-9 shows the Parkhurst theodolite. It was designed by D. L. Parkhurst of the U.S. Coast and Geodetic Survey and is used by that Bureau for first- and second-order triangulation and traverse. It has many distinctive features. It is introduced here to demonstrate the method of reading a circle with a certain type of micrometer microscope.

The circle, divided into 5-minute divisions, is read with two micrometer microscopes placed 180 degrees apart on the alidade. Each microscope has an objective lens close to the circle graduations. It forms an enlarged image of the circle near the micrometer eyepiece. The eyepiece further enlarges the image.

At the image plane is a fixed scale in the form of a comb. Two pairs of wires mounted on a movable frame are also in the image plane. The frame and wires can be moved left and right by a micrometer screw that carries a large drum. Figure 4-10 shows the appearance of the field of view and the drum. The lines on the circle shown are, from left to right, 126°55′, 127°00′, 127°05′, 127°10′, 127°15′, and 127°20′.

The approximate reading is determined from the position of the specially marked center notch on the comb. It is shown at about 127°07′. The bottoms of the notches on the comb are at 1-minute intervals. These show that the reading is about 127°06′40″.

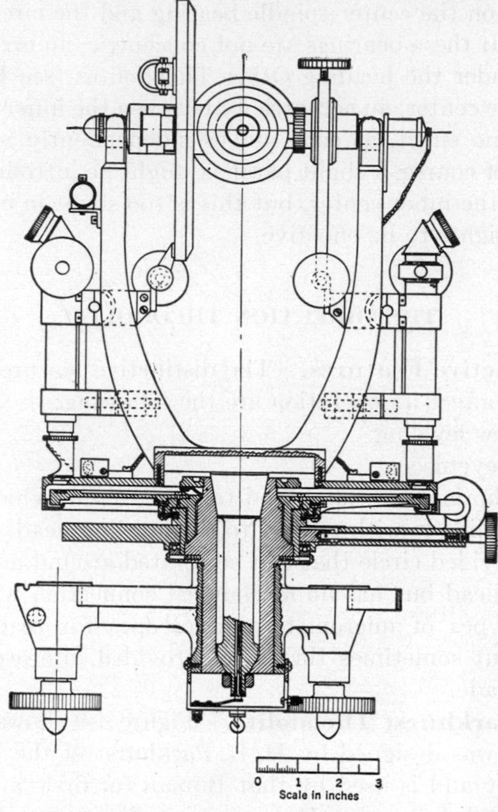

FIG. 4-9. A Parkhurst theodolite. (*U.S. Coast and Geodetic Survey.*)

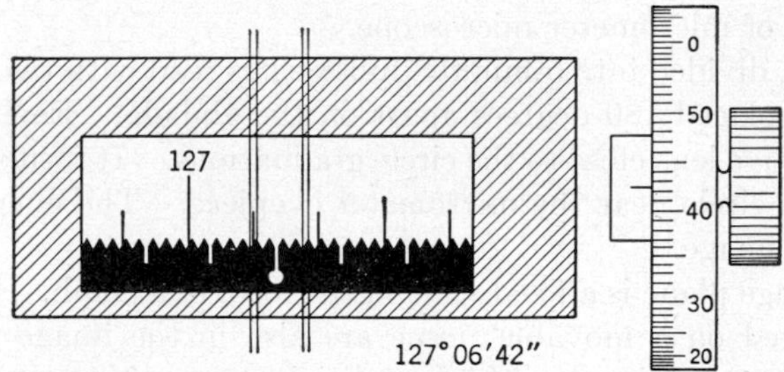

FIG. 4-10. Appearance of the field of view and the drum of a micrometer microscope for reading the circle.

The pitch of the micrometer screw is such that one revolution of the drum moves the wires exactly 2 minutes. The drum is divided into 120 parts, and each division represents 1 second. The drum is set so that it reads zero when either pair of wires is centered over a notch. The pairs of wires are mounted at an interval of 4 minutes. A single pair of wires could have been used; the second pair speeds up the work. When a pair of wires is centered over a line on the circle (the left-hand pair is centered over 127°05′ in the figure) the drum will read the number of seconds through which the wires would have had to be moved from the notch to the right of the line back to the line. In this case it is 42″. The complete reading is therefore 127°06′42″.

The right-hand pair is then centered over the nearest line (in this case 127°10′). Presumably exactly ½ revolution of the drum will do this. The drum is read again. If this reading were 44″, the average, 43″, would be recorded.

The accuracy of the device depends entirely on the accuracy of the micrometer screw. The comb is made so that the notches have the same pitch as the micrometer screw. Small variations in the notches are not important. The distance between the two pairs of wires is approximately four times the pitch. The whole device is adjusted so that the objective of the microscope focuses the image on the plane of the wires and gives such magnification that exactly five turns of the drum will move the wires a distance equal to the average separation of the lines on the circle. Corrections are applied when this result cannot be exactly obtained. A zero reading would occur when the zero line on the circle was in the center notch, in such a position that the average of the drum readings was zero when the pairs of wires were successively centered over it.

4-19. The Modern Theodolite. The most popular type of instrument in use today for triangulation, and often for traverse, is a small, accurate type of direction theodolite; it has many novel features that make it so different from others that it requires separate treatment. This type has so many small, precise parts and optical elements that the handwork required prevents economical manufacture in the United States. It is manufactured by many European firms. The first of this type to become well known in the United States is the **Wild,** manufactured by the Henry Wild Surveying Instruments Supply Co., Ltd., Heerbrugg, Switzerland. The particular instrument known as the T-2 will be described as an example.

4-20. The Wild Theodolite. Figure 4-11 shows the Wild T-2. It is roughly 12 in. high and 6 in. in diameter. The telescope is less than 6 in. long. The center (Fig. 4-12) has a plain, cylindrical bearing at the base and a conical ball bearing at the top. The horizontal bearings are cylindrical, with the bottom of the circle cut out (Fig. 4-13). Both circles are made of glass. The diameter of the horizontal circle is about 3½ in. and that of the vertical circle about 2¾ in. Both are illuminated through the glass by reflected daylight or by small electric lights. The instrument has a circular level for approximate leveling and a single plate level. The index of the vertical circle can be set by a vertical tangent screw according to an independent level of the coincidence type, so that the measurement of vertical angles is independent of any errors in leveling the instrument.

4-21. Circle Reading Device in Wild T-2. The circles in the Wild instrument are read by a very unusual optical micrometer. The eyepiece

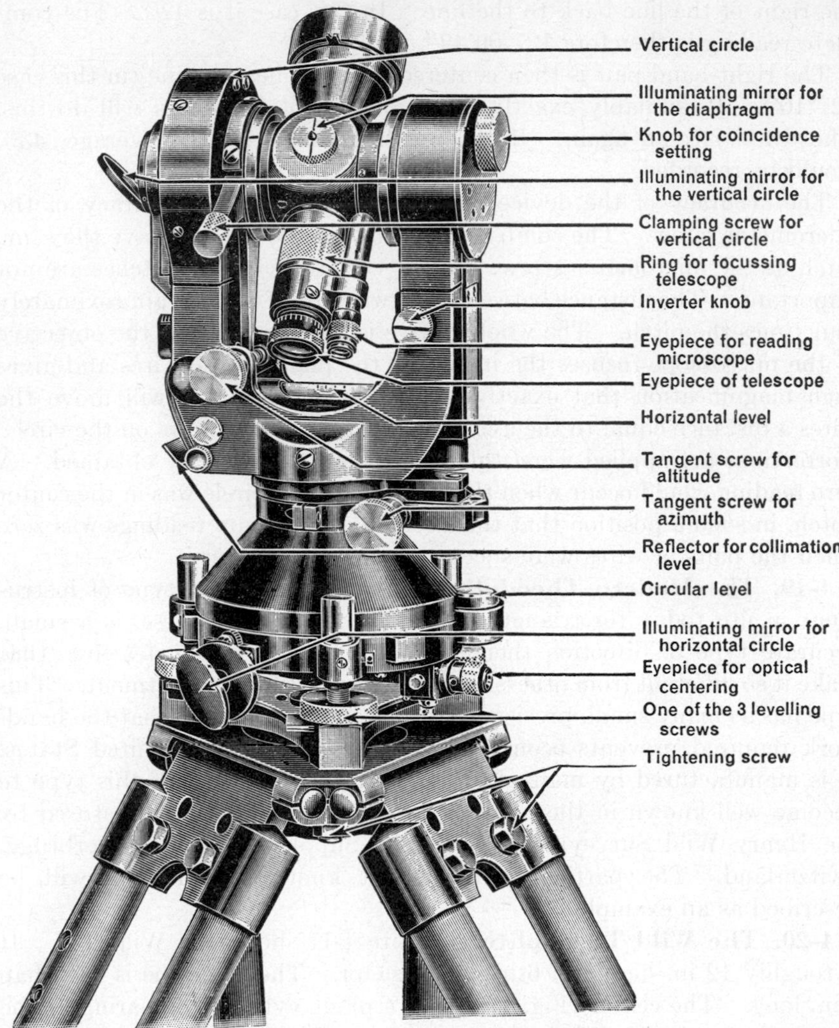

FIG. 4-11. Universal-theodolite Wild T-2. (*Henry Wild Surveying Instruments Supply Co. of America, Inc.*)

is beside the telescope eyepiece and moves with the telescope. In the field of view of the micrometer appear the circle graduations from two parts of the circle 180° apart. The appearance of the field is shown in each of the rectangles in Fig. 4-14. The circle is divided in 20-minute intervals.

The vertical line at the center of the lower half of each rectangle is the approximate index. In Fig. 4-14*a*, the readings of the circle at the index are about 13°45″ and 193°45″, respectively.

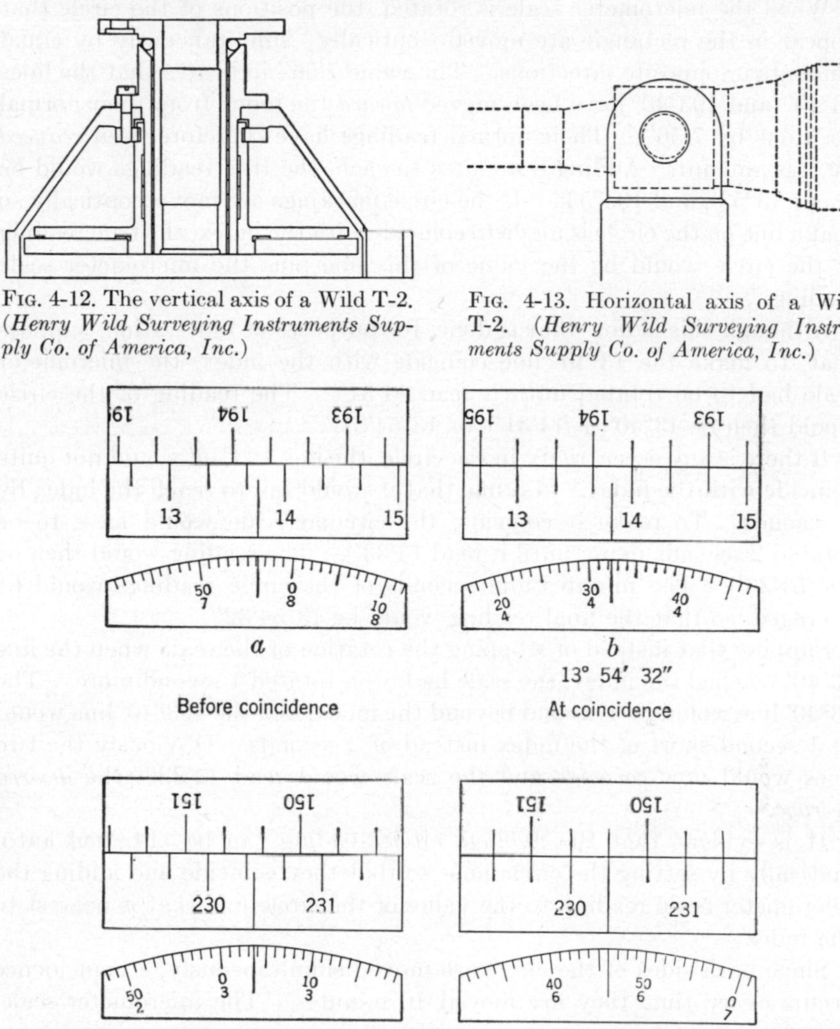

FIG. 4-12. The vertical axis of a Wild T-2. (*Henry Wild Surveying Instruments Supply Co. of America, Inc.*)

FIG. 4-13. Horizontal axis of a Wild T-2. (*Henry Wild Surveying Instruments Supply Co. of America, Inc.*)

FIG. 4-14. Field of view in Wild T-2 optical micrometer for reading the circles.

Near the top of the right-hand standard of the theodolite is a knob called the **knob for coincidence setting** (see Fig. 4-11). When this knob is turned, the micrometer scale which can be seen underneath the rectangle is rotated. The scale is graduated in seconds over a range from

0 to 10 minutes. The lower numbers give the minutes and the upper numbers give the seconds. In Fig. 4-14*a* the micrometer scale reads 7′56″.

When the micrometer scale is rotated, the positions of the circle that appear in the rectangle are moved—optically—simultaneously by equal amounts in opposite directions. The value 7′56″ indicates that the lines 13°40″ and 193°40′ have been moved *toward* the index from their normal positions by 7′56″. Their normal readings have therefore been *reduced* by this amount. Adding 8 minutes to each, the true readings would be about 13°53′, and 193°53′. If the circle markings are moved optically so that a line on the circle is made to coincide with the index, the true reading of the circle would be the value of this line plus the micrometer scale reading.

Although this is not what is done, for purposes of illustration, suppose that, to make the 13°40′ line coincide with the index, the micrometer scale had to be rotated until it read 14′31″. The reading of the circle would then be 13°40′ + 14′31″, or 13°54′31″.

If there is any eccentricity in the circle, the line 193°40′ would not quite coincide with the index. Assume that it would fail to reach the index by 2 seconds. To make it coincide, the circular scale would have to be rotated 2 seconds more, until it read 14′33″. The reading would then be 193°54′33″. The minutes and seconds of the circle readings would be averaged, so that the final reading would be 13°54′32″.

Suppose that instead of stopping the rotation of the scale when the line 13°40′ reached the index, the scale had been rotated 1 second more. The 13°40′ line would be 1 second beyond the index, and the 193°40′ line would be 1 second short of the index instead of 2 seconds. Obviously the two lines would now *coincide* and the scale would read 14′32″, the *desired average*.

It is evident that the average circle reading can be obtained automatically by setting the circle lines so that they coincide and adding the micrometer scale reading to the value of the circle line that is nearest to the index.

Since both sides of the circle are moved simultaneously, a coincidence occurs every time they are moved 10 minutes. The micrometer scale, therefore, need have a range of only 10 minutes to ensure reaching a coincidence. When a coincidence occurs, one of the 20-minute lines on the circle may be either at the index line or placed so that the index line falls halfway between that line and the next. The 10-minute value nearest the index is used.

Figure 4-14*b* shows the actual operation. The micrometer scale has been moved backward until the circle lines coincide. The index line is very near 13°50′. Applying the rule developed, to this value is added the reading of the micrometer scale. The final reading is 13°50′ + 4′32″ = 13°54′32″.

A better way to read the 50 minutes is to count the spaces from 13° to its opposite, 193°. Note that there are five spaces. Each space represents 10 minutes.

Since the index is read only to the nearest 10 minutes, it is not set by the manufacturer precisely with respect to the micrometer scale.

4-22. Procedure for Reading the Circle

1. Turn the coincidence setting knob until coincidence occurs.

2. Read the number to the left of the index. This gives the degrees (13°).

3. Count the spaces to the symmetrically placed number on the upper scale. This gives the tens of minutes (50′).

4. Read the remaining minutes and seconds on the micrometer scale (4′32″).

Note: Because of eccentricity, the size of the spaces may be slightly different on the two parts of the circle seen. It is then impossible to make all the lines exactly coincide at the same time. The best coincidence should be obtained nearest the index.

4-23. Recapitulation

1. The micrometer scale shows how far each side of the circle has been advanced optically toward the index from its normal position.

2. The true reading could be obtained at any time by adding the micrometer reading to the position of the theoretical index on the circle if it could be determined.

3. When coincidence occurs, the theoretical index must be at a 10-minute position plus one-half the error in eccentricity on one part of the circle and minus one-half the eccentricity on the other part of the circle. It is, therefore, at the position which averages that particular 10-minute position for both parts of the circle.

4. The final reading is that 10-minute value plus the reading of the micrometer scale.

Figure 4-14*c* shows another position of the circle. When coincidence is made, the line 230°20′ is to the left of the index, so that the reading is 230°26′46″.

Using the rules for taking the reading:

1. Set the coincidence Fig. 4-14*d*.

2. Read the number to the left of the index (230°).

3. Count the spaces from 230° to the symmetrically placed number on the upper scale (150°)—two tens of minutes.

4. Read the remaining minutes and seconds on the micrometer scale (6′46″).

The final reading is 230°26′46″.

OPERATION OF THE WILD T-2

4-24. To Set Up. Set up the tripod over the station. Attach the plumb-bob cord holder to the hold-down screw under the tripod head by its bayonet joint. Center the tripod over the point.

By pulling the leather straps outward, free the lever fastenings on the instrument case (Fig. 4-15). Remove the hood, remove the dust cap, unscrew the three hold-down screws, slide the hold-down fingers outward, remove the instrument, and place it on the tripod. In handling the instrument, avoid grasping the left-hand standard (the larger of the two), as this may affect the adjustment. Immediately thread the hold-down screw into the base plate. Never leave the theodolite on the tripod without threading this screw. If the screw is not threaded, the instrument will fall off when the tripod is disturbed.

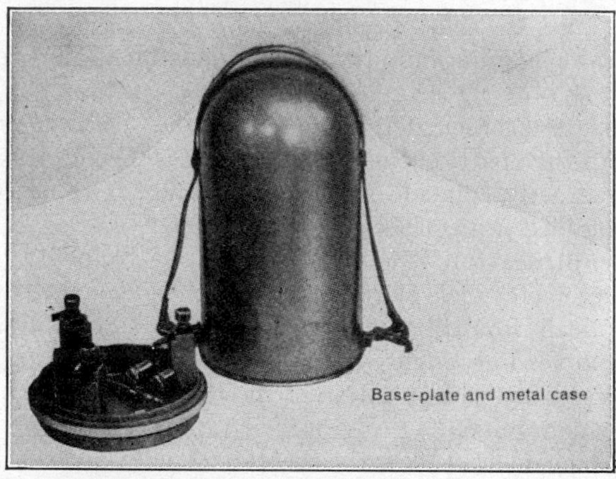

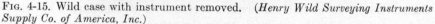

FIG. 4-15. Wild case with instrument removed. (*Henry Wild Surveying Instruments Supply Co. of America, Inc.*)

Slide the instrument laterally until the plumb bob is over the point and tighten the hold-down screw. Level the instrument according to the circular level. The bubble will move toward any leveling screw turned clockwise and vice versa.

Free the horizontal clamp (note Fig. 4-11). It is behind the instrument in Fig. 4-11 and at the level of its tangent screw (labeled "tangent screw for azimuth"). Turn the alidade so that the single level (labeled "horizontal level") is parallel with a line between two leveling screws. Center the bubble with these screws, turning them simultaneously in opposite directions. Turn the instrument to 90° in azimuth and center the bubble with the remaining screw alone. Check in the first azimuth position and repeat the procedure if necessary.

Remove the plumb-bob holder and the plumb bob. Focus the eyepiece of the optical centering device until the small black circle appears sharp. The circle marks the prolongation of the vertical axis of the instrument. When the instrument is level the circle should be on the point. If it is not centered, loosen the hold-down screw and slide the instrument on the tripod head to center it. Be careful to avoid rotating the base as this will throw the instrument out of level, and the position of the circle of the optical centering device will have no meaning. When the instrument is centered, tighten the hold-down screw, relevel, and check the centering.

4-25. To Measure Horizontal Angles. Adjust the eyepiece by rotating it. Turn the inverter knob so that the line on it is horizontal. Look through the eyepiece of the reading microscope and adjust the illuminating mirror for the horizontal circle until the scales are brightly illuminated. Focus the microscope eyepiece by turning it until the scales are sharp. Set the micrometer scale at the desired initial reading in minutes and seconds.

Aim the telescope at the initial signal. The open sights will aid in this operation. Focus by turning the knurled ring for focusing the telescope. Perfect the pointing with the horizontal and vertical clamps and their tangent screws. In Fig. 4-11 the vertical clamp is labeled "clamping screw for vertical circle" and the tangent screw, "tangent screw for altitude."

At the base of the instrument is a round, hinged cover about 1 in. in diameter. When it is opened, a knurled head is exposed; this turns the horizontal circle. Turn the circle to the desired initial circle reading and close the cover. Perfect the coincidence and record the reading.

Point at the other signals in clockwise order. Take the reading for each. The differences between the readings are the values of the angles. The circle is numbered from 0 to 360° clockwise.

4-26. To Measure a Vertical Angle. Turn the inverter knob so that the line on it is vertical. This turns a prism that brings the vertical circle instead of the horizontal circle into the rectangle in the field of the microscope. Adjust the illuminating mirror for the vertical circle until the scales are bright.

Open the reflector for the vertical index level (labeled "reflector for collimation level") by turning the knurled knob at the lower end of the left-hand standard. Adjust the prism so that the bubble can be seen. The prism is shown in Fig. 4-11 at the left of the left-hand standard about one-third of the way up from the bottom.

Point the instrument at the signal. Center the coincidence bubble by turning the tangent screw which is at the back of the left-hand standard. It is out of sight in Fig. 4-11. It has large smooth knurling.

Read the vertical angle in the same manner as a horizontal angle. The rectangle will be inclined in the field of view in accordance with the size of the vertical angle. The vertical circle is numbered from 0 to 360°. It reads 0 at the zenith, 90° when the telescope is direct and horizontal, and 270° when reversed and horizontal. Accordingly, it reads the zenith distance when the telescope is direct and reads the explement of the zenith distance (360° minus the zenith distance) when the telescope is inverted.

4-27. Night Operation. To change over for night operation, remove the two circle illuminating mirrors and replace by the lights found in the base of the instrument case. Both the mirrors and the lights snap on and off the instrument. Plug in the battery at the base of the instrument.

When observing a star or a light with a black background, turn the illuminating mirror for the diaphram until the line on it is at 45° to the telescope. This gives the maximum illumination on the cross hairs. By regulating the angle, the amount of illumination is controlled.

4-28. To Move to Another Station. The theodolite can be left on the tripod when the instrument is moved a short distance. Before picking up the tripod be sure the hold-down screw is tight.

4-29. To Replace the Instrument in the Case. Replace the dust cap, close the mirrors, unscrew the hold-down knob, and place the instrument in the case so that the three lugs at the base fit in their sockets. Slide the hold-down fingers over the lugs and tighten the screws. Place the hood in position, hook the fasteners, and snap them tight by pressing them against the hood.

PROCEDURE AT A TRIANGULATION STATION

4-30. Stations Observed. At each station on a triangulation system, as has been stated, the directions of all the triangle sides that meet at that point are observed with the circle in one position. The circle is then turned to another setting and the process repeated. Usually five directions must be observed, so five directions are used as an example.

Assume the station occupied is station 0 and that stations A, B, C, D, and E are situated clockwise around station 0 in the order stated.

4-31. Initial Station. One station is arbitrarily chosen as the *initial station.* Assume this to be station A. Each circle position is regulated by setting a precomputed value, called **initial setting,** on the circle when line of sight is pointing at A.

4-32. Initial Settings. The initial settings are chosen so that the entire circle and the entire range of the micrometer are used as uniformly as possible. Assume that the circle is read simultaneously at two points 180° apart and that the accuracy desired requires four circle positions. The interval between settings would be 180° ÷ 4 = 45°. The formula for the interval I is

$$I = \frac{360°}{mn} \tag{4-2}$$

where m is the number of places the circle is read and n is the number of circle positions required.

If the range of the micrometer is 10 minutes, the interval would be 2½ minutes for four circle positions.

The initial settings are therefore

0°0′00″
45°2′30″
90°5′00″
135°7′30″

4-33. Measuring the Directions (see the form of notes in Fig. 4-16). The line of sight is pointed at A. The micrometer scale is set to read 0′0″. The circle is set as nearly as possible at the coincidence for 0°0′. Coincidence is then accurately established with the coincidence setting knob. This changes the scale reading by a small quantity. The direction is then read, giving 0°0′32″. It is recorded at the bottom of the page in line A, position 1.

The directions to the other stations B, C, D, and E for that circle position are observed and recorded up the page in position 1. When these are complete, A is observed to make sure the circle has not moved. It should check the original reading within 2 seconds. If it checks, the original direction plus 360° is recorded in the top line. If it fails to check, the set of observations should be repeated.

While the observations are being made, the recorder computes the angles AB, BC, etc., by finding the differences between adjacent directions. All the angles are computed, even though they may not be part of the triangulation system, so that a check on the arithmetic can be obtained. The sum of all the angles for one position must be exactly 360°.

As soon as the recorder is satisfied with position 1 and while the line of sight is pointed at the initial station, the next initial setting is set on the micrometer scale and the circle. In this case the micrometer scale is set at 2′30″ and the circle is turned to approximate coincidence at 45°00′. Accurate coincidence is established with the micrometer, and the reading is taken. The directions for all the circle positions are observed similarly. Half the circle positions should be observed with the telescope direct and half with it reversed.

The values of each angle obtained with the various circle positions are compared. If one value is obviously in error, the circle position is repeated. When the values

	Station Occupied *O*				
Sta.	Pos. 1 Direct	Pos. 2 Direct	Pos. 3 Reversed	Pos. 4 Reversed	Av. angle
A | 360°	360 00 32	405 02 15	150–05–22	195–07–56	
E–A	172–25–51	172–25–42	172–25–45	172–25–43	172– 25– 45.2
E	187–34–41	232–36–33	277–39–37	322–42–13	
D–E	45–12–25	45–12–22	45–12–28	45–12–26	45– 12– 25.2
D	142–22–16	187–24–11	232–27–09	277–29–47	
C–D	47–11–54	47–12–00	47–12–01	47–11–58	47– 11– 58.2
C	95–10–22	140–12–11	185–15–08	230–17–49	
B–C	69–57–13	69–57–18	69–57–15	69–57–17	69– 57– 15.8
B	25–13–09	70–14–53	115–17–53	160–20–32	
A–B	25–12–37	25–12–38	25–12–31	25–12–36	25– 12– 35.5
A	0– 0–32	45–02–15	90–05–22	135–07–56	
				Check	358–117–17

FIG. 4-16. Form of notes for direction-theodolite observations.

agree within reasonable limits, the average of each angle is computed and recorded in the last column. The sum of the averages should be exactly 360° except for the effect of rounding off the tenths of seconds.

OTHER THEODOLITES

4-34. The Watts Theodolite. The Watts Microptic Theodolite No. 2 (Fig. 4-17) is an instrument of the type just described. It is made by Hilger and Watts, Ltd., London. The method of reading the circle is similar to that used in the Wild T-2 with some important differences.

The circle is graduated at 10-minute intervals. It is viewed in a rectangle as shown in Fig. 4-18. Each 20-minute line is extended outward a short distance and then continued a short distance farther as a double line. The single position of the 20-minute lines is seen in a small window above the rectangles. Imaged with them are the double portions

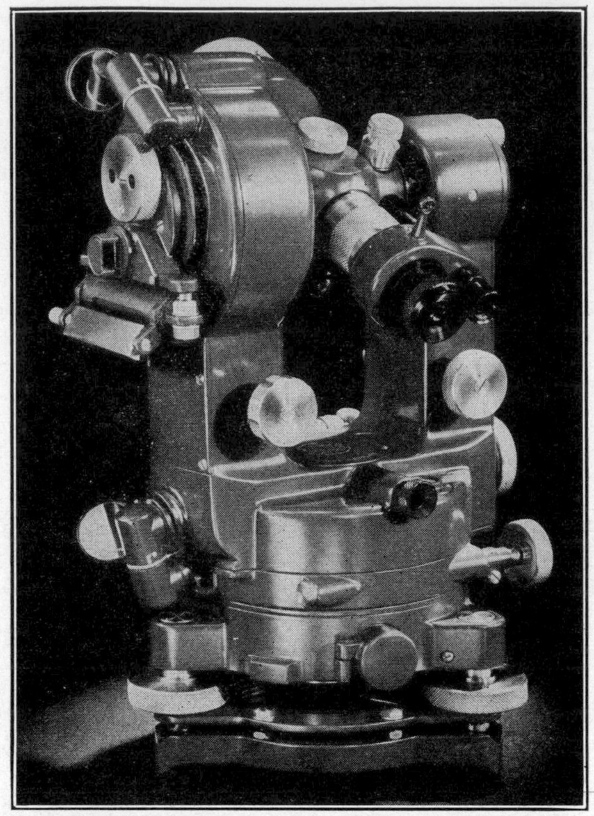

FIG. 4-17. Watts Microptic Theodolite No. 2. (*Jarrell-Ash Company.*)

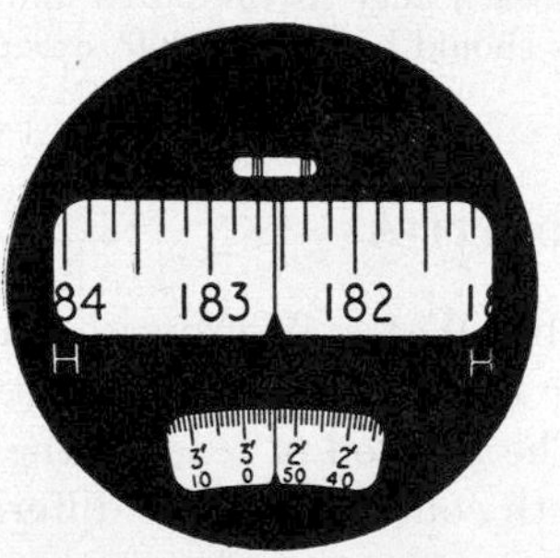

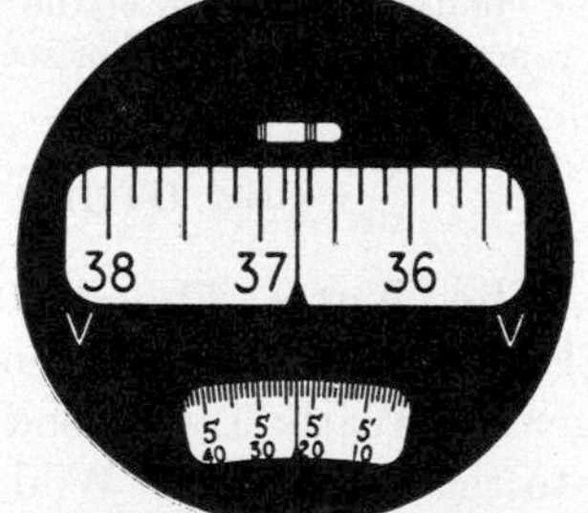

HORIZONTAL CIRCLE READING		VERTICAL CIRCLE READING	
Main scale................	182°30′	Main scale................	36°40′
Micrometer...............	2 54″	Micrometer...............	5 23″
Total....................	182°32′54″	Total....................	36°45′23″

FIG. 4-18. Circle reading device on Watts Theodolite No. 2. (*Jarrell-Ash Company.*)

of the 20-minute lines from the other side of the circle, arranged to move in the opposite direction as the circle is turned.

The micrometer moves only the double lines, that is, only the lines from the opposite side of the circle. The index marker therefore always gives the value of the angle as closely as it may be read. Coincidence is established by turning the micrometer knob until the single line appears in the center of the pair of double lines. The index is read to the nearest 10 minutes and the micrometer scale is read for the remaining minutes and seconds as shown in Fig. 4-18.

4-35. A Fourth Method for Circle Reading. Another system for reading the circle is often used. The ends of the lines from the opposite sides of the circle are imaged slightly offset beside the ends of the lines on the near side. When the circle is turned, both lines move in the same direction. Any eccentricity will cause the offset to change slightly in width. The microscope moves the pair of lines until an index line bisects the space between them.

4-36. Repeating Feature. Any of these instruments can be designed so that they will repeat angles by a proper arrangement of the clamps. However, this arrangement will not give precisely accurate results.

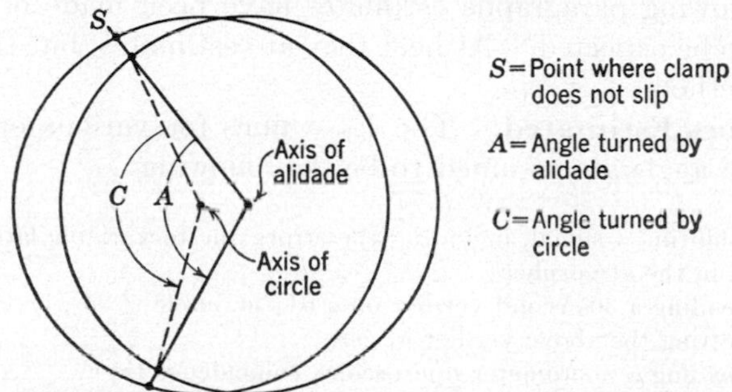

FIG. 4-19. Error in repeating angles when a double center is not used.

In repeating an angle, when the alidade is returned to the first direction, the circle must be clamped to the alidade so that the reading will not change. The axis of rotation of the alidade is determined by its bearing in the leveling head, and the axis of rotation of the circle is determined by another bearing in the leveling head. Eccentricity between these two bearings introduces an eccentric movement between the alidade and the circle. This changes the reading of the circle when no change should occur. An eccentricity of 0.0001 in. may change the reading 1 second on a 3½ in. circle. How much the reading is changed depends on the azimuth and extent of the eccentricity, the size of the angle measured, and where the clamp is forced to slip to adjust for the eccentric movement.

Figure 4-19 shows schematically how the change in the reading may occur. If the clamp between the alidade and the circle holds at point S, when the alidade is turned through the angle A the circle will be turned through the larger angle C.

4-37. Eccentricity of Glass Circles. Since a glass circle usually has a different coefficient of thermal expansion than the metal on which it is mounted, its position on its mounting will change from time to time when large temperature changes occur. It is, therefore, difficult if not impossible to mount a glass circle so that it stays permanently in its original position. It follows that larger eccentricities are to be expected with glass circles and, accordingly, they must always be read simultaneously at two or more equally spaced points on the circle.

ESTIMATES OF ACCURACIES

4-38. Accuracies to Be Expected. By making reasonable estimates of the errors involved in instrumental operation, proper procedures may be planned for any particular instrument to obtain the accuracies desired. When a survey is actually in operation, modifications should be made according to experience.

In the following paragraphs estimates have been made of the 90 per cent errors to be expected. At best they are estimates, but they demonstrate the method.

4-39. Errors Estimated. The E_{90} values for various operations in measuring an angle are assumed to be the following:

p = error of pointing a signal, including the errors due to variable lateral refraction in the atmosphere .... $p = 2''$
r = error of reading a 30-second vernier on a 6¼ in. circle .... $r = 15''$
s = error of setting the above vernier at zero .... $s = 10''$
m = error of reading a micrometer microscope, coincidence type .... $m = 2''$
c = error in circle graduation .... $c = 3''$

With these values it is possible to estimate the E_{90} values for measuring angles exclusive of errors in positioning the instrument or the signals over the stations. Four examples are given to illustrate.

Example 1. Engineer's transit, both verniers read, angle turned 3 D.R. The error i of the combination of setting the A vernier at zero, reading the B vernier, and taking the average can be computed as follows:

$$i = \frac{\sqrt{r^2 + s^2}}{2} = \frac{\sqrt{15^2 + 10^2}}{2} = 9''$$

The error f of the average of reading two verniers can be computed as follows:

$$f = \frac{\sqrt{2r^2}}{2} = \frac{\sqrt{2(15)^2}}{2} = 11''$$

The error of the final angle 3 D.R. which includes 12 pointings is

$$E_{90} = \frac{\sqrt{i^2 + f^2 + c^2 + 12p^2}}{6}$$
$$= \frac{\sqrt{9^2 + 11^2 + 3^2 + 12(2)^2}}{6}$$
$$= 2.7''$$

Example 2. Engineer's transit, both verniers read, angle turned 6 D.R.

$$E_{90} = \frac{\sqrt{i^2 + f^2 + c^2 + 24p^2}}{12}$$
$$= \frac{\sqrt{9^2 + 11^2 + 3^2 + 24(2)^2}}{12}$$
$$= 1.5''$$

Example 3. Engineer's transit, both verniers read, angle turned 6 D.R. but the system of "unwinding" the angle used. According to this system, the angle is turned 3 D.R., the reading taken, and then without disturbing the upper motion, a set of 3 D.R. is started for measuring the explement of the angle. When this set is completed, the vernier should read very nearly 360°, that is, very nearly zero. The zero reading and the near-zero reading are averaged and the result subtracted from the 3 D.R. reading to find the angles to be divided by 6.

$$E_{90} = \frac{\sqrt{(2c)^2 + (2f)^2 + i^2 + f^2 + 24p^2}}{12}$$

Note that the circle error and the error of the average reading of two verniers are both contained in the 3 D.R. reading. This reading is used with both the initial reading and the final reading. The error of the reading is therefore twice as effective as either of the others.

The formula may be developed as follows: Let R represent the final reading, d the reading at 3 D.R., and the symbols for the errors represent the actual values with which they are associated; then

$$R = \frac{d - \dfrac{i+f}{2}}{6} + \frac{24p}{12}$$

but

$$d = c + f$$
$$R = \frac{c+f}{6} - \frac{i+f}{12} + \frac{24p}{12}$$

Then by Eq. (21-22),

$$E_{90} = \sqrt{\left(\frac{\sqrt{c^2+f^2}}{6}\right)^2 + \left(\frac{\sqrt{i^2+f^2}}{12}\right)^2 + \left(\frac{\sqrt{24p^2}}{12}\right)^2}$$
$$= \sqrt{\frac{c^2+f^2}{36} + \frac{i^2+f^2}{144} + \frac{24p^2}{144}}$$
$$= \sqrt{\frac{4c^2+4f^2}{144} + \frac{i^2+f^2}{144} + \frac{24p^2}{144}}$$
$$= \frac{\sqrt{(2c)^2 + (2f)^2 + i^2 + f^2 + 24p^2}}{12}$$

Substituting values,

$$E_{90} = \frac{\sqrt{(6)^2 + (22)^2 + (9)^2 + (11)^2 + 24(2)^2}}{12} = 2.4''$$

Example 4. Four circle positions with a direction theodolite.

$$E_{90} = \frac{\sqrt{8m^2 + 4c^2 + 8p^2}}{4} = \frac{\sqrt{8(2)^2 + 4(3)^2 + 8(2)^2}}{4} = 2.5''$$

4-41. Application of Estimates of Errors. Tabulating the results, the values for E_{90} are

Engineer's transit, 3 D.R.	= 2.7″
Engineer's transit, 6 D.R.	= 1.5″
Engineer's transit, 6 D.R. (unwinding system)	= 2.4″
Theodolite (4 positions)	= 2.5″

It is evident that the unwinding system should not be used. It requires as much field work as the regular 6 D.R. method, but the precision is less.

To find the error in the sum of the three angles of a triangle, these values should be multiplied by $\sqrt{3}$. This gives

Engineer's transit, 3 D.R.	= 4.7″
Engineer's transit, 6 D.R.	= 2.6″
Theodolite (four positions)	= 4.3″

It is evident that 3 D.R. with an engineer's transit is about equivalent to four positions on the theodolite. Neither should be used for second-order accuracy, as the limit set for second-order accuracy is an average triangle closure of 3 seconds, and for triangulation it is safer to choose an operation that would seldom exceed the desired average.

The 6 D.R. with an engineer's transit would be sufficient for second-order work but the setups and signals should be established with care—particularly if the sights are short—if errors are to be kept within the requirements.

PROBLEMS

4-1. From the data in this chapter, compute the number of circle positions required for second-order accuracy.

4-2. What would be the probable errors for the four examples given?

4-3. Estimate the time saved by a direction instrument over a repeating instrument when five directions must be observed.

4-4. Would it be more accurate to measure an angle twice 2 D.R. and take the average of the values or 3 D.R. alone?

4-5. Compute E_{90} for four angles that close the horizon measured 3 D.R. with an engineer's transit and then adjusted for station adjustment.

CHAPTER 5

THE COMPASS AND THE EARTH'S MAGNETIC FIELD

5-1. The Compass as a Surveying Instrument. The magnetic compass has many uses in surveying. Its chief advantage is its ability to determine bearings that are entirely independent of any other measurements. Just as a level determines the horizontal at each setup, the compass determines magnetic north each time it is used.

In a compass traverse, the accidental errors of the determination of direction do not accumulate, since the direction is independently determined at each setup. Backsights are not usually necessary. An obstacle can be passed by offsetting a few feet without setting a foresight or a backsight, or it may be passed merely by placing the instrument beyond the obstacle as nearly on line as can be estimated.

The error of a compass determination is usually less than 15 minutes. It can sometimes be held to 5 minutes. Under most conditions when the distances involved are short this is within mapping accuracy.

The compass is used to determine true north, to check transit angles, to determine direction in transit-stadia surveys, plane-table surveys, preliminary-route surveys, and in many forms of rapid mapping. In the past, the surveyor's compass was used extensively in property surveys. A thorough understanding of the compass is often necessary to retrace property lines so surveyed.

5-2. The Earth's Magnetic Field. When a bar magnet is supported so that it is free to turn in azimuth, it will point approximately north and south. The end that points north is called the N pole of the magnet. Since opposite poles attract, the earth's north magnetic pole must be its S pole when it is considered as a magnet.

The earth, like any magnet, forms a field of magnetic force. The lines of force in a magnetic field can be defined as the paths along which a free N pole would move. A free S pole would follow the same paths in the opposite direction. A magnetic needle supported at its center of gravity and free to rotate will align itself with the lines of force.

Figure 5-1 shows a cross section of the lines of force in the earth's magnetic field. The cross section is taken in the plane of the meridians 80°E and 100°W. The lines of force run generally from south to north, but they are parallel to the earth's surface only near the equator. The angle they make with the earth's surface is called the **dip,** and a magnetic needle free to rotate will dip in accordance with these lines. In the Northern Hemisphere the needle will dip so that its north end is downward. When it is brought to an area near 70° north latitude, 96° west

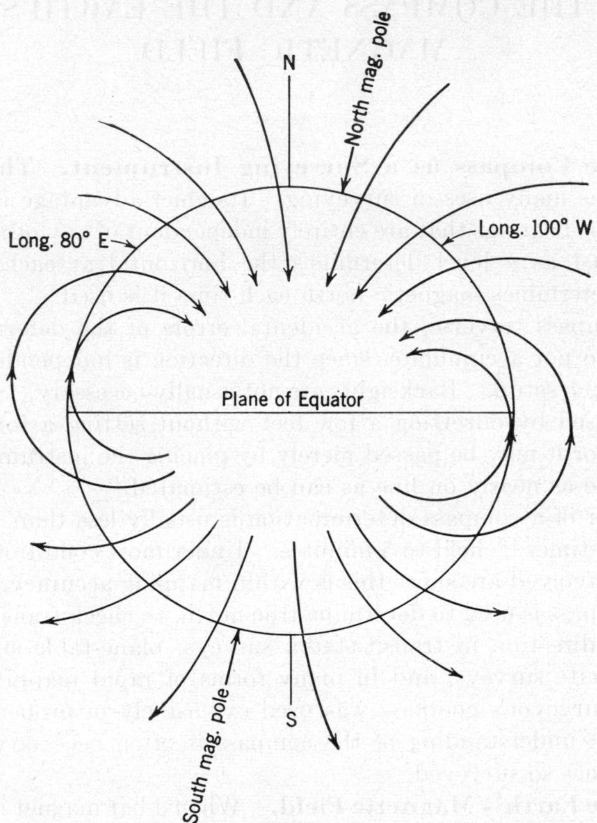

FIG. 5-1. Cross section of the earth's magnetic field.

longitude, it will dip 90°. This area is called the north magnetic pole. A similar area in the Southern Hemisphere is called the south magnetic pole. A free magnetic needle does not necessarily point toward the north magnetic pole but takes up the dip and azimuth of the lines of force where it happens to be located. Since the earth's field is quite irregular, the dip and azimuth of the lines of force are also irregular. The projections of the lines on the earth's surface are not straight nor do they converge exactly as would be expected.

5-3. A Compass Needle. A compass needle is hung from a conical jewel bearing supported on a sharp, hardened-steel pivot (see Fig. 5-2). The apex of the cone is above the center of gravity of the needle, so that before being magnetized it will hang in a horizontal position. A small coil of brass wire is wrapped around the needle to balance the force tending to make the needle dip. The position of the coil is adjusted for the dip in the locality in which the compass is to be used. The most practical means of determining the north end of the needle is to note where the brass wire is placed. North of the magnetic equator it is on the south end, and vice versa.

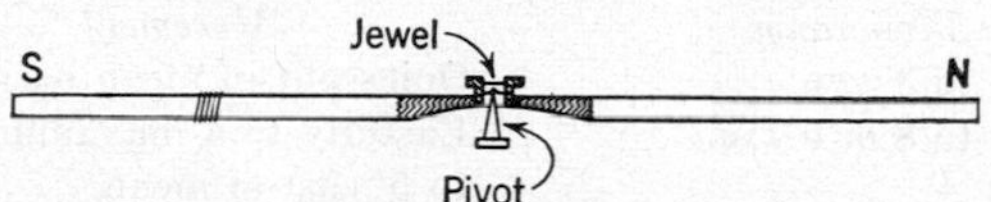

FIG. 5-2. A compass needle.

The pivot is so hard, and must be sharpened to such a small point, that the slightest jar will break its tip and destroy the accuracy of the compass. All surveying compasses have devices for raising the needle free of the point when not in use. If this is once forgotten, the pivot may be damaged and the compass will be too sluggish to use.

5-4. Declination. The horizontal angle measured from the true north to the direction of the north end of a perfect compass needle is called the declination. When the north end of the needle is east of true north the declination is said to be east, and vice versa (see Fig. 5-3). A more perfect definition would be: the declination at any place is the horizontal angle from the true north measured to the projection on the earth's surface of the lines of force of the earth's magnetic field. The direction of these lines is often called **magnetic north.** Because of irregularities of the field, this is not necessarily the direction of the north magnetic pole. Mariners call declination by the name **variation.**

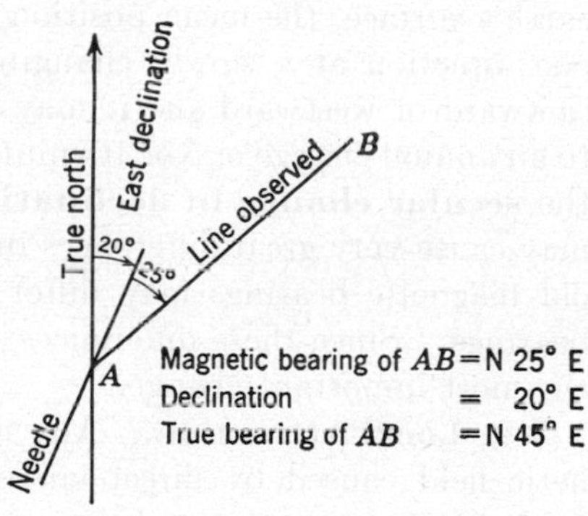

FIG. 5-3. Effect of declination.

5-5. Changes in Declination. The declination at any one place is never constant. The needle swings back and forth during the day and its mean position moves progressively from day to day. During magnetic storms the needle moves irregularly over a considerable range. These changes are usually called variations in declination, but here they will be called "changes" to avoid confusion with the mariners' term for declination itself.

5-6. Diurnal Change in Declination. The daily swing in declination is called the diurnal change. Its extent is different in different localities and is greater in summer than in winter. In the Northern Hemisphere the change can be visualized by imagining the north end of the needle to be attracted by the sun during the daytime. The motion in the United States can be outlined as follows:

Time range	*Movement*
9 P.M. to 3 A.M.	Quiescent in mean position
3 A.M. to 8 or 9 A.M.	Easterly to a maximum of 2 to 5′ east of mean
8 or 9 A.M. to 1 or 2 P.M.	Westerly to a maximum of 2 to 5′ west of mean
1 or 2 P.M. to 9 P.M.	Easterly to mean

The maxima given are often exceeded and, as noted, are greater in summer than in winter.

5-7. Annual Change in Declination. The declination has a yearly swing of about 1 minute in amplitude. It varies in different localities. The name should not be confused with the annual rate of the secular change in declination described in the next paragraph.

5-8. Secular Change in Declination. Nearly everywhere on the earth's surface, the mean position of the needle is moving constantly in one direction at a slowly changing, yearly rate. The motion may be eastward or westward and it may reverse itself. The rate may vary up to an annual change of 5 or 10 minutes. This progressive change is called the **secular change in declination.** As the years go by, this change may cause very great differences in the declination in any locality so that old magnetic bearings may differ radically from more recent magnetic bearings. Since these differences may be so large, the secular change is the most important change.

5-9. Local Attraction. Any mass of magnetic material, or the magnetic field caused by direct current, will deform the earth's magnetic field. Differences from the expected direction of the needle in these localities are said to be caused by **local attraction.** Mariners call differences caused by the ship itself deviations. The extent of local attraction can be measured by measuring the bearing of a line whose magnetic bearing has been determined elsewhere. The true angle between two lines can be measured where local attraction exists, however, by computing the difference between the bearings. Often, in a compass traverse, in addition to the foresights, backsights are taken at each station. When the bearings at the two ends of each line check, it can be assumed that no local attraction exists at either station. When a back-

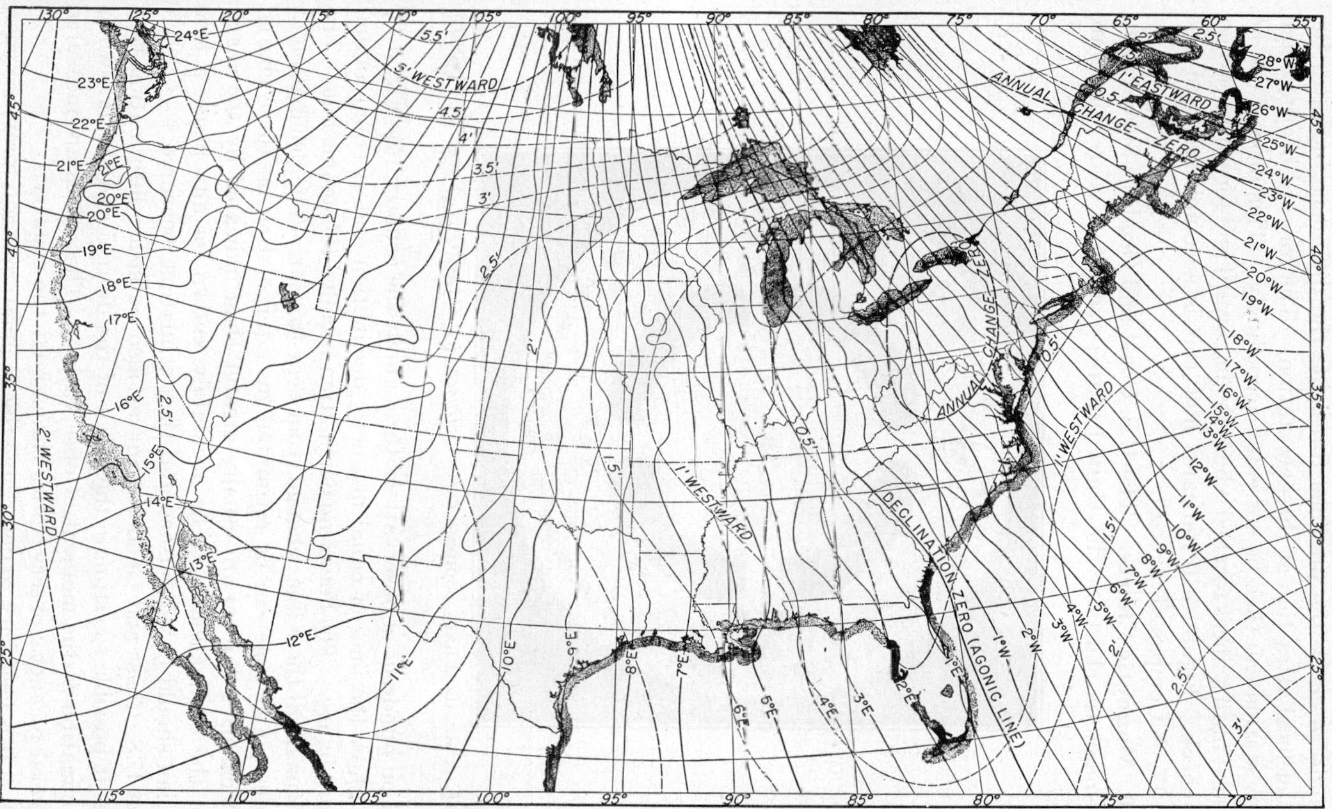

FIG. 5-4. Isogonic chart of the United States for Jan. 1, 1955. The solid lines are lines of equal magnetic declination. The broken lines are lines of equal annual rates of change in declination. The U.S. Coast and Geodetic Survey publishes such a chart every five years. (*U.S. Coast and Geodetic Survey.*)

sight fails to check the previous foresight, the difference indicates the extent of local attraction at the new station. The forward bearing from the new station can then be corrected accordingly. This process can be continued until a line is reached where the backward and forward bearings again check. Figure 5-10 indicates how this is accomplished (see also Sec. 5-17, A Compass Traverse).

5-10. Magnetic Charts. The U.S. Coast and Geodetic Survey publishes magnetic charts of the United States every five years; these are

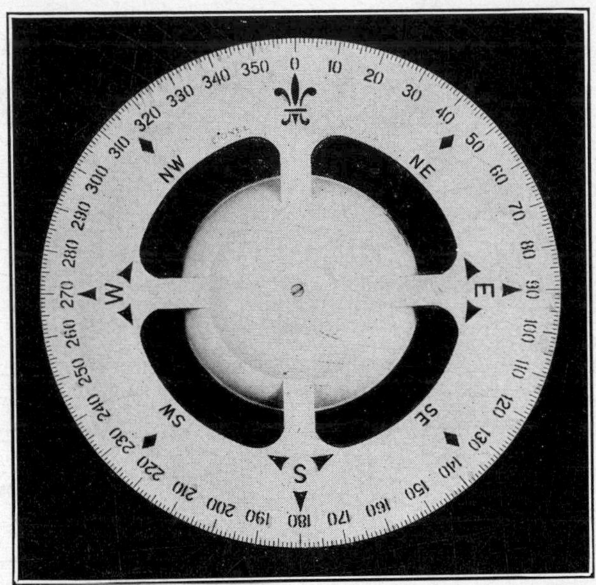

FIG. 5-5. A standard navy compass card. (*Bureau of Ships, U.S. Navy Department.*)

based on hundreds of observations throughout the country (see Fig. 5-4). They show the lines of equal magnetic declination and of equal annual secular change. The line of no declination is called the agonic line. The positions of all these lines change from year to year. The declination at any place and time can be estimated by interpolation on these charts. The local irregularities are usually greater than indicated by the chart. When the declination must be accurately known for any place, a true meridian should be established and the declination measured.

The U.S. Coast and Geodetic Survey also publishes information that makes it possible to estimate the effect of the diurnal and annual swings in declination at the moment of observation. These data are measured at several magnetic observatories throughout the country.

When it is necessary to determine the declination in any previous year, the difference between the present declination and the previous declination should be determined from the charts and applied to the present, measured declination.

5-11. Types of Compasses. There are two basic types of compasses, the card compass and the needle compass. The card compass (Fig. 5-5) is used universally by mariners. Usually a pair of magnetic bars are attached to the under side of the card (see Fig. 5-6). The card turns with the magnets and, when at rest, the magnetic bearing of a sight taken over the card can be read directly on the card. An index called a **lubber line** is placed in line with the ship's keel to mark the ship's heading.

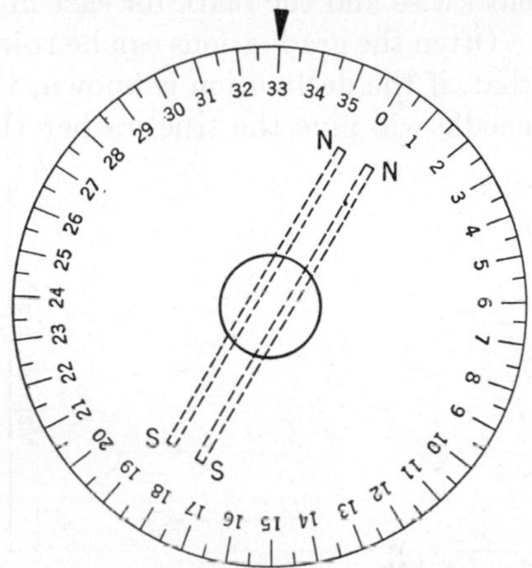

FIG. 5-6. Compass card showing a magnetic azimuth of 330°. The magnets are under the chart.

5-12. The Prismatic Compass. This is a hand card compass used in surveying. A prism is arranged so that the observer can aim the instrument along a line and at the same time read the azimuth of the line by the position of the card under an index.

FIG. 5-7. A surveyor's compass. (*W. & L. E. Gurley.*)

5-13. Needle-type Compasses. The most familiar form of the needle compass is the compass mounted on the upper plate of a transit. The needle travels over the graduations which are laid out from the direction of the line of sight. To make it possible to read the bearing of the line of sight directly at the needle position, the graduations are numbered in the reverse order. Clockwise azimuths must be numbered counter-clockwise and the mark for east must be on the left (see Fig. 5-8).

Often the graduations can be rotated according to a **declination arc** so that, if the declination is known, the graduations may be set so that the needle will give the true, rather than the magnetic, bearing.

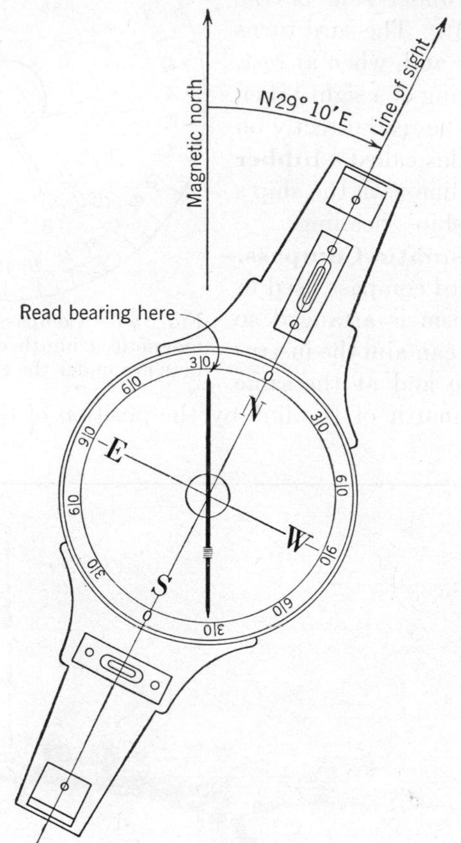

FIG. 5-8. Measuring a bearing with a surveyor's compass.

The surveyor's compass is illustrated in Figs. 5-7 and 5-8. The sights at each end are vertical slits about 7 in. high. The surveyor's compass is the type of compass that was used extensively for property surveys in days gone by and is still used where the value of the land is not very high. It is an excellent instrument for certain types of mapping.

Many hand surveying instruments are needle compasses, such as the mining compass, the forester's compass, and the Brunton compass.

5-14. Trough Compasses. These are mounted on plane-table alidades, plane tables, and some transits. A trough compass is a needle compass mounted in a narrow box that limits the needle to a few degrees of movement (see Fig. 7-17). When the needle is centered, the instrument points north. Other magnetic directions can then be measured.

5-15. Adjustments of the Compass. For the best results, both ends of the needle should be read and the average used. This eliminates the eccentricity of the pivot point and the effect of a bent needle. The following procedure is used to adjust the compass (see Fig. 5-9):

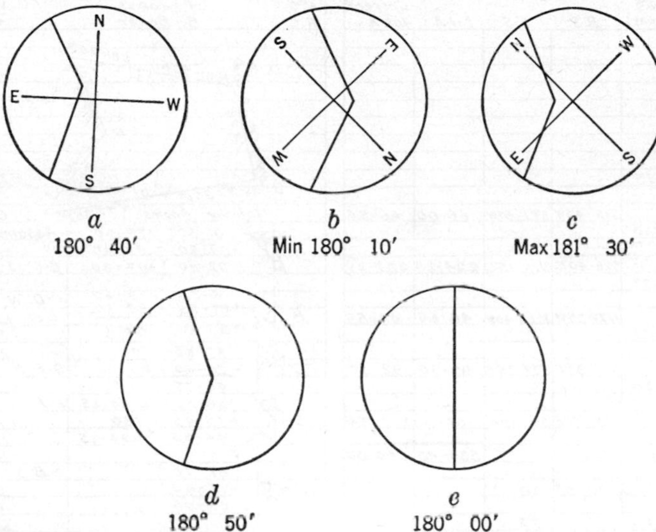

FIG. 5-9. Steps in centering the pivot point and straightening the needle. The angle to the right of the needle is stated under each figure. The maximum angle is found in *c*, and the pivot is then bent toward its bisector. Finally the needle is straightened.

First Adjustment

Object. To center the pivot point.

Test. Set up the compass, turn it until the north end of the compass reads exactly north. Read the south end. Repeat for the other three cardinal points. If the reading of the south end differs from that of the north end by a constant angle, the pivot is centered. If not, adjustment is necessary.

Adjustment. By trial, find the position that gives the greatest angular difference on one side of the needle between the two ends (Fig. 5-9*c*), and bend the pivot along the bisector of this angle. Repeat test.

Second Adjustment

Object. To make the readings of the two ends of the needle differ by 180°.

Test. Read both ends of the needle, after the first adjustment has been completed.

Adjustment. Bend needle as required and repeat test. The center of the needle is usually soft so that it will bend. Avoid touching the needle with the hands, as this tends to demagnetize it (see Figs. 5-9*d* and *e*).

5-16. The Sensitivity of the Compass. The sensitivity of the compass may be tested by noting how accurately it returns to the same reading when it is momentarily drawn off with a knife blade or other

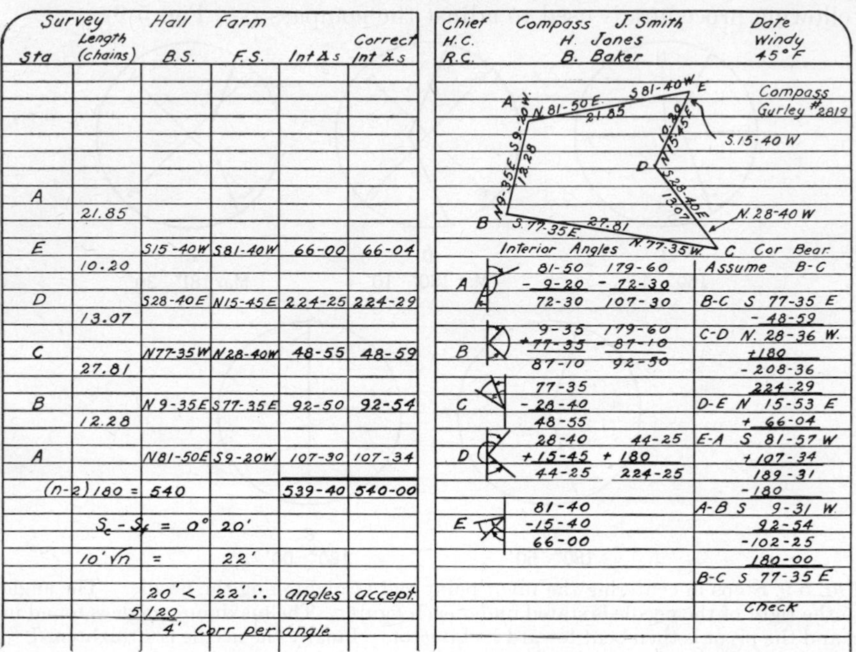

Survey	Hall Farm				
Sta	Length (chains)	B.S.	F.S.	Int ∡s	Correct Int ∡s
A					
	21.85				
E		S15-40W	S81-40W	66-00	66-04
	10.20				
D		S28-40E	N15-45E	224-25	224-29
	13.07				
C		N77-35W	N28-40W	48-55	48-59
	27.81				
B		N9-35E	S77-35E	92-50	92-54
	12.28				
A		N81-50E	S9-20W	107-30	107-34
(n-2) 180 =	540			539-40	540-00
	$S_c - S_f$ = 0°	20′			
	10′√n =	22′			
	20′ <	22′ ∴	angles	accept.	
	5/120	4′ Corr	per	angle	

FIG. 5-10. Compass-traverse field notes.

small piece of magnetic material. If it swings freely over small amplitudes and comes to rest at the same reading as before—as closely as can be determined—it is in good condition. If not, in almost every case the pivot point has been damaged. The pivot may be resharpened with a stone, but it is better practice to replace it, as it is very difficult to give it the proper point.

Sometimes under very unusual circumstances the needle will lose its magnetism. It can be remagnetized in the field by drawing the proper pole of a permanent magnet from the center of the needle outward to the end. This should be repeated a number of times on both ends. The proper pole attracts the end of the needle on which it is used.

5-17. A Compass Traverse. Figure 5-10 shows a compass traverse. The sketch shows the observed values. The interior angles are computed from the two bearings observed at each station. The computed angles are independent of local attraction. The interior angles are adjusted. The bearing *BC* is assumed to be correct, because the bearings at the two ends check. Starting with the bearing of *BC*, the other bearings are computed by successively applying the adjusted interior angles. Thereafter the computations are the same as those of a transit traverse. The error of closure should not exceed 1 part in 300.

PROBLEMS

5-1. The following are the field notes of a compass traverse. Draw a sketch approximately to the scale 1 in. = 2 chains. Compute the interior angles and the adjusted bearings.

Station	Length chains	Backsight	Foresight
A			
	5.10		
I		S64°40′W	S79°50′E
	0.96		
II		N22°10′E	N65°00′E
	0.57		
G		S82°20′W	S23°50′W
	3.52		
F		N43°50′W	N80°20′E
	1.20		
E		S 9°05′E	S44°50′E
	7.74		
D		S28°20′W	N 7°25′W
	4.02		
C		N88°50′E	N27°10′E
	6.11		
B		N26°20′E	N89°30′W
	10.88		
A		N79°50′W	S25°10′W

5-2. From the survey notes in Prob. 5-1, find the local attraction at each station.

5-3. What length stadia sights could you take, using magnetic bearings for direction? Assume that the maximum error desired in the map was 2 ft and that a stadia error was 1 ft in 500 ft.

5-4. Assume that the survey in Prob. 5-1 was made in June, 1893. What would be the bearing of the line *BC* measured at *B* in 1952? Given: the declination in the locality of the survey in 1890 was 5°10′ east and the annual change was 1°20′ west. The same values in 1950 were 11°20′ west and 0°32′ west.

5-5. Show, by a sketch indicating quantities, what is wrong with a compass that gave the following set of readings for the two ends of the needle:

	N end	*S end*
Max. difference..............	N20°10′E	S22°20′W
Min. difference...............	S21°00′W	N21°30′E

CHAPTER 6

SIMPLE CONTROL SURVEYS

6-1. Importance of Control. As has been stated frequently, a most important element in the economy and accuracy of any survey is the adoption, at the start, of a sound system of control.

6-2. Arrangement. Except in route surveys, both vertical and horizontal control take the form of a net covering the area to be surveyed. The control points are placed where they can be surveyed accurately and where other work can be conveniently tied to them. There should be just enough control positions to hold the accuracy desired.

6-3. Adjustment. The methods of running level lines and of measuring angles and lengths for triangulation and traverse are covered elsewhere. The chief subject of this chapter is the method of adjusting results.

From a theoretical standpoint the ideal adjustment would be a simultaneous adjustment of the entire net. This would include all control measurements of entirely different precisions, each weighted according to the precision used.

Such a plan is almost never practical. The accuracy gained is slight and since no elevations or positions would be available until the entire field work was complete, no progress could be made in any of the other surveying operations.

Instead, a simple primary system of control covering the whole area should be laid out at once, surveyed with comparatively high accuracy, and adjusted simultaneously at the earliest possible moment. All elevations or coordinates that result should be held fixed thereafter, and subsequent work should be based on them. Any secondary control nets that may be required will then cover a comparatively small area and each can be adjusted simultaneously as an independent unit, holding the original net fixed.

Working survey lines (in contrast to control surveys) are tied to the control in such a way that their measurements may be checked against the control values and, if the desired accuracy requires it, they can be adjusted to fit.

ADJUSTMENT OF A LEVEL NET

6-4. Methods of Adjustment. Three methods of adjustment of a level net are described here: consecutive adjustment by inspection, simultaneous adjustment by estimation, and the Dell method.[1] The circuit-reduction method devised by Norman F. Baaten of the U.S. Coast and Geodetic Survey and the method of least squares are discussed in Chaps. 17 and 18, respectively.

Consecutive Adjustment by Inspection

6-5. Advantages. The advantages of the method of consecutive adjustment by inspection are chiefly that it is practical and rapid and that it gives usable results. It is a "common-sense" method though devoid of much mathematical justification. In one form or another it is used extensively. The example given is only one type of solution.

6-6. Example. Figure 6-1 is a plan drawn approximately to scale showing a level net. The net consists of eight **junctions** *ABCDEFGH*

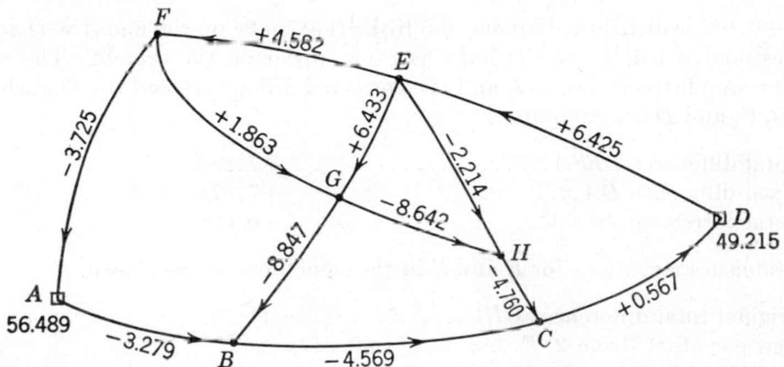

FIG. 6-1. Plan of level net. *A* and *D* have fixed elevations. The difference in elevation observed between junction points is stated. The arrows show the direction in which the differences in elevation apply.

where the level lines meet. The elevations given for *A* and *D* are to be held fixed. The direction in which the lines were run is shown by arrows and the resulting differences in elevation between the junctions are given for each line according to the direction each line was run. For example, the line *AB* was run from *A* to *B*, and *B* is found to be lower than *A* by 3.279 ft.

First, the shortest or most reliable connections between fixed elevations are adjusted. These are then held fixed and the next most reliable lines are adjusted to fit them. When more than one elevation results for a

[1] H. G. Dell, The Adjustment of a Level Net, *Proc. ASCE*, vol. 61, no. 4, pt. 1, April, 1935.

junction, the average is used. Finally, the corrections to the original field data are inspected and, if any corrections are obviously too large, changes in the corrections are introduced to decrease them.

It is planned to adjust the lines in the following order:

1. Line *ABCD*
2. Line *DEFA*
3. Line *FGHC*
4. Line *EGB*
5. Line *EH*

In Fig. 6-2 are listed the field elevations and the differences in elevation in accordance with the direction the lines were run. The last difference listed, −7.274, is a fixed difference. The differences are then entered on the sketch with arrows to show the directions to which they refer.

6-7. Adjustment.

1. Total difference *ABCD*		−7.281
Fixed difference *AD*		−7.274
Total correction *ABCD*		+0.007

The +0.007 is distributed among the **links** that make up the line *ABCD* according to the estimated lengths of the links as they appear on the sketch. The resulting corrections are listed in Fig. 6-2, and the corrected differences and the final elevations for *A*, *B*, *C*, and *D* are computed.

2. Total difference *DEFA*		+7.282
Fixed difference *DA*		+7.274
Total correction *DEFA*		−0.008

This establishes elevations for *E* and *F* in the same manner, as shown.

3. Original total difference *FGHC*		−11.539
New elevation (from 2) *F*	60.217	
New elevation (from 1) *C*	48.646	
New fixed difference *FC*		−11.571
Total correction *FGHC*		− 0.032

This establishes elevations for *G* and *H*, as shown.

4. Original total difference *EGB*		−2.414
New elevation (from 2) *E*	55.637	
New elevation (from 1) *B*	53.212	
New fixed difference *EB*		−2.425
Total correction *EGB*		−0.011

This establishes another elevation for *G*.

5. Original total difference *EH*		−2.214
New elevation (from 2) *E*	55.637	
New elevation (from 3) *H*	53.411	
New fixed difference *EH*		−2.226
Total correction *EH*		−0.012

It is noted that this correction is not too large.

There are now two elevations for G. They are nearly the same so the average is established. The changes are indicated in the column of final elevations.

B.M.	Field elevations	Differences	Corrections to lines	Corrections to links	Corrected differences	Final elevations
A	56.489					56.489
		−3.279		+ 2	−3.277	
B	53.210					53.212
		−4.569	+0.007	+ 3	−4.566	
C	48.641					48.646
		+0.567		+ 2	+0.569	
D	49.208					49.215
		+6.425		− 3	+6.422	
E	55.633					55.637
		+4.582	−0.008	− 2	+4.580	
F	60.215					60.217
		−3.725		− 3	−3.728	
A	56.490					56.489
F	60.215					60.217
		+1.863		−17	+1.846	
G	62.078					62.063
		−8.642	−0.032	−10	−8.652	
H	53.436					53.411
		−4.760		− 5	−4.765	
C	48.676					48.646
E	55.633					55.637
		+6.433		− 5	+6.428	
G	62.066		−0.011			62.065
		−8.847		− 6	−8.853	
B	53.219					53.212
E	55.633					
		−2.214		−12		
H	53.419					
A	56.489					
		−7.274		Fixed		
D	49.215					

FIG. 6-2. Consecutive adjustments by inspection.

When the final elevations of the junctions have been determined, the elevations of any intermediate benchmarks on a link are adjusted so that the corrections are in proportion to the positions of the benchmarks on the link.

Example. Assume that on link FG were benchmarks J at about the half point and K at about the two-thirds point (see Fig. 6-2a).

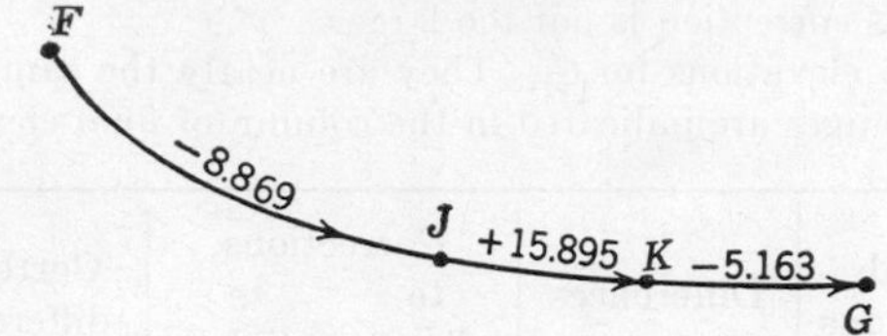

B.M.	Field elevation	Difference	Correction to sections	Corrected difference	Final elevation
F	60.215				60.217
		−8.869	−0.008	−8.877	
J	51.346				51.340
		+15.895	−0.003	+15.892	
K	67.241				67.237
		−5.163	−0.005	−5.168	
G	62.078				62.064

FIG. 6-2*a*. Adjustment of intermediate benchmarks between junctions.

Established elevation *F*.	60.217
Established elevation *G*.	62.064
Total difference *FG*.	+1.847
Original difference *FG*.	+1.863
Total correction *FG*.	0.016
Correction *FJ*, ½(−0.016).	−0.008
Correction *JK*, ⅙(−0.016).	−0.003
Correction *KG*, ⅓(−0.016).	−0.005
Check total.	−0.016

These differences are entered in Fig. 6-2*a* and the final elevations computed.

Simultaneous Adjustment by Estimation

6-8. Advantages. The method of simultaneous adjustment by estimation gives results that have a greater probability of being correct than the previous method. It therefore produces better results and no unusually large corrections. It takes longer to compute than the previous method but it is far quicker than any mathematically correct solution.

Example. The same level net will be adjusted as in Example 1. The differences in elevation are computed (Fig. 6-3) and the values are shown on a sketch (Fig. 6-4) in the same way as in the previous example.

The net can be considered to be built up of adjacent circuits. *F*, *E*, and *G* form circuit I. *F*, *G*, *B*, and *A* form circuit II, etc. By beginning at

B.M.	Field elevations	Differences	Corrections to links	Corrected differences	Corrected elevations
A	56.489				56.489
		−3.279	+0.006	−3.273	
B	53.210				53.216
		−4.569	+0.004	−4.565	
C	48.641				48.651
		+0.567	−0.003	+0.564	
D	49.208				49.215
		+6.425	−0.008	+6.417	
E	55.633				55.632
		+4.582	−0.003	+4.579	
F	60.215				60.211
		−3.725	+0.003	−3.722	
A	56.490				56.489✓
F	60.215				60.211
		+1.863	−0.008	+1.855	
G	62.078				62.066
		−8.642	−0.007	−8.649	
H	53.436				53.417
		−4.760	−0.006	−4.766	
C	48.676				48.651✓
E	55.633				55.632
		+6.433	+0.001	+6.434	
G	62.066				62.066✓
		−8.847	−0.003	−8.850	
B	53.219				53.216✓
E	55.633				55.632
		−2.214	−0.001	−2.215	
H	53.419				53.417✓
	Fixed elevations				
A	56.489				56.489
		−7.274	0.000	−7.274	
D	49.215				49.215✓

FIG. 6-3. Simultaneous adjustment by estimation.

any point in a circuit and applying the differences successively around the circuit to the point of beginning, the error or *closure* of the circuit may be computed. In order to obtain the proper sign of the closure all circuits must be computed in the same circular direction. Here the clockwise direction is used. Obviously the differences are applied with the sign they bear only when the arrow is in the direction of computation. When

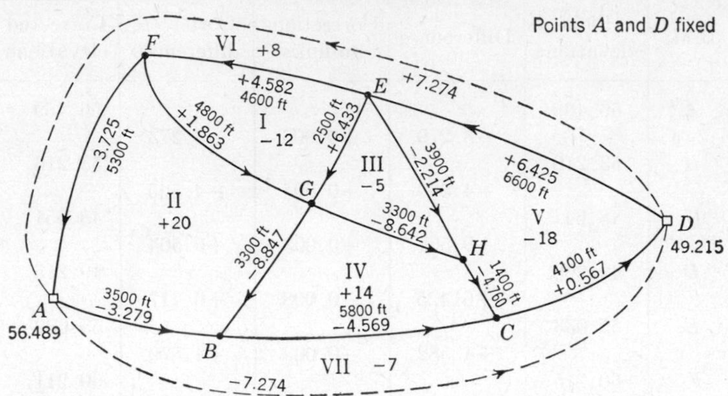

The differences in elevation given are for the directions shown by the arrows.

The dotted lines represent the loops formed by other leveling which is held fixed in elevation.

The error of each circuit is shown in the center and is stated in thousandths of a foot. It is computed clockwise. The computation is shown below.

COMPUTATION OF CIRCUIT ERRORS

Circuit	Line	+	−	Error
I	F–E		−4.582	
	E–G	+6.433		
	G–F		−1.863	
		+6.433	−6.445	= −0.012
II	F–G	+1.863		
	G–B		−8.847	
	B–A	+3.279		
	A–F	+3.725		
		+8.867	−8.847	= +0.020
III	E–H		−2.214	
	H–G	+8.642		
	G–E		−6.433	
		+8.642	−8.647	= −0.005
IV	B–G	+8.847		
	G–H		−8.642	
	H–C		−4.760	
	C–B	+4.569		
		+13.416	−13.402	= +0.014
V	C–H	+4.760		
	H–E	+2.214		
	E–D		−6.425	
	D–C		−0.567	
		+6.974	−6.992	= −0.018
VI	D–E	+6.425		
	E–F	+4.582		
	F–A		−3.725	
	A–D		−7.274	
		+11.007	−10.999	= +0.008
VII	A–B		−3.279	
	B–C		−4.569	
	C–D	+0.567		
	D–A	+7.274		
		+7.841	−7.848	= −0.007

Summary in thousandths of a foot

Circuit	+	−	
I		−12	
II	+20		
III		−5	
IV	+14		
V		−18	
VI	+8		
VII		−7	
	+42	−42	= 0

FIG. 6-4. Plan of level net showing field data and errors.

the arrow is contrary, the sign is reversed. The computed closure for each circuit is shown on the sketch near the center of the circuit.

The sum of the errors of two or more adjacent circuits is equal to the error of the loop formed by the exterior lines of these circuits. This statement can be proved as follows: When adjacent circuit closures are added, the common or interior lines (differences) enter the sum twice, once with a plus sign and once with a minus sign. Their effect is thus cancelled, and only the exterior loop closure is left in the sum.

It is convenient to indicate the differences between the fixed bench-marks by one or more dotted lines, each forming one side of a circuit. The differences to be shown on these lines are computed from the fixed elevations. When such dotted lines are arranged to completely surround the net, an exterior line is formed with a closure of zero; therefore, the sum of the circuit closures of all the circuits within these lines must be zero.

The circuit closures are computed and their sum checked against zero; they are then placed within the circuit to which they refer. The result is shown in Fig. 6-4.

6-9. Adjustment by Estimation. When a correction is applied to any line between two circuits, it will affect one closure additively and the other closure subtractively, by the same amount. Corrections are estimated according to this rule (see Fig. 6-5). Each correction is indicated by an arrow, flying across the line to be corrected, from the closure to be reduced toward the closure to be increased. A number is placed on each arrow to indicate, in thousandths of a foot, the amount of the correction. By trial and error an arrangement of corrections is obtained that cancels the error in every circuit. Larger corrections are applied to longer lines.

Begin with the largest closure error, viz., circuit II. All corrections probably must be outward from this circuit. The largest correction can be absorbed in circuit I because its error is large and of opposite sign. Also the line *FG* is rather long and a rather large error may be expected. An error of 8 thousandths is a good maximum to adopt for this kind of leveling so apply a correction of 8 out of II into I. Closure II now becomes +12 and closure I, −4. Continue to try corrections, using larger corrections where the lines are long and the differences between the closures greatest, until all circuit closures become zero. If possible, do not exceed a limit of 8; in general, keep the corrections as small as possible. Of course, the dotted lines cannot be corrected.

Obviously many solutions can be found. The ideal solution occurs when the sum of the squares of the corrections is minimum. Since all the corrections are interrelated each can be expressed as a function of some variable. The first derivative of the sum of the squares of these expressions can be equated to zero and solved for the value of the variable.

This can be substituted to give the values of the expression for the corrections (see Chap. 22).

The corrections found by estimation are shown on the work sketch, Fig. 6-5. They are listed in the computation table in Fig. 6-3 and given signs according to the following rule: Looking along the link in the direction of the arrow on the link, a correction arrow flying from left to right is plus.

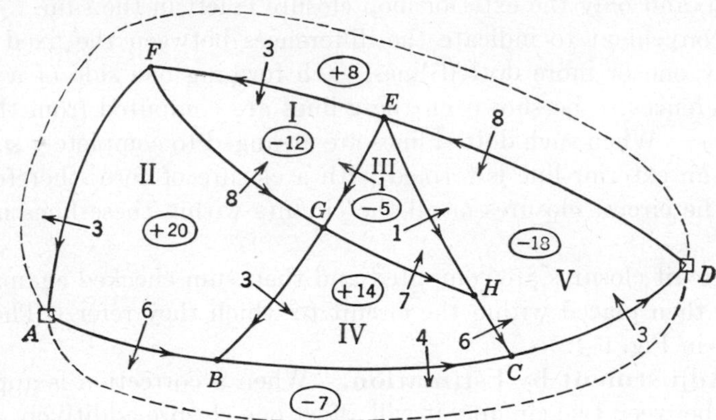

FIG. 6-5. Work sketch for computing estimated corrections. The values are given in thousandths of a foot.

When a correction is applied to the difference in elevation of a line of levels, it changes the circuit closures on both sides of the line by the amount of the correction. One circuit closure is affected additively, and the other, subtractively; therefore the correction can be represented by an arrow flying across the line toward the circuit affected additively, and away from the circuit affected subtractively. Systems of arrows of various values (corrections) can always be arranged which reduce all circuit closures to zero. The smaller these corrections are made, the better the system devised.

The corrected differences can be computed from the arrows (corrections) using their values and the following rule of signs.

When circuit closures are computed clockwise, arrows flying from left to right with respect to the direction of the differences in elevation are plus corrections to those differences.

These corrections are added algebraically to the differences to give the corrected differences. The corrected differences are applied successively to the elevations to obtain the corrected elevations. Elevations computed by all routes must agree. Checks are obtained at A, C, G, B, H, and D as shown in Fig. 6-3. Intermediate benchmarks are adjusted as in the previous example.

The Dell Method

6-10. The Dell Method. The Dell method has the great advantage that the results are the same as would be obtained by the method of least

	l	%	I	C	II	C	III	C	IV	C	V	C	VII	C	IV	C	II	C	I	C	VI	C	II	C	E	C	F	
FE	46	39	+ 5																−1								+ 4	✓
EG	25	21	+ 2				−2												−1						−1		− 2	✓
GF	48	40	+ 5		+ 4												+1		−1				+1				+10	✓
	119	100																										
I	−12		+12	0	+ 4	+ 4	−2	+ 2		+ 2		+2		+2		+2	+1	+3	−3	0		0	+1	+1	−1	0	+12	
FG	48	28	− 5		− 4												−1		+1				−1				−10	✓
GB	33	20			− 3				+ 4						−1												0	✓
BA	35	21			− 3								−1														− 4	✓
AF	53	31			− 5												−1				−1		−1				− 6	✓
	169	100																										
II	+20		− 5	+15	−15	0		0	+ 4	+ 4		+4	−1	+3	−1	+2	−2	0	+1	+1	−1	+2	−2	0		0	−20	
EH	39	40					+3				− 5																− 2	✓
HG	33	34					+2		+ 4						−1												+ 5	✓
GE	25	26	− 2				+2												−1						+1		+ 2	✓
	97	100																										
III	−5		− 2	− 7		− 7	+7	0	+ 4	+ 4	− 5	−1		−1	−1	−2		−2	+1	−1		−1		−1	+1	0	+ 5	
BG	33	24			+ 3				− 4						+1												0	✓
GH	33	24					−2		− 4						+1												− 5	✓
HC	14	10							− 1		− 2																− 3	✓
CB	50	42							− 6				−1		+1												− 6	✓
	138	100																										
IV	+14			+14	+ 3	+17	−2	+15	−15	0	− 2	−2	−1	−3	+3	0		0		0		0		0		0	−14	
CH	14	9							+ 1		− 2																+ 3	✓
HE	39	24					−3				+ 5																+ 2	✓
ED	66	41									+ 8										+1				+1		+10	✓
DC	41	26									+ 5		−1												−1		+ 3	✓
	160	100																										
V	−18			−18		−18	−3	−21	+ 1	−20	+20	0	−1	−1		−1		−1		−1	+1	0		0	0	0	+18	
DE	66	40									− 8										−1				−1		−10	✓
EF	46	28	− 5																+1								− 4	✓
FA	53	32			+ 5												+1				−1		+1				+ 6	✓
AD	0	0																									0	✓
	105	100																										
VI	+8		− 5	+ 3	+ 5	+ 8		+ 8		+ 8	− 8	0		0		0	+1	−1	−1	+2	−2	0	+1	+1	−1	0	− 8	
AB	35	26			+ 3								+1														+ 4	✓
BC	58	43							+ 6				+1		−1												+ 6	✓
CD	41	31									− 5		+1												+1		− 3	✓
DA	0	0																									0	✓
	134	100																										
VII	−7			− 7	+ 3	− 4		− 4	+ 6	+ 2	− 5	−3	+3	0	−1	−1		−1		−1		−1		−1	+1	0	+ 7	
				0		0		0		0		0		0		0		0		0		0		0		0		

		Final elev.
	A	56.489
−3.279 + 0.004 =		−3.275
	B	53.214
−4.569 + 0.006 =		−4.563
	C	48.651
+0.567 − 0.003 =		+0.564
	D	49.215
+6.425 − 0.010 =		+6.415
	E	55.630
+4.582 − 0.004 =		+4.578
	F	60.208
−3.725 + 0.006 =		−3.719
	A	56.489
	F	60.208
+1.863 − 0.010 =		+1.853
	G	62.061
−8.642 − 0.005 =		−8.647
	H	53.414
−4.760 − 0.003 =		−4.763
	C	48.651
	E	55.630
+6.433 − 0.002 =		+6.431
	G	62.061
−8.847 + 0 =		−8.847
	B	53.214
	E	55.630
−2.214 − 0.002 =		−2.216
	H	53.414

FIG. 6-6. Adjustment by the Dell method.

squares. It consists of successive approximations. Any circuit is chosen to start the work. The total correction for the circuit is distributed among all the lines that compose the circuit in proportion to their lengths. The closures for each of the adjacent circuits are changed in accordance

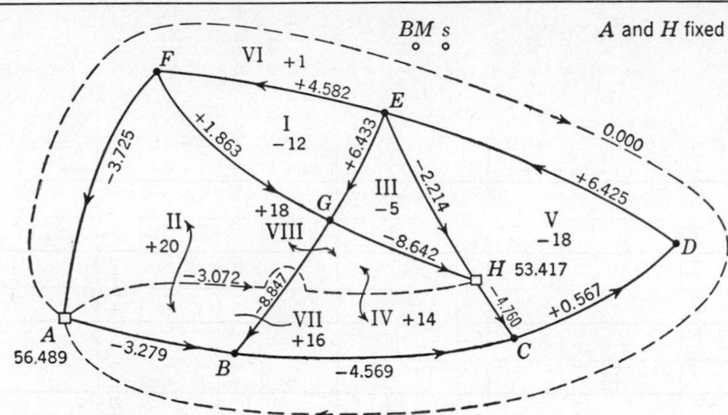

Dotted lines may be drawn anywhere as long as all points of fixed elevation are joined by them and the figure is surrounded by them.

Compute loops I, II, III, IV, and V as before.

Computation of Circuit Errors (Closures) in Clockwise Direction

	Line	+	−	
VI	*A–B*		−3.279	
	B–C		−4.569	
	C–D	+0.567		
	D–E	+6.425		
	E–F	+4.582		
	F–A		−3.725	
		+11.574	−11.573	= +0.001
VII	*A–H*		−3.072	
	H–C		−4.760	
	C–B	+4.569		
	B–A	+3.279		
		+7.848	−7.832	= +0.016

	Line	+	−	
VIII	*A–F*	+3.725		
	F–G	+1.863		
	G–H		−8.642	
	H–A	+3.072		
		+8.660	−8.642	= +0.018

Summary in thousandths of a foot

	+	−	
I		−12	
II + IV = VII + VIII	+34		
III		− 5	
V		−18	
VI	+ 1		
	+35	−35	= 0✓

Partial summary

II	+20	VII	+16
IV	+14	VIII	+18
	34		34✓

Fig. 6-7. Plan of level net showing field data and errors.

with the effect of these corrections. The corrections for the lines of another circuit are computed in the same way and all the affected circuit closures are changed accordingly. This process is continued until the closures for all the circuits becomes zero.

Figure 6-6 shows a form for this computation. The level net in Example I in the figure is adjusted. The lines and their lengths in hundreds of feet are listed by circuits and the percentage of the total length of circuit is computed for each line. The error of closure for each circuit in thousandths of a foot is shown under each set of lines.

The computation is carried out in pairs of columns. At the head of the first column of each pair is the name of the circuit adjusted. The second columns are all marked *C*. In these are placed the resulting closure for every circuit after each circuit is adjusted.

Circuit I is adjusted first. A total correction of +12 must be applied. This requires a correction to *FE* of +5; *EG*, +2; and *GF*, +5. The total, +12, is added to the error, −12, giving 0 in the *C* column. The corrections, with their signs changed, are then applied to the same set of lines in the circuits where they appear. The *C* column is completed. The sum of the values in each *C* column must be 0. The process is continued until all the circuit closures are reduced to 0 or to 0.001 ft. In the *E* column, these small values are eliminated by correcting the most convenient lines so that all closures are 0.

In the *F* column, the total correction to each line is computed. The sum of the corrections for the lines of each circuit must be equal to the original error of closure with its sign changed. Each line must appear in the *F* column twice with equal corrections but with opposite signs.

At the right is shown the computation of the corrected differences and the final elevations.

6-11. Confusing Arrangements. Sometimes the fixed elevations are within the level net. Dotted lines must be introduced so that all fixed elevations are connected. This sometimes forms superimposed loops which, though confusing, are not difficult to handle. Figure 6-7 shows how such an arrangement is handled. An arrow flying across a line like *BC* will affect three loops, IV, VI, and VII. The same treatment is used in the rare case when lines of levels that cross each other are not connected.

TRIANGULATION

6-12. Value of Triangulation. Horizontal control usually consists of a combination of triangulation and traverse. Triangulation is always cheaper when the distances are large, because horizontal measurement by tape is expensive. Triangulation requires many lines of sight from each station so that in areas where long sights are difficult to obtain, high observation points are required. If suitable existing positions for stations are not available, high towers must be built. As the distances become shorter, the expense of the towers overbalances any savings, and traverse becomes cheaper. On the other hand, once the proper triangulation stations have been established, the increase in cost for very precise measurements is comparatively small.

Control stations must be placed so that ties can be made conveniently. Usually triangulation stations must be located at inaccessible positions where connections are difficult, whereas traverse is so flexible that stations can usually be placed anywhere desired.

The reader is referred to Chap. 18 for further discussion of triangulation and to Chap. 22 for the adjustment of triangulation by the method of least squares.

It follows that usually the best plan is to use a simple triangulation system for primary control covering the whole area, and traverse (perhaps with supplementary triangulation) for secondary control.

6-13. Reconnaissance for Triangulation. In laying out a triangulation system, the first consideration is the shape of the triangles. The accuracy of the system depends on how accurately the length of the base can be carried through the angles to each length in the system. The basic formula for computation is, of course,

$$a = \frac{\sin A}{\sin B} b$$

which may be written

$$a = (\sin A \operatorname{cosec} B)b$$

but

$$\frac{da}{dB} = -\csc B \cot B(\sin A \cdot b)$$

This means that a given error in measuring B will create an error in a that is proportional to the function $\csc B \cot B$. The values of this function for several angles can be computed thus:

Angle	csc	cot	− csc cot
0°	$+\infty$	$+\infty$	$-\infty$
10°	+5.76	+5.67	−32.66
20°	+2.92	+2.75	−8.03
30°	+2.00	+1.73	−3.46
45°	+1.41	+1.00	−1.41
51°50′	+1.27	+0.79	−1.00
90°	+1.00	0	0
128°10′	+1.27	−0.79	+1.00
135°	+1.41	−1.00	+1.41
150°	+2.00	−1.73	+3.46
160°	+2.92	−2.75	+8.03
170°	+5.76	−5.67	+32.66
180°	$+\infty$	$-\infty$	$+\infty$

This table shows the relative errors of the determination of a side a for a given error in angle B for various sizes of B. Two examples will serve to illustrate:

Example 1

$$A = 30° \qquad \sin 30° = 0.5$$
$$B = 45° \qquad -\csc \cot 45° = -1.41$$
$$C = 105°$$

If there is an error of 1 second ($1'' = 0.00000485$) in B, the error in a caused by it would be

$$\text{Error} = (0.00000485)(-1.41)(0.5)b = -0.0000034b$$

Example 2

$$A = 30° \qquad \sin 30° = 0.5$$
$$B = 20° \qquad -\csc \cot 20° = -8.03$$
$$C = 130°$$

For an error of 1 second in B, the error in a caused by it would be

$$\text{Error} = (0.00000485)(-8.03)(0.5)b = -0.0000195b$$

When errors in angle measurement change the lengths by relatively small amounts the triangle is said to be **strong.** The **strength** of a triangle is a measure of this quality.

The function $-\csc \cot$ of angles smaller than 30° and larger than 150° approaches infinity very rapidly so that these values make good practical limits. In laying out a system, therefore, no angle opposite a side which is used in the computation to carry the lengths through the system (known as distance angle) should be much smaller than 30° or larger than 150°.

More precise methods of evaluating the strength of figure are given in Chap. 18.

Figures 6-8a and 6-8b show examples of good and very poor arrangements, respectively. In Fig. 6-8a, the length of the base is carried to the check base by two routes. For the first route, side a is computed by distance angles 1 and 2. Then b is computed by 3 and 4, and so on, using angles 5, 6, 7, and 8. All are well over 30°.

In the other route, angles 1, 12, 13, 14, 15, 16, 17, and 8 are used. All are over 30°.

In Fig. 6-8b only one route is available. The angles involved are 1, 2, 3, 4, 5, 6, 7, and 8; here angle 3 is larger than 150° and 4 and 8 are less than 30°.

The two routes in Fig. 6-8a alone make the system desirable, since an extra check is obtained at the check base. Only 12 sides are observed in Fig. 6-8*a* while 13 are necessary in Fig. 6-8*b*.

Usually the system can be arranged to form two or more simple **independent figures.** An independent figure is one that is connected to each adjacent figure by but one common side. The figures are independent only in the sense that if only one side is fixed, the angles of each independent figure can be adjusted independently.

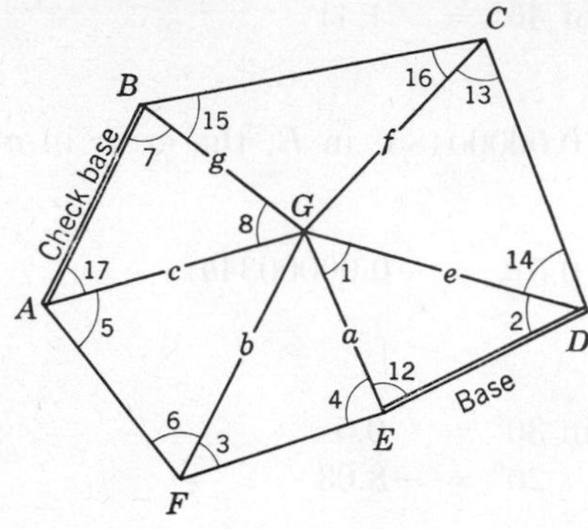

(*a*) Good system

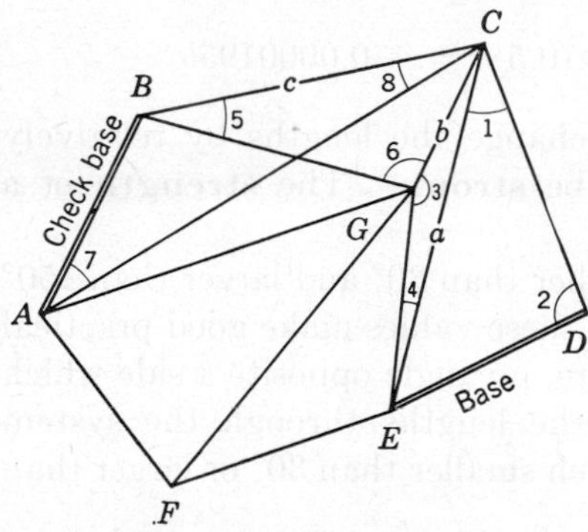

(*b*) Very poor system

Fig. 6-8. Examples of triangulation systems.

A line that joins two otherwise independent figures should be avoided in a primary system.

Figure 6-9 shows the four best types of figures to use. There is only one route through the system of simple triangles shown in Fig. 6-9*a*, but in each of the others two routes are available. In general, two routes should be possible. More than two add little to the accuracy unless expensive computation is performed. The system of single triangles should be used, therefore, only when it is impossible to arrange as strong a system by the other method.

Figure 6-9*b* shows the most common arrangement, a chain of quadrilaterals.

In Fig. 6-9*c* the central-point figure is very strong and often quite easy to arrange.

When, in a central point, the four-sided figure is skewed, it is greatly strengthened by a diagonal of the type shown (see Fig. 6-9*d*).

In general, not more than seven stations should be included in an independent figure.

6-14. Bases. The site for a base should be smooth, open, level ground where measurement is easily accomplished. Usually a base and a check base will suffice. They should be arranged so that all the triangles are used to carry the length from one base to the other. Sometimes a central base is used with two check bases at the two limits of the net (see Fig. 6-10).

The longer the base the more accurate will be all the lengths in the net.

When a short base must be used, a strong figure must be selected to expand it to fit the net.

Methods suitable for measuring bases more accurately than by the distance measurement given in Chap. 3 are covered in Chap. 18.

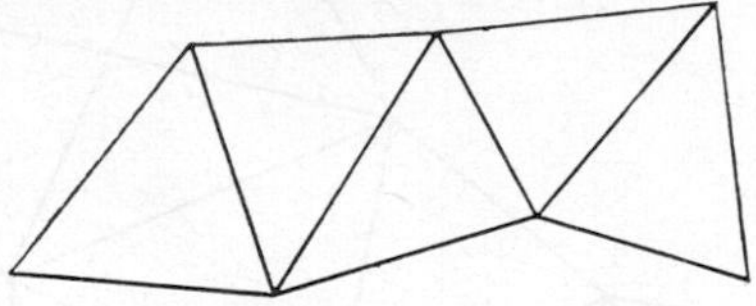

(*a*) Chain 5 independent triangles

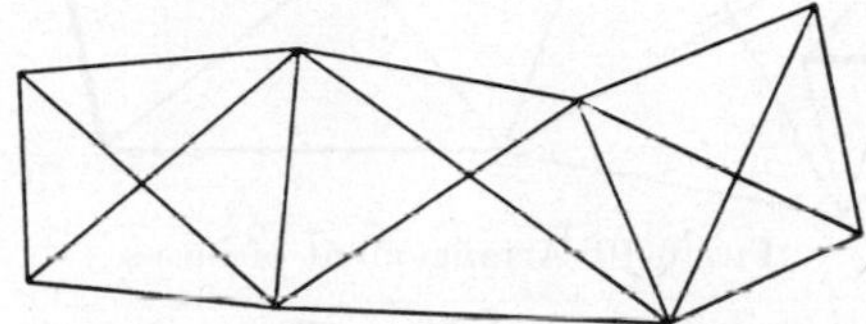

(*b*) Three independent quadrilaterals or quads

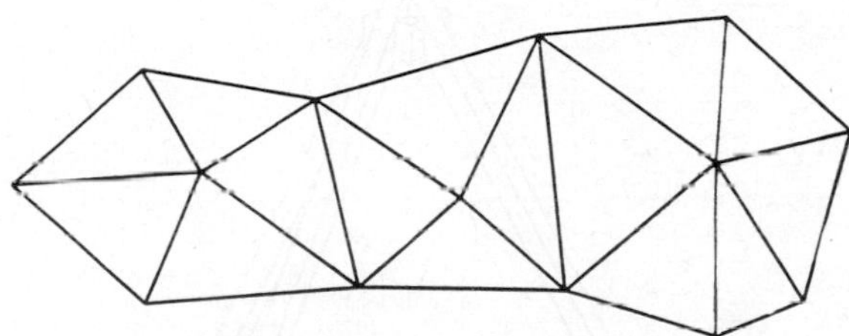

(*c*) Three independent central point figures

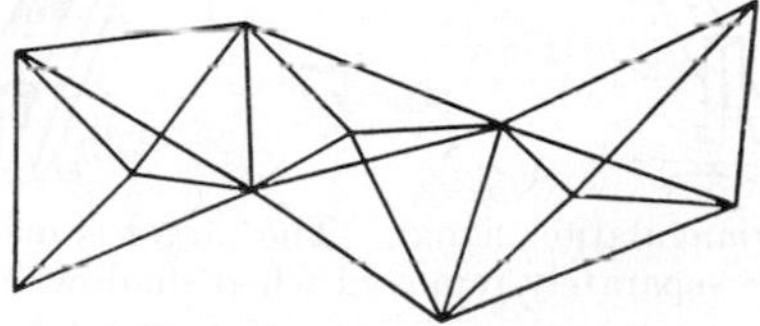

(*d*) Three independent central point figures each with extra diagonal

FIG. 6-9. Best types of triangulation figures.

6-15. Triangulation Stations. Triangulation stations should be permanently marked and witnessed by permanent reference marks properly tied to the station. The stations should be on the ground, if possible, for accessibility. Each station should have a permanent signal and

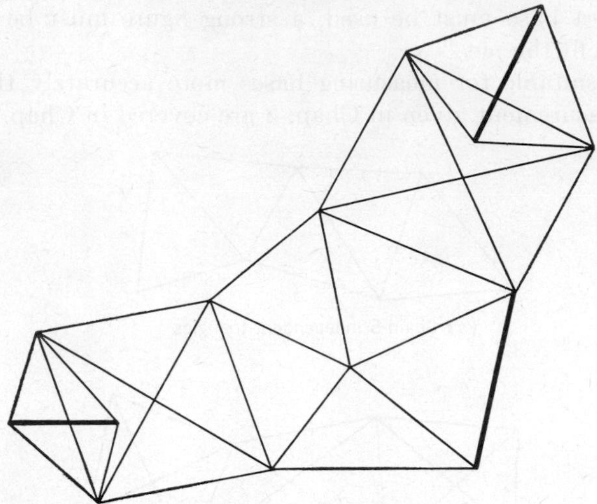

FIG. 6-10. Arrangement of bases.

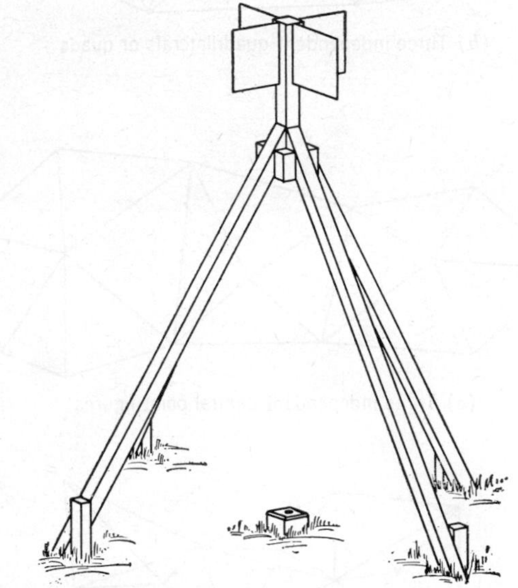

FIG. 6-11. Permanent triangulation signal. The target is made of heavily galvanized iron sheets which can be separately removed when shadows may create phase.

a permanent instrument position, both arranged to last throughout the survey (see Fig. 6-11). Often a church steeple or the center of a water tank can be used. They are excellent, since they can be seen so easily from a distance. Usually observations from such a station must be made at an eccentric position.

After the primary control is complete, the stations on the control will have to be occupied and observed frequently for other work. When, at any station, it is difficult to see any other triangulation station, a signal should be established or chosen that is clearly visible, to serve as an **azimuth mark.** Its direction is determined in the course of the work, for later use in making azimuth ties.

Often it is necessary to place stations on the roofs of buildings. Boards must usually be installed to support the observer so that his movements will not disturb the instrument. Boards extending from the walls or supported near the walls are effective. When such a station is occupied, the directions to two or three ground marks near the station should be observed to be used later for connecting subsequent surveys.

FIG. 6-12. Vertical collimator directly on station marker. (*U.S. Coast and Geodetic Survey.*)

6-16. Towers. Special towers for observing are sometimes necessary. They should be avoided if possible. They must be built so that the structural members that support the instrument are not used to support the observer. Two towers are necessary, therefore, one within the other. The signal can be placed on either. The positions for the instrument and the signal must be established on the tower from the ground mark by cross observations with a transit or a special device called a **vertical collimator,** which has a vertical telescope that may be set over the ground station mark (see Fig. 6-12).

6-17. Reduction to Center. When a triangulation station is established on a church steeple or any position where the instrument cannot be placed on a vertical line with it, the angle measurements at that station must be made from an eccentric position. Each direction observed from the eccentric position must be corrected to determine the values that would have been obtained had the observations been made at the station.

In the triangulation system shown in Fig. 6-13, T represents a station on a water tank and E the eccentric position used for observations. The angle observations to $PQRSU$ are made in the usual manner and at least two shots to T are taken so that the angles a measured from the direction ET to the direction of each station will be available.

The distance d is measured. If this cannot be accomplished directly, two short bases should be established extending from E as shown in the figure. EF, EG, and the angles shown are measured; thus d is determined by two independent triangles.

The lengths TP, TQ, TR, TS, and TU are computed from the field angles that were measured at the other stations or scaled from the layout drawing.

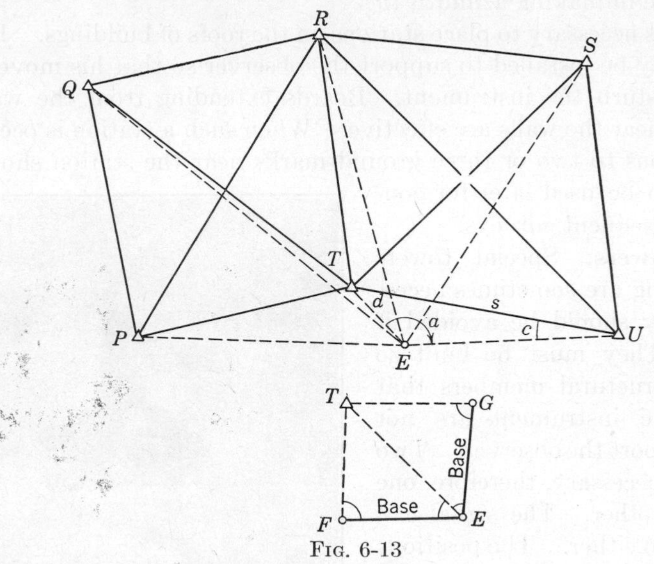

FIG. 6-13

The correction c to each direction is computed by the sine formula; s is the length from T to the station in question.

$$\sin c = \frac{d \sin a}{s} \tag{6-1}$$

or, when c is small,

$$c(\sin 1'') = \frac{d \sin a}{s} \tag{6-1a}$$

where c is expressed in seconds.

If a is always taken as the clockwise angle from T to the station in question, the sign of the correction will be correct.

If it is decided to use E as the station instead of T, that is, if the signal is assumed to be eccentric to the station rather than the other way around, then all the directions observed toward T must be corrected. These are UT, PT, QT, RT, and ST. The same formula will give the correction and the correct sign.

When s is large and d is quite small, the length of s can be obtained by scaling a layout drawing. The error in the correction for a given error in s can be computed as follows:

$$c_e = -s_e \frac{206{,}000d \sin a}{s^2}$$

where c_e = error in seconds in c

s_e = error in s

s_e, d, and s are in the same units

Example. When $s = 10{,}000'$, $d = 20'$, $a = 30°$, to find the error in the correction caused by an error of +5 ft in determining s,

$$c_e = -5\,\frac{(206{,}000)(20)(0.5)}{10^8}$$

$$c_e = -0.1''$$

This formula can be developed as follows:

$$\sin c = \frac{d \sin a}{s}$$

$$\frac{d \sin c}{ds} = -\frac{d \sin a}{s^2}$$

When the values are small, $dc = \Delta c$, $ds = \Delta s$.

$$\Delta \sin c = -\Delta s\,\frac{d \sin a}{s^2}$$

or

$$\sin c_e = -s_e\,\frac{d \sin a}{s^2}$$

If c_e is expressed in seconds, since it is so small that $\sin c_e = (\sin 1'')c_e = 0.00000485c_e$,

$$c_e = -s_e\,\frac{d \sin a}{0.00000485s^2}$$

$$c_e = -s_e\,\frac{206000d \sin a}{s^2}$$

ADJUSTMENT OF TRIANGULATION

6-18. Need for Adjustment. When the station adjustments have been made and any reductions to center are complete, the net must be adjusted to give geometric consistency.

6-19. Principles. Two angles completely determine the shape of a triangle. The measurement of the third angle is superfluous. It would be only by chance that the three measured angles should add up to 180°; yet if they do not, they cannot be used, since the computed results would be different depending on which angles were chosen. The corrections applied must be such that the final angles of each triangle add to 180°. This is the first set of conditions imposed on the corrections.

When two routes are available in an independent figure, if the angles of each triangle are adjusted independently, the angles of the figure as a whole would be geometrically consistent only by chance. This imposes a second set of conditions on the corrections used. Both sets described are called angle conditions.

When length can be computed through an independent figure by more than one route, the angles must also be adjusted so that computations by each route will give the same final length. This constitutes a third set of conditions. Other requirements are imposed on the corrections when the net includes other fixed values. It is logical to use corrections that have the greatest probability of being the true corrections. The method of least squares will give the most probable corrections that satisfy the conditions imposed. This method is covered in Chap. 22. The method of least squares involves so much computation that approximate methods are desirable. To be acceptable, an approximate method must result in geometric consistency and must give corrections that approach as nearly as possible the results of a least-squares solution. An approximate solution is given below that satisfies these requirements. Corrections are computed to satisfy each condition in turn so that several corrections are applied to each angle. The computations are arranged so that corrections computed to satisfy one condition will not disturb the corrections computed to satisfy the other conditions.

APPROXIMATE METHOD OF ADJUSTMENT

6-20. Types of Figures Covered. In the approximate method each independent figure is first adjusted by itself. In the following paragraphs the method of adjustment of each of the first three types of figures shown in Fig. 6-9 is covered. The fourth type is best handled by least squares.

6-21. Adjustment of a Single Triangle. Subtract the sum of the measured angles from 180° and divide by 3. The result is the correction to be applied to each angle.

6-22. Adjustment of a Quadrilateral. Although all corrections are applied to the angles, the term **angle conditions** refers to the requirements imposed on the angles by the directions of their sides, and the term **length conditions** refers to the requirements imposed by the lengths of the sides. In each quadrilateral, seven angle conditions must be satisfied. From Fig. 6-14 these are as follows:

$$a + b + c + h = 180° \qquad (6\text{-}2)$$

$$b + c + d + e = 180° \qquad (6\text{-}3)$$

$$d + e + f + g = 180° \qquad (6\text{-}4)$$

$$a + f + g + h = 180° \qquad (6\text{-}5)$$

$$a + b + c + d + e + f + g + h = 360° \qquad (6\text{-}6)$$

$$a + h - (d + e) = 0 \qquad (6\text{-}7)$$

$$b + c - (f + g) = 0 \qquad (6\text{-}8)$$

Equations (6-7) and (6-8) are correct because there is an angle at the intersection of the two diagonals that is the common exterior angle for both triangles involved.

If the sum and difference of Eqs. (6-7) and (6-8) are each added to and subtracted from Eq. (6-6), Eqs. (6-2) to (6-5) will be formed. It is evident, then, that if Eqs. (6-6) to (6-8) are satisfied, Eqs. (6-2) to (6-5) will also be satisfied. Accordingly, only three equations are required. The ones chosen are called the **angle condition equations.** The number of condition equations required can be computed for any figure by the methods given in Chap. 22. In a quadrilateral only three are required, and Eqs. (6-6) to (6-8) are the most convenient.

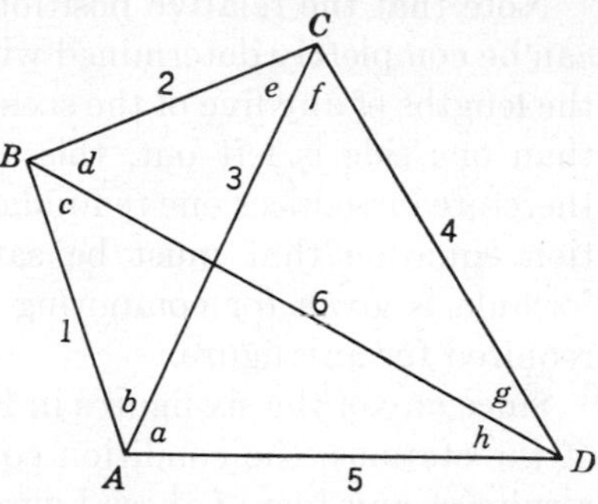

Fig. 6-14. Quadrilateral.

First, Eq. (6-6) is satisfied by correcting all the angles equally. Then Eq. (6-7) is satisfied by applying to each of the angles a set of corrections that are equal in size. The corrections are added to the first two angles and subtracted from the last two. Since the total effect on the sum of the angles is zero, Eq. (6-6) will still be satisfied. Equation (6-8) is treated in the same manner as Eq. (6-7).

6-23. Side Condition Equations. Suppose that the line 4 in Fig. 6-14 is rotated by an angle α as shown in Fig. 6-15a. Angle f will be

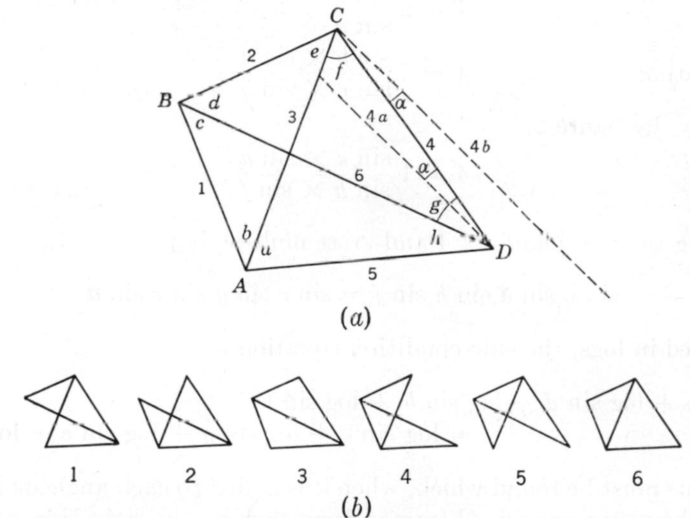

Fig. 6-15. Reason for side equations.

increased by α and angle g will be decreased by α so that the angle condition equations will still be satisfied. Obviously such a figure cannot exist, since D or C would have several positions. It will be found that any line in the figure may be rotated without interfering with the angle condition equations. It follows that the satisfaction of these equations does not ensure a geometric possibility. A side condition must be applied.

Note that the relative positions of all the stations in the quadrilateral can be completely determined without any knowledge of the angles, when the lengths of any five of the six sides are known (see Fig. 6-15*b*). If more than one side is left out, this cannot be accomplished. The final side therefore introduces one redundancy. This may be expressed by a condition equation that must be satisfied by adjustment. In Chap. 22 a formula is given for computing the number of side condition equations required for any figure.

Since any of the six figures in Fig. 6-15 establishes the relative positions of the stations, the condition equation can be made up by equating the results of any two of these figures. Figures 5 and 2 are usually chosen. For a given length of side 1, side 4 should have the same length when computed through both figures.

6-24. Computation of the Side Condition Equation. In figure 5 the computation would be as follows:

$$2 = 1\,\frac{\sin b}{\sin e}$$

$$4 = 2\,\frac{\sin d}{\sin g}$$

Substituting,

$$4 = 1\,\frac{\sin b \times \sin d}{\sin e \times \sin g}$$

Similarly, by figure 2,

$$4 = 1\,\frac{\sin c \times \sin a}{\sin h \times \sin f}$$

Equating the two values of 4 and cross-multiplying,

$$\sin b \sin d \sin h \sin f = \sin e \sin g \sin c \sin a$$

Expressed in logs, the side condition equation is

$$\log \sin b + \log \sin d + \log \sin h + \log \sin f$$
$$= \log \sin e + \log \sin g + \log \sin c + \log \sin a$$

A value must be found which, when it is added to each angle on the left side of the equation and subtracted from each on the right side, will give log sines that satisfy the equation. Note that the effect of such a system of corrections will cancel in the angle condition equations.

While looking up the logs, note the difference per second of each log sin. Find the sum of these differences per second (Σd). One second applied to

TABLE 6-1. ADJUSTMENT OF QUADRILATERAL

Angle	Angles, log sines	d 1″	Angle eqs.		Side eqs.	Adjusted angles, log sines
a	65°06′18″ 9.957628 18 9.957646	0.98	−0.7	−2.3	+0.8	65°06′15.8″ 9.957628 15 9.957643
b	44°03′38″ 9.842163 83 9.842246	2.18	−0.7	+3.0	−0.8	44°03′39.5″ 9.842163 86 9.842249
c	42°26′47″ 9.829131 108 9.829239	2.30	−0.6	+3.0	+0.9	42°26′50.3″ 9.829131 116 9.829247
d	50°44′05″ 9.888858 9 9.888867	1.72	−0.6	+2.3	−0.9	50°44′05.8″ 9.888858 10 9.888868
e	42°45′22″ 9.831742 50 9.831792	2.28	−0.6	+2.2	+0.9	42°45′24.5″ 9.831742 56 9.831798
f	55°51′40″ 9.917805 57 9.917862	1.43	−0.6	−3.0	−0.9	55°51′35.5″ 9.917805 51 9.917856
g	30°38′57″ 9.707180 202 9.707382	3.55	−0.6	−3.0	+0.9	30°38′54.3″ 9.707180 193 9.707373
h	28°23′18″ 9.677030 70 9.677100	3.90	−0.6	−2.2	−0.9	28°23′14.3″ 9.677030 56 9.677086
Σ	360°00′05″	18.34	−5.0	0	0	360°00′00.0″

		Field	*Adjusted*
	$a + h - (d + e)$	36″ − 27″ = +9″	30.1″ − 30.3″ = −0.2″
	$b + c - (f + g)$	25″ − 37″ = −12″	89.8″ − 89.8″ = 0″
Σ log sines	a, c, e, g	39.326059	39.326061
Σ log sines	b, d, f, h	39.326075	39.326059
		−16	+2

every angle would create this difference Σd in the sum of the log sines. Divide the total correction required by Σd. This will be the number of seconds to correct each angle. Apply this correction to each angle using the proper sign in accordance with the rule stated in the previous paragraph.

Example. Table 6-1 shows the adjustment of the quadrilateral shown in Fig. 6-14. The station adjustments and reductions to center have been completed. Under the angles are their log sines, the interpolations, and the differences per second.

The sum of the angles is 5 seconds too large.

$$\frac{-5}{8} = -0.6$$

with a remainder of −0.2. Two angles receive a correction of −0.7 and the other six, −0.6.

The computations for the other angle condition equations are at the bottom of the page. The second correction column shows the distribution.

The error in the logs of the side condition equation is computed at the bottom of the page. It is −16 in the sixth place of the logs. The sum of the log differences per second is 18.34. Each log sine must be corrected by

$$\frac{16}{18.34} = 0.87'' \qquad \text{correction per angle}$$

The corrected angles are shown in the last column. Under them are their log sines. The log sines to the previous minute are copied from the second column and the same differences per second are used for interpolation. In the lower right-hand corner are the various checks on the computation. Slight errors due to rounding off appear.

6-25. Computation of the Sides. The sides are computed as shown in Table 6-2. The order of computation is the following:

In each triangle the data are written down in the following order:

Known side.................................... log
Angle opposite known side........................ colog sin
Angle opposite first unknown side.................. log sin
Angle opposite second unknown side.............. log sin

The sum of the first two logs added to each of the next two logs gives the logs of the two unknown sides.

The logs of three sides are used in computation. The logs of the other three sides are computed twice. The check obtained is indicated by a correction to the log that was obtained through the weakest triangles. The error is due to rounding off.

6-26. Computation of Coordinates. The known bearing of AB and the corrected angles are used to compute the bearings. These will fail to check in the last place because of rounding off in the adjustment. The resulting bearings are shown in the coordinate computation in Table 6-3.

The coordinates of C are computed both from the known coordinates of A and from the known coordinates of B. The bearings just computed and the sides from

Table 6-2 are used. The two sets of coordinates for C should check closely. The coordinates for D are similarly handled.

Table 6-3 shows the form for machine computation. Table 22-23 shows the logarithmic form. Table 6-3 gives the interpolation for the determining of the lengths of the sides from their logarithms in Table 6-2 and the interpolation of the natural cosines and sines.

TABLE 6-2. COMPUTATION OF SIDES
A–B = 9304.10 (from coord.)

A–B	9304.10	3.968675
h		0.322914
$a + b$	109°09′55.3″	9.975236
c		9.829247
B–D		4.266825
A–D		4.120836
A–D		4.120836
f		0.082144
$g + h$	59°02′08.6″	9.933228
a		9.957643
A–C		4.136208 (−2)
C D		4.160623
B–D		4.266825
$e + f$	98°37′00.0″	0.004930
g		9.707373
d		9.888868
B–C		3.979128 (−2)
C–D		4.160623 (+0)
A–B		3.968675
e		0.168202
b		9.842249
$c + d$	93°10′56.1″	9.999329
B–C		3.979126
A–C		4.136206

Triangle	*Known side*	*Sides computed*
ABD	AB	BD, AD
ADC	AD	AC, CD
BCD	BD	BC, CD
ABC	AB	BC, AC

The lengths are multiplied first by the cosines of the bearings to find the latitudes and then by the sines to find the departures. The latitudes are applied to the y coordinates and the departures to the x coordinates. The box at the bottom of the table shows how these are arranged.

6-27. Adjustment of a Central-point Figure. Figure 6-16 shows a typical central-point figure. The angle conditions to be satisfied are

TABLE 6-3. COMPUTATION OF COORDINATES

A–C N21°06′28.5″E				*B–C* N63°51′52.9″E			
4.136206			13,683.77	3.979126			9,530.73
0.136086				0.979093			
120/318				33/45			
0.932954	−104.7			0.440723	−261.2		
−50		*A*	1,432.56	−230		*B*	10,000.00
0.932904			+12,765.64	0.440490			+ 4,198.19
		C	14,198.20 (−1)			*C*	14,198.19
0.359997	+271.4			0.897643	+128.2		
+129		*A*	4,428.39	+113		*B*	800.00
0.360126			+ 4,927.88	0.897756			+ 8,556.27
		C	9,356.27			*C*	9,356.27
A–D N86°12′44.3″E				*C–D* S34°45′07.1″E			
4.120836			13,207.96	4.160623			14,475.13
0.120574				0.160469			
262/329				154/300			
0.066274	−290.3			0.821647	−165.8		
−215		*A*	1,432.56	−20		*C*	14,198.19
0.066060			+ 872.52	0.821627			−11,893.16
		D	2,305.08			*D*	2,305.03 (+5)
0.997802	+19.2			0.569997	+238.9		
+14		*A*	4,428.39	+28		*C*	9,356.27
0.997816			+13,179.11	0.570025			+ 8,251.19
		D	17,607.50			*D*	17,607.46 (+4)

	Name of side	*Bearing*
Log of side		Length
Nearest log		
Diff./tab. diff.		
Nat. cos. deg. min. bearing	Tab. diff.	
Tab. diff. × seconds/60		Sta. *y*-coord.
Nat. cos. bearing		latitude
		Sta. *y*-coord.
Nat. sin. deg. min. bearing	Tab. diff.	
Tab. diff. × seconds/60		Sta. *x*-coord.
Nat. sin. bearing		departure
		Sta. *y*-coord.

the following (there are also many polygons composed of triangles whose angle conditions will obviously be satisfied if the triangles are satisfied):

$$a + b = 180° - k \tag{6-9}$$
$$c + d = 180° - l \tag{6-10}$$
$$e + f = 180° - m \tag{6-11}$$
$$g + h = 180° - n \tag{6-12}$$
$$i + j = 180° - o \tag{6-13}$$
$$a + b + c + d + e + f + g + h + i + j = 540° \tag{6-14}$$
$$k + l + m + n + o = 360° \tag{6-15}$$

Equation (6-15) is already satisfied by the station adjustment at the central station. Adding Eqs. (6-9) to (6-13),

$$a + b + c + d + e + f + g + h + i + j = 900° - (k + l + m + n + o) \tag{6-16}$$

Substituting the values from Eq. (6-15),

$$a + b + c + d + e + f + g + h + i + j = 540°$$

Thus, if Eqs. (6-9) to (6-13) are satisfied as well as Eq. (6-15), all the conditions are satisfied.

Find the correction required for each triangle. These are shown within each triangle in Fig. 6-16.

Assign a set of balanced corrections to the angles about the central station. If their sum is zero the station adjustment will remain satisfied. The corrections should be small and not greater than one-third the largest triangle correction. This operation is shown in the table in Fig. 6-16, Columns 1 to 5.

Column 1 gives the required triangle corrections. The values in Column 1 are divided by 3 to find the values in Column 2. Their sum is at the bottom of the column. This sum is made as nearly zero as possible by equal adjustments as shown in Column 3, giving the results in Column 4. Column 5 gives the final values for the central correction. It is the same as Column 4 except for the change in the correction for angle in order to make the sum exactly equal to zero. In Columns 6 and 7 are the corrections for the other angles. They are made as nearly equal as possible in each triangle and large enough to complete the required triangle corrections.

6-28. Side Equations. In a central-point figure, note that the relative positions of the stations can be determined if one side is omitted. This requires one side equation. It is developed and satisfied by the same methods that were used for a quadrilateral.

$$\Sigma \log \sin a, c, e, g, i = \Sigma \log \sin b, d, f, h, j$$

6-29. Adjustment to Check Bases. In large triangulation nets, the angles must be adjusted so that all measured base lengths are held fixed. The method of least squares is usually required. In small nets, lengths do not deteriorate. Once the angles are adjusted for angle and side conditions, the actual lengths can be made to depend on an adjusted value of all the bases.

The longest base is used in the side computation. This will give a computed value for each check base. The lengths of all the sides are then proportionately changed in accordance with the mean of the bases

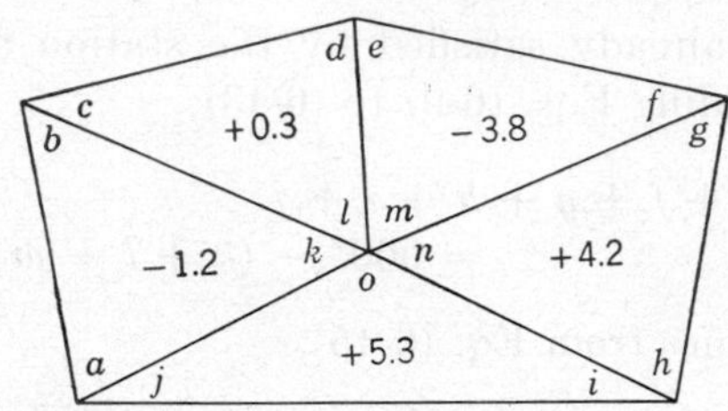

Required triangle corrections	1st trial, central corrections	Adjustment to 1st trial	2nd trial, central corrections	Final corrections		
(1)	(2)	(3)	(4)	(5)	(6)	(7)
−1.2	k −0.4	−0.3	−0.7	k −0.7	a −0.3	b −0.2
+0.3	l +0.1	−0.3	−0.2	l −0.2	c +0.3	d +0.2
−3.8	m −1.3	−0.3	−1.6	m −1.5	e −1.2	f −1.1
+4.2	n +1.4	−0.3	+1.1	n +1.1	g +1.6	h +1.5
+5.3	o +1.6	−0.3	+1.3	o +1.3	i +2.0	j +2.0
	+1.4		−0.1	0		

FIG. 6-16. Adjustment of a central-point figure.

weighted in proportion to the square root of their lengths. This can be expressed as follows:

$$P = 1 + \frac{\dfrac{1}{m_2\sqrt{m_2}}C_2 + \dfrac{1}{m_3\sqrt{m_3}}C_3}{\dfrac{1}{\sqrt{m_1}} + \dfrac{1}{\sqrt{m_2}} + \dfrac{1}{\sqrt{m_3}}}$$

where P = value by which each length is multiplied

m_1 = measured length of base used in the original computation

m_2, m_3 = measured lengths of check bases

C_2, C_3 = measured lengths of check bases minus computed lengths

Log P is added to the log of each side before the coordinate computation is begun.

Example. Assume $m_1 = 3600.00$, $m_2 = 1600.20$, $m_3 = 2500.10$, and the computed lengths of the check bases are the following: 1600.00, 2500.00.

$$P = 1 + \frac{\dfrac{0.20}{(16)(4)} + \dfrac{0.10}{(25)(5)}}{\frac{1}{6} + \frac{1}{4} + \frac{1}{5}} 10^{-2}$$

$$P = 1.0000636$$

$$\log P = 0.0000277$$

The modification of the formula for other than two check bases is obvious.

6-30. Supplementary Stations. While observations are in progress on the main scheme, directions should be observed to objects that are visible from ground points in the area. These include church steeples, water tanks, beacons, and the like. Each should be observed from at least three stations so that a position check is possible. A traverse terminal can usually be established wherever several of these points are visible.

As stated previously, at each main scheme station, arrangements must be made for traverse connections. Supplementary points must often be observed for this purpose. The computations for supplementary positions are made after the main scheme has been adjusted.

TRAVERSE

6-31. Connections with Triangulation Stations. Traverse terminals should always be connected to superior control by both position and azimuth ties. It is not always easy to plan the simplest and strongest ties with a triangulation station. The following procedure is recommended.

6-32. Types of Connections with Triangulation Stations.[1] In making a reconnaissance for a traverse tie to a triangulation station, try the following methods in the order given.

Legend: In Figs. 6-18 to 6-27, which illustrate the various cases, the symbols in Fig. 6-17 apply.

Case 1. *Connection by Superimposing Traverse on Azimuth Line.* This method represents the best form of tie to a triangulation station (see Fig. 6-18). The traverse line is run through the azimuth mark, and the distance from the azimuth mark to the triangulation station is measured. Coordinates can be computed for the azimuth mark, and then the triangulation station and azimuth mark become companion monuments on the traverse line.

[1] From *ASCE Manuals of Engineering Practice*, no. 20, Horizontal Control Surveys to Supplement the Fundamental Set, prepared by the Committee of the Surveying and Mapping Division on Control Surveys under the chairmanship of the author.

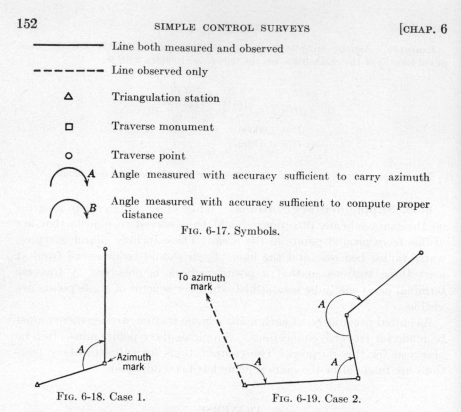

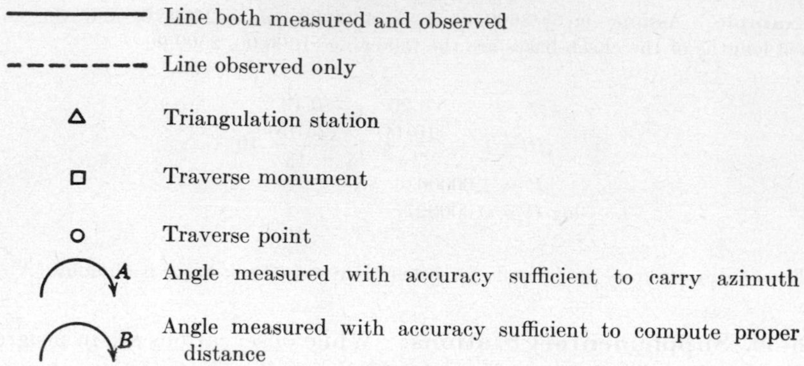

——— Line both measured and observed

- - - - - - - Line observed only

△ Triangulation station

□ Traverse monument

○ Traverse point

A Angle measured with accuracy sufficient to carry azimuth

B Angle measured with accuracy sufficient to compute proper distance

FIG. 6-17. Symbols.

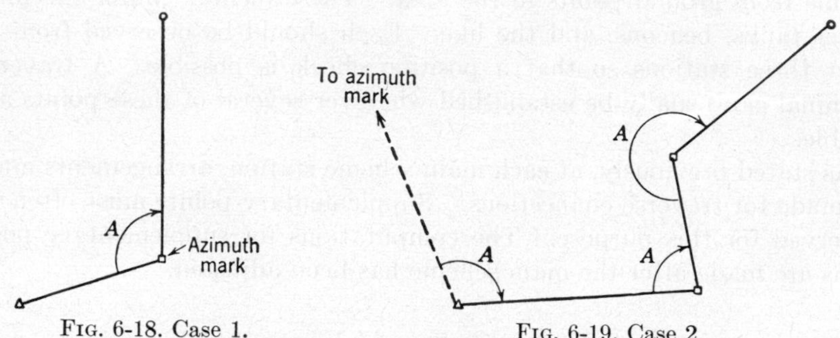

FIG. 6-18. Case 1. FIG. 6-19. Case 2.

Case 2. Standard Connection. Case 2 (see Fig. 6-19) should be used if Case 1 cannot be adopted because the azimuth mark is inaccessible.

Case 3. Separate Azimuth and Position Tie for Ground Station. Case 3 (see Fig. 6-20) is to be used when it is impossible to tape directly from the monument to the triangulation station.

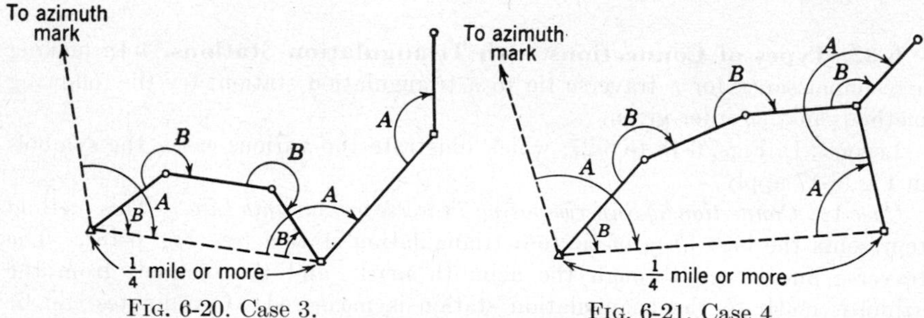

FIG. 6-20. Case 3. FIG. 6-21. Case 4.

Case 4. Special Case of Position Tie and Separate Azimuth Tie. Case 4 (see Fig. 6-21) is a variation of Case 3.

Case 5. Connection for a Triangulation Station on a Building, or on Ground, When Inaccessible. In this case (see Fig. 6-22) monuments should be set so that an approximately equilateral triangle is formed. No angle should be less than 45°.

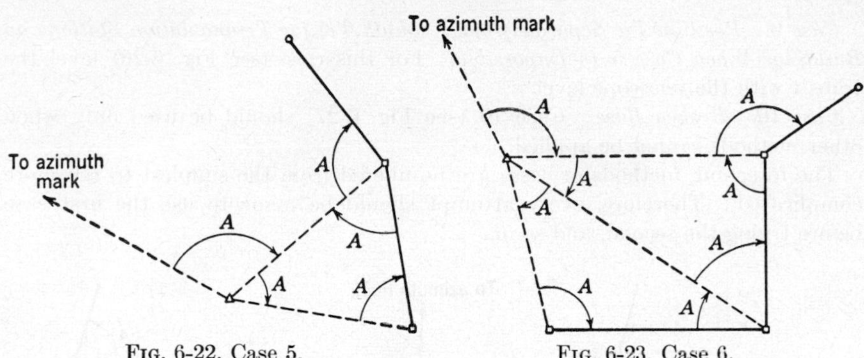

FIG. 6-22. Case 5. FIG. 6-23. Case 6.

Case 6. *Check Connection for a Triangulation Station on a Building, or on Ground, When Inaccessible.* For Case 6 (see Fig. 6-23) monuments should be set so that no angle is less than 30°. The other diagonal may be measured and the minimum angle may then be reduced to 20°.

Case 7. *For Ground Triangulation Station When Traverse Must Be Close to Station and When It Is Impossible to Measure Directly from Monument B to Station.* In Case 7 (see Fig. 6-24), keep the angle at Monument *A* close to a right angle, and keep the distance from Monument *A* to the triangulation station 100 ft or less.

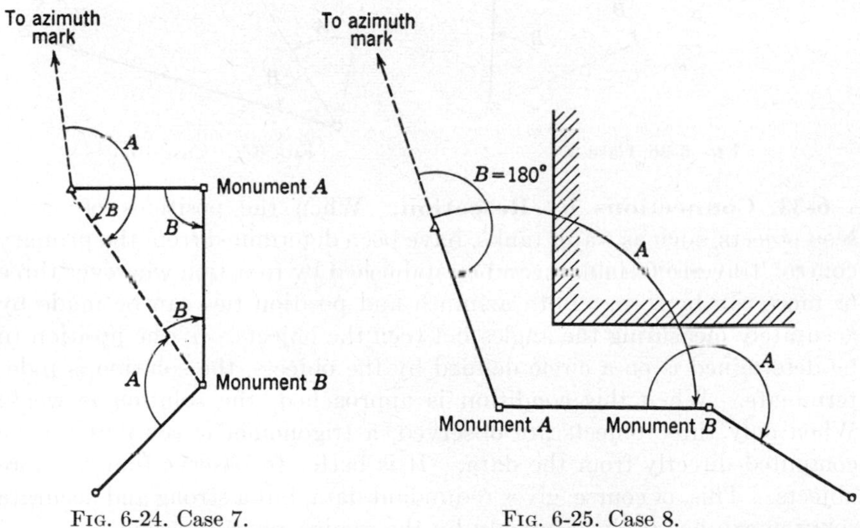

FIG. 6-24. Case 7. FIG. 6-25. Case 8.

Case 8. *Superimposing Traverse on Azimuth, When Azimuth Mark Is on Distant Spire or Tank.* In Case 8 (see Fig. 6-25) it is important: (1) that no direct method be possible; (2) that Monument *A* be located by three sets of double centering and placed near triangulation station; and (3) that the angle at *A* be turned from the azimuth mark to Monument *B*.

Case 9. *Position Tie Separate from Azimuth Tie for Triangulation Stations on Buildings When Case* 5 *Is Impossible.* For this case (see Fig. 6-26) level the transit with the telescope level.

Case 10. *Broken Base.* Case 10 (see Fig. 6-27) should be used only when other methods cannot be applied.

The foregoing methods or cases are numbered from the simplest to the more complicated. Therefore, every attempt should be made to use the first case before trying the second, and so on.

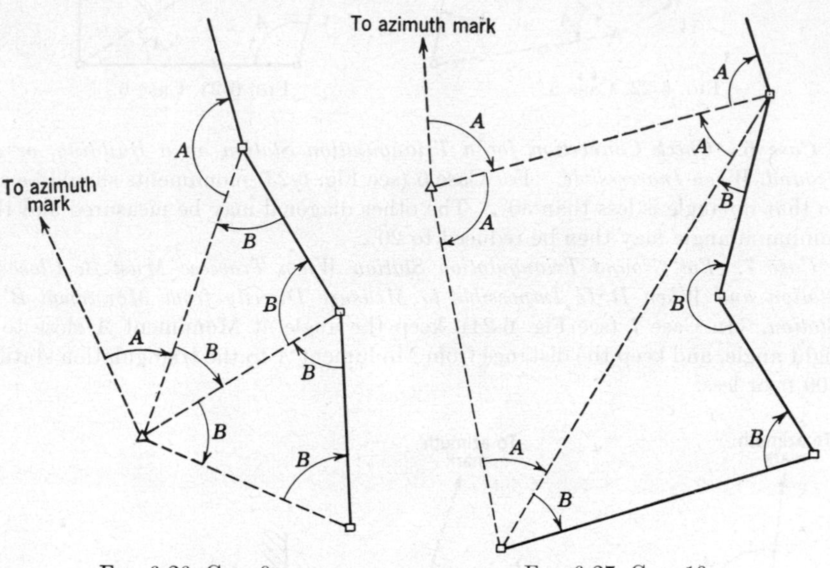

FIG. 6-26. Case 9. FIG. 6-27. Case 10.

6-33. Connections by Resection. When the positions of easily seen objects, such as water tanks, have been determined from the primary control, traverse terminals can be established by resection wherever three or more can be seen. Both azimuth and position ties can be made by accurately measuring the angles between the objects. If the position to be determined is on a circle defined by the objects, the solution is indeterminate. When this condition is approached, the solution is weak. When only three objects are observed, a trigonometric solution can be computed directly from the data. It is better to observe four or more objects. This, of course, gives redundant data, but a strong and accurate solution can be obtained quickly by the semigraphical method.

6-34. The Trigonometric Solution for Three Points. Figure 6-28 shows the three possible arrangements. A, B, and C are objects of known position. P is the position to be determined from the measured angles α and β. The sum of i plus j is known. The problem is to find the angles i, j, g, and h. When these are known, the lengths and bearings of AP, BP, and CP can be computed and the position P determined from each of these lines.

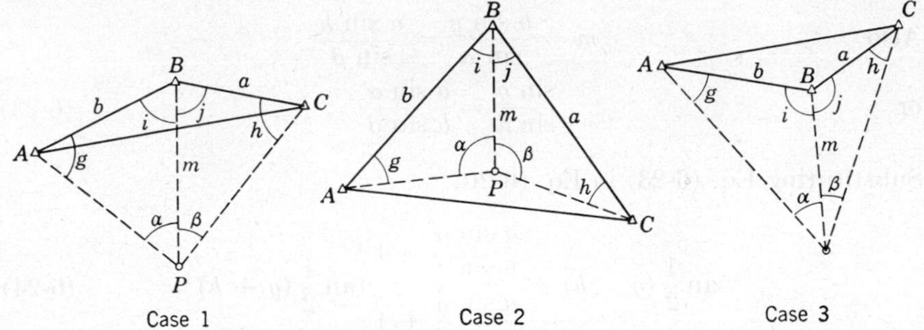

FIG. 6-28. The three-point problem of resection.

6-35. Theory.[1] The following three trigonometric formulas are used in this demonstration:

$$\sin g - \sin h = 2 \sin \tfrac{1}{2}(g - h) \cos \tfrac{1}{2}(g + h) \tag{6-17}$$

$$\sin g + \sin h = 2 \cos \tfrac{1}{2}(g - h) \sin \tfrac{1}{2}(g + h) \tag{6-18}$$

$$\tan (45° + \theta) = \frac{\tan 45° + \tan \theta}{1 - \tan 45° \tan \theta} \tag{6-19}$$

Dividing Eq. (6-17) by Eq. (6-18),

$$\frac{\sin g - \sin h}{\sin g + \sin h} = \frac{\sin \frac{1}{2}(g - h)}{\cos \frac{1}{2}(g - h)} \times \frac{\cos \frac{1}{2}(g + h)}{\sin \frac{1}{2}(g + h)}$$

or

$$\frac{\dfrac{\sin g}{\sin h} - 1}{\dfrac{\sin g}{\sin h} + 1} = \tan \frac{1}{2}(g - h) \cot \frac{1}{2}(g + h)$$

and

$$\tan \frac{1}{2}(g - h) = \frac{\dfrac{\sin g}{\sin h} - 1}{\dfrac{\sin g}{\sin h} + 1} \tan \frac{1}{2}(g + h) \tag{6-20}$$

From Eq. (6-19) (since $\tan 45° = 1$),

$$\tan (45° + \theta) = \frac{1 + \tan \theta}{1 - \tan \theta}$$

Substituting for $\tan \theta$ its equal, $1/\cot \theta$, and clearing,

$$\tan (45° + \theta) = \frac{\cot \theta + 1}{\cot \theta - 1} \tag{6-21}$$

[1] This method is the one published by the U.S. Coast and Geodetic Survey in various publications of that bureau.

From the figure, in each case,

$$g + h = 360 - (\alpha + \beta + i + j) \tag{6-22}$$

Also

$$m = \frac{b \sin g}{\sin \alpha} = \frac{a \sin h}{\sin \beta}$$

or

$$\frac{\sin g}{\sin h} = \frac{a \sin \alpha}{b \sin \beta} \tag{6-23}$$

Substituting Eq. (6-23) in Eq. (6-20),

$$\tan \frac{1}{2}(g - h) = \frac{\dfrac{a \sin \alpha}{b \sin \beta} - 1}{\dfrac{a \sin \alpha}{b \sin \beta} + 1} \tan \frac{1}{2}(g + h) \tag{6-24}$$

Equation (6-22) can be solved for $(g + h)$ and Eq. (6-24) can be solved for $\frac{1}{2}(g - h)$. These will give the required values. A solution of Eq. (6-24) that may be computed more easily is the following:

Let

$$\theta = \text{arc tan} \frac{b \sin \beta}{a \sin \alpha}$$

$$\cot \theta = \frac{a \sin \alpha}{b \sin \beta}$$

Substituting in Eq. (6-21),

$$\cot (45° + \theta) = \frac{\dfrac{a \sin \alpha}{b \sin \beta} - 1}{\dfrac{a \sin \alpha}{b \sin \beta} + 1}$$

Substituting in Eq. (6-24),

$$\tan \tfrac{1}{2}(g - h) = \cot (45° + \theta) \tan \tfrac{1}{2}(g + h) \tag{6-25}$$

where θ is defined as arc tan $\dfrac{b \sin \beta}{a \sin \alpha}$.

Example

$$a = 2545.19$$
$$b = 4285.38$$
$$i + j = 205°36'14''$$
$$\alpha = 45°55'02''$$
$$\beta = 19°40'08''$$

By Eq. (6-22),

$\alpha =$	45°55′02″	$\sin \alpha =$	0.718335
$\beta =$	19 40 08	$\sin \beta =$	0.336584
$i + j =$	205 36 14		
Explement of $(g + h) =$	271°11′24″		
	359 59 60		
$g + h =$	88°48′36″		
$\frac{1}{2}(g + h) =$	44°24′18″	$\tan \frac{1}{2}(g + h) =$	0.979443

By definition,

$$\tan \theta = \frac{b \sin \beta}{a \sin \alpha} = \frac{1442.39}{1828.30} = 0.788924$$

$$\begin{aligned} \theta &= 38°16'15'' \\ &\quad +45 \\ 45° + \theta &= 83°16'15'' \\ \cot (45° + \theta) &= 0.117989 \end{aligned}$$

From Eq. (6-25),

$$\begin{aligned} \tan \tfrac{1}{2}(g - h) &= (0.117989)(0.979443) \\ &= 0.115564 \\ \tfrac{1}{2}(g - h) &= 6°35'32'' \\ \tfrac{1}{2}(g + h) &= 44°24'18'' \\ \tfrac{1}{2}(g - h) &= 6\ 35\ 32 \\ g &= 50°59'50'' \qquad \sin g = 0.777115 \\ h &= 37°48'46'' \qquad \sin h = 0.613083 \end{aligned}$$

Check by Eq. (6-23):

$$\frac{\sin h}{\sin g} = 0.788922 \qquad \text{cf. } 0.788924$$

Final angles:

$$\begin{aligned} g &= 50°59'50'' & h &= 37°48'46'' \\ +\alpha &= 45\ 55\ 02 & +\beta &= 19\ 40\ 08 \\ & \ 96°54'52'' & & \ 57°28'54'' \\ & \ 179\ 59\ 60 & &= 179\ 59\ 60 \\ i &= 83°05'08'' & j &= 122°31'06'' \end{aligned}$$

6-36. Semigraphical Resection. In the illustration given here, it is assumed that a plane coordinate system has been established and that the coordinates of the points observed are known. The observations are made at station X.

Points observed	y	x
A	5,087.53	975.48
B	11,534.65	6,694.24
C	4,501.16	6,923.33
D	3,178.46	2,943.80

Angles at X after station adjustment:

AB	94°06′37″
BC	58 12 29
CD	79 14 27
DA	128 26 27

Plot $ABCD$ as shown in Fig. 6-29 and plot the measured angles on tracing paper as shown in Fig. 6-30. Move the tracing paper over the coordinate system until the lines pass through the proper points. Record the resulting coordinates of station X. Call this position X_1.

	y	x
X_1	4270	2700

Compute the bearing of the longest line, X_1B, from these coordinates:

	y	x
B	11,534.65	6,694.24
$-X_1$	4,270	2,700
	7,264.65	3,994.24

$$\tan \text{bearing } X_1B = \frac{3994.24}{7264.65} = 0.5498186$$

$$\text{Bearing } X_1B = \text{N}28°48'10''\text{E}$$

Starting with this bearing, compute the bearings of the other lines from the angles measured at X. Determine the tangents of all bearings less than 45° and the cotangents of those over 45°.

Course	*Bearings*	*Functions*
X_1B	N28°48′10″E	tan 0.5498178
X_1C	N87 00 39 E	cot 0.0522182
X_1D	S13 44 54 E	tan 0.2446676
X_1A	N65 18 27 W	cot 0.4597900

Choose two vertical and two horizontal lines on the coordinate system as close together as possible but so spaced that the point X will surely fall within them. These lines will form a rectangle.

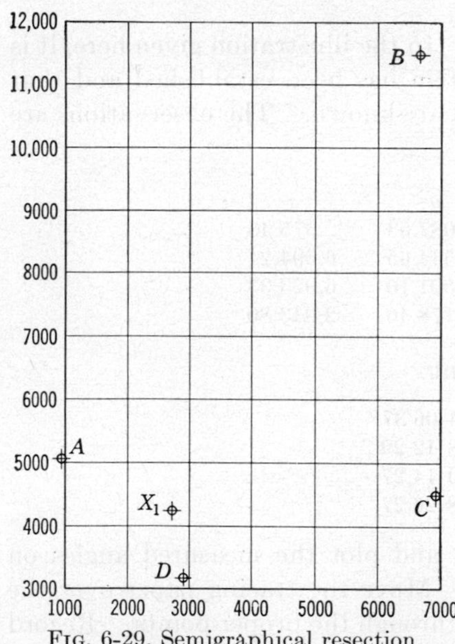

FIG. 6-29. Semigraphical resection.

	y	x
Lines	4330	2660
X_1	4270	2700
Lines	4230	2730

Plot this rectangle at large scale as shown in Fig. 6-31.

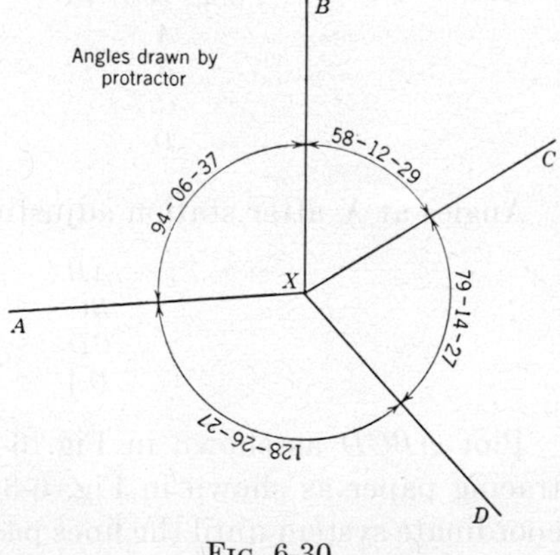

FIG. 6-30

Compute where the boundaries of the rectangle chosen will be intersected by lines originating at the known coordinate positions $ABCD$ and having the directions BX_1, CX_1, etc. These are the back values of the

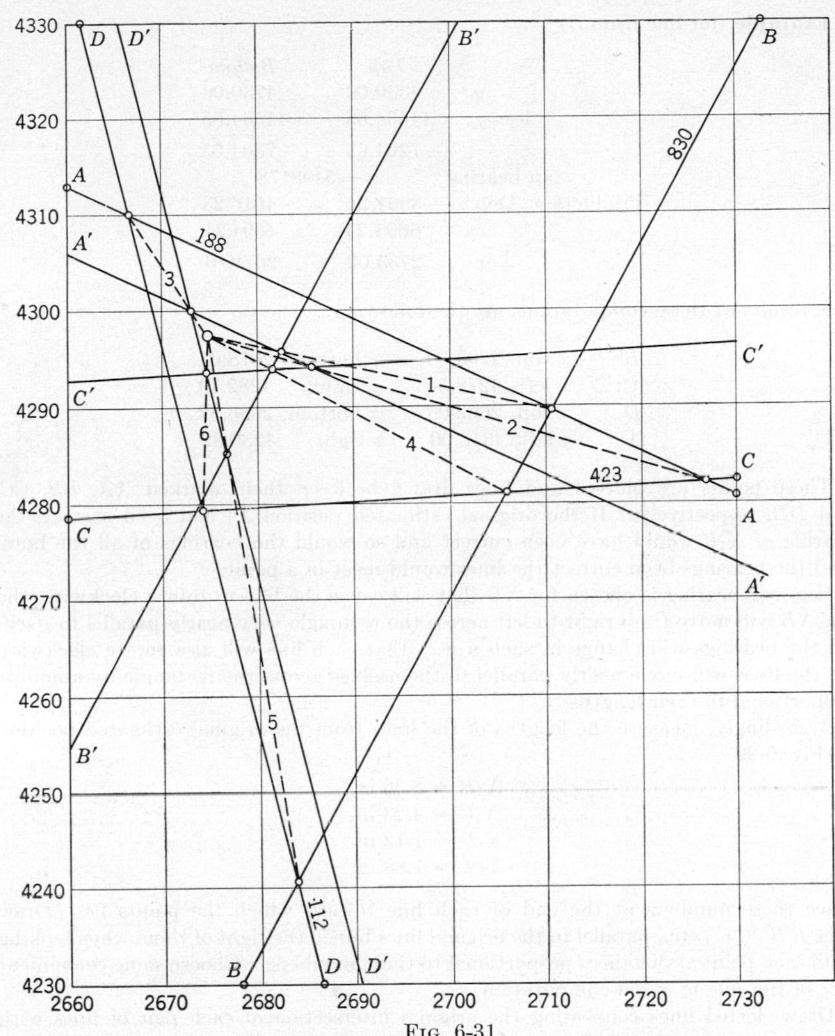

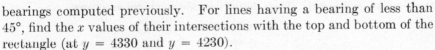

FIG. 6-31

bearings computed previously. For lines having a bearing of less than 45°, find the x values of their intersections with the top and bottom of the rectangle (at $y = 4330$ and $y = 4230$).

$$x = x_p + (y_r - y_p) \tan \text{bearing}$$

where x = value sought
x_p, y_p = coordinates of known station
y_r = y coordinate of top or bottom of rectangle

For lines having a bearing greater than 45° compute the intersections with the sides of the rectangle.

$$y = y_p + (x_r - x_p) \cot \text{bearing}$$

Example (for line from B)

	Top	*Bottom*
y_r	4330.00	4230.00
less y_p	11534.65	11534.65
	−7204.65	−7304.65
tan bearing	+.5498178	
Products = Dep.	−3961.24	−4016.23
x_p	6694.24	6694.24
x	2733.00	2678.01

The results of these computations are the following:

B:	x top, 2733.00	x bottom, 2678.01
C:	y left, 4278.54	y right, 4282.19
D:	x top, 2662.06	x bottom, 2686.52
A:	y left, 4313.00	y right, 4280.82

These points are plotted and lines drawn between them marked AA, BB, CC, and DD, respectively. If the original estimated position X_1 had been correct, the bearing of X_1B would have been correct and so would the bearings of all the lines. Had the bearings been correct the lines would meet in a point.

If a new bearing is chosen for XB that will cause the line to rotate clockwise, the line XB will move from right to left across the rectangle very nearly parallel to itself. All the bearings will change in such a way that each line will also rotate clockwise. All the lines will move nearly parallel to themselves across the rectangle by amounts proportional to their lengths.

Accordingly, measure the lengths of the lines from the original estimated position on Fig. 6-29.

$$X_1B = 8.30 \text{ in.}$$
$$X_1C = 4.23 \text{ in.}$$
$$X_1D = 1.12 \text{ in.}$$
$$X_1A = 1.88 \text{ in.}$$

Place these numbers at the end of each line toward which the points lie. Draw lines $B'B'$, $C'C'$, etc., parallel to the original lines but to the right of them when looking from each point at distances proportional to these numbers. Choose some convenient face of the rule to avoid computation.

Draw dotted lines connecting the original intersection of each pair of lines with their new intersection. These represent the movement of each intersection when the bearings are changed. These dotted lines are numbered as follows:

A and B:	1
A and C:	2
A and D:	3
B and C:	4
B and D:	5
C and D:	6

Since all the lines should meet at a point, these dotted lines should all intersect at that point. Extend them, if necessary, to the intersection. This occurs at $y = 4296.7$ and $x = 2675.0$. This is X_2, the second estimate of the position of X. It should be correct within a foot.

This whole operation is repeated to a much larger scale as shown in Fig. 6-32. The scale is so large that hundredths of a foot can be estimated. The first operation is to compute a new starting bearing for the longest line, X_2B, based on the new position X_2. Thereafter the procedure is the same as before. The final result is

	y	x
X_3	4297.35	2675.05

The position can be checked by computing the bearings of the various lines and checking the angles between them against the original angles. The position of X will now be known and the bearing of any of the lines can be used for an azimuth tie.

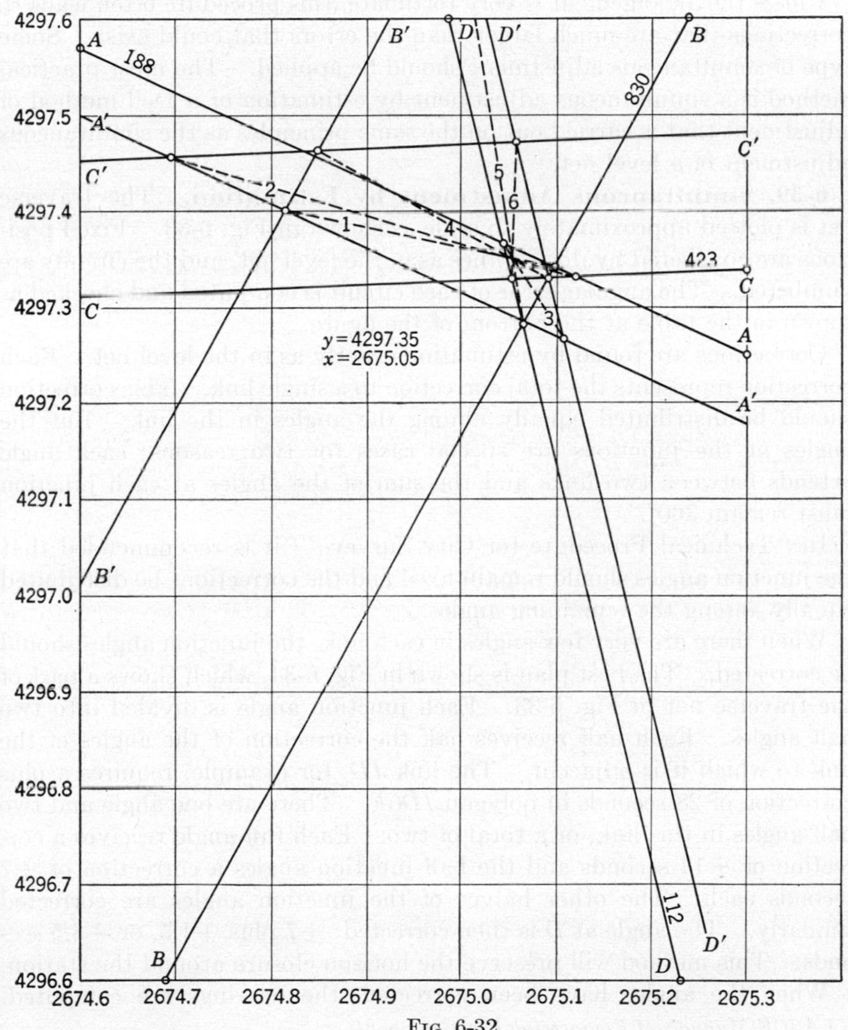

FIG. 6-32

6-37. Traverse Adjustment. Traverses are usually adjusted as follows: (1) azimuth closure is obtained by correcting all the angles by equal amounts; (2) the latitudes and departures are computed from the bearings that result after angle adjustment and each is corrected in proportion to the lengths of the course to which it applies.

6-38. Traverse Net Adjustment. When the arrangement of the traverses forms a net, it is usual practice to adjust each traverse separately in order of size, beginning with the shortest traverse that connects with superior control. Each adjusted traverse is held fixed after adjustment and the next traverse is adjusted to it.

Unless the arrangement is very fortunate, this procedure often leads to corrections that are much larger than the errors that could exist. Some type of simultaneous adjustment should be applied. The most practical method is a simultaneous adjustment by estimation or a Dell method of adjustment that is carried out on the same principles as the simultaneous adjustment of a level net.

6-39. Simultaneous Adjustment by Estimation. The traverse net is plotted approximately to scale as shown in Fig. 6-33. Fixed positions are connected by dotted lines as in the level net, and the circuits are numbered. The angular error of each circuit is computed and checked as shown in the table at the bottom of the figure.

Corrections are found by estimation exactly as in the level net. Each correction represents the total correction to a single link. This correction should be distributed equally among the angles in the link. But the angles at the junctions are special cases for two reasons; each angle extends between two links and the sum of the angles at each junction must remain 360°.

In "Technical Procedure for City Surveys"[1] it is recommended that the junction angles should remain fixed and the corrections be distributed equally among the remaining angles.

When there are very few angles in each link, the junction angles should be corrected. The best plan is shown in Fig. 6-34, which shows a part of the traverse net in Fig. 6-33. Each junction angle is divided into two half angles. Each half receives half the correction of the angles of the link to which it is adjacent. The link *JD*, for example, requires a plus correction of 28 seconds in polygon *JDGF*. There are one angle and two half angles in this link, or a total of two. Each full angle receives a correction of +14 seconds and the half-junction angles a correction of +7 seconds each. The other halves of the junction angles are corrected similarly. The angle at *D* is then corrected: +7 plus +1.5, or +8.5 seconds. This method will preserve the horizon closure around the station.

When the angles have been corrected, the bearings are computed.

[1] *ASCE Manuals of Engineering Practice*, no. 10.

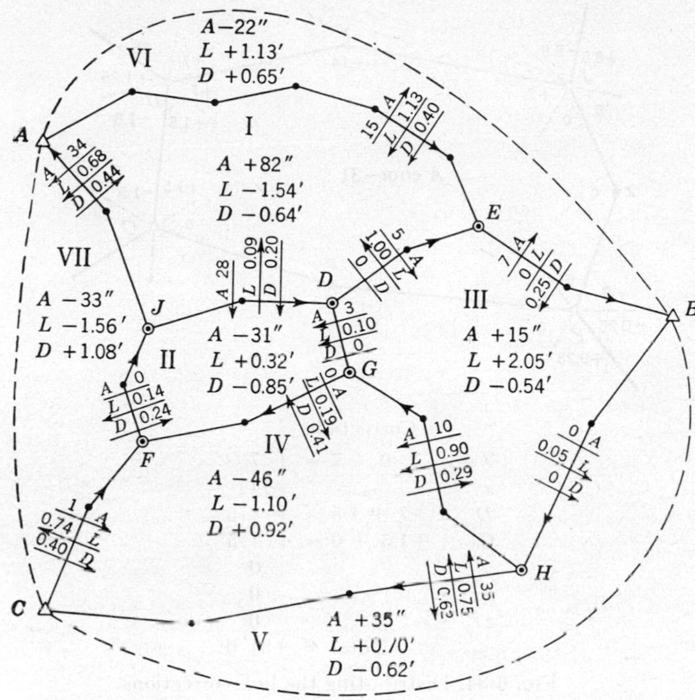

ERRORS OF CLOSURE

	Angles		Lat.		Dep.	
	+	−	+	−	+	−
I	82″			1.54		0.64
II		31″	0.32			0.85
III	15		2.05			0.54
IV		46		1.10	0.92	
V	35		0.70			0.62
VI		22	1.13		0.65	
VII		33		1.56	1.08	
Σ	132	132	4.20	4.20	2.65	2.65

FIG. 6-33. Adjustment of a traverse net.

Each should be computed by two routes. The results by each route should be the same. The latitudes and departures are then computed with the new bearings and the measured lengths. The latitude and the departure error for each circuit are computed clockwise. These are adjusted exactly as for the level net. Finally the coordinates are computed. These should agree by every route chosen.

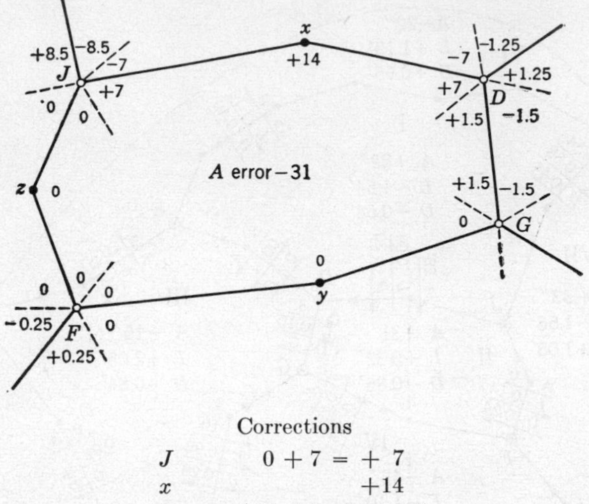

	Corrections
J	$0 + 7 = +\ 7$
x	$+14$
D	$7 + 1.5 = +\ 8.5$
G	$+1.5 + 0 = +\ 1.5$
y	0
F	0
z	0
	Total $= +31.0$

FIG. 6-34. Distributing the link corrections.

PROBLEMS

6-1. Adjust the following level net by the method of simultaneous adjustment by inspection. Fixed elevations: $A = 90.23$, $F = 82.73$, $I = 99.81$.

Link	*Elevation differences in directions given*	*Lengths, thousands of ft*
AB	−10.31	2
BC	− 5.24	2
CD	− 2.40	1
DE	+ 4.82	3
EF	+ 5.66	2
FG	+ 2.28	4
GH	+ 8.27	3
HI	+ 6.50	2
BH	+13.42	3
HJ	− 8.22	2
JE	− 8.07	3
GJ	+ 0.08	2
JC	−10.63	3

6-2 to 6-4. Adjust to the nearest second the quadrilateral whose angles are given below, according to Fig. 6-13.

	6-2	**6-3**	**6-4**
a	31°49′02″	41°13′06″	38°44′06″
b	61 12 27	56 41 22	23 44 38
c	48 04 01	49 19 12	42 19 09
d	34 24 30	37 10 20	44 52 01
e	36 19 00	36 49 11	69 04 21
f	43 06 09	56 28 15	39 37 48
g	66 10 25	49 32 20	26 25 51
h	38 54 20	32 46 15	75 12 14

6-5. Adjust to the nearest second the center-point figure whose angles are given below, according to Fig. 6-15.

a	17°48′30″	*f*	45°32′35″	*k*	47°44′05″
b	60 27 44	*g*	53 11 22	*l*	71 02 49
c	51 02 21	*h*	55 19 41	*m*	94 02 09
d	57 54 36	*i*	56 34 20	*n*	71 29 06
e	40 25 22	*j*	47 43 42	*o*	75 41 51

6-6 to 6-8. Determine by semigraphical resection the coordinates of station *O* where the angles shown were measured.

Station	Angle	Coordinates	
		North	East
6-6. *A*		2,250.10	1,357.00
	20°40′19″		
B		1,823.13	1,213.52
	69 29 39		
C		1,034.72	1,216.69
	42 34 04		
D		832.57	699.02
	227 16 00		
A			
6-7. *A*		561,793.57	1,990,183.56
	7°55′28.2″		
B		553,644.08	2,000,547.75
	65 39 35.6		
C		555,046.79	2,002,889.33
	31 13 42.1		
D		561,960.04	2,015,012.37
	255 11 14.1		
A			

Station	Angle	Coordinates	
		North	East
6-8. *A*		5,087.52	975.48
	94°06′37″		
B		11,534.65	6,694.24
	58 12 29		
C		4,501.16	6,923.33
	79 14 27		
D		3,178.46	2,943.80
	128 26 27		
A			

6-9. Given a traverse net of the same shape as that shown in Fig. 6-32; the closures are as given below. Show the corrections determined by inspection.

Errors of Closure

	Angles		Lat.		Dep.	
	+	−	+	−	+	−
I	76″		1.05			0.81
II	18			0.90		0.47
III		24″	1.50		1.08	
IV	10		0.42			0.22
V		52		0.61	0.90	
VI		20		0.17		1.50
VII		8		1.29	1.02	
Σ	104	104	2.97	2.97	3.00	3.00

CHAPTER 7

TOPOGRAPHIC SURVEYS

7-1. Introduction. The methods of establishing the control necessary for a topographic survey have been described in other chapters. This chapter deals with the methods of making the numerous minor measurements that tie the various features to be shown on the map to the control systems.

7-2. Methods for Minor Measurements. There are four important methods in general use for making these measurements: **transit stadia, plus and offset, photogrammetry,** and **plane-table surveying.** Each has its advantages and its limitations. Sometimes all four methods are used on the same survey.

7-3. Transit Stadia. Except for route surveys which are very long and narrow, transit stadia requires less time in the field than any other ground method. The field notes can be plotted to any scale desired or to several scales, no extra equipment is required, and the method will bridge distances that cannot be taped.

On the other hand, it requires more time in the office, and objects and contour lines must be plotted when neither the ground nor the positions of the points observed can be seen by the draftsman.

The method is especially useful when field time is at a premium because of bad weather, difficult field conditions, the allocation of personnel and their duties, or other causes. It is an excellent means of augmenting other surveys and a means of making unusual measurements without previous preparation.

7-4. The Plus-and-Offset Method. The plus-and-offset method is excellent for route surveys and for any survey in which the degree of accuracy required becomes less as the distance from the survey line decreases. Since the values of all the pluses are determined by a single measurement along the traverse, the amount of distance measurement, when the offsets are short, is reduced to a minimum. The data obtained are simple, usually error-free, and can be plotted quickly and easily without mistakes.

7-5. Photogrammetry. Large areas can usually be mapped by photogrammetry to any degree of accuracy desired at less than half the cost of equivalent ground surveys. The photographs themselves are an added asset. They nearly always contain information and details that cannot otherwise be discovered or recorded.

The necessary delays and overhead costs inherent in air mapping usually prevent its use for small surveys. The photographs can be taken very rapidly and large areas can be covered quite quickly, but the proper conditions for photography occur only on a comparatively few days during the year. The leaves must be off the trees, the ground must be nearly free of snow, and the day must be clear and sparkling—absolutely free of clouds or haze. In order to take advantage of such weather, the entire field equipment and personnel must be kept ready to start at a moment's notice.

It is evident that a small area can be mapped economically only when the photographs can be taken as part of a flight required for other work. In mapping a large area the costs of standing by are absorbed in the savings over ground methods, and the delays are not important since ground methods would take at least equally as long. In mapping a small area neither of these conditions obtain.

In certain terrain, photogrammetry is impossible. It cannot be used where high conifers or tropical rain forests conceal the ground or in areas such as grass-covered plains, that contain no discernible objects. It is difficult to use in areas of rugged topography or where high buildings are close together. The ground will be concealed over large areas of the photographs. Sometimes special flights can be arranged to augment the coverage. Often in a large area that is mapped from the air, small sections are in the categories described. Ground methods are used to map these sections.

The subject of photogrammetry is taken up in detail in Chap. 20.

7-6. Plane-table Surveys. The plane-table survey is almost always the best method, next to photogrammetry. It is the only means of mapping directly from the ground without notes or intermediate steps. Since the map is made in the field, omissions are reduced to a minimum and only the barest number of observations are required. The ground forms and small variations in contours can be sketched accurately by eye and the exact relationships between the points observed and the surrounding area are always before the topographer as he works.

The method requires a certain amount of previous planning. The scale must be decided in advance and any existing control must be plotted on the plane-table sheets in advance.

TRANSIT STADIA

7-7. Variations in Transit Stadia Methods. The usual method of transit stadia is covered in "Surveying" and is assumed to be familiar to the reader. Variations in this method that have been developed in Europe are described here.

7-8. Horizontal Stadia. Some instruments have vertical stadia wires so that the rod may be read in a horizontal position. The rod must be held exactly level, at right angles to the line of sight, and at the h.i. (the height of the instrument above its station). To fulfill these requirements, the rod is supported on a stand that is adjustable in height, and a level and sights are provided to establish the alignment. The method eliminates the need to apply a cosine correction to stadia intercepts on inclined sights. The formulas for horizontal stadia are the following:

Internal focusing:

$$H = 100\,S \cos \alpha$$
$$V = 100\,S \sin \alpha$$

External focusing:

$$H = (100\,S + f + c) \cos \alpha$$
$$V = (100\,S + f + c) \sin \alpha$$

The method is not affected by the errors caused by differential refraction of stratified atmosphere.

7-9. The Distance Wedge. Figure 7-1 shows the operation of a device called a distance wedge. It consists of two separate glass wedges

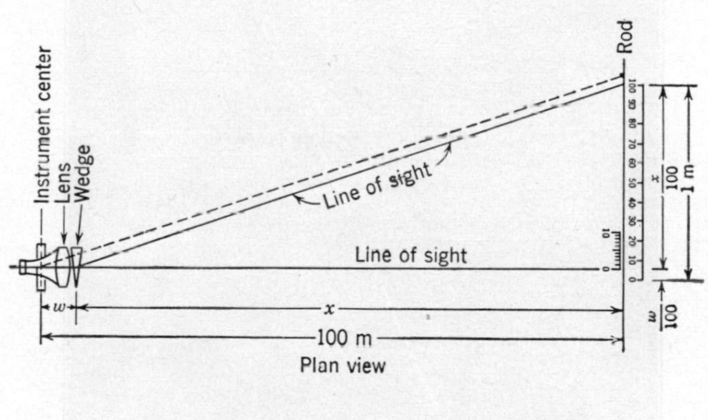

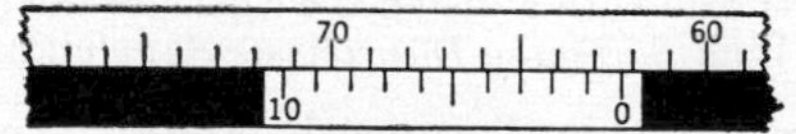

Appearance of rod with vernier as seen through telescope
Reading 62.3 m

FIG. 7-1. Operation of a distance wedge.

that together are achromatic. They cover a narrow, horizontal strip of the objective lens and superimpose on the normal image the image of objects at an angle to the left or right of the line of sight. The angle between any two superimposed points is equal to the angle of deflection caused by the wedge. This is set at a ratio of 1:100. The stadia rod is held in a horizontal position. It has a vernier at the right end (Fig. 7-2). The wedge superimposes a certain part of the rod scale on the vernier.

The actual length of rod from the index (the zero) of the vernier to the part of the scale superimposed on it is $\frac{1}{100}$ of the distance from the wedge

Fig. 7-2. How the rod is read with a distance wedge and the rod itself. Stadia reading, 56.78 m. (*Henry Wild Surveying Instruments Supply Co. of America, Inc.*)

to the rod. In Fig. 7-1 this distance is $x/100$. In order to make the scale read the distance to the center of the instrument instead of the distance x, the scale is shifted toward the index a distance of $w/100$ (w is the distance from the wedge to the instrument center). Thus the actual distance from the center of the instrument to the rod is read directly.

The distance wedge is superior to stadia wires for four reasons. It provides a vernier reading instead of an estimated reading. Both ends of the rod can be read simultaneously. Since the rod is horizontal, no cosine corrections need be applied to the stadia intercept on inclined shots and there is no additive constant in the stadia formulas. The formulas are

$$H = 100\,S\cos\alpha$$
$$V = 100\,S\sin\alpha$$

On the other hand, a stand and a special rod are required. The rod must be leveled and turned at right angles to the instrument with a special sighting device mounted on the rod each time it is used.

In use, the rod is set at the h.i. and the vertical support is set over the station. When pointing the instrument for vertical and horizontal

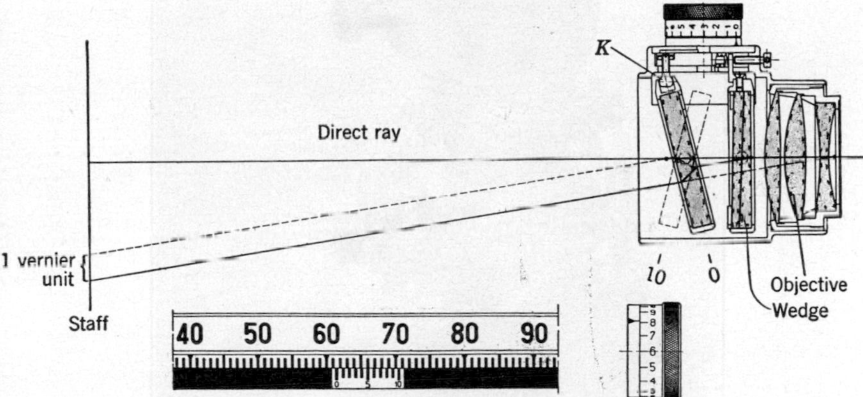

FIG. 7-3. A distance wedge with a plano-parallel. *(Henry Wild Surveying Instruments Supply Co. of America, Inc.)*

angles, the image from the wedge is blocked out by covering the wedge, and a sight taken on the target on the vertical support.

Figure 7-3 shows a distance wedge with a plano parallel attached so that an exact vernier coincidence can be obtained and the final reading taken from the drum. Figure 7-4 shows an instrument equipped with this device.

7-10. Automatic Stadia Compensation. Instead of stadia wires, some instruments have a device that practically eliminates computation. Replacing one half of the usual field of view is the image of a glass arc mounted on one of the instrument standards. The arc is engraved with a system of curves as shown in Fig. 7-5. As the telescope is inclined, different parts of the glass arc are seen through an optical train. The arc apparently slides under the vertical edge of the field. The telescope is aimed so that the edge of the field strikes the rod, and the horizontal wire strikes the target or any predetermined point on the rod in the usual way. In Fig. 7-6 a special rod is illustrated. Curve N is concentric with the

Fig. 7-4. A Wild T2 theodolite with a distance wedge and a plano-parallel. (*Henry Wild Surveying Instruments Supply Co. of America, Inc.*)

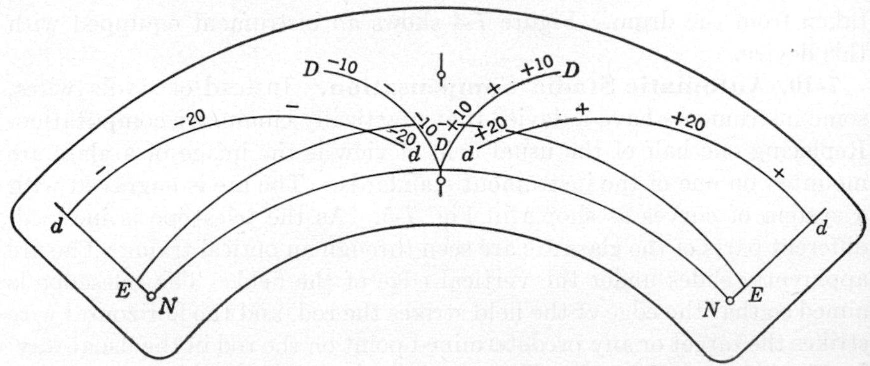

Fig. 7-5. (*Fennel Instrument Corp. of America.*)

horizontal axis and at the proper radius, so that as the telescope is tilted it remains in contact with the horizontal cross hair. The intercept on the rod between the points where curves N and E intersect the edge of the field is read for the horizontal distance. A similar intercept between N and D, or d, is read for the vertical distance. The curves are designed so that the actual values are obtained by multiplying by simple constants. In this case, 100 (N to E) is the horizontal distance and 10 (N to D) or

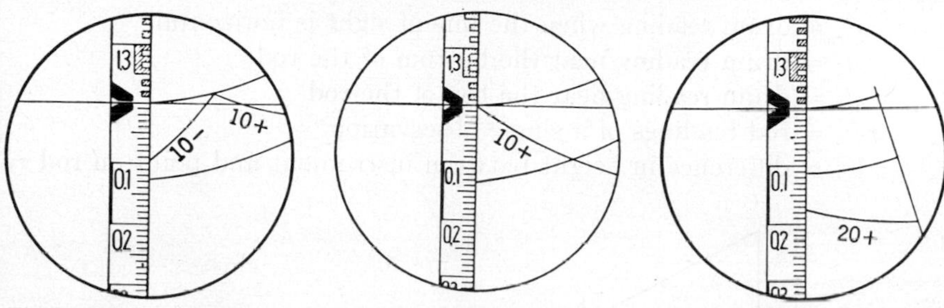

Field of view with telescope **depressed** Field of view with telescope **horizontal** Field of view with telescope **elevated**

$H = 12.7$ m $H = 13.4$ m $H = 11.3$ m

$V = 10(-0.095) = -0.95$ m $V = 20(+0.175) = 3.50$ m

FIG. 7-6. (*Fennel Instrument Corp. of America.*)

20 (N to d) is the vertical distance. The actual separation of the lines are the following:

$$N \text{ to } E = i \cos^2 \alpha'$$
$$N \text{ to } D = 10\, i \sin \alpha \cos \alpha$$
$$N \text{ to } d = 5\, i \sin \alpha \cos \alpha$$

where i = distance between stadia hairs when $f/i = 100$; thus $i = f/100$

α = vertical angle

7-11. The Gradienter. The vertical tangent screw on some instruments is arranged to tilt the line of sight a difference of grade of 1 per cent for exactly one revolution. If this is to be strictly true, the tangent screw must bear on a plane surface which, if extended, would contain the horizontal axis; the axis of the screw must be perpendicular to this surface when the line of sight is horizontal; and the distance from the horizontal axis to the center line of the screw C must be one hundred times its pitch. Slight errors in any but the last requirement have little effect. The screw has a graduated drum for reading partial revolutions and a scale along which the drum advances to read the number of revolutions.

To use this device, set the line of sight at a foot mark near the bottom of the rod. Read the drum. Raise the line of sight to a point near the top of the rod. Set the drum at the same partial revolution as before but advanced by several whole revolutions. Read the rod. The difference in the drum readings will be in feet per hundred feet without fractions. Then

$$H = \frac{100(r_2 - r_1)}{d_2 - d_1}$$

$$V_1 = \frac{(d_1 - d_0)(r_2 - r_1)}{d_2 - d_1}$$

where H = horizontal distance
d_0 = drum reading when the line of sight is horizontal
d_1 = drum reading near the bottom of the rod
d_2 = drum reading near the top of the rod
$r_1 r_2$ = rod readings of a single observation
V_1 = difference in height between instrument and point on rod r_1

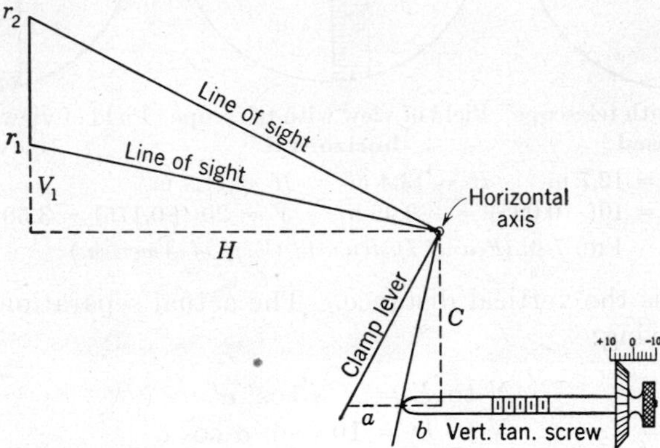

FIG. 7-7. Theory of gradimeter.

In Fig. 7-7, from similar triangles,

$$\frac{H}{r_2 - r_1} = \frac{C}{a}$$

but
$$a = \frac{C}{100}(d_2 - d_1)$$

Substituting,
$$H = 100\,\frac{r_2 - r_1}{d_2 - d_1}$$

Also,
$$\frac{V_1}{H} = \frac{b}{C}$$

but
$$b = \frac{C}{100}(d_1 - d_0)$$

Substituting,
$$V_1 = (d_1 - d_0)\,\frac{H}{100}$$

$$= (d_1 - d_0)\,\frac{r_2 - r_1}{d_2 - d_1}$$

Example. $r_1 = 2.00$; $d_1 = 3.58$; $d_2 = 8.58$; $r_2 = 10.76$; $d_0 = 0$.

$$H = 100\,\frac{8.76}{5} = 175.2$$

$$V_1 = (3.58)(1.752) = 6.27$$

7-12. Subtense Bar. A subtense bar is a device that holds two targets at a precise horizontal separation (see Fig. 7-8). The separation is usually maintained by a strip or wire of invar, under tension. It has a sighting device and a level to set it perpendicular to the line of sight and horizontal.

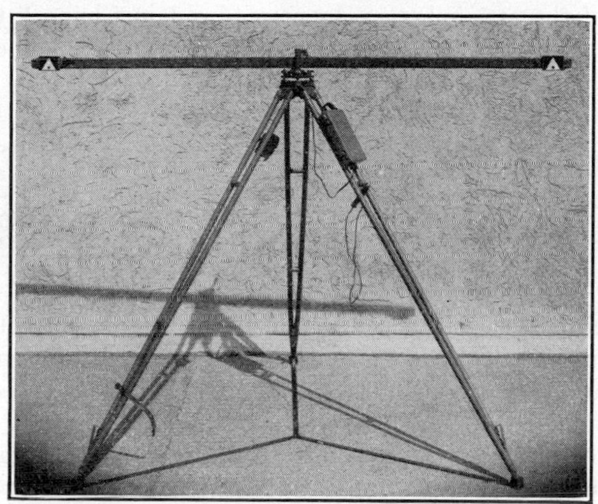

FIG. 7-8. A subtense bar. *(Henry Wild Surveying Instruments Supply Co. of America, Inc.)*

The angle subtended by the targets is measured with a theodolite or by a special tangent screw attached to an ordinary transit:

$$H = \frac{S}{2 \tan \frac{1}{2}\theta}$$

where S = separation between the targets
θ = angle measured

No correction for slope is required since the angle measured is a horizontal angle.

The vertical angle is measured as usual, and

$$V = H \tan \alpha$$

The accuracies obtained with subtense bars up to reasonable distances compare very favorably with ordinary steel-tape measurements. However, the error increases with the square of the distance, and therefore distances over 300 or 400 ft must be measured in sections to attain second-order accuracy.

The same principle is sometimes used with a 100-ft steel tape. Targets are placed at the 0 and 100-ft marks. Temperature corrections are applied. A sighting device must be used to keep the tape at 90° to the line of sight and a level to keep it horizontal. The angle measured subtends the whole 100 ft. Second-order accuracy and better can be attained with sights ¼ mile long.

7-13. Errors in Stadia Measurement. In addition to the instrumental and observational errors that are evident from the foregoing paragraphs, stadia measurements are affected by variations in the refractive index of the atmosphere. As pointed out in Chap. 2, the air near the ground is often stratified in layers of different temperature and

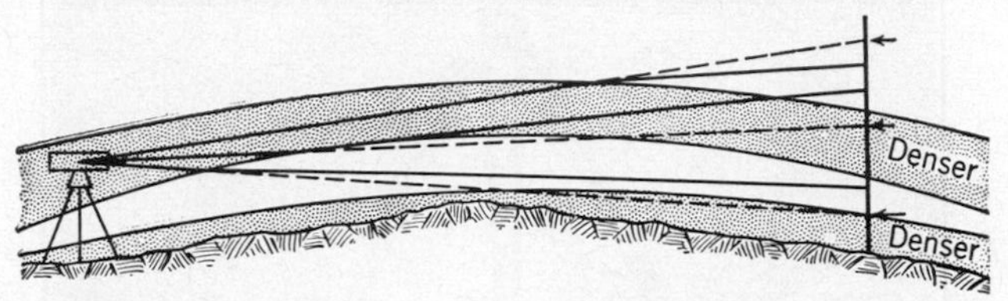

FIG. 7-9. Effect of stratification on stadia. Dashes show true lines.

moisture contents which, accordingly, have different refractive indices. These layers can have almost any arrangement of warm or cold, or moist or dry, air and they can change rapidly. They bend the lines of sight by unknown amounts and hence introduce unknown errors into the vertical angles and, except in horizontal stadia, into the values of stadia intercepts (see Fig. 7-9). Under constant conditions, the effect on vertical angles would be canceled, because the values will always tend to be too large or always too small, and therefore the errors in the backsights would cancel the errors in the foresights. In practice the effect has a tendency to cancel and the residual errors are somewhat accidental, so that levels carried by stadia are quite accurate. The stadia intercept determined depends on lines of sight that pass through different sets of air layers. Under constant conditions they will always be bent together or always bent apart. The errors in backsights will have the same sign as errors in foresights and therefore there is no tendency to cancel.

Accordingly, stadia surveys have a tendency to change in scale but not in shape. They should be fitted to control as much as possible by introducing proportional changes in the lengths of lines. Errors of 1 part in 300 can be expected. The results can be adjusted by simple graphical or slide-rule methods.

PLUS AND OFFSET

7-14. Right-angle Instruments. The plus-and-offset method supplemented by the engineer's level and the hand level is too well known to be covered here. The method is very fast for certain types of work and unusually simple to plot in the office.

The chief disadvantage of the method is the difficulty of estimating accurately the right angle required for each plus. In Europe, right-angle prisms and mirrors, designed as hand instruments, are used extensively and they are rapidly coming into use in the United States. Two of the several types are shown in Fig. 7-10. They all work on the same principle: when a ray of light is reflected by two reflecting surfaces held stationary with respect to each other, it will be changed in direction by twice the

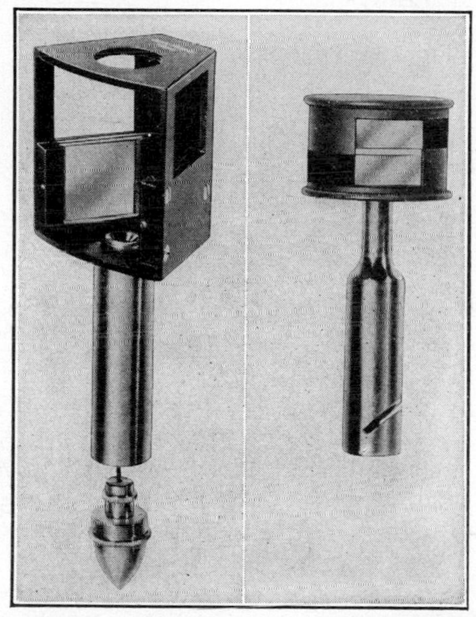

FIG. 7-10. Right-angle instruments. (*Keuffel & Esser Co.*)

angle between the surfaces no matter in what direction the mirrors are turned.[1] The two reflecting surfaces are set at 45° so that the line of sight is bent 90°. When an object is seen in the instrument, it will be at

[1] The general principle is as follows. The trace of a light ray reflected by two plane surfaces (in a plane perpendicular to both surfaces) is always deflected by twice the angle of opening between them, to the left if the left side of the angle is first encountered, and conversely.

In Fig. 7-11, m is the angle of opening between two reflecting surfaces. AB is a light ray striking the left side of the angle at any angle a, and r is the angle of deflection to the left.

$$a = a' = a'' \qquad b = b' = b''$$
$$2b = r + 2a \text{ (exterior angle)}$$
$$b = m + a \text{ (exterior angle)}$$
$$r = 2m$$

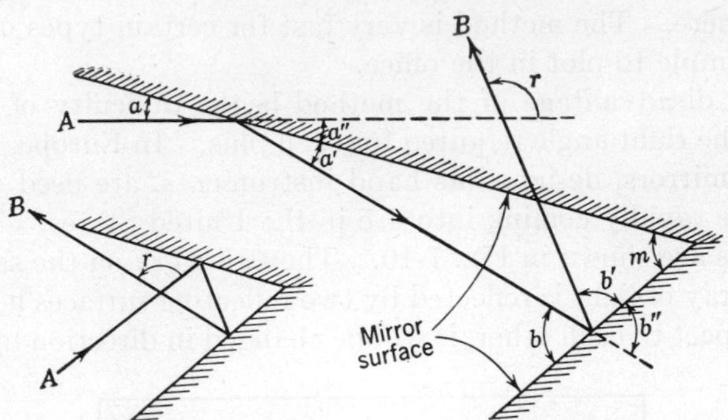

FIG. 7-11. Principle of double reflection.

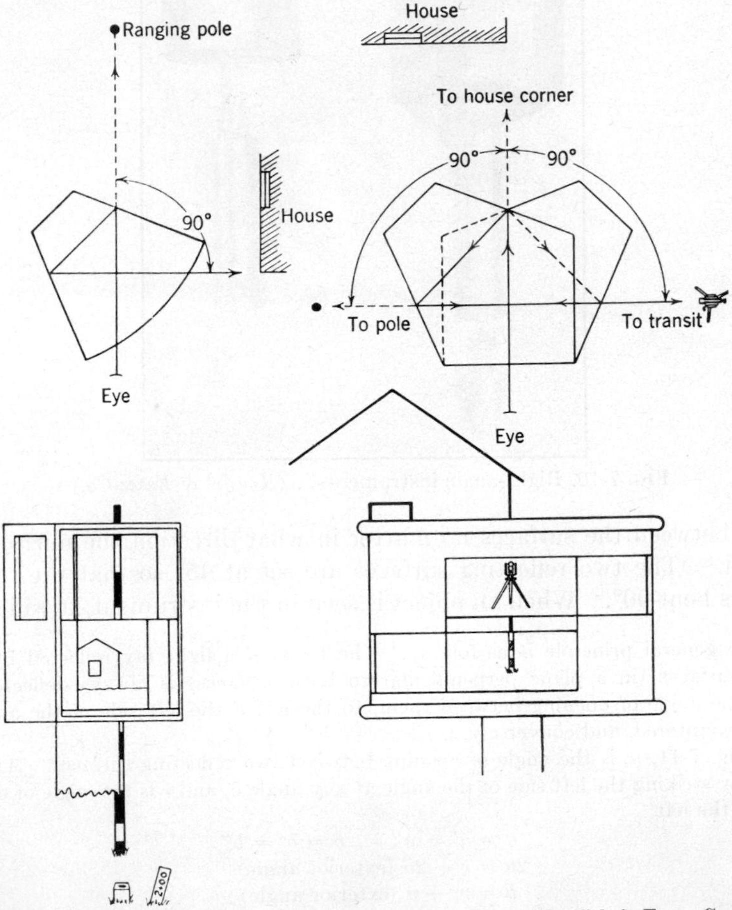

FIG. 7-12. How the right-angle instruments work. (*Keuffel & Esser Co.*)

90° to any object viewed directly that is apparently in line (see Fig. 7-12). The vertex of the right angle is very near the center of the instrument.

The instrument shown in Fig. 7-10*a* is held on line by sighting two marks beyond the instrument, the stake and the range pole (Fig. 7-12*a*). The observer moves the instrument along the line until the image of the object whose plus is to be determined, the house corner, appears in the instrument in line with the two marks. A plumb bob attached to the instrument facilitates marking the position found.

The instrument in Fig. 7-10*b* is a double prism that can be placed on line *between* two marks, the transit and the range pole (Fig. 7-12*b*). When the images of two marks at each end of the line are lined up in the instrument, the instrument is on line. The observer moves the mirror along the line until the object, the house corner, viewed directly (between the prisms or above or below them) is in line with the two images.

PLANE-TABLE SURVEYING

7-15. The Plane-table Method. As previously stated, plane-table surveying is a means of making a manuscript map in the field while the ground can be seen by the topographer and without the intermediate steps of recording and transcribing field notes. It can be used to tie topography to existing control and to carry its own control systems by triangulation or traverse and by lines of levels. Horizontal position can be determined to about 1 part in 300 and elevations to about 1 ft $\sqrt{\text{miles}}$.

7-16. Plane-table Equipment. A plane table consists of a drawing board mounted on a tripod and arranged for field use (see Fig. 7-13). The table can be leveled and oriented to any desired azimuth. A **plane-table alidade** is a straightedge with some form of sighting device. For accurate mapping the sighting device is a telescope equipped for stadia. The alidade rests on the plane table and can be moved wherever desired.

The **plane-table sheet** or **map sheet** used on the plane table is made of paper that is as dimensionally stable as possible. Fiberglas sheets or paper backed with sheet aluminum are often used. Any horizontal control points that are available within the area to be mapped on the sheet are plotted previous to the field work. Notes are added giving elevation data. Sometimes topographic details taken from air photographs or other sources are also placed on the sheet.

7-17. Principle of Operation. Levels are carried exactly as in transit stadia. An area is usually mapped by short plane-table traverses run from the existing control (see Fig. 7-14). Let A and E represent two horizontal control points plotted on the sheet. The plane table is set up and leveled at field station A and the alidade is placed along the map line AE. The table is turned until the alidade points to E. This orients the

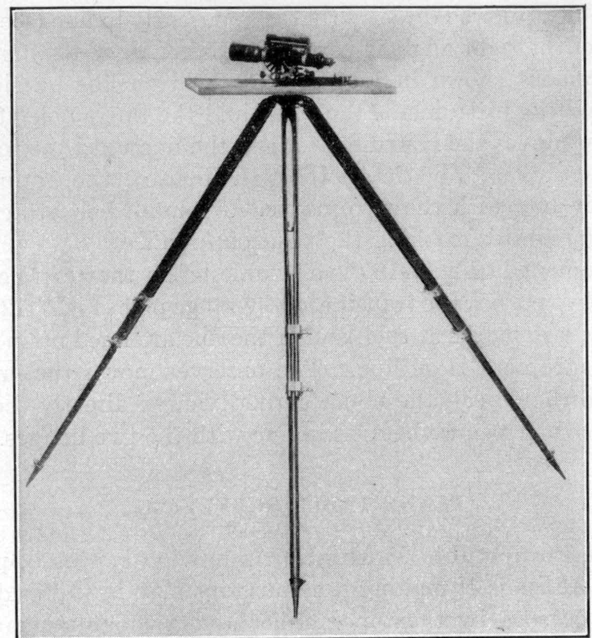

FIG. 7-13. A plane table and alidade. (*W. & L. E. Gurley.*)

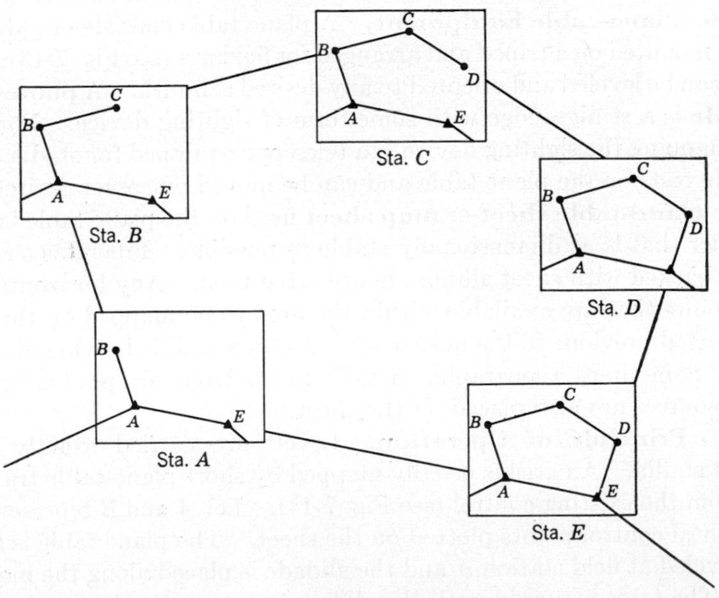

FIG. 7-14. Plane-table traverse.

table. The alidade is then pointed to the next plane-table station B with the straightedge passing through A. A line is drawn along the straightedge representing AB, and when the distance and difference in elevation have been determined by stadia or otherwise, the position of B is plotted to scale and its elevation noted. Topographic features are plotted in the same manner as B until the work at A is complete. The plane table is then moved to B and oriented by placing the alidade along the line from B to A and turning the table until A is sighted. The traverse is continued in this way, preferably until it is closed on control as at E.

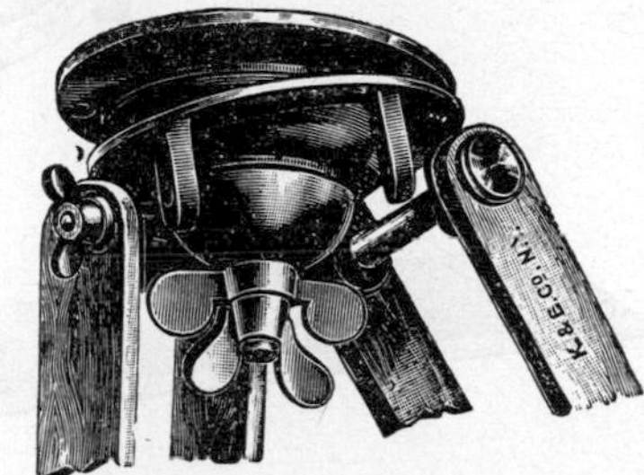

FIG. 7-15. A Johnson head. (*Keuffel & Esser Co.*)

The Plane Table. There are several devices for leveling the plane table and controlling its orientation. One consists of a three- or four-screw leveling head and a clamp-and-tangent screw to turn the table. The most generally used device is the **Johnson head** (see Fig. 7-15). It has two thumbscrews on the underside. When the upper screw is free, the table may be tilted for leveling. When the lower screw is loosened, the table may be turned in azimuth.

7-18. The Alidade. Figures 7-16 to 7-18 show two types of alidades. The standards attached to the straightedge support the telescope trunions. The inclination of the telescope is controlled by a clamp-and-tangent

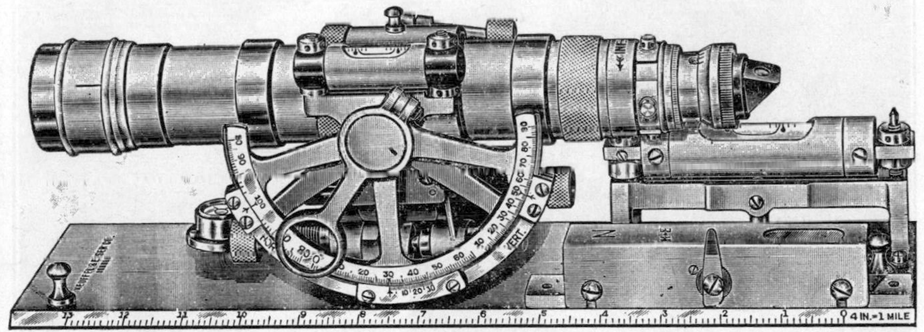

FIG. 7-16. An expedition alidade. (*Keuffel & Esser Co.*)

screw. The telescope is usually 12 to 16 power. In addition to the usual stadia hairs, a third hair is usually placed between the center hair and the upper hair to indicate a 1/4 intercept for long shots (see Fig. 7-19). The smaller instrument has a prismatic eyepiece.

The vertical arc carries the two Beaman's arc graduations for computing vertical and horizontal distances, respectively, as well as the usual angular graduations. It is customary to number vertical angles 30°

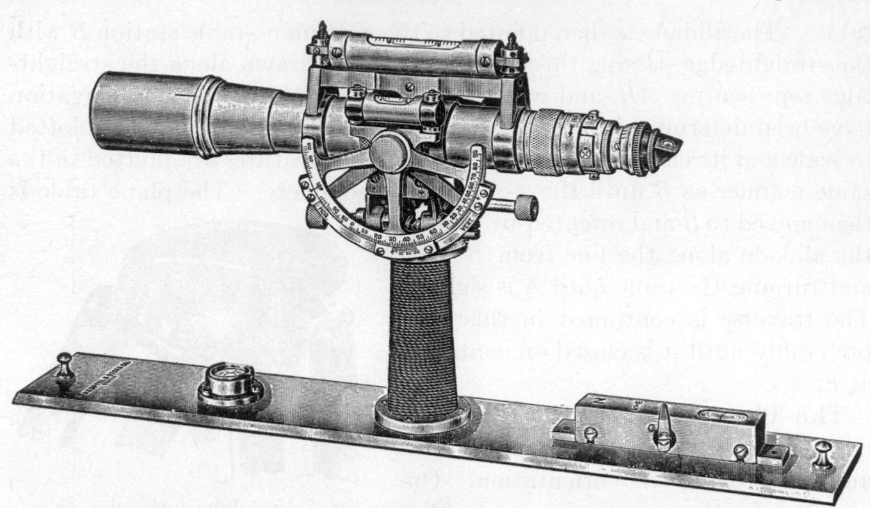

FIG. 7-17. A U.S. Geological Survey type alidade. (*Keuffel & Esser Co.*)

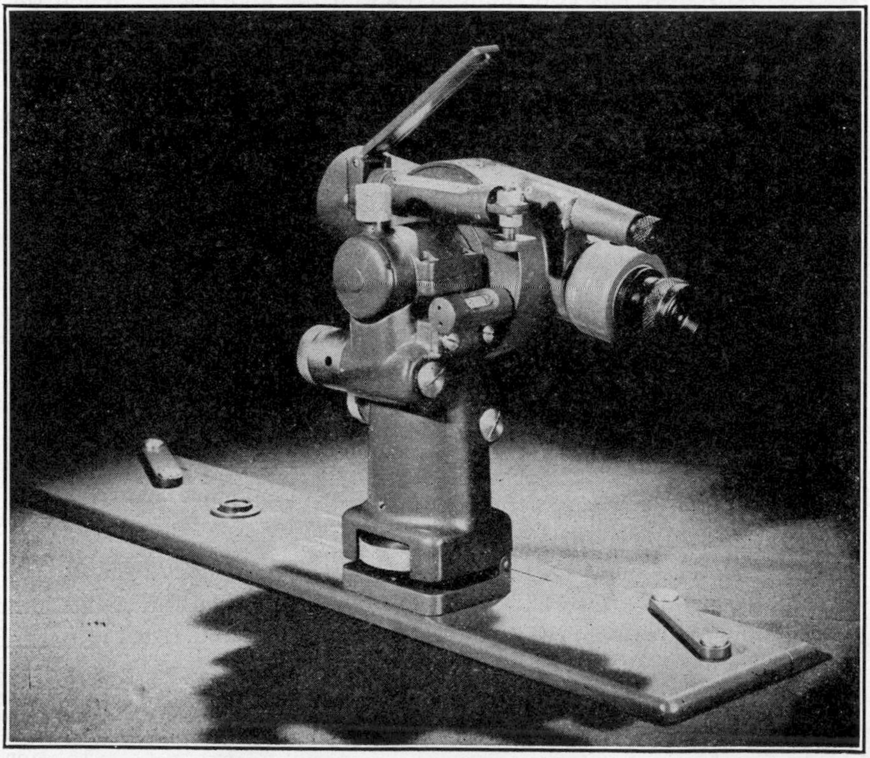

FIG. 7-18. A Watts microptic plane-table alidade. (*Jarrell-Ash Co.*)

larger than their actual value and to add 50 to the indications for vertical height on Beaman's arc. The values read are thus always plus and always read in one direction, whether the vertical angle is plus or minus. The values stated are subtracted from the readings before the computations are made.

As on the transit, the Beaman's arc indications are multiplied by the stadia intercept to give the horizontal and vertical distances, respectively.

The vernier and the two indices for Beaman's arc are mounted on a circular member that carries a level vial (the **index level vial**). The whole assembly may be rotated a few degrees around the horizontal axis by a tangent screw. The level is centered before the readings are taken. It is adjusted so that the readings obtained are the correct values for the actual inclination of the line of sight with respect to gravity. These readings are therefore independent of any irregularities of the surface of the plane table or the accuracy to which it is leveled.

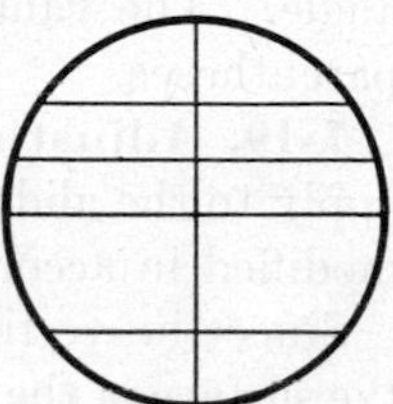

FIG. 7-19. Quarter stadia intercept.

An alidade has been designed in which the vertical arcs are mounted on a pendulum so that they automatically compensate for variations of level.

The alidade has a small circular level for leveling the plane table, a striding level that may be placed on collars on the telescope for accurate level shots and to adjust the instrument, and a box compass to be used when the plane table is oriented by compass instead of by backsight as described in the previous section.

The stadia formulas are the same for the alidade as for a transit. They are developed in "Surveying." The working formulas are the following:

$$H = \left(S + \frac{c + f}{100}\right) 100 \cos^2 \alpha$$

$$V = \left(S + \frac{c + f}{100}\right) 100 \sin \alpha \cos \alpha$$

where H = horizontal distance
V = difference in elevation
S = stadia intercept
α = vertical angle
$c + f$ = additive stadia constant

The value $c + f$ varies with different alidades. It is usually given by the maker. For internal-focusing instruments it can be taken as zero. For external-focusing instruments it can be determined as described in "Surveying." In large-scale surveys, to this constant must also be added the distance from the plotted position of the station occupied to the center of the instrument. Unless the scale of the map is very great, this is negligible. Usually the additive constant is taken as 1 ft for external-focusing instruments and 0 for internal-focusing instruments on medium-scale maps.

Tables, stadia, slide rules, or the Beaman's arc give the values of the parts of the formulas not in parentheses for the various values of α. A Beaman's arc consists of two specially graduated parts of the circle that give these values directly. This makes it unnecessary to read the vertical angle. The values thus determined are multiplied by the value of the parentheses.

7-19. Adjustments of the Alidade. The same geometric principles apply to the alidade as to the transit, but the adjustments are somewhat modified in accordance with the lower degree of accuracy required.

Since the vertical axis about which the alidade turns is controlled by the surface of the plane table, its direction varies more than the tolerances that can be held in manufacture. The following relationships can be built into the instrument without providing a means of adjustment: (1) the circular level will center when the base of the alidade is horizontal; (2) the horizontal axis is parallel to the base; and (3) the axis of the telescope collars is perpendicular to the horizontal axis.

The only important adjustment of the alidade is to set the index level so that when the bubble is centered, the index (the vernier) will read the actual inclination of the line of sight. This can be accomplished directly by the peg method.

Find two supports for the alidade about 100 to 200 ft apart. They should support the alidade firmly with the circular bubble centered. Two plane tables are best. Make a small target that stands above the table at the height of the horizontal axis of the alidade. Read the vertical angle from each table to the target on the other. The angles should be equal but of opposite sign. If they are not, set the index at their mean and adjust the level until the bubble centers.

Most alidades are arranged so that they may be adjusted without extra equipment and without removing them from the plane table. In these instruments the telescope can be rotated approximately about the line of sight. In some instruments the telescope can be lifted out of its bearings and thus rotated. In most instruments the telescope can be rotated within a bearing in the horizontal axis. Usually a knurled ring around the telescope must be backed off to allow this rotation. The process of adjustment is described in the following paragraphs.

1. *To make the plane of the striding level coincide with the axis of the collars.* Place the striding level on the collars and center the bubble with the vertical tangent screw. Tilt the striding level about 20° to one side.

If the bubble moves off center bring it all the way back with the lateral adjustment.

2. *To make the striding level center when the axis of the collars is horizontal.* Center the bubble with the tangent screw. Reverse the level. If it fails to center, bring it halfway back with the tangent screw and adjust it until it centers.

3. *To make the vertical cross hairs lie in a plane perpendicular to the horizontal axis.* Loosen the knurled ring. Make sure the telescope is rotated against the stop that establishes its normal position. Tighten the knurled ring. Proceed as for the transit.

4. *To make the line of sight coincide with the axis of the collars.* Arrange a scale or a level rod so that it is held vertical about 100 to 200 ft from the position of the instrument. Back off the knurled ring. Center the striding level. Read the rod. Rotate the telescope 180°. Center the striding level. If the rod reading is not the same, bring the line of sight halfway back with the cross-hair adjusting screws. See "Surveying" under Y-level Adjustments. Check and note the reading.

Rotate the telescope 90°, center the striding level, and adjust the vertical cross hair (which will now be horizontal) until it strikes the rod at the final reading previously determined. Very little movement should be required. As it is easy to disturb either adjustment while making the other, both should be checked if either one has been changed.

5. *To make the index level center when the index position on the vertical arc reads the inclination of the axis of the collars.* Center the striding level. With the index tangent screw set the index at 30° (zero). Adjust the bubble until it centers.

7-20. Plane-table Rods. In very large-scale mapping, a level rod can be used. The operation of reading the rod is exactly like that in transit stadia. Small-scale maps are made more frequently. In these the distances are greater; this, in addition to the lower-power telescopes, makes it necessary to use a rod that is more easily read.

Three types of rods are shown in Fig. 7-20. No target is used, so that the method of handling the h.i. is not the same as with a level rod.

The pennant rod in Fig. 7-20 is excellent for medium distances. The E rod is very accurate for long shots but there is no means of recognizing which foot is observed except by counting. The reconnaissance type is excellent for long distances and can be used successfully for medium distances.

The reconnaissance rod shown has only one or two **divided feet** and no numbers. It will be noticed that when any two adjacent foot marks are observed, the number of feet from the bottom of the rod is known, because no two pairs are the same.

In using any of these rods, the stadia intercept is taken first. Either hair is placed on a divided foot; the other hair is then placed on any foot division and the stadia intercept is read. The center hair is then moved to the nearest foot mark and the vertical angle determined. The particular foot mark is noted and used in the computations. It is called the **rod reading.**

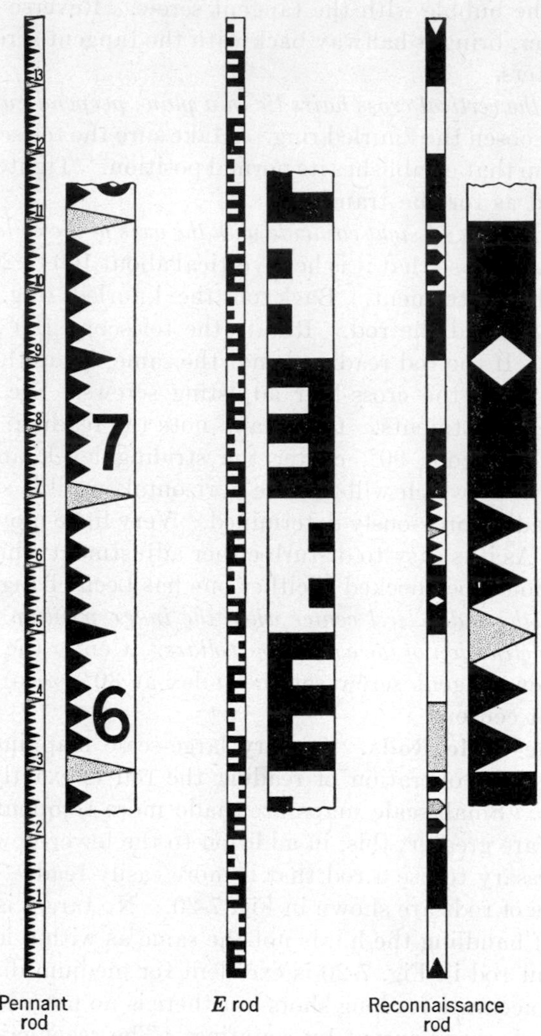

FIG. 7-20. Stadia rods. Light shading indicates red.

7-21. To Take a Backsight. This is the first operation at each plane-table station. It orients the table and determines the elevation of the instrument (H.I.). The point observed may be a previous plane-table station or a horizontal control station. When the elevation of the horizontal control station is not known it is used only for orientation and position determination. A stadia shot to a benchmark must be taken separately to determine the H.I.

Assume that the plane table is set up at *B* in Fig. 7-14 and that station *A* is to be observed. The line *AB* will have been drawn on the plane-table sheet and the elevation of *A* (105.4 ft) will have been noted. If the scale is large, the plane table should be placed so that the plotted point *B* on the sheet is over the point *B* on the ground. The device shown in Fig. 7-21 facilitates this operation, or, instead, a pebble can be dropped from the point.

Place the straightedge on the line *BA* with the instrument facing toward *A*. When the map scale is large, a convenient index mark on the straightedge should be chosen or marked. The index should be placed at *A*. When the computations are made, the additive stadia correction should be increased by the distance on the straightedge measured forward from the index to a point opposite the horizontal axis.

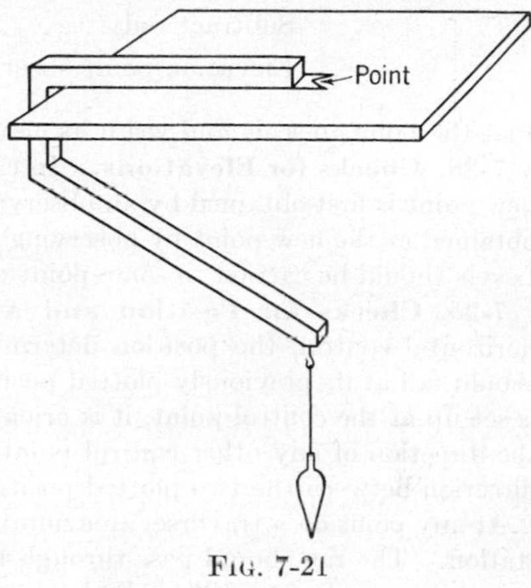

FIG. 7-21

Turn the table until the line of sight is in line with the rod at *A*. Determine the stadia intercept (6.24). Set the horizontal cross hair on the nearest foot mark (4.0). Wave "all right" to the rodman. Center the index bubble and read the vertical angle or the indications of the Beaman's arc (hor. 99.8, vert. 45.1). Measure the height of the instrument above the station (3.6). This is known as the h.i. Make the following computations.

7-22. Computations for a Backsight.

Vertical angle		*Beaman's arc*	
Angle =	27°12′	Hor. = 99.8 Vert. =	45.1
Minus Index cor. =	−30 00		−50.0
Vertical angle =	− 2°48′		− 4.9
By stadia slide rule or tables		*By Beaman's arc*	
$H = (6.24)(99.76) =$	-622.5	$H = (6.24)(99.8) =$	623
$V = (6.24)(-4.88) =$	-30.45	$V = (6.24)(-4.9) =$	-30.6

Elevation station observed........	105.4
Add rod........................	+ 4.0
	109.4
Subtract V.....................	−(−30.4)
H.I...........................	139.8
Subtract h.i....................	− 3.6
Elevation station occupied.........	136.2

Check the plotted distance *AB* and write the elevation at *B*.

7-23. To Take a Foresight. A foresight is used to observe another station or to observe any point to be located for topography. With the straightedge in contact with *B*, aim the instrument at the rod at *C*. A pin is often put in the map point *B* to facilitate keeping the straightedge in position. Some alidades have a parallel rule or a parallel motion to simplify this operation.

Draw a light line along the straightedge. Take the readings as described for a foresight. Determine the values of V (+10.62) and H (422.3) as before. Assume the rod is 6.0.

7-24. Computations for a Foresight.

H.I. (at station occupied)..........	139.8
Add V..............................	+(+10.6)
Total..............................	150.4
Subtract rod........................	− 6.0
Elevation point observed..........	144.4

Plot the point to scale and write its elevation near it.

7-25. Checks for Elevations. In carrying a traverse forward, the elevation of a new point is first obtained by an observation from a previous point. It can be again obtained at the new point by observing the old point. This provides a useful check. Levels should be carried to some point of known elevation as a final check.

7-26. Checks for Position and Azimuth. When a traverse is returned to horizontal control, the position determined by the traverse for the control station should fall at its previously plotted position. To check the azimuth, the plane table is set up at the control point, it is oriented by observing the last traverse point, and the direction of any other control point is observed. This should coincide with the direction between the two plotted positions.

At any point on a traverse, an azimuth check can be made by sighting any control station. The ray should pass through the plotted position of the station. Sighting two stations is better still, and when rays to three stations check, both position and orientation are assured (see Three-point Resection, Sec. 7-30).

7-27. Use of Compass. The plane table can be oriented by compass under the following conditions:

1. When there is no second point available for orientation
2. When speed is more important than accuracy
3. When the traverse is so long that accumulated errors in carrying the azimuth forward might be greater than orientation by compass
4. When a few short shots are to be made from a station as an isolated operation and it would take too long to send a rodman to another station
5. In certain resection problems

To make use of the compass, first draw a line on the map in the direction of the magnetic meridian. This direction can be assumed or calculated, or placed on the map with the alidade when the plane table has been previously oriented. The last method can be carried out as follows. After the table is oriented by sighting some other station, the alidade is turned so that the needle comes to rest at its zero indication. A line is then drawn along the straightedge to represent the magnetic meridian. Thereafter, the table can be oriented at any station by placing the alidade along the magnetic meridian and turning the table until the needle comes to rest at zero.

7-28. Traverses without Control. The plane table alone can be used for mapping without other control. A direction and an elevation must be assumed at the start. Usually the work is arranged to form a closed traverse.

7-29. Location by Intersection. When a point is to be mapped that can be observed from two stations, it can be plotted by triangulation without sending a rodman to it. A ray is drawn toward it at each station (Fig. 7-22) and the vertical angles are observed. The intersection of the two rays gives the position of the point, and the horizontal distance H

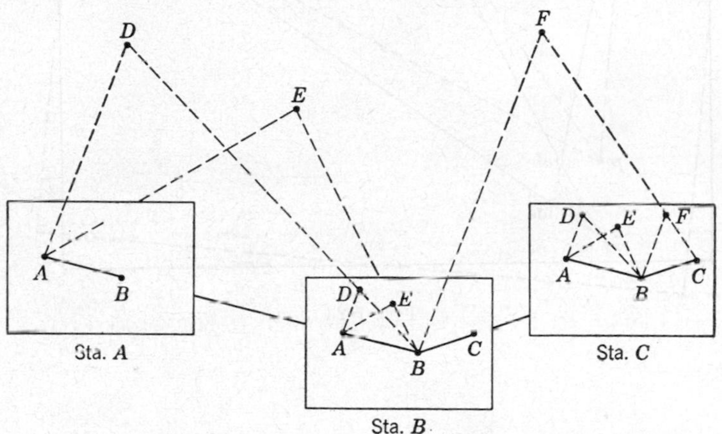

FIG. 7-22. Locating points D, E, and F by intersection.

from each station to the point is scaled off. The difference in elevation from each station to the point is computed by the following formula:

$$\text{Diff. in elev.} = H \tan \alpha$$

The elevations computed from each station should be the same.

7-30. Three-point Resection. A plane-table station can be established wherever three stations that are already plotted on the map sheet can be observed. The position on the sheet and the orientation of the plane table can be obtained by a graphical solution of the three-point problem. Many methods have been used but the method of successive approximations is generally accepted as the most practical. The procedure is described in the next paragraphs.

7-31. Procedure for Three-point Resection. A, B, and C are stations plotted on the sheet (see Figs. 7-23 and 7-24). The plane table is set up anywhere that A, B, and C can be observed. The table is oriented by eye or by compass. The alidade is aimed at each of the stations in turn and a ray is drawn through each plotted position in the direction of the station represented. The rays are shown by heavy lines. If the table has been properly oriented, these rays should meet in a point P representing

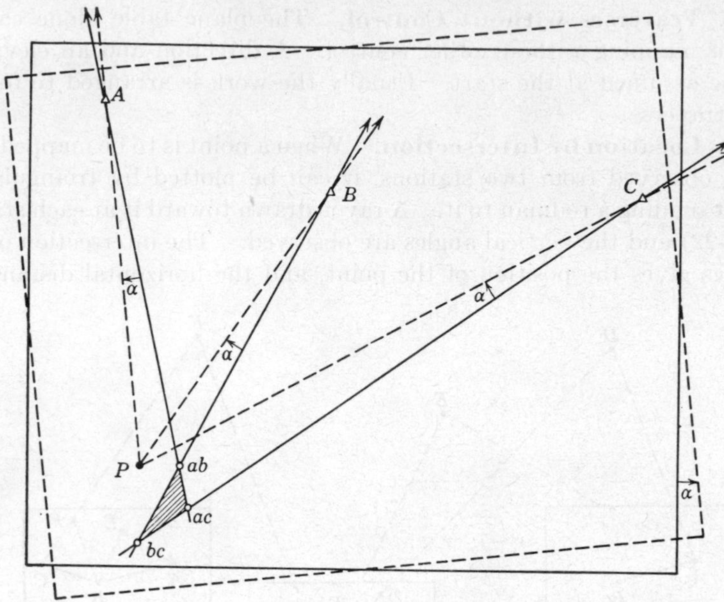

FIG. 7-23

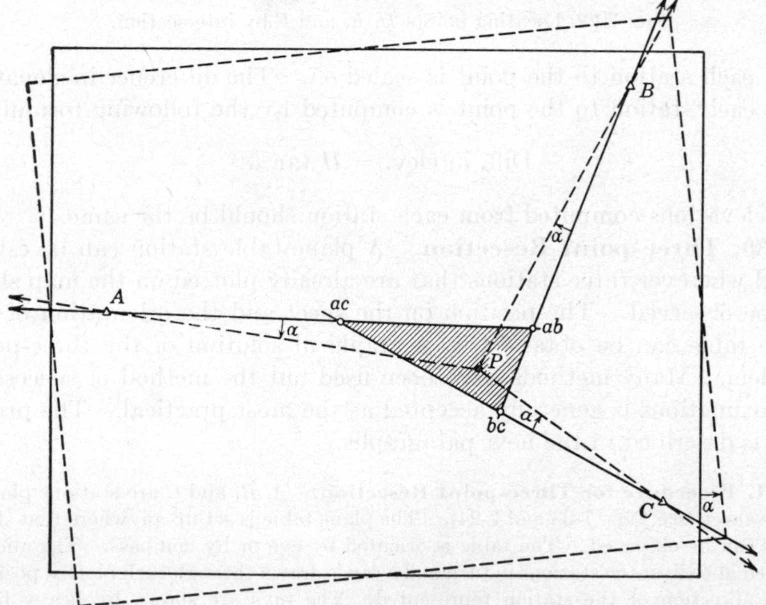

FIG. 7-24

the plotted position of the station occupied. The position of P is estimated from the triangle of error (shaded area in the figure) by **Lehmann's method.**

If the table should be turned counterclockwise through the angle α, as shown by the dotted outline, the effect on the map would be the same as if the ground had been turned clockwise through the angle α. All the rays would rotate clockwise through the angle α to the positions shown by the dotted rays. In the vicinity of the triangle of error they will all move to the left or all to the right, each by an amount proportional to its length. A point is chosen by estimation through which all the rays will pass according to this rule. When the point is marked, the table is oriented by a backsight along the line from the point to the most distant station. The two other rays are drawn as before. If the estimate was accurate they will meet in a point. If not, a second estimate is made.

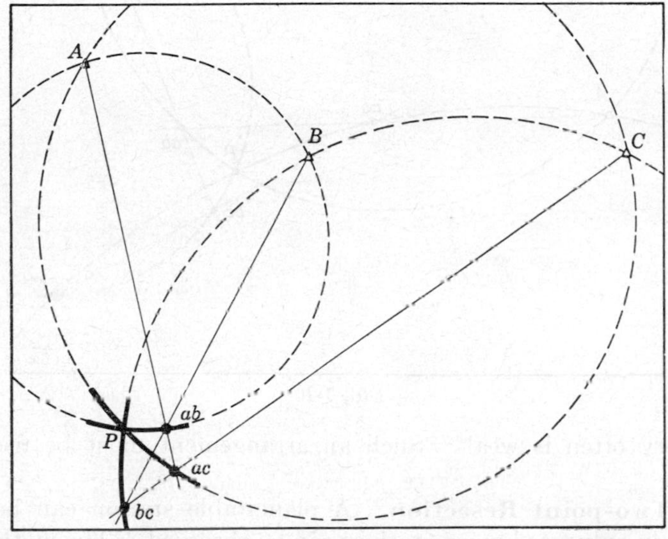

FIG. 7-25

The author has used the following method for estimating the position of P. The angles APB, APC, and BPC will not change when the table is turned. The locus of the intersection of each pair of rays—ab, ac, and bc—therefore, is a circle through the two points observed and the intersection of the two rays. An arc is sketched by eye through each intersection that would be on such a circle (see Figs. 7-25 and 7-26). It is not difficult to imagine the positions of the circles well enough to sketch these arcs. All three arcs should meet at the true position of P.

A third method is often used. A piece of tracing paper is fastened on the table. Rays are drawn with the alidade toward each station from any convenient point. The paper is shifted until the rays pass through the plotted positions of the three stations. The intersection of the rays will then be at the point P. It is pricked through, and the table is oriented by the line from P to the most distant station.

7-32. The Strength of Figure of the Three-point Solution. If the station to be located falls on the circle determined by the three known stations, the solution is indeterminate. When the station is near the circle, the solution is weak.

To judge the strength of the figure, estimate the position of the arcs used in the author's solution. The strongest solution results when the three arcs intersect at 60°. The strength decreases as the angles of intersection become less. At least one angle should be 30° or larger. When the central fixed point is farther from the station than the others, the figure very often is weak. Such an arrangement must be used with caution.

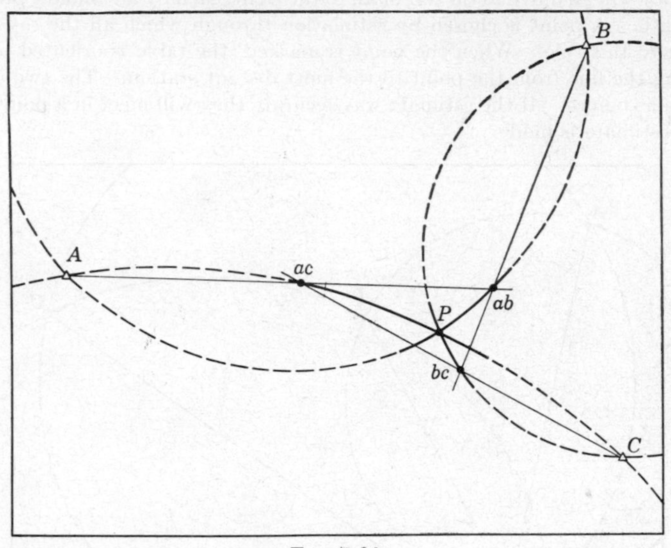

Fig. 7-26

7-33. Two-point Resection. A plane-table station can be established when only two known stations can be observed. The method does not provide a convenient check, and sometimes an unnecessary, auxiliary plane-table station must be established.

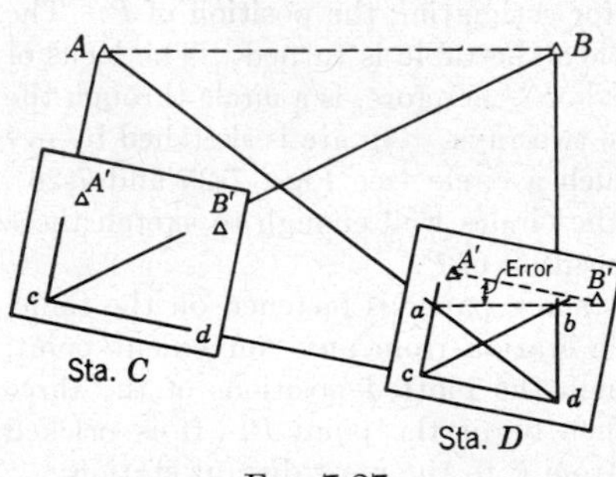

Fig. 7-27

The procedure can be explained as follows. Assume that the known stations are *A* and *B* (Fig. 7-27). Two plane-table stations *C* and *D* can be established anywhere, and any line of any length *cd* can be drawn on the sheet to represent the line between them. Then *A* and *B* can be located on the sheet by intersection (*a* and *b*). The figure *abcd* would then be similar to *ABCD*. If the plane table had been properly oriented at the start, the line *ab* would be parallel to the plotted direction of the line AB. The angle between them represents the error in orientation. The plane table is oriented by turning it through this angle. The position of D' (Fig. 7-28) can then be found by plotting the intersection of the new rays $A'D'$ and $B'D'$ drawn backward with this new orientation.

To correct the orientation, lay out a line from A' or B' (in this case B') at the proper direction for the line $B'D'$. The line is drawn so that it makes an angle with bd equal to the angle between ab and $A'B'$. The alidade is placed on this line and the table is oriented by sighting B. The ray from A' is then drawn; its intersection with the first line gives D'.

Sometimes C can be placed where it can usefully be made part of the plane-table traverse. In that case a traverse line is run to C in the usual way and its true position is plotted. When the table is again set up at C and oriented by a sight on D, the points A' and B' should be in line with their respective ground points A and B.

7-34. Hand Topographic Instruments. There are a number of hand instruments not previously mentioned that can be used successfully to tie topographic features to control, or to make approximate maps. They are all various forms of hand compasses to determine direction and **clinometers** to determine slope. The three most important are the **Brunton compass, the prismatic compass,** and the **Abney level.**

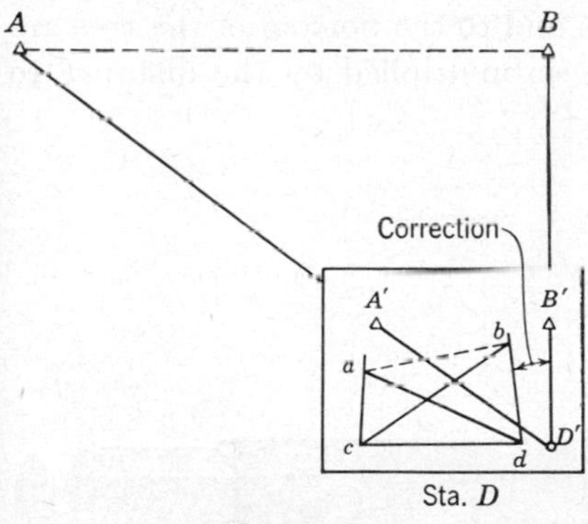

FIG. 7-28

FIG. 7-29. A Brunton compass. (*Keuffel & Esser Co.*)

The Brunton compass (Fig. 7-29) will give compass bearings and measure vertical angles.

The prismatic compass (Fig. 7-30) gives quite accurate compass azimuths. It is arranged so that the observer can see the azimuth reading and the pointing of the line of sight simultaneously.

The Abney level (Fig. 7-31) is a form of clinometer that gives the value of the slope in terms of arc measure, per cent grade, or rise—or fall—per

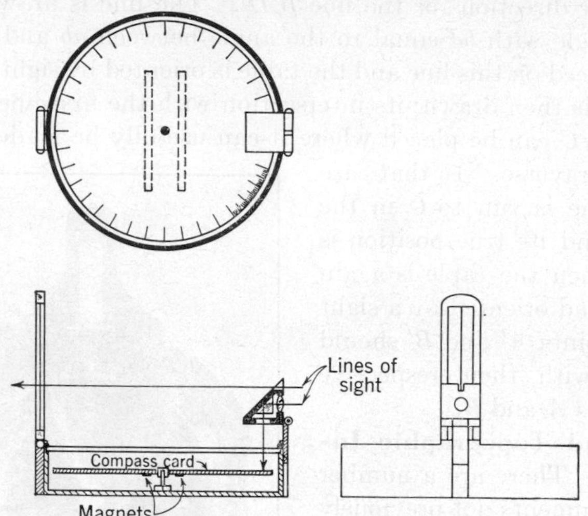

FIG. 7-30. The principle of the prismatic compass.

chain, as desired. Foresters use this instrument to determine the height of trees. The rates of grade to the top and to the bottom of the tree are measured. The arithmetic sum of these, multiplied by the distance to the tree, is the height.

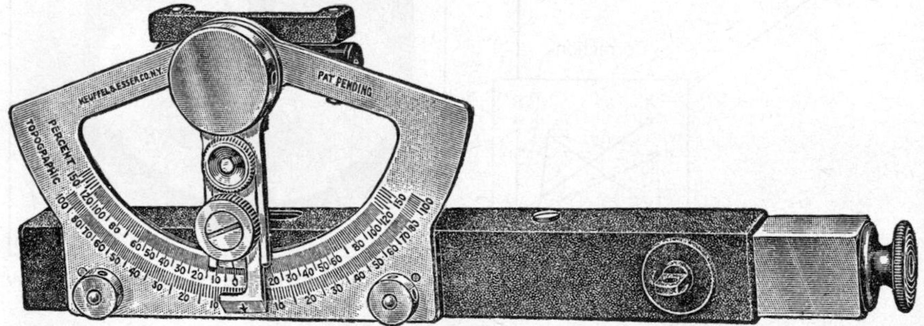

FIG. 7-31. The Abney level. The instrument can be made to read in degrees of slope by reversing the plate.

7-35. The Art of Topographic Mapping. The art of taking topography can be perfected only by long field practice. The topographer must be able to choose points to be observed that give the maximum information, he must be sure never to omit a necessary point, and he must be able to sketch contours and topographic features with maximum fidelity. The key points that give the greatest information are:

1. Summits
2. Saddles (low points in ridges or high points in valleys)
3. Depressions
4. Valley profiles
5. Ridge profiles
6. Boundary and building corners
7. Profiles along boundaries and buildings
8. Profiles along toes of slopes
9. Profiles along brows of hills (tops of slopes)
10. Profiles along shoulders

To decide correctly whether or not to observe any particular point is one of the most difficult and important arts to acquire. In general, the point should not be observed if the topographer is confident that he can sketch the contours within one-half a contour interval of their correct positions without taking an observation on it.

The essential points most likely to be omitted are saddles and midpoints on gently changing slopes. Saddles are particularly difficult to notice when they are in ridges or valleys that are on a line with the topographer. Necessary points on slopes are often missed when an upward slope faces the topographer. Omissions of this kind can best be avoided by observing a point that is apparently unnecessary near the center of any large area.

The accuracy of the contours sketched depends on the topographer's ability to visualize the actual contours on the ground. Of course, without a sufficient number of observed elevations he cannot avoid sketching contours that run gradually uphill or downhill; but the ability can be developed to sketch land forms between observed points so that the contours will be accurate to far greater detail than could be possibly obtained by direct interpolation between observed elevations.

Careful thought should be given to the choice of positions for the plane-table stations. The stations should be placed at points of vantage where the largest ground area can be most easily seen. However, the most common mistake is to attempt to use too few stations rather than too many. The time necessary to set up and take the required control observations is often overestimated, while the delay caused by overlong shots is underestimated. When the shots are long, the rodmen take longer to reach the points, it is more difficult to direct them, observation is more difficult and slower, and the topographer is often too far from the ground he is mapping to choose proper points or to sketch accurately.

The moment there is delay due to long shots in any area, the plane table should be moved into the area.

The economical length of shot depends on the required accuracy of detail. Shots can be as long as 3000 ft when the scale is very small and the accuracy requirements are very low. When mapping at 20 ft to the inch, shots of much more than 200 ft are usually uneconomical.

It goes without saying that the topographer must be sure that each observation is properly computed. So many small computations are involved that the work should be carefully systematized. It is best to record the observed data and the computations in a field book according to some standard form. This practice reduces the possibilities for error or omission. It usually increases the speed of computation and provides a permanent record so that the work can be checked and errors traced.

7-36. Special Terms. The term **tachymeter,** tacheometer, or tachometer is often applied to any instrument, except tapes and similar instruments, that can be used to measure distance. Stadia transits, transits equipped with a distance wedge, and similar devices are tachymeters.

7-37. The Anallactic Point. The anallactic point has several applications in surveying but it is best understood when applied to stadia. Due to the effect of focusing, the stadia multiplier constant (usually 100) can give the distance to the rod from a point at the instrument center *exactly* only at one rod position. When the position of the rod, and hence the focus, is changed, the distance given by the stadia constant is measured from a point that is not quite at the instrument center. The point, so described, is called the anallactic point. It changes position with changes in focus.

This principle is illustrated for an internal-focusing instrument in Fig. 7-32. H_1 and H_2 are the principal planes of the combined objective and focusing lens. By the definition of these principal planes, in air the angle α equals the angle β (Fig. 7-32*a*). Therefore,

$$\frac{u_0}{S} = \frac{v_0}{i} \qquad \text{or} \qquad u_0 = 1.00\,\frac{297}{3} = 99 \text{ ft}$$

In Fig. 7-32*b*, the rod has been moved until the intercept S equals 0.5 ft. When the telescope is focused at the new position, v_0 changes from 297 mm to v' at 303 mm. This reduces the size of the equal angles to α' and β', and H_1 and H_2 move to their new positions H_1 and H_2. Thus

$$u' = 0.50\,\frac{303}{3} = 50.5 \text{ ft}$$

In practice, after the telescope is designed, the stadia hairs are placed so that when D_0 has a chosen value—usually 100 ft—the stadia intercept will be 1.00 ft (or some other convenient value), as shown in Fig. 7-32*a*. In this case the stadia multiplier constant is said to be 100; that is,

$$D = 100S$$

The anallactic point A is then at the center of the instrument.

When the rod is held elsewhere, as shown in Fig. 7-32*b*, the anallactic point will move to A' where the distance determined by stadia would be correct. Since the intercept is 0.5, this would be 50 ft from the rod.

Thus the anallactic point can be defined as the point on the optical axis where for any given focus, the stadia multiplier constant holds true.

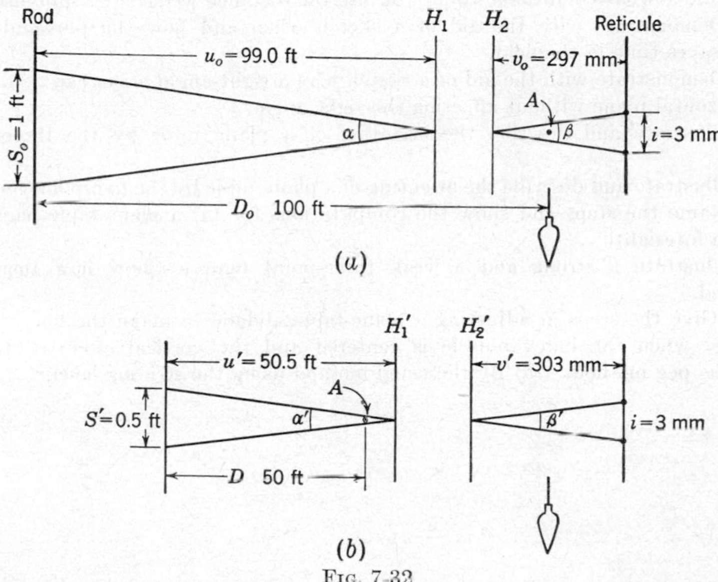

FIG. 7-32

Since v changes very little except when the rod is close to the instrument, the anallactic point changes in position very little under the same circumstances. When an instrument is so designed that, on shots of ordinary length, the anallactic point remains very nearly at the instrument center, the instrument is often called an **anallactic instrument.** Only internal-focusing telescopes can be so designed and therefore the term anallactic instrument is often used to mean an internal-focusing instrument.

In the Zeiss automatic level, the anallactic point is that point on the optical axis from which the angular change in the line of sight, caused by the movement of the pendulum, is subtended on the rod. It is therefore at the elevation of the point sighted on the rod and represents the height of instrument. It is at the instrument center when the rod is at 66 ft, and 0.3 ft in front of the instrument center when the rod is at the minimum focus of 11 ft. If the telescope azimuth axis is tilted, the height of instrument will change very slightly when the telescope is turned in azimuth, especially on short sights.

PROBLEMS

7-1. Name the comparative advantages of the following mapping procedures: transit stadia, plus and offset, photogrammetry, plane-table surveying.

7-2. Name the various types of distance-measuring devices that are used for mapping, excluding chains and tapes.

7-3. Under what conditions would you use the distance wedge; the subtense bar?

7-4. Demonstrate with the aid of a sketch when and how the plus-and-offset method saves time in the field.

7-5. Demonstrate with the aid of a sketch why a right-angle prism can be rotated in a horizontal plane without affecting the right angle.

7-6. Illustrate and describe the orienting of a plane table by the three-point method.

7-7. Illustrate and describe the orienting of a plane table by the two-point method.

7-8. Name the steps and show the computations for (*a*) a plane-table backsight and (*b*) a foresight.

7-9. Illustrate a strong and a weak three-point figure. Show how these are recognized.

7-10. Give the steps in adjusting a plane-table alidade to make the line of sight horizontal when the index bubble is centered and the vertical circle reads 30°. (*a*) By the peg method. (*b*) By the usual method using the striding level.

CHAPTER 8
PLANNING AND ESTIMATING FROM TOPOGRAPHIC MAPS

8-1. The Purpose of Topographic Maps. A topographic map has many uses, but by and large its chief purpose is to provide data in their most convenient form so that construction can be planned. Unless a project is planned so that it conforms closely to the existing topography, the results may not be satisfactory and the costs may be prohibitive. The cost of earthwork is the item most affected by the relationship maintained between the plan and the topography. For this reason, careful estimates of the required earthwork must be made before a plan can be safely accepted. Usually cost estimates of earthwork for several proposed plans must be compared before the best solution can be selected. Certain methods have been developed for using the contours themselves for quickly selecting the most promising locations and for making rapid estimates of earthwork and other quantities involved. While these estimates are too approximate for a basis of payment, they are excellent for comparisons between the costs of various possible solutions and for preliminary estimates for budgeting and financing.

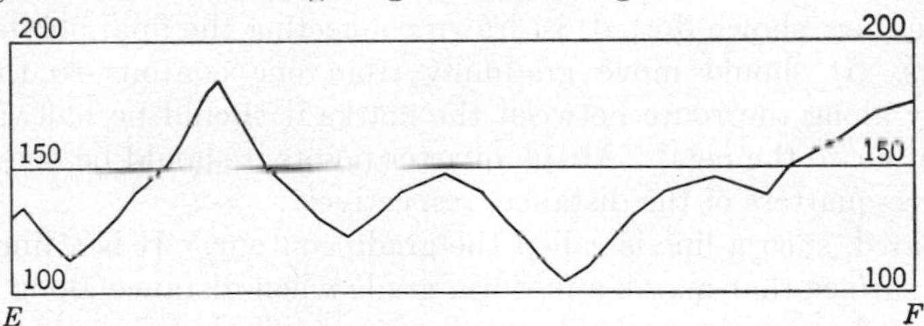

FIG. 8-1. A profile of the line *EF* in Fig. 8-2.

8-2. Profiles. A profile of the ground surface along the line of proposed construction will give an immediate indication of the probable costs and the probable rates of grade required. Figure 8-1 shows the profile of the line *EF* in Fig. 8-2 constructed from the contours. The plus of each contour line is scaled along the line *EF*. It is then plotted on the profile according to the elevation it represents.

8-3. Grade Contours. It is often necessary to choose a route between two points where a road or other facilities can be constructed at a uniform grade. For example, it might be necessary to build a road from *B* in Fig. 8-2 through the saddle at *C* and on to the north. Any grade but a uniform grade must, at some point, be steeper than the uniform grade. Accordingly, a uniform grade should be established if possible.

A glance at the map shows that only two routes are feasible, one passing through the saddle *D* and the other around the hill *G*. The total rise from *B* at elevation 100 ft to *C* at elevation 142 ft is 42 ft. The route through *D* at elevation 122 ft would rise 22 ft—or over half the total rise—in less than half the distance. This would prevent a uniform grade through *D*; so the route west of *G* is chosen.

8-4. To Draw the Grade Contour. A **grade contour** is a line of uniform grade along the route of the proposed project. It is constructed as follows. The probable route is measured. In the example it will be found to be about 6 in. long. A uniform grade therefore would rise 10 ft in about 1.4 in. A point is marked on the 110-ft contour that is 1.4 in. from *B*, measured along the *estimated direction of the road.* The 120-ft contour is then marked 1.4 in. farther along the estimated route, and so on. It is evident that the road will not run up and then down the valleys as the grade contour is shown but will cut across each valley in a curve. The distance along the route is therefore estimated along the probable curves.

The process is continued until the 140-ft contour has been marked. The distance from this point to *C* should be just the distance required for the uniform grade to rise the remaining 2 ft, or about 0.3 in. If an extension of 0.3 in. does not fall at *C*, all the points are moved proportionally along the direction of the route.

A line (here shown dotted) is drawn connecting the final points on the contours. It should move gradually from one contour to the next. Halfway along the route between the marks it should be halfway from one contour to the next. At the quarter points it should be one-quarter and three-quarters of the distance, respectively.

As stated, such a line is called the grade contour. It is a line on the ground surface that marks a uniform grade when distance along the line is measured along the probable location of the final route. If the route could be laid out along the grade contour, no cut or fill would be required to obtain a uniform grade except to level the road transversely.

With the grade contour as a guide, the location for the center line can be established so that it will require a minimum of earthwork.

8-5. Drainage Area. The first consideration in making plans for the utilization, control, or almost any type of economic development of a stream is to determine the size of its drainage area. For example, it might be planned to form a reservoir by building a dam at *AB* as shown

FIG. 8-2. Solutions of problems by contour lines.

in Fig. 8-2. The drainage area of the stream above the dam would be required.

A drainage area for a given point in a stream can be defined as the area that forms the source of all water that passes that point. A line that marks the limits of a drainage area has the following characteristics:

1. It begins and ends at the point in the stream to which it applies.
2. It passes through every saddle that divides the drainage area in question from others.
3. It is always perpendicular to the contour lines.
4. It often follows ridges.
5. It often turns sharply at a ridge and seldom turns sharply elsewhere.

To draw such a line, first lay out sections of the line extending from the point in the stream in question (in this case the line AB) and from every saddle that forms a divide. Draw them perpendicular to the contours that they cross. Extend them until they reach a ridge. There they must be turned to follow the ridge until they connect with another section. The final line is shown in Fig. 8-2 by dot and dash.

8-6. The Planimeter. The area contained in a drainage area is usually measured with a planimeter. A planimeter is an instrument that will measure the area of a plane figure when its tracing point is moved around the perimeter of the figure (see Figs. 8-6 and 8-11). The theory and operation of a planimeter are given at the end of this chapter.

8-7. Reservoir Capacity. An estimate of the capacity of a proposed reservoir is always of great importance. If, as illustrated in Fig. 8-2, the spillway of the dam at AB is to have an elevation of 100 ft, the water in the reservoir will cover the area enclosed by the 100-ft contour as shown by crosshatching. The solid figure represented by the water can be thought of as composed of layers 10 ft thick separated by level surfaces each at the elevation of a contour line.

Each layer—except the layer below the lowest contour line—would then be a solid figure bounded at the top and bottom by parallel plane surfaces. The area at each of these surfaces would be the area bounded by the contour line in question. If the ground sloped everywhere uniformly from one contour to the next, each solid would be bounded on the sides by a surface generated by a straight line moving in contact with the two contour lines. The volume of such a solid can be found exactly by **prismoidal formula,** or approximately by the **end-area formula.** The total volume is the sum of these layers plus the layer below the last contour.

8-8. The Prismoidal Formula. A demonstration of the prismoidal formula can be found in any text on solid geometry. It gives the volume of a solid bounded by two plane parallel figures and a surface generated by a straight line moving in contact with the perimeters of the two figures. It is stated as follows:

$$V = \frac{h}{6}(A_1 + 4M + A_2) \tag{8-1}$$

where V = volume of the solid defined
h = perpendicular distance between the two bounding planes or bases
A_1 = area of one base
A_2 = area of the other base
M = area of the **midsection,** a section of the solid cut by a plane parallel to and halfway between the bases

8-9. The End-area Formula. The end-area formula is an approximate formula often used for simplicity:

$$V = h\left(\frac{A_1}{2} + \frac{A_2}{2}\right) \tag{8-2}$$

where the symbols are the same as in Eq. (8-1). When several volumes are added, this becomes

$$V = h\left(\frac{A_1}{2} + A_2 + A_3 + \cdots + A_{n-1} + \frac{A_n}{2}\right) \tag{8-3}$$

8-10. Application of the Prismoidal Formula. The areas A_1 and A_2 can be found for each layer by tracing the contour with a planimeter. The value h is the contour interval, in this case 10 ft. But, unfortunately, the area of the midsection cannot be found. It can be estimated, however, if it is assumed that areas bounded by adjacent contours are similar figures and that the midsection is similar to both. The linear dimensions of the midsection would then be equal to the average of those of the bases. Then, since linear dimensions of similar figures are to each other as the square roots of the areas of the figures,

$$\sqrt{M} = \frac{\sqrt{A_1} + \sqrt{A_2}}{2}$$

$$M = \frac{A_1 + 2\sqrt{A_1A_2} + A_2}{4}$$

Substituting in the prismoidal formula,

$$V = \frac{h}{6}\left(A_1 + 4\,\frac{A_1 + 2\sqrt{A_1A_2} + A_2}{4} + A_2\right)$$

$$= \frac{h}{3}(A_1 + \sqrt{A_1A_2} + A_2) \tag{8-4}$$

When several layers are added together, the areas of each contour except the first and last are added twice. Then the simplest method is to divide the first and last by 2, add them to the others, and finally multiply the sum by 2. The middle term is handled separately.

The volume of the last solid must be computed by the formula that applies to the nearest approach to its shape. Since all the areas are obtained by planimeter, no accuracy is lost by using a slide rule.

Example. In the example in Fig. 8-2:

Contour	Sq ft	$\sqrt{A_1A_2}$		
100	108,400		×½	54,200
		85,000		
90	66,800		×1	66,800
		48,500		
80	35,200		×1	35,200
		16,400		
70	7,600		×½	3,800
Σ		149,900		160,000
				×2
			$\Sigma A_1 + A_2 =$	320,000
			$\Sigma \sqrt{A_1A_2} =$	149,900
			$\Sigma(A_1 + \sqrt{A_1A_2} + A_2) =$	469,900 sq ft
			$\frac{h}{3} =$	3.333
			Volume = product =	1,566,000 cu ft

The last layer is the volume between the 70-ft contour and the ground. It is nearly a wedge, with a base of 7600 sq ft and an altitude of about 4 ft.

$$\tfrac{1}{2} \times 4 \times 7600 = 15{,}200 \text{ cu ft}$$

The capacity of the reservoir is then 1,581,000 cu ft.

8-11. Application of the End-area Formula. The same volume computed by the end-area formula would be treated as follows:

Contour	Sq ft		
100	108,400	×½	54,200
90	66,800	×1	66,800
80	35,200	×1	35,200
70	7,600	×½	3,800
		Σ =	160,000
		h =	×10
		Volume = product =	1,600,000 cu ft
Volume of the last solid = ½ × 4 × 7600 =			15,200 cu ft
		Total capacity =	1,620,000 cu ft

Note that the volume computed is 40,000 cu ft larger by the end-area formula than by the prismoidal formula.

8-12. Earthwork Quantities from Contours. Once a plan has been developed, the earthwork quantities it involves can be determined approximately from the contours. The process is described in five steps.

1. Plot the plan in its correct horizontal position on the topographic map.

2. From the elevation data on the plan, draw the new contours that will appear on the surface of the earthwork called for by the plan.

3. Draw the new contours on the side slopes at the cuts and fills called for by the plan. The horizontal separation between them will depend on the values of the slopes that are to be used. They may be drawn without reference to the existing contours but later they are terminated at the existing contours.

4. Draw the line at the limits of the slopes (the line of no cut or fill) and the lines of 1 ft cut, 2-ft cut or similar lines having the same vertical interval as the contours. These new lines are called **quantity contours.**

5. Measure the areas of the quantity contours and compute the volume by either Eq. (8-1) or Eq. (8-2).

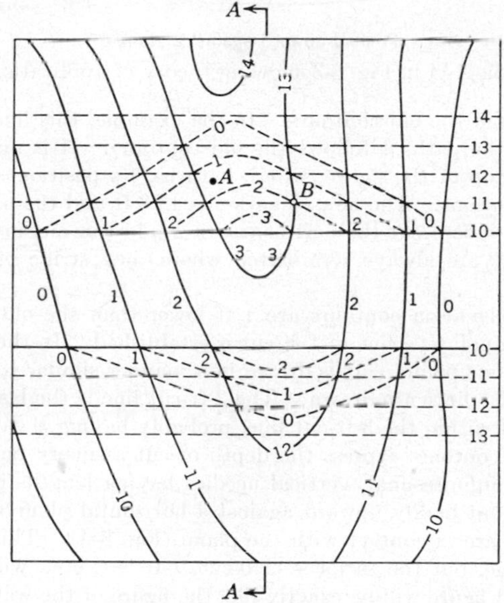

FIG. 8-3. Earthwork quantities from contours.

Example 1. Figure 8-3 shows the spur of a hill. A level path 8 ft wide is to be cut through the spur at elevation 10 ft. Beyond the spur it will follow the ground surface. The side slopes are 2:1 (2 horizontal to 1 vertical).

1. The plan of the path is drawn as shown.

2. The new 10-ft contour will run along the sides of the path and cross the path at both ends along the existing 10-ft contour.

3. The new slopes will slope upward from the path, rising 1 ft in every 2 ft on the horizontal. The new 11-ft contour will be parallel to the edges of the path and 2 ft distant. The 12-ft contour will be parallel and 2 ft from the 11-ft contour, and so on. When the new contours (as shown by the straight dashed lines) have been drawn, the cut at any point can be found by subtracting the new elevation determined from the new contours from the old elevation determined by the old contours. At point A, for example, the new elevation is 11.8 ft and the old elevation is 13.4 ft. This shows a cut of 1.6 ft. Where new and old contours cross, the cut is expressed in exact feet. At B, for example, the new elevation is 11 ft and the old one 13 ft, giving a cut of 2 ft.

4. Where like contours cross, there is no cut or fill. By connecting such points, a line of zero cut is established. When the position of such a line between intersections is in doubt, its exact position can be found from a comparison of elevations

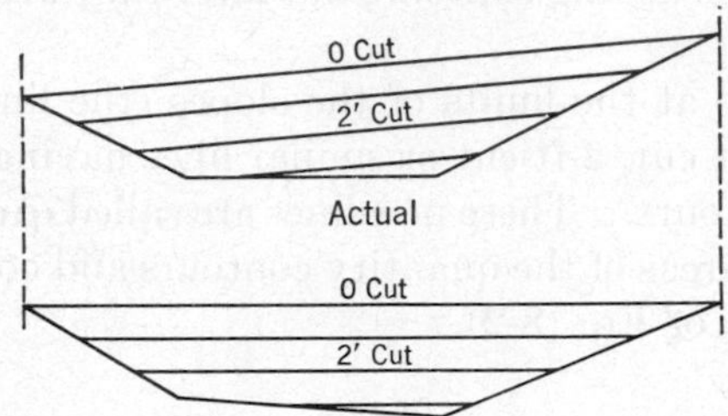

FIG. 8-4. Section A-A in Fig. 8-3 showing theory of application of formulas.

given by the new and the old contours. In the example, this line is dotted (except where, in crossing the path, it follows the old contour). It is marked 0 at various points. Near the top of the figure, where it turns, its position can be checked by comparing the elevations. The new elevation is 13.8 ft and the old is 13.8 ft.

Beyond the line of zero cut the new contours would run out into space. Accordingly, new contours are always terminated where they strike old contours of the same elevation.

Where the new elevation contours are 1 ft lower than the old, there is 1-ft cut. By connecting such points, a line of 1-ft cut is established. In the example, this line is marked 1 at various points. It is drawn in a manner similar to the 0-cut line.

The other quantity lines are drawn. The 3-ft-cut line is the last. The maximum cut would be found within the 3-ft-cut line, probably having a value of 3.2-ft cut.

5. The quantity contours express the depth of cut at every point. Consider the cut as made up of infinitesimal, vertical needles having lengths equal to the depth of cut. Move the cut bodily upward against a horizontal plane surface so that the tops of the needles are in contact with the plane (Fig. 8-4). This operation has no effect on the volume, but the surfaces of 0-cut, 1-ft cut, etc., will all be horizontal planes and the solid figure will be exactly like the figure of the water in the reservoir described previously.

Accordingly, the areas of the various quantity contours are measured with a planimeter and the volumes computed by either Eq. (8-1) or Eq. (8-2).

Example 2. Figure 8-5 shows the surface of the earthwork—the subgrade—to be prepared for a width of 40 ft for a highway. The surface has a crown of 1 ft. The slopes are at 4:1. The plans call for a center elevation of 65 ft (at the bottom of the map) and a uniform grade to 120 ft (at the top of the map) as shown.

The new 5-ft contours are drawn across the subgrade curving with the crown. The

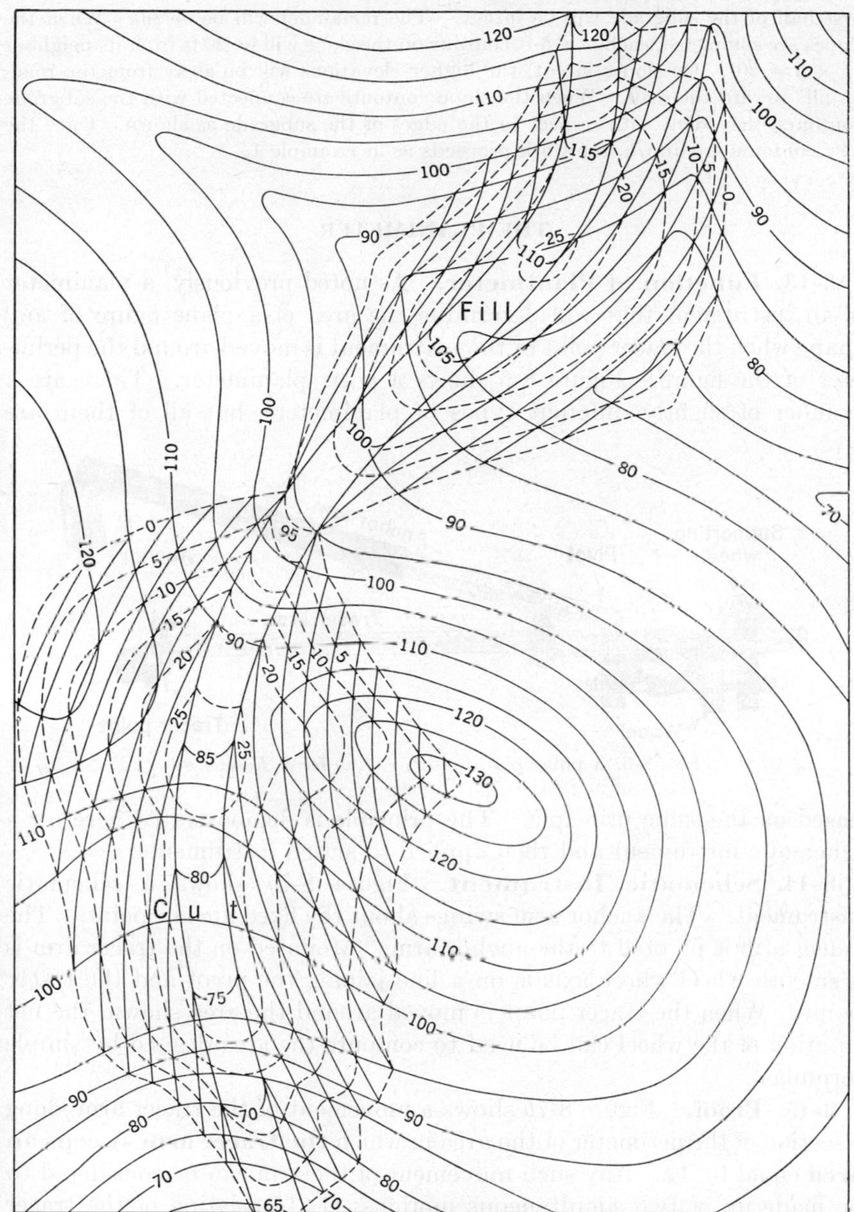

FIG. 8-5. Quantity contours for a highway.

first half of the subgrade will be in cut. The remainder will be on fill. When the slopes are completed, each new 5-ft contour on the slope will be 20 ft from its neighbor ($4 \times 5 = 20$). On slopes in cut, the higher elevations will be away from the road, on fills toward the road. When the slope contours are connected with the subgrade contours, they slant with respect to the edges of the subgrade as shown. Once the new contours are drawn, the work proceeds as in Example 1.

THE PLANIMETER

8-13. Function of Planimeter. As noted previously, a planimeter is an instrument that will determine the area of a plane figure of any shape when the tracer point of the instrument is moved around the perimeter of the figure. Figure 8-6 shows a polar planimeter. There are a number of slightly different types of planimeters, but all of them are

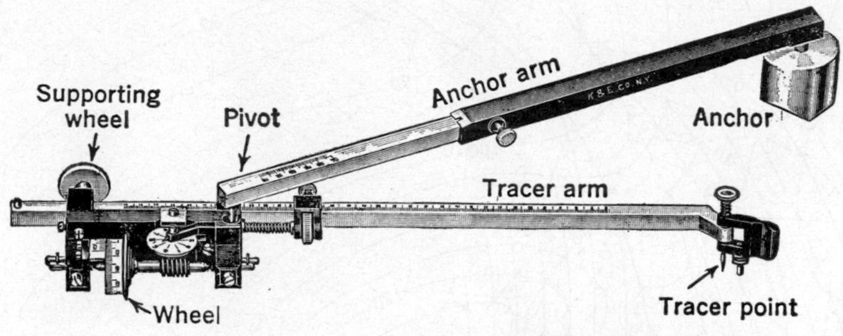

FIG. 8-6. A polar planimeter. (*Keuffel & Esser Co.*)

based on the same principle. The principle is demonstrated here for a schematic instrument and then applied to actual instruments.

8-14. Schematic Instrument. Figure 8-7*a* shows a schematic instrument. The anchor arm swings about the fixed anchor point. The tracer arm is pivoted to the anchor arm. Mounted on the tracer arm is a smooth wheel whose axis is on a line joining the pivot and the tracer point. When the tracer point is moved around the area shown, the net rotation of the wheel can be used to compute the area desired by simple formulas.

8-15. Proof. Figure 8-7*b* shows a movement of the tracer arm along a section of the perimeter of the area in which **the tracer arm sweeps an area** equal to A_S. Any such movement of the arm can be considered to be made up of two simultaneous motions: (1) translation of the tracer arm in parallel motion, and (2) rotation of the arm about the pivot. Figure 8-7*c* shows a differential movement of the arm in which these two motions are shown separately. From Fig. 8-7*c*,

$$dA_S = L\,dh + \tfrac{1}{2}L^2\,d\theta \tag{8-5}$$

It is a basic principle of a wheel that when it is moved over a surface, it will roll out a length along its circumference equal to its total movement perpendicular to its axis. Accordingly, if dw equals the distance rolled

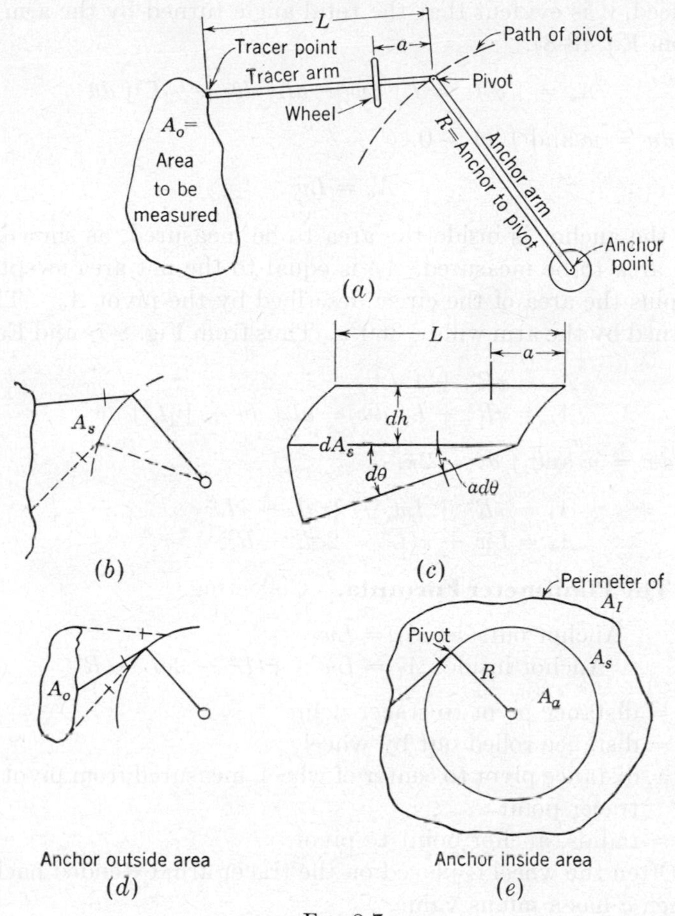

FIG. 8-7

out by the wheel,

$$dw = dh + a\ d\theta \tag{8-6}$$

or

$$dh = dw - a\ d\theta \tag{8-7}$$

The area given by Eq. (8-5) can be expressed in terms of the distance rolled out by the wheel as given by Eq. (8-6), by substituting the value of dh from Eq. (8-7). Thus

$$dA_S = L\ dw - aL\ d\theta + \tfrac{1}{2}L^2\ d\theta \tag{8-8}$$

When the anchor is outside the area to be measured, as shown in Fig. 8-7d, the net area swept by the arm is equal to the area to be measured.

In this example, the arm moves downward along one side of the area and upward along the other side so that the net area swept, A_S, is equal to A_o. Since the position of the arm is the same before and after the perimeter is traced, it is evident that the total angle turned by the arm is zero. Thus from Eq. (8-8),

$$A_o = \int dA_S = L\int dw - aL\int d\theta + \tfrac{1}{2}L^2\int d\theta$$

where $\int dw = w$ and $\int d\theta = 0$.

Hence $$A_o = Lw \tag{8-9}$$

When the anchor is inside the area to be measured, as shown in Fig. 8-7*e*, the area to be measured, A_I, is equal to the net area swept by the arm A_S plus the area of the circle described by the pivot A_a. The total angle turned by the arm will be 360°. Thus from Fig. 8-7*e* and Eq. (8-8),

$$A_I = \pi R^2 + \int dA_S$$
$$A_I = \pi R^2 + L\int dw - aL\int d\theta + \tfrac{1}{2}L^2\int d\theta$$

where $\int dw = w$ and $\int d\theta = 2\pi$.

Hence $$A_I = \pi R^2 + Lw - 2\pi aL + \pi L^2$$
or $$A_I = Lw + \pi(L^2 - 2aL + R^2) \tag{8-10}$$

8-16. The Planimeter Formula. Collecting,

$$\text{Anchor outside: } A_o = Lw \tag{8-9}$$
$$\text{Anchor inside: } A_I = Lw + \pi(L^2 - 2aL + R^2) \tag{8-10}$$

where L = distance pivot to tracer point
w = distance rolled out by wheel
a = distance pivot to center of wheel, measured from pivot toward tracer point
R = radius, anchor point to pivot

Note: Often the wheel is placed on the tracer arm extended back of the pivot; then a has a minus value.

8-17. Circle of Correction. It will be noted that the formula for the **anchor outside** is the same as the formula for the **anchor inside** with an area added. The expression for this area is the second term and may be checked by geometry. If the tracer point is moved in a circle around the anchor so that the wheel moves axially, the wheel will not turn and the reading will be zero (see Fig. 8-8). Therefore the area of the circle traced is the area that must be added to Lw. It is called the circle of correction.

In Fig. 8-8 are shown the two arrangements for the wheel, *a* between the pivot and the tracer point and *b* back of the pivot. In each case the tracer arm is held so that the wheel moves axially and hence the tracer point traces the circle of correction. From Fig. 8-8,

$$r^2 = (L - a)^2 + x^2$$

but

$$x^2 = R^2 - a^2$$

Substituting,

$$r^2 = L^2 - 2aL + R^2$$

$$\text{Correction area} = \pi r^2 = \pi(L^2 - 2aL + R^2)$$

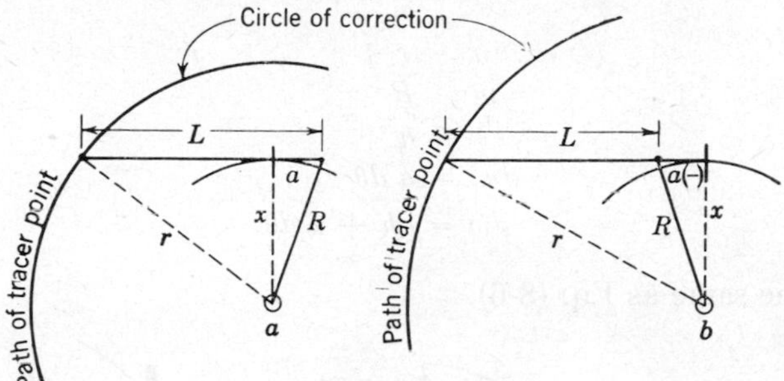

FIG. 8-8

8-18. Offset Wheel Axis. In most planimeters the wheel is offset from the tracer arm as shown in Figs. 8-9*a* and *b*. It will be shown that,

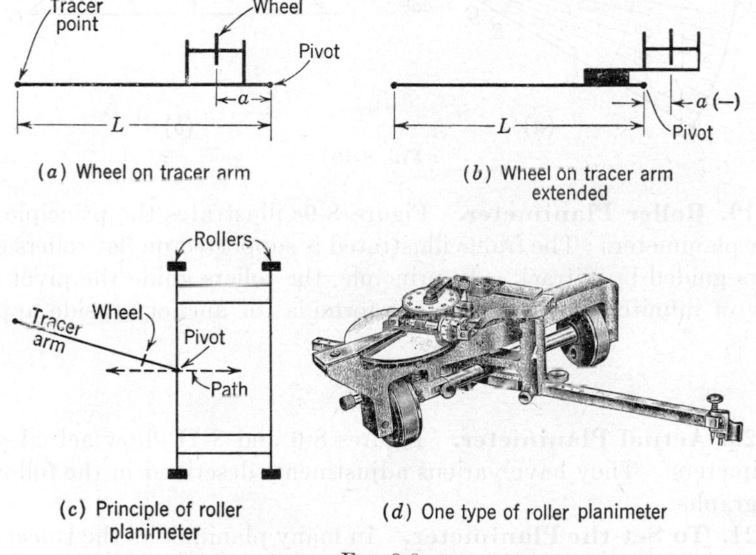

(*a*) Wheel on tracer arm

(*b*) Wheel on tracer arm extended

(*c*) Principle of roller planimeter

(*d*) One type of roller planimeter

FIG. 8-9

as long as the axis of the wheel is parallel to the line joining the tracer point and the pivot, the planimeter will operate properly and the formulas previously derived will apply.

From Fig. 8-7*c* it is evident that in translation the wheel rolls out the distance *dh*. Figure 8-10 illustrates the distance rolled out in rotation.

The wheel is moved along the circumference of the circle R in the direction of m. Since it records only motion perpendicular to its axis, it records only motion in the direction of n. Thus for a movement of $d\theta$, the distance rolled out in rotation is expressed as follows:

$$dw_r = \frac{n}{m} R\, d\theta$$

but

$$\frac{n}{m} = \frac{r}{R}$$

and

$$r = a$$

Thus

$$dw_r = a\, d\theta$$

Therefore

$$dw = dh + a\, d\theta$$

which is the same as Eq. (8-6).

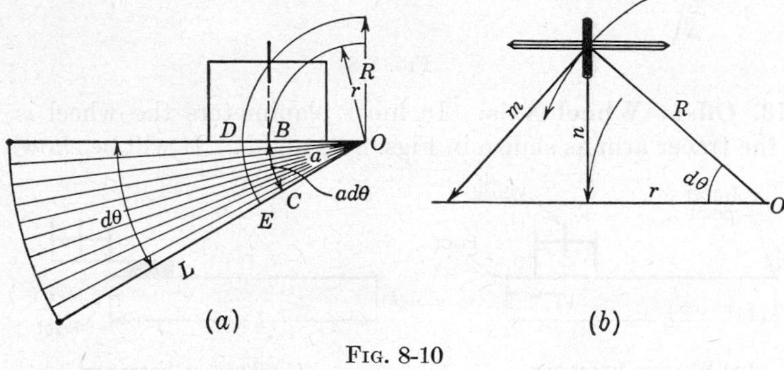

FIG. 8-10

8-19. Roller Planimeter. Figure 8-9c illustrates the principle of a roller planimeter. The frame illustrated is supported on flat rollers or on rollers guided by a track. In principle, the rollers guide the pivot on a circle of infinite radius so that the formula for anchor outside applies; that is,

$$A_o = L\omega$$

8-20. Actual Planimeter. Figures 8-6 and 8-11 show actual polar planimeters. They have various adjustments described in the following paragraphs.

8-21. To Set the Planimeter. In many planimeters the tracer arm is adjustable. This makes it possible to set the distance L so that the readings of the wheel will be related to the area by a convenient ratio. The anchor arm is also sometimes adjustable so that the area of the correction circle can be given by a round value once the distance L is established. The adjustable arms have scales and verniers that give their lengths, usually in units of half centimeters.

A common value for the circumference of the wheel is 2.5 in. The wheel carries a scale that turns against a vernier. One thousand vernier units usually equal one revolution so that one vernier unit represents 0.0025 in. A counter keeps track of the number of whole revolutions. The value of a varies. The Keuffel and Esser Co. uses $a = -1.151$ in. on many of their instruments. The above values are used in all the examples given in this book.

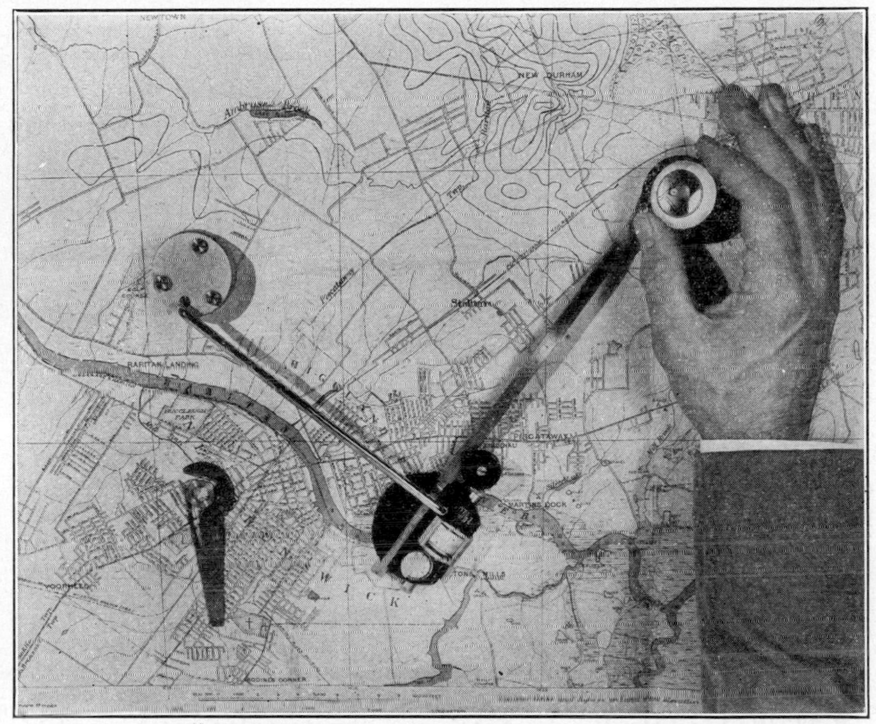

FIG. 8-11. A modern type of polar planimeter. (*Trans-Gobal Co.*)

Values of L, R, w, and a are given in the table below in inches and centimeters, where s is the reading of the arm setting and u is the number of vernier units read on the wheel, for planimeters made as described above.

Distance	*Inches*	*Centimeters*
L and R	$0.1969s$	$0.5s$
w	$0.0025u$	$0.00635u$
a	-1.151	-2.924

For example, if the setting of an arm scale s is 20, the length of L is $0.1969 \times 20 = 3.938$ in. If the wheel reads u equals 1.248, the distance rolled out is $0.0025 \times 1.248 = 0.00312$ in.

Example 1. Required to compute the tracer-arm setting so that one vernier unit equals 0.01 sq in. Substitute in Eq. (8-9)

$$A = 0.01 \qquad L = 0.1969s \qquad w = (0.0025)(\text{unity})$$
$$0.01 = (0.1969s)(0.0025)$$
$$s = 20.31$$

Example 2. Required to compute the tracer setting so that one vernier unit equals 0.1 sq cm. Substitute in Eq. (8-9)

$$A = 0.1 \qquad L = 0.5s \qquad w = (0.00635)(\text{unity})$$
$$0.1 = (0.5s)(0.00635)$$
$$s = 31.50$$

Example 3. Required to compute the tracer-arm setting so that one vernier unit equals 0.01 acre for a map scale of 200 ft to the inch.

$$1 \text{ sq in.} = 40{,}000 \text{ sq ft}$$
$$1 \text{ acre} = 43{,}560 \text{ sq ft}$$
$$0.01 \text{ acre} = \frac{435.60}{40{,}000} \text{ sq in.} = 0.01089 \text{ sq in.}$$

Substitute in Eq. (8-9)

$$A = 0.01089 \qquad L = 0.1969s \qquad w = (0.0025)(\text{unity})$$
$$0.01089 = (0.1969s)(0.0025)$$
$$s = 22.12$$

Example 4. Required to compute the setting of the anchor arm so that the area of the correction circle is 300 acres for a map scale of 200 ft to the inch. From the previous example, 300 acres = 326.7 sq in., $s = 22.12$, $L = 4.355$ in. Substitute in Eq. (8-10)

$$326.7 = \pi[(4.355)^2 - 2(4.355)(-1.151) + R^2]$$
$$R = 8.660 \text{ in.}$$
$$R = 0.1969s$$
$$8.660 = 0.1969s$$
$$s = 43.98$$

8-22. To Find the Precise Settings. Since no planimeter is perfect, the computed settings can only be approximate. The instrument should be set at these values and then checked by tracing a carefully drawn square or circle whose area is known. Usually the instrument will be supplied with a small metal strip that has a pin near one end on the underside and two or three pinholes along its length on the upper side. When the pin is pressed into the paper, each pinhole will guide the tracer arm around a circle of known area and thus check the setting.

If the setting of L gives too large an area reading, the length of L should be *increased* in proportion to the error. Once the L setting is established, the anchor-arm setting should be tested by measuring a large circle with the anchor at the center. If the setting gives too large an area, the anchor arm should be *increased* in length.

8-23. To Use the Planimeter. The work should be performed on a horizontal surface. Place the anchor with two requirements in mind.

1. The tracing point must reach the entire perimeter of the figure without causing the tracer arm and the anchor arm to come within 15° of alignment with each other.

2. Neither the wheel nor other supporting parts should leave the paper.

By circling a small area, determine which direction causes the readings to increase in value. Usually tracing a figure clockwise increases the readings.

Mark a point to start tracing the perimeter of the figure. A fine line across the perimeter is excellent. The work must be started and ended *exactly* at this point. A discrepancy in any direction will introduce an error. It is much more important to start and stop at the *same* point than to follow the perimeter exactly.

Adjust the tracer point until it is near, but clears, the paper. Move it to the starting point; read the wheel vernier and the rotation counter. Some engineers prefer to set the wheel at zero. Trace the perimeter. If the pointer wanders off one side of the line, balance it by moving off the other side of the line so as to make the areas of the errors equal. Finally take the last reading and find the difference. Make sure the subtraction is made in the proper direction. If the counter has passed zero, make the proper correction.

If it is suspected that the axis of the wheel is not parallel with the line from the tracer point to the pivot, the resulting error can be eliminated by *compensation.* To compensate, first trace the area with the pivot so placed that the smaller angle between the tracer arm and the anchor arm is above the pivot; then trace a second time with the smaller angle below. The average of the results is error-free.

PROBLEMS

Use the data for planimeters given in this chapter. The limit for the setting for the tracer arm is between 12 and 35 and for the anchor arm between 12 and 20.

8-1. For a map scale of 20 ft per inch, what should be the tracer-arm setting so that one vernier unit equals 5 sq ft?

8-2. For a map scale of 1 in. = 1 mile, what should be the tracer-arm setting so that one vernier unit equals 10 acres?

8-3. If the tracer arm is set at 16.3 and the map scale is 1 in. = 50 ft, what is the value of one vernier unit?

8-4. For a scale ratio of 1:5, what is the tracer-arm setting so that the value of one vernier unit is 2 sq cm?

8-5. For a scale ratio of 1:8 and a tracer-arm setting of 19.5, what is the value of one vernier unit?

8-6. Determine a convenient setting and value for a vernier unit when measuring acres with a map scale of 100 ft per inch.

8-7. Compute a convenient setting for the anchor arm for each of the above problems.

Part II. Operations

CHAPTER 9

CELESTIAL OBSERVATIONS[1]

INTRODUCTION

9-1. Celestial Observations Applied to Surveys. The usual purpose of a celestial observation in surveying operations is *to establish a direction.* A direction so established is called an **astronomic direction.** Astronomic directions are used chiefly for the following purposes: to give permanence to the directions of property lines, to correlate surveys, to check the angles of long traverses, and to orient important maps or charts. They are also used to orient directional radio and radar antennae, the polar axes of astronomical instruments, phototheodolite test installations and similar devices, and they are an important element in geodetic surveys and in the determinations of magnetic declinations.

To establish direction it is often necessary also to determine geographic position, that is, latitude and longitude. Frequently, geographic position must be determined for other purposes.

9-2. The Terms "Astronomic" and "Geodetic." As will be explained in Chap. 12, direction and position determined astronomically are not quite the same as those determined by a geodetic survey. The adjectives **astronomic** and **geodetic** are used to distinguish them. Both types of directions are often called true directions to distinguish them from magnetic directions. However, the word **true** is usually confined to astronomic directions, as it is in this text.

[1] The following publications are frequently mentioned in this chapter: (1) "The American Nautical Almanac," referred to in this text as the Nautical Almanac. (2) "The American Ephemeris and Nautical Almanac." This is an entirely different publication. It is referred to in this text as The American Ephemeris. Both are issued each year by the U.S. Naval Observatory and are for sale by the Superintendent of Documents, U.S. Printing Office, Washington 25, D.C. (3) "K & E Solar Ephemeris and Other Data," published each year by Keuffel & Esser Co., Hoboken, N.J. and available on request. It is referred to in this text as the K & E Ephemeris.

GENERAL CONCEPTS[1]

9-3. The Celestial Sphere. To simplify the computations necessary for astronomical determinations, certain arbitrary concepts of the heavens have been generally adopted. They are the following:

1. The earth is stationary.

2. The heavenly bodies have been projected outward, along lines which extend from the center of the earth, to a sphere of infinite radius called the **celestial sphere.** The celestial sphere has the following characteristics:

a. Its center is at the center of the earth.

b. Its equator is on the projection of the earth's equator.

c. With respect to the earth, the celestial sphere rotates from east to west about a line which coincides with the earth's axis. Accordingly, the poles of the celestial sphere are at the prolongations of the earth's poles.

d. The speed of rotation of the celestial sphere is constant. It is that of the stars. Its value can be taken as 360°59.13884′ per 24 hr. Note that it makes slightly more than one revolution per day.

e. With the important exception of bodies in the solar system, which change position slowly, all heavenly bodies remain practically fixed in their positions on the celestial sphere, never changing more than almost negligible amounts in 24 hr, and, accordingly, they are often called *fixed stars.*

9-4. The Position of a Heavenly Body. The position of a heavenly body on the celestial sphere is given by its **declination *d*** and its **sidereal hour angle S.H.A.** (see Fig. 9-1).

A plus declination of a body is its angular distance measured north from the celestial equator. A minus declination is measured south. A declination therefore is equivalent to a latitude and is always *equal to the earth latitude* of the body.

The sidereal hour angle of a body is the angle measured westward on the celestial equator from the vernal equinox to the point where the body's celestial meridian intersects the celestial equator. The **right ascension (R.A.)** of a body is 360° minus the S.H.A. It is usually expressed in hours rather than in degrees and it is not as easy to use as the S.H.A.

9-5. The Vernal Equinox ♈. A reference point on the equator of the celestial sphere is known as the **vernal equinox,** the symbol for which is ♈. It is the mean point where the sun crosses the celestial equator in the spring. It is sometimes called the **first point of Aries.**

9-6. Tables. The declination and the sidereal hour angle of about 64 easily identified stars, the four planets visible to the naked eye, and the

[1] Many passages in this chapter are taken from "Celestial Observations," written by the author and published in the "K & E Solar Ephemeris" (copyright by Keuffel & Esser Co., Hoboken, N.J.), and used by permission.

sun and the moon can be quickly computed for any moment of time from data in the Nautical Almanac. These stars and others are covered in many other publications. The word *apparent* is often found preceding the names of data in tables. It means that the data are those which would result if an observation were made at the center of the earth. Since all computations are based on this premise and all observations are corrected accordingly, these are the data to use. When the data are not otherwise defined, they can be assumed to be apparent.

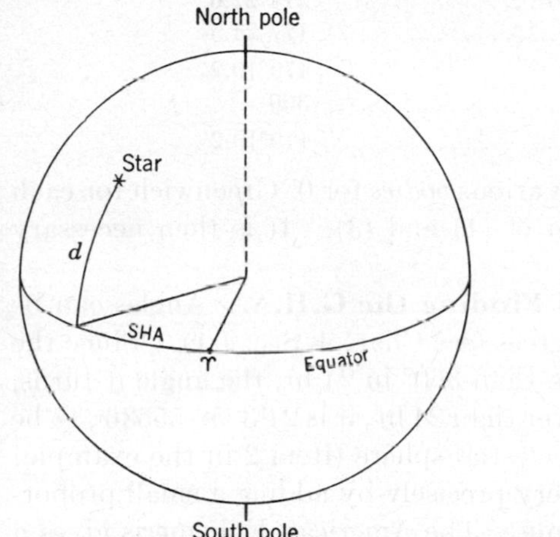

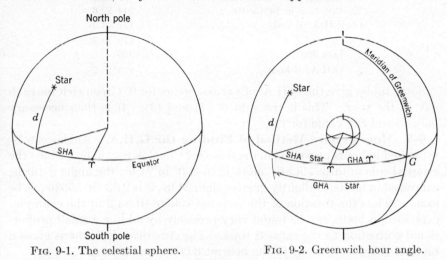

FIG. 9-1. The celestial sphere.

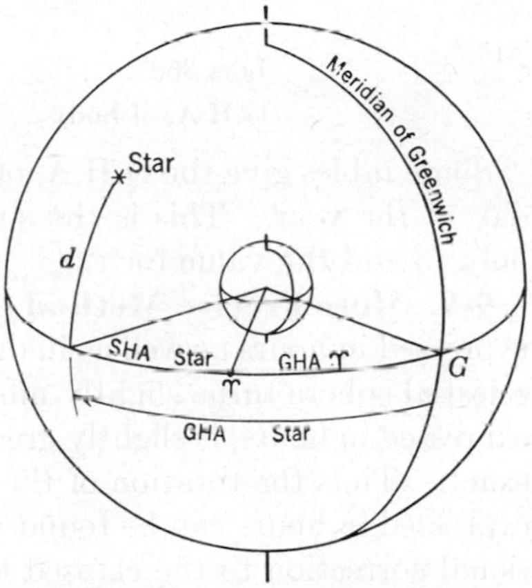

FIG. 9-2. Greenwich hour angle.

9-7. Greenwich Hour Angle G.H.A. The Greenwich hour angle **G.H.A.** of a body (or a point) on the celestial sphere at any moment is the angle measured westward from the meridian of Greenwich (projected on the celestial sphere) to the meridian of the body (see Fig. 9-2). Up to 180°, the G.H.A. is measured like west longitude. Thereafter, it continues up to 360°. Thus the G.H.A. of every body or point on the celestial sphere is always increasing as the heavens move toward the west at very nearly the rate of the rotation of the celestial sphere (360°59.13884′ per 24 hr). This rate of rotation can be expressed as follows:

$$15°2.46412' \text{ per hour}$$
$$15.04107' \text{ per minute}$$
$$0.250684' \text{ per second}$$

Publications giving celestial data often give a multiplication table by which the angular movement of the celestial sphere can be quickly computed for any time interval. Table 9-1 is such a table.

9-8. To Find the G.H.A. of a Body. The Nautical Almanac lists the G.H.A. of ♈ for each hour of the year. Some publications list it only for the moment of Greenwich midnight (0^h) for each day of the year. Since the angular speed of rotation of the celestial sphere is known, the

G.H.A. of ♈ can thus be computed for any moment of time. The G.H.A. for any body is then found by adding the body's S.H.A. to the G.H.A. of ♈ (see Fig. 9-2).

Thus the G.H.A. of a body at any moment of time is found by adding three quantities: (1) the G.H.A. of ♈ for a certain moment, (2) the rotation of the celestial sphere since that moment, and (3) the S.H.A. of the body. Thus:

1. G.H.A. ♈ 0^h Greenwich	87°32.1′
2. Rotation in $14^h15^m10^s$	214 22.6
3. S.H.A. of body	177 24.5
	479°19.2′
Less 360°	−360
G.H.A. of body	119°19.2′

Some tables give the G.H.A. of various *bodies* for 0^h Greenwich for each day of the year. This is the sum of (1) and (3). It is then necessary only to add the value for (2).

9-9. More Precise Method of Finding the G.H.A. Angles can be expressed in hours as well as in degrees (see Chap. 4, Sec. 4-1). Since the celestial sphere turns slightly more than 360° in 24 hr, the angle it turns, expressed in hours, is slightly greater than 24 hr; it is $24^h3^m56.55536^s$, to be exact. Thus the rotation of the celestial sphere (item 2 in the example) expressed in hours can be found very precisely by adding a small proportional correction to the elapsed time. The American Ephemeris gives a table of these corrections to the nearest 0.001 sec of time.

This publication also gives the G.H.A. of ♈ for 0^h Greenwich (item 1) for each day of the year expressed in hours rather than in degrees. Item 3 in the computation is usually expressed in R.A. rather than in S.H.A., as R.A. is expressed in hours. The R.A. must be subtracted, as it is equal to 360° − S.H.A. The R.A. and the declination of 212 stars are given in the American Ephemeris.

Such a computation would be as follows:

1. G.H.A. ♈ at 0^h Greenwich[1]	5^h50^m 8.32^s	
2. Elapsed time	+14 15 10.05	
Correction[2]	+ 2 20.48	
G.H.A. ♈ at the moment of time desired	$20^h07^m38.85^s$	
3. R.A. of body[3]	−12 10 22.01	
G.H.A. of body	$7^h57^m16.84^s$	in hours
	119°19′12.6″	in degrees

[1] In "The American Ephemeris" this value is found in the table labeled "Sun" under column labeled "Sidereal Time (Right Ascension of Mean Sun + 12^h)."

[2] In "The American Ephemeris" this value is found in Table III, labeled "Mean Solar into Sidereal Time."

[3] In "The American Ephemeris" this value is found in the table labeled "Apparent Places of the Stars."

9-10. Time. The word *time* has two meanings that are often confused, **elapsed time** and **moment of time.**

The measure of elapsed time used throughout this discussion is the familiar hour of which there are 24 in a day. Elapsed time, so measured, is more accurately called **mean solar time** (MST), **mean time** (MT), or **civil time** (CT).

A moment of time is given by the year, the day of the month, and the elapsed time since midnight (0^h) at the beginning of the day named. It must be further defined by the meridian from which it is reckoned. Accordingly, **Greenwich civil time** (GCT), often called **universal time,** is civil time reckoned from the moment of midnight at the Greenwich meridian; and 75th meridian time [Eastern standard time (EST)], for example, is civil time reckoned from midnight at the 75th meridian. **Local civil time** (LCT) is civil time reckoned from the precise meridian of longitude where an observation is taken. To convert a moment of time, reckoned from any meridian, to GCT, 1 hr is added for every 15° of west longitude and 1 hr is subtracted for every 15° of east longitude. Obviously, when the value of the longitude is not evenly divisible by 15°, fractional hours will result.

The following are a few examples:

1 A.M. EST	=	6^h GCT
2 P.M. EST	=	19^h GCT
10 P.M. EST	=	3^h GCT (the next day)
4 A.M. 76°W LCT	=	9^h4^m GCT
10 A.M. 30°E LCT	=	8^h GCT

9-11. Sidereal Time. The G.H.A. ♈ at any moment is often called sidereal time. Local sidereal time is the same kind of angle, but measured from a local meridian instead of from the Greenwich meridian. Thus sidereal time can be used to express a moment of time, but it is best to think of it as an angle as defined above.

The angular rotation of the celestial sphere in a given **elapsed** time is also called sidereal time. It would be better called **elapsed sidereal time.** Since the angular rotation of the celestial sphere in 24 hr is greater than 360°, more than 24 sidereal hours occur in 24 hr. Accordingly, a sidereal hour passes more quickly than a mean solar hour. Thus, elapsed time can be measured in two different units. The sidereal hour is a convenient unit for the astronomer who uses an instrument in a permanent location with the main axis (the polar axis) inclined and oriented so that it is parallel to the earth's axis. When he knows the sidereal time he can set his instrument on the vernal equinox or, more practically, on any desired body whose S.H.A. is known. For the engineer and for the navigator two different time units only lead to confusion.

Accordingly, in this text both elapsed time and moments of time will be expressed only in units of mean solar time, the familiar units. The rotation of the celestial sphere and position of the celestial sphere will be treated only in terms of angles.

Angles expressed in hours rather than in degrees are usually said to be expressed in time. This expression will also be avoided. The conversion of the value of a given angle from hours (^{h}), minutes of hours (^{m}), and seconds of hours (^{s}) to degrees (°), minutes of degrees (′), and seconds of degrees (″) will be called *hours to degrees* instead of the more usual *time to arc*.

This section on sidereal time would be unnecessary to the reader were it not for the fact that he may be obliged to find data in astronomical publications.

9-12. The Sun. As the earth traverses its yearly path, the background of the celestial sphere behind the sun changes slightly from day to day. Viewed from the earth, the sun apparently completes a circuit around the celestial sphere once in the course of a year. Accordingly, the rate of increase in the G.H.A. of the sun differs slightly from the rate of increase of the G.H.A. of the fixed stars, which move very nearly with the celestial sphere.

Since the earth's axis slants with respect to the plane of the earth's path around the sun, the sun moves north and south on the celestial sphere as the earth moves along its path. This causes the sun's declination to change more rapidly than do the nearly constant declinations of fixed stars. In the course of a year the sun's declination ranges almost 23½° each way north and south of the equator. For these reasons, special tables are necessary for the sun.

The sun does not make its daily passage around the earth at a constant rate. For this reason it is impossible to base elapsed time on the daily passage of the sun. The 24-hr unit used for the day is based on the *average* rate of movement of the sun. In fact the sun is sometimes ahead of and sometimes behind noon L.C.T. by an amount which varies from 0 to about 16 min (see the next section).

The variation in the sun's rate can be explained as follows. The earth's path is an ellipse with the sun at one of the foci. The angular speed of the earth on its path is greater when the earth is nearer the sun. At this period the sun apparently moves faster through the heavens. The apparent change in rate is exaggerated in accordance with the size of the sun's declination. When the sun is further from the equator, more earth meridians are crossed for the same angular change in the sun's position. Thus, in 24 hr the sun may apparently move more or less than 360°.

9-13. The Mechanics of the Sun's Movement. The apparent movement of the sun on the celestial sphere is not easy to visualize. Figure 9-3 is a schematic drawing arranged to illustrate the two causes of the chief irregularities of its motion. Minor irregularities are neglected in this discussion.

The path or **orbit** of the earth revolving around the sun is called the **ecliptic.** It defines a plane called the **plane of the ecliptic.** Figure 9-3 shows the earth half immersed in this plane with the north pole at the top. The earth's axis is inclined at about 23½° from the normal to this plane. About December 21 and June 21 the projection of the earth's axis on the plane of the ecliptic passes through the sun.

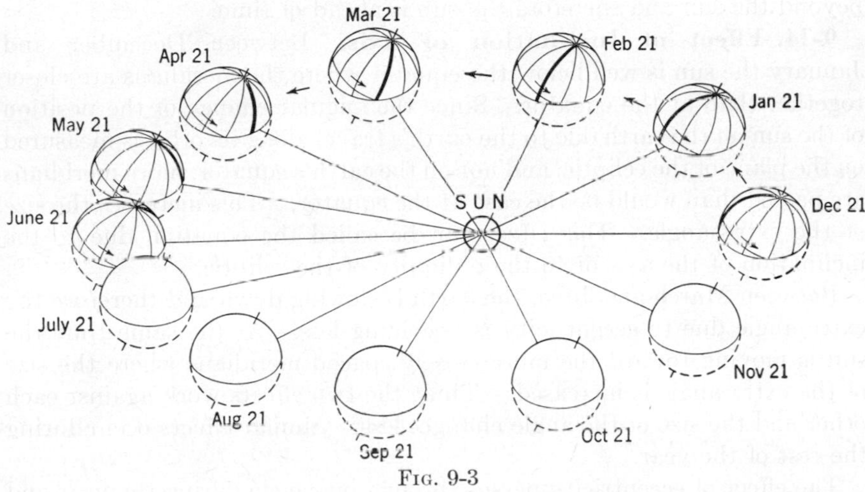

FIG. 9-3

As previously mentioned, the earth's orbit is an ellipse with the sun at one of the foci. The earth is at a point nearest the sun (called the **perihelion** of the earth's orbit) on about January 4 and at a point farthest from the sun (at **aphelion**) about July 6. The earth's rate of angular movement around the sun is, of course, greatest at perihelion and least at aphelion. The movement is counterclockwise when viewed from the north side of the ecliptic.

The earth rotates on its own axis at a very constant rate. This is the rate of the apparent rotation of the celestial sphere. The rotation of the earth is counterclockwise when viewed from the north.

Since the earth rotates and also moves around the sun in the same direction, a given meridian will come under the sun each time the earth rotates 360° *plus a small extra angle* caused by the earth's movement along its orbit and by the inclination of its axis. If the earth's axis were normal to the ecliptic, this extra angle would be equal to the angular movement of the earth around the sun in the time elapsed. The extra angle would then be greatest at perihelion and least at aphelion. This effect can be called the equation due to eccentricity. The effect due to the inclination of the earth's axis is covered in the next section.

Figure 9-3 shows the actual positions of the earth at about the twenty-first-day of each month. The average extra angle for a month would be one-twelfth of 360°, or 30°. The meridian shown by a heavy line in each position is advanced 30° for each month. In December the sun is nearly over this meridian. By January it can be seen that the meridian will have to turn farther to reach the sun so that the sun is behind time. By February the difference is still greater. By May the meridian is slightly beyond the sun and therefore the sun is ahead of time.

9-14. Effect of Inclination of Axis. Between December and January the sun is well below the equator where the meridians are closer together than at the equator. Since the angular change of the position of the sun on the earth due to the earth's travel along its orbit is measured on the plane of the ecliptic, and not on the earth's equator, more meridians are passed than would be the case at the equator. This increases the size of the extra angle. This effect can be called the equation due to the inclination of the axis or to the obliquity of the ecliptic.

Between March and June, the earth is slowing down and therefore the extra angle due to eccentricity is becoming less. At the same time the sun is moving toward the more closely spaced meridians where the size of the extra angle is increased. Thus, the two effects work against each other and the size of the angle changes less. Similar effects occur during the rest of the year.

The effect of eccentricity passes through one cycle during the year and the effect of obliquity passes through two, since the sun crosses the equator twice. The combination of these two effects gives the sun quite an irregular motion.

In order to keep the length of the days the same, elapsed time is based on the average rate of the sun, and the term **mean sun** is used to represent the sun's average position. Local noon (i.e., noon LCT) occurs at any locality when the mean sun crosses the meridian of that place. The **equation of time** is the angle by which the true sun precedes the mean sun. It is defined in the next section. A minus value indicates that the true sun follows the mean sun.

9-15. The Equation of Time. The equation of time is the G.H.A. of the sun minus the G.H.A. of noon LCT. Noon LCT here is the same thing as the mean sun. But for all practical purposes the equation of time may be defined as the elapsed time by which the sun precedes noon LCT. A minus value indicates that the sun follows noon LCT. For example, the following data can be true:

Moment of time	*Position of sun*	*Equation of time*
12^h GCT	G.H.A. 3° 8.4′	$+12^m33.6^s$
12^h GCT	G.H.A. 356°37.1′	$-13^m31.6^s$

Tables are available that give the value of the equation of time for any moment of time throughout the year.

9-16. True Solar Time. True solar time is, for all practical purposes, a moment of time measured in civil hours but based on the sun's passage for the day in question. Noon (12^h) true solar time on any day and at any meridian is the moment the sun crosses that meridian that day. Accordingly, at any given meridian:

$$\text{True solar time} = \text{local civil time} + \text{equation of time}$$

or

$$\text{Local civil time} = \text{true solar time} - \text{equation of time} \qquad (9\text{-}1)$$

THE TRIGONOMETRY INVOLVED

9-17. A Spherical Triangle. A **great circle** is the trace on a sphere of a plane which passes through the center of the sphere. A **spherical triangle** is the figure on a sphere bounded by the arcs of three great circles. It has six parts, three sides and three angles. When any three parts are known, the other three can be found. Both sides and angles are measured in angular units, usually in degrees and minutes. The length of a side is measured by the angle at the center of the sphere between the radii extended to its ends. The size of an angle is measured by the dihedral angle between the planes of the great circles which form it. It may also be measured by the angle between the tangents to the great circles at their intersection.

Any three points on a sphere may be joined by great circles to form a triangle in which no part is greater than 180°. Such a triangle is the one always considered in this text.

9-18. Principles. Figure 9-4 illustrates the spherical trigonometry involved in every observation for position or for true north. It represents the conditions that exist at the moment of observation. P is a pole of the celestial sphere (in this case the north pole),[1] S is the celestial body observed (the arrows represent the path of the body), and Z is the observer's zenith. The lines joining them are arcs of great circles.

9-19. The Zenith. The observer's zenith is a point on the celestial sphere found by projecting the center of the instrument at the time of observation upward along the direction opposite to that of gravity.

9-20. The Astronomical Triangle. The triangle PZS is known as the **astronomical triangle** (see Fig. 9-4). It may be formed west of the

[1] The south pole may be used, but the signs of the latitude and the declination must then be reversed. To avoid confusion, the symbol P in this text is always taken as the north pole.

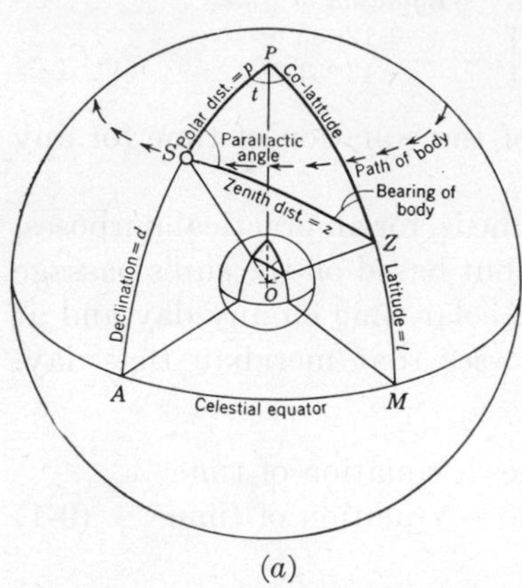

(a)

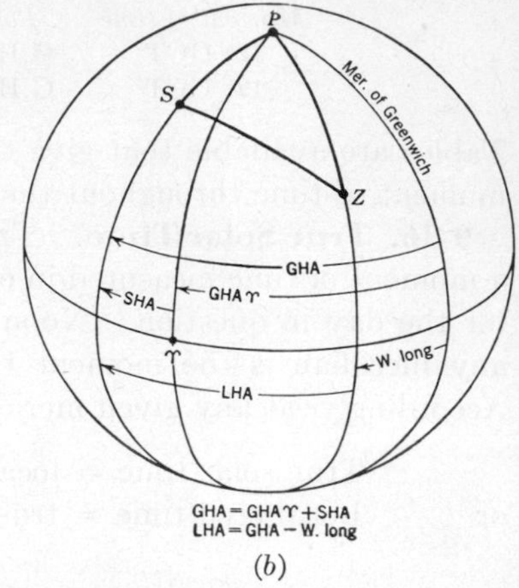

(b)

FIG. 9-4

meridian as shown, or east of the meridian if the body is so located. It is a true spherical triangle formed by great circles, and spherical trigonometric formulas apply. When Z or S, or both, are in the Southern Hemisphere, other arrangements are created. Figure 9-5 shows the twelve possibilities. Note that no angle or side is greater than 180°. All forms of the triangle are solved by the same formulas but the results of the solutions do not indicate whether the body is east or west of the meridian. This can be determined from the L.H.A. described in the next paragraph. The six parts of the triangle are named and described below.

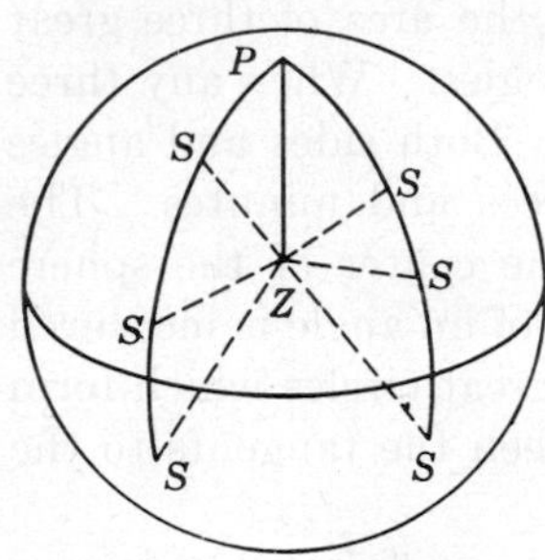

Observer in northern hemisphere

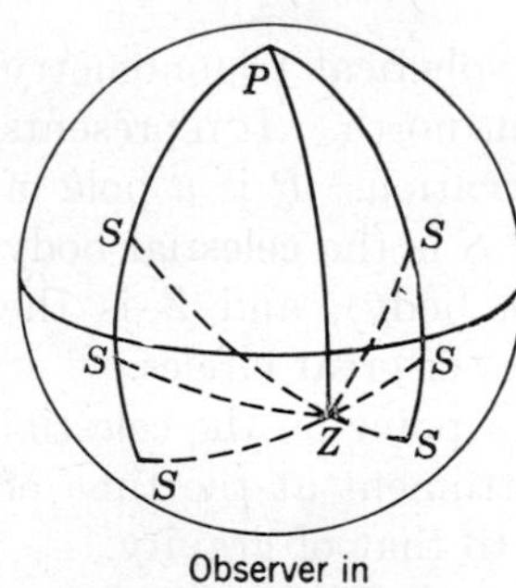

Observer in southern hemisphere

FIG. 9-5

Angle t in Fig. 9-4a is known as the **meridian angle.** The **local hour angle** (L.H.A.) of a body is the angle measured westward around the axis of the celestial sphere from the meridian of the observation to the meridian of the body. The arc MA represents the L.H.A. Obviously from Fig. 9-4b

$$\text{L.H.A.} = \text{G.H.A.} - \text{west longitude} \quad (9\text{-}2a)$$
$$\text{L.H.A.} = \text{G.H.A.} + \text{east longitude} \quad (9\text{-}2b)$$

When the L.H.A. is less than 180°, the body is west of north and $t = \text{L.H.A.}$

When the L.H.A. is greater than 180°, the body is east of north and $t = 360° - \text{L.H.A.}$

Angle Z is the **bearing**[1] of the body S, since it is equal to the horizontal angle between the north direction of the observer's meridian and the direction of the body. It is measured east or west of north according to the position of the PZS triangle.

Angle S is the **parallactic angle.** It is usually unnecessary to use the value of this angle.

Side PS (p) is the **polar distance.** It is equal to 90° minus the declination d of the body S. In the formulas used for many observations sin d is substituted for cos p, etc.

Side PZ is the **colatitude** of Z. It is equal to 90° minus the latitude l of the observer. The formulas are often written using sin l substituted for cos colatitude, etc.

Side ZS (z) is the **zenith distance** of the body S. It is equal to 90° minus true altitude h. The true altitude can be found in the field by observing the altitude of a body and correcting the result for *refraction* and *parallax,* as described in the next section.

9-21. Refraction and Parallax. Figure 9-6 shows the effect of refraction and of parallax. The true altitude is obtained by the following formula:

$$\text{True altitude} = h = \text{measured altitude} - \text{refraction} + \text{parallax} \qquad (9\text{-}3)$$

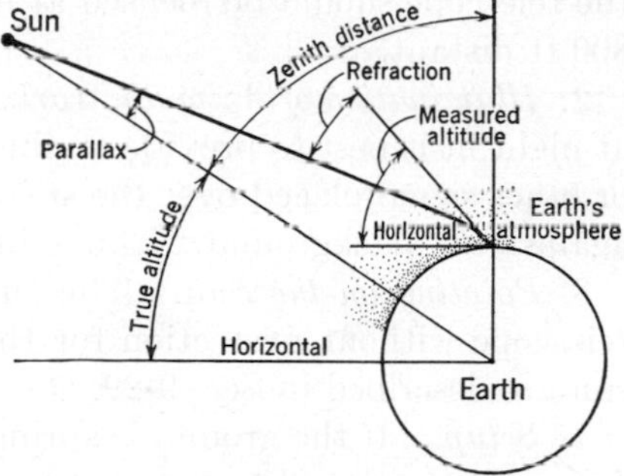

True altitude = measured altitude − refraction + parallax

FIG. 9-6

The value of the refraction is affected by the altitude and to a lesser degree by the air density. Its effect can be estimated for the measured altitude and the ground temperature. The sun's parallax is very small (never more than 0.2′) and the parallax of the stars is infinitesimal. Table 9-2 gives values for refraction and Table 9-3 gives values for the sun's parallax.

FIELD METHODS

9-22. Operations. The field operations consist of measuring vertical and horizontal angles to celestial bodies and recording the precise time that a body is sighted. Horizontal angles must, of course, be measured from an *azimuth mark* on the ground.

Usually several observations are made in a group, the angles and times

[1] The term *bearing* is used here in preference to *azimuth* since the angle measured from the north may be either east or west. It may exceed 90° but it is always less than 180°.

averaged, and only the averages used in computation. At least two groups should be averaged and computed separately for a check. As a rule no single group should include observations made more than 10 min apart, because of the curvature of the path of the body. Corrections for the effect of curvature can be computed as shown in Sec. 9-56. The methods of observing are identical to ordinary surveying methods with the following additional requirements:

1. *Pointing on a Star.* At dusk a star can be observed without special equipment. When the sky is so dark that the cross hairs cannot be seen, a flashlight should be pointed diagonally into the objective lens. The position of the flashlight can be regulated so that the stray light gives sufficient background to show the cross lines but is not so bright that the star is obliterated. Instruments equipped with illuminated cross hairs can of course be used at any time. Before attempting to point at a star, the telescope should be focused at infinity by focusing on a point at least 800 ft distant.

2. *Illumination of Azimuth Mark.* While any light will serve as a mark at night, it is best to provide an illuminated background for a plumb bob or other signal placed over the station. The cross lines can then be seen against the background without illuminating the instrument.

3. *Pointing on the Sun.* The sun should not be observed through a telescope without protection for the eye. Methods for pointing on the sun are described in Sec. 9-22.

4. *Setup.* If the ground is springy, the tripod should be supported on firm stakes. Leveling must be extremely accurate to avoid serious errors. If the instrument is equipped with a telescope level, proceed as follows, after leveling with the plate bubbles:

Step 1. Set the vernier at zero, turn the instrument in azimuth until the telescope is in line with a pair of opposite leveling screws and center the telescope bubble with the vertical motion.

Step 2. Turn 180° in azimuth. If the bubble fails to center, correct half the error with the leveling screws, and the other half with the tangent screw.

Step 3. Repeat steps 1 and 2 until the bubble remains in the center.

Step 4. Turn 90° in azimuth. Center the bubble with the leveling screws.

Step 5. Test the leveling at 0, 90, 180, and 270°.

If the instrument is without a telescope level, perfect the leveling until the plate level that is parallel to the telescope remains in the same position in the tube at all azimuths.

5. *Recording Time.* For certain observations, time must be recorded within two or three seconds. For these observations the watch correction should be determined to the nearest second before, and preferably also

after, the field work. Before comparison, the minute hand should be set so that it coincides exactly with a minute mark when the second hand is at zero. Accurate time is best obtained from radio time signals.

During the observation the transitman calls "tip" when his pointing is perfected. The recorder first reads the second, then the minute, and then the hour. When no recorder is available, a stop watch is of great assistance. The observer starts the stop watch when the pointing on the body is correct and stops it when the second hand of the timing watch is on a 10-seconds mark. The reading of the stop watch is subtracted from that of the timing watch.

Example. If the stop watch was stopped at 6 sec when the timing watch read $8^h42^m30^s$, the stop watch was evidently started at $8^h42^m24^s$. Therefore, the time of observation was $8^h42^m24^s$.

9-23. Star Identification. The ability to identify stars can be acquired best by memorizing a star chart and then finding the stars in

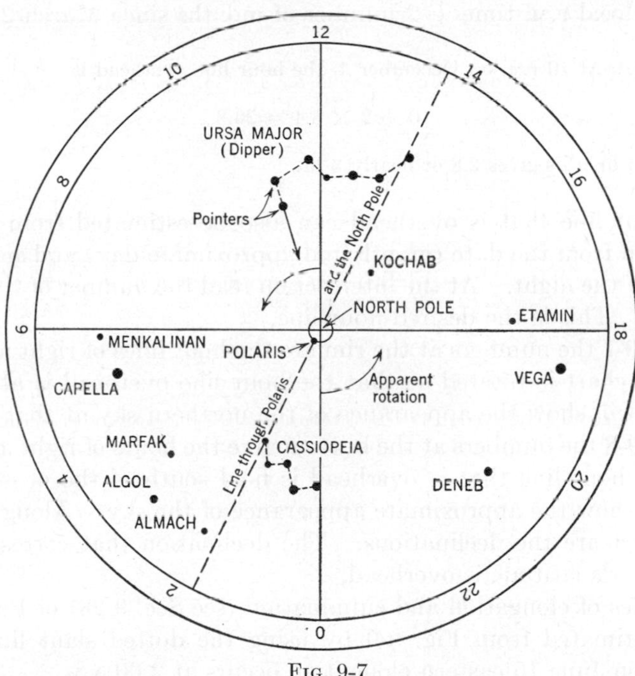

FIG. 9-7

the sky. If possible, the stars shown in the northern sky (Fig. 9-7) should be identified first. The Dipper, Cassiopeia, and Polaris can be found at once and the others then found by their relation to them.

The rest of the heavens are well illustrated in Fig. 9-8. They are shown as they appear when looking south. It will be seen that the curve of the Dipper handle extended passes through Arcturus, a red star, and then through Spica. Spica is in line with the gaff of the mainsail that the constellation Corvus resembles.

The constellations along the equator can be identified next. In the early evening in winter, Orion is instantly recognizable; in summer, Cygnus can easily be found. Others can be picked out by their relationship with these two. As soon as a constellation is recognized, the various stars can be identified.

It is important to be familiar with the stars listed in the Nautical Almanac, as computations for observations on these are simplified by the tables given.

9-24. To Use the Charts. The hour line of right ascension (at the bottom of Fig. 9-8) that is overhead at any time can be estimated by the following formula:

Hour line overhead

$$= \text{local P.M. time} + 2(\text{number of months since March 22}) \qquad (9\text{-}4)$$

Example. At 10 P.M. on December 4, the hour line overhead is

$$10 + 2 \times 8.4 = 26.8$$

Rejecting 24 hr, this gives 2.8 or nearly 3 hr.

The hour line that is overhead can also be estimated from Fig. 9-9. Read down from the date (month and approximate day) and across from the hour of the night. At the intersection read the number of the nearest slant line. This is the desired hour line.

In Fig. 9-7 the numbers at the rim are the hour lines of right ascension. When the chart is rotated so that the hour line overhead is at the top, the chart will show the appearance of the northern sky at that time.

In Fig. 9-8 the numbers at the bottom give the hours of right ascension. When the hour line that is overhead is held south of the observer, the chart will show the approximate appearance of the sky. Along the edge of the chart are the declinations. The declination that corresponds to the observer's latitude is overhead.

The times of elongation and culmination (see Sec. 9-28) of Polaris can also be estimated from Fig. 9-9 by using the dotted slant lines. For example, on June 15 eastern elongation occurs at 2:00 A.M.

METHODS OF DETERMINING DIRECTION

9-25. Basic Types of Observations. Two types of observations are used to determine direction:

1. Observations on a circumpolar star
2. Observations on the sun, an east-west star, or on a planet

9-26. Observations on a Circumpolar Star. A **circumpolar** star is a star near either the north or the south pole. The horizontal angle from a mark to the star is measured and then the direction Z of the star is computed from the star's position at the time of the observation.

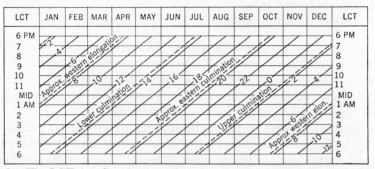

FIG. 9-9. The LCT (see Sec. 9-10) used in the chart is substantially correct for any year. To convert watch time to LCT, add to the watch time 4 min for every degree of longitude that the observation is east of the meridian from which the watch time is reckoned. Subtract if west. Example: watch 10:00 CST (90th meridian); observation 87° west longitude; 10:00 + 3 × 4 = 10:12 (LCT). This conversion is necessary only when watch time and LCT differ substantially.

From the known angle to the star and its direction, the direction of the mark is computed. The known quantities are:

- t (from the time of observation, the S.H.A. from the tables, and the longitude taken from a map)
- l (latitude from a map)
- p (from the tables)

Since the direction of a circumpolar star changes so little, small errors in the determination of time only slightly affect the result.

The procedure is changed in detail depending on what data are available. When no map can be had so that latitude and longitude are both missing, a special observation is required to determine latitude. The four most useful procedures for determining a direction from a circumpolar are given in detail in later paragraphs.

When a circumpolar is observed at **elongation,** that is, when the parallactic angle is 90°, the *time* need not be recorded.

9-27. Observations on an East or West Body. When a circumpolar star cannot be used, the next best method is to observe an east or west body. Such a body is moving comparatively slowly in azimuth but rapidly in altitude. The horizontal angle from the mark to body is measured and either accurate time or the altitude of the body is determined. The direction Z is computed from the following known quantities:

t (from the time, the S.H.A. from the tables, and the longitude taken from a map)
or z (computed from the altitude corrected for refraction, and, if a body in the solar system, corrected for parallax)
l (taken from a map)
p (taken from the tables)

Since the azimuth is changing slowly, the small errors in t or z affect the result only slightly. However, it must be remembered that the determination of z depends, at best, on an estimated value of the refraction. Latitude must be known accurately. When a map is not available, a special observation is necessary to determine latitude.

The direction of true north can be determined without any data whatever by finding the direction of the same star at any altitude when it rises and at the same altitude when it sets. The meridian is halfway between two corresponding directions. The method is not very accurate and it requires two observations on the same night. However, several altitudes can be obtained each time and the average of the pairs of directions determined.

The method of determining direction from an east-west body is given in detail for the sun, in later paragraphs.

Method 1. By Observation of Polaris at Elongation

9-28. Description. True north can be determined by a Polaris observation at elongation when accurate time is not available. The method is simple and, with care, should give results within ± 0.5 minute. However, it often must be performed at inconvenient hours.

Polaris follows a circular path around the pole similar to that shown by the arrows in Fig. 9-4. Viewed from the earth, the path is circular and the motion is counterclockwise. The points in the path where Polaris is farthest west or farthest east are called *western* and *eastern elongation* respectively. When the parallactic angle S is 90°, the star is at elongation.

9-29. Directions for Observation. Required to determine the bearing of a mark B from transit station A (see Sec. 9-22). The instrument must be in perfect adjustment unless the star can be observed with the telescope both direct and reversed as mentioned below. From observation of the sky or by means of Fig. 9-7, estimate the time of that elongation which occurs during dark. Set up at A. Set the horizontal vernier at zero. Sight B with the lower motion. Point on Polaris with the upper motion.

Before western (eastern) elongations the star will be moving down (up) and also moving slightly toward the west (east).

Follow the star until the westward (eastward) movement ceases and the only movement is vertical. Read the horizontal vernier. The motion east or west is imperceptible for about 10 min. It therefore may be possible to repeat the angle with the telescope reversed, thus increasing the accuracy and eliminating errors of instrument adjustment.

Example. Date, Sept. 10, 1950; latitude (from map), 40°20′; clockwise angle from mark B to Polaris, 75°20′. Use the following formula:

$$\text{Bearing of Polaris (in minutes)} = \frac{\text{polar distance (in minutes)}}{\text{cos latitude}} \tag{9-5}$$

$$\text{Polar distance Polaris, Sept. 10, 1950} = 0°58.20'$$

$$Z = \frac{58.20}{0.762} = 76.4' = 1°16.4'$$

$$\text{Bearing } AB = \text{N}76°36'\text{W} \quad \text{(if elongation was western)}$$
$$= \text{N}74°04'\text{W} \quad \text{(if elongation was eastern)}$$

Method 2. By Observation of Polaris at Any Time

9-30. Description. The method described below is designed for accurate observation. It is quick and convenient, and the computations are simple. It can be made at any time from dusk to dawn. Evening twilight is recommended. Pointings can be made on the star at that time while the instrument can be set up and the angles read by daylight. Polaris can easily be seen through the telescope an hour before sunset. To find the star, estimate its place in the sky with respect to the pole to ±¼° by means of Fig. 9-9. The telescope can be pointed to the pole by setting the vertical circle at the latitude and the compass needle at the magnetic bearing of the pole. By correcting these settings, according to its estimated position, the star can be brought into the field of view. A signal should be placed on the ground under the star about 30 ft from the instrument, to aid in finding the star again.

9-31. Directions for Observation. Required to determine the bearing of mark B from transit station A. Determine the watch correction to the nearest second (see Sec. 9-22). Set up at A. Turn the angle from B to the star using six repetitions of the angle, three with the telescope in its normal position and three in its reversed position, i.e., 3 D.R. Record the time to the nearest second each time the cross line is brought on the star. Average the time and divide the total angle by 6. At least two separate observations should be made.

Example. Time of observation, May 5, 1950; watch reading, $8^h28^m23^s$ P.M., 90th meridian time; watch known to be 2^m03^s slow; latitude (from map) N42°22.6′;

longitude (from map) W92°58.3′; clockwise angle, mark to star 25°53.0′. Find the L.H.A. and t (see Sec. 9-20) as shown below.

Watch time	$8^h28^m23^s$	P.M.
Watch correction (slow is plus)	2 03	slow
Standard time (90th meridian)	$8^h30^m26^s$	P.M.
Correction to 24-hr basis	+12	
90th meridian time	$20^h30^m26^s$	
Correction for time zone	+ 6	
GCT (Sec. 9-10)	$26^h30^m26^s$	
GCT May 6, 1950	$2^h30^m26^s$	
G.H.A. (Secs. 9-7 and 9-8) 0^h May 6, 1950	196°28.0′	
Correction for 2^h30^m (Table 9-1)	+ 37 36.2	
Correction for 26^s (Table 9-1)	+ 6.5	
G.H.A.	234°10.7′	
Less west longitude (from map)	− 92 58.3	
L.H.A. (Sec. 9-20)	141°12.4′	
t = L.H.A. or 360 − L.H.A. (use smaller)	141°12.4′	

9-32. Computation. Three optional procedures for the computation are given below. Procedure A is the simplest but it requires the use of the K & E Ephemeris. Procedures B and C are added to give a solution under different conditions of latitude and instrumentation.

Procedure A. For a precise computation, if the station is between N. Lat. 25° and N.Lat. 50° (latitudes of the United States), use the formula

$$Z = \frac{\sin t}{\cos h} p \qquad (9\text{-}6)$$

where Z = bearing of Polaris expressed in minutes of arc with a computational accuracy of $\pm 0.02'$ (see Sec. 9-20)

t = meridian angle to the nearest minute (see Sec. 9-20)

h = true altitude to the nearest minute from Table 6 in K & E Ephemeris[1] (see Sec. 9-20)

p = polar distance expressed in minutes to the nearest 0.02′ (see Sec. 9-20)

Example

p (Sec. 9-20) for Polaris May 6, 1950	58.24′
l (from map)	42°22.6′
Correction (K & E Table 6, L.H.A. = 141.2°)	−45.4
h (Sec. 9-20)	41°37.2′

$$Z = \frac{\sin 141°12'}{\cos 41°37'} \times 58.24' = \frac{0.6266}{0.7476} \times 58.24' = 48.81'$$

Procedure B. For latitudes below 25° or above 50° where Polaris can be observed, greater accuracy can be achieved by measuring h directly,

[1] If not available, use Procedure B or C.

rather than by using a correction of l, provided the instrument has a full vertical circle. Use the same formula Eq. (9-6) as above, but observe the altitude and the temperature and obtain the true altitude h by correcting for refraction. The altitude should be observed once direct at the first pointing on Polaris and once reversed at the last pointing on Polaris. The average is used. Assume the average measured altitude is 21°10.0′ and the temperature is 70°.

Measured altitude..........................	21°10.0′
Less refraction correction (Table 9-2)..........	−2.4
h..	21°07.6′

Procedure C. When procedure *B* cannot be followed because the instrument has no vertical circle, use the following formula:

$$Z \text{ (minutes)} = \frac{3438 \sin t}{\cos l \cot p - \sin l \cos t} \tag{9-7}$$

Example

$$Z \text{ (minutes)} = \frac{3438 \sin 141°12'}{\cos 42°23' \cot 58.21' - \sin 42°23' \cos 141°12'}$$

$$= \frac{(3438)(0.6266)}{(0.7386)(59.02) - (0.67)(-0.78)}$$

$$Z = 48.83'$$

9-33. To Compute the Bearing of the Mark. Having computed Z by any one of the above three procedures, continue as follows (see Fig.

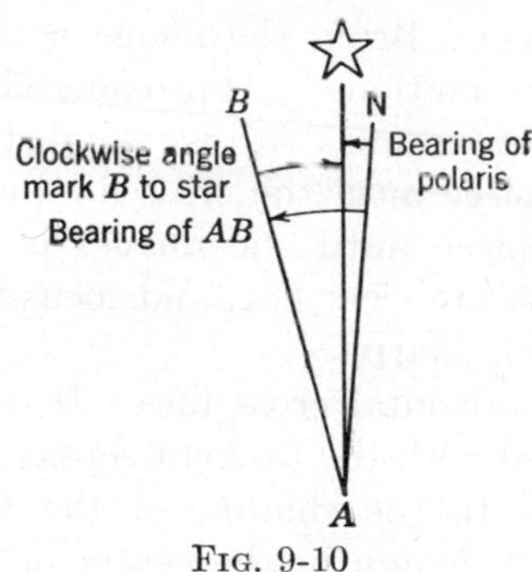

FIG. 9-10

9-10): Since L.H.A. is less than 180°, the star is west of north. Using the value 0°48.8′, the bearing of mark *B* is computed as follows (sec Sec. 9-20):

Clockwise angle mark *B* to star..........	25°53.0′
Bearing of Polaris......................	+N0 48.8 W
Bearing of *AB*........................	N26°41.8′W

Note: When neither accurate time nor longitude are available, the G.H.A. and the *t* of Polaris can be obtained by the method outlined in Sec. 9-61.

Method 3. By the Altitude of the Sun

9-34. Description. The bearing of a line can be determined to within $\pm 2'$ by an observation on the sun. The average of several such observations will, of course, give a higher accuracy.

9-35. Directions for Observation. Required to find the bearing of mark *B* from transit station *A* (Sec. 9-22). The observation should be made when the sun's altitude is at least 20° and at least 2 hr before or after noon. Correct time within ± 10 min must be observed. The watch should be checked accordingly.

Set up at *A* using the telescope level as described in Sec. 9-22, paragraph 4. Set the *A* vernier at zero and, with the telescope direct, sight *B* using the lower motion.

For direct observation of the sun a dark filter must be used. A method preferred by many, however, is based on using the image of the sun that is formed on any white surface held behind the eyepiece. The latter method is described below, but the instructions can also be followed for direct observations through the telescope. The difference in the procedure is obvious.

9-36. To Sight the Sun. Have the recorder hold the back (white) page of the field book about 6 in. behind the eyepiece. With the upper and vertical clamps free, turn the telescope until its shadow on the page is circular. As the telescope is moved into this position, the sun's image will flash across the page. Bring the image within the shadow of the telescope and clamp both motions. Approximately center the sun in the shadow with the tangent screws. Adjust the distance between the field book page and the eyepiece until the cross lines can be seen most effectively. Focus the eyepiece until the images of the cross lines which appear on the sun's disk are clear cut, and focus the objective until the edge of the sun's image is sharp.

Identify the center horizontal cross line. It will be noticed that, as the sun's image is moved with the tangent screws, the cross lines remain stationary with respect to the shadow of the telescope. The center horizontal cross line will remain in the center of the shadow, while the stadia lines will be found at the top and bottom respectively.

To make the observations, the sun's image may be centered on the cross lines or brought tangent to them. The first method is simpler to perform; the second method is more accurate. If the second method is used, half of the observations should be made with the sun in one quadrant and the other half with the sun in the opposite quadrant. The mean of the results will then be the same as that which would have been obtained had the cross lines been centered (see Fig. 9-11).

While either pair of opposite quadrants may be used, the scheme illustrated in Fig. 9-11 is recommended. When using an erecting instrument, the sun's image moves in the same direction as a shadow that is cast by the sun. Figure 9-11 shows this motion. If it is turned upside down Fig. 9-11 will illustrate an observation with an inverting instrument. In the Southern Hemisphere the horizontal components of the sun's motion will be in the opposite direction from those shown in Fig. 9-11. To make the observation, a quadrant is chosen in which the sun's image is moving toward one cross line and away from the other. The disk is kept tangent to the line towards which it is moving but allowed to make its own tangency with the other line. When the moment of tangency occurs, stop moving the telescope and read the circles.

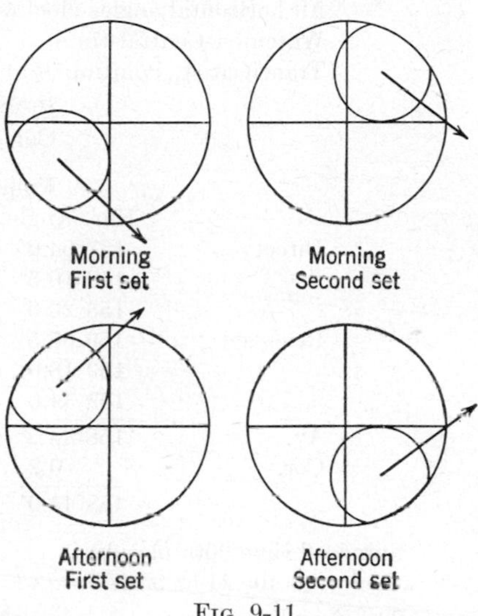

FIG. 9-11

Make three pointings with the telescope direct and three with the telescope reversed, without moving the circle and therefore without using the lower motion. Read the A vernier and the vertical angle vernier both to the nearest vernier division at each pointing. Record the time of the first and last pointings. Finally sight mark B with the telescope reversed and read the A vernier. It should read 180° within ±1 minute. Subtract 180° from all reversed pointings.

Find the averages, respectively, of the horizontal readings, the vertical readings, and the time. Correct the horizontal readings by the average of the two readings on the mark B. Take the air temperature or estimate it.

This method is also applicable to star observations as described in Sec. 9-57. When a star is observed, the barometric pressure should be observed for greater accuracy.

9-37. Computation. Use the following formulas:

Form for machine computation:

$$\cos Z = \frac{\sin d - \sin h \sin l}{\cos h \cos l} \qquad (9\text{-}8a)$$

Form for logarithmic computation:

$$\cos Z = \frac{\sin d}{\cos h \cos l} - \tan h \tan l \qquad (9\text{-}8b)$$

where Z (Sec. 9-20) is the sun's bearing. It may be greater than 90°. A minus value of cos Z indicates a value of Z greater than 90°. If the observation was taken in the morning, the bearing is east of north and vice versa. d (Sec 9-20) is the declination taken from Table 1, to the nearest 0.1′. h (Sec. 9-21) is the measured altitude to the nearest 0.1′ corrected for refraction and parallax. l is the latitude to the nearest 0.1′ taken from the map. Five-place tables give sufficient accuracy.

Example. Assume latitude (from map) N38°10.1′, temperature 70°F; the field notes are:

Date: May 6, 1950

All horizontal angles clockwise
Watch on Central time

Transit at A, Pointing B at start	0°00′00″
B at end	0 00 30
Cor.	0°00′15″

Pointing Sun

	Hor. Angle	Vert. Angle	Time
Direct	157°54.0′	33°48.0′	3:40 P.M.
	158 10.5	33 26.5	
	158 23.0	33 12.0	
Reversed	159 05.5	32 50.0	
	159 18.0	32 31.0	
	159 34.0	32 09.0	3:50 P.M.
Av.	158°44.2′	32°59.4′	3:45 P.M.
Cor.	0.2		
	158°44.0′		

Standard time 90th meridian	3:45 P.M.
Correction for 24-hr basis	+12
90th meridian time	15:45
Correction for time zone	+ 6
GCT (Sec. 9-10)	21:45
Sun's d 0^h May 6, 1950	+16°18.8′
Change since 0^h; 21.75 × 0.71	+ 15.4
Sun's d	+16°34.2′
Measured altitude	32°59.4′
Refraction and parallax (Tables 9-2 and 9-3)	−1.3
True altitude h	32°58.1′

$$\cos Z = \frac{\sin 16°34.2' - \sin 32°58.1' \sin 38°10.1'}{\cos 32°58.1' \cos 38°10.1'}$$

$$\cos Z = \frac{0.28519 - (0.54418)(0.61797)}{(0.83897)(0.78620)}$$

$$\cos Z = -0.07747$$
$$Z = 94°26.6'$$

See Fig. 9-4.

Sun's bearing	S85°33.4′W
Clockwise angle B to sun	158 44.0
Bearing of mark B	S73°10.6′E

9-38. Use in Southern Hemisphere. The formula can be used without change for observations in the Southern Hemisphere. South latitude as well as south declination must be taken as minus. Z will be measured from the north, as before.

Method 4. By Use of the Solar Attachment

9-39. Description. A **solar attachment** is a special device which can be mounted on a transit that can be made to solve the astronomical triangle automatically. There are several forms of this device. Essentially it consists of a supplementary telescope sight mounted on three

FIG. 9-12. A solar attachment on a transit. (*W. & L. E. Gurley.*)

axes: a **latitude axis,** a **polar axis,** and a **declination axis.** The instrument is set according to the latitude of the place of observation and the sun's declination. Then when it is aimed at the sun, the polar axis will point north. Figure 9-12 shows one type of solar attachment.

9-40. Theory. Figure 9-13 shows schematically the operation of one type of solar attachment. The latitude axis is parallel with the horizontal axis of the transit. The observer looks down through the supplementary telescope mounted on the latitude axis at a slanted mirror. The mirror can be rotated around the axis of the telescope which is the polar axis. The slant of the mirror can be changed around the declination axis.

For purposes of demonstration, assume that the supplementary telescope is set so that its inclination equals the latitude, as shown, and the main telescope is turned to the north. The polar axis will then be parallel to the earth's axis. The mirror is attached to a large arc (not shown) so that its slant can be accurately established. Assume that it is set so that the line of sight makes an angle with the polar axis of 90° less the sun's declination, as shown. The figure shows the sun on the meridian

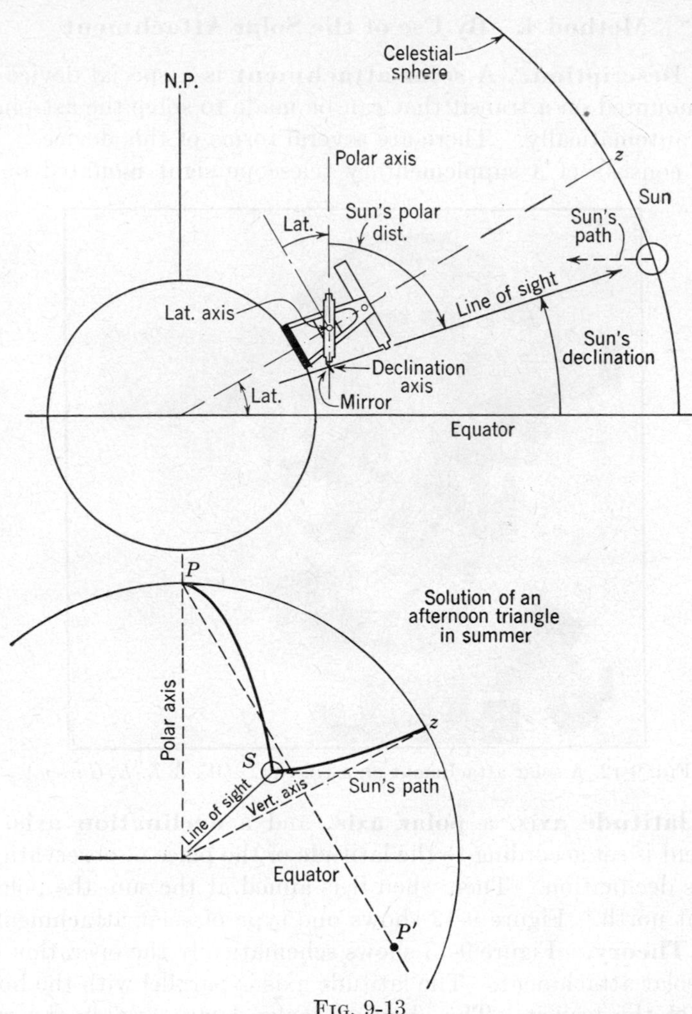

FIG. 9-13

of observation, that is, at local noon. As the sun moves westward, it follows a path parallel to the equator. The line of sight could be kept on the sun by rotating the mirror on the polar axis. After the sun has moved (assume a reasonable distance; 30° or more), any movement in azimuth of the main telescope left or right of true north would immediately move the line of sight of the supplementary telescope off the sun. As the main telescope is moved in azimuth, the polar axis would describe an arc PP' on the celestial sphere with the zenith as its center (note the lower part of Fig. 9-13). The radius of this arc would be the angular distance PZ as shown on the lower part of Fig. 9-13. Since the angle between the polar axis and the line of sight is set equal to the sun's polar distance PS, the line of sight could only be brought to bear on the sun when the polar axis is pointed at the pole P or at a single other point P' such that the arc $P'S = PS$. P' could never be mistaken for the pole and, moreover, if P' were used by mistake, the sun could not be followed by rotating the mirror on the polar axis, as the polar axis would not be parallel to the earth's axis.

In other words, considering the astronomical triangle, the side PZ is set in the instrument when the latitude is set, PS is set in when the sun's declination is set. Consider PZ and PS, then, as two fixed arcs pivoted at P and with Z fixed on the celestial sphere. The only way that PS can be made to reach the sun is by placing P at the pole.

9-41. Procedure for Observation. The observation should be made when the sun's altitude is at least 20° and at least 2 hr before or after noon. Correct time within ±10 min must be observed. The watch must be checked accordingly.

Compute the sun's declination as demonstrated in the previous method. The declination must be corrected for the effect upon it of refraction and parallax. Table 9-4 has been designed to give this special correction. The following example shows the use of this table.

Example. Latitude N28° (from map), sun's $d = -20°$; time about 8 A.M. In Table 9-4 under latitude 28°, "Hours from noon" line 1; column −20°; take out +2.2′. The declination to set on the instrument is then $-20° + 2.2' = -19°57.8'$.

Set the latitude and the declination on the instrument according to the directions for the particular solar attachment. Set the azimuth vernier at zero. Turn the supplementary line of sight to its approximate direction (sometimes hours are marked to aid this operation). Sight the sun by using the azimuth motion and rotating the supplementary line of sight around the polar axis. When the sight is correct, the sun can be followed for a few minutes by rotating the supplementary line of sight around the polar axis.

The main telescope will now point north. Any desired bearing can be laid off.

THE DETERMINATION OF LATITUDE

9-42. Description. The latitude of **the point of observation** is required for several methods of determining a true bearing. For these observations, when it cannot be accurately determined from a map, latitude must be determined in the field.

Latitude can be determined, without accurate time and independent of other observations, by measuring the altitude of the sun or a star when it reaches its culmination.

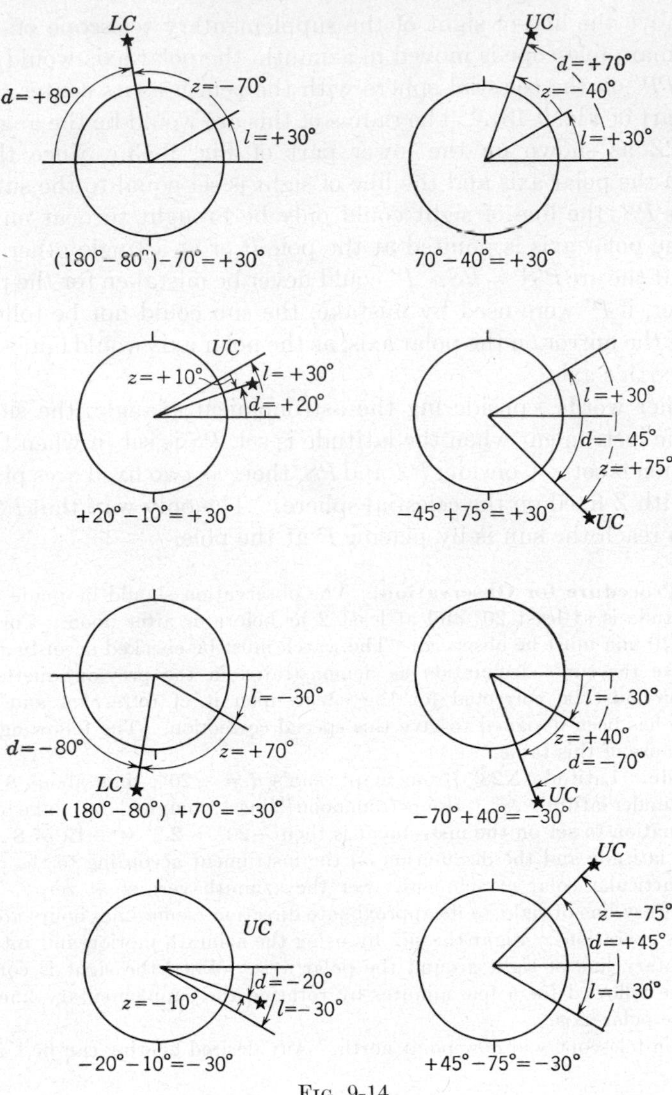

FIG. 9-14

A heavenly body is at upper culmination when it is on the meridian of the observation. This is the highest point that it reaches in the sky. At upper culmination it is moving west. It is at lower culmination when it is on the meridian 180° from the observer. This is the lowest point it reaches in the sky. At lower culmination it is moving east.

9-43. Directions for Observation. The instrument must be in good adjustment. Choose a time when any star, or the sun, is approaching a point north or south of the point of observation. Follow the body with the cross lines to the point of maximum (or minimum) altitude.[1] When this point is reached, stop moving the cross lines and record the altitude.

9-44. Computation. The observed altitude is corrected for refraction (and for semidiameter and parallax if the sun is observed) and the zenith distance (Sec. 9-20) is derived from it. The formulas used are the following (see also Fig. 9-14):

$$z = 90° - h \tag{9-9}$$

Then at upper culmination

$$l = d + z \tag{9-10}$$

and at lower culmination

$$l = (180° - d) + z$$

where l is the latitude, d is the declination (Sec. 9-4), and z is the zenith distance. The signs of each of these terms must be carefully taken into account as follows: (1) south latitudes are minus; (2) south declinations are minus; and (3) zenith distances of bodies north of the observer are minus in both the Northern and Southern hemispheres.

Note: In Eq. (9-10) the parenthetical term $(180° - d)$ is computed as though d were plus. The result is then given the actual sign of d.

One of the 64 stars for which data are available passes the meridian nearly every few minutes (see Fig. 9-8). It is unnecessary, therefore, to predetermine when the observation must be made. If it is desired to observe Polaris or the sun, it is convenient to estimate the proper time to make the observation. The time of upper or lower culmination of Polaris can be estimated from Fig. 9-7. The time of upper culmination of the sun is noon, true solar time. It is estimated as follows, when approximate longitude is available:

$$\text{GCT of true solar noon} = 12 \text{ hr} + \text{west longitude} - \text{equation of time} \tag{9-11}$$

Example 1. Estimating the Time of the Sun's Upper Culmination. Assume longitude approximately W102°; observation to be made May 6, 1950; temperature 40°.

Noon, true solar time		12.0 hr
Add west longitude converted to hours	$^{102}/_{15}$ =	+ 6.8
Less equation of time (Sec. 9-15)		− 0.1
GCT (Sec. 9-15) of true solar noon at W102°		18.7 hr
Zone correction to mountain time (105th meridian)		− 7.0
Mountain time		11.7 hr

[1] A maximum occurs at *upper culmination*, a minimum at *lower culmination*. Lower culmination can be observed only on bodies having a declination which, when added to the latitude of the observer, gives a value numerically larger than 90°.

Example 2. Sun Observation. Assume that the observation described above resulted in the following observed altitude of the sun's upper limb (top):

Observed altitude		58°29.8′
Less sun's semidiameter May 6, 1950		− 15.9
Observed altitude of sun's center		58°13.9′
Less refraction and parallax (Tables 9-2 and 9-3)		− 0.6
True altitude h		58°13.3′
$z = 90 - h$ (z is taken as plus when body is south of observer)		+31°46.7′
Sun's d 0^h May 6, 1950	+16°18.8′	
Change since 0^h; 18.7 × .71	+ 13.3	
Sun's d		+16 32.1
Latitude		+48°18.8′

Example 3. Polaris Observation. Assume observation on Polaris at lower culmination May 6, 1950; temperature 50°; observed altitude 35°21.9′.

Observed altitude		35°21.9′
Less refraction (Table 9-2)		− 1.3
True altitude, h		35°20.6′
$z = 90 - h$ (z is taken as minus when body is north of observer)		−54°39.4′
Polaris d May 6, 1950	+89°1.8′	
180° − d (lower culmination)		+90 58.2
Latitude		36°18.8′

DETERMINATION OF POSITION

9-45. Line of Position. Latitude and longitude can be determined separately as described in Secs. 9-42 and 9-61. However, when both are required the method described here should be used. The principle involved is based on **Sumner's line of position** and the procedure and method of computation is known as the **method of Marq St. Hilaire.** The method is used almost always for navigation and very often on land. A single complete observation with a transit will give position within a mile and, when a special land instrument called an **astrolabe** is used, astronomic position can be determined within 100 ft.

9-46. Observations. The altitude and time of observation is observed for two bodies differing in azimuth, preferably by as nearly 90° as possible.

9-47. Computations. The latitude and longitude are assumed at the best values that can be estimated. If the error in assumption is too great, the computations will give only an approximation of the position. With this approximation, however, a second computation will usually give the correct results. The altitude and azimuth of each of the two bodies are then computed using the *assumed position* and the *time of observation.* The known quantities are the following:

t (from the time, the S.H.A., and the assumed longitude)
l (the assumed latitude)
p (from the tables)

If the computed altitudes should agree with the observed altitudes of the bodies, the assumed position would be correct. If the computed altitude of a body is too small, the true position must be more nearly under that body, that is, **toward** that body from the assumed position and vice versa.

9-48. Chart. A chart like that shown in Fig. 9-15 is made up. The lines of minutes of latitude are straight and uniformly spaced at any

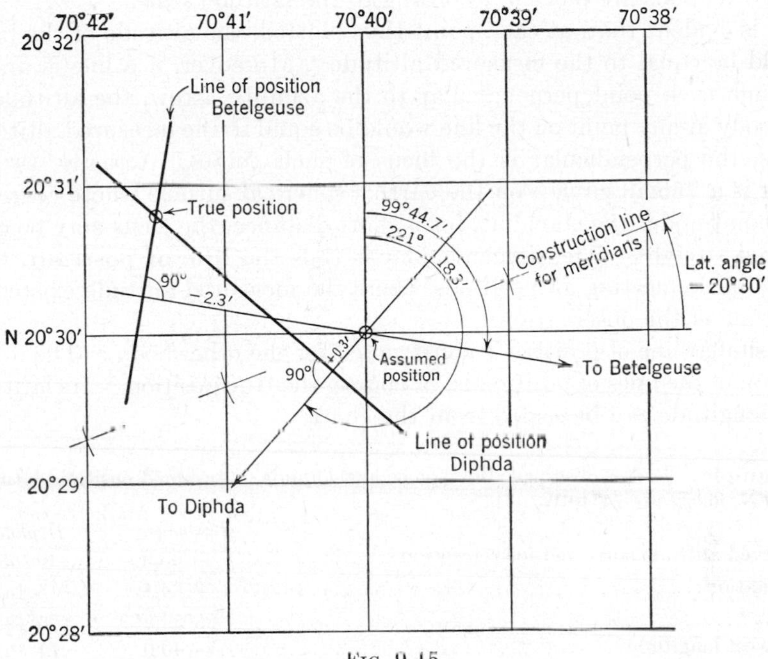

FIG. 9-15

convenient scale. The lines of minutes of longitude are drawn straight and at right angles to the lines of latitude. They are spaced at intervals equal to the intervals used for latitude multiplied by the cosine of the assumed latitude. Such a chart is a very nearly correct representation of any small part of the earth's surface. The spacing for the meridians can be established graphically. A line is drawn that slopes with respect to the horizontal at the angle of the latitude and it is marked off in units of latitude minutes. The longitude lines are drawn through these points. The latitude and longitude lines are given convenient values so that the assumed position comes near the center of the chart.

9-49. Graphical Solution. A long arrow is drawn through the assumed position, pointing along the computed azimuth of each of the bodies. Each measured altitude is corrected for refraction and parallax to obtain h. The angular position correction is computed for each body thus:

$$\text{Correction} = \text{observed } h - \text{computed } h \qquad (9\text{-}12)$$

A plus correction is plotted along the arrow toward the body from the assumed position. A minus correction is plotted from the assumed position away from the body. Since the space between minute lines of *latitude* correctly represents the lengths of minutes on the earth's surface, the corrections are plotted according to the latitude scale.

It is evident that, at each point thus plotted, the altitude of the body would be equal to the measured altitude. Moreover, if a line is drawn through each point perpendicular to the azimuth arrow, the altitude of the body at any point on the line would be equal to the measured altitude. Thus, the perpendicular is the locus of such points. Actually, such a locus is a "small circle" on the earth's spherical surface whose center is the point under the star; but, for a short distance, the locus may be considered straight. The perpendicular is thus the **line of position,** that is, a line connecting all positions where the measured altitude existed at the time of the observation.

A similar line of position is constructed for the other body. The intersection of the lines of position is, of course, the true position. Its latitude and longitude can be scaled from the chart.

Example. Bodies observed, Betelgeuse and Diphda. Assumed position: latitude 20°30′N, longitude 74°40′W.

	Betelgeuse	*Diphda*
Observed altitude corrected for refraction	43°53.9′	40°02.8′
Declination	+ 7 24.1	−18 15.8
G.H.A.	28°58.0′	106°49.0′
Less west longitude	−74 40.0	−74 40.0
t	−45°42.0′	+32°09.0′

COMPUTATION

Use the following formulas:

$$\tan \tfrac{1}{2}(Z + S) = \frac{\cos \frac{1}{2}(l - d)}{\sin \frac{1}{2}(l + d)} \cot \frac{t}{2}$$

$$\tan \tfrac{1}{2}(Z - S) = \frac{\sin \frac{1}{2}(l - d)}{\cos \frac{1}{2}(l + d)} \cot \frac{t}{2}$$

$$\tan (45 - \tfrac{1}{2}h) = \frac{\sin \frac{1}{2}(Z + S)}{\sin \frac{1}{2}(Z - S)} \tan (l - d)$$

Check:
$$\frac{\cos h}{\sin t} = \frac{\cos d}{\sin Z} = \frac{\cos l}{\sin S}$$

Formula	Betelgeuse	Diphda
$\tan \frac{1}{2}(Z+S)$	$= \frac{0.993474}{0.241089} 2.37311$	$= \frac{0.943329}{0.019517} 3.47026$
	$= 9.77906$	$= 167.731$
$\frac{1}{2}(Z+S)$	$= 84°09'41''$	$= 89°39'30''$
$\tan \frac{1}{2}(Z-S)$	$= \frac{0.114056}{0.970503} 2.37311$	$= \frac{0.331859}{0.999810} 3.47026$
	$= 0.278894$	$= 1.15186$
$\frac{1}{2}(Z-S)$	$= 15°35'01''$	$= 49°02'12''$
Z	$= 99°44'42''$	$= 138°41'42''$
Azim.	$= 99°44.7'$ (t is $-$)	$= 221°18.3'$ (t is $+$)
S	$= 68°34'40''$	$= 40°37'18''$
$\tan(45° - \frac{1}{2}h)$	$= \frac{0.994812}{0.268644} 0.114805$	$= \frac{0.999982}{0.755129} 0.351796$
	$= 0.425133$	$= 0.465867$
$45° - \frac{1}{2}h$	$= 23°01'55''$	$= 24°58'45''$
h	$= 43°56'10''$	$= 40°02'30''$
$h_o - h_c$	$= -0°2.3'$	$= +0°00.3'$
$\frac{\cos h}{\sin t}$	$= \frac{0.720114}{0.715693} = 1.00618$	$= \frac{0.765577}{0.532138} = 1.43868$
$\frac{\cos d}{\sin Z}$	$= \frac{0.991667}{0.985571} = 1.00619$	$= \frac{0.949626}{0.660067} = 1.43868$
$\frac{\cos l}{\sin S}$	$= \frac{0.936672}{0.930914} = 1.00619$	$= \frac{0.936672}{0.651061} = 1.43869$

The intersection of the lines of position is at Lat. 20°30.8′N, Long. 70°41.5′W.

9-50. Azimuth. It might be said that since the azimuth computations are based on the assumed rather than the true position, the azimuth will be in error. However, quite a large error in azimuth will have little effect on the result. The azimuth is computed accurately only to obtain a check.

9-51. Other Methods of Computation. Many methods have been devised to reduce the length of these computations. Chief among these are the following:

1. United States Navy Department Hydrographic Office H.O. No. 214, Tables of Computed Altitude and Azimuth, available from the Superintendent of Documents.

2. United States Navy Department Hydrographic Office publication H.O. No. 211, Dead Reckoning Altitude and Azimuth Table by Ageton. For sale by the Hydrographic Office, Washington, D.C.

3. Line of Position Book, by Philip Van Horn Weems. U.S. Naval Institute, Annapolis, Md.

9-52. More Bodies Observed. Usually at least three and sometimes many more bodies are observed. The mean of the intersection of the lines of position is used and any erroneous value can be discovered and rejected. The bodies should be chosen so that they are uniformly distributed in azimuth. This gives a stronger solution, and since refraction is nearly the same in every direction, the effects of incorrect estimates of its values tend to cancel each other. In Chap. 21 is a method of determining position from observations on a number of stars by the method of least squares.

9-53. The Astrolabe. An astrolabe, as the term is used today, is a device that indicates very precisely when a heavenly body has an observed altitude of exactly 60°. Two types of astrolabes are in general use, the **prismatic astrolabe** and the **pendulum astrolabe.**

Figure 9-16 shows a schematic view of the prismatic astrolabe. The instrument consists of a telescope with a 60° prism and a mercury pool mounted in front of the objective lens, as shown. Rays A and A' come from a body whose vertical angle is greater than 60°. The rays that strike the prism directly form an image at a' and those that are reflected

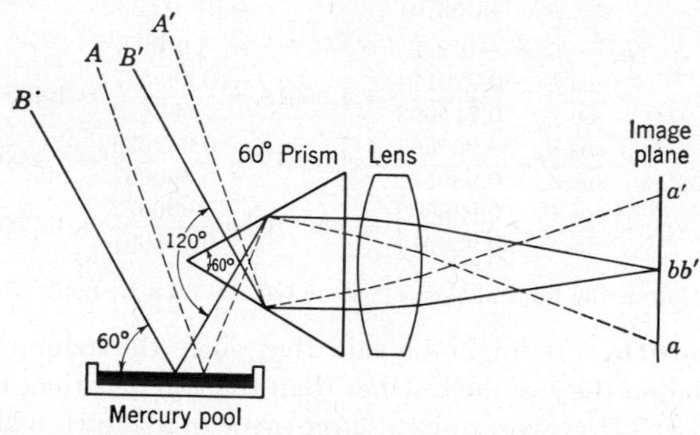

FIG. 9-16

from the mercury pool form an image at a. As the vertical angle decreases, the images move toward each other until, when the vertical angle is 60°, the images will coincide at bb'.

The horizontality of the telescope does not affect the observation. It only changes the position of the point of coincidence on the image plane. The images will still coincide when the vertical angle is 60°. Rotating the prism in the plane of the paper or in a horizontal plane merely moves the position of coincidence. Rotating it in a plane perpendicular to the paper makes the images pass each other to one side or the other. When the geometry is completely worked out it will be found that any triangular prism (except for slight effects of refraction) will emit rays parallel to each other if they enter the prism at twice the angle between the faces through which they enter. In a 60° prism the effects of refraction are also eliminated.

Accordingly, even if the prism is not made with a forward angle of exactly 60°, the angle between two entering rays is *always the same* (though not 120°) when the images are superimposed. Since the surface of the mercury pool is always exactly horizontal, the ray that is reflected from the pool makes an angle with the direct ray from the body equal to exactly twice the vertical angle. Thus, with this device, when the two images of a star are superimposed, the vertical angle is *always the same* for a given prism and very nearly equal to 60°.

The instrument is mounted so that it can be turned in azimuth and a graduated circle is provided so that it can be pointed at any desired azimuth.

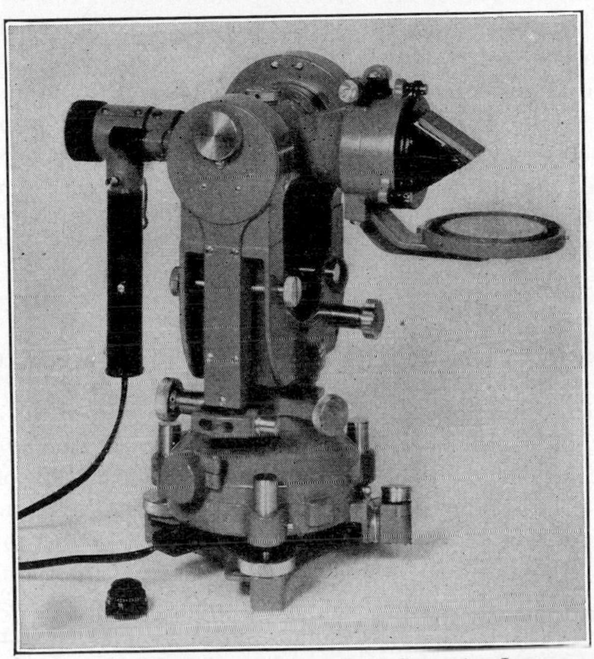

FIG. 9-17. A Wild T2 with astrolabe. *(Henry Wild Surveying Instruments Supply Co. of America, Inc.)*

9-54. To Use the Instrument. First, approximate latitude and longitude must be determined. Then a list is made out of the stars that will pass within 30° of the zenith during the hours of observation. The azimuth and time when each of these stars will have a vertical angle of 60° are computed. Before the appearance of each star the instrument is turned to the approximate azimuth. When the two images of that star coincide, a key is pressed that records the time on a chronograph that is carefully checked against accurate time signals. The National Bureau of Standards' wireless station WLWL sends out a time signal once per second on several frequencies. These should be used if possible.

The final solution is usually made by plotting as described previously. Since stars are observed that are fairly uniformly distributed in azimuth, any errors in estimating refraction and any error in the angle of the prism are automatically canceled.

It is beyond the scope of this book to describe in detail the techniques involved in the use of the astrolabe. It is enough if the principles have been demonstrated and if it is evident that the device gives results that are entirely free from primary instrumental errors. It is true that if the relative position of the prism and the mirror should change or if the focus is changed during the observations, certain shifts in the images will occur that will introduce secondary errors in the results. But as long as no changes in the optics are made during the observations, no instrumental errors can occur.

9-55. The Pendulum Astrolabe. While a pendulum astrolabe is quite different in appearance, it is based on the same principle. The mercury pool is replaced by an optically flat mirror that is held horizontal by a damped pendulum that can swing in any direction. Any error in the horizontality of the mirror when the pendulum comes to rest is eliminated by the practice described of observing stars at well distributed azimuths.

CURVATURE OF THE PATH OF A CIRCUMPOLAR

9-56. Principles. It is evident that if Polaris passed through elongation during the middle of an observation for azimuth, the average of the times of the various pointings would be very nearly the time of elongation and, since Polaris moves very slowly in azimuth near elongation, the computed azimuth would be practically that at elongation. On the other hand, the average direction of the instrument pointings would always be less than the azimuth of elongation. When the angle from the mark to the average of the instrument pointing is applied to the azimuth of the star thus computed, a small error will result.

This is an example of the error caused by the "curvature of the path of the star." The error is greatest at elongation and practically nil at culmination.

Equation (9-13) expresses the relation between the azimuth and the other quantities.

$$Z\ (\text{minutes}) = \frac{3438 \sin t}{\cos l \cot p - \sin l \cos t} \tag{9-13}$$

Cot p is always so large that small changes in the second term of the denominator hardly affect the result. It is therefore evident that Z changes very nearly with sin t. Thus, instead of using a value of t computed from the average of the times of the pointings, we should find the value of sin t for the time of each pointing and use the average of these sines for the expression sin t in the equation.

To be sure of keeping this error negligible, no group of time records that extend over a range of more than three minutes should be used.

Each group should be averaged and the sine of the resulting t found. The weighted average of the sin t should then be used in the equation. The cosine for the corresponding t can be used in the denominator of Eq. 9-13.

Example

Time of Observations

$9^h32^m41^s$ P.M.
9 33 46
9 34 50
9 35 24
9 40 03
9 41 06

Group 1		*Group 2*
$9^h32^m41^s$		$9^h40^m03^s$
9 33 46		9 41 06
9 34 50	Av. =	$9^h40^m34.5^s$
9 35 24		
Av. = $9^h34^m10.2^s$		
t = 82°13′12″		t = 83°49′30″
sin t = 0.990795		sin t = 0.994198
×4		×2
3.963180		1.988396

3.963180
1.988396
6|5.951576

Final sin t = 0.991929 Final t = 82°42′57″

By the usual method sin t would be 0.992027.

TRUE BEARING WITHOUT POLARIS

9-57. Description. In high northern latitudes, when Polaris is too near the zenith to be observed, and in southern latitudes where Polaris cannot be seen, the most accurate method of determining a true bearing is by observations of the altitudes of stars.

The method consists of observing simultaneously the horizontal angle from a mark to a star and the star's altitude. The procedure for this operation is the same as that for finding true north by the altitude of the sun as described in Secs. 9-34ff. Each observation consists of six direct and six reversed pointings with the horizontal circle in the same position. Time does not have to be recorded, as the declinations of the stars do not change rapidly enough to make it necessary. The method is not very accurate, especially in high latitudes, and therefore many observations must be taken. The stars chosen should be nearly on the **prime vertical,** that is, nearly east or west from the observer and preferably 25 to 30° above the horizon. They should be observed in pairs, one east and one west, at about the same altitude. This process eliminates the effect of errors in estimating refraction.

9-58. Directions for Observation. Choose two bright stars, one in the east and one in the west, at altitudes of about 25 to 30°. Unless they can be immediately recognized, make sketches showing the positions of the stars that surround them, to aid in later identification. Complete several observations of each star, alternating between the two.

9-59. Computations. Use the formula given on page 237 (see Eq. 9-8*a*). In this case Z (Sec. 9-20) is the star's true bearing from the north. Assume, star observed Nunki (in east), June 3, 1950; latitude S40°20′ (from map); clockwise horizontal angle from mark B to star, 125°10′; vertical angle 35°55.7′; temperature 40°.

d Nunki (50) June 3, 1950	−26°21.7′
Measured altitude	35°55.7′
Refraction correction (Sec. 9-21, Table 9-2)	− 1.3
True altitude, h (Sec. 9-21)	35°54.4′

$$\cos Z = \frac{\sin(-26°21.7') - \sin 35°54.4' \sin(-40°20.0')}{\cos(35°54.4')\cos(-40°20.0')}$$

$$\cos Z = \frac{-0.44404 - (0.58647)(-0.64723)}{(0.80997)(0.76229)}$$

$$\cos Z = -0.10440$$

$$Z = 95°59.6'$$

Star's bearing = S84°00.4′E (Sec. 9-20)
Clockwise angle B to star = 125°10.0′
Bearing mark B = N29°10.4′W

9-60. True Bearing by a Southern Circumpolar Star. Stars that are near the poles are called circumpolar stars. In the Southern Hemisphere the nearest bright star to the pole is about 30° distant from it. Several such stars exist, and an observation for true bearing can be made on one of them in exactly the same way as the observation on Polaris, which is made in the Northern Hemisphere. The field work is identical. The computation must be based on the following formula:

$$\tan Z = \frac{\sin t}{\cos l \cot p - \sin l \cos t} \tag{9-14}$$

Whichever star is nearest elongation (Sec. 9-28) should be chosen. A star that is on an hour line about 6 hr from the overhead position is nearly at elongation.

OBSERVATION TO DETERMINE *t*

9-61. Principle of "Time Shots." When an observation is made on Polaris or on a southern circumpolar star, if either longitude, accurate time, or both, are unavailable, the angle t of the circumpolar can be found by an observation on a star near the prime vertical, as follows:

Observe the altitude and take the time of each pointing. The horizontal angle is not required. Using the average altitude corrected for refraction and the average time, compute angle t for the east-west star by the following formula:

$$\cos t = \frac{\sin h - \sin d \sin l}{\cos d \cos l} \tag{9-15}$$

Since the S.H.A. of each star is given in the tables, the difference between S.H.A.'s can be computed. The difference in t is the same as the difference in S.H.A. Thus if the value of t for the east-west star at the time of the observation is known, the value of t for the circumpolar at that moment can be computed by applying the difference of the S.H.A.'s. The change in t that occurred between the times of the two observations can be computed by Table 9-1.

Example. Assume that neither longitude nor time had been available for the example given in Sec. 9-31, and that instead, the star Pollux had been observed in the west. *Required to obtain t of Polaris.*

Time of observation of Polaris, May 5, 1950, $8^h49^m47^s$ P.M. (watch not correct); latitude N42°22.6′ from map.

Time of observation of Pollux, $9^h10^m04^s$ (by the same watch); measured altitude 40°11.0′; temperature 60°.

In the formula given above, h is the altitude corrected for refraction, d is the declination of east or west star, l is the latitude. All are taken to the nearest 0.1. Five places of decimals are sufficient.

Measured altitude..........................	40°11.0′
Less refraction correction (Table 9-2)..........	1.1
True altitude h..........................	40°09.9′

$$\cos t = \frac{\sin 40°09.9' - \sin 28°09.1' \sin 42°22.6'}{\cos 28°09.1' \cos 42°22.6'}$$

$$\cos t = \frac{0.64499 - (0.47181)(0.67400)}{(0.88170)(0.73873)}$$

$$\cos t = 0.50203$$

$$t = 59°51.9' \qquad \text{(Pollux)}$$

L.H.A. Pollux 59°5.19′ (see Sec. 9-20).

1. G.H.A. Polaris 0^h G May 6, 1950..............................	196°28.0′
	360 00.0
	556°28.0′
2. G.H.A. ♈ 0^h G May 6, 1950..............................	223 16.8
3. S.H.A. Polaris all day May 6, 1950, (1) − (2)....................	333°11.2′
4. S.H.A. Pollux all day May 6, 1950..........................	244 26.0
5. Cor. to be added to L.H.A. Pollux to obtain L.H.A. Polaris (3) − (4)	88°45.2′
6. L.H.A. Pollux at $9^h10^m04^s$ by watch (above)....................	59 51.9
7. L.H.A. Polaris at $9^h10^m04^s$ by watch (5) + (6)..................	148°37.1′
8. Elapsed time between observations $8^h49^m47^s - 9^h10^m04^s = -0^h20^m17^s$	
9. Correction for elapsed time (Table 9-1)........................	− 5 05.1
10. t for Polaris at $8^h49^m47^s$ by watch..........................	143°32.0′

Having obtained t for Polaris, the computation for the bearing can be made as before.

TO FIND LONGITUDE

To find longitude except from a map, accurate time must be available. When it is available, an observation such as described for Pollux above will give longitude if the time of each pointing on the star is recorded.

Example. Assume that the correct GCT (see Sec. 9-31 for computation GCT) for the observation of the altitude of Pollux was $2^h53^m39^s$, May 6, 1950.

S.H.A. Pollux May 6, 1950	244°26.0′
G.H.A. ♈ 0^h *G* May 6, 1950	223 16.8
Correction for 2^h53^m (Table 9-2)	43 22.1
Correction for 39 (Table 9-2)	9.8
G.H.A. Pollux (add above four)	511°14.7′
Rejecting 360°	151°14.7′
Less L.H.A. as computed	59 51.9
West longitude	91°22.8′

PROBLEMS

9-1 to 9-4. Compute for longitude 74°39′W at 8^h30^m P.M. EST. the day this assignment is due, the value of t for:

9-1. Polaris.

9-2. Sirius.

9-3. Vega.

9-4. Fomalhaut.

9-5 to 9-8. Find the EST. of culmination for these stars:

9-5. Polaris.

9-6. Sirius.

9-7. Vega.

9-8. Fomalhaut.

9-9 to 9-12. At what latitudes can lower culmination be seen? Assume that you can see not less than 5° above horizon.

9-9. Polaris.

9-10. Sirius.

9-11. Vega.

9-12. Fomalhaut.

9-13. Given latitude 40°20′N and previous data (above), compute the bearing of Polaris.

9-14. Given for an observation on the sun, temperature 80°F, latitude 40°20′N, average time 3^h40^m P.M. EST, May 21 of the present year, average vertical angle 39°09′, horizontal angle from mark 94°21′; compute the azimuth of the mark.

9-15. Given measured noon altitude of sun's lower limb, 26°38.4′, January 2 of the present year at longitude 74°39′W; find the latitude.

9-16. Given time of observation $9^h32^m15^s$ P.M. EST, January 2 of the present year; star observed, Spica in the west, observed altitude 17°18′, latitude 40°20′N; find the longitude.

9-17. Given assumed position latitude 40°20′N, longitude 74°39′W, date April 8 of the present year; observed Sirius, altitude 20°46′ at $8^h25^m07^s$ EST, Arcturus, altitude 29°17′ at $8^h35^m46^s$ EST; find the latitude and longitude.

Table 9-1. Increase in G.H.A. for Elapsed Time

Min	Hours of Greenwich civil time								Sec	Cor.
	0h	1h	2h	3h	4h	5h	6h	7h		
	° ′	° ′	° ′	° ′	° ′	° ′	° ′	° ′		′
0	0 0.0	15 2.5	30 4.9	45 7.4	60 9.9	75 12.3	90 14.8	105 17.2	0	0.0
1	0 15.0	15 17.5	30 19.9	45 22.4	60 24.9	75 27.4	90 29.8	105 32.3	1	0.3
2	0 30.1	15 32.5	30 35.0	45 37.5	60 39.9	75 42.4	90 44.9	105 47.3	2	0.5
3	0 45.1	15 47.6	30 50.1	45 52.5	60 55.0	75 57.4	90 59.9	106 2.4	3	0.8
4	1 0.2	16 2.6	31 5.1	46 7.6	61 10.0	76 12.5	91 14.9	106 17.4	4	1.0
5	1 15.2	16 17.7	31 20.1	46 22.6	61 25.1	76 27.5	91 30.0	106 32.5	5	1.3
6	1 30.2	16 32.7	31 35.2	46 37.6	61 40.1	76 42.6	91 45.0	106 47.5	6	1.5
7	1 45.3	16 47.8	31 50.2	46 52.7	61 55.1	76 57.6	92 0.1	107 2.5	7	1.8
8	2 0.3	17 2.8	32 5.3	47 7.7	62 10.2	77 12.6	92 15.1	107 17.6	8	2.0
9	2 15.4	17 17.8	32 20.3	47 22.8	62 25.2	77 27.7	92 30.2	107 32.6	9	2.3
10	2 30.4	17 32.9	32 35.3	47 37.8	62 40.3	77 42.7	92 45.2	107 47.7	10	2.5
11	2 45.5	17 47.9	32 50.4	47 52.8	62 55.3	77 57.8	93 0.2	108 2.7	11	2.8
12	3 0.5	18 3.0	33 5.4	48 7.9	63 10.3	78 12.8	93 15.3	108 17.7	12	3.0
13	3 15.5	18 18.0	33 20.5	48 22.9	63 25.4	78 27.9	93 30.3	108 32.8	13	3.3
14	3 30.6	18 33.0	33 35.5	48 38.0	63 40.4	78 42.9	93 45.4	108 47.8	14	3.5
15	3 45.6	18 48.1	33 50.5	48 53.0	63 55.5	78 57.9	94 0.4	109 2.9	15	3.8
16	4 0.7	19 3.1	34 5.6	49 8.0	64 10.5	79 13.0	94 15.4	109 17.9	16	4.0
17	4 15.7	19 18.2	34 20.6	49 23.1	64 25.6	79 28.0	94 30.5	109 32.9	17	4.3
18	4 30.7	19 33.2	34 35.7	49 38.1	64 40.6	79 43.1	94 45.5	109 48.0	18	4.5
19	4 45.8	19 48.2	34 50.7	49 53.2	64 55.6	79 58.1	95 0.6	110 3.0	19	4.8
20	5 0.8	20 3.3	35 5.7	50 8.2	65 10.7	80 13.1	95 15.6	110 18.1	20	5.0
21	5 15.9	20 18.3	35 20.8	50 23.3	65 25.7	80 28.2	95 30.6	110 33.1	21	5.3
22	5 30.9	20 33.4	35 35.8	50 38.3	65 40.8	80 43.2	95 45.7	110 48.2	22	5.5
23	5 45.9	20 48.4	35 50.9	50 53.3	65 55.8	80 58.3	96 0.7	111 3.2	23	5.8
24	6 1.0	21 3.4	36 5.9	51 8.4	66 10.8	81 13.3	96 15.8	111 18.2	24	6.0
25	6 16.0	21 18.5	36 21.0	51 23.4	66 25.9	81 28.3	96 30.8	111 33.3	25	6.3
26	6 31.1	21 33.5	36 36.0	51 38.5	66 40.9	81 43.4	96 45.8	111 48.3	26	6.5
27	6 46.1	21 48.6	36 51.0	51 53.5	66 56.0	81 58.4	97 0.9	112 3.4	27	6.8
28	7 1.1	22 3.6	37 6.1	52 8.5	67 11.0	82 13.5	97 15.9	112 18.4	28	7.0
29	7 16.2	22 18.7	37 21.1	52 23.6	67 26.0	82 28.5	97 31.0	112 33.4	29	7.3
30	7 31.2	22 33.7	37 36.2	52 38.6	67 41.1	82 43.6	97 46.0	112 48.5	30	7.5
31	7 46.3	22 48.7	37 51.2	52 53.7	67 56.1	82 58.6	98 1.1	113 3.5	31	7.8
32	8 1.3	23 3.8	38 6.2	53 8.7	68 11.2	83 13.6	98 16.1	113 18.6	32	8.0
33	8 16.4	23 18.8	38 21.3	53 23.7	68 26.2	83 28.7	98 31.1	113 33.6	33	8.3
34	8 31.4	23 33.9	38 36.3	53 38.8	68 41.2	83 43.7	98 46.2	113 48.6	34	8.5
35	8 46.4	23 48.9	38 51.4	53 53.8	68 56.3	83 58.8	99 1.2	114 3.7	35	8.8
36	9 1.5	24 3.9	39 6.4	54 8.9	69 11.3	84 13.8	99 16.3	114 18.7	36	9.0
37	9 16.5	24 19.0	39 21.4	54 23.9	69 26.4	84 28.8	99 31.3	114 33.8	37	9.3
38	9 31.6	24 34.0	39 36.5	54 39.0	69 41.4	84 43.9	99 46.3	114 48.8	38	9.5
39	9 46.6	24 49.1	39 51.5	54 54.0	69 56.5	84 58.9	100 1.4	115 3.9	39	9.8
40	10 1.6	25 4.1	40 6.6	55 9.0	70 11.5	85 14.0	100 16.4	115 18.9	40	10.0
41	10 16.7	25 19.1	40 21.6	55 24.1	70 26.5	85 29.0	100 31.5	115 33.9	41	10.3
42	10 31.7	25 34.2	40 36.7	55 39.1	70 41.6	85 44.0	100 46.5	115 49.0	42	10.5
43	10 46.8	25 49.2	40 51.7	55 54.2	70 56.6	85 59.1	101 1.5	116 4.0	43	10.8
44	11 1.8	26 4.3	41 6.7	56 9.2	71 11.7	86 14.1	101 16.6	116 19.1	44	11.0
45	11 16.8	26 19.3	41 21.8	56 24.2	71 26.7	86 29.2	101 31.6	116 34.1	45	11.3
46	11 31.9	26 34.4	41 36.8	56 39.3	71 41.7	86 44.2	101 46.7	116 49.1	46	11.5
47	11 46.9	26 49.4	41 51.9	56 54.3	71 56.8	86 59.2	102 1.7	117 4.2	47	11.8
48	12 2.0	27 4.4	42 6.9	57 9.4	72 11.8	87 14.3	102 16.8	117 19.2	48	12.0
49	12 17.0	27 19.5	42 21.9	57 24.4	72 26.9	87 29.3	102 31.8	117 34.3	49	12.3
50	12 32.1	27 34.5	42 37.0	57 39.4	72 41.9	87 44.4	102 46.8	117 49.3	50	12.5
51	12 47.1	27 49.6	42 52.0	57 54.5	72 56.9	87 59.4	103 1.9	118 4.3	51	12.8
52	13 2.1	28 4.6	43 7.1	58 9.5	73 12.0	88 14.5	103 16.9	118 19.4	52	13.0
53	13 17.2	28 19.6	43 22.1	58 24.6	73 27.0	88 29.5	103 32.0	118 34.4	53	13.3
54	13 32.2	28 34.7	43 37.1	58 39.6	73 42.1	88 44.5	103 47.0	118 49.5	54	13.5
55	13 47.3	28 49.7	43 52.2	58 54.6	73 57.1	88 59.6	104 2.0	119 4.5	55	13.8
56	14 2.3	29 4.8	44 7.2	59 9.7	74 12.2	89 14.6	104 17.1	119 19.5	56	14.0
57	14 17.3	29 19.8	44 22.3	59 24.7	74 27.2	89 29.7	104 32.1	119 34.6	57	14.3
58	14 32.4	29 34.8	44 37.3	59 39.8	74 42.2	89 44.7	104 47.2	119 49.6	58	14.5
59	14 47.4	29 49.9	44 52.4	59 54.8	74 57.3	89 59.7	105 2.2	120 4.7	59	14.8
60	15 2.5	30 4.9	45 7.4	60 9.9	75 12.3	90 14.8	105 17.2	120 19.7	60	15.0

TABLE 9-1. INCREASE IN G.H.A. FOR ELAPSED TIME (*Continued*)

Min	Hours of Greenwich civil time								Sec	Cor.
	8^h	9^h	10^h	11^h	12^h	13^h	14^h	15^h		
	° ′	° ′	° ′	° ′	° ′	° ′	° ′	° ′		′
0	120 19.7	135 22.2	150 24.6	165 27.1	180 29.6	195 32.0	210 34.5	225 37.0	0	0.0
1	120 34.8	135 37.2	150 39.7	165 42.1	180 44.6	195 47.1	210 49.5	225 52.0	1	0.3
2	120 49.8	135 52.3	150 54.7	165 57.2	180 59.7	196 2.1	211 4.6	226 7.0	2	0.5
3	121 4.8	136 7.3	151 9.8	166 12.2	181 14.7	196 17.2	211 19.6	226 22.1	3	0.8
4	121 19.9	136 22.3	151 24.8	166 27.3	181 29.7	196 32.2	211 34.7	226 37.1	4	1.0
5	121 34.9	136 37.4	151 39.8	166 42.3	181 44.8	196 47.2	211 49.7	226 52.2	5	1.3
6	121 50.0	136 52.4	151 54.9	166 57.3	181 59.8	197 2.3	212 4.7	227 7.2	6	1.5
7	122 5.0	137 7.5	152 9.9	167 12.4	182 14.9	197 17.3	212 19.8	227 22.2	7	1.8
8	122 20.0	137 22.5	152 25.0	167 27.4	182 29.9	197 32.4	212 34.8	227 37.3	8	2.0
9	122 35.1	137 37.5	152 40.0	167 42.5	182 44.9	197 47.4	212 49.9	227 52.3	9	2.3
10	122 50.1	137 52.6	152 55.1	167 57.5	183 0.0	198 2.4	213 4.9	228 7.4	10	2.5
11	123 5.2	138 7.6	153 10.1	168 12.6	183 15.0	198 17.5	213 19.9	228 22.4	11	2.8
12	123 20.2	138 22.7	153 25.1	168 27.6	183 30.1	198 32.5	213 35.0	228 37.5	12	3.0
13	123 35.2	138 37.7	153 40.2	168 42.6	183 45.1	198 47.6	213 50.0	228 52.5	13	3.3
14	123 50.3	138 52.8	153 55.2	168 57.7	184 0.1	199 2.6	214 5.1	229 7.5	14	3.5
15	124 5.3	139 7.8	154 10.3	169 12.7	184 15.2	199 17.6	214 20.1	229 22.6	15	3.8
16	124 20.4	139 22.8	154 25.3	169 27.8	184 30.2	199 32.7	214 35.2	229 37.6	16	4.0
17	124 35.4	139 37.9	154 40.3	169 42.8	184 45.3	199 47.7	214 50.2	229 52.7	17	4.3
18	124 50.4	139 52.9	154 55.4	169 57.8	185 0.3	200 2.8	215 5.2	230 7.7	18	4.5
19	125 5.5	140 8.0	155 10.4	170 12.9	185 15.3	200 17.8	215 20.3	230 22.7	19	4.8
20	125 20.5	140 23.0	155 25.5	170 27.9	185 30.4	200 32.9	215 35.3	230 37.8	20	5.0
21	125 35.6	140 38.0	155 40.5	170 43.0	185 45.4	200 47.9	215 50.4	230 52.8	21	5.3
22	125 50.6	140 53.1	155 55.5	170 58.0	186 0.5	201 2.9	216 5.4	231 7.9	22	5.5
23	126 5.7	141 8.1	156 10.6	171 13.0	186 15.5	201 18.0	216 20.4	231 22.9	23	5.8
24	126 20.7	141 23.2	156 25.6	171 28.1	186 30.6	201 33.0	216 35.5	231 37.9	24	6.0
25	126 35.7	141 38.2	156 40.7	171 43.1	186 45.6	201 48.1	216 50.5	231 53.0	25	6.3
26	126 50.8	141 53.2	156 55.7	171 58.2	187 0.6	202 3.1	217 5.6	232 8.0	26	6.5
27	127 5.8	142 8.3	157 10.7	172 13.2	187 15.7	202 18.1	217 20.6	232 23.1	27	6.8
28	127 20.9	142 23.3	157 25.8	172 28.2	187 30.7	202 33.2	217 35.6	232 38.1	28	7.0
29	127 35.9	142 38.4	157 40.8	172 43.3	187 45.8	202 48.2	217 50.7	232 53.2	29	7.3
30	127 50.9	142 53.4	157 55.9	172 58.3	188 0.8	203 3.3	218 5.7	233 8.2	30	7.5
31	128 6.0	143 8.4	158 10.9	173 13.4	188 15.8	203 18.3	218 20.8	233 23.2	31	7.8
32	128 21.0	143 23.5	158 26.0	173 28.4	188 30.9	203 33.3	218 35.8	233 38.3	32	8.0
33	128 36.1	143 38.5	158 41.0	173 43.5	188 45.9	203 48.4	218 50.8	233 53.3	33	8.3
34	128 51.1	143 53.6	158 56.0	173 58.5	189 1.0	204 3.4	219 5.9	234 8.4	34	8.5
35	129 6.1	144 8.6	159 11.1	174 13.5	189 16.0	204 18.5	219 20.9	234 23.4	35	8.8
36	129 21.2	144 23.7	159 26.1	174 28.6	189 31.0	204 33.5	219 36.0	234 38.4	36	9.0
37	129 36.2	144 38.7	159 41.2	174 43.6	189 46.1	204 48.6	219 51.0	234 53.5	37	9.3
38	129 51.3	144 53.7	159 56.2	174 58.7	190 1.1	205 3.6	220 6.1	235 8.5	38	9.5
39	130 6.3	145 8.8	160 11.2	175 13.7	190 16.2	205 18.6	220 21.1	235 23.6	39	9.8
40	130 21.4	145 23.8	160 26.3	175 28.7	190 31.2	205 33.7	220 36.1	235 38.6	40	10.0
41	130 36.4	145 38.9	160 41.3	175 43.8	190 46.2	205 48.7	220 51.2	235 53.6	41	10.3
42	130 51.4	145 53.9	160 56.4	175 58.8	191 1.3	206 3.8	221 6.2	236 8.7	42	10.5
43	131 6.5	146 8.9	161 11.4	176 13.9	191 16.3	206 18.8	221 21.3	236 23.7	43	10.8
44	131 21.5	146 24.0	161 26.4	176 28.9	191 31.4	206 33.8	221 36.3	236 38.8	44	11.0
45	131 36.6	146 39.0	161 41.5	176 43.9	191 46.4	206 48.9	221 51.3	236 53.8	45	11.3
46	131 51.6	146 54.1	161 56.5	176 59.0	192 1.5	207 3.9	222 6.4	237 8.8	46	11.5
47	132 6.6	147 9.1	162 11.6	177 14.0	192 16.5	207 19.0	222 21.4	237 23.9	47	11.8
48	132 21.7	147 24.1	162 26.6	177 29.1	192 31.5	207 34.0	222 36.5	237 38.9	48	12.0
49	132 36.7	147 39.2	162 41.6	177 44.1	192 46.6	207 49.0	222 51.5	237 54.0	49	12.3
50	132 51.8	147 54.2	162 56.7	177 59.2	193 1.6	208 4.1	223 6.5	238 9.0	50	12.5
51	133 6.8	148 9.3	163 11.7	178 14.2	193 16.7	208 19.1	223 21.6	238 24.1	51	12.8
52	133 21.8	148 24.3	163 26.8	178 29.2	193 31.7	208 34.2	223 36.6	238 39.1	52	13.0
53	133 36.9	148 39.3	163 41.8	178 44.3	193 46.7	208 49.2	223 51.7	238 54.1	53	13.3
54	133 51.9	148 54.4	163 56.9	178 59.3	194 1.8	209 4.2	224 6.7	239 9.2	54	13.5
55	134 7.0	149 9.4	164 11.9	179 14.4	194 16.8	209 19.3	224 21.8	239 24.2	55	13.8
56	134 22.0	149 24.5	164 26.9	179 29.4	194 31.9	209 34.3	224 36.8	239 39.3	56	14.0
57	134 37.0	149 39.5	164 42.0	179 44.4	194 46.9	209 49.4	224 51.8	239 54.3	57	14.3
58	134 52.1	149 54.6	164 57.0	179 59.5	195 1.9	210 4.4	225 6.9	240 9.3	58	14.5
59	135 7.1	150 9.6	165 12.1	180 14.5	195 17.0	210 19.5	225 21.9	240 24.4	59	14.8
60	135 22.2	150 24.6	165 27.1	180 29.6	195 32.0	210 34.5	225 37.0	240 39.4	60	15.0

TABLE 9-1. INCREASE IN G.H.A. FOR ELAPSED TIME (*Continued*)

Min	Hours of Greenwich civil time								Sec	Cor.
	16^h	17^h	18^h	19^h	20^h	21^h	22^h	23^h		
	° ′	° ′	° ′	° ′	° ′	° ′	° ′	° ′		′
0	240 39.4	255 41.9	270 44.4	285 46.8	300 49.3	315 51.7	330 54.2	345 56.7	0	0.0
1	240 54.5	255 56.9	270 59.4	286 1.9	301 4.3	316 6.8	331 9.3	346 11.7	1	0.3
2	241 9.5	256 12.0	271 14.4	286 16.9	301 19.4	316 21.8	331 24.3	346 26.8	2	0.5
3	241 24.5	256 27.0	271 29.5	286 31.9	301 34.4	316 36.9	331 39.3	346 41.8	3	0.8
4	241 39.6	256 42.1	271 44.5	286 47.0	301 49.4	316 51.9	331 54.4	346 56.8	4	1.0
5	241 54.6	256 57.1	271 59.6	287 2.0	302 4.5	317 7.0	332 9.4	347 11.9	5	1.3
6	242 9.7	257 12.1	272 14.6	287 17.1	302 19.5	317 22.0	332 24.5	347 26.9	6	1.5
7	242 24.7	257 27.2	272 29.6	287 32.1	302 34.6	317 37.0	332 39.5	347 42.0	7	1.8
8	242 39.8	257 42.2	272 44.7	287 47.1	302 49.6	317 52.1	332 54.5	347 57.0	8	2.0
9	242 54.8	257 57.3	272 59.7	288 2.2	303 4.7	318 7.1	333 9.6	348 12.0	9	2.3
10	243 9.8	258 12.3	273 14.8	288 17.2	303 19.7	318 22.2	333 24.6	348 27.1	10	2.5
11	243 24.9	258 27.3	273 29.8	288 32.3	303 34.7	318 37.2	333 39.7	348 42.1	11	2.8
12	243 39.9	258 42.4	273 44.8	288 47.3	303 49.8	318 52.2	333 54.7	348 57.2	12	3.0
13	243 55.0	258 57.4	273 59.9	289 2.4	304 4.8	319 7.3	334 9.7	349 12.2	13	3.3
14	244 10.0	259 12.5	274 14.9	289 17.4	304 19.9	319 22.3	334 24.8	349 27.2	14	3.5
15	244 25.0	259 27.5	274 30.0	289 32.4	304 34.9	319 37.4	334 39.8	349 42.3	15	3.8
16	244 40.1	259 42.5	274 45.0	289 47.5	304 49.9	319 52.4	334 54.9	349 57.3	16	4.0
17	244 55.1	259 57.6	275 0.0	290 2.5	305 5.0	320 7.4	335 9.9	350 12.4	17	4.3
18	245 10.2	260 12.6	275 15.1	290 17.6	305 20.0	320 22.5	335 24.9	350 27.4	18	4.5
19	245 25.2	260 27.7	275 30.1	290 32.6	305 35.1	320 37.5	335 40.0	350 42.5	19	4.8
20	245 40.2	260 42.7	275 45.2	290 47.6	305 50.1	320 52.6	335 55.0	350 57.5	20	5.0
21	245 55.3	260 57.8	276 0.2	291 2.7	306 5.1	321 7.6	336 10.1	351 12.5	21	5.3
22	246 10.3	261 12.8	276 15.3	291 17.7	306 20.2	321 22.6	336 25.1	351 27.6	22	5.5
23	246 25.4	261 27.8	276 30.3	291 32.8	306 35.2	321 37.7	336 40.2	351 42.6	23	5.8
24	246 40.4	261 42.9	276 45.3	291 47.8	306 50.3	321 52.7	336 55.2	351 57.7	24	6.0
25	246 55.4	261 57.9	277 0.4	292 2.8	307 5.3	322 7.8	337 10.2	352 12.7	25	6.3
26	247 10.5	262 13.0	277 15.4	292 17.9	307 20.3	322 22.8	337 25.3	352 27.7	26	6.5
27	247 25.5	262 28.0	277 30.5	292 32.9	307 35.4	322 37.9	337 40.3	352 42.8	27	6.8
28	247 40.6	262 43.0	277 45.5	292 48.0	307 50.4	322 52.9	337 55.4	352 57.8	28	7.0
29	247 55.6	262 58.1	278 0.5	293 3.0	308 5.5	323 7.9	338 10.4	353 12.9	29	7.3
30	248 10.7	263 13.1	278 15.6	293 18.0	308 20.5	323 23.0	338 25.4	353 27.9	30	7.5
31	248 25.7	263 28.2	278 30.6	293 33.1	308 35.6	323 38.0	338 40.5	353 42.9	31	7.8
32	248 40.7	263 43.2	278 45.7	293 48.1	308 50.6	323 53.1	338 55.5	353 58.0	32	8.0
33	248 55.8	263 58.2	279 0.7	294 3.2	309 5.6	324 8.1	339 10.6	354 13.0	33	8.3
34	249 10.8	264 13.3	279 15.7	294 18.2	309 20.7	324 23.1	339 25.6	354 28.1	34	8.5
35	249 25.9	264 28.3	279 30.8	294 33.3	309 35.7	324 38.2	339 40.6	354 43.1	35	8.8
36	249 40.9	264 43.4	279 45.8	294 48.3	309 50.8	324 53.2	339 55.7	354 58.2	36	9.0
37	249 55.9	264 58.4	280 0.9	295 3.3	310 5.8	325 8.3	340 10.7	355 13.2	37	9.3
38	250 11.0	265 13.4	280 15.9	295 18.4	310 20.8	325 23.3	340 25.8	355 28.2	38	9.5
39	250 26.0	265 28.5	280 31.0	295 33.4	310 35.9	325 38.3	340 40.8	355 43.3	39	9.8
40	250 41.1	265 43.5	280 46.0	295 48.5	310 50.9	325 53.4	340 55.8	355 58.3	40	10.0
41	250 56.1	265 58.6	281 1.0	296 3.5	311 6.0	326 8.4	341 10.9	356 13.4	41	10.3
42	251 11.1	266 13.6	281 16.1	296 18.5	311 21.0	326 23.5	341 25.9	356 28.4	42	10.5
43	251 26.2	266 28.7	281 31.1	296 33.6	311 36.0	326 38.5	341 41.0	356 43.4	43	10.8
44	251 41.2	266 43.7	281 46.2	296 48.6	311 51.1	326 53.6	341 56.0	356 58.5	44	11.0
45	251 56.3	266 58.7	282 1.2	297 3.7	312 6.1	327 8.6	342 11.1	357 13.5	45	11.3
46	252 11.3	267 13.8	282 16.2	297 18.7	312 21.2	327 23.6	342 26.1	357 28.6	46	11.5
47	252 26.4	267 28.8	282 31.3	297 33.7	312 36.2	327 38.7	342 41.1	357 43.6	47	11.8
48	252 41.4	267 43.9	282 46.3	297 48.8	312 51.3	327 53.7	342 56.2	357 58.6	48	12.0
49	252 56.4	267 58.9	283 1.4	298 3.8	313 6.3	328 8.8	343 11.2	358 13.7	49	12.3
50	253 11.5	268 13.9	283 16.4	298 18.9	313 21.3	328 23.8	343 26.3	358 28.7	50	12.5
51	253 26.5	268 29.0	283 31.4	298 33.9	313 36.4	328 38.8	343 41.3	358 43.8	51	12.8
52	253 41.6	268 44.0	283 46.5	298 48.9	313 51.4	328 53.9	343 56.3	358 58.8	52	13.0
53	253 56.6	268 59.1	284 1.5	299 4.0	314 6.5	329 8.9	344 11.4	359 13.8	53	13.3
54	254 11.6	269 14.1	284 16.6	299 19.0	314 21.5	329 24.0	344 26.4	359 28.9	54	13.5
55	254 26.7	269 29.1	284 31.6	299 34.1	314 36.5	329 39.0	344 41.5	359 43.9	55	13.8
56	254 41.7	269 44.2	284 46.6	299 49.1	314 51.6	329 54.0	344 56.5	359 59.0	56	14.0
57	254 56.8	269 59.2	285 1.7	300 4.2	315 6.6	330 9.1	345 11.5	0 14.0	57	14.3
58	255 11.8	270 14.3	285 16.7	300 19.2	315 21.7	330 24.1	345 26.6	0 29.1	58	14.5
59	255 26.8	270 29.3	285 31.8	300 34.2	315 36.7	330 39.2	345 41.6	0 44.1	59	14.8
60	255 41.9	270 44.4	285 46.8	300 49.3	315 51.7	330 54.2	345 56.7	0 59.1	60	15.0

TABLE 9-2. MEAN REFRACTION r_m

Pressure = 760 mm; temperature = 10°C; relative humidity = 60%

z	00′	10′	20′	30′	40′	50′	60′
°	″	″	″	″	″	″	″
0	0.0	0.2	0.3	0.5	0.7	0.8	1.0
1	1.0	1.2	1.3	1.5	1.7	1.9	2.0
2	2.0	2.2	2.4	2.5	2.7	2.9	3.0
3	3.0	3.2	3.4	3.5	3.7	3.9	4.0
4	4.0	4.2	4.4	4.6	4.7	4.9	5.1
5	5.1	5.2	5.4	5.6	5.7	5.9	6.1
6	6.1	6.3	6.4	6.6	6.8	6.9	7.1
7	7.1	7.3	7.5	7.6	7.8	8.0	8.1
8	8.1	8.3	8.5	8.7	8.8	9.0	9.2
9	9.2	9.3	9.5	9.7	9.9	10.0	10.2
10	10.2	10.4	10.6	10.7	10.9	11.1	11.3
11	11.3	11.4	11.6	11.8	12.0	12.1	12.3
12	12.3	12.5	12.7	12.8	13.0	13.2	13.4
13	13.4	13.5	13.7	13.9	14.1	14.3	14.4
14	14.4	14.6	14.8	15.0	15.2	15.3	15.5
15	15.5	15.7	15.9	16.1	16.2	16.4	16.6
16	16.6	16.8	17.0	17.2	17.3	17.5	17.7
17	17.7	17.9	18.1	18.3	18.4	18.6	18.8
18	18.8	19.0	19.2	19.4	19.6	19.8	19.9
19	19.9	20.1	20.3	20.5	20.7	20.9	21.1
20	21.1	21.3	21.5	21.7	21.8	22.0	22.2
21	22.2	22.4	22.6	22.8	23.0	23.2	23.4
22	23.4	23.6	23.8	24.0	24.2	24.4	24.6
23	24.6	24.8	25.0	25.2	25.4	25.6	25.8
24	25.8	26.0	26.2	26.4	26.6	26.8	27.0
25	27.0	27.2	27.4	27.6	27.8	28.0	28.2
26	28.2	28.4	28.7	28.9	29.1	29.3	29.5
27	29.5	29.7	29.9	30.1	30.4	30.6	30.8
28	30.8	31.0	31.2	31.4	31.7	31.9	32.1
29	32.1	32.3	32.5	32.8	33.0	33.2	33.4
30	33.4	33.6	33.9	34.1	34.3	34.6	34.8
31	34.8	35.0	35.2	35.5	35.7	35.9	36.2
32	36.2	36.4	36.6	36.9	37.1	37.4	37.6
33	37.6	37.8	38.1	38.3	38.6	38.8	39.0
34	39.0	39.3	39.5	39.8	40.0	40.3	40.5
35	40.5	40.8	41.0	41.3	41.5	41.8	42.1
36	42.1	42.3	42.6	42.8	43.1	43.3	43.6
37	43.6	43.9	44.1	44.4	44.7	44.9	45.2
38	45.2	45.5	45.8	46.0	46.3	46.6	46.9
39	46.9	47.1	47.4	47.7	48.0	48.3	48.6
40	48.6	48.8	49.1	49.4	49.7	50.0	50.3
41	50.3	50.6	50.9	51.2	51.5	51.8	52.1
42	52.1	52.4	52.7	53.0	53.3	53.6	54.0
43	54.0	54.3	54.6	54.9	55.2	55.5	55.9
44	55.9	56.2	56.5	56.9	57.2	57.5	57.9

TABLE 9-2. MEAN REFRACTION r_m (*Continued*)

z	00′	10′	20′	30′	40′	50′	60′
°	″	″	″	″	″	″	″
45	57.9	58.2	58.5	58.9	59.2	59.6	59.9
46	59.9	60.2	60.6	61.0	61.3	61.7	62.0
47	62.0	62.4	62.8	63.1	63.5	63.9	64.2
48	64.2	64.6	65.0	65.4	65.7	66.1	66.5
49	66.5	66.9	67.3	67.7	68.1	68.5	68.9
50	68.9	69.3	69.7	70.1	70.6	71.0	71.4
51	71.4	71.8	72.2	72.7	73.1	73.6	74.0
52	74.0	74.4	74.9	75.3	75.8	76.2	76.7
53	76.7	77.2	77.6	78.1	78.6	79.1	79.5
54	79.5	80.0	80.5	81.0	81.5	82.0	82.5
55	82.5	83.0	83.5	84.1	84.6	85.1	85.6
56	85.6	86.2	86.7	87.3	87.8	88.4	88.9
57	88.9	89.5	90.1	90.7	91.2	91.8	92.4
58	92.4	93.0	93.6	94.2	94.8	95.5	96.1
59	96.1	96.7	97.4	98.0	98.6	99.3	100.0
60	100.0	100.6	101.3	102.0	102.7	103.4	104.1
61	104.1	104.8	105.5	106.3	107.0	107.7	108.5
62	108.5	109.2	110.0	110.8	111.6	112.4	113.2
63	113.2	114.0	114.8	115.6	116.5	117.3	118.2
64	118.2	119.0	119.9	120.8	121.7	122.6	123.5
65	123.5	124.5	125.4	126.4	127.4	128.3	129.3
66	129.3	130.3	131.4	132.4	133.4	134.5	135.6
67	135.6	136.7	137.8	138.9	140.0	141.2	142.3
68	142.3	143.5	144.7	145.9	147.2	148.4	149.7
69	149.7	151.0	152.3	153.6	155.0	156.4	157.8
70	157.8	159.2	160.6	162.1	163.6	165.1	166.6
71	166.6	168.2	169.7	171.4	173.0	174.7	176.3
72	176.3	178.1	179.8	181.6	183.4	185.3	187.2
73	187.2	189.1	191.0	193.0	195.1	197.1	199.2
74	199.2	201.4	203.6	205.8	208.1	210.4	212.8
75	212.8	215.2	217.7	220.2	222.8	225.5	228.2
76	228.2	230.9	233.7	236.6	239.6	242.6	245.7
77	245.7	248.9	252.1	255.4	258.9	262.3	265.9
78	265.9	269.6	273.4	277.2	281.2	285.3	289.5
79	289.5	293.8	298.2	302.8	307.5	312.3	317.3
80	317.3	322.4	327.7	333.2	338.8	344.6	350.6
81	350.6	356.8	363.2	369.8	376.6	383.7	391.1
82	391.1	398.7	406.6	414.8	423.3	432.1	441.3
83	441.3	450.9	460.9	471.2	482.0	493.3	505.1
84	505.1	517.4	530.3	543.8	558.0	572.8	588.4
85	588.4	604.4	621.6	639.7	658.8	678.9	700.2
86	700.2	722.7	746.6	771.8	798.7	827.2	857.6
87	857.6	890.0	924.7	961.6	1001.3	1043.9	1089.7
88	1089.7	1138.9	1192.0	1249.2	1311.4	1378.6	1452.0
89	1452.0	1531.7	1618.8	1714.0	1818.4	1933.1	2059.5

TABLE 9-2a. PRESSURE CORRECTION FACTOR C_B

Apply to mean refraction in Table 9-2. $r = (r_m)(C_B)(C_T)$

Barometer In.	Barometer Mm	C_B	Barometer In.	Barometer Mm	C_B	Barometer In.	Barometer Mm	C_B	Barometer In.	Barometer Mm	C_B	Barometer In.	Barometer Mm	C_B
20.0	508	0.670	22.4	569	0.749	24.8	630	0.829	27.2	691	0.909	29.6	752	0.989
20.1	511	0.673	22.5	572	0.752	24.9	632	0.832	27.3	693	0.912	29.7	754	0.992
20.2	513	0.676	22.6	574	0.755	25.0	635	0.835	27.4	696	0.916	29.8	757	0.996
20.3	516	0.679	22.7	576	0.759	25.1	637	0.838	27.5	699	0.920	29.9	759	0.999
20.4	518	0.682	22.8	579	0.762	25.2	640	0.842	27.6	701	0.923	30.0	762	1.003
20.5	521	0.685	22.9	582	0.766	25.3	643	0.846	27.7	704	0.926	30.1	765	1.007
20.6	523	0.688	23.0	584	0.770	25.4	645	0.849	27.8	706	0.929	30.2	767	1.010
20.7	526	0.692	23.1	587	0.773	25.5	648	0.853	27.9	709	0.933	30.3	770	1.013
20.8	528	0.696	23.2	589	0.776	25.6	650	0.856	28.0	711	0.936	30.4	772	1.016
20.9	531	0.699	23.3	592	0.779	25.7	653	0.859	28.1	714	0.939	30.5	775	1.020
21.0	533	0.703	23.4	594	0.783	25.8	655	0.862	28.2	716	0.942	30.6	777	1.023
21.1	536	0.706	23.5	597	0.786	25.9	658	0.866	28.3	719	0.946	30.7	780	1.026
21.2	538	0.709	23.6	599	0.789	26.0	660	0.869	28.4	721	0.949	30.8	782	1.029
21.3	541	0.712	23.7	602	0.792	26.1	663	0.872	28.5	724	0.953	30.9	785	1.033
21.4	544	0.716	23.8	605	0.796	26.2	665	0.875	28.6	726	0.956	31.0	787	1.036
21.5	546	0.719	23.9	607	0.799	26.3	668	0.879	28.7	729	0.959			
21.6	549	0.722	24.0	610	0.803	26.4	671	0.882	28.8	732	0.963			
21.7	551	0.725	24.1	612	0.806	26.5	673	0.885	28.9	734	0.966			
21.8	554	0.729	24.2	615	0.809	26.6	676	0.889	29.0	737	0.970			
21.9	556	0.732	24.3	617	0.813	26.7	678	0.892	29.1	739	0.973			
22.0	559	0.735	24.4	620	0.816	26.8	681	0.896	29.2	742	0.976			
22.1	561	0.739	24.5	622	0.820	26.9	683	0.899	29.3	744	0.979			
22.2	564	0.742	24.6	625	0.823	27.0	686	0.902	29.4	747	0.983			
22.3	566	0.746	24.7	627	0.826	27.1	688	0.905	29.5	749	0.986			

TABLE 9-2b. TEMPERATURE CORRECTION FACTOR C_T

Apply to mean refraction in Table 9-2. $r = (r_m)(C_B)(C_T)$

Temp. °F	Temp. °C	C_T	Temp. °F	Temp. °C	C_T	Temp. °F	Temp. °C	C_T	Temp. °F	Temp. °C	C_T	Temp. °F	Temp. °C	C_T
−25	−31.7	1.172	8	−13.3	1.089	41	5.0	1.018	74	23.3	0.955	107	41.7	0.900
−24	−31.1	1.169	9	−12.8	1.087	42	5.6	1.016	75	23.9	0.953	108	42.2	0.899
−23	−30.6	1.166	10	−12.2	1.085	43	6.1	1.014	76	24.4	0.952	109	42.8	0.897
−22	−30.0	1.164	11	−11.7	1.082	44	6.7	1.012	77	25.0	0.950	110	43.3	0.895
−21	−29.4	1.161	12	−11.1	1.080	45	7.2	1.010	78	25.6	0.948	111	43.9	0.894
−20	−28.9	1.158	13	−10.6	1.078	46	7.8	1.008	79	26.1	0.946	112	44.4	0.892
−19	−28.3	1.156	14	−10.0	1.076	47	8.3	1.006	80	26.7	0.945	113	45.0	0.891
−18	−27.8	1.153	15	− 9.4	1.073	48	8.9	1.004	81	27.2	0.943	114	45.6	0.890
−17	−27.2	1.151	16	− 8.9	1.071	49	9.4	1.002	82	27.8	0.941	115	46.1	0.888
−16	−26.7	1.148	17	− 8.3	1.069	50	10.0	1.000	83	28.3	0.939	116	46.7	0.886
−15	−26.1	1.145	18	− 7.8	1.067	51	10.6	0.998	84	28.9	0.938	117	47.2	0.885
−14	−25.6	1.143	19	− 7.2	1.064	52	11.1	0.996	85	29.4	0.936	118	47.8	0.884
−13	−25.0	1.140	20	− 6.7	1.062	53	11.7	0.994	86	30.0	0.934	119	48.3	0.882
−12	−24.4	1.138	21	− 6.1	1.060	54	12.2	0.992	87	30.6	0.933	120	48.9	0.881
−11	−23.9	1.135	22	− 5.6	1.058	55	12.8	0.990	88	31.1	0.931	121	49.4	0.880
−10	−23.3	1.133	23	− 5.0	1.056	56	13.3	0.988	89	31.7	0.929	122	50.0	0.878
− 9	−22.8	1.130	24	− 4.4	1.054	57	13.9	0.986	90	32.2	0.928	123	50.6	0.877
− 8	−22.2	1.128	25	− 3.9	1.051	58	14.4	0.985	91	32.8	0.926	124	51.1	0.876
− 7	−21.7	1.125	26	− 3.3	1.049	59	15.0	0.983	92	33.3	0.924	125	51.7	0.874
− 6	−21.1	1.123	27	− 2.8	1.047	60	15.6	0.981	93	33.9	0.923	126	52.2	0.873
− 5	−20.6	1.120	28	− 2.2	1.045	61	16.1	0.979	94	34.4	0.921	127	52.8	0.871
− 4	−20.0	1.118	29	− 1.7	1.043	62	16.7	0.977	95	35.0	0.919	128	53.3	0.870
− 3	−19.4	1.115	30	− 1.1	1.041	63	17.2	0.975	96	35.6	0.917	129	53.9	0.868
− 2	−18.9	1.113	31	− 0.6	1.039	64	17.8	0.973	97	36.1	0.916	130	54.4	0.867
− 1	−18.3	1.111	32	0.0	1.036	65	18.3	0.972	98	36.7	0.914			
0	−17.8	1.108	33	+ 0.6	1.034	66	18.9	0.970	99	37.2	0.912			
+ 1	−17.2	1.106	34	1.1	1.032	67	19.4	0.968	100	37.8	0.911			
2	−16.7	1.103	35	1.7	1.030	68	20.0	0.966	101	38.3	0.909			
3	−16.1	1.101	36	2.2	1.028	69	20.6	0.964	102	38.9	0.908			
4	−15.6	1.099	37	2.8	1.026	70	21.1	0.962	103	39.4	0.906			
5	−15.0	1.096	38	3.3	1.024	71	21.7	0.961	104	40.0	0.905			
6	−14.4	1.094	39	3.9	1.022	72	22.2	0.959	105	40.6	0.903			
7	−13.9	1.092	40	4.4	1.020	73	22.8	0.957	106	41.1	0.902			

TABLE 9-3. PARALLAX OF THE SUN FOR THE FIRST DAY OF EACH MONTH
Based on solar parallax = 8.80″

Altitude	Jan. 1	Feb. 1	Mar. 1	Apr. 1	May 1	June 1	July 1	Aug. 1	Sept. 1	Oct. 1	Nov. 1	Dec. 1
0°	8″.95	8″.93	8″.88	8″.81	8″.73	8″.68	8″.66	8″.67	8″.72	8″.79	8″.87	8″.92
3	8.94	8.92	8.87	8.80	8.72	8.67	8.65	8.66	8.71	8.78	8.86	8.91
6	8.90	8.88	8.83	8.76	8.68	8.63	8.61	8.62	8.67	8.74	8.82	8.87
9	8.84	8.82	8.77	8.70	8.62	8.57	8.55	8.56	8.61	8.68	8.76	8.81
12	8.75	8.73	8.69	8.62	8.54	8.49	8.47	8.48	8.53	8.60	8.68	8.72
15	8.64	8.63	8.58	8.51	8.43	8.38	8.36	8.37	8.42	8.49	8.57	8.62
18	8.51	8.49	8.45	8.38	8.30	8.26	8.24	8.25	8.29	8.36	8.44	8.48
21	8.36	8.34	8.29	8.23	8.15	8.10	8.08	8.09	8.14	8.21	8.28	8.33
24	8.18	8.16	8.11	8.05	7.97	7.93	7.91	7.92	7.97	8.03	8.10	8.15
27	7.97	7.96	7.91	7.85	7.78	7.73	7.72	7.72	7.77	7.83	7.90	7.95
30	7.75	7.73	7.69	7.63	7.56	7.52	7.50	7.51	7.55	7.61	7.68	7.72
33	7.51	7.49	7.45	7.39	7.32	7.28	7.26	7.27	7.31	7.37	7.44	7.48
36	7.24	7.22	7.18	7.13	7.06	7.02	7.01	7.01	7.05	7.11	7.18	7.22
39	6.96	6.94	6.90	6.85	6.78	6.75	6.73	6.74	6.78	6.83	6.89	6.93
42	6.65	6.64	6.60	6.55	6.49	6.45	6.44	6.44	6.48	6.53	6.59	6.63
44	6.44	6.42	6.39	6.34	6.28	6.24	6.23	6.24	6.27	6.32	6.38	6.42
46	6.22	6.20	6.17	6.12	6.06	6.03	6.02	6.02	6.06	6.11	6.16	6.20
48	5.99	5.98	5.94	5.89	5.84	5.81	5.79	5.80	5.83	5.88	5.93	5.97
50	5.75	5.74	5.71	5.66	5.61	5.58	5.57	5.57	5.61	5.65	5.70	5.73
52	5.51	5.50	5.47	5.42	5.38	5.34	5.33	5.34	5.37	5.41	5.46	5.49
54	5.26	5.25	5.22	5.18	5.13	5.10	5.09	5.10	5.13	5.17	5.21	5.24
56	5.00	4.99	4.97	4.93	4.88	4.85	4.84	4.85	4.88	4.92	4.96	4.99
58	4.74	4.73	4.71	4.67	4.63	4.60	4.59	4.59	4.62	4.66	4.70	4.73
60	4.48	4.46	4.44	4.40	4.36	4.34	4.33	4.34	4.36	4.40	4.44	4.46
62	4.20	4.19	4.17	4.14	4.10	4.08	4.07	4.07	4.09	4.13	4.16	4.19
64	3.92	3.91	3.89	3.86	3.83	3.81	3.80	3.80	3.82	3.85	3.89	3.91
66	3.64	3.63	3.61	3.58	3.55	3.53	3.52	3.53	3.55	3.57	3.61	3.63
68	3.35	3.35	3.33	3.30	3.27	3.25	3.24	3.25	3.27	3.29	3.32	3.34
70	3.06	3.05	3.04	3.01	2.99	2.97	2.96	2.97	2.98	3.01	3.03	3.05
72	2.77	2.76	2.74	2.72	2.70	2.68	2.68	2.68	2.69	2.72	2.74	2.76
74	2.47	2.46	2.45	2.43	2.41	2.39	2.39	2.39	2.40	2.42	2.44	2.46
76	2.17	2.16	2.15	2.13	2.11	2.10	2.09	2.10	2.11	2.13	2.15	2.16
78	1.86	1.86	1.85	1.83	1.81	1.80	1.80	1.80	1.81	1.83	1.84	1.85
80	1.55	1.55	1.54	1.53	1.52	1.51	1.50	1.51	1.51	1.53	1.54	1.55
82	1.25	1.24	1.24	1.23	1.22	1.21	1.21	1.21	1.21	1.22	1.23	1.24
84	0.94	0.93	0.93	0.92	0.91	0.91	0.90	0.91	0.91	0.92	0.93	0.93
86	0.62	0.62	0.62	0.61	0.61	0.61	0.60	0.61	0.61	0.61	0.62	0.62
88	0.31	0.31	0.31	0.31	0.30	0.30	0.30	0.30	0.30	0.31	0.31	0.31
90	0	0	0	0	0	0	0	0	0	0	0	0

Table 9-4. Solar Attachment Declination Corrections*

Barometer 29.6 in., temperature 50°F

Hr from noon	The sun's apparent declination										
	+25°	+20°	+15°	+10°	+5°	0°	−5°	−10°	−15°	−20°	−25°
					Latitude 2°						
	′	′	′	′	′	′	′	′	′	′	′
0	−0.4	−0.3	−0.2	−0.1	0.0	0.0	0.1	0.2	0.3	0.4	0.5
1	−0.4	−0.3	−0.2	−0.1	0.0	0.0	0.1	0.2	0.3	0.4	0.5
2	−0.4	−0.3	−0.2	−0.1	0.0	0.0	0.1	0.2	0.3	0.4	0.5
3	−0.4	−0.3	−0.2	−0.1	0.0	0.0	0.1	0.2	0.3	0.4	0.5
4	−0.4	−0.3	−0.2	−0.1	0.0	+0.1	0.2	0.2	0.3	0.4	0.5
5	−0.3	−0.2	−0.1	0.0	0.0	0.1	0.2	0.3	0.4	0.5	0.6
					Latitude 4°						
	′	′	′	′	′	′	′	′	′	′	′
0	−0.4	−0.3	−0.2	−0.1	0.0	0.1	0.2	0.2	0.3	0.4	0.5
1	−0.4	−0.3	−0.2	−0.1	0.0	0.1	0.2	0.2	0.3	0.4	0.5
2	−0.4	−0.3	−0.2	−0.1	0.0	0.1	0.2	0.2	0.3	0.4	0.5
3	−0.3	−0.2	−0.2	−0.1	0.0	0.1	0.2	0.3	0.4	0.5	0.6
4	−0.3	−0.2	−0.1	0.0	0.0	0.1	0.2	0.3	0.4	0.5	0.6
5	−0.2	−0.1	0.0	0.1	0.2	0.3	0.3	0.4	0.6	0.7	0.8
					Latitude 6°						
	′	′	′	′	′	′	′	′	′	′	′
0	−0.3	−0.2	0.3	0.1	0.0	0.1	0.2	0.3	0.4	0.5	0.6
1	−0.3	−0.2	−0.1	−0.1	0.0	0.1	0.2	0.3	0.4	0.5	0.6
2	−0.3	−0.2	−0.1	−0.1	0.0	0.1	0.2	0.3	0.4	0.5	0.6
3	−0.3	−0.2	−0.1	0.0	0.1	0.1	0.2	0.3	0.4	0.5	0.6
4	−0.2	−0.1	−0.1	0.0	0.1	0.2	0.3	0.4	0.5	0.6	0.7
5	0.0	0.0	+0.1	0.2	0.3	0.4	0.5	0.6	0.7	0.9	1.0
					Latitude 8°						
	′	′	′	′	′	′	′	′	′	′	′
0	−0.3	−0.2	−0.1	−0.0	0.0	0.1	0.2	0.3	0.4	0.5	0.6
1	−0.3	−0.2	−0.1	0.0	0.1	0.1	0.2	0.3	0.4	0.5	0.6
2	−0.3	−0.2	−0.1	0.0	0.1	0.1	0.2	0.3	0.4	0.5	0.6
3	−0.2	−0.1	−0.1	0.0	0.1	0.2	0.3	0.4	0.5	0.6	0.7
4	−0.2	−0.1	0.0	0.1	0.2	0.3	0.4	0.5	0.6	0.7	0.8
5	0.1	0.1	0.2	0.3	0.4	0.5	0.6	0.8	0.9	1.1	1.3

* The correction taken from the table is added algebraically to the sun's declination, taking account of the sign (+ or −) of the correction and of the sign of the declination. When the latitude is south, the sign of the declination is changed before entering the tables and thereafter used with the new sign.

TABLE 9-4. SOLAR ATTACHMENT DECLINATION CORRECTIONS (*Continued*)

Hr from noon	The sun's apparent declination										
	+25°	+20°	+15°	+10°	+5°	0°	−5°	−10°	−15°	−20°	−25°
	Latitude 10°										
	′	′	′	′	′	′	′	′	′	′	′
0	−0.3	−0.2	−0.1	0.0	0.1	0.2	0.3	0.3	0.4	0.5	0.7
1	−0.2	−0.2	−0.1	0.0	0.1	0.2	0.3	0.4	0.5	0.6	0.7
2	−0.2	−0.2	−0.1	0.0	0.1	0.2	0.3	0.4	0.5	0.6	0.7
3	−0.2	−0.1	0.0	0.1	0.2	0.2	0.3	0.4	0.5	0.6	0.8
4	−0.1	0.0	0.1	0.2	0.2	0.3	0.4	0.5	0.7	0.8	0.9
5	0.2	0.2	0.3	0.4	0.5	0.6	0.8	0.9	1.1	1.3	1.6
	Latitude 12°										
	′	′	′	′	′	′	′	′	′	′	′
0	−0.2	−0.1	0.0	0.0	0.1	0.2	0.3	0.4	0.5	0.6	0.7
1	−0.2	−0.1	0.0	0.0	0.1	0.2	0.3	0.4	0.5	0.6	0.7
2	−0.2	−0.1	0.0	0.1	0.1	0.2	0.3	0.4	0.5	0.6	0.8
3	−0.1	−0.1	0.0	0.1	0.2	0.3	0.4	0.5	0.6	0.7	0.8
4	0.0	0.1	0.1	0.2	0.3	0.4	0.5	0.6	0.7	0.9	1.1
5	0.2	0.3	0.4	0.5	0.7	0.8	0.9	1.1	1.3	1.6	2.0
	Latitude 14°										
	′	′	′	′	′	′	′	′	′	′	′
0	−0.2	−0.1	0.0	0.1	0.2	0.2	0.3	0.4	0.5	0.6	0.8
1	−0.2	−0.1	0.0	0.1	0.2	0.2	0.3	0.4	0.5	0.7	0.8
2	−0.1	−0.1	0.0	0.1	0.2	0.3	0.4	0.5	0.6	0.7	0.8
3	−0.1	0.0	0.1	0.2	0.2	0.3	0.4	0.5	0.7	0.8	0.9
4	0.0	0.1	0.2	0.3	0.4	0.5	0.6	0.7	0.8	1.0	1.2
5	0.3	0.4	0.5	0.6	0.8	0.9	1.1	1.3	1.6	2.0	2.5
	Latitude 16°										
	′	′	′	′	′	′	′	′	′	′	′
0	−0.2	−0.1	0.0	0.1	0.2	0.3	0.4	0.5	0.6	0.7	0.8
1	−0.1	−0.1	0.0	0.1	0.2	0.3	0.4	0.5	0.6	0.7	0.8
2	−0.1	0.0	0.1	0.1	0.2	0.3	0.4	0.5	0.6	0.8	0.9
3	0.0	0.0	0.1	0.2	0.3	0.4	0.5	0.6	0.7	0.9	1.0
4	0.1	0.2	0.3	0.3	0.4	0.5	0.7	0.8	0.9	1.1	1.3
5	0.4	0.5	0.6	0.7	0.9	1.1	1.3	1.5	1.9	2.4	3.1
	Latitude 18°										
	′	′	′	′	′	′	′	′	′	′	′
0	−0.1	0.0	0.0	0.1	0.2	0.3	0.4	0.5	0.6	0.7	0.9
1	−0.1	0.0	0.1	0.1	0.2	0.3	0.4	0.5	0.6	0.8	0.9
2	−0.1	0 0	0.1	0.2	0.3	0.4	0.5	0.6	0.7	0.8	1.0
3	0.0	0.1	0.2	0.2	0.3	0.4	0.5	0.7	0.8	0.9	1.1
4	0.1	0.2	0.3	0.4	0.5	0.6	0 7	0.9	1.1	1.3	1.5
5	0.5	0.6	0.7	0 8	1.0	1.2	1.4	1.8	2.2	2.8	3.9

TABLE 9-1. SOLAR ATTACHMENT DECLINATION CORRECTIONS (*Continued*)

Hr from noon	The sun's apparent declination										
	+25°	+20°	+15°	+10°	+5°	0°	−5°	−10°	−15°	−20°	−25°
Latitude 20°											
	′	′	′	′	′	′	′	′	′	′	′
0	−0.1	0.0	0.1	0.2	0.3	0.3	0.4	0.5	0.7	0.8	0.9
1	−0.1	0.0	0.1	0.2	0.3	0.4	0.5	0.6	0.7	0.8	1.0
2	0.0	0.0	0.1	0.2	0.3	0.4	0.5	0.6	0.7	0.9	1.0
3	0.0	0.1	0.2	0.3	0.4	0.5	0.6	0.7	0.9	1.0	1.2
4	0.2	0.3	0.4	0.5	0.6	0.7	0.8	1.0	1.2	1.4	1.7
5	0.5	0.7	0.8	0.9	1.1	1.3	1.6	2.0	2.6	3.5	5.2
Latitude 22°											
	′	′	′	′	′	′	′	′	′	′	′
0	0.0	0.0	0.1	0.2	0.3	0.4	0.5	0.6	0.7	0.9	1.0
1	0.0	0.0	0.1	0.2	0.3	0.4	0.5	0.6	0.7	0.9	1.0
2	0.0	0.1	0.2	0.3	0.3	0.4	0.5	0.7	0.8	1.0	1.1
3	0.1	0.2	0.3	0.3	0.4	0.5	0.7	0.8	0.9	1.1	1.3
4	0.2	0.3	0.4	0.5	0.6	0.8	0.9	1.1	1.3	1.6	1.9
5	0.6	0.7	0.9	1.0	1.2	1.5	1.8	2.3	3.0	4.3	7.1
Latitude 24°											
	′	′	′	′	′	′	′	′	′	′	′
0	0.0	0.1	0.2	0.2	0.3	0.4	0.5	0.6	0.8	0.9	1.1
1	0.0	0.1	0.2	0.3	0.3	0.4	0.5	0.7	0.8	0.9	1.1
2	0.0	0.1	0.2	0.3	0.4	0.5	0.6	0.7	0.9	1.0	1.2
3	0.1	0.2	0.3	0.4	0.5	0.0	0.7	0.9	1.0	1.2	1.5
4	0.3	0.4	0.5	0.6	0.7	0.8	1.0	1.2	1.5	1.8	2.2
5	0.7	0.8	0.9	1.1	1.3	1.6	2.0	2.5	3.5	5.3	10.5
Latitude 26°											
	′	′	′	′	′	′	′	′	′	′	′
0	0.0	0.1	0.2	0.3	0.4	0.5	0.6	0.7	0.8	1.0	1.2
1	0.0	0.1	0.2	0.3	0.4	0.5	0.6	0.7	0.8	1.0	1.2
2	0.1	0.2	0.2	0.3	0.4	0.5	0.7	0.8	0.9	1.1	1.3
3	0.2	0.2	0.3	0.4	0.5	0.7	0.8	0.9	1.1	1.3	1.6
4	0.3	0.4	0.5	0.6	0.8	0.9	1.1	1.3	1.6	2.0	2.5
5	0.7	0.9	1.0	1.2	1.4	1.8	2.2	2.9	4.2	6.8	—
Latitude 28°											
	′	′	′	′	′	′	′	′	′	′	′
0	0.0	0.1	0.2	0.3	0.4	0.5	0.6	0.7	0.9	1.1	1.3
1	0.1	0.1	0.2	0.3	0.4	0.5	0.6	0.8	0.9	1.1	1.3
2	0.1	0.2	0.3	0.4	0.5	0.6	0.7	0.8	1.0	1.2	1.4
3	0.2	0.3	0.4	0.5	0.6	0.7	0.9	1.0	1.2	1.5	1.8
4	0.4	0.5	0.6	0.7	0.8	1.0	1.2	1.4	1.8	2.2	2.9
5	0.8	0.9	1.1	1.3	1.5	2.0	2.5	3.2	4.9	9.2	—

TABLE 9-4. SOLAR ATTACHMENT DECLINATION CORRECTIONS (*Continued*)

Hr from noon	The sun's apparent declination										
	+25°	+20°	+15°	+10°	+5°	0°	−5°	−10°	−15°	−20°	−25°
	Latitude 30°										
	′	′	′	′	′	′	′	′	′	′	′
0	0.1	0.2	0.3	0.3	0.4	0.5	0.7	0.8	1.0	1.1	1.4
1	0.1	0.2	0.3	0.4	0.5	0.6	0.7	0.8	1.0	1.2	1.4
2	0.1	0.2	0.3	0.4	0.5	0.6	0.8	0.9	1.1	1.3	1.6
3	0.2	0.3	0.4	0.5	0.6	0.8	0.9	1.1	1.3	1.6	2.0
4	0.4	0.5	0.6	0.8	0.9	1.1	1.3	1.6	2.0	2.5	3.3
5	0.8	1.0	1.2	1.4	1.7	2.1	2.7	3.6	5.9	—	—
	Latitude 32°										
	′	′	′	′	′	′	′	′	′	′	′
0	0.1	0.2	0.3	0.4	0.5	0.6	0.7	0.9	1.0	1.2	1.5
1	0.1	0.2	0.3	0.4	0.5	0.6	0.7	0.9	1.1	1.3	1.5
2	0.2	0.3	0.4	0.5	0.6	0.7	0.8	1.0	1.2	1.4	1.7
3	0.3	0.4	0.5	0.6	0.7	0.8	1.0	1.2	1.4	1.8	2.2
4	0.5	0.6	0.7	0.8	1.0	1.2	1.4	1.7	2.2	2.8	3.9
5	0.9	1.0	1.2	1.5	1.8	2.3	3.0	4.3	7.3	—	—
	Latitude 34°										
	′	′	′	′	′	′	′	′	′	′	′
0	0.2	0.2	0.3	0.4	0.5	0.6	0.8	0.9	1.1	1.3	1.6
1	0.2	0.3	0.3	0.4	0.5	0.7	0.8	0.9	1.1	1.4	1.6
2	0.2	0.3	0.4	0.5	0.6	0.7	0.9	1.1	1.3	1.5	1.8
3	0.3	0.4	0.5	0.6	0.8	0.9	1.1	1.3	1.6	1.9	2.4
4	0.5	0.6	0.8	0.9	1.1	1.3	1.6	1.9	2.4	3.2	4.6
5	0.9	1.1	1.3	1.6	2.0	2.5	3.3	4.9	9.1	—	—
	Latitude 36°										
	′	′	′	′	′	′	′	′	′	′	′
0	0.2	0.3	0.4	0.5	0.6	0.7	0.8	1.0	1.2	1.4	1.7
1	0.2	0.3	0.4	0.5	0.6	0.7	0.9	1.0	1.2	1.5	1.8
2	0.3	0.3	0.4	0.5	0.7	0.8	0.9	1.1	1.4	1.7	2.0
3	0.4	0.5	0.6	0.7	0.8	1.0	1.2	1.4	1.7	2.2	2.8
4	0.6	0.7	0.8	1.0	1.2	1.4	1.7	2.1	2.7	3.7	5.7
5	1.0	1.1	1.4	1.7	2.1	2.7	3.7	5.7	—	—	—
	Latitude 38°										
	′	′	′	′	′	′	′	′	′	′	′
0	0.2	0.3	0.4	0.5	0.6	0.7	0.9	1.1	1.3	1.5	1.9
1	0.2	0.3	0.4	0.5	0.6	0.8	0.9	1.1	1.3	1.6	1.9
2	0.3	0.4	0.5	0.6	0.7	0.9	1.0	1.2	1.5	1.8	2.2
3	0.4	0.5	0.6	0.7	0.9	1.1	1.3	1.5	1.9	2.3	3.1
4	0.6	0.7	0.9	1.0	1.2	1.5	1.8	2.3	3.0	4.1	7.1
5	1.0	1.2	1.5	1.8	2.2	2.9	4.0	6.6	—	—	—

TABLE 9-4. SOLAR ATTACHMENT DECLINATION CORRECTIONS (*Continued*)

Hr from noon	The sun's apparent declination										
	+25°	+20°	+15°	+10°	+5°	0°	−5°	−10°	−15°	−20°	−25°
	Latitude 40°										
	′	′	′	′	′	′	′	′	′	′	′
0	0.3	0.3	0.4	0.5	0.7	0.8	1.0	1.1	1.4	1.7	2.0
1	0.3	0.4	0.5	0.6	0.7	0.8	1.0	1.2	1.4	1.7	2.1
2	0.3	0.4	0.5	0.6	0.8	0.9	1.1	1.3	1.6	2.0	2.5
3	0.4	0.5	0.7	0.8	0.9	1.1	1.4	1.6	2.0	2.6	3.5
4	0.7	0.8	0.9	1.1	1.3	1.6	2.0	2.5	3.3	4.8	9.4
5	1.0	1.3	1.5	1.9	2.3	3.0	4.3	7.6	—	—	—
	Latitude 42°										
	′	′	′	′	′	′	′	′	′	′	′
0	0.3	0.4	0.5	0.6	0.7	0.9	1.0	1.2	1.5	1.8	2.2
1	0.3	0.4	0.5	0.6	0.7	0.9	1.1	1.3	1.5	1.9	2.4
2	0.4	0.5	0.6	0.7	0.8	1.0	1.2	1.4	1.7	2.1	2.8
3	0.5	0.6	0.7	0.9	1.0	1.2	1.5	1.8	2.2	2.9	4.1
4	0.7	0.8	1.0	1.2	1.4	1.7	2.1	2.8	3.8	6.0	—
5	1.1	1.3	1.6	1.9	2.5	3.3	4.9	9.1	—	—	—
	Latitude 44°										
	′	′	′	′	′	′	′	′	′	′	′
0	0.3	0.4	0.5	0.6	0.8	0.9	1.1	1.3	1.6	1.9	2.5
1	0.3	0.4	0.5	0.7	0.8	0.9	1.1	1.4	1.6	2.0	2.6
2	0.4	0.5	0.6	0.7	0.9	1.1	1.3	1.5	1.9	2.4	3.1
3	0.5	0.6	0.8	0.9	1.1	1.3	1.6	1.9	2.4	3.3	4.8
4	0.7	0.9	1.0	1.2	1.5	1.7	2.3	3.1	4.2	7.4	—
5	1.1	1.4	1.6	2.0	2.6	3.5	5.2	10.9	—	—	—
	Latitude 46°										
	′	′	′	′	′	′	′	′	′	′	′
0	0.4	0.5	0.6	0.7	0.8	1.0	1.2	1.4	1.7	2.2	2.8
1	0.4	0.5	0.6	0.7	0.9	1.0	1.2	1.5	1.8	2.2	2.9
2	0.4	0.6	0.7	0.8	1.0	1.1	1.3	1.7	2.0	2.6	3.6
3	0.6	0.7	0.8	1.0	1.2	1.4	1.7	2.1	2.7	3.7	5.8
4	0.8	0.9	1.1	1.3	1.6	2.0	2.5	3.4	4.8	9.5	—
5	1.2	1.4	1.7	2.1	2.8	3.8	5.6	—	—	—	—
	Latitude 48°										
	′	′	′	′	′	′	′	′	′	′	′
0	0.4	0.5	0.6	0.7	0.9	1.1	1.3	1.5	1.9	2.4	3.1
1	0.4	0.5	0.6	0.8	0.9	1.1	1.3	1.6	1.9	2.5	3.3
2	0.5	0.6	0.7	0.9	1.0	1.2	1.5	1.8	2.2	2.9	4.1
3	0.6	0.7	0.9	1.0	1.2	1.5	1.8	2.3	3.0	4.3	7.2
4	0.8	1.0	1.2	1.4	1.7	2.1	2.7	3.8	5.5	12.8	—
5	1.2	1.5	1.8	2.2	2.9	4.1	6.1	—	—	—	—

TABLE 9-4. SOLAR ATTACHMENT DECLINATION CORRECTIONS (*Continued*)

Hr from noon	The sun's apparent declination										
	+25°	+20°	+15°	+10°	+5°	0°	−5°	−10°	−15°	−20°	−25°
						Latitude 50°					
	′	′	′	′	′	′	′	′	′	′	′
0	0.4	0.5	0.7	0.8	1.0	1.1	1.4	1.6	2.1	2.6	3.5
1	0.5	0.6	0.7	0.8	1.0	1.2	1.4	1.7	2.1	2.8	3.8
2	0.5	0.6	0.8	0.9	1.1	1.3	1.6	1.9	2.4	3.3	4.9
3	0.6	0.8	0.9	1.1	1.3	1.6	2.0	2.5	3.3	4.9	9.5
4	0.9	1.0	1.2	1.5	1.8	2.3	3.0	4.1	6.5	—	—
5	1.2	1.5	1.9	2.3	3.0	4.3	6.9	—	—	—	—
						Latitude 52°					
	′	′	′	′	′	′	′	′	′	′	′
0	0.5	0.6	0.7	0.9	1.0	1.2	1.5	1.8	2.2	2.9	4.1
1	0.5	0.6	0.7	0.9	1.1	1.3	1.5	1.9	2.6	3.1	4.5
2	0.6	0.7	0.8	1.0	1.2	1.4	1.7	2.1	2.7	3.8	5.9
3	0.7	0.8	1.0	1.2	1.4	1.7	2.1	2.8	3.8	6.1	—
4	0.9	1.1	1.3	1.6	1.9	2.4	3.2	4.7	7.8	—	—
5	1.3	1.6	1.9	2.4	3.2	4.6	8.4	—	—	—	—
						Latitude 54°					
	′	′	′	′	′	′	′	′	′	′	′
0	0.5	0.6	0.8	0.9	1.1	1.3	1.6	1.9	2.5	3.3	4.9
1	0.5	0.7	0.8	0.9	1.1	1.4	1.6	2.0	2.6	3.5	5.4
2	0.6	0.7	0.9	1.0	1.3	1.5	1.9	2.3	3.1	4.4	7.5
3	0.7	0.9	1.0	1.3	1.5	1.7	2.3	3.1	4.4	7.6	—
4	1.0	1.1	1.4	1.6	2.1	2.6	3.9	5.4	10.9	—	—
5	1.3	1.6	2.0	2.5	3.4	5.1	9.6	—	—	—	—
						Latitude 56°					
	′	′	′	′	′	′	′	′	′	′	′
0	0.6	0.7	0.8	1.0	1.2	1.4	1.7	2.1	2.8	3.9	6.0
1	0.6	0.7	0.9	1.0	1.2	1.5	1.8	2.2	2.9	4.1	6.7
2	0.7	0.8	0.9	1.1	1.3	1.6	2.0	2.6	3.5	5.3	10.3
3	0.8	0.9	1.1	1.3	1.6	2.0	2.6	3.5	5.2	9.9	—
4	1.0	1.2	1.4	1.7	2.2	2.8	3.9	6.3	—	—	—
5	1.4	1.7	2.0	2.6	3.6	5.4	11.1	—	—	—	—
						Latitude 58°					
	′	′	′	′	′	′	′	′	′	′	′
0	0.6	0.7	0.9	1.1	1.3	1.5	1.9	2.4	3.1	4.5	7.7
1	0.6	0.8	0.9	1.1	1.3	1.6	1.9	2.5	3.3	4.8	8.9
2	0.7	0.8	1.0	1.2	1.4	1.8	2.2	2.9	4.0	6.5	—
3	0.8	1.0	1.2	1.4	1.7	2.2	2.8	3.9	6.2	—	—
4	1.0	1.2	1.5	1.8	2.3	3.0	4.3	7.4	—	—	—
5	1.4	1.7	2.1	2.7	3.8	5.7	—	—	—	—	—

TABLE 9-4. SOLAR ATTACHMENT DECLINATION CORRECTIONS (*Continued*)

Hr from noon	The sun's apparent declination										
	+25°	+20°	+15°	+10°	+5°	0°	−5°	−10°	−15°	−20°	−25°
Latitude 60°											
	′	′	′	′	′	′	′	′	′	′	′
0	0.7	0.8	1.0	1.1	1.4	1.6	2.0	2.6	3.5	5.4	—
1	0.7	0.8	1.0	1.2	1.4	1.7	2.1	2.7	3.8	5.9	—
2	0.8	0.9	1.1	1.3	1.5	1.9	2.4	3.2	4.6	8.3	—
3	0.9	1.0	1.3	1.5	1.8	2.3	3.1	4.4	7.6	—	—
4	1.1	1.3	1.6	1.9	2.5	3.3	4.8	8.9	—	—	—
5	1.4	1.7	2.2	2.8	3.9	6.0	—	—	—	—	—
Latitude 62°											
	′	′	′	′	′	′	′	′	′	′	′
0	0.7	0.9	1.0	1.2	1.5	1.8	2.2	2.9	4.1	6.8	—
1	0.7	0.9	1.0	1.3	1.5	1.8	2.3	3.1	4.4	7.5	—
2	0.8	1.0	1.1	1.4	1.7	2.1	2.6	3.6	5.6	—	—
3	0.9	1.1	1.3	1.6	2.0	2.5	3.4	5.1	9.8	—	—
4	1.1	1.4	1.7	2.0	2.6	3.6	5.5	—	—	—	—
5	1.5	1.8	2.3	3.0	4.2	6.9	—	—	—	—	—
Latitude 64°											
	′	′	′	′	′	′	′	′	′	′	′
0	0.8	0.9	1.1	1.3	1.6	1.9	2.5	3.3	4.9	9.1	—
1	0.8	0.9	1.1	1.3	1.6	2.0	2.6	3.5	5.3	—	—
2	0.9	1.0	1.2	1.5	1.8	2.2	2.9	4.2	6.9	—	—
3	1.0	1.2	1.4	1.7	2.1	2.8	3.8	6.0	—	—	—
4	1.2	1.4	1.7	2.2	2.8	3.9	6.3	—	—	—	—
5	1.5	1.8	2.3	3.1	4.4	7.5	—	—	—	—	—
Latitude 66°											
	′	′	′	′	′	′	′	′	′	′	′
0	0.8	1.0	1.2	1.4	1.7	2.1	2.8	3.8	6.0	—	—
1	0.8	1.0	1.2	1.4	1.8	2.2	2.9	4.0	6.5	—	—
2	0.9	1.1	1.3	1.6	1.9	2.5	3.3	4.9	9.0	—	—
3	1.0	1.2	1.5	1.8	2.2	3.0	4.3	7.3	—	—	—
4	1.2	1.5	1.8	2.3	3.0	4.3	7.2	—	—	—	—
5	1.5	1.9	2.4	3.2	4.5	8.2	—	—	—	—	—
Latitude 68°											
	′	′	′	′	′	′	′	′	′	′	′
0	0.9	1.1	1.3	1.5	1.9	2.4	3.1	4.5	7.8	—	—
1	0.9	1.1	1.3	1.6	1.9	2.4	3.2	4.7	8.6	—	—
2	1.0	1.2	1.4	1.7	2.1	2.7	3.7	5.8	—	—	—
3	1.1	1.3	1.6	2.0	2.5	3.4	5.0	9.2	—	—	—
4	1.3	1.6	1.9	2.4	3.2	4.7	8.5	—	—	—	—
5	1.6	2.0	2.5	3.3	4.9	9.1	—	—	—	—	—

TABLE 9-4. SOLAR ATTACHMENT DECLINATION CORRECTIONS (*Continued*)

Hr from noon	The sun's apparent declination										
	+25°	+20°	+15°	+10°	+5°	0°	−5°	−10°	−15°	−20°	−25°
	Latitude 70°										
	′	′	′	′	′	′	′	′	′	′	′
0	1.0	1.1	1.4	1.6	2.0	2.6	3.5	5.4	10.9	—	—
1	1.0	1.2	1.4	1.7	2.1	2.7	3.7	5.8	12.4	—	—
2	1.0	1.2	1.5	1.8	2.3	3.0	4.3	7.3	—	—	—
3	1.2	1.4	1.7	2.1	2.7	3.7	5.0	—	—	—	—
4	1.3	1.6	2.0	2.6	3.5	5.3	10.3	—	—	—	—
5	1.6	2.0	2.6	3.4	5.0	10.1	—	—	—	—	—

TABLE 9-5. CONVERSION OF TIME TO ARC

T.	A.	T.	A.	T.	A.	T.	A.	T.	A.	T.	A.
Hr	°	Hr	°	Hr	°	Hr	°	Hr	°	Hr	°
1	15	5	75	9	135	13	195	17	225	21	315
2	30	6	90	10	150	14	210	18	270	22	330
3	45	7	105	11	165	15	225	19	285	23	345
4	60	8	120	12	180	16	240	20	300	24	360

Min Sec	° ′ ′ ″	Min Sec	° ′ ′ ″	Min Sec	° ′ ′ ″
1	0 15	21	5 15	41	10 15
2	0 30	22	5 30	42	10 30
3	0 45	23	5 45	43	10 45
4	1 0	24	6 0	44	11 0
5	1 15	25	6 15	45	11 15
6	1 30	26	6 30	46	11 30
7	1 45	27	6 45	47	11 45
8	2 0	28	7 0	48	12 0
9	2 15	29	7 15	49	12 15
10	2 30	30	7 30	50	12 30
11	2 45	31	7 45	51	12 45
12	3 0	32	8 0	52	13 0
13	3 15	33	8 15	53	13 15
14	3 30	34	8 30	54	13 30
15	3 45	35	8 45	55	13 45
16	4 0	36	9 0	56	14 0
17	4 15	37	9 15	57	14 15
18	4 30	38	9 30	58	14 30
19	4 45	39	9 45	59	14 45
20	5 0	40	10 0	60	15 0

100ths of sec of time	0.00	0.01	0.02	0.03	0.04	0.05	0.06	0.07	0.08	0.09
Sec	″	″	″	″	″	″	″	″	″	″
0.00	0.00	0.15	0.30	0.45	0.60	0.75	0.90	1.05	1.20	1.35
.10	1.50	1.65	1.80	1.95	2.10	2.25	2.40	2.55	2.70	2.85
.20	3.00	3.15	3.30	3.45	3.60	3.75	3.90	4.05	4.20	4.35
.30	4.50	4.65	4.80	4.95	5.10	5.25	5.40	5.55	5.70	5.85
.40	6.00	6.15	6.30	6.45	6.60	6.75	6.90	7.05	7.20	7.35
0.50	7.50	7.65	7.80	7.95	8.10	8.25	8.40	8.55	8.70	8.85
.60	9.00	9.15	9.30	9.45	9.60	9.75	9.90	10.05	10.20	10.35
.70	10.50	10.65	10.80	10.95	11.10	11.25	11.40	11.55	11.70	11.85
.80	12.00	12.15	12.30	12.45	12.60	12.75	12.90	13.05	13.20	13.35
.90	13.50	13.65	13.80	13.95	14.10	14.25	14.40	14.55	14.70	14.85

TABLE 9-6. CONVERSION OF ARC TO TIME

°	h m	°	h m	°	h m	°	h m	°	h m	°	h m	′	m s	″	s
0	0 0	60	4 0	120	8 0	180	12 0	240	16 0	300	20 0	0	0 0	0	0.00
1	0 4	61	4 4	121	8 4	181	12 4	241	16 4	301	20 4	1	0 4	1	0.07
2	0 8	62	4 8	122	8 8	182	12 8	242	16 8	302	20 8	2	0 8	2	0.13
3	0 12	63	4 12	123	8 12	183	12 12	243	16 12	303	20 12	3	0 12	3	0.20
4	0 16	64	4 16	124	8 16	184	12 16	244	16 16	304	20 16	4	0 16	4	0.27
5	0 20	65	4 20	125	8 20	185	12 20	245	16 20	305	20 20	5	0 20	5	0.33
6	0 24	66	4 24	126	8 24	186	12 24	246	16 24	306	20 24	6	0 24	6	0.40
7	0 28	67	4 28	127	8 28	187	12 28	247	16 28	307	20 28	7	0 28	7	0.47
8	0 32	68	4 32	128	8 32	188	12 32	248	16 32	308	20 32	8	0 32	8	0.53
9	0 36	69	4 36	129	8 36	189	12 36	249	16 36	309	20 36	9	0 36	9	0.60
10	0 40	70	4 40	130	8 40	190	12 40	250	16 40	310	20 40	10	0 40	10	0.67
11	0 44	71	4 44	131	8 44	191	12 44	251	16 44	311	20 44	11	0 44	11	0.73
12	0 48	72	4 48	132	8 48	192	12 48	252	16 48	312	20 48	12	0 48	12	0.80
13	0 52	73	4 52	133	8 52	193	12 52	253	16 52	313	20 52	13	0 52	13	0.87
14	0 56	74	4 56	134	8 56	194	12 56	254	16 56	314	20 56	14	0 56	14	0.93
15	1 0	75	5 0	135	9 0	195	13 0	255	17 0	315	21 0	15	1 0	15	1.00
16	1 4	76	5 4	136	9 4	196	13 4	256	17 4	316	21 4	16	1 4	16	1.07
17	1 8	77	5 8	137	9 8	197	13 8	257	17 8	317	21 8	17	1 8	17	1.13
18	1 12	78	5 12	138	9 12	198	13 12	258	17 12	318	21 12	18	1 12	18	1.20
19	1 16	79	5 16	139	9 16	199	13 16	259	17 16	319	21 16	19	1 16	19	1.27
20	1 20	80	5 20	140	9 20	200	13 20	260	17 20	320	21 20	20	1 20	20	1.33
21	1 24	81	5 24	141	9 24	201	13 24	261	17 24	321	21 24	21	1 24	21	1.40
22	1 28	82	5 28	142	9 28	202	13 28	262	17 28	322	21 28	22	1 28	22	1.47
23	1 32	83	5 32	143	9 32	203	13 32	263	17 32	323	21 32	23	1 32	23	1.53
24	1 36	84	5 36	144	9 36	204	13 36	264	17 36	324	21 36	24	1 36	24	1.60
25	1 40	85	5 40	145	9 40	205	13 40	265	17 40	325	21 40	25	1 40	25	1.67
26	1 44	86	5 44	146	9 44	206	13 44	266	17 44	326	21 44	26	1 44	26	1.73
27	1 48	87	5 48	147	9 48	207	13 48	267	17 48	327	21 48	27	1 48	27	1.80
28	1 52	88	5 52	148	9 52	208	13 52	268	17 52	328	21 52	28	1 52	28	1.87
29	1 56	89	5 56	149	9 56	209	13 56	269	17 56	329	21 56	29	1 56	29	1.93
30	2 0	90	6 0	150	10 0	210	14 0	270	18 0	330	22 0	30	2 0	30	2.00
31	2 4	91	6 4	151	10 4	211	14 4	271	18 4	331	22 4	31	2 4	31	2.07
32	2 8	92	6 8	152	10 8	212	14 8	272	18 8	332	22 8	32	2 8	32	2.13
33	2 12	93	6 12	153	10 12	213	14 12	273	18 12	333	12 12	33	2 12	33	2.20
34	2 16	94	6 16	154	10 16	214	14 16	274	18 16	334	22 16	34	2 16	34	2.27
35	2 20	95	6 20	155	10 20	215	14 20	275	18 20	335	22 20	35	2 20	35	2.33
36	2 24	96	6 24	156	10 24	216	14 24	276	18 24	336	22 24	36	2 24	36	2.40
37	2 28	97	6 28	157	10 28	217	14 28	277	18 28	337	22 28	37	2 28	37	2.47
38	2 32	98	6 32	158	10 32	218	14 32	278	18 32	338	22 32	38	2 32	38	2.53
39	2 36	99	6 36	159	10 36	219	14 36	279	18 36	339	22 36	39	2 36	39	2.60
40	2 40	100	6 40	160	10 40	220	14 40	280	18 40	340	22 40	40	2 40	40	2.67
41	2 44	101	6 44	161	10 44	221	14 44	281	18 44	341	22 44	41	2 44	41	2.73
42	2 48	102	6 48	162	10 48	222	14 48	282	18 48	342	22 48	42	2 48	42	2.80
43	2 52	103	6 52	163	10 52	223	14 52	283	18 52	343	22 52	43	2 52	43	2.87
44	2 56	104	6 56	164	10 56	224	14 56	284	18 56	344	22 56	44	2 56	44	2.93
45	3 0	105	7 0	165	11 0	225	15 0	285	19 0	345	23 0	45	3 0	45	3.00
46	3 4	106	7 4	166	11 4	226	15 4	286	19 4	346	23 4	46	3 4	46	3.07
47	3 8	107	7 8	167	11 8	227	15 8	287	19 8	347	23 8	47	3 8	47	3.13
48	3 12	108	7 12	168	11 12	228	15 12	288	19 12	348	23 12	48	3 12	48	3.20
49	3 16	109	7 16	169	11 16	229	15 16	289	19 16	349	23 16	49	3 16	49	3.27
50	3 20	110	7 20	170	11 20	230	15 20	290	19 20	350	23 20	50	3 20	50	3.33
51	3 24	111	7 24	171	11 24	231	15 24	291	19 24	351	23 24	51	3 24	51	3.40
52	3 28	112	7 28	172	11 28	232	15 28	292	19 28	352	23 28	52	3 28	52	3.47
53	3 32	113	7 32	173	11 32	233	15 32	293	19 32	353	23 32	53	3 32	53	3.53
54	3 36	114	7 36	174	11 36	234	13 36	294	19 36	354	23 36	54	3 36	54	3.60
55	3 40	115	7 40	175	11 40	235	15 40	295	19 40	355	23 40	55	3 40	55	3.67
56	3 44	116	7 44	176	11 44	236	15 44	296	19 44	356	23 44	56	3 44	56	3.73
57	3 48	117	7 48	177	11 48	237	15 48	297	19 48	357	23 48	57	3 48	57	3.80
58	3 52	118	7 52	178	11 52	238	15 52	298	19 52	358	23 52	58	3 52	58	3.87
59	3 56	119	7 56	179	11 56	239	15 56	299	19 56	359	23 56	59	3 56	59	3.93
60	4 0	120	8 0	180	12 0	240	16 0	300	20 0	360	24 0	60	4 0	60	4.00

CHAPTER 10
BOUNDARY SURVEYS

10-1. Definition. The art of measuring, laying out, and describing land boundaries is called **land surveying.** It is a highly specialized branch of civil engineering that can be mastered only through long experience. To become proficient at land surveying, a civil engineer must acquire a high type of surveying skill, a fundamental understanding of the law relating to land, a personal knowledge and the records of the survey lines where he works, and a familiarity with the customs and practices of land surveying in his community.

He must have the will to persevere in the accumulation of all the evidence that might affect the problem in hand and the imagination and analytical ability to correlate conflicting evidence.

LEGAL ASPECTS

10-2. The Genesis of the Position of Boundary Lines. Boundaries are both indicated and created by acts on the ground or by legal instruments. When these acts or instruments designate a new tract that is part, not the whole, of an old tract of land, new boundaries are created. The boundary lines of the new tract are then made up of new lines and usually parts of the old lines that bounded the old tract. The positions of the new lines depend on the present acts or instruments but the positions of the old lines depend on those acts or instruments, executed in the past, that in each case designated a new tract.

It is therefore usually the case that the positions of the boundary lines of a parcel of land depend on surveys and descriptions made at various times in the past as well as the present survey and description of the parcel. Two principles of law regulate the positions of boundaries.

10-3. Principle No. 1. Position is determined by the *intent* of the parties that establish the new boundary. Their intent is judged by the evidence of their acts, their written instruments, and the circumstances involved.

With the passage of time, this evidence becomes more difficult to secure.

When incomplete, it frequently leads to absurdities. Consequently more positive evidence must be sought. This has lead to Principle No. 2.

10-4. Principle No. 2. A number of rules of jurisprudence have been developed, all of which tend to eliminate old evidence in favor of present conditions. These rules may be grouped together and loosely expressed as Principle No. 2 as follows: The basic evidence of the position of an old boundary is the *acceptance* of that position over a period of years. The longer the period of acceptance, the stronger the evidence becomes. The principle is more particularly applied when improvements have been built that would be affected by any change in position of the boundary.

This principle does not cover minor changes in position that are too unimportant to be considered, since something that is not considered cannot be accepted as evidence.

The rules that govern these principles are expressed in the two types of law, **common law** and **statutory law.**

10-5. Common Law. The greater proportion of law relating to land ownership is common law. Common law is the body of rules and principles that have been accepted by immemorial usage. In the written decisions of the courts down through the years, these principles have become clear and definite. A great body of the common law of the United States follows the jurisprudence developed originally in England.

10-6. Statutory Law. Statutory law is composed of the body of law enacted by governing bodies. Many statutory laws relate to land. In the following discussion any law that is statutory is so described.

10-7. Sources of Evidence. As stated previously, the position of a boundary must be decided on evidence of intent or evidence of long acceptance. Such evidence is found in transfers of ownership (**title transfer**), in transfer of rights (the establishment of **easements** especially for **rights of way**), in acts leading to **adverse possession,** and in aquiescence, agreement, and thus especially **marks on the ground.**

10-8. Title Transfer. Title can be transferred only by deeds, wills, inheritance without a will, or by adverse possession. Deeds and wills are legal instruments and they are said to **convey** title. Adverse possession is covered in another section of this chapter. Deeds must contain some sort of **description** of the boundaries of the land conveyed. The description must be tied directly or indirectly to marks on the ground and the evidence given must be interpreted in accordance with these marks. Deeds are usually recorded in the office of the proper public official where they are open to public inspection. A surveyor uses these descriptions more than any other form of evidence.

10-9. Partitions and Seniority. To clarify the legal principles involved, it is well to consider the proper procedure for creating a boundary. It is the following:

1. An owner decides to sell *part* of a tract of land. He and the prospective purchaser agree on the boundaries that will separate the land to be sold from the land to be retained. They have agreed on a **partition,** a separation of part of the land.

2. A surveyor is instructed to mark these boundaries and write a description of them. The court assumes that the grantor and the grantee inspect the marked boundaries and are satisfied that the surveyor has followed their instructions and that the description properly describes the new boundary. Accordingly, the boundary is forever fixed by the surveyor, and the more nearly any future surveyor can "follow in the footsteps" of the original surveyor the more truly will the boundary be located.

3. The method of "following the footsteps" is laid down by the many rules that are indicated in this chapter.

The deed for the first parcel to be sold off from a tract of land is called the **senior deed.** Further deeds take seniority in the order in which they are executed. If the owner gives deeds that call for more than he owns, one or more grantees will get less than they expect. Between two grantees, the senior gets all that is called for, the less senior (the inferior) gets the remnant. This, of course, makes sense. The owner sells what he stipulates until a time comes when he sells (unknowingly) more than he owns. This grantee (the inferior) gets less than called for. In deciding on the location of a boundary this principle must be remembered.

Sometimes, when two (or more) deeds are executed on the same day, the land is divided proportionately. When a plan of subdivisions is used, particularly when it is filed in the public records, the boundaries are usually assumed to have been established simultaneously and therefore they should be marked on the ground by proportional measurement. This rule is applied nearly always when the description merely calls for a lot by numbers according to the plan that has been filed.

10-10. Easements and Rights of Way. An **easement** is a right to use the land of another for some specific purpose. When it is no longer used, the right terminates. A **right of way** is an easement giving the right to pass across the land. Easements can be created by the owner, by the public, or by the state. An owner can create an easement by a deed or by **dedication.** He dedicates land to the public when he makes a public record of his dedication or when he allows it to be used by the public for a statutory period. In the latter case, the public is said to acquire the right by **prescription.** The filing of a plan of streets in the public records is assumed to be a dedication of the streets. In many jurisdictions it must be accepted by a special procedure.

The state has the right of **eminent domain.** This is the right to use private property for specific purposes provided the owner receives just compensation. The chief use of the right of eminent domain is to establish rights of way.

Counties, municipalities, and other branches of the state have the same right. The state can grant the right of eminent domain to railroad and to other public utilities that need such a right to operate.

The right to use the land is taken by **condemnation proceedings** and the records are made part of the public record. Taking land for a street or highway is usually called a **street** or **road opening** and almost always includes a description, often called the **return.** When the right of way is no longer used, the right ceases. The right can be terminated by a **street** or **road closing.** Frequently, statutes are enacted that give the state the right to acquire title for right of way by actual purchase.

Abutting property owners have the right of access to public rights of way. In some jurisdictions, statutes have been enacted that allow the state to take property for rights of way of **limited access.** Most parkways, throughways, etc., are of this nature. In these cases access is barred to the abutter.

Abutting property owners usually own to the center of the right of way, especially when it has been condemned. Often they own part of it. The descriptions of parcels sometimes include the area in the public right of way and sometimes they do not, even when the land in the right of way is really part of the parcel. Adjacent parcels that actually have the same depth may then have a common side line that is given a different length in each description. One length runs to the center line, the other to the side of the right of way.

10-11. Adverse Possession. It is evident that title to land depends on evidence. When a man believes that he has acquired title to a tract of land, he is merely convinced by the available evidence that it is his. After he has taken possession of the land and built improvements on it, he would be done an injury if some remote evidence were advanced that proved that he did not own the land after all. Since all title is subject to such an attack, no title would be good were it not for rules to prevent such an occurrence.

The courts provide that should such a happening occur, the presumable owner can claim title by **adverse possession.** He must prove his case according to certain, very strict rules. In some jurisdictions these rules are slightly modified by statutes. The rules are outlined here.

1. He must have **color of title;** that is to say, he must prove that he had reason to believe that he had acquired title. Almost any instrument in writing purporting to convey title is sufficient. If the instrument contains a description, upon winning the case he owns to the boundaries described. Otherwise, he owns only to the boundaries of his occupancy.

2. He must be in **actual, open, notorious,** and **exclusive possession.**

This means that he must perform visible acts on the ground that accrue to his own benefit, such as cultivating the fields, building a house, or renting the property to a tenant who performs visible acts. He must prevent others from using the property. Actions of this kind further prove his belief in his title and give notice of his belief so that anyone having claim to title may be warned that he must advance his claim.

3. He must be in continuous possession for a statutory period.

4. His possession must be **hostile.** This means that there cannot be the slightest implication of **permission** from the title claimant who is not in possession.

Usually the principle of adverse possession is applied to a separate tract whose boundaries are accepted and do not interfere with other tracts. Sometimes, however, when two tracts overlap, one owner will take title from the other by adverse possession and thus establish the position of the boundary.

10-12. Encroachments. The gradual taking possession of land not owned is an encroachment. Usually it consists of a fence, building, or the eaves of a building built over the line. The owner can remove it up to the boundary and collect the cost of removal from the encroacher. However, if it remains in position for a statutory period, the encroacher can claim the land actually occupied by adverse possession. However, if the owner is aware that his adjoiner is building on his (the owner's) land he must notify the adjoiner at once. If he fails to notify him he cannot claim title to the land occupied. In effect, the owner loses the land occupied. The legal bar to his making such a claim is called an estoppel.

10-13. Adverse Possession Against the State. Adverse possession cannot be claimed against the state. This is very important to the surveyor, as no owner can obtain any right in a right of way by owning a structure standing in the right of way, merely because it has been standing for a long period of time.

However, this is somewhat modified by either of two conditions:

1. If the owner has been led to believe that his structure was built outside the right of way by an *overt act* of a responsible official, the state is barred from taking action against him by the principle of estoppel.

2. If the position of the right-of-way boundary cannot be determined, it must be assumed that a structure of long standing was built outside the right of way.

10-14. Acquiescence. When two persons owning adjoining land (adjoiners) use the land up to a certain line for a reasonable period and show by their actions that they believe it to be the boundary, they establish the position of the boundary at that line by acquiescence or **practical location.**

10-15. Agreement. When two adjoiners agree upon where a boundary is located and either record the agreement or show by their actions where the line is located, the location is legally acceptable.

10-16. Records of Boundary Positions. It is obvious that the evidence of boundary locations is either written or visual. **Written evidence** of the position of boundaries is found in the land descriptions included in the following instruments: deeds, wills, dedications, condemnation proceedings, and agreements. As a protection to the parties involved, these instruments are usually recorded and thus made part of the public records where they are available to anyone who wishes to see them. The place of record varies in different jurisdictions. Usually the deeds, dedications, condemnations, and agreements are recorded or filed in the county or town clerk's office or the recorder's office, and the wills in the surrogate's office, if such office exists.

10-17. Visual evidence of boundary positions are the marks on the ground. They may be structures, fences, or surveyor's landmarks, like monuments, irons, or stakes. Although underground structures are difficult to note, they are assumed to be visible. They include cellars, sewers, or drains.

10-18. The Selection of Proper Evidence. It is evident from the foregoing that when a boundary was created by a recent partition, the intent of the parties involved is usually found in the original description of the partition and the original marks on the ground. When the partition was created a long time ago, such evidence is often not so strong as later descriptions and marks based on more recent surveys. The record of the position of the boundary as found in these later descriptions and later marks becomes, perhaps not the best evidence of the "footsteps of the surveyor," but certainly the best evidence of the accepted position established by long acquiescence, agreement, easement, encroachment, or adverse possession that occurred during the years but left no record other than the more recent surveys and landmarks.

RIPARIAN RIGHTS

10-19. Introduction. Bodies of water provide economic advantages that are so different from those provided by land that a special body of law has been developed to regulate their use. Chief among these advantages is transportation, that is, navigation in its broadest sense. Accessories to navigation include harbor improvements, docks, wharves, warehouses, canals, basins, and dams. Other economic considerations are fish, ice, water-power mills, irrigation, water supply, underwater plants, stranded natural objects, oyster beds, increased land areas built or uncovered by the water, and artificial fills over flats.

The laws regulating these economic possibilities or *rights* are quite different in the various states, and court decisions within each state follow different precedents in apparently similar cases. In this text it is impossible to do more than to outline the major principles and to warn the surveyor that he must carefully study the law in the jurisdiction in which he works.

10-20. Principles. The right to maintain and regulate navigation as it affects commerce remains in the United States. The ownership of bodies of water is in the state subject to the rights of the United States. The state can transfer by law such ownership and the rights associated with this ownership.

When a body of water forms the boundary of a parcel of land, the owner of the parcel is called the **upland owner.** Unless specifically excepted in the grant, he has the **riparian rights** to the water that are granted to such owners by state law. He owns to the water, but the exact position of the water boundary is usually defined by state law.

10-21. Definition of Water Boundaries. The courts are not consistent in the interpretations given or used to describe water boundaries. Those most generally accepted are given below.

10-22. High-water Mark. This is the line reached by the water when the body of water is full but not in flood. It is marked by the line of vegetation that shows no sign of alternate wetting and drying. In tidal waters it is the line of the neap high tides (see Chap. 15).

10-23. Low-water Mark. This is the line reached when the body of water is at its minimum flow, if the minimum flow is due to normal conditions and is not caused by a temporary drought. In tidal water it is the line of the neap low tides.

10-24. Thread of Stream. This is the line where the greatest volume of water flows. It may be considered to be the channel, or a line midway between (1) the high-water lines, (2) the low-water lines, (3) averages of these two, or (4) edges of the water at normal flow. It is sometimes called the center line. In a lake or pond it is the line extending throughout the greatest dimension equidistance from one of the types of lines described above.

10-25. Lines of Ownership. The line of ownership along a water boundary may be established by state law. It may be the high-water mark, the lower-water mark, or the thread. Often one of these lines is used for navigable waters and another for nonnavigable waters. When either high- or low-water mark is used, the state retains title to the stream between the lines of ownership on the banks.

The area covered by water is the **bed.** It usually is assumed to be bounded by the limits of the state ownership if such is retained.

10-26. Shore. The words **shore of a stream, bank of a stream,** or **by the stream** are usually interpreted to mean the line of ownership. When one of these expressions is used, it is interpreted to mean a riparian boundary.

10-27. Changes of Position of Water Boundaries. When the body of water changes position suddenly and obviously so that it no longer touches its old bed, as may happen in a storm or in a flood, the boundaries remain in the position they occupied before the change. The riparian rights of the owner whose boundary is thus suddenly left without water are lost by what is called **avulsion** or **revulsion.** Other parcels may suddenly acquire riparian boundaries by the same process.

When the movement is sudden but the water does not entirely leave its old bed, the courts are inclined to disagree. If the state by law retains the bed of the stream, it presumably loses title to the part of the bed no

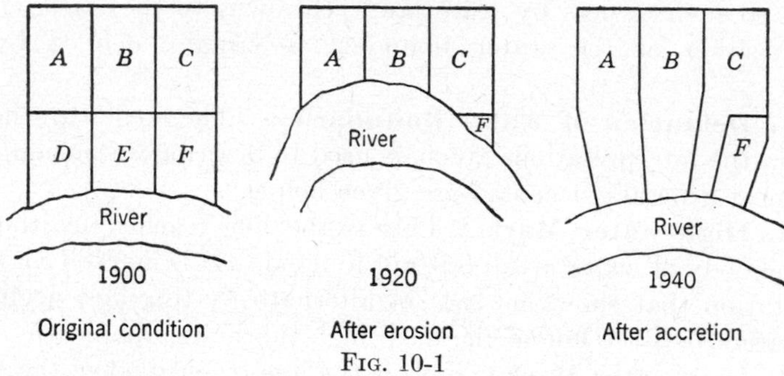

FIG. 10-1

longer occupied by the water. The riparian owner, it may be argued, advances to the water over unclaimed land.

When the body of water changes position slowly, and imperceptibly, the riparian owner's boundary moves with it. The land area may be **increased** by **accretion** when the water deposits materials (**alluvium**) on the banks. It may also be **decreased** by erosion.

When the water recedes, as when a lake dries up, the land is increased by what is called **reliction** or **dereliction.**

If an entire parcel is eliminated by erosion, title and riparian rights are extinguished. If the water comes in contact with a parcel that previously did not have a water boundary, the newly made water boundary carries with it all riparian rights. If, subsequently, the water moves slowly back to its original position, the new riparian boundary moves with it. The original owner cannot claim title. For example, Fig. 10-1 shows three positions for a river, one in 1900, one in 1920, and one in 1940, representing the result of gradual changes. In 1900 the riparian owners were *D*, *E*, and *F*. By 1920, *D* and *E* had lost their property but *F* still owned a small portion of his original tract. The riparian owners are then *A*, *B*, *C*, and *F*. When the river moved back to the position shown for 1940, the riparian boundaries moved with it so that *A*, *B*, *C*, and *F* remain the riparian owners and the new land is divided by various principles but usually as shown. *D* and *E* cannot claim title. The same principles apply to ocean boundaries.

10-28. Side Boundaries. When land is acquired by accretion or by reliction as illustrated in Fig. 10-1, the court must establish the side

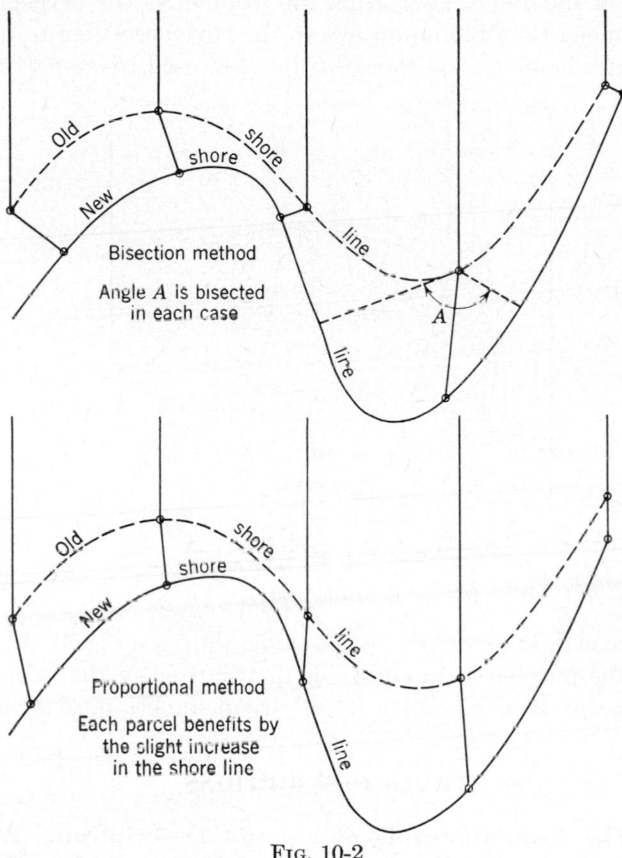

FIG. 10-2

boundaries. The court attempts to arrange them so that an equitable division results. When water frontage is of paramount importance, the new water boundary is sometimes prorated and the new corners are joined with the old by straight lines. Often the side lines are run perpendicular to the general shore line. This is sometimes called the **bisection method** because the court will direct that the line be run at the bisector of an angle formed by perpendiculars to the shore line on each side of the old corner (see Fig. 10-2).

Side boundaries must often be established under water to divide certain riparian rights among the riparian owners. Chief among these is the right to build wharves over shoal water to the line of navigation. These boundaries are also established by the courts in much the same manner as the side boundaries of land acquired by accretion or reliction. Sometimes the side boundaries are merely prolonged but often perpendiculars to the shore line are established by the bisection method at headlands and the length of the line of navigation intercepted by the perpendiculars is prorated among the riparian owners in the cove according to the lengths of their water boundaries. Straight lines are used to connect the points

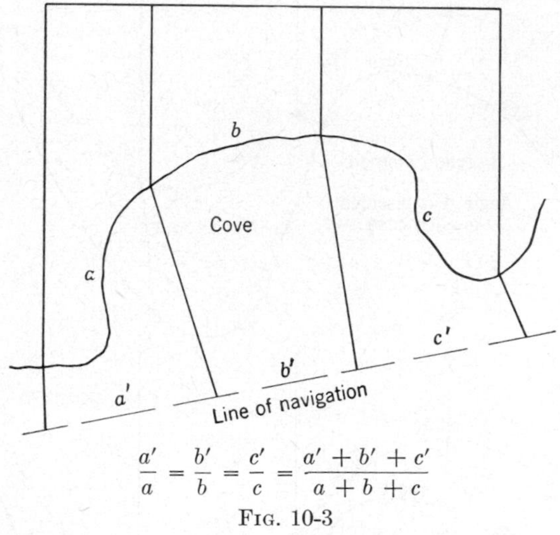

$$\frac{a'}{a} = \frac{b'}{b} = \frac{c'}{c} = \frac{a' + b' + c'}{a + b + c}$$

FIG. 10-3

on the shore with those on the line of navigation (see Fig. 10-3). Since in each case the purpose is an equitable distribution of the major riparian rights in the particular locality, local circumstances must control.

LAND DESCRIPTIONS

10-29. The Fundamentals of a Land Description. As pointed out, a land description is a necessary and important part of any legal document that refers to a parcel of land. The description has two purposes: (1) to identify the land and (2) to state its size, shape, and location. The identification should be clear and definite, and the statement of size, shape, and location should give complete survey directions for marking the boundaries by measurements from precise, durable landmarks. In legal phraseology, landmarks are called **monuments** and they must have visibility, permanence, stability, and definite location independent of measurements.

Unfortunately, few land descriptions entirely comply with these requirements. The surveyor is therefore frequently faced with many difficulties inherent in the descriptions themselves.

10-30. Forms of Descriptions in Use. There are two basic forms of descriptions, the description by plat and the description by words. A description by plat consists of a drawing upon which all the survey data is placed (see Fig. 10-8). The legal instrument that utilizes this description must contain an exact reference to the plat. The plat may be a drawing for a single parcel or for many parcels. When several parcels are shown, as in a subdivision plan, each city block and each parcel is given a distinguishing letter or number (see Fig. 10-15). The form of description used in the U.S. Public Land System (Chap. 11) is essentially a description by plat.

Verbal descriptions are of two types, bounding descriptions and descriptions by courses and distances. Bounding descriptions, often called **descriptions by adjoiners,** merely state the names of the adjoiners and indicate the boundary on which they adjoin. They are usually almost worthless. Descriptions by courses and distances, often called **running descriptions,** have developed a conventional form. In these descriptions it is customary first to describe a point of beginning and then to describe the directions and lengths of the boundary lines in order, as one would traverse the boundaries around the parcel, beginning with the boundary along the most important public right of way. When the adjoiners are also stated, the description is said to be by **metes and bounds.**

10-31. Relative Directions. The relative directions of lines used in a description are usually stated in terms of bearings or azimuths. In old descriptions the bearings used are usually magnetic bearings determined when the original survey was made. The absolute direction of these bearings seldom can be reestablished today because of the uncertainty of the date of the observations and the questionable accuracy of any determination of the magnetic declination at the time. The angles between lines, however, can be correctly determined from the bearings stated, to an accuracy equal to the original work.

It is customary today to measure the angles with a transit and to assume the bearing often used in previous deeds of some important line, frequently a public right-of-way line. The line used for the assumed bearing is called the **bearing base.** With this assumption, the other bearings are computed from the measured angles. Thus the bearings serve as a means of defining angles but have little meaning in themselves. Angles can be computed from bearings in the same survey but bearings from two different surveys must never be combined (see Fig. 10-4). Thus a line common to two parcels may be given different bearings in the two deeds. Sometimes persons ignorant of what is meant by bearings will make up a description using the bearing appearing in descriptions of adjoining parcels. The result, of course, is a geometric absurdity. The error, however, does not show in the bearings used. It can be discovered only by computing the closure.

Sometimes an astronomical determination is used as a basis for the bearing of one line. The bearings of the other lines are then computed from it using the measured angles, and the statement is made that the bearings are "true" bearings. This procedure is helpful, as the only uncertainty in absolute direction is the convergence of meridians between

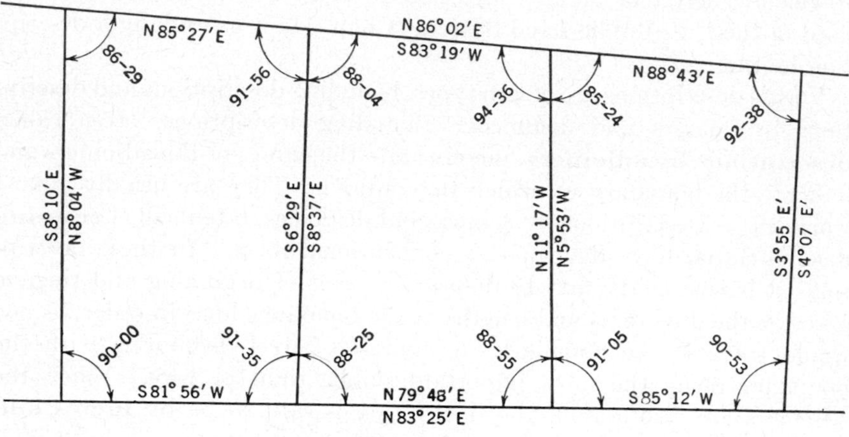

FIG. 10-4. Illustration of the usual use of bearings to give values of angles not absolute directions. All bearings are constant within each parcel, and they give correct values for the angles. The bearings in each parcel are obviously based on different assumed directions.

the unknown point where the observation was made and the point where a direction is to be established. When the State System of Plane Coordinates is used, "grid azimuths" or "grid bearings" should always be used (see Chap. 12). Such directions can be independently established anywhere with certainty and, therefore, provide another and a more reliable evidence of the locations of the lines.

10-32. Examples of a Typical Metes and Bounds Description. Given below is a typical metes and bounds description (see Fig. 10-5). It is not a model description, as it lacks certain requirements. It is, however, the usual type that is accepted as a good description. It is given here to illustrate the usual source of written evidence.

It is assumed that Richard Roe has decided to sell to Robert Smith 300 ft of frontage along Maple Street. The metes and bounds description might be as follows:

Beginning at a stone bound in the northerly line of Maple Street and marking the southwesterly corner of the lands hereby conveyed and running;

1. Thence, along the northerly line of Maple Street N84°15′47″E 300.00 ft to a concrete monument in the boundary of the lands of the grantor;

2. Thence, along the lands of the grantor N5°44′13″W 556.44 ft to a concrete monument in the boundary of the lands of John Doe;

3. Thence, along the lands of John Doe S81°46′34″W 252.54 ft to a concrete monument in the boundary of the lands of James Jones;

4. Thence, along the lands of James Jones S0°44′19″E 547.56 ft to the point of beginning.

The important desirable elements in this description are the following: All the monuments are connected by survey ties and they become wit-

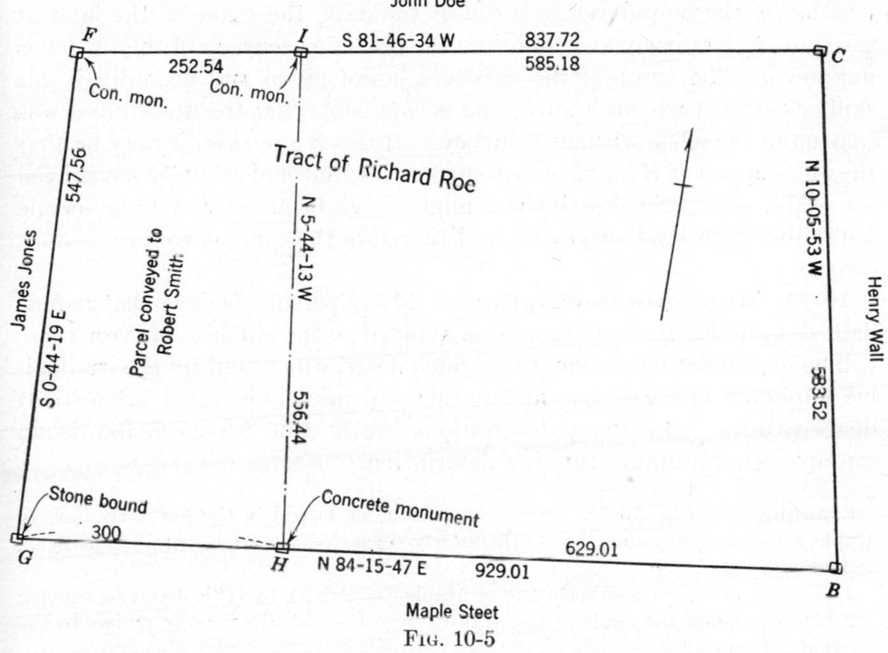

FIG. 10-5

nesses to each other. Should all be destroyed but two, the line HI can be reestablished. Obviously, many of the data in this description are redundant but this has many advantages. For example, the bearing and length of the line FG is determined geometrically from the description of the other lines without the necessity of using bearing and length of the line itself. The bearing and length of IF and the length of HI are determined by the positions of the boundaries of John Doe and James Jones, and the monuments at each corner define the lines without other considerations. The purpose of this description is, of course, only to provide an identification of the property and a means of establishing the location of the new boundary *HI*. Although, in effect, it gives information for locating the lines *IF*, *FG*, and *GH*, this information can be used for locating these lines only when other means fail. It therefore indicates where the surveyor believed these lines to be when he made his survey, and thus it can be used as indirect evidence to establish the position of *HI* even though an error had been made.

On the other hand, the description fails to refer to the record of the deed that describes the land of the grantor and fails to give any indication of where in the north line of Maple Street the point of beginning can be found. It therefore lacks the proper quality of identification. The date of the survey is not given. It is impossible to tell whether the survey was made before or after any possible events that might have changed the position of the boundaries. Without the date, the value of the land at the time of the survey and therefore the possible accuracy of the survey is unknown. The name of the surveyor is not given and accordingly his skill or customs are not known, and it is possible that the description was written in the office without a survey. If this is the case, it may be that the monuments at *H* and *I* were to be set later but had actually never been set. Moreover, the descriptions might have been written by someone unfamiliar with land surveying and therefore they might contain serious errors.

10-33. Minimum Descriptions. Many persons believe that redundant data in descriptions should be avoided so that in case of error there will be no conflicting statements. Such descriptions contain the irreducible minimum of necessary information and might be called **minimum descriptions.** Bounding descriptions are of this type. Below is an example of a minimum running description.

Example (See Fig. 10-5). Beginning at a stone bound in the northerly lines of Maple Street and marking the southwesterly corner of the lands hereby conveyed, and running:

1. Thence, along the northerly line of Maple Street N84°15′47″E 300 ft to a point;
2. Thence, along the lands of the grantor N5°44′13″W 556 ft more or less to the lands of John Doe;
3. Thence, westerly along the lands of John Doe to the lands of James Jones;
4. Thence, southerly along the lands of James Jones to the point of beginning.

This description creates and describes the new boundary *HI*. But it describes the Smith parcel in such a way that the boundaries *GH*, *IF*, and *FG* are clearly dependent on earlier legal instruments, presumably those that created them. There is no conflicting evidence and there is little difficulty in searching the title. But if the boundaries of the adjoiners are lost, there is no evidence available to locate the parcel. Actually, no survey measurements or computations are necessary for such a description and there is no assurance that the ground was even visited.

10-35. Parol Evidence (Evidence from Outside Sources). Unfortunately, after a few years have elapsed, it is seldom possible to locate a boundary using only the description in the instrument that created it, however perfect the description may be. There are four causes for this, as follows:

1. Blunders in surveys and descriptions
2. Disturbed, lost, or indefinite landmarks
3. Surveys of low accuracy made when the land was cheap and high accuracy was not justified
4. The operation of the principle of acceptance

Thus, to determine the location of a boundary, the instrument which created it must be used and in addition *all other evidence that is available* must be given careful consideration. Obviously, the more landmarks and survey ties mentioned without sacrifice of clarity, the more surely will each boundary be fixed. Minimum descriptions, therefore, are of little use as evidence or as an aid in marking the lines in the field. They provide little protection to the grantee and are not recommended.

Actually there is very little chance that conflicting statements will cause difficulties. The courts have developed rules of weighing evidence which are based on a common-sense method of establishing the intent of the parties involved. The order of importance of evidence is usually taken as follows:

1. Evidence of occupancy
2. Landmarks mentioned in the description (that can be properly identified in the field)
3. Other landmarks (that can be properly identified in the field)
4. Adjoiners
5. Lengths
6. Directions
7. Area

(items 4–7: that are mentioned in the description)

This order is sometimes changed when special conditions indicate that it should be. It might be pointed out that one of the best identifications of landmarks is reasonable consistency with lengths and bearings. Of course, none of these items are considered unless they are specifically stated in the description. (Such a statement in a deed is known as a **call.**)

10-36. Multiple Boundary Lines, Subdivisions. When a tract is to be sold off in building lots, i.e., **subdivided,** it is customary to place on public record a plat of the development showing the survey monuments that have been placed on the ground and the survey data necessary to locate all the lots and streets from them (Fig. 10-15). Each lot and block is numbered or lettered. The description need then refer only to the lot and block and the name and place of record of the plan. All the boundary lines are thus created simultaneously by the plat rather than separately by the individual deeds. Monuments are placed at the intersections of the center or the side lines of the streets or at the intersections of parallel offset lines. They thus indicate the physical location of the streets. Usually the actual distances between the streets vary slightly from the distances recorded on the plat. Such variations are called overages or shortages. When the sum of the lot dimensions given on the plat is equal to the distance between the streets as given on the plat, it is obvious that the overages or shortages should be apportioned to the lots in proportion to the lot dimensions. By showing on the plan that a block is 600 ft long and made up of 12 50-ft lots, the owner has in effect said that each lot shall occupy one-twelfth of the block. In most jurisdictions the courts have held that when lots are described by lot and block numbers, the boundary lines shall be proportionately increased or decreased according to the actual monumentation. Lots described by other means, even if lot and block are also mentioned, shall be located according to all the evidence available and not *necessarily* according to the above rule.

10-37. Principle of Proportion Otherwise Applied. Sometimes a verbal description is written so that the principle of proportion applies. For example, the first course in the parcel shown in Fig. 10-5 might have been described:

1. Thence, along the northerly line of Maple Street N84°15′47″ 300.00 ft of the 929.01 ft of frontage of the grantor.

In this case the courts might hold that an overage or shortage of the 929.01 ft would be applied proportionately to the 300.00 ft.

10-38. Recommended Descriptions. The foregoing descriptions are given as examples of the type the surveyor must work with. Recommended descriptions are covered later.

PROCEDURE FOR A BOUNDARY SURVEY

10-39. The Role of the Surveyor. The land surveyor is called upon to perform the following functions:

1. Write descriptions
2. Survey land so that descriptions can be written
3. Stake out property corners
4. Determine the positions of various structures or lines with respect to boundary lines

To write descriptions he must have a thorough knowledge of the methods of locating boundaries. To perform the other functions he must first stake out the boundary lines themselves. Thus, all his work hinges on his ability to determine the actual positions of boundaries. He has no legal power to establish the positions of boundaries. When boundaries are in dispute this is a function of the court and can only be accomplished by a court decree. The surveyor's function is to mark on the ground where, in his *opinion*, the lines are located. His opinion must be based on the evidence that exists, interpreted in the light of the law. When he is not reasonably convinced that the view that a court would take will coincide with his opinion, he should recommend that the owner petition the court to issue a decree which will establish the boundaries. The data he has collected and his expert opinion will usually be used by the court in making the decree.

10-40. Procedure for a Property Survey. It is apparent from the foregoing discussion that a considerable body of evidence must be collected before boundaries can be marked in the field. The following routine for making property surveys is therefore recommended:

1. Obtain a copy of the deed from the owner. This will contain a description of the parcel which will at least identify it.

2. Obtain the abstract of title from the owner if one exists. This will contain a list of legal instruments by which the title to the land has been successively transferred until it has reached the present owner. It will thus indicate the various instruments that have created the boundaries up to the time the present owner took title. The proper references are given in the abstract by which the instruments may be found in the public records. Deeds, agreements, and condemnations are usually found at the county clerk's office, wills at the surrogate's office, and subdivision plans at the county clerk's office or at the city hall, although these arrangements differ widely in different jurisdictions.

3. Abstract the descriptions in these instruments by using sketches. By a study of these sketches, the tract or city block which affects the parcel can be determined.

4. Abstract, using sketches, the latest descriptions of all the parcels in this tract or city block. They are most easily found through the names of the present owners. The city tax map is the best assistance in finding the owners' names in metropolitan areas. In the country, inquiry on the ground is usually necessary.

5. By inquiry, discover the marks that show the accepted location of the public streets, roads, or other public properties adjacent to the block or tract. The engineer of the city, township, county, or state may be able to provide this information. Often no marks exist and the original descriptions made at the time the streets or roads were opened are inadequate. The positions of boundaries of the public land must then be determined from the physical location.

6. If the parcel is in the city, establish the street lines which bound the entire block from the best available marks or indications on the ground. The street lines are the boundaries of the rights of way. The sidewalk and parkway areas are included between them (Fig. 10-6). When no monuments exist, the following must be used to locate these lines: fences, faces of buildings, and private property corners marked by monuments, iron pins, iron pipes, or stakes. The faces of buildings are usually laid out along the *building line*. This line is usually parallel to, or coincides with, the street lines. By averaging these marks the line can be established. Usually, in older parts of the city, the lines on opposite sides of the street are not parallel and the lines are not continuous from block to block, so that each street line in one block can and should be determined separately.

In the country, in the absence of monuments or other marks, old fence lines and stones marking corners must be relied upon. Sometimes there exist at the office of record, documents that give a description of the rights of way. Usually the only information that can be used is the width of the right of way and the extent of the straight portions. When the width of the right of way is known, fences on both sides can be averaged to determine the line. Fences beyond the straight portions should not be used in the average, of course.

7. Having determined the right-of-way lines, a control traverse should be established around the entire block or tract (Fig. 10-7). Traverse lines should also often be established, running near interior property corners. It usually reduces computation to establish these control courses parallel to the right-of-way lines and property

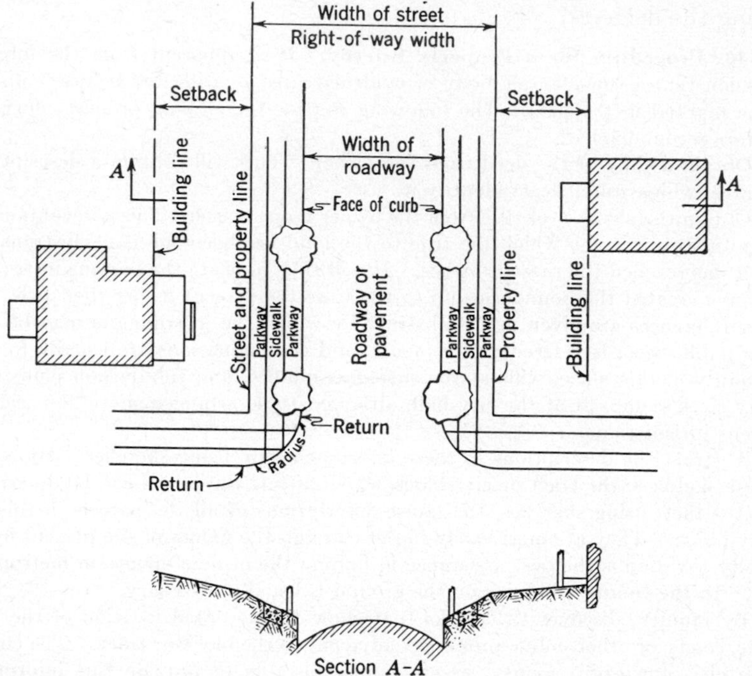

FIG. 10-6. Nomenclature of streets (when there is no setback, building lines and street lines coincide).

lines. All traverse angle points should be permanently marked and ties measured to permanent objects to ensure reestablishment of the traverse. In the country, it is seldom possible to make a survey of much more than the original parcel.

Measure accurate ties from the control traverse to the following:

a. All marked property corners
b. All indications of occupancy or possession
c. Buildings or other permanent structures

8. The resulting data constitute a preliminary survey. It should be plotted to a large scale and the survey data placed upon the drawing. Often it is well to establish a coordinate system and to compute the coordinates of important points.

9. Using the data in the descriptions and the physical information, determine a

solution that will most nearly satisfy all the evidence both written and physical. This step is the most difficult and requires skill, knowledge, and experience. Due regard must be given to possible mistakes, to the meaning of words used in old instruments, to customs in the locality, to the reputation of the early surveyors, and particularly to the law and the weights of evidence. This solution results in the determination of the positions of all the property corners in the area with respect to the control traverse.

Obviously, the program outlined is very costly. The client cannot be expected to pay the cost of all of it. When the surveyor is employed to stake out any other parcel in the same block or tract, the same solution can be used and a charge can be made accordingly. It is, therefore, most important that a complete record should be kept of the solution and of the marks by which the original traverse can be regained.

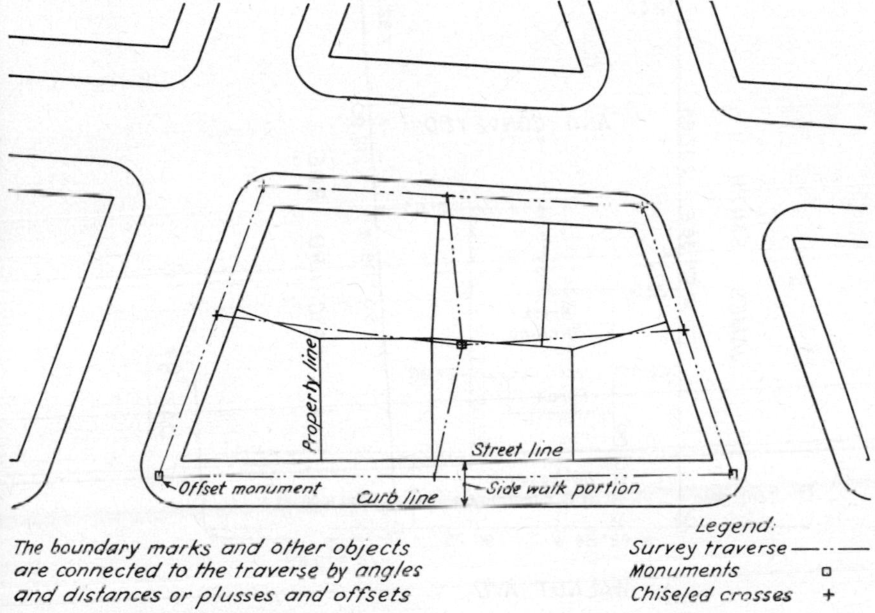

FIG. 10-7. Typical control traverses for survey in city block.

10. Stake out the parcel according to the solution, measuring from the control traverse.

11. Construct a plat of the parcel (Fig. 10-8) on which should be shown the boundaries, the physical features located, and any easements. The bearings or azimuths and lengths of all boundaries should be stated, and all other dimensions should be given in terms of rectangular offsets from the property lines.

Any structure extending into the parcel from outside the boundaries is an *encroachment*. Eaves or other projections above or below ground are encroachments. They should be especially noted.

In closely built-up city areas, small encroachments are so common that it is customary to show the boundaries as they would be located from the deed description if the only physical marks on the ground were the street lines. These are usually called *lines by deed*. The lines of the building, fences, etc., are also shown and called *lines of occupancy* (see Fig. 10-9).

These small encroachments are seldom considered to constitute a basis for adverse possession on the principle that they are not *visible*. Accordingly, the description remains the same from conveyance to conveyance.

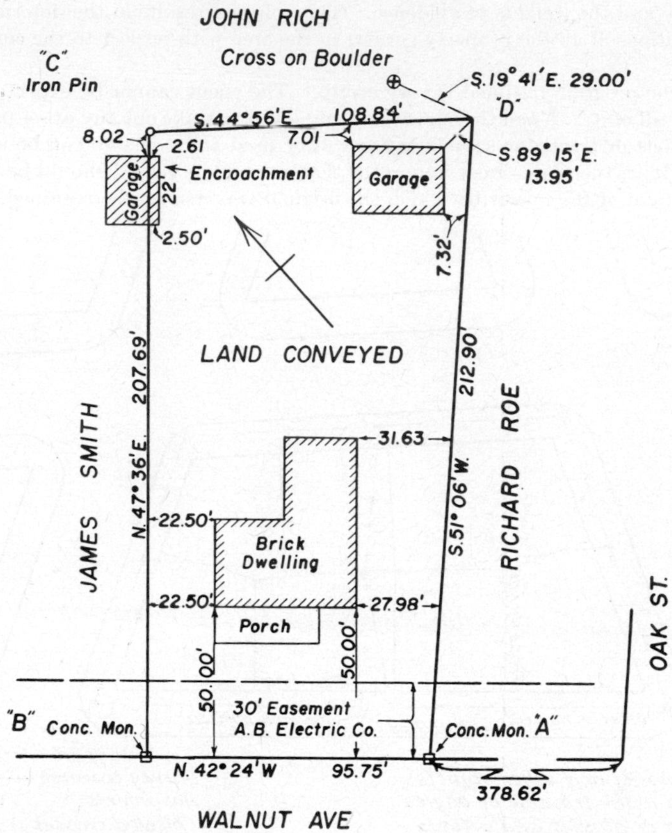

FIG. 10-8. Form of description by plat which is filed with deed. When plat is filed elsewhere, slight changes are required.

Survey lines for this type of lot must be actually established inside the house by running through windows and doors. Measurements to the party walls are made from these lines.

12. Prepare a short written report for the owner which includes a print of the plat. The report should state what assurance the surveyor may have in his location and

whether, in his opinion, a petition should be made to the court for a decree establishing the boundaries for a writ of ejectment to force the removal of encroachments.

10-41. Accuracy. The accuracy required for the various survey operations depends on value of the land surveyed. In the country, errors of 1 to 3 parts in 1000 might be allowable. In some city work errors should not be greater than 1 part in 50,000.

10-42. Boundary Disputes. When a boundary dispute is imminent, the surveyor should attempt an equitable settlement. He can point out

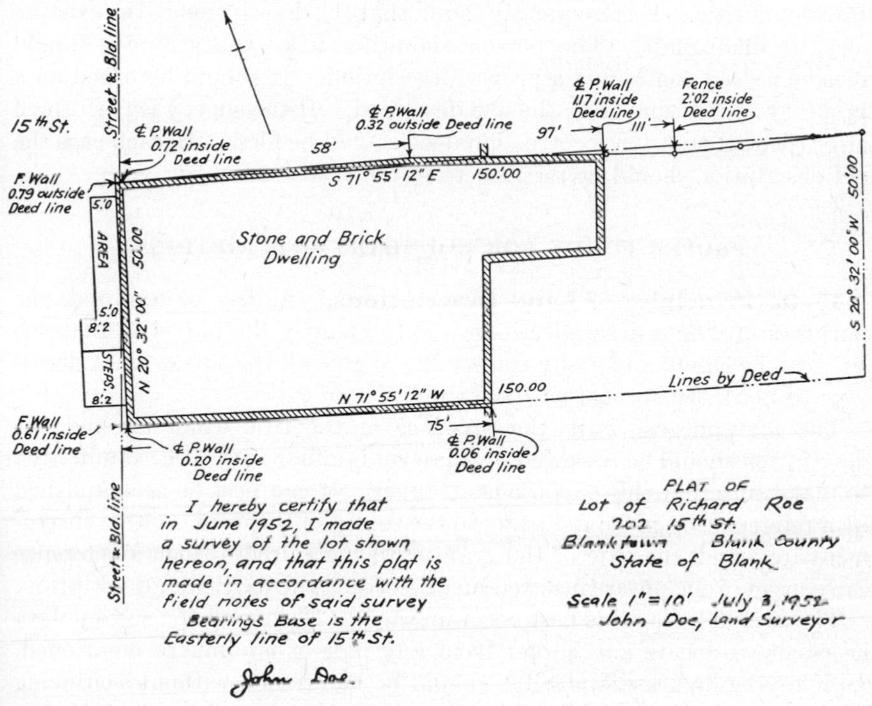

FIG. 10-9

that an agreement can be made out of court by the adjoiners, and he can explain to both parties the basis of his opinion concerning the location of the boundary.

10-43. Testimony in Court. When the case must be argued before the court, the surveyor is called upon to testify as an expert witness to the existing facts and to interpret the results of survey measurements in readily understandable terms. The location he believes to be correct is invariably challenged by the surveyor representing the other side. It is essential that he shall have compiled sufficient evidence to support his opinion if he is to be of service.

10-44. Monumenting the Corners. When the client has accepted the location, the surveyor should point out to him the importance of monumenting the corners. He should remind the owner that they will be of use to him when he builds or otherwise uses the land. He should explain how much weight is given to monuments in determining a boundary and the possibility that the position of the line may be changed by adverse possession if the corners are not monumented.

10-45. Writing Descriptions. In the usual case, the survey is made after a contract or an **agreement of sale** has been executed but preceding transfer of title. It is frequently found that the description in the existing deed is inadequate. The person acquiring title (the grantee) should receive a deed containing a proper description. It should be based on a property survey similar to the one described. If the survey is performed after the title transfer, a correction deed should be filed. In each case the old description should be referred to in the new deed.

PROPER FORMS FOR BOUNDARY DESCRIPTIONS

10-46. Principles of Land Descriptions. As has been stated, the purposes of a land description are: (1) to identify the parcel so that the title can be surely and easily traced; (2) to give all the survey data necessary to mark the corners on the ground.

The first purpose is in the province of the title examiner and the description should be reviewed by a lawyer familiar with title examination to make sure that this purpose is attained. It can best be accomplished by a reference, by book and page, to the record of the deed or other instrument by which the title of the grantor was acquired. Such a reference can consist of one phrase inserted just preceding the particular description.

The second purpose is best accomplished by giving all the survey data necessary to locate any corner from any pair of landmarks mentioned. As many landmarks as possible should be mentioned without sacrificing clarity, provided the landmarks are reasonably durable and precise in location.

10-47. Recommended Form of Description by Plat. The best form of description consists chiefly of a plat. This is filed in a place of public record either with the deed or separately. All survey data should be shown on the plat. An example of a plat to be filed with the recorded deed is shown in Fig. 10-8. If not filed with the deed, it should be filed in the same office as the deed for ready reference. When filed separately, the title should be changed to read "Plat of Parcel of Land of Richard Roe."

In every case the plat must be incorporated in the deed by a carefully drawn reference which identifies the plat and names the place of filing (see the plat reference used in Sec. 10-50). Without such a reference, the plat does not serve as a description in the deed. The plat may be a subdivision plan or the Survey of the Public Lands of the United States (see Chap. 11).

A metes-and-bounds description is the best substitute for a plat. As no drafting is required, it is the customary form.

10-48. Recommended Form of Metes-and-bounds Description (Fig. 10-8). ________________, situated in the City of Blankville, County of Blank, State of Blank, being a part of the same tract conveyed by Leslie Ware to Richard Roe by warranty deed dated June 15, 1907, and recorded in Book 100, page 100, June 20, 1907, at the Blank County Clerk's Office, and bounded as follows:

Beginning at a concrete monument in the northeasterly line of Walnut Avenue, at the southerly corner of the land hereby conveyed, said monument bearing N42°24′W, 378.62 ft along the northeasterly line of Walnut Avenue from the intersection of said northeasterly line of Walnut Avenue and the northwesterly line of Oak Street, and running:

1. Thence, N42°24′W, 95.75 ft along the northeasterly line of Walnut Avenue, to a concrete monument at the southerly corner of the land of James Smith and the westerly corner of the land hereby conveyed;

2. Thence, N47°36′E, 207.69 ft along the southeasterly line of the land of James Smith to an iron pin at the northerly corner of the land hereby conveyed;

3. Thence, S44°56′E, 108.84 ft along the southwesterly line of the land of John Rich to a point at the easterly corner of the land hereby conveyed, said point bearing S19°41′E, 29.00 ft from a cross chiseled on a boulder on the land of John Rich and also bearing S89°15′E, 13.95 ft from the easterly corner of the face of the foundation of the garage on the land hereby conveyed;

4. Thence, S51°06′W, 212.90 ft along the northwesterly boundary hereby established of the land of Richard Roe to the point of beginning.

All bearings are based on the stated direction of the northwesterly line of Walnut Avenue.

This description was written on May 5, 1942, by John Doe, Civil Engineer, from data secured by a survey made by said John Doe in March and April, 1942.

10-49. Description of a Lot Having a Curved Boundary (Fig. 10-15). Lot *B*-5, ________________, situated in Blankville, Blank County, State of Blank, and bounded as follows:

Beginning at a point in the northerly line of Somerset Street at the southwesterly corner of the land hereby conveyed, said point bearing N72°04′E, 72.58 ft from a concrete monument in the northerly line of Somerset Street, said monument bearing N58°04′E, 302.28 ft measured along the northerly line of Somerset Street from a concrete monument at the intersection of the northerly line of Somerset Street and the northerly line of Overville Street and running:

1. Thence, easterly on the arc of a circle 150 ft in radius curving to the right an arc distance of 72.08 ft, along the northerly line of Somerset Street, the chord of said arc running S30°10′E, 71.39 ft, to a point at the southeasterly corner of the land hereby conveyed;

2. Thence, N23°36′E, 212.60 ft along the westerly line of the land of (here insert the name of the owner of lot *B*-6) to a point at the easterly corner of the land hereby conveyed;

3. Thence, N44°56′W, 107.56 ft along the southerly line of the land of John Stout to a concrete monument at the northerly corner of the land hereby conveyed;

4. Thence, S58°04′W, 109.92 ft along the southeasterly line of the land of Harry King to a point at the westerly corner of the land hereby conveyed;

5. Thence, S3°56′E, 201.10 ft along the easterly line of the land of (here insert the name of the owner of lot *B*-4) to the point of beginning.

All bearings are based on the stated direction of the northerly line of Somerset Street.

This description was written on June 10, 1942, by John Doe, Civil Engineer, from data secured by a survey by said John Doe during March and April, 1942.

10-50. Description by Lot and Block (Fig. 10-15). Lot *B*-5, ______ ______, situated in the city of Blankville, County of Blank, State of Blank, shown as lot 5 in block *B* upon a plat filed in book 253, page 121, in the Blank County Clerk's Office and entitled, "Plan of Lots, Somerset Development, Blankville, Blank County, State of Blank, A. B. Realty Company, Owners, Scale 1″ = 100′, March 10, 1942, John Doe, Civil Engineer," to which plat reference is hereby made for more particular description.

COMPUTATION OF AREA

10-51. Purpose. When a tract is larger than a city lot, the area should be computed. Most purchasers want to know the area, and a knowledge of the area is essential when the land is to be used for agriculture. Usually a statement of the area is added at the end of a metes-and-bounds description, thus: "being an area of 46.24 acres more or less." Sometimes this statement aids in interpreting the description.

10-52. Method. There are a number of convenient methods of computing the area of a parcel, once the latitudes and departures have been determined. The method described here is the best known and the most generally used. It is called the method of **double meridian distances.**

Assume that it is required to find the area of any parcel such as *ABCDE* in Fig. 10-10.

Through the most westerly point *A* construct a meridian *C′E′*. The desired area is then equal to the area of the whole figure less the shaded area.

$$\text{Area } ABCDE = \text{area } C'CDEE' - \text{area } C'CBAEE'$$

From each corner, drop a perpendicular to the meridian: *BB′*, *CC′*, *DD′*, and *EE′*. These form either a trapezoid or a triangle between each course and the meridian. In each figure thus formed, the altitude is the latitude of the course, and the average of the bases is the perpendicular from the midpoint dropped to the meridian. The latter are shown by the dotted lines *FF′*, *GG′*, *HH′*, *II′*, and *JJ′*. They are called the **meridian distances** of the various courses.

The area of each figure is the product of the latitude of the course multiplied by its meridian distance.

When the meridian *C′E′* is placed as shown, the meridian distances will always be plus. It will be noticed that the latitudes of the shaded areas are minus while all other latitudes are plus. Thus the area desired is the algebraic sum of the areas of the trapezoids (and triangles). If the survey had been carried out in the opposite direction (clockwise), all the latitudes would have opposite signs and the desired area would have a minus sign. It will be found that the meridian can be passed through any corner desired as long as the signs are carefully followed.

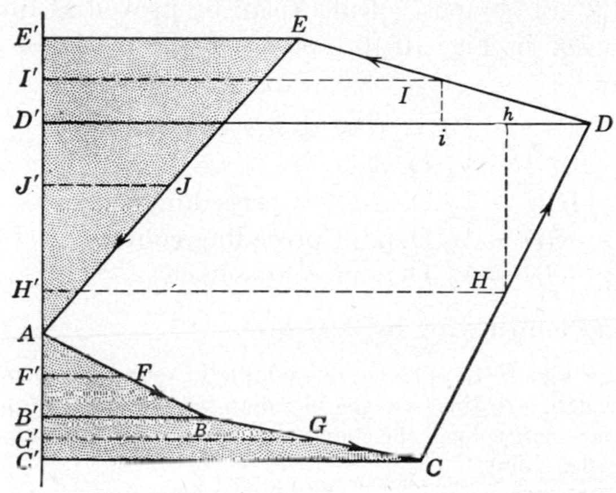

Course	Chains			Double areas	
	Lat.	Dep.	D.M.D.	+	−
AB	−2	+4	+ 4		8
BC	−1	+5	+13		13
CD	+8	+4	+22	176	
DE	+2	−7	+19	38	
EA	−7	−6	+ 6		42
				+214	−63
				− 63	
				151	

Area = 151 × ½ = 75.5 sq chains
= 7.55 acres

FIG. 10-10

10-53. To Find the Double Meridian Distances. Less arithmetic is required to find twice the meridian distance (the double meridian distance, the D.M.D.) for each course than the meridian distance itself. Accordingly, the latitudes are multiplied by the D.M.D.'s and the resulting area is divided by two.

The D.M.D.'s are found by the following rules:

1. The D.M.D. of the first course (the course that begins at the meridian) is equal to the departure of that course.

2. The D.M.D.'s of the remaining courses are each equal to the D.M.D. of the preceding course plus the departure of that course plus the departure of the course itself.

3. The D.M.D. of the last course is equal to the departure of that course with its sign changed.

Rules 1 and 3 are obvious. Rule 2 can be proved as follows: In any course such as DE in Fig. 10-10, construct the meridian distance II'. From the figure

$$I'I = H'H + hD - Di$$

But

$$\begin{aligned} I'I &= \tfrac{1}{2}\text{ D.M.D. of course desired} \\ H'H &= \tfrac{1}{2}\text{ D.M.D. of preceding course} \\ hD &= \tfrac{1}{2}\text{ Dep. of preceding course} \\ -Di &= \tfrac{1}{2}\text{ Dep. of course itself} \end{aligned}$$

Substituting and multiplying by 2, *Q.E.D.*

Example 1. In Fig. 10-10 is given an example in round numbers for computing the area in the sketch. In this case the meridian was assumed through A. In the column headed *course* write down the courses in order, beginning with the course that starts at the meridian point.

Compute the D.M.D.'s as follows:

D.M.D. $AB = +4$	D.M.D. $CD = +22$	
Dep. $AB = +4$	Dep. $CD = +4$	
Dep. $BC = +5$	Dep. $DE = -7$	
D.M.D. $BC = +13$	D.M.D. $DE = +19$	
Dep. $BC = +5$	Dep. $DE = -7$	
Dep. $CD = +4$	Dep. $EA = -6$	
D.M.D. $CD = +22$	D.M.D. $DE = +6$	*Check*

Compute the double areas by multiplying the D.M.D.'s by the latitudes. Half the algebraic sum of the areas is 75.5 sq chains, which is 7.55 acres.

When the area is found in square feet, the number of acres is found by dividing by 43,560.

Example 2. Using the same values, pass the meridian through E. The first course would be EA; then

Course	Chains			Double areas	
	Lat.	Dep.	D.M.D.	+	−
EA	−7	−6	− 6	+42	
AB	−2	+4	− 8	+16	
BC	−1	+5	+ 1		−1
CD	+8	+4	+10	+80	
DE	+2	−7	+ 7	+14	
				+152	−1

Area $= 151 \times \tfrac{1}{2} = 75.5$ as before

D.M.D. $EA = -6$	D.M.D. $BC = +1$
Dep. $EA = -6$	Dep. $BC = +5$
Dep. $AB = \underline{+4}$	Dep. $CD = \underline{+4}$
D.M.D. $AB = -8$	D.M.D. $CD = +10$
Dep. $AB = +4$	Dep. $CD = +4$
Dep. $BC = \underline{+5}$	Dep. $DE = \underline{-7}$
D.M.D. $BC = +1$	D.M.D. $DE = +7$ *Check*

Example 3. This illustrates the computation of the area of the whole tract shown in Fig. 10-15.

Course	Lat.	Dep.	D.M.D.	Double areas
A–J	+653.68	−596.88	−596.88	− 390,169
			−596.88	
			+514.79	
J–O	+320.84	+514.79	−678.97	− 217,841
			+514.79	
			+575.25	
O–W	−576.59	+575.25	+411.07	− 237,019
			+575.25	
			−493.16	
W–A	−397.93	−493.16	+493.16	− 196,243

Double area sq ft = −1,041,272
Area sq ft = 520,636

$$\frac{520{,}636}{43{,}560} = 11.95 \text{ acres}$$

SUBDIVISION PLANS

10-54. Importance. As a metropolitan area expands, a demand is created for residential building lots in the surrounding neighborhood. This land is usually occupied by farms or country estates and therefore divided into comparatively large tracts. When an owner of one of these tracts decides to sell it for building lots, he should adopt a definite arrangement of lots and streets called a **subdivision plan.** To follow such a plan avoids wasting space and makes the lots more attractive to purchasers. A few random sales, not based on a plan, might prevent further development of the property. Usually a land surveyor is requested to construct a plat of such a plan and he is thus called upon to advise the owner of the conditions which affect such a project.

10-55. Approval of Municipal Authorities. While often there is nothing to prevent the owner from adopting any plan he desires, it is usually advantageous and sometimes necessary to obtain the approval of the municipal government. Various zoning laws and subdivision regulations may often control the layout. Permits must usually be obtained for street entrances, sewer construction, and water and sewer connections. These are difficult to obtain without approval of the plan. The owner will want the municipality to take over the maintenance of the streets, the removal of snow, and the maintenance of the various city services. Without approval, this can be indefinitely postponed. If the plan is approved, it should be filed in the proper office as a protection to purchasers.

10-56. Requirements for Approval of the Plan. Many municipalities have ordinances regulating the approval of subdivision plans. The usual requirements are the following. The development must contribute to value of adjacent tracts and if a city plan or a master plan has been adopted, it must be closely in accordance with it. All proposed street lines must be monumented, the monuments must be shown on the plat, and all bearings, lengths, and ties must be set forth. Frequently street profiles must be shown.

10-57. Filing Subdivision Plans. The owner may *file* the subdivision plat at the offices of certain government officials. In many jurisdictions this is required. Preferably, such plats should be filed where the deeds are recorded, which is usually at the county clerk's office. When this is contemplated each block and lot should be identified on the plat by a number or letter. Each lot can then be described in the various deeds merely by stating the lot and block identification and giving reference to the plat and place of filing, as pointed out previously. This clarifies the description and reduces the cost of recording the various deeds. The act of filing dedicates the streets as mentioned previously.

10-58. Advantages to the Purchaser of Obtaining City Approval and of Filing. When municipal approval has been secured and the plan filed, it is difficult for the municipality to close the streets shown or to refuse to maintain the streets and services that are constructed. This protection makes the lots more saleable. However, it must be remembered that as soon as the plan is filed, the owner is usually required to pay taxes based on the higher assessments that are applied to building lots instead of those applied to farm land.

10-59. Arrangement of Streets and Lots. The size and arrangement of streets and lots are primarily based on the nature of the neighborhood. When the lots must be small, the layout usually must be rectangular, since lots of this shape can be more effectively utilized (Fig. 10-11). When the frontages are greater than 100 ft or so, the streets can be curved (Figs. 10-12 and 10-13). Curved streets give greater privacy and can be better adapted to the topography. The side lines of lots should be nearly at right angles to the street. As modern traffic is detrimental to residential lots, dead-end streets, with loops for turn-arounds called **cul-de-sacs,**

and U-shaped streets, are often used to eliminate through traffic. Where two cul-de-sacs approach each other, an easement connecting them should be set aside for service connections and sometimes for pedestrians and emergency vehicles.

10-60. Building Lines. In heavily built-up sections the faces of the buildings are placed on the street lines. Where space is not at such a premium they are set back from the streets for greater privacy, more light and air, and better appearance. The line of the front faces of the buildings is called the *building line.* It is good practice and is often required to establish building lines on the plat.

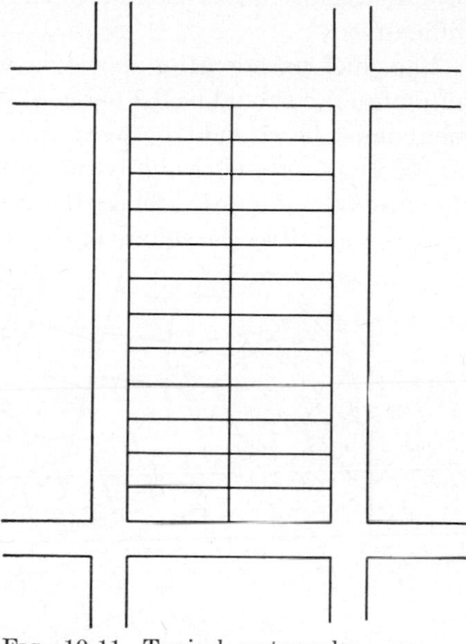

FIG. 10-11. Typical rectangular arrangement.

10-61. Restrictions or Protective Covenants. Frequently, the type of development is controlled by introducing restrictions in deeds. These refer to the cost, type, use, and location of the buildings that may be erected. Often zoning laws regulate the type and location of the buildings.

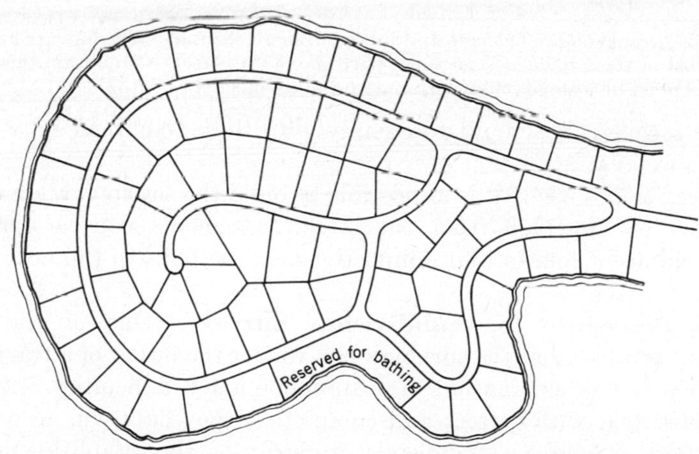

FIG. 10-12. Curved layout development of a headland.

10-62. Adaptation of Subdivisions to Topography. In general, streets should be placed at a lower elevation than the abutting lots,

The streets should have a longitudinal fall of at least 0.4 per cent for drainage and not more than a 10 per cent grade for safety. Sags in profile within blocks should be avoided, as these introduce drainage difficulties.

The chief consideration, however, is the sanitary sewage system. The minimum fall should be 0.4 per cent for small sewers, the grade and alignment must be straight between manholes, and the flow line should preferably be lower and certainly not be much higher than the cellar floors of the proposed houses. Since the unit cost of excavation for a sewer increases about as the square of the depth, the streets must be arranged so

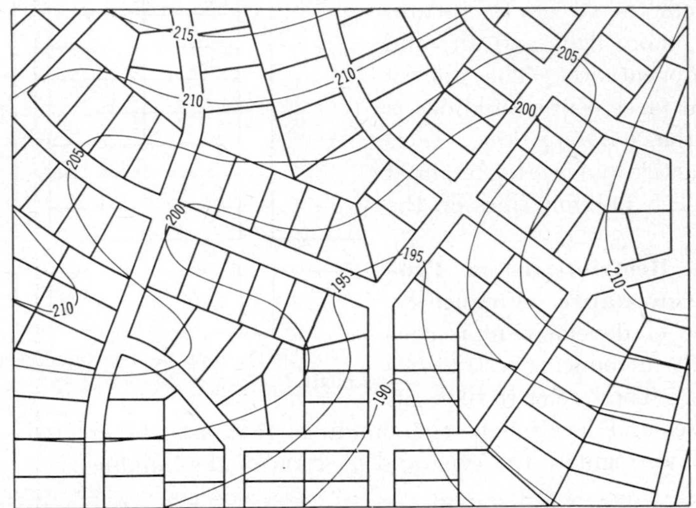

FIG. 10-13. Arrangement of streets and lots in difficult terrain. The lots are arranged so that most of them have at least some part above the street. To obtain this result, the streets must be located on low ground and must meet the contours at right angles.

that the sewers can be built to comply with these requirements without excessive excavation.

On steep slopes where it is impossible to build the sewers low enough to serve the lots downhill from the street, easements can be arranged through adjacent lots so that connections can be built to the next street below.

10-63. Procedure for Subdivision Surveys. One of the most important services that the surveyor can render the owner of a tract to be subdivided is the elimination of future boundary difficulties. A well-constructed plat with correct and complete survey data will accomplish this purpose. Such a plat depends primarily on the establishment of a precise control traverse. Such a traverse should be established at the outset and utilized thereafter for several purposes. The traverse is arranged so that a traverse station is located within the tract near each right-of-way corner of the tract to be subdivided. The stations should be intervisible and should provide as long courses as practicable. However, sufficient stations should be established to provide control within 100 ft of any course in the boundaries. The stations should be marked by iron pins or pipes. The traverse should be surveyed with second-order accuracy. Figure 10-14 shows such a control traverse later used to lay out the street monuments for the plan shown in Fig. 10-14.

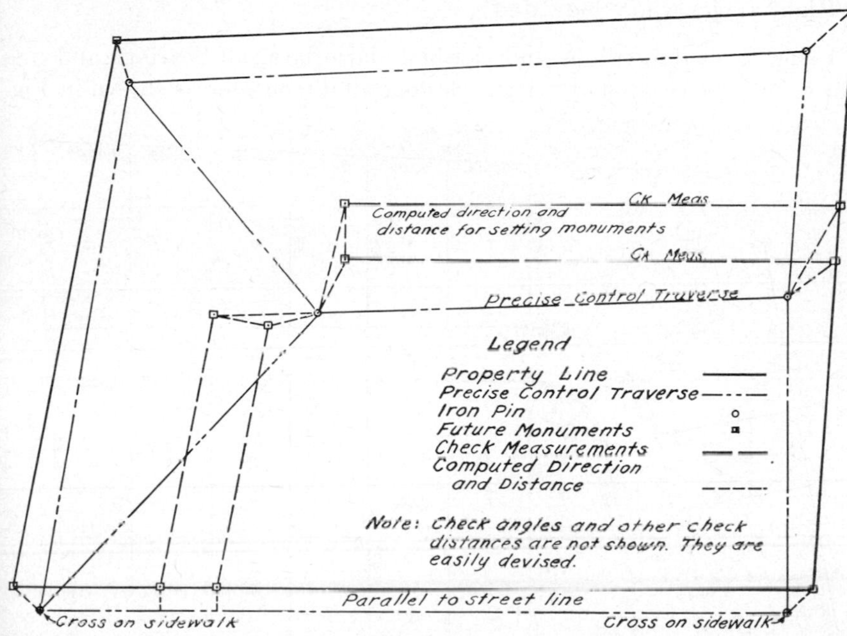

FIG. 10-14. Control traverse for subdivision.

The boundaries of the tract should be determined by a property survey based on the control traverse and the property corners thus determined should be monumented.

A system of benchmarks should be established based on city datum, and a contour map should be made of the tract having a contour interval of 2 to 5 ft depending on the topography and at a scale of 30 to 100 ft to the in.

Using tracing paper, possible arrangements of streets and lots should be sketched in pencil. When a particular arrangement is adopted, certain basic dimensions are thereby established. For example, in the subdivision shown in Fig. 10-15 the following dimensions were established:

I–17, 160 ft from, and parallel to, *JO*
Curve, 17–16–15–14, radius 150 ft
Street, 60 ft wide

3–Y, 360 ft from, and parallel to, AJ
1–Z, 180 ft from, and parallel to, AJ
G–Q, through PT of curve and perpendicular to AJ
Basic lot frontage, 90 ft
Angle between radial 17–M and 16–N, 28°
Angle between radial 15–P and 14–Q, 24°
KL, 100 ft
Right angles where indicated

Using these dimensions and the boundary data, all bearings and distances may be computed. By including all dimensions as shown in Fig.

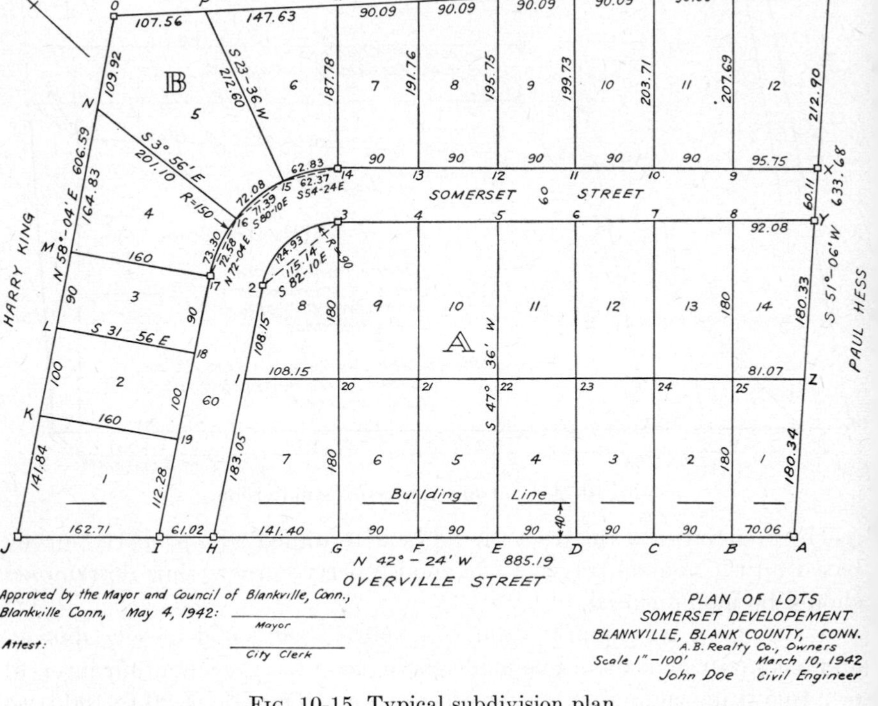

FIG. 10-15. Typical subdivision plan.

10-15, all monuments are connected by survey ties so that if any monument is disturbed it can be replaced by measurements from others. The completeness of the data protects the purchaser and permanently establishes the lines.

10-64. Principle of Minimum Description Applied to Subdivision Plans. By using the plat shown in Fig. 10-16, the precise traverse, the boundary survey, and all computations can be avoided. Such a plan can be made up from the original description of the tract, and when the monuments shown are set, the lots can be staked out. This plan complies with the theory which advocates minimum descriptions and has all the advantages and disadvantages inherent in them. It is not recommended.

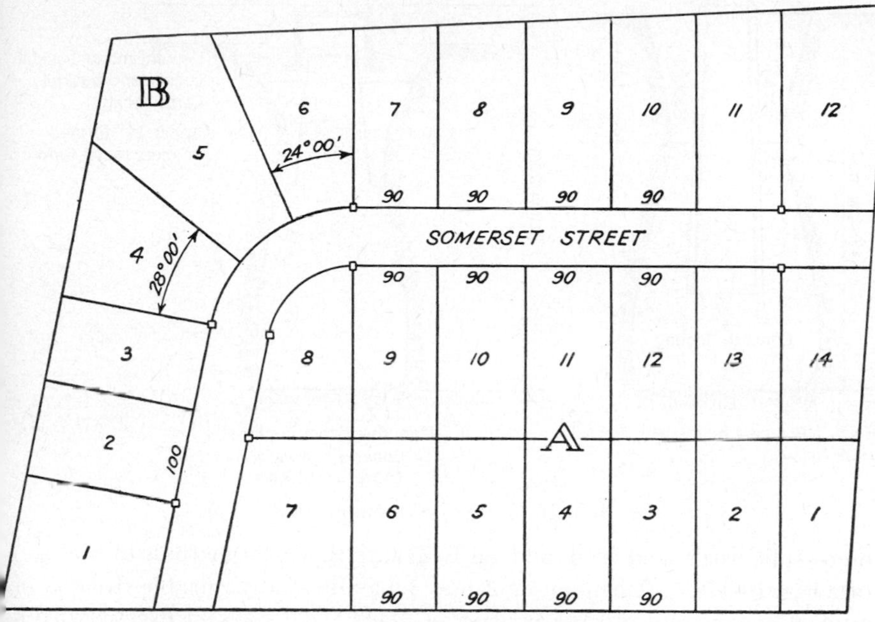

FIG. 10-16. Example of minimum dimensions for a subdivision plan. Not recommended. Such a plan can be constructed without a survey and without computations. The monuments shown can be placed by estimation, and thereafter all corners are fixed.

10-65. Establishing the Monuments. When the plan is complete and preliminary approval is obtained from the municipal authorities, the street line monuments should be set in the ground. This is accomplished by measurements from the precise preliminary traverse. The arrangement of the proposed plan dictates the method (see Fig. 10-14). Angles between long lines must be established by repetition. Frequently, the position for a monument is best determined by computing a direction and a distance from a preliminary traverse station. The positions for the monuments should be marked with stakes and tacks, and traverses should be run over the monument positions along the street lines and tied to the preliminary traverse as a check. In general, the monuments should be set on the actual street lines, either on the side lines or the center lines. This prevents later confusion. In areas where the building line coincides with the street line, the monuments are set at the intersection of offset lines within the sidewalk areas.

10-66. Street Monuments. Monuments must be set so that they will not be moved by settlement or frost action. The hole for the monument should be as small as possible, it should reach below the frost line

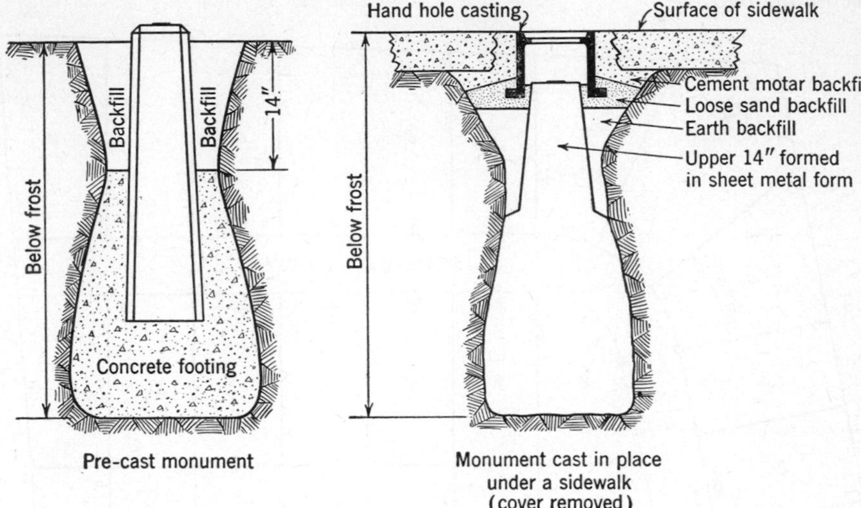

FIG. 10-17. Typical monuments.

or to solid rock, and it should be larger at the bottom than at the top (see Fig. 10-17). The monument may be made of any durable stone or of reinforced concrete precast or cast in place. All backfill except within 14 in. of the surface should be concrete. If earth is used, settlement will move the monument. At least the top 14 in. of the monument should be smooth and tapering upward. A galvanized iron or wooden form may be used for the top section when the monument is cast in place. The exact point is marked with a drill hole, or with a copper plug or a bronze disk with extended prongs. The plug or disk should be set flush with the surface so that it cannot be removed.

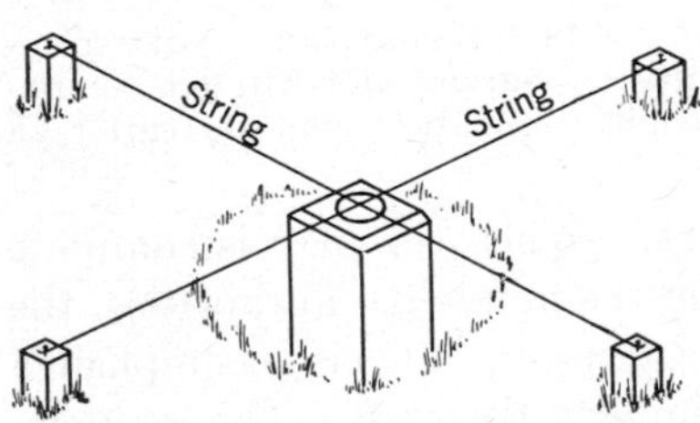

FIG. 10-18. Setting a monument. Monument should be set plumb.

In order to replace a stake with a monument, strings between two pairs of stakes are arranged to intersect at the monument position, Fig. 10-18. The strings can be removed while the hole is dug and replaced whenever the exact position is desired.

10-67. Surveys for Building Streets. When the streets are actually to be constructed, preliminary profiles are surveyed and typical cross sections and grades are designed from them for the road pavement, sidewalks, curbs, drainages, and other services, according to the principles described elsewhere. The lines for these improvements are established from the monuments and the grades from the established system of benchmarks (see Chap. 13).

COMPUTATIONS FOR SUBDIVISION PLANS

10-68. Steps for a Real-estate Subdivision. The various operations hitherto described, which are required to subdivide a tract into building lots, are collected here. The first five steps should be completed at the outset. The remainder can be put off until a street must be built or a lot is sold. Then only the necessary parts of the tract need be treated.

1. Establish a second-order **control traverse** near the boundaries of the tract.
2. Complete a **boundary survey** based on the control traverse, and monument the property corners thus established.
3. Establish a system of **benchmarks.**
4. Make a topographic map.
5. Determine the best plan for the streets and lots and draw a sketch to scale showing the street and lot lines. The basic dimensions thus established should be given. These dimensions are enclosed in small rectangles in Fig. 10-19. Prints of this plan can be used in selling the lots. However, more protection is given the purchaser if the plan is completed, approved, and filed.
6. Compute the bearings and lengths of the lot boundaries and street lines.
7. Set the **monuments** that establish street lines. Any necessary streets and utilities can then be built as required.
8. Mark the lot corners as required. Descriptions of each lot can be supplied the owner as required.

10-69. Computation. The computations referred to in step 6, above, usually impose mathematical problems that require considerable skill and experience to solve. It is impossible to cover all the types of problems that arise.

The most efficient method of computation usually follows these steps:

1. Compute the necessary angles from the bearings.
2. Compute the missing lengths and bearings by trigonometric solution of triangles. Avoid simultaneous equations and coordinate computations.
3. Compute the final bearings from the angles.
4. Compute the coordinates by latitudes and departures.

The procedure is illustrated by demonstrating the computations necessary for the subdivision shown in Fig. 10-15. The values required are illustrated in Fig. 10-19.

It is assumed that the boundary survey has been completed and adjusted so that the dimensions shown form a closed figure. These dimensions and the basic subdivision dimensions are the fixed values. Each is enclosed in a small rectangle in Fig. 10-19.

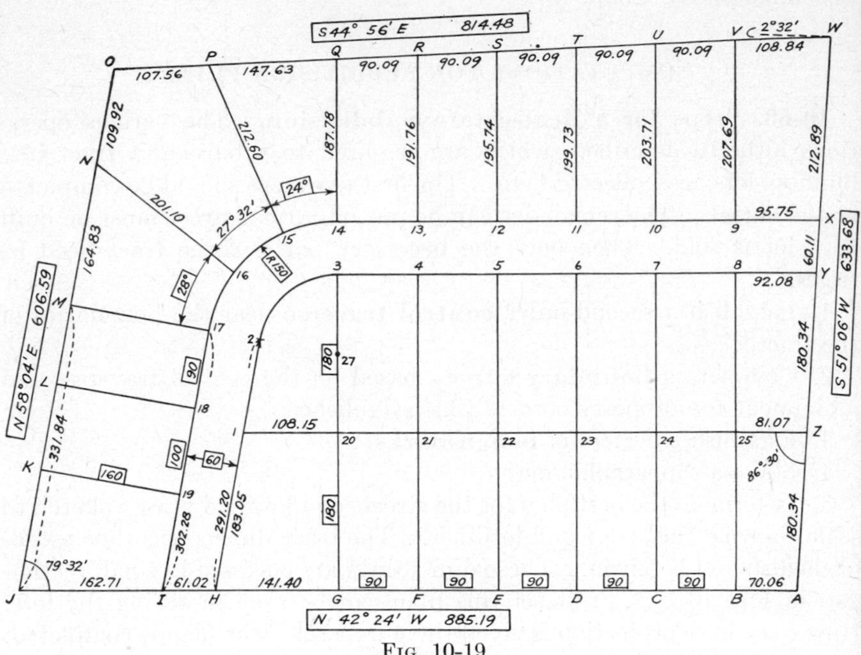

Fig. 10-19

Unless dimensions indicate to the contrary, lines are exactly parallel or perpendicular as they appear.

First compute these angles from the bearings.

$$J = 180° - 58°04' - 42°24' = 79°32'$$
$$Z = 180° - 51°06' - 42°24' = 86°30'$$
$$W = 44°56' - 42°24' = 2°32'$$

Compute the following lengths (note that 27 is a center):

$$G\text{–}27 = 360 - 90 = 270$$
$$\perp \text{ from } M \text{ to } JA = (G\text{–}27) + (M\text{–}27)\cos 79°32'$$
$$JM = \frac{270 + 310 \cos 79°32'}{\sin 79°32'} = 331.84$$
$$I\text{–}17 = \frac{270 + 150 \cos 79°32'}{\sin 79°32'} = 302.28$$
$$H\text{–}2 = \frac{270 + 90 \cos 79°32'}{\sin 79°32'} = 291.20$$

Then

$$GH = (2\text{–}27) \sin 79°32' + (H\text{–}2) \cos 79°32'$$
$$GH = 90 \sin 79°32' + 291.20 \cos 79°32' = 141.40$$
$$HI = \frac{60}{\sin 79°32'} = 61.02$$
$$IJ = \frac{160}{\sin 79°32'} = 162.71$$
$$AB = 885.19 - 815.13 = 70.06$$
$$Z\text{–}25 = 70.06 + 180 \cot 86°30' = 81.07$$
$$Y\text{–}8 = 70.06 + 360 \cot 86°30' = 92.08$$
$$X\text{–}9 = 70.06 + 420 \cot 86°30' = 95.75$$
$$\perp \text{ from } W \text{ to } VB = AB + AW \cos 86°30'$$
$$\perp \text{ from } W \text{ to } VB = 70.06 + 633.68 \cos 86°30' = 108.74$$
$$WV = \frac{108.74}{\cos 2°32'} = 108.85$$
$$AZ = \frac{180}{\sin 86°30'} = 180.34$$
$$YX = \frac{60}{\sin 86°30'} = 60.11$$
$$XW = 633.68 - 420.79 = 212.89$$
$$UV = \frac{90}{\cos 2°32'} = 90.09$$
$$\perp \text{ from } W \text{ to } X\text{–}14 = 212.89 \sin 86°30' = 212.50$$
$$9\text{–}V = 212.50 - 108.74 \tan 2°32' = 207.69$$
$$10\text{–}U = 212.50 - 198.74 \tan 2°32' = 203.71$$

etc.

$$14\text{–}Q = 212.50 - 558.74 \tan 2°32' = 187.78$$
$$\text{Angle } N\text{–}27\text{–}P = 79°32' - 28° - 24° = 27°32'$$
$$27\text{–}N = \frac{310}{\cos 28°} = 351.10$$
$$16\text{–}N = 351.10 - 150 = 201.10$$
$$MN = 310 \tan 28° = 164.83$$

In the triangle P–Q–27: $P = 68°32'$, $Q = 87°28'$, $27 = 24°$, and $27\text{–}Q = 337.78$.

$$PQ = \frac{\sin 24°(337.78)}{\sin 68°32'} = 147.63$$
$$27\text{–}P = \frac{\sin 87°28'(337.78)}{\sin 68°32'} = 362.60$$
$$15\text{–}P = 362.60 - 150 = 212.60$$
$$\text{Arc } 17\text{–}16 = \frac{28°}{360°} 2\pi 150 = 73.30$$
$$\text{Arc } 16\text{–}15 = \frac{27°32'}{360°} 2\pi 150 = 72.08$$
$$\text{Arc } 15\text{–}14 = \frac{24°}{360°} 2\pi 150 = 62.83$$

Then by Eq. (13-3) chords for these arcs and arc 2–3.

Similarly:

$$H\text{-}1 = \frac{180}{\sin 79°32'} = 183.05$$

$$1\text{–}20 = 141.40 - 180 \cot 79°32' = 108.15$$

All other lengths can be computed by addition and subtraction. The bearings are then computed through the angles.

The closure of each lot should be checked by bearings and lengths.

STATE COORDINATE SYSTEMS

10-70. The Use of State Plane Coordinates. The U.S. Coast and Geodetic Survey has established a system of plane coordinates for each state. These are described more fully in Chaps. 12 and 23. The state coordinates have been determined for U.S. Coast and Geodetic Survey triangulation stations and, in some states, for monumented points throughout the state by first- and second-order triangulation and traverse. Since the position of the coordinate system is marked by many monuments in any given area, state coordinate positions are permanently fixed with considerable precision. All the monuments connected with the system serve as witnesses to each other so that should any monument be displaced, it could be replaced or a coordinate position monument established nearby. Since the accuracy of the basic surveys is 1 part in 10,000 or better, the probable error of replacement is comparatively small. These monuments thus provide the two requirements desired in landmarks used as a basis for boundary positions. They are both permanent and precise. Every effort should therefore be made to include these monuments as landmarks in boundary descriptions, and survey ties should always be made to them whenever possible. The monuments are usually found in pairs so that direction as well as position ties can be made. Using such ties the state coordinates of each of the property corners can be computed, and their inclusion in the description has many advantages. It establishes the identity of the property without possible error, it states the relative position of the property with respect to any other property so described, it permanently establishes the lines, and it facilitates the relocation of the corners from any position for which the state coordinates are known. The use of state coordinates in boundary descriptions is recommended by the American Bar Association and the American Society of Civil Engineers in Joint Committee Reports[1] and by action of the governing bodies of the two organizations.

The second of these reports recommends three types of descriptions for utilizing the state coordinate systems. These are given here.

[1] "Land Surveys and Titles," First and Second Progress Report of the Real Property Division, American Bar Association and the Surveying and Mapping Division, American Society of Civil Engineers, November, 1938, and June, 1941, respectively.

10-71. Descriptions Utilizing State Coordinates Recommended by the ABA and the ASCE. Here are quoted parts of the Second Report noted above.[1]

1. *Sample Description, Including a Plat as Part of the Description.*

"* * * situated in the Town of __________, County of __________, and State of __________, shown (____)[2] upon a plat drawn by Fred L. Connor, Civil Engineer, dated May 1, 1929, and filed (state here the place of filing) to which plat reference is hereby made for more particular description."

It is to be noted that the use of this form of description requires in every instance the use of a plat [see Fig. 10-20] as an essential part thereof, which plat must be made a matter of public record.

The use of this form is not recommended when the omission from the body of the deed of words identifying the tract causes extra labor in title search (for example, when the place of storage of the plats is separate from that of the deeds). In such cases, the form Sample 2 should be used.

2. *Sample Description, Including a Plat as Part of the Description.* Where further identification in the body of the deed is desired than is provided by Description 1, the following is suggested:

"* * * situated in the Town of __________, County of __________, State of __________, and bounded and described as follows:

Southerly by Farm Road, 123.39 ft.
Westerly by land of Arthur C. Hicks, 145.82 ft.
Northerly by land of Peter L. Prince, 62.04 ft.
Easterly by land of Peter L. Prince, 133.09 ft.

Being the same premises shown on a plat of the land hereby conveyed, drawn by Fred L. Connor, Civil Engineer, dated May 1, 1929, and filed (state here the place of filing) to which plat reference is hereby made for more particular description."

The verbal part of Description 2 furnishes positive identification of the tract for purposes of tracing title, especially if the brief description is used, verbatim, over and over in successive transfers, and, of course, much more conveniently when read with the plat at hand—as should always be the case. The plat (see Fig. 10-20), incorporated by reference, permits the clear showing of all data which are essential for recovering the boundaries on the ground. These data include, among other items, physical marks of boundary, a position and directions stated in terms of the state coordinate system, and names of adjoining owners. It will be noticed that the

[1] Copyright, American Society of Civil Engineers.

[2] If the parcel has a name, lot number or other designation such as "Blackacre," or "containing ____ sq ft" that should be here stated. A sample plat is shown in Fig. 10-20. The foregoing description is to be used where only the one parcel described is shown on the plat. If the plat includes more area than that to be included in the description, there should appear in the place marked (____)[2] such a qualification as "see Lot 5 of Block 7."

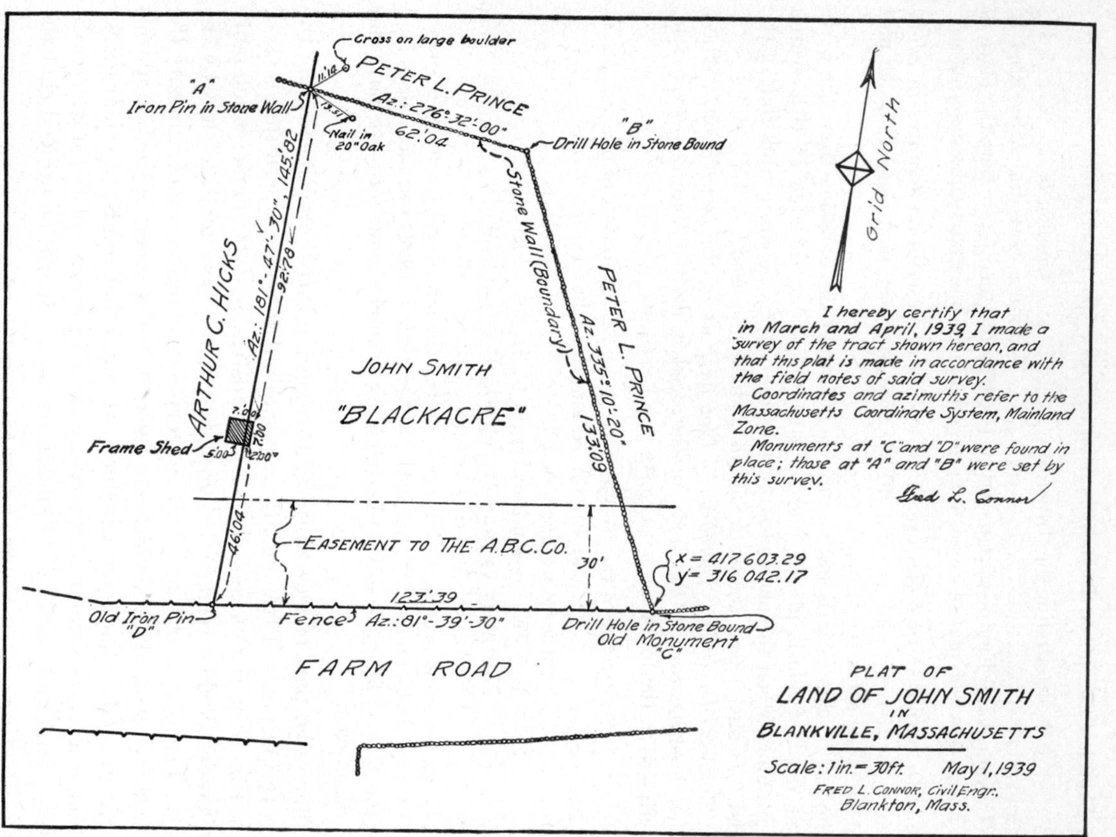
Cross on large boulder
"A"
Iron Pin in Stone Wall
PETER L. PRINCE
Az.: 276°-32'-00"
62.04
Nail in 20" Oak
"B"
Drill Hole in Stone Bound
Grid North
Stone Wall (Boundary)
PETER L. PRINCE
Az. 335°-10'-20"
133.09
ARTHUR C. HICKS
Az.: 181°-47'-30", 145.82
92.78
Frame Shed
JOHN SMITH
"BLACKACRE"
EASEMENT TO THE A.B.C. Co.
30'
46.04
x = 417 603.29
y = 316 042.17
123.39
Old Iron Pin "D"
Fence
Az.: 81°-39'-30"
Drill Hole in Stone Bound
Old Monument "C"
FARM ROAD
I hereby certify that in March and April, 1939, I made a survey of the tract shown hereon, and that this plat is made in accordance with the field notes of said survey.
Coordinates and azimuths refer to the Massachusetts Coordinate System, Mainland Zone.
Monuments at "C" and "D" were found in place; those at "A" and "B" were set by this survey.
Fred L. Connor
PLAT OF
LAND OF JOHN SMITH
IN
BLANKVILLE, MASSACHUSETTS
Scale: 1 in. = 30 ft.
May 1, 1939
FRED L. CONNOR, Civil Engr.
Blankton, Mass.

FIG. 10-20

boundary on the highway is stated first, this being for the purpose of rapid identification.

3. *Sample Description by Metes and Bounds, Where No Plat is Available.*

"* * * situated in the Town of __________, County of __________, State of __________, and bounded as follows:

"Beginning at a drill hole in a stone bound which is set in the corner of a stone wall on the north line of Farm Road at the southwest corner of land of Peter L. Prince and at the southeast corner of land hereby conveyed, the coordinates of which monument referred to the Massachusetts Coordinate System, Mainland Zone, are: x = 417,603.29, y = 316,042.17.

"Thence, on an azimuth of 81°39′30″, 123.39 feet along the northerly line of Farm Road to an iron pin at the southwest corner of the tract hereby conveyed;

"Thence, on an azimuth of 181°47′30″, 145.82 feet along the easterly line of land of Arthur C. Hicks to an iron pin in a stone wall at the northwest corner of the tract hereby conveyed;

"Thence, on an azimuth of 276°32′00″, 62.04 feet along a stone wall on the southerly line of land of Peter L. Prince to a drill hole in a stone bound in the wall at the northeast corner of land hereby conveyed;

"Thence, on an azimuth of 335°10′20″, 133.09 feet along a stone wall on the westerly line of land of said Prince, to the point of beginning.

"Zero azimuth is grid south in the Massachusetts Coordinate System, Mainland Zone.

"This description was written June 1, 1939 from data secured by survey made by Fred L. Connor, Civil Engineer, in March and April, 1939.

"Together with all right, title, and interest in and to all roads and ways adjoining the above-described premises."

In Description 3 the location of the point of beginning is first definitely described, both as to physical monument and as to location with respect to adjoining land and in the Massachusetts Coordinate System. Each clause defining one course of the boundary contains the four essential elements of (*a*) direction, (*b*) distance, (*c*) along a physical monument if one exists, such as a wall, and along the boundary of the adjoining tract, and (*d*) physical terminus. The manner of writing the directions and distances indicates the precision with which they are stated (number of decimal places, for example). Intentionally, there is no statement of "more or less." The bounds return the description to the point of beginning.

The meaning of the directions is definitely indicated by the sentence beginning "Zero azimuth is grid south . . ."; that is, there is no uncertainty as to whether directions are referred to the true meridian, to the magnetic needle, or to the direction of the central meridian of the state coordinate system.

However, this description, even for such a simple tract of land, is long and involved, and its use is not recommended. It is susceptible to error in transcription. Moreover, it is difficult to understand and visualize until a sketch showing the major features has been made up by the reader. For these reasons, the forms shown by Descriptions 1 and 2, involving the use of a plat, should always be required.

PROBLEMS

10-1. What is the connection between surveying and land surveying?

10-2. Name and describe the two legal principles that regulate the positions of boundaries.

10-3. Describe common law and statutory law.

10-4. What is a right of way?

10-5. Define an easement.

10-6. How are rights of way established?

10-7. Define a partition and seniority of title.

10-8. Define adverse possession. What are the advantages of this rule?

10-9. Define an encroachment.

10-10. Define acquiescence and agreement.

10-11. Where is the evidence of boundary position found?

10-12. What evidence is used for boundary location?

10-13. What are riparian rights?

10-14. Who has riparian rights?

10-15. Define high-water mark, low-water mark, and thread of stream.

10-16. Where is a riparian boundary?

10-17. What are the two basic functions of a description?

10-18. How are bearings used in a description?

10-19. What are the arguments for and against plus or minus added to a value in a description?

10-20. When are lengths prorated?

10-21. What is the fundamental function of a land surveyor?

10-22. Outline the procedure for a property survey.

Assume that the surveyor's client is a prospective purchaser, the agreement of sale has been executed, the survey has been completed, but the deed has not been written. What advice should the surveyor give the client if the survey indicates the following conditions concerning the parcel?

10-23. A building apparently encroaches 0.3 ft into the street.

10-24. The eaves of a roof encroach on the property.

10-25. The adjoining descriptions and the physical location are in agreement but the description of the property indicates a slightly different location.

10-26. A path is worn across the property.

10-27. There is no public record of the street in front of the property.

10-28. It is necessary to change a dimension from 83.5 to 85.3 ft to obtain consistency.

10-29. A monument is found that disturbs an otherwise consistent location.

10-30. One of the boundaries is described "by the pond" and the pond has long since drained away.

10-31. A fence line cuts diagonally across the property but fulfills the requirements of the adjoiner's deed.

CHAPTER 11

THE UNITED STATES PUBLIC LAND SURVEYS[1]

11-1. Establishment. In order to dispose of the great land areas acquired from time to time by the United States, it was necessary to develop a system to divide these areas into reasonably small holdings and to mark the boundaries of these holdings on the ground. It was important that the parcels should have a useful shape, be easily described and marked on the ground, and that data concerning them should be available to the public.

In 1785, Congress passed the first ordinance regulating the division of public lands. This ordinance established the basic principle, in force today, that the public lands should be divided into townships six miles square containing 36 sections each one mile square.

Shortly afterward, surveys were initiated north of the Ohio River under the supervision of the geographer of the United States. In 1796 a surveyor general was appointed who directed surveys northwest of the Ohio River and above the mouth of the Kentucky River. Thereafter, these surveys were continued under various auspices.

In 1812 the General Land Office was established in the Department of the Treasury, and the Public Land Surveys were assigned to this bureau. In 1849 the Department of the Interior was established and the General Land Office was transferred to the new Department. In 1946 the General Land Office was abolished and its functions transferred to a newly established Bureau of Land Management, also in the Department of the Interior.

Various acts of Congress and instructions issued by the officials in charge have established definite rules for the surveys. While the methods have changed throughout the years and different methods have been used in different localities, by and large the basic system is the same everywhere.

The system is generally known and referred to as "The System of Rectangular Surveys."

[1] The material in this chapter is taken almost exclusively from the "Manual of Instructions for the Survey of the Public Lands of the United States, 1947," prepared and published by the Bureau of Land Management, U.S. Department of the Interior. It can be procured from the Superintendent of Documents.

11-2. Operation. The United States government marks the section corners and those quarter-section corners that lie on the section lines. When the title to the land passes from the United States, these marks become final, despite any errors. The written evidence (record evidence) of their position consists of the survey field notes and the plats made from the field notes.

When holdings smaller than sections are disposed of by the United States, any necessary corners are almost always marked in the field by local surveyors. They must be marked in accordance with the regulations of the Bureau of Land Management, as these regulations represent the intent of the parties involved. When the corners are lost or obliterated they must be restored to their original positions whenever possible. Any actions or decisions that may involve changes in the location of established boundaries are subject to review by the state courts.

The usual minimum parcel that the United States will dispose of is the quarter quarter section. This is 1/4 mile square and contains 40 acres. Some fractional lots limited by natural barriers or other surveys are smaller.

The records of the surveys are kept in Washington, D.C., but when the surveys by the United States are practically complete in any state, the originals are transferred to that state (with the exception of Oklahoma), and usually duplicates are retained in Washington. The records are open to public inspection and copies can be procured upon payment of a certain fee.

After the lands are disposed of by the United States, subsequent land surveys are conducted according to the same principles as elsewhere. Obviously the determination of land boundaries will depend to a very large degree on the positions of the public land monuments. It is evident that every land surveyor who practices in an area originally subdivided by the Public Land Surveys must be familiar with the basic system and with the special conditions that exist in his territory. The general outline is given in this text. For the intricate specific rules that govern these surveys, the reader is referred to the manual mentioned at the head of this chapter and to its various supplements.

The initial surveys in Ohio were made before the standard system had been developed and accordingly contain many exceptions to the general rules. Careful study of the development, execution, and map of the public land survey in that state is necessary to clarify the meaning of descriptions and to conduct properly land surveys in that area.

11-3. Standard Township. Throughout the public land surveys, a township is approximately 6 miles square and bounded by due north-south and due east-west lines. In Ohio the lines are not exactly north-south. When a township is limited by a natural boundary or an old survey, part of it may be eliminated. Due to the convergence of meridians and other factors that are explained later, no township is **exactly** 6 miles square. A township contains 36 sections approximately one mile square. They are arranged and numbered as shown in Fig. 11-1 and each section contains approximately 640 acres. When part of a township is cut off by natural barriers or old surveys, the existing sections are numbered as though the whole township could have been laid out. Quarter sections and quarter quarter sections ($\frac{1}{16}$ of a section) are arranged as shown in Fig. 11-1 and are designated by their compass positions. The

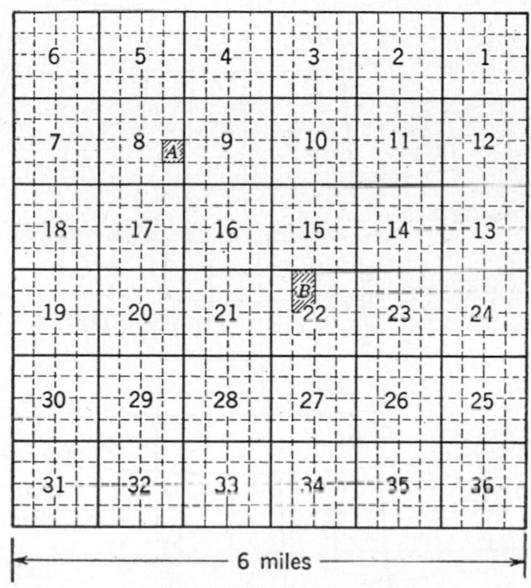

FIG. 11-1. A township.

quarter quarter section indicated by the letter *A* in Fig. 11-1 is described as the NE $\frac{1}{4}$ of SE $\frac{1}{4}$ of Sec. 8. Sometimes two adjacent quarters are described as the N $\frac{1}{2}$ or the E $\frac{1}{2}$, etc. Letter *B* indicates the E $\frac{1}{2}$ of NW $\frac{1}{4}$ of Sec. 22.

11-4. The Initial Points. The Public Land Surveys within the United States are based on 35 separate surveys. Each survey is a separate entity. A survey is begun with the establishment of an **initial point** (see Fig. 11-2). A line is run true north and south from the initial point. It is known as the **principal meridian.** The principal meridian is named, its name is used to identify the survey, and the name is included in descriptions of all parcels based on that survey. An initial point is established by setting a monument and reference marks according to special instructions. The astronomical position is then determined by field observations.

STANDARD LINES

11-5. Survey Methods. A framework of **standard lines** is laid out from the initial point (see Fig. 11-2). These run due north and south and due east and west. They are usually spaced at 24-mile intervals, forming **quadrangles** approximately 24 miles square, and they are established in the order given here. They consist of the principal meridian, the base

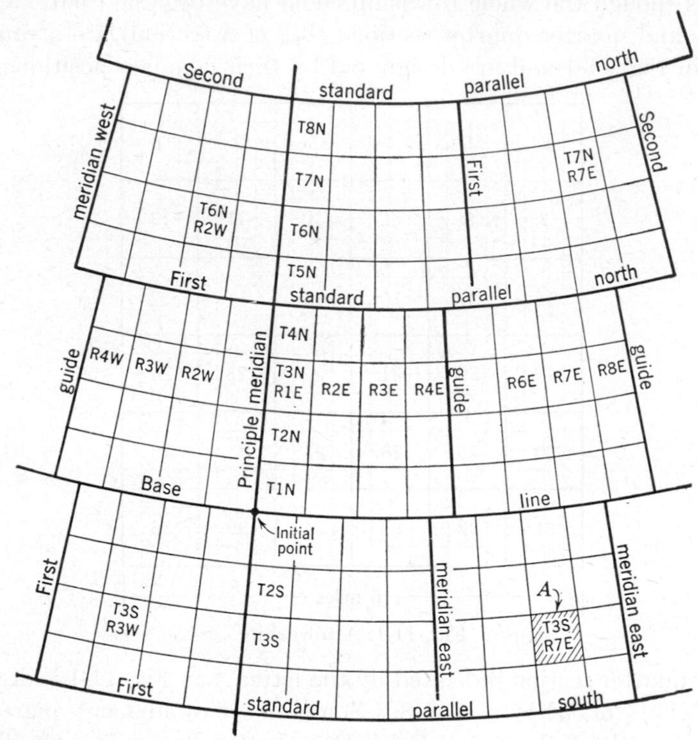

FIG. 11-2

line, the standard parallels, and the guide meridians. All are marked at intervals of 40 chains (one-half mile), that is to say, at the section-and quarter-section corners. These corners on the standard lines are called **standard corners.** Direction is established by observations on Polaris or other stars, or the sun. Observations on the sun are usually made with the solar attachment (the solar transit). In recent surveys directions are held within ± 3 minutes, and distances are chained twice and held to a maximum difference of 14 links in 80 chains, or 1 part in 571.

The directions of the lines are maintained either by backsight (the **transit method**) or by orienting the transit with the solar attachment at frequent setups (the **solar method**). When the **transit method** is employed for east-west lines, either the **tangent method** or the **secant method** is used to establish the curvature. Direction need not be checked more often than every 20 chains, but it should be checked at least every 40 chains to keep the line in its proper position. When the solar method is used the curvature is established automatically.

11-6. The Tangent Method. Beginning at any township corner, a transit line is established at 90 degrees from true north and run straight for 6 miles (see Fig. 11-3). At each 40-chain point the corner is established by measuring due north the proper distance. The direction is established by turning the proper angle from the transit line. The angle and the distance measured for any corner depend on the latitude. The

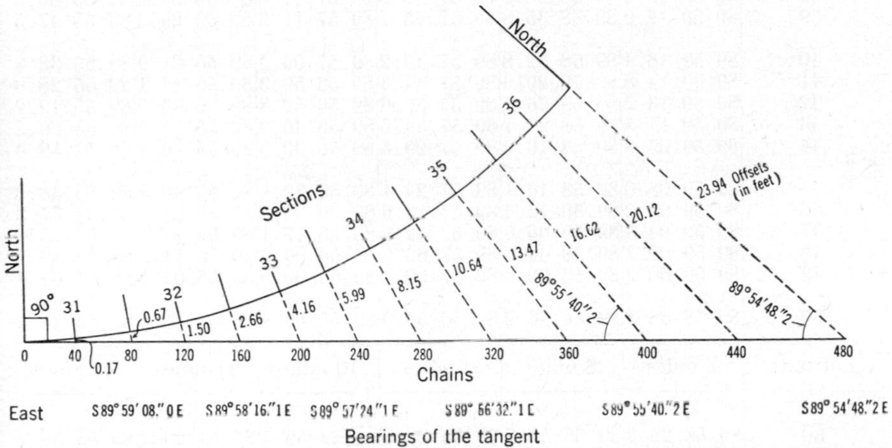

FIG. 11-3. Tangent method (latitude 45°).

values necessary can be found in Tables 11-1 and 11-2.[1] Those in the figure are for latitude 45°. The angles turned are derived from the stated bearings of the tangent line.

11-7. The Secant Method. The secant method (see Fig. 11-4) is usually preferred to the tangent method, as it requires shorter offsets. In this method the transit line is arranged to run through the first and fifth section corner. The transit line begins at a point due south of a township corner. True north is established at this point and the proper angle turned off. Points are set at 40-chain intervals and the corners are set due north as shown in the figure. The necessary offsets and bearings depend on the latitude and are given in Table 11-3. The values given in the figure are for latitude 45°. The angles turned are derived from the stated bearings of the secant line.

[1] Values for the half-mile points can be found by noting the following: angles vary directly as the distance; offsets vary as the square of the distance.

TABLE 11-1. AZIMUTHS OF THE TANGENT TO THE PARALLEL

The azimuth is the smaller angle the tangent makes with the true meridian; it is always measured from the north and toward the tangential point

Latitude	1 mile	2 miles	3 miles	4 miles	5 miles	6 miles
°	° ′ ″	° ′ ″	° ′ ″	° ′ ″	° ′ ″	° ′ ″
30	89 59 30.0	89 58 59.9	89 58 29.9	89 57 59.9	89 57 29.9	89 56 59.8
31	89 59 28.8	89 58 57.5	89 58 26.3	89 57 55.0	89 57 23.8	89 56 52.5
32	89 59 27.5	89 58 55.0	89 58 22.5	89 57 50.0	89 57 17.5	89 56 45.0
33	89 59 26.2	89 58 52.5	89 58 18.7	89 57 44.9	89 57 11.2	89 56 37.4
34	89 59 24.9	89 58 49.9	89 58 14.8	89 57 39.7	89 57 04.6	89 56 29.6
35	89 59 23.6	89 58 47.2	89 58 10.8	89 57 34.4	89 56 58.0	89 56 21.6
36	89 59 22.2	89 58 44.4	89 58 06.8	89 57 28.9	89 56 51.1	89 56 13.4
37	89 59 20.8	89 58 41.6	89 58 02.5	89 57 23.3	89 56 44.1	89 56 05.0
38	89 59 19.4	89 58 38.8	89 57 58.2	89 57 17.5	89 56 36.9	89 55 56.3
39	89 59 17.9	89 58 35.8	89 57 53.7	89 57 11.6	89 56 29.6	89 55 47.5
40	89 59 16.4	89 58 32.8	89 57 49.2	89 57 05.5	89 56 21.9	89 55 38.3
41	89 59 14.8	89 58 29.6	89 57 44.4	89 56 59.3	89 56 14.1	89 55 28.9
42	89 59 13.2	89 58 26.4	89 57 39.6	89 56 52.8	89 56 06.0	89 55 19.2
43	89 59 11.5	89 58 23.1	89 57 34.6	89 56 46.2	89 55 57.7	89 55 09.2
44	89 59 09.8	89 58 19.6	89 57 29.5	89 56 39.3	89 55 49.1	89 54 58.9
45	89 59 08.0	89 58 16.1	89 57 24.1	89 56 32.1	89 55 40.2	89 54 48.2
46	89 59 06.2	89 58 12.4	89 57 18.6	89 56 24.8	89 55 31.0	89 54 37.2
47	89 59 04.3	89 58 08.6	89 57 12.9	89 56 17.1	89 55 21.4	89 54 25.7
48	89 59 02.3	89 58 04.6	89 57 06.9	89 56 09.2	89 55 11.5	89 54 13.8
49	89 59 00.2	89 58 00.5	89 57 00.7	89 56 00.9	89 55 01.2	89 54 01.4
50	89 58 58.1	89 57 56.2	89 56 54.3	89 55 52.6	89 54 50.5	89 53 48.5

Latitude	7 miles	8 miles	9 miles	10 miles	11 miles	12 miles
°	° ′ ″	° ′ ″	° ′ ″	° ′ ″	° ′ ″	° ′ ″
30	89 56 29.8	89 55 59.8	89 55 29.8	89 54 59.7	89 54 29.7	89 53 59.7
31	89 56 21.3	89 55 50.0	89 55 18.8	89 54 47.6	89 54 16.3	89 53 45.1
32	89 56 12.5	89 55 40.0	89 55 07.6	89 54 35.1	89 54 02.6	89 53 30.1
33	89 56 03.6	89 55 29.9	89 54 56.1	89 54 22.3	89 53 48.5	89 53 14.8
34	89 55 54.5	89 55 19.4	89 54 44.4	89 54 09.3	89 53 34.2	89 52 59.1
35	89 55 45.2	89 55 08.8	89 54 32.3	89 53 55.9	89 53 19.5	89 52 43.1
36	89 55 35.6	89 54 57.8	89 54 20.0	89 53 42.3	89 53 04.5	89 52 26.7
37	89 55 25.8	89 54 46.6	89 54 07.4	89 53 28.2	89 52 49.1	89 52 09.9
38	89 55 15.7	89 54 35.1	89 53 54.5	89 53 13.9	89 52 33.2	89 51 52.6
39	89 55 05.4	89 54 23.3	89 53 41.2	89 52 59.1	89 52 17.0	89 51 34.9
40	89 54 54.7	89 54 11.1	89 53 27.5	89 52 43.8	89 52 00.2	89 51 16.6
41	89 54 43.7	89 53 58.5	89 53 13.4	89 52 28.2	89 51 43.0	89 50 57.8
42	89 54 32.4	89 53 45.6	89 52 58.8	89 52 12.0	89 51 25.2	89 50 38.4
43	89 54 20.8	89 53 32.3	89 52 43.8	89 51 55.4	89 51 06.9	89 50 18.5
44	89 54 08.7	89 53 18.5	89 52 28.4	89 51 38.2	89 50 48.0	89 49 57.8
45	89 53 56.3	89 53 04.3	89 52 12.3	89 51 20.4	89 50 28.4	89 49 36.4
46	89 53 43.4	89 52 49.5	89 51 55.7	89 51 01.9	89 50 08.1	89 49 14.3
47	89 53 30.0	89 52 34.3	89 51 38.6	89 50 42.9	89 49 47.2	89 48 51.4
48	89 53 16.1	89 52 18.4	89 51 20.7	89 50 23.0	89 49 25.3	89 48 27.6
49	89 53 01.7	89 52 01.9	89 51 02.1	89 50 02.4	89 49 02.6	89 48 02.8
50	89 52 46.6	89 51 44.7	89 50 42.8	89 49 40.9	89 48 39.0	89 47 37.1

TABLE 11-2. OFFSETS, IN FEET, FROM TANGENT TO PARALLEL

Latitude	1 mile	2 miles	3 miles	4 miles	5 miles	6 miles
30°	0.38	1.54	3.46	6.15	9.61	13.83
31	0.40	1.60	3.60	6.40	10.00	14.40
32	0.42	1.66	3.74	6.65	10.40	14.97
33	0.43	1.73	3.89	6.91	10.80	15.56
34	0.45	1.80	4.04	7.18	11.22	16.16
35	0.47	1.86	4.19	7.45	11.65	16.77
36	0.48	1.93	4.35	7.73	12.09	17.40
37	0.50	2.01	4.51	8.02	12.53	18.05
38	0.52	2.08	4.68	8.32	12.99	18.71
39	0.54	2.15	4.85	8.62	13.47	19.39
40	0.56	2.23	5.02	8.93	13.95	20.09
41	0.58	2.31	5.20	9.25	14.46	20.81
42	0.60	2.40	5.39	9.58	14.97	21.56
43	0.62	2.48	5.58	9.92	15.50	22.33
44	0.64	2.57	5.78	10.28	16.06	23.12
45	0.67	2.66	5.99	10.64	16.62	23.94
46	0.69	2.75	6.20	11.02	17.21	24.79
47	0.71	2.85	6.42	11.41	17.83	25.67
48	0.74	2.95	6.65	11.81	18.46	26.58
49	0.76	3.06	6.88	12.24	19.12	27.53
50	0.79	3.17	7.13	12.68	19.81	28.52

Latitude	7 miles	8 miles	9 miles	10 miles	11 miles	12 miles
30°	18.83	24.50	31.13	38.43	46.50	55.33
31	19.59	25.59	32.39	39.99	48.39	57.58
32	20.38	26.61	33.68	41.58	50.32	59.88
33	21.18	27.66	35.00	43.22	52.29	62.23
34	21.99	28.73	36.36	44.88	54.31	64.63
35	22.83	29.82	37.74	46.59	56.38	67.00
36	23.69	30.04	39.16	48.34	58.49	69.61
37	24.57	32.09	40.61	50.13	60.66	72.19
38	25.47	33.27	42.10	51.98	62.89	74.85
39	26.40	34.48	43.63	53.87	65.18	77.57
40	27.35	35.72	45.21	55.82	67.54	80.38
41	28.33	37.01	46.83	57.82	69.96	83.26
42	29.34	38.33	48.51	59.89	72.46	86.24
43	30.39	39.69	50.24	62.02	75.04	89.31
44	31.47	41.10	52.02	64.22	77.71	92.48
45	32.58	42.56	53.86	66.50	80.46	95.76
46	33.74	44.07	55.78	68.86	83.32	99.16
47	34.94	45.63	57.76	71.30	86.28	102.68
48	36.18	47.26	59.81	73.84	89.35	106.33
49	37.48	48.95	61.95	76.48	92.54	110.13
50	38.82	50.71	64.17	79.23	95.87	114.09

TABLE 11-3. AZIMUTHS OF THE SECANT, AND OFFSETS, IN FEET, TO THE PARALLEL

Latitude	Azimuths and offsets at							Deflection angle* and nat. tan. to rad. 66 ft
	0 miles	½ mile	1 mile	1½ miles	2 miles	2½ miles	3 miles	
°								
30	89°58.5′ 1.93 N	89°58.7′ 0.87 N	89°59.0′ 0.00	89°59.2′ 0.67 S	89°59.5′ 1.15 S	89°59.7′ 1.44 S	90° (E or W) 1.54 S	3′00.2″ 0.69 in.
31	89°58.4′ 2.01 N	89°58.6′ 0.91 N	89°58.9′ 0.00	89°59.2′ 0.70 S	89°59.5′ 1.20 S	89°59.7′ 1.50 S	90° (E or W) 1.60 S	3′07.4″ 0.72 in.
32	89°58.4′ 2.09 N	89°58.6′ 0.94 N	89°58.9′ 0.00	89°59.2′ 0.73 S	89°59.5′ 1.25 S	89°59.7′ 1.56 S	90° (E or W) 1.67 S	3′15.0″ 0.75 in.
33	89°58.3′ 2.17 N	89°58.5′ 0.97 N	89°58.8′ 0.00	89°59.1′ 0.76 S	89°59.4′ 1.30 S	89°59.7′ 1.62 S	90° (E or W) 1.73 S	3′22.6″ 0.78 in.
34	89°58.2′ 2.25 N	89°58.5′ 1.01 N	89°58.8′ 0.00	89°59.1′ 0.79 S	89°59.4′ 1.35 S	89°59.7′ 1.69 S	90° (E or W) 1.80 S	3′30.4″ 0.81 in.
35	89°58.2′ 2.33 N	89°58.5′ 1.05 N	89°58.8′ 0.00	89°59.1′ 0.82 S	89°59.4′ 1.40 S	89°59.7′ 1.75 S	90° (E or W) 1.87 S	3′38.4″ 0.84 in.
36	89°58.1′ 2.42 N	89°58.4′ 1.09 N	89°58.7′ 0.00	89°59.0′ 0.85 S	89°59.4′ 1.46 S	89°59.7′ 1.82 S	90° (E or W) 1.94 S	3′46.4″ 0.87 in.
37	89°58.0′ 2.51 N	89°58.3′ 1.13 N	89°58.6′ 0.00	89°58.9′ 0.88 S	89°59.3′ 1.51 S	89°59.7′ 1.89 S	90° (E or W) 2.01 S	3′55.0″ 0.90 in.
38	89°58.0′ 2.61 N	89°58.3′ 1.17 N	89°58.6′ 0.00	89°58.9′ 0.91 S	89°59.3′ 1.56 S	89°59.7′ 1.95 S	90° (E or W) 2.08 S	4′03.6″ 0.93 in.
39	89°57.9′ 2.70 N	89°58.2′ 1.21 N	89°58.6′ 0.00	89°58.9′ 0.94 S	89°59.3′ 1.62 S	89°59.7′ 2.02 S	90° (E or W) 2.16 S	4′12.6″ 0.97 in.
40	89°57.8′ 2.79 N	89°58.1′ 1.25 N	89°58.5′ 0.00	89°58.9′ 0.98 S	89°59.3′ 1.68 S	89°59.7′ 2.10 S	90° (E or W) 2.24 S	4′21.6″ 1.00 in.
41	89°57.7′ 2.89 N	89°58.0′ 1.30 N	89°58.4′ 0.00	89°58.8′ 1.02 S	89°59.2′ 1.74 S	89°59.6′ 2.17 S	90° (E or W) 2.32 S	4′31.2″ 1.04 in.
42	89°57.7′ 3.00 N	89°58.0′ 1.35 N	89°58.4′ 0.00	89°58.8′ 1.05 S	89°59.2′ 1.80 S	89°59.6′ 2.25 S	90° (E or W) 2.40 S	4′40.8″ 1.08 in.
43	89°57.6′ 3.11 N	89°58.0′ 1.40 N	89°58.4′ 0.00	89°58.8′ 1.08 S	89°59.2′ 1.86 S	89°59.6′ 2.33 S	90° (E or W) 2.48 S	4′50.8″ 1.12 in.
44	89°57.5′ 3.22 N	89°57.9′ 1.45 N	89°58.3′ 0.00	89°58.7′ 1.12 S	89°59.2′ 1.93 S	89°59.6′ 2.41 S	90° (E or W) 2.57 S	5′01.0″ 1.16 in.
45	89°57.4′ 3.33 N	89°57.8′ 1.50 N	89°58.3′ 0.00	89°58.7′ 1.16 S	89°59.1′ 2.00 S	89°59.5′ 2.49 S	90° (E or W) 2.66 S	5′11.8″ 1.20 in.
46	89°57.3′ 3.44 N	89°57.7′ 1.55 N	89°58.2′ 0.00	89°58.6′ 1.21 S	89°59.1′ 2.07 S	89°59.5′ 2.59 S	90° (E or W) 2.76 S	5′22.8″ 1.24 in.
47	89°57.2′ 3.57 N	89°57.6′ 1.61 N	89°58.1′ 0.00	89°58.6′ 1.25 S	89°59.1′ 2.14 S	89°59.5′ 2.67 S	90° (E or W) 2.86 S	5′34.2″ 1.28 in.
48	89°57.1′ 3.70 N	89°57.5′ 1.66 N	89°58.0′ 0.00	89°58.5′ 1.30 S	89°59.0′ 2.22 S	89°59.5′ 2.78 S	90° (E or W) 2.96 S	5′46.2″ 1.33 in.
49	89°57.0′ 3.82 N	89°57.5′ 1.72 N	89°58.0′ 0.00	89°58.5′ 1.34 S	89°59.0′ 2.30 S	89°59.5′ 2.87 S	90° (E or W) 3.06 S	5′58.6″ 1.38 in.
50	89°56.9′ 3.96 N	89°57.4′ 1.78 N	89°57.9′ 0.00	89°58.4′ 1.39 S	89°59.0′ 2.38 S	89°59.5′ 2.97 S	90° (E or W) 3.17 S	6′11.4″ 1.43 in.

* Between adjacent secants.

TABLE 11-4. LENGTHS OF ARCS OF THE EARTH'S SURFACE*

Lengths of degrees of the parallel				Lengths of degrees of the meridian	
Lat.	Statute miles	Lat.	Statute miles	Lat.	Statute miles
° ′		° ′		°	
25 0	62.729	47 30	46.818	25	68.829
30	62.473	48 0	46.372	26	68.839
26 0	62.212	30	45.922	27	68.848
30	61.946	49 0	45.469	28	68.858
27 0	61.676	30	45.012	29	68.869
30	61.401	50 0	44.552	30	68.879
28 0	61.122	30	44.088	31	68.890
30	60.837	51 0	43.621	32	68.901
29 0	60.548	30	43.150	33	68.912
30	60.254	52 0	42.676	34	68.923
30 0	59.956	30	42.199	35	68.935
30	59.653	53 0	41.719	36	68.946
31 0	59.345	30	41.235	37	68.958
30	59.033	54 0	40.749	38	68.969
32 0	58.716	30	40.259	39	68.981
30	58.396	55 0	39.766	40	68.993
33 0	58.071	30	39.270	41	69.006
30	57.741	56 0	38.771	42	69.018
34 0	57.407	30	38.269	43	69.030
30	57.068	57 0	37.764	44	69.042
35 0	56.725	30	37.256	45	69.054
30	56.378	58 0	36.745	46	69.066
36 0	56.027	30	36.232	47	69.079
30	55.671	59 0	35.716	48	69.091
37 0	55.311	30	35.196	49	69.103
30	54.947	60 0	34.674	50	69.115
38 0	54.579	30	34.150	51	69.127
30	54.206	61 0	33.623	52	69.139
39 0	53.829	30	33.093	53	69.151
30	53.448	62 0	32.560	54	69.163
40 0	53.063	30	32.025	55	69.175
30	52.674	63 0	31.488	56	69.186
41 0	52.281	30	30.948	57	69.197
30	51.884	64 0	30.406	58	69.209
42 0	51.483	30	29.862	59	69.220
30	51.078	65 0	29.315	60	69.230
43 0	50.669	30	28.766	61	69.241
30	50.257	66 0	28.215	62	69.251
44 0	49.840	30	27.661	63	69.261
30	49.419	67 0	27.106	64	69.271
45 0	48.995	30	26.548	65	69.281
30	48.567	68 0	25.988	66	69.290
46 0	48.136	30	25.426	67	69.299
30	47.700	69 0	24.862	68	69.308
47 0	47.261	30	24.297	69	69.316
47 30	46.818	70 0	23.729	70	69.324

* The lengths of degrees of the meridian are tabulated to correspond to the length of the arc of which the tabulated latitude is the middle; thus, the quantity 68,993, opposite latitude 40°0′, is the number of miles between latitudes 39°30′ and 40°30′.

The above table is an abridgment of a table published by the U.S. Coast and Geodetic Survey based on the values of the Clarke spheroid.

TABLE 11-5. CONVERGENCY OF MERIDIANS, 6 MILES LONG AND 6 MILES APART, AND DIFFERENCES OF LATITUDE AND LONGITUDE

Lat.	Convergency		Difference of longitude per range		Difference of latitude for	
	On the parallel	Angle	In arc	In time	1 mile	1 Tp.
°	Lks.	′ ″	′ ″	Sec		
25	33.9	2 25	5 44.34	22.96		
26	35.4	2 32	5 47.20	23.15		
27	37.0	2 39	5 50.22	23.35	0.871	5.229
28	38.6	2 46	5 53.40	23.56		
29	40.2	2 53	5 56.74	23.78		
30	41.9	3 0	6 0.26	24.02		
31	43.6	3 7	6 3.97	24.26		
32	45.4	3 15	6 7.87	24.52	0.871	5.225
33	47.2	3 23	6 11.96	24.80		
34	49.1	3 30	6 16.26	25.08		
35	50.9	3 38	6 20.78	25.39		
36	52.7	3 46	6 25.53	25.70		
37	54.7	3 55	6 30.52	26.03	0.870	5.221
38	56.8	4 4	6 35.76	26.38		
39	58.8	4 13	6 41.27	26.75		
40	60.9	4 22	6 47.06	27.14		
41	63.1	4 31	6 53.15	27.54		
42	65.4	4 41	6 59.56	27.97	0.869	5.216
43	67.7	4 51	7 6.29	28.42		
44	70.1	5 1	7 13.39	28.89		
45	72.6	5 12	7 20.86	29.39		
46	75.2	5 23	7 28.74	29.92		
47	77.8	5 34	7 37.04	30.47	0.869	5.211
48	80.6	5 46	7 45.80	31.05		
49	83.5	5 59	7 55.05	31.67		
50	86.4	6 12	8 4.83	32.32		
51	89.6	6 25	8 15.17	33.03		
52	92.8	6 39	8 26.13	33.74	0.868	5.207
53	96.2	6 54	8 37.75	34.52		
54	99.8	7 9	8 50.07	35.34		
55	103.5	7 25	9 3.18	36.22		
56	107.5	7 42	9 17.12	37.14		
57	111.6	8 0	9 31.97	38.13	0.867	5.202
58	116.0	8 19	9 47.83	39.19		
59	120.6	8 38	10 4.78	40.32		
60	125.5	8 59	10 22.94	41.52		
61	130.8	9 22	10 42.42	42.83		
62	136.3	9 46	11 3.38	44.22	0.866	5.198
63	142.2	10 11	11 25.97	45.73		
64	148.6	10 38	11 50.37	47.36		
65	155.0	11 8	12 16.82	49.12		
66	162.8	11 39	12 45.55	51.04		
67	170.7	12 13	13 16.88	53.12	0.866	5.195
68	179.3	12 51	13 51.15	55.41		
69	188.7	13 31	14 28.77	57.92		
70	199.1	14 15	15 10.26	60.68	0.866	5.193

11-8. Order of Survey. The procedure is as follows:

1. *Principal meridian.* The principal meridian is established. Measurements are north and south from the initial point.

2. *Base Line.* The base line is established. Measurements are east and west from the initial point.

3. *Standard Parallels.* The standard parallels, also called **correction lines,** are run. The measurements are east and west from the principal meridian. They begin at corners previously set on the principal meridian at 24-mile intervals north and south from the initial point. In some of the older surveys they are 36 miles apart. They are numbered north and south from the base line.

4. *Guide Meridians.* The guide meridians are run due *north* from the standard parallel (or from the base line). They begin at 24-mile intervals at corners previously established. They are terminated at the next standard parallel (or the base line). Due to the convergence of the meridians, each will strike the parallel where they terminate at a point nearer the principal meridian than the 24-mile corner. The intersection is called a **closing corner.** It is found by retracing the parallel and it is marked wherever it falls. It is not corrected to agree with its theoretical location. The distance from the closing corner to the nearest standard corner on the parallel is recorded. Any excess or deficiency in the length of the guide meridian is allowed to remain in the last 40 chains.

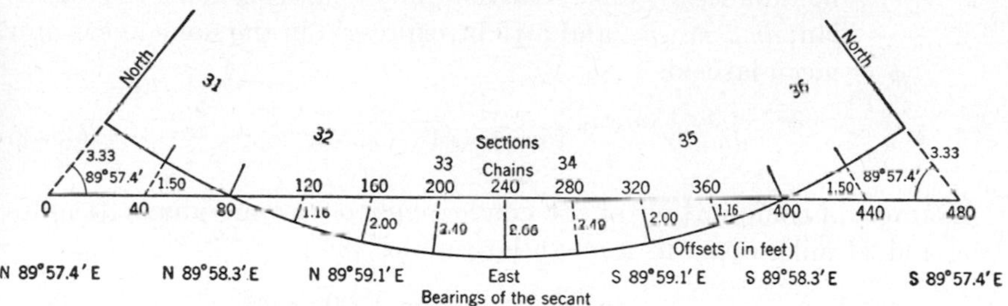

FIG. 11-4. Secant method (latitude 45°N).

In some of the older surveys, the guide meridians are spaced at greater intervals.

Guide meridians are numbered east and west from the principal meridian.

When existing conditions require, a guide meridian may be run *south.* It is begun at a closing corner set at its theoretical position, which is computed from the convergence of the meridians.

CONVERGENCE OF MERIDIANS

11-9. The linear amount of the convergency of two meridians is a function of their distance apart, of the length of the meridian between two reference parallels, of the latitude, and of the spheroidal form of the earth's surface.

The following equation is convenient for the analytical computation of the linear amount of the convergency on the parallel of two meridians any distance apart and of any length. The correction for convergency in any closed figure is proportional to the area and may be computed from an equivalent rectangular area:

m_λ = measurement along the parallel
m_ϕ = measurement along the meridian
a = equatorial radius of the earth = 3963.3 miles
e = factor of eccentricity, log e = 8.915 2515
dm_λ = linear amount of the convergency on the parallel, of two meridians, distance apart m_λ, and length m_ϕ along the meridian; dm_λ, m_λ, m_ϕ and a to be expressed in the same linear unit
ϕ = mean latitude

$$dm_\lambda = \frac{m_\lambda m_\phi}{a} \tan \phi \sqrt{1 - e^2 \sin^2 \phi} \tag{11-1}$$

Example of computation of the convergency of two meridians 24 miles long and 24 miles apart in a mean latitude of 43°20′:

nat 1		= 1.0000000	
log e	= 8.915 2515		
log e	= 8.915 2515		
log sin 43°20′	= 9.836 477		
log sin 43°20′	= 9.836 477		
log $e^2 \sin^2 \phi$	= 7.503 457		
nat $e^2 \sin^2 \phi$		= 0.0031875	
nat $(1 - e^2 \sin^2 \phi)$		= 0.9968125	
log $(1 - e^2 \sin^2 \phi)$	= 9.998 614		
log $\sqrt{1 - e^2 \sin^2 \phi}$	= 9.999 307		
log tan 43°20′	= 9.974 720		
log 24	= 1.380 211		
log 24	= 1.380 211		
log 5280	= 3.722 634		(converts miles to feet)
log product	= 6.457 083		
log 3963.3	= 3.598 057		
log dm_λ	= 2.859 026		
nat dm_λ	722.81 ft		

The convergency, measured on the parallel, of two meridians 24 miles apart and 24 miles long, in a mean latitude of 43°20′, is therefore found to be 722.81 ft. The convergency of the east and west boundaries of a regular township in the same latitude would be equal to $\frac{1}{16}$ of the convergency of the east and west boundaries of the quadrangle as computed above, or 45.18 ft.

The *angular* convergence of meridians very nearly equals the difference in longitude times the sine of the latitude. The convergency as described above, i.e., the *linear* convergence is then equal to the distance measured north times the sine of the angular convergence.

Table 11-4 gives the length of 1° of latitude and 1° of longitude at various latitudes. These values may be used for computations requiring transfer from angular to linear measure. For example, to determine the effect of convergence given in the previous example by the approximate rules:

$$\text{Degrees in 24 miles on a parallel} = \frac{24}{50.394} = 0.476247°$$

$$0.476247° \times \sin \text{lat} = 0.326821° = 19.6093' \text{ convergence}$$

$$24 \times 5280 \sin 19.6093' = 722.81 \text{ ft}$$

Table 11-5 gives useful values of meridian convergence.

TOWNSHIP EXTERIORS

11-10. Tiers and Ranges. A row of townships extending east and west is called a **tier.** Tiers are numbered north and south from the base line. A row of townships extending north and south is called a **range.** Ranges are numbered east and west from the principle meridian. The position of a township 3 tiers south and 7 ranges east is written T3S, R7E. But to simplify the expression, the word *tier* is omitted so that this is read, township 3 south, range 7 east. Ranges are separated by **range lines** but the east-west lines that separate townships have no names.

11-11. Township Boundaries. Range lines are established in exactly the same way as guide meridians.

The latitudinal lines are established in 6-mile lengths extending from one meridional line to the next. Each length is run by one of the methods described for curved east-west lines. First a random line is run from a corner in a meridional line to the next meridional line. Then the whole random line is adjusted in azimuth so that it will meet the proper corner. The shortage in length due to the convergence between the meridional lines and any accumulated errors are placed in the most westerly 40 chains.

The usual practice is to lay out the southerly tier of a quadrilateral first, then the next most southerly, and so on. Usually the random latitudinal is run from the new to the old meridian. In general, the random is run from east to west to simplify placing the shortage and the errors in the last 40 chains.

When defective old work is encountered, special correction lines are established. In general, however, double corners are avoided if possible.

SUBDIVISION OF TOWNSHIPS

11-12. Meridional Section Lines. The north-south lines that separate sections within township exteriors are *not run due north.* They are run *parallel to the east boundary* of the township. This is accomplished by running on the bearing of the east boundary (see Table 11-3). It results in a small north-west bearing. The last 80 chains are run first as a random line. It is then adjusted to meet the proper corner in the north boundary of the township. The accumulated errors in length measurement are placed in the last 40 chains. Of course, if the north boundary is a standard parallel or the base line, the meridional section line will not terminate on a corner and the last 80 chains are run as a permanent line at once, and not run as a random. The intersection with the parallel is a closing corner. It is placed where it falls and the distance to the nearest standard corner is recorded.

Latitudinal Section Lines. The latitudinal lines that separate sections within a township are straight lines approximately 80 chains long that join the proper corners on each pair of meridional lines. The quarter-section corner is set at the *midpoint* of each line except, of course, in the westerly sections. There the quarter corner is set 40 chains from the easterly meridional line. These latitudinals are first run as randoms and then adjusted.

Corners Established by the Local Surveyor. The corners for further subdivision are usually not established by the United States but by the local surveyor. The procedure required is described in the following paragraphs.

Quarter-section Corners. The quarter-section corner in the middle of each section is set at the intersection of straight lines that join pairs of opposite quarter-section corners in the section boundaries. When a quarter-section corner in a section boundary has not been set because of natural barriers or the like, the necessary straight line is run at the mean bearing of the adjacent section lines. If only one section line exists, the line is run parallel to it.

Quarter Quarter Sections (Sixteenth Sections). It is necessary to set five corners to divide a quarter section into four quarter quarter sections. First, a corner is set at the midpoint of each boundary of the quarter section. Then the central corner is set by intersection as in a quarter section.

When the quarter section is adjacent to the north or west boundary of the township, the sixteenth-section corner is not set at the midpoint of the quarter-section boundary that runs to the township line. Instead, the sixteenth-section corner is set 20 chains from the south or east quarter-section corner, whichever applies, so that any discrepancy is thrown in the last 20 chains.

When the sixteenth corners on the boundaries of the quarter sections cannot be set, the problem is handled in accordance with the principles employed for the quarter sections.

Meander Lines. Since the highwater mark of important bodies of water limits certain uses of the land by private owners, traverses called **meander lines** are run in the Public Land Surveys along these high-water marks in order to determine the approximate usable acreage at the time the survey was made. These lines are not boundaries nor do they mark the limits of the rights of the owner. These rights are limited by the actual position of high-water mark in accordance with the general principles of riparian rights.

A mark called a meander corner is set at the intersection of high-water mark with every standard line, township line, or section line. A witness corner should also be set, if necessary, on the same line far enough away to be safe from destruction. The meander is run along high-water mark from one meander corner to the next.

11-13. Official Plats. The official plat of a township shows the section lines, such topography as appears in the field notes, the boundaries of bodies of water as determined from the meanders, and the boundaries of old surveys with their survey dimensions (Fig. 11-5). The latter usually consist of Spanish grants, mining claims, and townsites. The dimensions are those obtained by the Public Land Surveys.

The lengths of section lines are given only if they differ from 80 chains. The areas of regular sections are given as 640 acres despite slight variations in dimensions.

Although the quarter-section lines are not surveyed, they are platted. Platting without ground measurement is called platting by **protraction.**

Irregular sections will always be found along the north and west boundaries of townships, along the boundaries of old surveys, and along natural barriers. In these sections, every quarter section that is irregular is broken down by protraction so that the entire irregular boundary is adjacent to parts of sixteenth sections, called **fractional lots.** Full dimensions and full acreage are assumed for all parcels not containing these fractional lots. The dimensions of the fractional lots are worked out from the field notes and noted on the plat. The resulting acreage for each lot is also computed and noted.

11-14. The Local Surveyor. It must be remembered that when any land comprising less than a section is disposed of by the United States, the boundaries must be marked for the entryman or the new owner by a local surveyor. The description of the parcel is that given by the official plat. The local surveyor, therefore, must carefully follow this plat and the regulations laid down by the Bureau of Land Management. In areas where the lots are irregular, special care must be exercised.

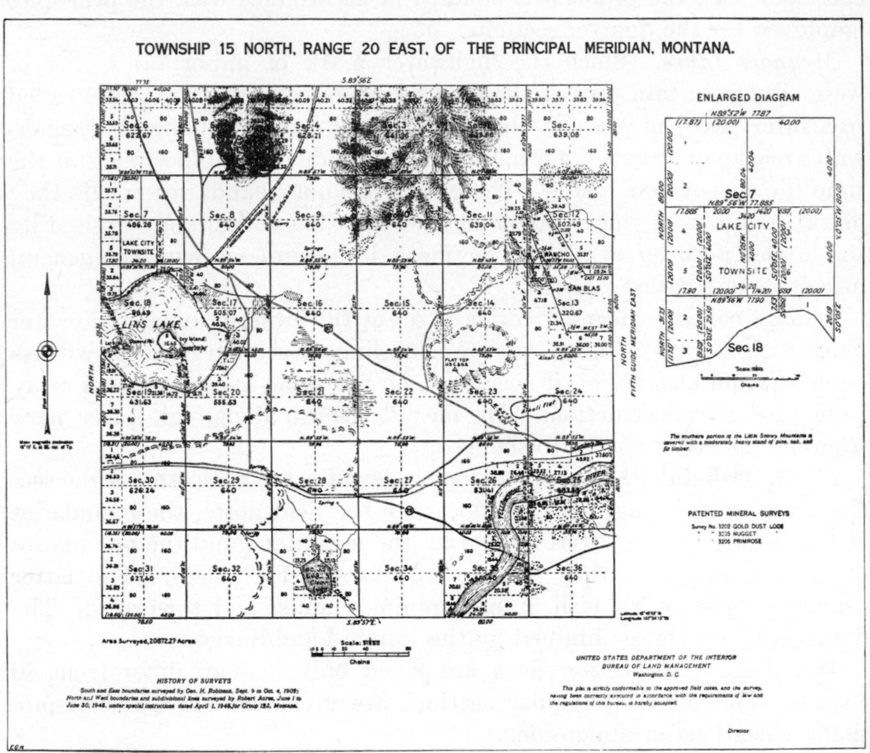

FIG. 11-5. Sample official plot of a township.

11-15. Proportionate Measurement. The first step in marking a parcel to be acquired from the United States always consists of establishing corners on lines surveyed by the Federal government. Each line must be retraced and remeasured between the two existing adjacent corners. Any excess or deficiency between the measured distance and the distance given on the plat is prorated in establishing the new corner. This is called **proportionate measurement.** The actual measured distances are reported.

When only one existing corner is available, judgment may be exercised but the decision must depend on the spirit of the principles of the Public

Land System. Sometimes the stated area of the parcel is all the evidence available to guide this decision.

11-16. State Jurisdiction Controls. After the corners have been established to the satisfaction of the United States and the prospective owner, the corner monuments, so established, mark the **final position** of the boundaries. Any later disagreements are resolved in accordance with the legal principles of land transfer under the jurisdiction of the state courts.

11-17. Lost or Obliterated Corners. "A lost corner is a point of a survey whose position cannot be determined, beyond reasonable doubt, either from acceptable evidence or testimony that bears upon the original position, and whose location can be restored only by reference to one or more independent corners."[1]

Since the positions of boundaries within the areas of the Public Land Surveys usually depend on the monuments established under this system long after the land has been disposed of by the United States, the local land surveyor is often called upon to reestablish a lost corner.

The most difficult and important problem to solve in this connection is whether or not the corner is actually lost. Another definition of a lost corner which is not so accurate but one that clarifies the problem is the following:

A lost corner is one whose position is evidenced only by the original field notes.

Since the original field measurements are not accurate enough nor were they meant to be accurate enough to give precise evidence of position, they therefore should be used only as a last resort.

These principles are in accordance with the principles of land surveying everywhere. The courts have long recognized that marks on the ground once identified are more trustworthy evidence than recorded measurements.

Accordingly, lost corners should be restored in accordance with the general principles of land surveying. The evidence of all boundary marks, previous surveys that substantiate each other, local knowledge, and a keen analysis of possible mistakes and errors that, when corrected, give consistency to the results, must be brought to bear.

If local evidence of this kind fails, the surveyor should retrace and remeasure those lines of the Public Land Surveys that affect the location, beginning with the two nearest corners that can be found on each line. These measurements may result in as many as four positions for the lost corner. The same analysis must be applied of possible errors and mistakes in the original survey. Finally, if it is impossible to obtain

[1] From the "Manual" described at the head of this chapter.

consistent results, the corner must be established by proportionate measurement.

11-18. Double Proportionate Measurement. When both the distance and alignment of the original surveys are in doubt, the corner should be restored by **double proportionate measurement** between the nearest existing adjacent corners on the two lines that intersect at the lost corner. This method eliminates bearings and other alignment data from consideration (see Fig. 11-6).

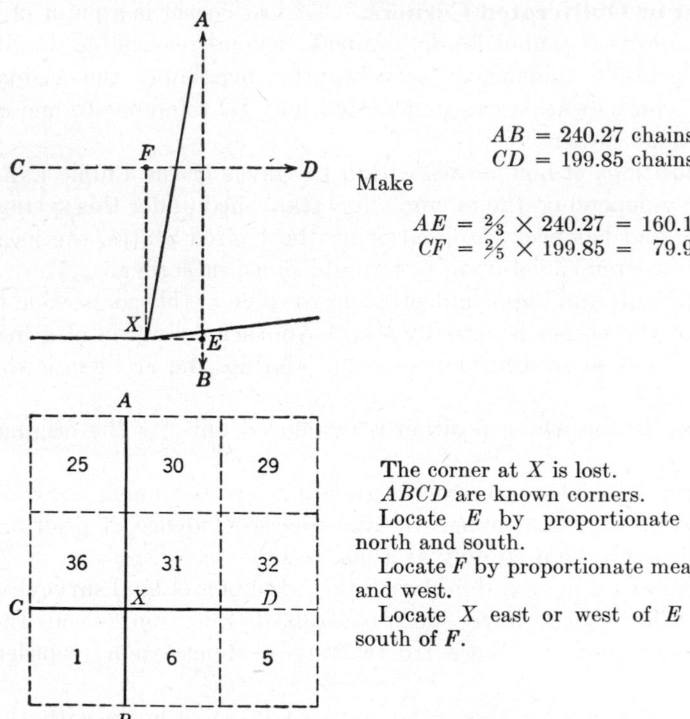

AB = 240.27 chains
CD = 199.85 chains

Make

$AE = \frac{2}{3} \times 240.27 = 160.18$ chains
$CF = \frac{2}{5} \times 199.85 = 79.94$ chains

The corner at X is lost.

$ABCD$ are known corners.

Locate E by proportionate measurement north and south.

Locate F by proportionate measurement east and west.

Locate X east or west of E and north or south of F.

FIG. 11-6. Double proportionate measurement.

11-19. Methods of Marking a Corner.[1] The position of a corner is marked by the objects described in the following list. The object used for any particular course is recorded in the original field notes.

1. A monument at the corner
2. A witness monument called a witness corner
3. Corner accessories
   a. Reference monuments
   b. Bearing objects
   c. Pits and mounds

[1] In this discussion, the word "monument" is used not in the legal sense but in the sense employed in land surveying, that is, a physical artificial marker of durable material placed for the specific purpose of marking a point.

The monuments used can be any recognizable object. They usually are one of the following:

1. A wrought-iron pipe, zinc coated, 2 in. inside diameter, 30 in. long. The lower end is split for 4 or 5 in. and spread. A brass cap is fastened to the top. The pipe is filled with concrete. It is set with ¾ of its length in the ground.
2. A durable native stone at least 20 by 6 by 6 in., set with ¾ of its length in the ground.
3. A cross mark on surface rock.
4. A tablet set in surface rock.
5. A living tree when it occupies the position of the corner.
6. A steel rod.
7. A wooden post.
8. A deposit of charcoal or glass or any durable artificial material (called a **memorial**).

Often the mark is protected by a nearly flat stone mound 3 to 6 ft in diameter.

Witness Corner. A witness corner is used when it is impractical to occupy the site of the corner. If possible, it is placed as near the corner as is practical, on one of the lines running to the corner. It must not be more than 10 chains distant. If this is impossible, it is placed anywhere within 5 chains.

Reference Monuments. A reference monument is placed within a short distance of the corner. At least two and sometimes four are set. They are used when the corner mark is liable to destruction and no trees are available that can be used.

Bearing Objects. Any tree or other natural object near the corner (ordinarily within 3 chains) can be used as a bearing object. The bearing and distance *from* the corner to the object is recorded. The distance to a tree is measured to the center of the tree just above the root bole. Bearings only are often recorded to distant landmarks.

Pits and Mounds. Pits are rectangular and placed on lines that run to the corner. The excavated material is placed in a mound at the corner or on one of the lines. New and old specifications differ.

Numbers and Letters Marked on Monuments and Bearing Objects. Various systems are employed for marks of identification. The 1947 "Manual"[1] gives very specific rules. So many kinds of corners must be marked that the rules cannot be listed here. The meaning of the marks should be decipherable from the following list of abbreviations.

AM	Amended (new corner position when old remains)
AMC	Auxiliary meander corner
AP	Angle point
BO	Bearing object
BT	Bearing tree
C	Center
CC	Closing corner
E	East
LM	Location monument (for U.S. Survey not connected with Public Land System)
M	Mile
MC	Meander corner
N	North
NE	Northeast
NW	Northwest

[1] See head of this chapter.

PL	Public land (unsurveyed)
R	Range
RM	Reference monument
S	Section
S	South
SC	Standard corner
SE	Southeast
SMC	Special meander corner
SW	Southwest
T	Township
TR	Tract
W	West
WC	Witness corner
WP	Witness point
¼	Quarter section
1/16	Sixteenth section

Grooves or notches are used on stone monuments marking closing corners. Grooves are cut in the face, notches in the corner. They are placed toward each of the three township corners of the township to which the corner belongs. The number of grooves or notches indicate the number of miles to that corner.

11-20. Marking Lines. The final lines (**true lines**) are blazed through timber. Ordinary blazes and **hack marks** are used. They are placed about breast high. A blaze should be 5 to 6 in. high and 2 to 4 in. wide, cut into the wood tissue. A hack is a horizontal V-shaped notch cut through the bark and well into the wood. Two hacks are used, one above the other.

Trees on line are marked by a pair of hacks on each side facing the line. A sufficient number of trees within 50 links of the line are blazed so that the line may be easily retraced. The blazes are placed on two sides quartering toward the line. On nearby trees, the blazes are nearly opposite. The further the tree is out from the line, the nearer the blazes are brought together.

Witness Points. A **witness point** is a monumented station on the line of the survey and is employed to perpetuate an important location more or less remote from, and without special relation to, any regular corner.

Description of Parcels. The parcel indicated by the letter *A* in Figs. 11-1 and 11-2 is described as follows: NE ¼ of SE ¼, Sec. 8, T3S, R7E, Salt Lake Mer.

The fractional lot just south of Lins Lake in Fig. 11-5 is described as follows: Lot 5, Sec. 19, T15N, R20E, Montana Mer.

Descriptions of small parcels for private transfer usually recite the Public Lands description in the general statement of the locality that is written previous to the metes and bounds description, thus:

Situated in the State of Montana, Count of Blank, NE ¼ of SE ¼ Section 22, T15N, R20E, Montana, Mer. and more particularly described as follows:

Beginning at a . . . etc.

11-21. Accuracy. It is evident that the methods used to lay out the Public Land System are not designed to give accurate position to the various corners. There is no real system of control nor are the measurements made with very great accuracy. In the older surveys, the measurements are still less accurate. However, the system has been wisely designed to conform with the best principles of land surveying practice that, through the years, have proved to be the safest and the most practical. The positions of boundaries depend on actual landmarks. The recorded measurements are accurate enough to find the monuments and to identify them and to give the areas with an error of less than 1 per cent. The measurements, however, cannot be used *by themselves* to hold the position of a corner.

This, of course, has certain disadvantages particularly when a corner is lost without a trace. When such an event occurs, the methods described are applied to place a new corner as nearly in the old position as possible, but these methods can never give complete assurance that it is in its original position.

Wherever possible, recognized corners should be tied to the U.S. Coast and Geodetic Survey triangulation by the state coordinate system as explained in Chap. 12. When such a tie has been made, the corner can always be replaced very accurately in its original position.

11-22. Identification. The Public Land System provides a perfect method for identifying parcels. No two parcels can have the same description nor can there be any doubt as to the location of the parcel described.

This quality makes it possible to file records of deeds and other instruments according to the parcels involved rather than by the persons involved. The entire record of title of any parcel is found in one place in the file of each type of instrument. It is really an all-embracing system of filing by lot and block. Title can be traced rapidly and surely and any information concerning a parcel can be found with a minimum loss of time.

PROBLEMS

11-1. *a.* Using sketches, name and show the three methods of laying out latitudinal lines.

b. Place on the sketches estimates of the bearings and lengths of the various lines.

c. Name the lines for which these methods are used.

11-2. Describe the controlling factors for the bearings and final measurements for the following:

a. Principal meridian
b. Base line
c. Standard parallels
d. Guide meridians
e. Range lines
f. Latitudinal township boundaries
g. Meridional section lines
h. Latitudinal section lines
i. Quarter-section lines
j. Sixteenth-section lines

11-3. Using two sketches show the location of NE 1/4 of SW 1/4 T5N R3E fifth principal meridian. Name all the lines shown. Indicate tiers and ranges.

11-4. Define:

a. Standard corner
b. Meander corner
c. Witness corner
d. Closing corner
e. Reference monument
f. Bearing object
g. Amended corner
h. Location monument

11-5. Describe the method of finding a corner and restoring a lost corner.

11-6. What is the survey practice when a private holding like a Spanish grant or a mining claim is found within the public lands?

11-7. Describe the functions performed by a local surveyor working in the areas of the public land surveys.

11-8. How would use of state plane coordinates affect the Public Land System?

11-9. What are the advantages of the Public Land System for the lawyer?

11-10. Under what conditions can a corner be moved?

CHAPTER 12

FUNDAMENTAL SURVEY CONTROL AND THE STATE COORDINATE SYSTEMS

FUNDAMENTAL CONTROL

12-1. Introduction. The two fundamental survey control systems of the United States are the triangulation net and the level net established by the U.S. Coast and Geodetic Survey. These nets cover the entire United States. The triangulation net is connected with the triangulation nets of Canada and Mexico.

The level net is used extensively throughout the United States for surveys of every character but the triangulation net has been used in the past only for large and important surveys as its use required geodetic reduction. Geodetic reduction is unnecessary for most surveys and since it increases the cost of the computations, it is seldom used or understood by most engineers. To make this net available for surveys of every type, the U.S. Coast and Geodetic Survey in 1933 introduced the **State Coordinate Systems.** Under these systems each state is divided into from one to seven zones, generally bounded by county lines. A system of plane rectangular coordinates has been designed for each zone which makes it possible to express positions interchangeably in terms of geodetic latitude and longitude or in terms of the plane coordinates of the particular system established for that zone. The one or more zones covering each state are called collectively the State Coordinate System. The systems thus provide, within each zone, a standard horizontal datum expressed in plane coordinates that extends over an exceptionally large area.

The U.S. Coast and Geodetic Survey publishes special tables for each zone that simplify the necessary transformation from geodetic position to plane coordinates or the reverse.

When the geodetic position (given in latitude and longitude) of any triangulation station is transformed to plane coordinates on a state system, the coordinates thus obtained can be used exactly like those of any plane coordinate system except for surveys of exceptionally high accuracy. This makes the entire fundamental net readily available for the control of any local survey. Moreover, once a survey is tied to the net, it is automatically tied to every other survey that is also tied to the net.

In order to use these fundamental control nets (the level net and the triangulation net) to the greatest advantage, and particularly to make the best use of the state systems of coordinates in surveys of higher accuracy, certain geodetic concepts must be understood. The first part of this chapter is devoted to this subject.

THE SHAPE OF THE EARTH

12-2. The Forces Involved. Each particle of matter in the earth is attracted toward every other particle by the force of **gravitation.** On

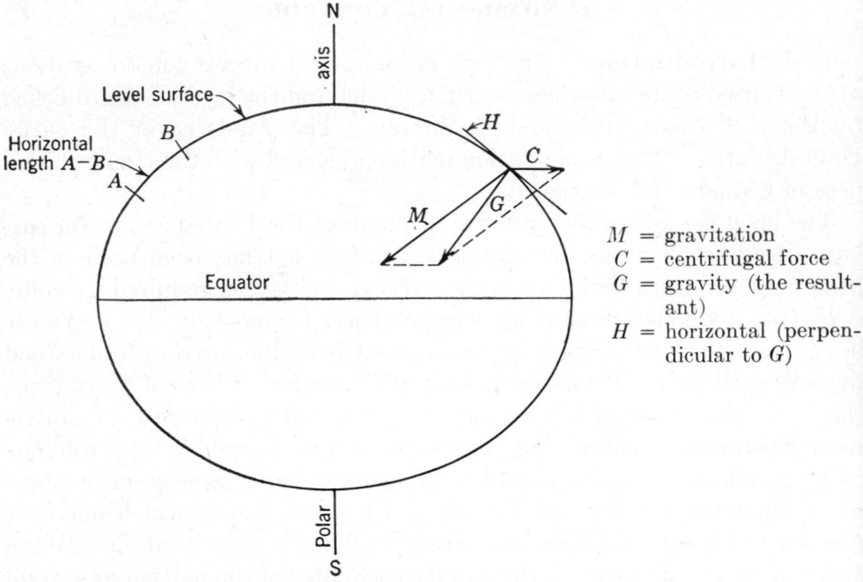

FIG. 12-1. Direction of gravity.

the earth's surface the direction of the resultant of all these forces is more or less toward the center of the earth's mass. Variations in the density of the earth affect the direction of this resultant. These effects are especially noticeable near mountains or islands that rise from ocean depths where the resultant is deflected toward these great masses. Each particle of matter not on the earth's axis is also pulled outward perpendicular to the earth's axis by the centrifugal force of the earth's rotation. The resultant of gravitation and centrifugal force combined is known as the force of **gravity** (see Fig. 12-1). Due to the component of centrifugal force, the direction of gravity is not toward the center of the earth's mass but is deflected slightly toward the plane of the equator except, of course, at the poles and at the equator itself.

Since the centrifugal force is greatest at the equator, the earth, being fluid, takes the general shape of an ellipsoid of revolution. The stiffness and varying densities of the crust prevent the formation of a perfect mathematical figure.

12-3. A Level Surface. A level surface is a continuous surface everywhere perpendicular to gravity. Such a surface is basically the shape of an ellipsoid of revolution, but since the direction of gravity is affected by

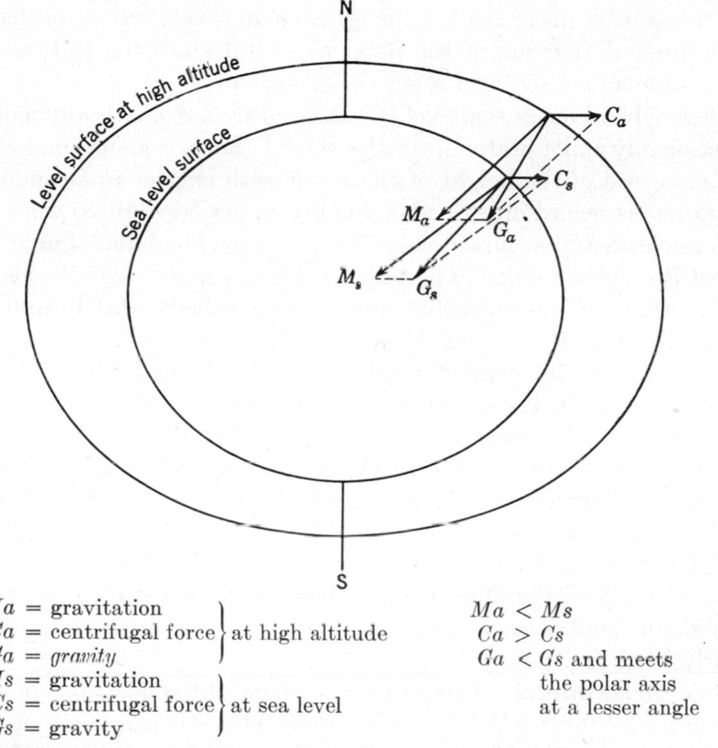

FIG. 12-2. Orthometric corrections. Force diagrams showing the effect of increased centrifugal force and reduced gravitation at high altitude. The direction of gravity is changed; hence the level surface, which is perpendicular to it, is not parallel with sea level but approaches it toward the poles.

local concentrations of mass, a level surface, although continuous, is somewhat irregular.

Level surfaces are not parallel. At a high altitude gravitation is less and since the velocity due to rotation is greater, the centrifugal force is greater. The resultant therefore points more toward the plane of the equator than the equivalent resultant at a lower altitude (see Fig. 12-2). Level surfaces at higher altitudes are therefore more elliptical than at lower altitudes and they accordingly tend to become closer together at the poles.

12-4. Mean Sea Level. The surface of still water must be everywhere perpendicular to the direction of gravity and is therefore, by definition, a level surface. The mean surface of the sea is very nearly a level surface, so nearly in fact that for surveying purposes it is assumed to be level. The slight departures from level are caused chiefly by ocean currents, variations in the density of the water, and variations in local mean barometric pressures.

The height of mean sea level is apparently slowly rising, probably due to the present thawing of the polar ice. But since the process is slow, mean sea level is used as a reference elevation.

The height of mean sea level is measured by automatic tide gauges set in reasonably quiet water near the shore. Such a gauge makes a continuous record of the height of the water with respect to an index mark. A continuous record of water heights over a period of three years usually gives a reasonably accurate value for mean sea level and thus the elevation of the index mark. The index mark is connected by levels, preferably to three or more nearby bench marks called **tidal bench marks** (see also Chap. 15).

The fundamental control level net in the United States covers the entire country. It is connected with many tidal bench marks at frequent intervals along the coast line. Although slight differences in mean sea level are evident from the results of this leveling, the entire net is adjusted to hold zero elevation at mean sea level at every tide gauge. A readjustment of the entire level net was made in 1929 and the resulting elevations are said to be based on the "Sea Level Datum of 1929."

Mean sea level therefore can be thought of as a surface that extends throughout land areas. Like all level surfaces it is continuous and slightly irregular.

12-5. Orthometric Correction. Since the operation of leveling automatically follows level surfaces, a line of levels run north or south at a high altitude will at first result in elevations referenced to surfaces that are not at a constant height above sea level as previously mentioned (see Fig. 12-2). Elevations so obtained must therefore be corrected by what is called the **orthometric correction.**

12-6. The Geoid. The figure of mean sea level as described is called the geoid (see Fig. 12-3). It is the figure to which levels and astronomical observations refer. This statement is explained in the next paragraphs.

An astronomical observation that is used for surveying purposes basically determines two things as follows:

1. The coordinates of the point on the celestial sphere where the local direction of gravity strikes it. It must be remembered that when an instrument is leveled, the vertical axis is made to coincide with the direction of gravity. The zenith, therefore, is the local direction of gravity

projected to the celestial sphere. The measurements made and the trigonometry applied give the celestial coordinates of the zenith. These are translated into the earth coordinates of latitude and longitude.

2. The azimuth of a mark measured in a plane perpendicular to gravity.

Thus, latitude, longitude, and azimuth, determined astronomically, depend on the local direction of gravity and hence are referred to the geoid. Figure 12-4 illustrates this graphically.

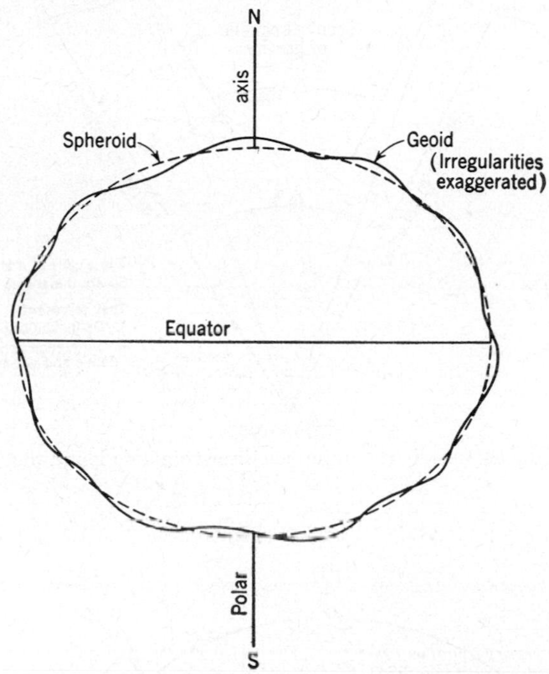

FIG. 12-3. Geoid and spheroid. The spheroid is a perfect mathematical figure which very closely approximates the geoid. The geoid, affected by local disturbances of gravity, is irregular.

Astronomical values may be defined as follows:

The **astronomic latitude** of a station is the angle between the **direction of gravity** at the station and the plane of the earth's equator (see Fig. 12-5).

The **astronomic longitude** of a station is the dihedral angle whose edge coincides with the earth's polar axis, whose initial side contains the meridian of origin of longitude (Greenwich), and whose final side is parallel to the direction of gravity at the station (see Fig. 12-4).

The **astronomic bearing** of a station B from any station A (see Fig. 12-6) is the dihedral angle whose edge coincides with the direction of gravity at station A, whose initial side contains the pole of the celestial sphere, and whose final side contains station B.

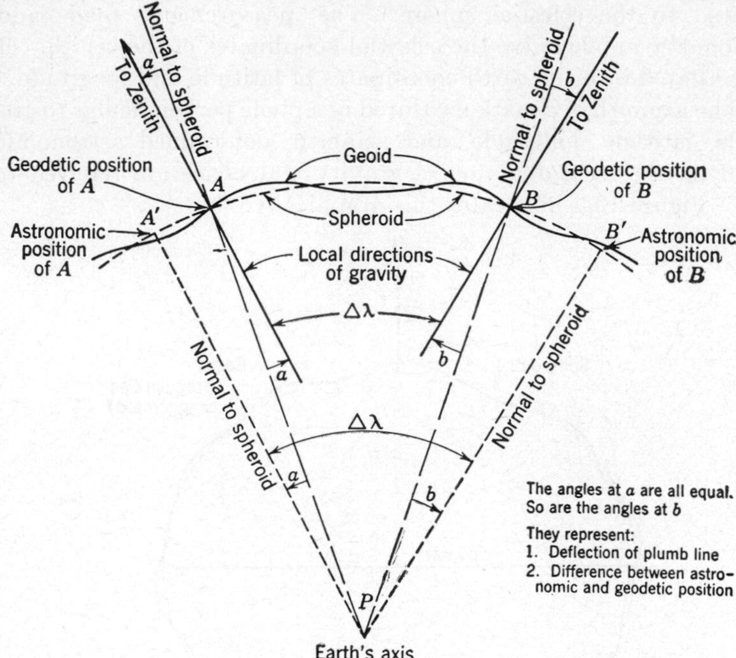

FIG. 12-4. Δλ is the difference in astronomic longitude.

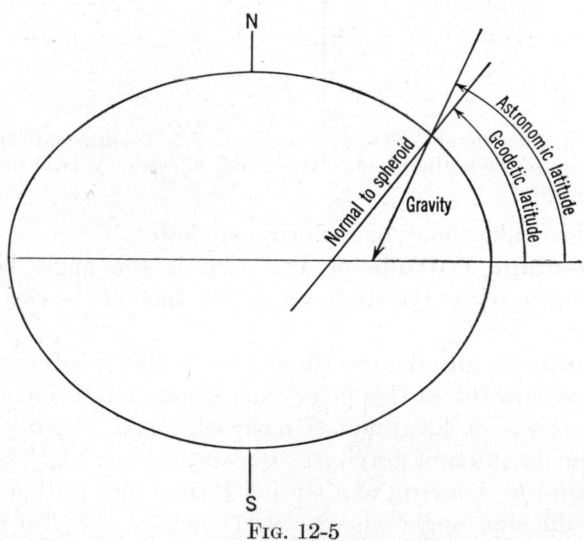

FIG. 12-5

12-7. The Spheroid. Horizontal survey measurements for fundamental control are always first reduced to their equivalents on the geoid. Horizontal angles are automatically measured in a horizontal plane; sloping length measurements are immediately reduced to their horizontal equivalents; and length measurements made at high altitudes are reduced

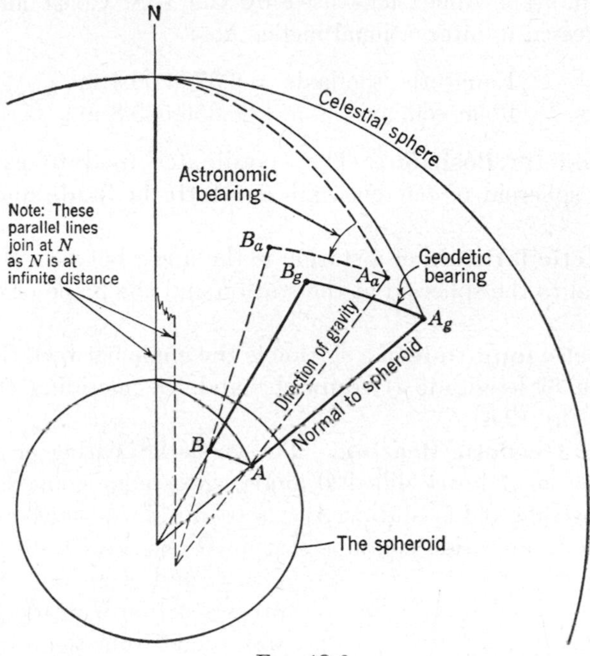

FIG. 12-6

to their sea level equivalents (see Fig. 12-7; length AB is reduced to length $A'B'$).

Since it is impossible to compute a survey closure or to establish a system of coordinates on an irregular surface like the geoid, a mathematical surface must be chosen that so nearly represents the shape of the geoid that it can be assumed, with negligible error, that the measurements were made on this surface rather than on the geoid.

A plane can be used for a small survey, but for a larger survey some surface that more nearly fits the geoid is necessary.

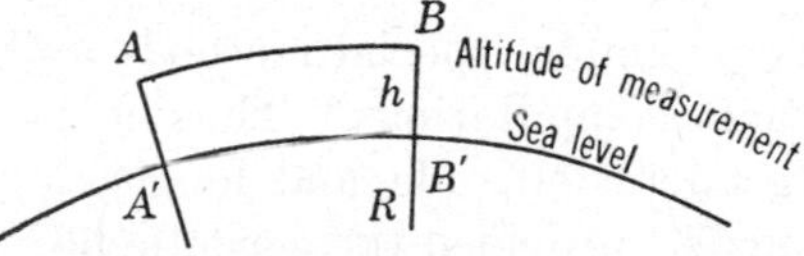

FIG. 12-7. Sea-level reduction.

Computations based on many arcs of triangulation have shown that an oblate spheroid can be made to fit the geoid very closely. This is the figure that would be generated by an ellipse rotated on its minor axis when the minor axis coincided with the axis of the earth. Such a figure is completely defined by the dimensions of its two axes. Many different spheroids have been suggested and used. The U.S. Coast and Geodetic Survey has selected Clarke's spheroid of 1866 for the standard surface of reference. This spheroid was devised by Captain A. R. Clarke, a British ordnance officer, after a very complete investigation of existing European triangulation. The dimensions used by the U.S. Coast and Geodetic Survey expressed in international meters are:

Equatorial semiaxis = 6,378,206.4 m
Polar semiaxis = 6,356,583.8 m

12-8. Geodetic Position. The coordinates used to express position on the spheroid of reference are **geodetic latitude** and **geodetic longitude.**

The **geodetic latitude** of a station is the angle between the direction of the normal to the spheroid at the station and the plane of the equator (see Fig. 12-5).

The **geodetic longitude** of a station is the angle between the meridian of the origin of longitude (Greenwich) and the meridian through the station (see Fig. 12-4).

12-9. The Geodetic Bearing. The geodetic bearing of a station B from any station A is the dihedral angle whose edge coincides with the normal to the spheroid at station A, whose initial side contains the polar axis, and whose final side contains station B (see Fig. 12-6). The U.S. Coast and Geodetic Survey expresses azimuth in terms of clockwise angle from the south.

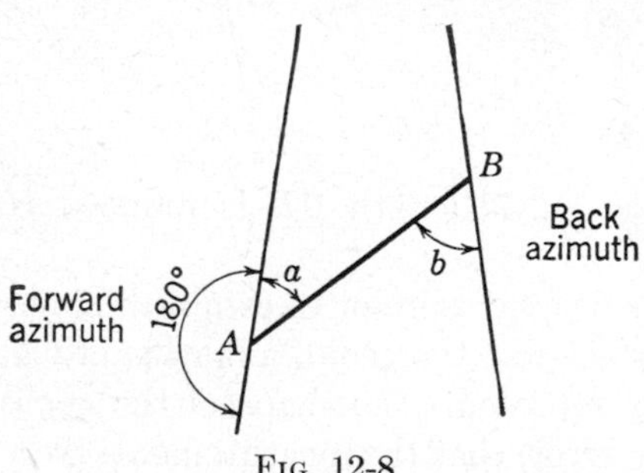

Fig. 12-8

12-10. Computation of Geodetic Position. Since the spheroid is a mathematical surface, when the direction and distance are measured from any station A of known position to any station B, the position of B can be computed. Since the spheroidal surface is not very simple, special methods and tables have been devised to simplify such computations.[1] Thus it is possible to compute the coordinates (geodetic latitude and longitude) of all the stations of any horizontal survey, provided the geodetic position and a geodetic direction is known at one station.

It should be pointed out that, due to the convergence of meridians, the back azimuth of a line on the spheroid will not be equal to the forward

[1] See U.S. Coast and Geodetic Survey Special Publication No. 8, "Formulas and Tables for Compution of Geodetic Positions," 7th ed., 1933.

azimuth of the line ±180° (except under unusual circumstances). In Fig. 12-8, angle *a* is not equal to angle *b*.

It should also be pointed out that, on a spheroid, the distance on the surface equivalent to a given unit of latitude is minimum at the equator and increases toward the pole (see Fig. 12-9, where $AB < BC < CD$ although each covers 30° of latitude). On a sphere, of course, each degree of latitude would represent the same distance.

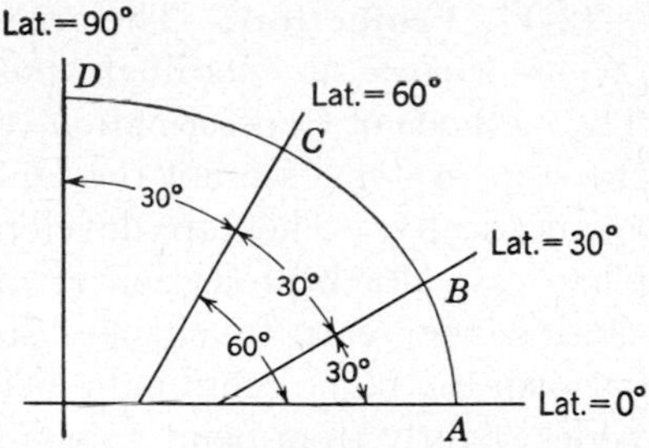

FIG. 12-9

12-11. To Establish a Geodetic Datum. Astronomic positions and azimuths are determined at many stations on any large triangulation net. One astronomic position and direction is assumed as a basis for geodetic computation. When sufficient geodetic positions have been computed, it is possible to shift the positions of the whole net so that the resulting geodetic positions agree closely with the majority of the astronomic positions, particularly those at stations where there is no evidence that the direction of gravity is deflected from the normal to the spheroid. In the United States such considerations led to the selection of a certain geographic position and azimuth for a triangulation station known as **Meades Ranch** in Kansas. Geodetic positions based on this position and azimuth computed from the most recent adjustment of the triangulation net are said to be in the North American Datum of 1927.

12-12. Deflection of the Plumb Line. The angular difference between the astronomical positions and the geodetic positions of a point is called the **deflection of the plumb line** or the **deflection of the vertical,** as it is very nearly equal to the difference in direction between the normal to the geoid and the normal to the spheroid. The deflection of the plumb line in certain localities may be as much as 1 minute of arc. It averages about 3 seconds.

It is evident that while the distance between two points is not precisely given by the differences of their astronomic positions, it is, however, given precisely by the differences in their geodetic positions (see Fig. 12-4).

STATE COORDINATE SYSTEMS

12-13. Requirements. As stated previously, the state coordinate systems provide a means of expressing geodetic positions in terms of plane rectangular coordinates. Since measurements on a spheroid cannot be expressed exactly in terms of plane coordinates, a process must be developed by which the plane coordinates used to express geodetic positions give very nearly the actual distances and angles between points. This process can be successfully devised if many plane coordinate systems are used and none covers too great an area.

12-14. Projection. The process of transforming coordinate positions on one surface to coordinate positions on another is called projection. The methods of transformation (that is, the kinds of projection) used for the state systems are not the familiar type that can easily be illustrated geometrically. They are developed mathematically as demonstrated in Chap. 23. Each projection or **grid** consists of a systematic method of causing the scale to change throughout. In other words, the ratio between the plane coordinate distances and their actual geodetic lengths varies slightly from point to point on the grid in accordance with a plan chosen to give the desired results as nearly as possible.

Three features are especially desirable:

1. The grid should be as large as possible.
2. The grid scale[1] should vary from unity as little as possible.
3. The angular relationships (shape) should be retained, that is to say, horizontal angles derived from the coordinates should agree with the corresponding measured angles.

12-15. Conformal Projections. Since the grid scale must change over the projection, angular relationships cannot be precisely retained. However, if at each point the scale is the same in all directions, the angular relationships will be retained as nearly as possible. For example, in Fig. 12-10 *ABC* represents any triangle on the grid that is small enough so that the scale change on it is negligible. If the three sides have the same scale, the angles will be the same as in the original triangle. Thus, if at each point in the projection, the scale is the same in every direction, the projection will retain angular relationships and it is called a **conformal projection.**

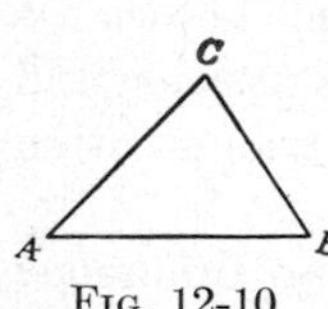

FIG. 12-10

12-16. Zones. In order to limit the size of the grids, and thus to keep the scale variations within the prescribed limits, it was thought best to assign a certain system of grids to the area covered by each state. Each state system consists of one or more **zones.** A zone is an area covered by a single coordinate system (grid). Most zones are about 158 miles wide and they extend in length to the state boundaries. They are oriented either north and south or east and west so that their greatest dimension corresponds to the greatest dimension of the state.

12-17. The Projections Used. In zones whose length is east and west, the projection used is known as the **Lambert conformal conic two-parallel projection,** hereafter called the Lambert projection or

[1] The grid scale is the distance between two points computed by their grid coordinates divided by the actual (geodetic) distance between the two points.

grid. In zones whose length is north and south, the projection used is known as the **Transverse Mercator projection** or **grid.** It is also conformal and in fact is a special form of the Lambert.

THE LAMBERT PROJECTION

12-18. The Lambert Projection. In the Lambert projections, the parallels of latitude are represented on the **plane** by concentric circles (see Fig. 12-11). Since the projection is conformal, the meridians must be represented by lines everywhere perpendicular to these circles and hence by straight lines that meet at the common center of the circles (the **central point**).

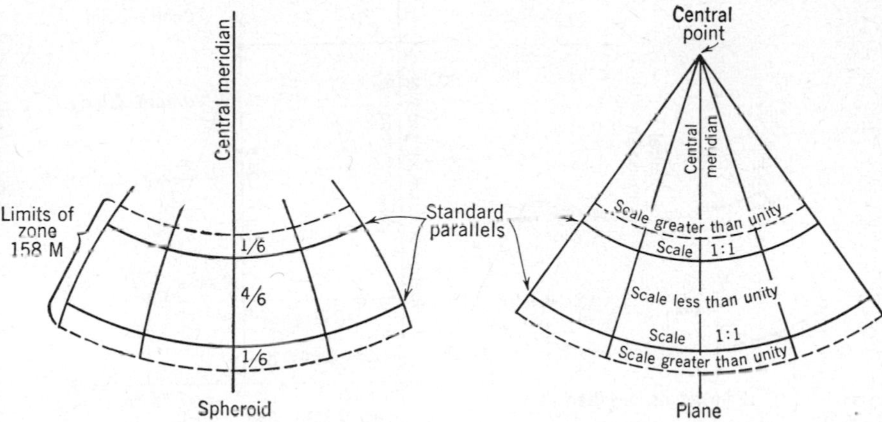

FIG. 12-11. The Lambert projection.

A Lambert projection for a given zone is established according to the following outline. First, a meridian near the center of the zone is chosen as the central meridian and hence the Y axis.

Next two parallels of latitude called **standard parallels** are chosen about ⅙ the width of the zone from its north and south limits respectively. The lengths of these parallels between meridians on the spheroid can be computed from the known shape of the spheroid. Both parallels are to be represented on the grid by their true lengths, that is to say, the scale along the two circular arcs representing them will be unity.

From a consideration of Fig. 12-11, it is evident that the arcs representing parallels that lie between the standard parallels must be slightly shorter than the actual lengths of the parallels they represent because of the bulge in the spheroid.

In other words, the scale for these arcs on the grid plane must be less than unity. The scale will vary from nearly unity near the standard parallels to a minimum near the line about halfway between.

In like manner, the scale of the arcs representing parallels outside the standard parallels will be greater than unity, increasing with their distance from the standard parallels. By selecting the positions of the standard parallels as described, the maximum scale at the north and south borders of the zone will be approximately as much greater than unity as the minimum scale near the center is less than unity. By limiting the width of the zone (north and south) to 158 miles, the grid scale is held so that it will differ from unity by no more than the limits of ± 1 part in 10,000.

In addition, in order to make the projection conformal, the scale of the distances along the lines representing meridians must be everywhere equal to the scale of the intersecting parallels. These two requirements

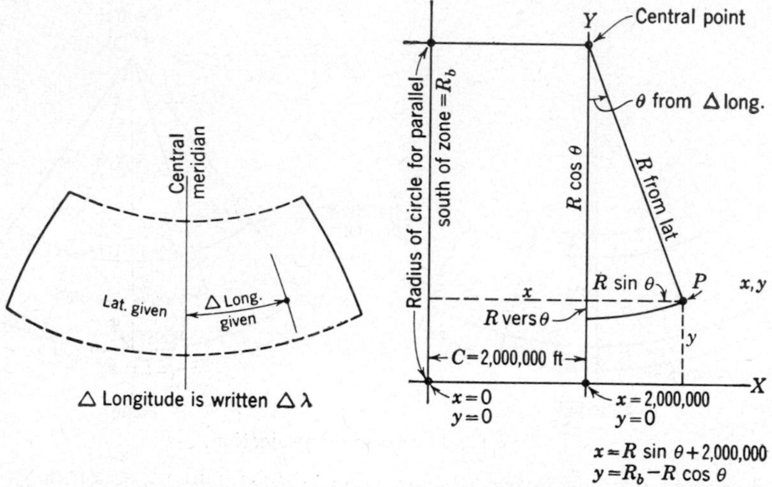

FIG. 12-12

applied simultaneously establish the radii of the standard parallels. They also establish the **radii of all other parallels** and the **angles between the meridians.**

Thus, once the projection has been established, if we know the latitude of a point on the spheroid, we know the radius of the circle that represents its latitude on the plane; and if we know the longitude, we know the direction of the straight line from the central point that locates it on the plane. That is, in Fig. 12-12 we know R and θ. The latter is sometimes called the **mapping angle.** The values for R and θ for each latitude and longitude are given in the special tables for the zone published by the U.S. Coast and Geodetic Survey.

In other words, when the latitude and longitude of a point are known, we can immediately determine the polar coordinates of the point on the plane with respect to the central meridian and the central point. The polar coordinates can be reduced to rectangular coordinates as shown in Fig. 12-12.

For any point (as P):

$$x = R \sin \theta + C \qquad (12\text{-}1)$$

$$y = R_b - R \cos \theta \qquad (12\text{-}2)$$

where C and R_b are constants for the zone. C is equal to 2,000,000 ft. Often the value $R \sin \theta$ is called x'. Figure 12-13 is a graphic representation of the result showing the coordinate lines.

It is possible to draw a map of the zone (Fig. 12-14) showing parallel circles through points of equal scale. When a precise survey is made in the zone, the scale that should be applied to length measurements of the survey to make the results agree with the coordinate distances can be

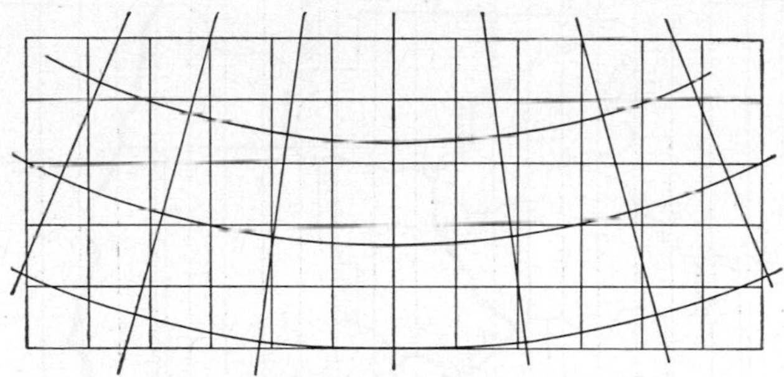

FIG. 12-13. Meridians and parallels on the Lambert grid.

found by locating the position of the survey on the map. It is also possible to find the exact scale from the special tables.

When Lambert coordinates are to be transformed into geodetic latitude and longitude, the following formulas are used. They can readily be derived from Fig. 12-12.

$$\tan \theta = \frac{x - C}{R_b - y} \qquad (12\text{-}3)$$

$$R = \frac{R_b - y}{\cos \theta} \qquad (12\text{-}4)$$

From tables, using R as an argument,

$$\Delta\lambda = \frac{\theta}{l} \qquad (12\text{-}5)$$

$$\lambda = \text{central meridian} - \Delta\lambda \qquad (12\text{-}6)$$

where l is the ratio $\theta/\Delta\lambda$ for any point. It is equal to the number of seconds in θ for 1 second in $\Delta\lambda$.[1]

[1] $\Delta\lambda = \Delta$ Long. = difference in longitude.

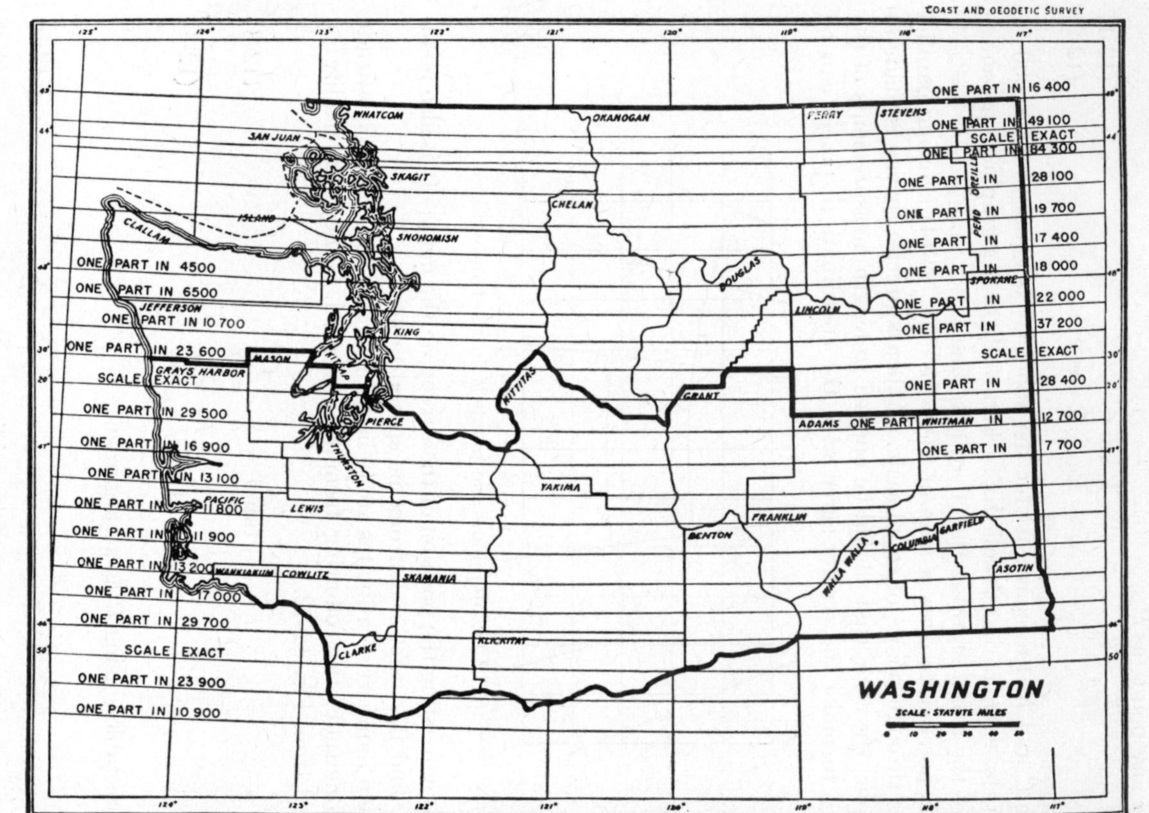

STATE PLANE-COORDINATE ZONES AND SCALE FACTORS

FIG. 12-14. Two zones on the Lambert grid. Lines of equal scale factors are shown. A scale factor is the difference between the grid scale and unity. (*U.S. Coast and Geodetic Survey.*)

Example. Figure 12-15 shows the computation of the Lambert coordinates of a station from its geodetic position. Extracts from the special tables for the zone are given. Once the data are taken from the tables, the formulas for x and y can easily be solved.

Mason Lat. = ϕ = 40°35′06.829″, Long. = λ = 98°28′10.329″

Lambert Projection for Nebraska (South)

R_b = 24,590,781.86 C = 2,000,000

TABLE I (extract)

Lat.	R, ft	Tabular difference for 1″ of lat., ft	Scale expressed as a ratio
40°35′	24,256,835.32	101.19400	0.9999485
40 36	24,250,763.68	101.19417	0.9999465

Lat.	*R*
40°35′	24,256,835.32
6.829 × 101.19400 =	691.05
40°35′6.829″	24,256,144.27

TABLE II (extract)

1″ of long. = 0.65607640″ of θ

Long.	θ
98°28′	+0°40′40.6042″
98 29	+0 40 01.2396

Long.	θ
98°28′	+0°40′40.6042″
10.329 × 0.65607640 =	6.7766
98°28′10.329″	+0°40′33.8276″

$\sin\theta = 0.0117992553$ $\cos\theta = 0.9999303862$

$x = R\sin\theta + C = 2{,}286{,}204.44$

$y = R_b - R\cos\theta = 336{,}326.15$

FIG. 12-15. Lambert grid coordinates computed from geodetic position.

THE TRANSVERSE MERCATOR PROJECTION

12-19. The Transverse Mercator Projections. In the Transverse Mercator projections, the Y axis represents the central meridian, as in the Lambert projection (see Fig. 12-16). Straight lines (not lines of latitude) perpendicular to the central meridian on the spheroid are made straight and perpendicular to the central meridian on the grid. Such lines obviously converge on the spheroid (for they are like great circles on a sphere), but they do not converge on the grid so that an expanding scale must be used to keep them equidistant on the grid. For conformality, this expanding scale must also be applied to lengths measured along these lines. To keep the expanding scale as near unity as possible, the scale at the central meridian is established at slightly less than unity, using such a value that it reaches unity for lines parallel to the central meridian at one-sixth the width of the zone from each boundary.

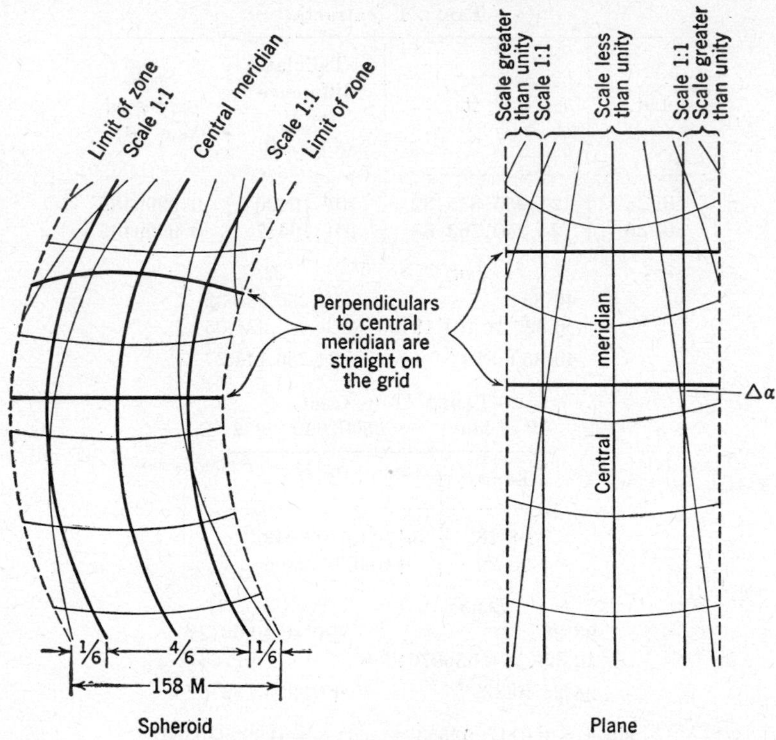

FIG. 12-16. The Transverse Mercator projection.

To find the grid coordinates of any geodetic position (see Fig. 12-17), the actual length on the spheroid of the perpendicular s from the central meridian to the point is first computed by a manipulation of ordinary geodetic computations. It is then multiplied by the integral of the expanding scale. The result is multiplied by the scale ratio at the central meridian to obtain x. Finally, the x value of the Y axis—usually 500,000 ft—is added to obtain the x coordinate of the geodetic position.

Next, the latitude at the central meridian of the foot of the perpendicular from the geodetic position is computed by an inverse geodetic computation. This is reduced to a, the distance from the latitude of the point

$y = 0$ on the central meridian. The result is multiplied by the scale ratio at the central meridian to obtain the y coordinate of the geodetic position.

As in the Lambert projection, by limiting the width—east and west—of the zone to 158 miles, the maximum variation in scale from unity is ± 1 part in 10,000.

Thus we have a plane, rectangular coordinate system in which the scale is constant along the central meridian and on all lines parallel to it but increases as a function of the distance from the central meridian.

It is possible to draw a map of the zone showing lines parallel to the central meridian through points of equal scale (see Fig. 12-18). By

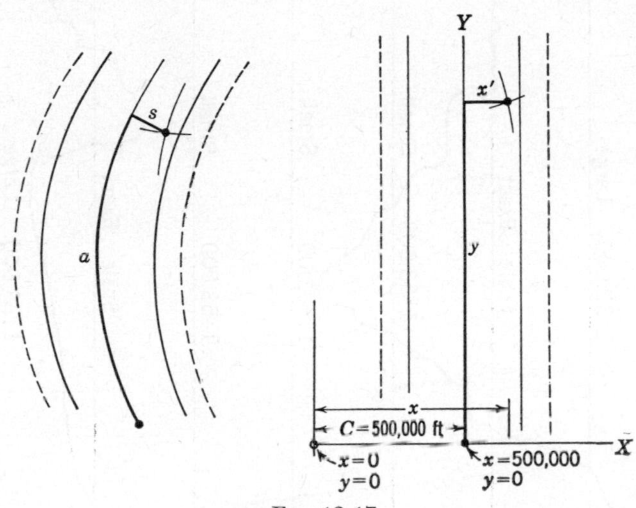

FIG. 12-17

selecting the scale value nearest the center of a survey, the scale applicable to that survey can be found. The scale can also be found from the special tables.

It should be noted that on this projection neither the meridians nor the parallels are straight lines. Both are somewhat complicated curves. The lines of equal scale, however, are straight and parallel to the central meridian.

12-20. Formulas for Computing Coordinates on the Transverse Mercator Grid.

$$x = x' + C \tag{12-7}$$

$$x' = H \times \Delta\lambda'' \pm ab \tag{12-8}$$

$$y = y_0 + V\left(\frac{\Delta\lambda''}{100}\right)^2 \pm c \tag{12-9}$$

where C is a constant for the zone, usually 500,000; y_0 (called tabular y),

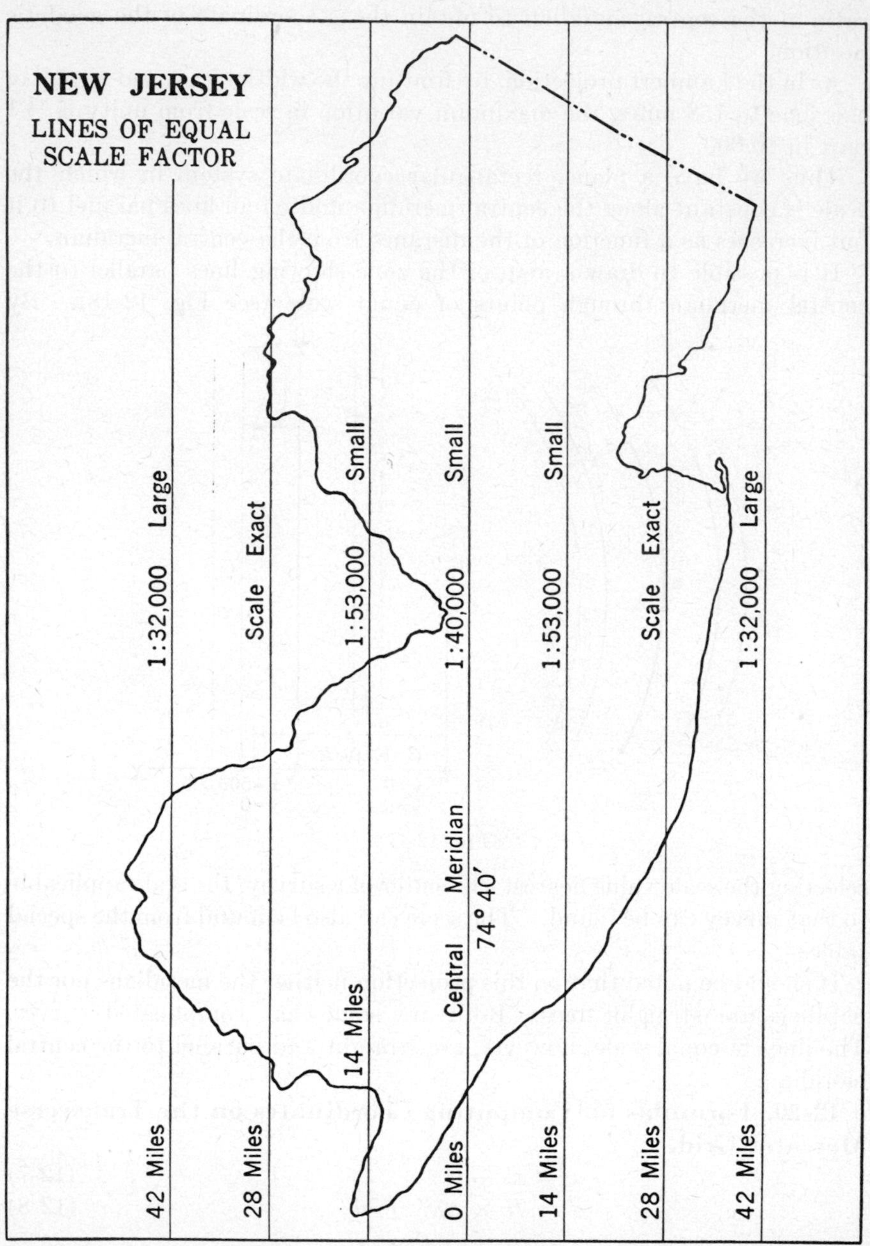

FIG. 12-18. Example of a Transverse Mercator projection. (*U.S. Coast and Geodetic Survey.*)

H, V, and a depend on the latitude and are found in the special tables for the zones; $\Delta\lambda''$ is the longitude of the central meridian minus the longitude of the point, expressed in seconds of arc; and b and c are based on $\Delta\lambda''$ and are given in the special tables for the zone.

The sign of ab and the sign of c are found in the table. When the sign of the term $H \times \Delta\lambda''$ is minus, the sign of ab is reversed from that given in the tables.

When Transverse Mercator coordinates are to be transformed into geodetic latitude and longitude, the following formulas are used:

$$x' = x - C \tag{12-10}$$

$$y_0 = P\left(\frac{x'}{10{,}000}\right)^2 + d \tag{12-11}$$

P is obtained from tables, given the y coordinate. d is obtained from tables, given the x coordinate.

In the tables, find latitude from y_0, H, and find a from latitude.

$$\text{Approx. } \Delta\lambda \text{ in seconds} = x' \div H \tag{12-12}$$

In the tables find b from approximate $\Delta\lambda$ in seconds.

$$\Delta\lambda \text{ in seconds} = (x' \mp ab) \div H \tag{12-13}$$

The sign of ab comes from the tables. When x' is minus, the sign from the tables is used. When x' is plus, the sign is reversed.

$$\text{Long.} = \text{central meridian} + \Delta\lambda \tag{12-14}$$

Example. Figure 12-19 shows the computation of the Transverse Mercator coordinates of a station from its geodetic position. Extracts from the special tables for the zone are given.

GENERAL CONSIDERATIONS

12-21. Scales. It should be remembered that the word **scale** means here the value obtained by dividing a distance computed from grid coordinates by the actual value of the distance computed from geodetic coordinates. In other words, it is the scale of grid lengths with respect to equivalent true lengths. In precise work a measured distance should be multiplied by the scale in the locality to obtain the coordinate distance.

Sometimes the scale, the grid scale, is said to be "too large." This is merely a substitute for the words "greater than unity." The term **scale factor** is often used instead of scale. This is the difference between the grid scale and unity. Measurements are usually corrected for scale by adding algebraically the **scale correction.** The scale correction is the

Flat Lat. = ϕ = 30°43′01.454″, Long. = λ = 88°26′04.338″
Transverse Mercator Projection for Mississippi (East)
Central mer. = 88°50′00.000″ C = 500,000 ft
TABLE (extract)

Lat.	y_0, ft	Δy_0 per second	H	ΔH per second	V	ΔV per second	a
30°43′	381,855.11	101.02833	87.295468	250.33	1.081083	5.72	−0.928
30 44	387,916.81	101.02850	87.280448	250.45	1.081426	5.72	−0.927

Central mer.		88°50′00.000″
$-\lambda$		−88 26 04.338
	+	23′55.662″
$\Delta\lambda''$ = $\Delta\lambda$ in seconds	+	1435.662
$\left(\dfrac{\Delta\lambda''}{100}\right)^2$		206.113
a (from table) =		−0.928

Lat.		*H*
30°43′00″		87.295468
1.454 × 250.33	−	364
30°43′01.454″		87.295104

Lat.		*V*
30°43′00″		1.081083
1.454 × 5.72	+	8
30°43′01.454″		1.081091

Lat.		y_0
30°43′00″		381,855.11
1.454 × 101.02833	+	146.90
30°43′01.454″		382,002.01

TABLE (extract)

$\Delta\lambda''$	b	Δb	c
1400	+5.216	+0.295	−0.043
1500	+5.511	+0.281	−0.049

$\Delta\lambda''$	b	$\Delta\lambda''$	c
1400	+5.216	1400	−0.043
35.662 × 0.295	+ 105	0.36 × 6	− 2
1435.662	+5.321	1435.662	−0.045

$H\,\Delta\lambda''$ = (87.295104)(+1435.662)	+125,326.26
ab = (−0.928)(+5.321) = −4.94*	− 4.94
$x' = H\,\Delta\lambda'' \pm ab$ =	+125,321.32
C =	500,000.00
$x = x' + C$ =	625,321.32
$V\left(\dfrac{\Delta\lambda''}{100}\right)^2$ = (1.081091)(206.113) =	+222.827
C =	− 0.045
	+222.782
y_0 = tabular y =	382,002.01
$y = y_0 + \left(\dfrac{\Delta\lambda''}{100}\right)^2 \pm C$ =	382,224.79

* When $H\,\Delta\lambda''$ is minus, reverse the sign of this value.

FIG. 12-19. Transverse Mercator grid coordinates computed from geodetic position.

product of the scale factor multiplied by the distance. Thus

$$D = (S - 1)M + M \tag{12-15}$$

where D = coordinate distance
S = scale
$S - 1$ = scale factor
M = measured distance

In logarithmic computation, the logarithm of the scale is added to the logarithm of the measured distance.

12-22. Azimuths. A straight line on the earth is a line prolonged by successive double centering with a transit, that is, it is a line that does not change direction in a surface perpendicular to gravity. If such a line were run without error for many miles and its grid coordinates determined at a number of points throughout its length by computation from the geodetic positions of these points, it would be found that these grid coordinates could not be related by a linear equation; if these coordinates were plotted on the grid, they would not be quite on a straight line. In other words, a straight line on the spheroid is not quite straight on the grid. This condition is always true of a conformal projection. A straight line always curves toward the area where the scale is smaller. Of course, straight lines perpendicular to lines of equal scale do not curve on any projection, nor do lines parallel to the straight lines of equal scale on the mercator grid.

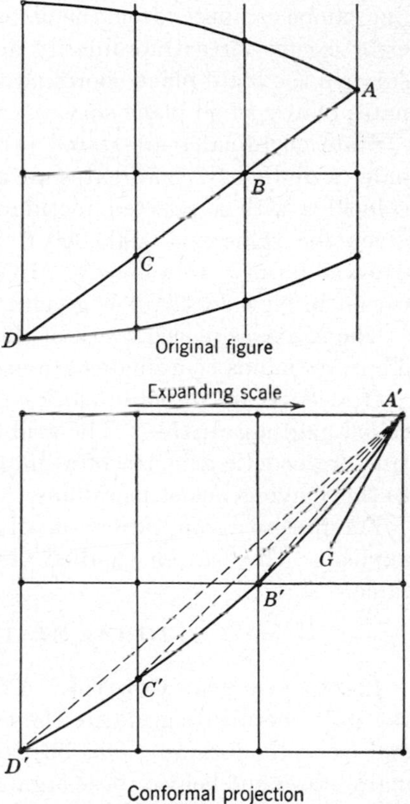

FIG. 12-20

Figure 12-20 shows this condition in an exaggerated form. The lower figure is a conformal projection of the upper figure similar to a Transverse Mercator projection. The scale expands the proper amount to straighten out the curved lines. $ABCD$ is a straight line on the upper figure. When these points are plotted at $A'B'C'D'$, the line curves toward the smaller scale. $A'G$ represents the tangent to $A'B'C'D'$ at A' and therefore is the observed direction of D, C, or B. $A'D'$ is the direction of D' computed from the coordinates. $A'C'$ and $A'B'$ are the same for the points B' and C'. The nearer the point to A', the more nearly the direction computed by coordinates will agree with the observed direction. The procedure for computing the direction between two points from their plane coordinates, when this small effect is taken into consideration, is given in later paragraphs.

12-23. Summation. In each zone (grid) in a state system, the scale factor is never greater than 1 part in 10,000[1] and in most parts of the grid it is much less. This means that the length of a line computed from the plane coordinates of its ends will be correct within this limit. Also, the error in azimuth is negligible. This means that the grid azimuth of a line can be computed from the plane coordinates of its ends with negligible error except for extraordinarily accurate work. Accordingly, for most surveys the state plane coordinates can be used exactly like the coordinates of any small plane survey.

State coordinates are stated in feet. East coordinates are given the name x and north coordinates are given the name y. The central Y axis coincides with a selected meridian called the central meridian. It is given the value $x = 2{,}000{,}000$ ft in all zones whose length is east and west, and also in New Jersey. In zones whose length is north and south, except in New Jersey, it is given the value $x = 500{,}000$ ft.

The X axis is placed south of the zone and is given the value $y = 0$ ft. Thus, no minus coordinates are used.

Directions are stated in clockwise azimuths from the *south*. They are called **grid azimuths.** The grid azimuth of a line is different from the true or geodetic azimuth of a line—except at the central meridian—due to the convergence of meridians.

For precise computations of distance, a **scale correction** should be applied. This can be handled very simply by referring to a map or a table.

TO USE STATE COORDINATES

12-24. Available Control. The chief source of control available for the state coordinate grids are the triangulation stations of the U.S. Coast and Geodetic Survey. This bureau also inaugurated geodetic surveys in many states and helped these organizations to establish systems of second-order traverse based on the triangulation scheme but computed directly by the coordinates of the state grids. As engineers become familiar with the advantages of these systems, more and more surveys are connected with these control systems and computed by grid coordinates. These surveys include state highway surveys, other state surveys, county and

[1] North Carolina is the one exception. In this state it reaches 1 part in 5000.

municipal surveys, surveys of large corporations, and many smaller surveys. The stations of these surveys also furnish a means of using the state grid coordinates.

12-25. U.S. Coast and Geodetic Survey Stations. Information concerning the fundamental control can be obtained from the Director, U.S. Coast and Geodetic Survey. In addition to completed publications of this bureau, the Director has available temporary publications covering work in progress. Usually the state coordinate values and grid azimuths are available. If geodetic values are completed, the grid data can be computed by the tables and the instructions issued by this bureau, as explained in previous paragraphs.

12-26. State Bureaus. In some states the data required can be obtained from a state bureau that maintains the geodetic survey records within the state.

12-27. To Base a Survey on the State Grid. In almost every case the state systems can be utilized exactly like an ordinary coordinate system. A survey to be based on a state system is connected with stations in the system by as many position and azimuth ties as possible. It is then adjusted by the usual methods. The positions of the stations on the state system are held fixed.

If the exterior limits of the survey have been connected to the control, this adjustment gives very accurate results. When exceptional accuracy is desired, or when the connections were available only at one part of the survey, certain refinements can be applied that will improve the results

12-28. Refinements. If the fixed coordinate positions and the survey itself were both perfect, to obtain perfect agreement the following corrections to the survey data would be necessary:

1. Reduction to sea level
2. Reduction for scale
3. Reduction for curvature

If these corrections are to be applied, they should be applied before adjustment. Any adjustment is designed to eliminate accidental errors, and it therefore introduces corrections in accordance with the likelihood of the occurrence of accidental errors in both length and angular measurements. The three reductions specified are values that can be exactly computed. Their distribution is precisely known, and it is not in accordance with the distribution of accidental errors. For example, the sea-level reduction affects length measurements. It is a fixed ratio of the length, depending on the altitude at which the length was measured. If a discrepancy in a traverse due to measurements at a high altitude were to be eliminated by ordinary traverse adjustment, angles as well as lengths would be changed, and the changes would be entirely independent of the altitude at which the lengths were measured. Accordingly, a slightly erroneous adjustment would result.

12-29. Reduction to Sea Level. From Fig. 12-7,

$$A'B' = \frac{R}{R + h} AB$$

or

$$A'B' = AB - \frac{h}{R + h} AB$$

in which $A'B'$ is the length reduced to its sea-level equivalent, AB is the measured length, and h is the elevation in feet.

Taking the mean radius of the earth within the United States as 20,906,000 ft[1] this becomes

$$A'B' = AB - \frac{h}{20{,}906{,}000 + h} AB \tag{12-16}$$

in which the expression $h/(20{,}906{,}000 + h)$ is the correction factor per foot of length for altitude expressed in feet. The correction factor per 1000 ft of altitude per 10,000 ft of length is very nearly 0.4781. Then

$$C' = 0.4781 \frac{h}{1000} \frac{l}{10{,}000} \tag{12-17}$$

where C' is the correction, h is the elevation, and l is the measured length.

Example. To reduce to sea level a length of 4,289.762 measured at an elevation of 2340 ft,

$$c = 0.4781 \times 2.34 \times 0.429 = 0.480$$
$$\text{Sea-level length} = 4289.762 - 0.480 = 4289.282$$

The value for the altitude in the formula is taken as the average altitude of the length measurement to be reduced. Often, the same average altitude can be used for the lengths in a large portion of the survey or for the entire survey.

12-30. Reduction for Scale. The value of the grid scale (grid length/true length) can be found from the location of the survey on a map showing the scale ratios, or it can be taken from the various U.S. Coast and Geodetic Survey special tables. In the Lambert projection the scale depends on the latitude. In the Transverse Mercator projection the scale depends on the x' value. Each measured length can be corrected according to the scale at its midpoint. The very slight error that this might cause can be eliminated by certain tables also published by the

[1] Oscar S. Adams and Charles N. Claire, U.S. Coast and Geodetic Survey Special Publications No. 194, "Manual of Traverse Computation on the Lambert Grid," p. 8, 1935, and No. 195, "Manual of Traverse Computation on the Transverse Mercator Grid," p. 12, 1935.

U.S. Coast and Geodetic Survey. Usually the lengths of the whole survey can be corrected by the same scale ratio.

Example. Measured distance 4289.282. From the tables, for the particular location: scale in units of the seventh place of logarithms, -105.5. Scale expressed as a ratio, 0.9999757. Then

$$\begin{aligned} \log 4289.282 &= 3.6323846 \\ \log \text{scale} &= \underline{-0.0000106} \\ \log \text{grid length} &= 3.6323740 \\ \text{Grid length} &= 4289.178 \end{aligned}$$

or

$$\text{Grid length} = 0.9999757 \times 4289.282 = 4289.178$$

12-31. Reduction for Curvature. The corrections for curvature introduce so small a difference that they need seldom be used for traverses, as the usual traverse adjustments will take care of them with sufficient accuracy. However, when the grid coordinates of a triangulation station are to be computed through the side of a triangle of over 5 miles in length in a survey of first-order accuracy, the computation of the necessary grid azimuth from the known geodetic azimuth must include this correction (see Fig. 12-20).

It is necessary also to use this correction when an azimuth tie for a first-order traverse is based on a sight of over 5 miles in length, as is often the case in making corrections with triangulation. The principles are explained in the next section.

12-32. Azimuth. Figure 12-21 (taken from Fig. 12-20) shows the various elements of the azimuth of a line on a grid. AD is the line represented by $A'D'$ on the grid. At A' the observer would point toward G in order to observe D'.

1. *Geodetic Azimuth.* This is the angle from geodetic south to the point observed. Its value on the grid is exactly equal to the true measured value.

2. *Terminal Grid Azimuth.* This is the angle on the grid from grid south to the tangent to the line observed, at the station occupied. This is the azimuth that must be used for azimuth ties, as its value on the grid is exactly equal to the true measured value.

3. *Grid Azimuth.* This is the angle from the grid south to a straight line on the grid running from the point of observation to the point observed. It can also be defined as follows:

$$\text{Grid azimuth} = \text{arc tan}\,\frac{x_1 - x_2}{y_1 - y_2} \qquad (12\text{-}18)$$

where x_1y_1 are the coordinates of the point of observation and x_2y_2 are those of the point observed. It is used to compute coordinates or, when the coordinates are known, it is computed from them as a first step in finding the terminal grid azimuth.

12-33. Angular Meridian Convergence on the Grid (with Respect to the Central Meridian), *M*. This is the difference between a geodetic azimuth and the terminal grid azimuth of the same line. It is the difference between geodetic north and grid north. On the Lambert projection it is equal to θ (see Fig. 12-12). On the Transverse Mercator projection it is the value $\Delta\alpha$ (see Fig. 12-16).

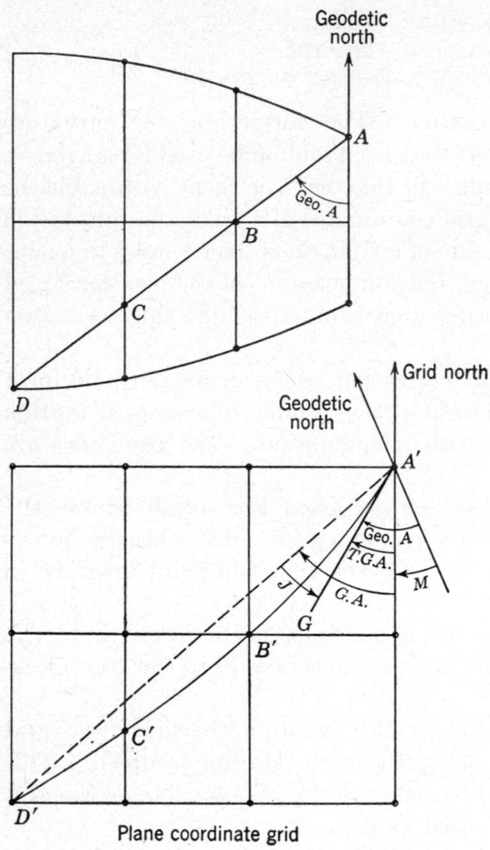

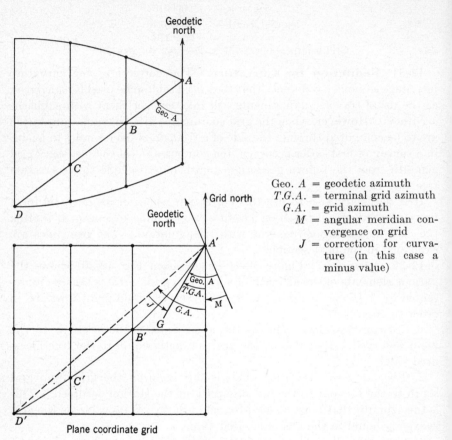

Geo. A = geodetic azimuth
$T.G.A.$ = terminal grid azimuth
$G.A.$ = grid azimuth
M = angular meridian convergence on grid
J = correction for curvature (in this case a minus value)

Fig. 12-21

12-34. Correction for Curvature, *J*. This is the difference between the terminal grid azimuth and the grid azimuth.

12-35. Equations. From Fig. 12-21,

$$\text{Terminal grid azimuth} = \text{grid azimuth} + J \qquad (12\text{-}19)$$

$$\text{Terminal grid azimuth} = \text{geodetic azimuth} - M \qquad (12\text{-}20)$$

$$\text{Grid azimuth} = \text{terminal grid azimuth} - J \qquad (12\text{-}21)$$

$$\text{Grid azimuth} = \text{geodetic azimuth} - M - J \qquad (12\text{-}22)$$

In the Lambert projection

$$M = \theta$$

$$J \text{ (seconds)} = -\frac{x_2 - x_1}{2\rho_0^2 \sin 1''}\left(y_1 - y_0 + \frac{y_2 - y_1}{3}\right) \qquad (12\text{-}23)$$

In the Mercator projection

$$M = \Delta\alpha$$

$$J \text{ (seconds)} = \frac{(y_2 - y_1)(2x_1' + x_2')}{(6\rho_0^2 \sin 1'')_g} \qquad (12\text{-}24)$$

where x_1y_1 are the coordinates of the beginning of the line where the observation is made, x_2y_2 are the coordinates of the end of the line, and x_1' and x_2' are the x coordinates of these points minus the x value of the central meridian.

The values of y_0, θ, $1/(2\rho_0^2 \sin 1'')$, and $1/(6\rho_0^2 \sin 1'')_g$ can be taken from the tables for the particular projection.

$$\Delta\alpha \text{ in seconds} = (\sin \phi)(\Delta\lambda \text{ in seconds}) + g$$

where ϕ is the latitude, $\Delta\lambda$ is the longitude of the central meridian minus the longitude of the point, and g is given in the projection tables. The sign of g is reversed when $(\sin \phi)(\Delta\lambda$ in seconds) has a minus value.

12-36. Example on the Lambert Grid. Figure 12-22 shows actual examples of azimuth computation on the Lambert grid for the two cases that arise in making an azimuth tie for a traverse. It is assumed that the traverse originates at Station Mason and that the azimuth tie consists of measuring the angle from the direction of Prosser to the direction of the first course of the traverse.

Since the *line* to Prosser will be curved on the grid, the *direction* to Prosser, when plotted on the grid, will be the line to A which is tangent to the curved line to Prosser at the station Mason. The direction of this line to A on the grid is expressed by the terminal grid azimuth. The terminal grid azimuth must therefore be used, with the measured angle, to compute the azimuth of the first course of the traverse. There are two ways of finding the terminal grid azimuth, depending on the data available:

1. If the geodetic azimuth, Mason to Prosser, and the longitude of Mason are known, Eq. (12-20) is used as shown. θ for Mason can be found from the special tables (see Fig. 12-15).

2. If the grid coordinates of Mason and Prosser are known, Eq. (12-19) is used. This involves J. The complete computation is shown in Fig. 12-22.

It is obvious that if the distance from Mason to Prosser were not very great, the value of J would be negligible and the computation of J would be unnecessary. This is the usual case.

12-37. Triangulation. If it were desired to compute the grid coordinates of Prosser when only the geodetic length and geodetic azimuth of Mason to Prosser were known (the typical problem in computing triangulation by grid coordinates), then the grid azimuth must be computed by Eq. (12-22).

12-38. Example on the Transverse Mercator Grid. Figure 12-23 shows an example of the computation of the terminal grid azimuth on a Transverse Mercator grid.

12-39. Example in Traverse Computation on the Lambert Grid. Figure 12-24 shows a traverse running from Mason to Prosser computed on a Lambert grid. The fixed data and the actual survey measurements are given. If the actual lines

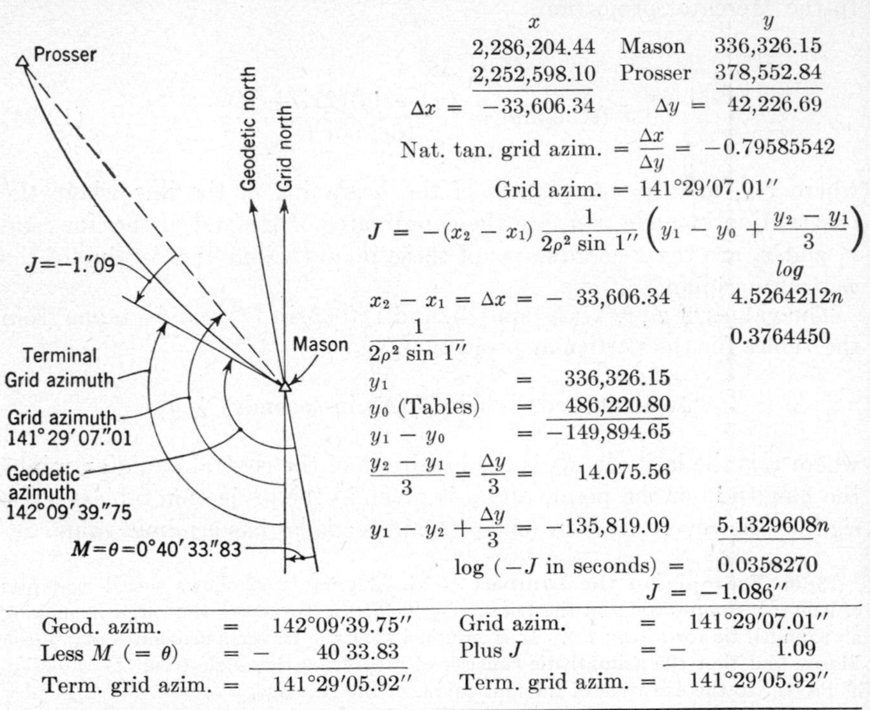

	x		y
	2,286,204.44	Mason	336,326.15
	2,252,598.10	Prosser	378,552.84
$\Delta x =$	−33,606.34	$\Delta y =$	42,226.69

$$\text{Nat. tan. grid azim.} = \frac{\Delta x}{\Delta y} = -0.79585542$$

$$\text{Grid azim.} = 141°29'07.01''$$

$$J = -(x_2 - x_1)\frac{1}{2\rho^2 \sin 1''}\left(y_1 - y_0 + \frac{y_2 - y_1}{3}\right)$$

			log
$x_2 - x_1 = \Delta x$	=	− 33,606.34	4.5264212n
$\dfrac{1}{2\rho^2 \sin 1''}$			0.3764450
y_1	=	336,326.15	
y_0 (Tables)	=	486,220.80	
$y_1 - y_0$	=	−149,894.65	
$\dfrac{y_2 - y_1}{3} = \dfrac{\Delta y}{3}$	=	14.075.56	
$y_1 - y_2 + \dfrac{\Delta y}{3}$	=	−135,819.09	5.1329608n
log ($-J$ in seconds)	=		0.0358270
		$J =$	−1.086″

Geod. azim.	= 142°09′39.75″		Grid azim.	= 141°29′07.01″
Less M (= θ)	= − 40 33.83		Plus J	= − 1.09
Term. grid azim.	= 141°29′05.92″		Term. grid azim.	= 141°29′05.92″

FIG. 12-22. Computation of the terminal grid azimuth on the Lambert grid.

between the stations were shown, they would be curved lines beginning and ending on the short lines shown at each station that give the directions of the terminal grid azimuth. Straight lines connecting the stations are drawn instead, as these lines are the lines used in the grid computations. They follow the grid azimuth.

12-40. Azimuth Marks. Usually at each triangulation station on the main scheme, a mark is set or chosen that can be seen from the ground at the station. This is called the azimuth mark as it can be used for an azimuth tie without elevating the instrument on a tower.

12-41. Fixed Data. The following data are usually available for any station on the main scheme of triangulation:

1. Geodetic position
2. Geodetic azimuth to at least one station on the main scheme
3. A list of directions

The list of directions is a method of stating the results, after the station adjustment is complete, of the angular measurements made at a triangulation station. The direction of one of the points observed is chosen as the *initial direction*. This direction is given the value 0°00′00″,

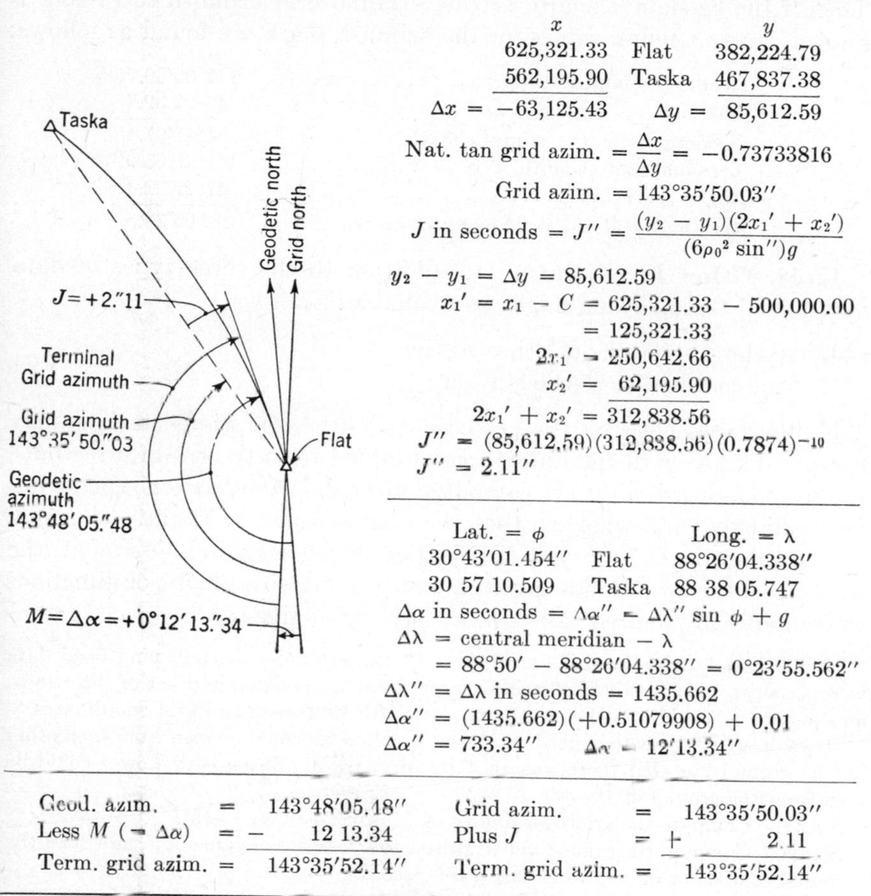

FIG. 12-23. Computation of the terminal grid azimuth on the Transverse Mercator grid.

without regard to its actual azimuth. The directions of all other points observed are then given by clockwise angles from this initial direction. From this list the angle between any two points observed can be found from difference in their listed directions. For example, in the list of directions at Mason might be found the following:

South base	00°00′00.00″
Azimuth mark	10°17′22.3″
Prosser	57°39′54.70″

The angle between azimuth mark and Prosser is found as follows:

Prosser	57°39′54.70″
− Azimuth mark	10 17 22.3
Angle	47°22′32.4″

Then if the geodetic azimuth or the terminal grid azimuth to Prosser is known, corresponding values for the azimuth mark are found as follows:

Geodetic azimuth, Prosser	142°09′39.76″
− Angle	47 22 32.4
Geodetic azimuth, azimuth mark	94°47′07.4″
Terminal grid azimuth, Prosser	141°29′05.9″
− Angle	47 22 32.4
Terminal grid azimuth, azimuth mark	94°06′33.5″

12-42. Other Fixed Data. In addition to the three types of data mentioned, the following might be available:

4. Geodetic azimuths of other stations
5. Grid coordinates of the station

The differences between the geodetic azimuths of various stations may not quite agree with the differences obtained from the list of directions, as the geodetic azimuths are computed after the triangulation scheme has been adjusted as a whole so that the angles would be slightly changed. As the azimuths of long sides of triangles change least in adjustment, the azimuth of the longest side given should be used with the list of directions to compute any desired azimuth not otherwise given.

12-43. Fixed Data in the Example. In the example, the following fixed data were assumed to be available: at Mason the geodetic position and list of directions; at Prosser the geodetic position, geodetic azimuth to Tower, and grid coordinates.

It was also assumed that neither Mason nor Prosser could be seen from each other on the ground but that there was an azimuth mark at Mason, and Tower could be seen from the ground at Prosser.

Step 1. Compute the grid coordinates of Mason (see Fig. 12-15).

Step 2. Compute the grid azimuth Mason to Prosser, the value of J, and then the terminal grid azimuth of Mason to Prosser (see Fig. 12-22).

Step 3. Compute the terminal grid azimuth Mason to azimuth mark (see Sec. 12-42).

Step 4. Compute the terminal grid azimuth Prosser to Tower from its geodetic azimuth by Eq. (12-20). Thus:

Geodetic azimuth	164°23′47.2″
− M	36 18.3
Terminal grid azimuth	163°47′28.9″

12-44. Traverse Computation on the Lambert Grid. An accurate scale drawing should first be made of the traverse, as in Fig. 12-24.

It is assumed that the angles a, b, c, d, e, and f in Fig. 12-24, which represent the difference between terminal grid azimuths and grid azimuths, are negligible. If it is thought that they should be computed, the method for computing J is shown in Fig. 12-22. The coordinate values for the computation of J can be scaled from the sketch.

The traverse computations are shown in Fig. 12-25.

12-45. Traverse Computation on the Transverse Mercator Grid. A traverse computation similar to that just described, but on the Transverse Mercator grid, is shown in Figs. 12-26 and 12-27.

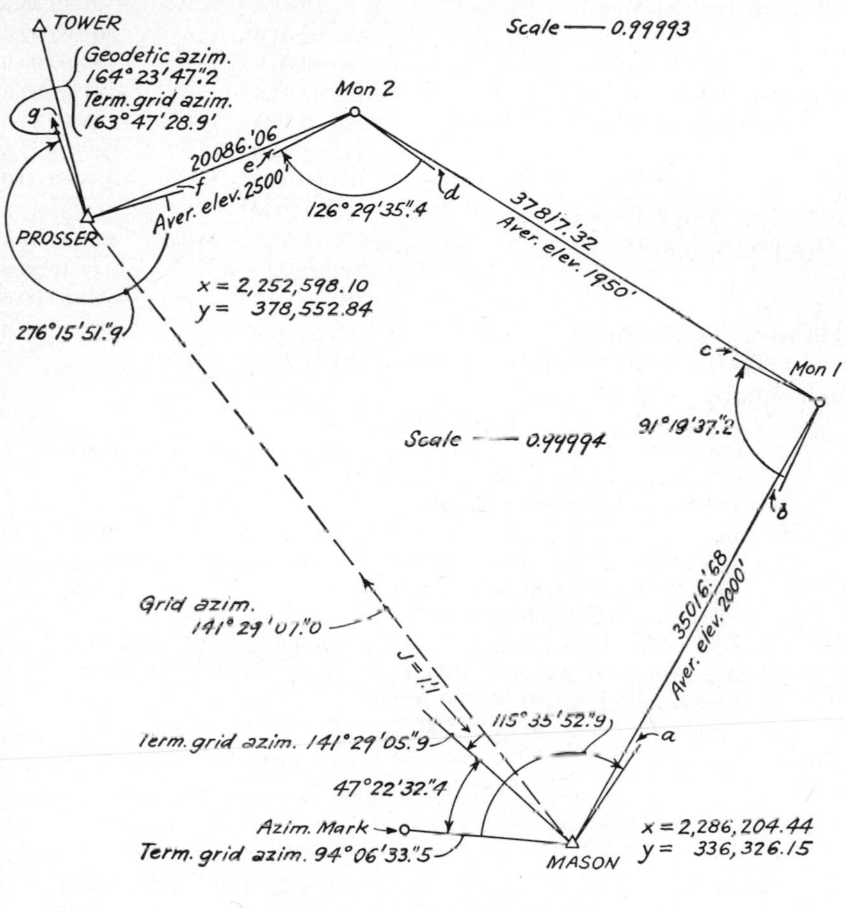

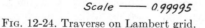

FIG. 12-24. Traverse on Lambert grid.

12-46. Practical Rules for the Use of State Coordinates. In conclusion, it should be pointed out that the use of state coordinate systems is extremely simple. Only the following two rules are necessary:

1. State coordinates can be used exactly like the coordinates of any plane rectangular system. Scale corrections can be applied to measured distances if desired.
2. All azimuths used or computed from coordinates are grid azimuths. They are measured from the grid and not from meridians. They therefore differ from such azimuths. This difference can be computed if desired.

Grid Azimuths on Traverse	*Measured*	*Cor.*	*Adjusted*
Mason–Prosser Term. grid azim.	141°29′05.9″		
Angle from list of directions	− 47 22 32.4		
Mason–azim. mark Term. grid azim.	94°06′33.5″		94°06′33.5″
Traverse angle at Mason	+115 35 52.9	−0.5	+115 35 52.4
Grid azim. Mason–Mon 1	209°42′26.4″		209°42′25.9″
Traverse angle at Mon 1	+ 91 19 37.2	−0.5	+ 91 19 36.7
	301°02′03.6″		301°02′02.6″
	−180 00 00.0		−180 00 00.0
Grid azim. Mon 1–Mon 2	121°02′03.6″		121°02′02.6″
Traverse angle at Mon 2	+126 29 35.4	−0.5	+126 29 34.9
	247°31′39.0″		247°31′37.5″
	−180 00 00.0		−180 00 00.0
Grid azim. Mon 2–Prosser	67°31′39.0″		67°31′37.5″
Traverse angle at Prosser	+276 15 51.9	−0.5	276 15 51.4
	343°47′30.9″		343°47′28.9″
	−180 00 00.0		−180 00 00.0
Grid azim. Prosser–Tower	163°47′30.9″		163°47′28.9″
Prosser–Tower Term. grid azim.	163 47 28.9		
Error in traverse angles	+2.0″		

Lengths on Traverse

Mason–Mon 1 Av. elev. 2000 ft	Meas. length	35,016.68
Sea-level cor. 0.4781 × 2.00 × 3.50		− 3.35
Scale cor. 0.0000562 × 35,000		− 1.97
	Grid length	35,011.36
Mon 1–Mon 2 Av. elev. 1950 ft	Meas. length	37,817.32
Sea-level cor. 0.4781 × 1.95 × 3.78		− 3.52
Scale cor. 0.0000636 × 37,817		− 2.41
	Grid length	37,811.39
Mon 2–Prosser Av. elev. 2500 ft	Meas. length	20,086.06
Sea-level cor. 0.4781 × 2.50 × 2.01		− 2.40
Scale cor. 0.0000652 × 20,100		− 1.31
	Grid length	20,082.35

Grid Coordinates

Stations	*Bearings* / *Lengths*	*cos* / *sin*	*y*	*x*
Mason			336,326.15	2,286,204.44
	N29°42′25.9″E	0.86856930	+ 30,409.79	+ 17,350.50
	35,011.36	0.49556774		
Mon 1			366,735.94	2,303,554.94
	N58°57′57.4″W	0.51554747	+ 19,493.57	− 32,399.11
	37,811.39	0.85686102		
Mon 2			386,229.51	2,271,155.83
	S67°31′37.5″W	0.38224668	− 7,676.41	− 18,557.30
	20,082.35	0.92406031		
Prosser			378.553.10	2,252,598.53
Prosser	Fixed		−378,552.84	2,252,598.10
		Error	+0.26	+0.43

FIG. 12-25. Traverse computation on Lambert grid.

Stations		Final Grid Coordinates		
		y		*x*
Mason		336,326.15		2,286,204.44
	+30,409.79 − 0.10	+ 30,409.69	+17,350.50 − 0.16	+ 17,350.34
Mon 1		366,735.84		2,303,554.78
	+19,493.57 − 0.11	+ 19,493.46	−32,399.11 − 0.18	− 32,399.29
Mon 2		386,229.30		2,271,155.49
	− 7,676.41 − 0.05	− 7,676.46	−18,557.30 − 0.09	− 18,557.39
Prosser		378,552.84		2,252,598.10

FIG. 12-25 (*Continued*)

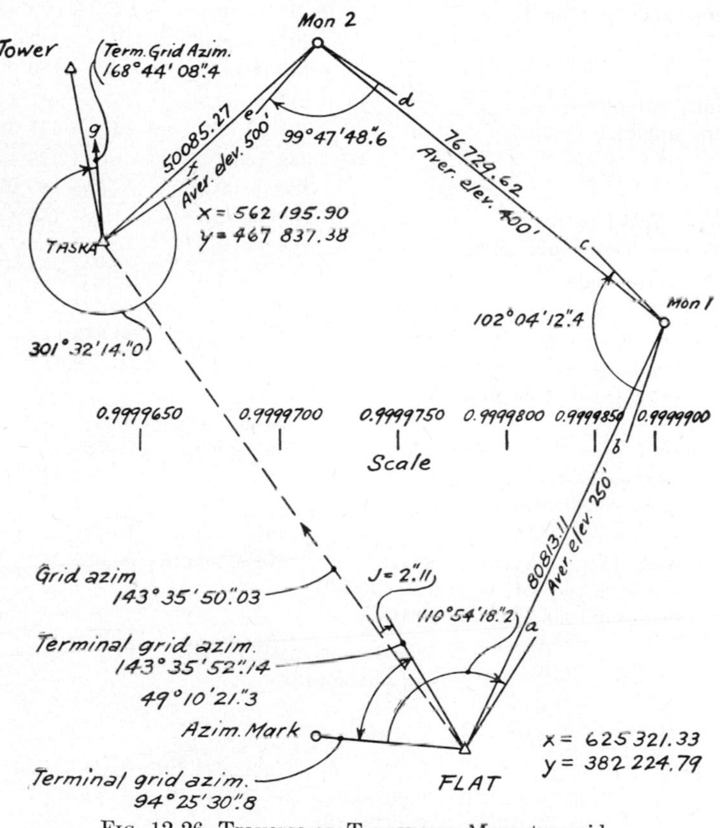

FIG. 12-26. Traverse on Transverse Mercator grid.

12-47. Advantages of State Systems of Plane Coordinates. By making the fundamental control system available, state coordinate systems provide a means of controlling local surveys and placing them on a single datum. Where monumented positions are available in sufficient density, the necessity of establishing a basic triangulation for surveys in a new area is eliminated, and route surveys can be checked and adjusted for greater accuracy. All surveys for which the state coordinates are used are automatically connected with each other so that data can be

Grid Azimuths on Traverse	*Measured*	*Cor.*	*Adjusted*
Flat–Taska Term. grid azim.	143°35′52.1″		
Angle from list of directions	− 49 10 21.3		
Flat–azim. mark Term. grid azim.	94°25′30.8″		94°25′30.8″
Traverse angle at Flat	+110 54 18.2	+1.1	+110 54 19.3
Grid azim. Flat–Mon 1	205°19′49.0″		205°19′50.1″
Traverse angle at Mon 1	+102 04 12.4	+1.1	+102 04 13.5
	307°24′01.4″		307°24′03.6″
	−180 00 00.0		−180 00 00.0
Grid azim. Mon 1–Mon 2	127°24′01.4″		127°24′03.6″
Traverse angle at Mon 2	+ 99 47 48.6	+1.1	+ 99 47 49.7
	227°11′50.0″		227°11′53.3″
	−180 00 00.0		−180 00 00.0
Grid azim. Mon 2 to Taska	47°11′50.0″		47°11′53.3″
Traverse angle at Taska	+301 32 14.0	+1.1	+301 32 15.1
	348°44′04.0″		348°44′08.4″
	−180 00 00.0		−180 00 00.0
Grid azim. Taska to Tower	168°44′04.0″		168°44′04.8″
Taska–Tower Term. grid azim.	168°44′08.4″		
Error in traverse angles	−4.4		

Lengths on Traverse

Flat–Mon 1 Av. elev. 250 ft	Meas. length	80,813.11
Sea-level cor. 0.4781 × 0.250 × 8.08		−0.97
Scale cor. 0.0000167 × 80,800		−1.35
	Grid length	80,810.79
Mon 1–Mon 2 Av. elev. 400 ft	Meas. length	76,729.62
Sea-level cor. 0.4781 × 0.400 × 7.67		−1.47
Scale cor. 0.0000209 × 76,700		−1.60
	Grid length	76,726.55
Mon 2–Taska Av. elev. 500 ft	Meas. length	50,085.27
Sea-level cor. 0.4781 × 0.500 × 5.01		−1.20
Scale cor. 0.0000326 × 50,100		−1.63
	Grid length	50,082.44

Grid Coordinates

Stations	*Bearings* *Lengths*	*cos* *sin*	*y*	*x*
Flat			382,224.79	625,321.34
	N25°19′50.1″E	0.90385430	+ 73,041.18	+ 34,574.12
	80,810.79	0.42784039		
Mon 1			455,265.97	659,895.46
	N52°35′56.4″W	0.60738971	+ 46,602.92	− 60,951.88
	76,726.55	0.79440402		
Mon 2			501,868.89	598,943.58
	S47°11′53.3″W	0.67946513	− 34,029.27	− 36,745.88
	50,082.44	0.73370780		
Taska			467,839.62	562,197.70
Taska	Fixed		467,837.38	562,195.90
		Error	+2.24	+1.80

FIG. 12-27. Traverse computation on Transverse Mercator grid.

Stations		*Final Grid Coordinates* *y*		*x*
Flat		382,224.79		625,321.34
	+73,041.18 − 0.87	+73,040.31	+34,574.12 − 0.71	+ 34,573.41
Mon 1		455,265.10		659,894.75
	+46,602.92 − 0.83	+46,602.09	−60,951.88 − 0.66	− 60,952.54
Mon 2		501,867.19		598,942.21
	−34,029.27 − 0.54	−34,029.81	−36,745.88 − 0.43	− 36,746.31
Taska		467,837.38		562,195.90

FIG. 12-27 (*Continued*)

interchanged. Such surveys are also given permanent position so that lost stations can be restored.

As the use of state coordinates grows, their usefulness increases. It is probable that ultimately all surveys of any importance will be computed in terms of state coordinates and thus become parts of one great survey. Many examples of advantageous use of State coordinates can be given. Three will be pointed out here.

State, county, and municipal boundary lines are of considerable importance, as they often mark the boundaries of districts for road maintenance, tax payment, court jurisdiction, and the recording of deeds and other documents. Few county lines are marked with sufficient monuments because of the high cost of such work. When the state coordinates of the angle points of a county line are known, it is possible to establish the line at any point desired by surveys from any marked coordinate positions.

Mapping of large areas today can be most successfully accomplished by aerial photography. For this work, a considerable density of marked positions must be available. Where state coordinates are in use, such marked positions are available. In such an area, maps can be made with maximum rapidity and economy.

The difficulties inherent in property surveys are chiefly caused by the lack of permanence of landmarks. When the state coordinates of a landmark have been determined, it can always be reestablished if lost. Every other monumented position serves as a witness to it.

When state coordinates are determined for the corners of land parcels, the relative positions of all such parcels are immediately evident. This is important for drawing tax maps, for the determination of title, and for the location of boundaries.

At present writing, 29 states have enacted statutes that define the state system of coordinates mathematically and give it a name. In any of these states, the state coordinates of a point can be used in a land description by merely stating the name of the system and giving the values of the coordinates. The first of these laws was written by the author of this book and enacted in New Jersey in 1935.

PROBLEMS

12-1. Explain the reason for (*a*) orthometric corrections, (*b*) sea-level corrections.

12-2. What is the difference between gravity and gravitation?

12-3. What is the difference between astronomic and geodetic (*a*) position, (*b*) bearing?

12-4. What is a conformal projection?

12-5. Describe the general procedure for establishing (*a*) a Lambert projection, (*b*) a Transverse Mercator projection.

12-6. Define (*a*) grid scale, (*b*) grid scale factor.

12-7. What lines have constant grid scales throughout their length on a Lambert projection and on a Transverse Mercator projection?

12-8. Which way does a straight line on the ground curve on a grid?

12-9. Define (*a*) geodetic azimuth, (*b*) grid azimuth, (*c*) terminal grid azimuth, (*d*) angular meridian convergence on a grid, (*e*) correction for curvature.

12-10. Draw a sketch illustrating the items in Prob. 12-9 above.

12-11. When an accurate survey is made between two pairs of marked points having known grid coordinates, what changes from ordinary computations are required?

12-12. What is a list of directions?

12-13. Compute an assigned problem on a Lambert grid.

12-14. Compute an assigned problem on a Transverse Mercator grid.

(U.S. Coast and Geodetic Survey tables must be available for Probs. 12-13 and 12-14.)

CHAPTER 13

CITY SURVEYS

13-1. The term **city surveying** has been used to cover many types of surveying. In this text it is applied to the great class of detailed surveys necessary for the types of construction usually performed in cities. It includes surveys for buildings, culverts, curbs and streets, small bridges, retaining walls, sewers, drains, etc., and is therefore the kind of surveying usually encountered by most engineers. Structures of this type must be measured and located with considerable accuracy, usually within two or three hundredths of a foot. But the extent of the work is seldom large so that an accuracy of 1 part in 3000 to 5000 is usually sufficient. Accurate property surveys must be performed in cities; this branch of surveying, which requires special treatment, was covered in Chap. 10.

In many cities a horizontal control system is used. High accuracies are required to establish such a system. It is described later in this chapter.

Most of the surveying methods required are familiar to readers of this text. However, in the location of lines and grades for streets which include those for pavements, curbs, gutters, sidewalks, and right of way, horizontal and vertical curve techniques not hitherto covered are necessary. Certain survey operations in connection with street improvements, new street layouts, real estate developments, and city engineering in general require the special treatment included in this chapter.

HORIZONTAL CURVES

13-2. Types of Horizontal Curves. There are four types of horizontal curves: simple curves, compound curves, reversed curves, and spirals. The first three are covered in this book. Spirals are very seldom used in city surveying; they are omitted here. The reader is referred to works on highway curves for this subject.

13-3. Simple Curves. A **simple curve** is a circular arc joining two straight lines known as **tangents** (see Fig. 13-1). It is assumed to have a forward direction according to the stationing or the direction used in property descriptions. Its various parts are named as shown in Fig. 13-1. R = radius, T = tangent distance, C = long chord, L = length of curve, E = external distance, M = middle ordinate, PC = point of curve, PT = point of tangency, Δ = central angle and also the deflection angle between the two tangents. The angle between the long chord and a tangent equals $\frac{1}{2}\Delta$ (from geometry). From Fig. 13-1,

$$T = R \tan \tfrac{1}{2}\Delta \tag{13-1}$$

$$L = \frac{\Delta,\ \text{deg}}{180^\circ} \pi R = 0.0174532925 R\Delta^\circ \tag{13-2}$$

$$C = 2R \sin \tfrac{1}{2}\Delta \tag{13-3}$$

$$E = R \operatorname{ex\,sec} \tfrac{1}{2}\Delta \tag{13-4}$$

$$M = R \operatorname{vers} \tfrac{1}{2}\Delta \tag{13-5}$$

13-4. Degree of Curvature. Instead of the radius, degree of curvature D is often used to designate the sharpness of a curve. It has two definitions which give slightly different results. The first is generally used in highway practice and the second in railroad practice. The highway definition is usually used in city construction.

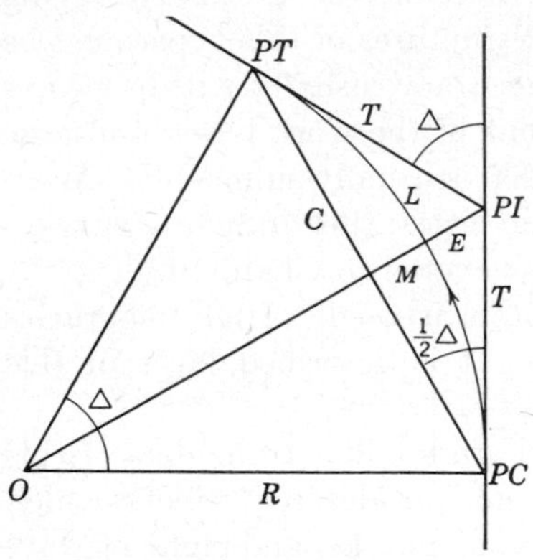

FIG. 13-1

Highway Practice. D equals the central angle in degrees subtended by an arc of 100 ft. This is called the **arc definition.** Then

$$\frac{D}{180^\circ} \pi R = 100$$

$$RD = 5729.578 \tag{13-6}$$

Railroad Practice. D equals the central angle subtended by a chord of 100 ft. This is called the **chord definition.** Then

$$2R \sin \tfrac{1}{2}D = 100$$

$$R = \frac{50}{\sin \frac{1}{2}D} \tag{13-7}$$

13-5. To Stake Out a Curve. Usually the two tangents are first established and run to intersection at the PI. Angle Δ is measured and T is computed by Eq. (13-1). The PC and the PT are then set by measuring T from the PI. If the radius is short, the center can be set and the curve marked by swinging an arc.

13-6. Method of Deflection Angles. When the radius is long, the curve is usually staked out by deflection angles. If the line is stationed, it is common practice to carry the stationing around the curve without interruption.

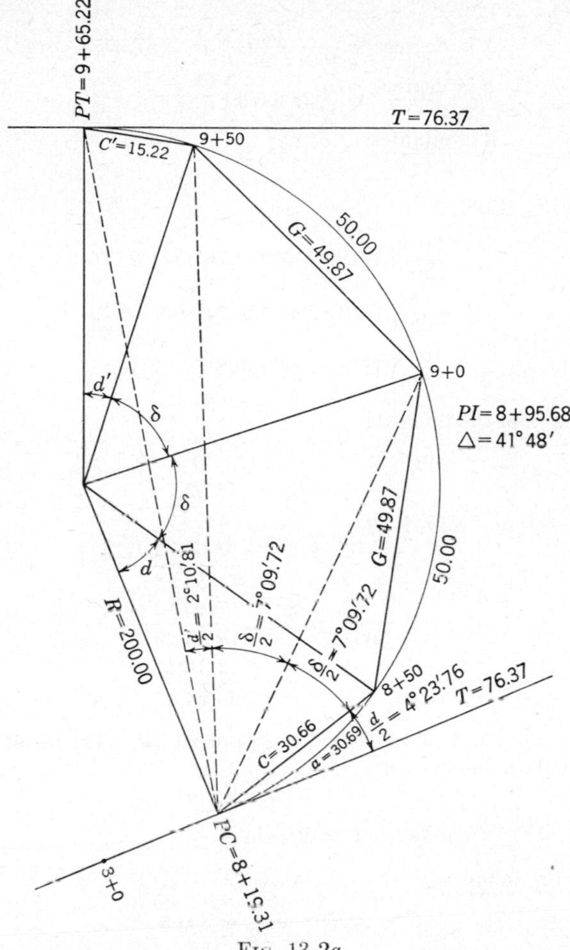

FIG. 13-2a

Figure 13-2a shows a 200-ft-radius curve staked out at the half-station points. The method is as follows:

Set PI

Measure plus of PI = 8 + 95.68

Measure Δ = 41°48′

From Eq. (13-1), T = (200)(0.381863) = 76.37

From Eq. (13-2), L = (0.017453)(200)(41.8) = 145.91

Plus PI =	8 + 95.68
Subtract T =	76.37
Plus PC =	8 + 19.31
Add L =	1 + 45.91
Plus PT =	9 + 65.22

Arc a = 8 + 50 − 8 + 19.31 = 30.69

Arc a' = 9 + 65.22 − 9 + 50 = 15.22

From Eq. (13-2),

$$\Delta \text{ in degrees} = \frac{L}{0.0174533R} = \frac{L}{R}57.2958$$

$$\frac{d \text{ in degrees}}{2} = \frac{a}{R}28.6479$$

$$\frac{d \text{ in minutes}}{2} = \frac{a}{R}1718.87 \qquad (13\text{-}8)$$

Substituting in Eq. (13-8),

$$\frac{d}{2} = \frac{30.69}{200}1718.87 = 263.76' = 4°23.76'$$

$$\frac{\delta}{2} = \frac{50}{200}1718.87 = 429.72' = 7°09.72'$$

$$\frac{d'}{2} = \frac{15.22}{200}1718.87 = 130.81' = 2°10.81'$$

Total deflections from the tangent:

PC		0
	$+ d/2$	4°23.76′
8 + 50		4 23.76
	$+ \delta/2$	7 09.72
9 + 0		11 33.48
	$+ \delta/2$	7 09.72
9 + 50		18 43.20
	$+ d'/2$	2 10.81
PT		20°54.01′

The deflection to the PT should be equal to ½Δ or 20°54′. The result checks within the limits of computational error.

From Eq. (13-3),

$$C = 2R \sin \frac{d}{2} \qquad (13\text{-}9)$$

Substituting in Eq. (13-9),

$$C = 2(200) \sin \frac{d}{2} = 30.66$$

$$G = 2(200) \sin \frac{\delta}{2} = 49.87$$

$$C' = 2(200) \sin \frac{d'}{2} = 15.22$$

The PT and the PC are set by measuring T from the PI. The PC can be set by measuring from the nearest station on the back tangent. The transit is set up at the PC and with the vernier set at zero, a sight is taken on the PI.

The deflection to 8 + 50 (4°24′) is turned off and the distance C (30.66 ft) is laid off on line. The deflection to 9 + 0 (11°33′) is next turned off and 9 + 0 is set by measuring G (49.87 ft) from 8 + 50, etc. Finally the PT set in this manner should check with the PT as set on the tangent.

13-7. Setup on the Curve. When necessary, the transit can be set up at any point on the curve. The instrument is then **oriented to the curve** by sighting any station on the curve. The vernier should be first set at the deflection angle for the station sighted. The telescope should be direct for a forward sight and reversed for a backward sight. When the instrument is then turned to the deflection angle of any other station, the line of sight will be pointing to the position of that station.

For example, if the transit were set up at 9 + 0, it could be oriented to the curve by taking a reversed sight on 8 + 50 with the vernier set at 4°24′, or on the PC with the vernier set at 0°. 9 + 50 could then be set by turning to 18°43′ and the PT by turning to 20°54′. This fortunate feature depends on the geometric principle that the intercept on the circumference between two chords or between a chord and a tangent is measured by half the intercepted arc.

Also, once the transit is oriented to the curve, the tangent at the station where the transit is set up can be established by turning to the deflection angle of the station itself.

13-8. Lines Offset to Curves. A curve parallel to the centerline curve can be staked out with the same deflection angles as the center line by using chord lengths that are increased or decreased in accordance with the increase or decrease of the radius for the offset curve (see Fig. 13-2*b*). The figure shows the chord lengths used for 50-ft offsets both left and right. In this case the center-line radius of 200 ft is increased and decreased by $^{50}\!/_{200} = 0.25$. The center-line chord lengths are therefore increased and decreased by 0.25 times their lengths.

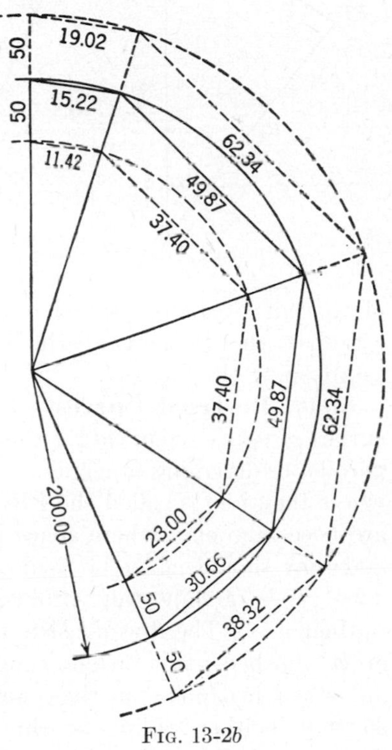

FIG. 13-2*b*

13-9. Compound Curves.

A compound curve consists of two or more simple curves that join at common tangent points and turn in the same direction. Each common tangent point is called a point of compound curve (PCC). At these points the adjacent curves have a common tangent and their radii coincide in direction. The case of a compound curve composed of two simple curves is taken up here. In the discussion, the subscript 1 is used for the curve of smaller radius and subscript 2 for the curve of larger radius. From Fig. 13-3,

$$\Delta = \Delta_1 + \Delta_2 \tag{13-10}$$

$$t_1 = R_1 \tan \tfrac{1}{2}\Delta_1 \tag{13-11}$$

$$t_2 = R_2 \tan \tfrac{1}{2}\Delta_2 \tag{13-12}$$

$$\frac{VG}{\sin \Delta_2} = \frac{VH}{\sin \Delta_1} = \frac{t_1 + t_2}{\sin \Delta} \tag{13-13}$$

$$T_1 = VG + t_1 \tag{13-14}$$

$$T_2 = VH + t_2 \tag{13-15}$$

Usually Δ is measured in the field or known from the plans and certain values have been established for R_1 and R_2. One other value is necessary to establish the curve. Either Δ_1 or Δ_2 can be scaled from the plans. The equations are then solved in the order given to find Δ_1 or Δ_2, T_1 and T_2.

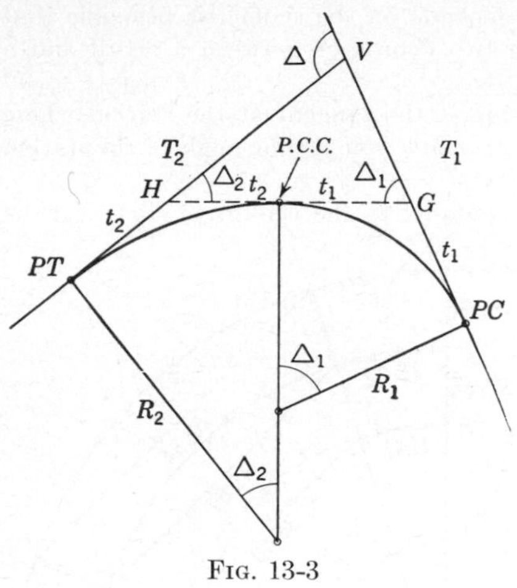

FIG. 13-3

Sometimes one of the T values must be held fixed. The most convenient method is then to scale Δ_1 or Δ_2 and hold fixed only the adjacent R. For example: Given Δ, Δ_2, T_1, R_1; to find Δ_1, T_2, and R_2:

Solve Eq. (13-10) for Δ_1
Eq. (13-11) for t_1
Eq. (13-14) for VG
Eq. (13-13) for VH and t_2
Eq. (13-15) for T_2
Eq. (13-12) for R_2

Field Procedure. When the values in the previous paragraphs have been computed, the two parts of the curve can be staked out like simple curves. It is necessary to establish the tangent at the PCC, either by deflection angles or by establishing the line *HG*.

13-10. Reversed Curves. A reversed curve consists of two simple curves of opposite curvature that join at a common tangent and connect parallel or diverging tangents. The point where they join is called the PRC. They are necessary only where space is limited and they should never be used elsewhere. Two types of important problems occur in their use. The first consists of joining an established point on one tangent with an established point on the other tangent so that neither radius is shorter than necessary. This purpose is attained only when the two radii are equal. The second type of problem consists of using the minimum allowable common radius to join two established tangents.

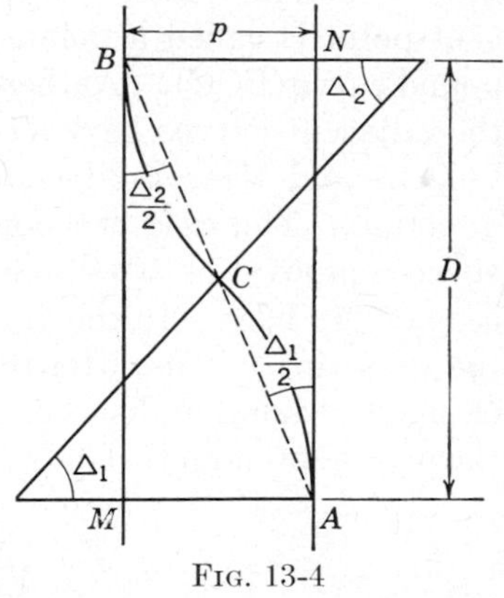

FIG. 13-4

13-11. Parallel Tangents (Case I). Given p and D (Fig. 13-4), to find the common R and Δ. From Fig. 13-4,

$$\Delta = \Delta_1 = \Delta_2$$

$$\text{Angle } CAN = \tfrac{1}{2}\Delta_1 = \text{angle } CBM = \tfrac{1}{2}\Delta_2$$

Therefore the line ACB is straight.

$$\tan\frac{\Delta}{2} = \frac{p}{D} \qquad (13\text{-}16)$$

$$p = R \text{ vers } \Delta_1 + R \text{ vers } \Delta_2$$

$$R = \frac{p}{2 \text{ vers } \Delta} \qquad (13\text{-}17)$$

13-12. Parallel Tangents (Case II). Given p and R (Fig. 13-4), to find Δ and D. From Eq. (13-17),

$$\text{vers } \Delta = \frac{p}{2R}$$

From Eq. (13-16), $\quad D = p \cot\frac{\Delta}{2}$

13-13. Diverging Tangents (Case I). Given length AB and angles α and β (Fig. 13-5), to find Δ_1, Δ_2, and the common R. From Fig. 13-5,

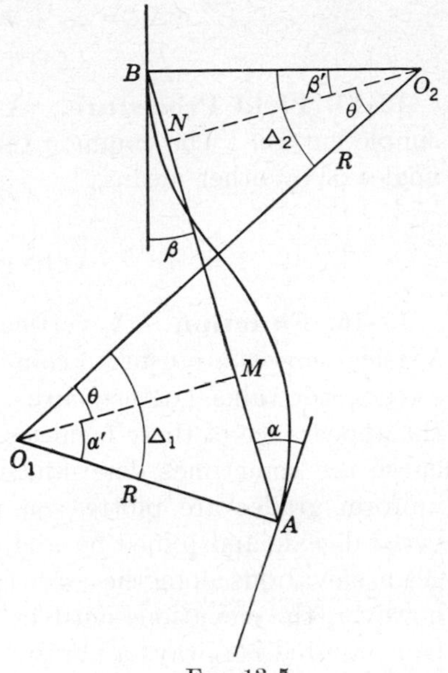

FIG. 13-5

$$\alpha' = \alpha \qquad \beta' = \beta$$

$$\cos\theta = \frac{O_1M + NO_2}{O_1O_2}$$

$$\cos\theta = \frac{R\cos\alpha + R\cos\beta}{2R}$$

$$\cos\theta = \tfrac{1}{2}(\cos\alpha + \cos\beta)$$

$$\Delta_1 = \alpha + \theta \qquad \Delta_2 = \beta + \theta$$

$$AB = R\sin\alpha + 2R\sin\theta + R\sin\beta \qquad (13\text{-}18)$$

$$R = \frac{AB}{\sin\alpha + 2\sin\theta + \sin\beta} \qquad (13\text{-}19)$$

13-14. Diverging Tangents (Case II). Given θ, T_1, and the common R (Fig. 13-6), to find Δ_1, Δ_2, and T_2. From Fig. 13-6,

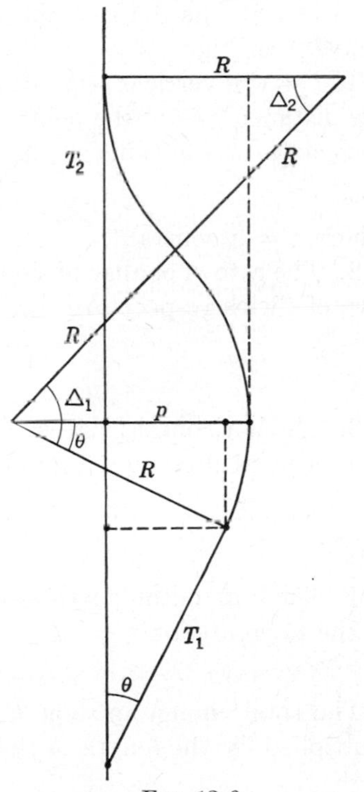

FIG. 13-6

$$p = T_1 \sin\theta + R \text{ vers } \theta$$

$$\text{vers } \Delta_2 = \frac{p}{2R} = \frac{T_1\sin\theta + R \text{ vers } \theta}{2R}$$

$$\text{vers } \Delta_2 = \tfrac{1}{2}\left(\frac{T_1}{R}\sin\theta + \text{vers } \theta\right) \qquad (13\text{-}20)$$

$$\Delta_1 = \Delta_2 + \theta \tag{13-21}$$

$$T_2 = T_1 \cos\theta + R\sin\theta + 2R\sin\Delta_2 \tag{13-22}$$

13-15. Field Procedure. A reversed curve is staked out like two simple curves. The common tangent must be established by deflection angles or by other means.

VERTICAL CURVES

13-16. Function. A vertical curve is a changing rate of grade. Vertical curves are used to connect uniform grades of different slopes in streets, sidewalks, gutters, curbs, etc., to improve the riding qualities and the appearance of these facilities. The grade elevations along a vertical curve are sometimes determined graphically. The profiles of the two uniform grades are plotted on profile paper at a greatly exaggerated vertical scale and joined by a circular curve of the desired length. The grade elevations along the curve can then be read off the profile. Usually, however, the elevations must be computed. A vertical curve that is to be computed is always a parabola whose axis is vertical.

13-17. The Properties of a Parabola. Certain properties of a parabola are used to compute the grade elevations. These are given below. In Fig. 13-7,

1. t = the vertical separation between any tangent and a parabola at the distance x from the point of tangency. It is often called the tangent correction or the tangent offset. In any parabola

$$t = ax^2 \tag{13-23}$$

where a = a constant.

2. The rate of change of slope of the parabola is then the second derivative of t with respect to x, or,

$$\text{Rate} = \frac{dt^2}{d^2x} = 2a \tag{13-24}$$

3. The horizontal lengths (T_1 and T_2) of tangents to a parabola drawn from a point are equal. From Eq. (13-23),

$$t = aT_1^2 = aT_2^2$$

$$T_1 = T_2 \tag{13-25}$$

4. The horizontal length of the parabola L is equal to twice that of one of the tangents, or

$$L = 2T \tag{13-26}$$

The total change in slope R is equal to the rate of change of slope $2a$ multiplied by the length of the parabola L. Thus

$$R = L(2a) \tag{13-27}$$

13-18. Choosing the Parabola. The value of $2a$—the rate of change of slope—determines the sharpness of the parabola, just as the radius determines the sharpness of a circular curve. A minimum value is often chosen for sags and another for summits. To simplify computation, horizontal lengths should be measured in larger units than feet. The most convenient unit is the minimum distance required between grade stakes. Such a unit can be called a **stage** and its length denoted by l.

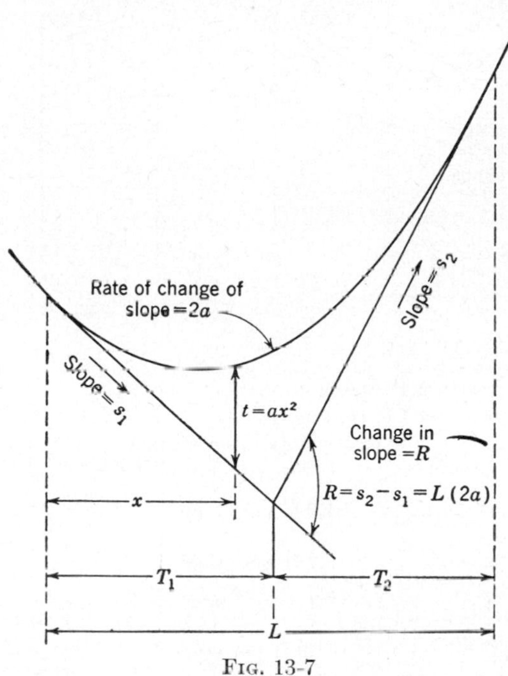

FIG. 13-7

Slopes can then be expressed in feet per stage and rates of change of slope in feet per stage per stage.

When $2a$ is stated in per cent per 100-ft station, it can be converted to feet per stage per stage by multiplying by $(l/100)^2$. Thus

$$2a_l = \left(\frac{l}{100}\right)^2 2a_{100} \tag{13-28}$$

13-19. To Compute a Vertical Curve. The steps required to compute a vertical curve are given below (see Fig. 13-8).

1. From the plans, take off

Plus PVI = 22 + 31.47 Elev. PVI = 83.62
Rate of grade first tangent, g_1 = +2.4%
Rate of grade second tangent, g_2 = −1.4%
Desired rate of change of slope = 1.44% per station
Length of stage = 50 ft

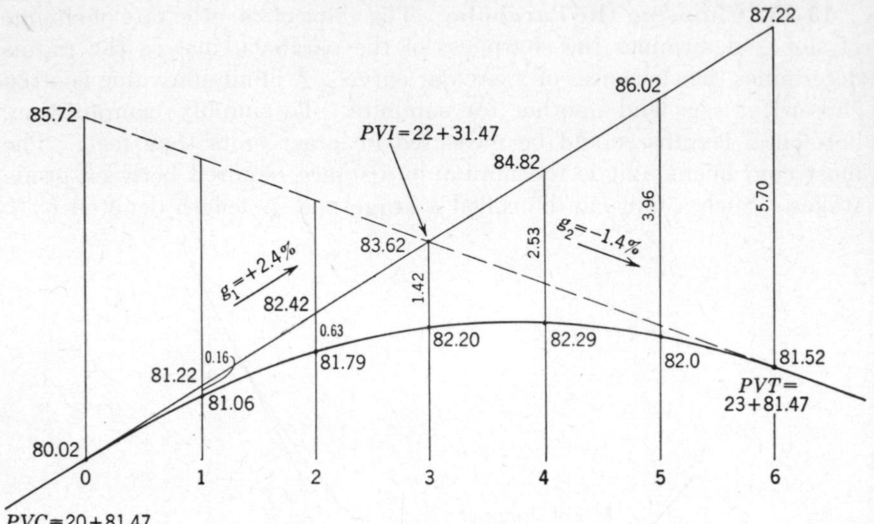

Fig. 13-8

2. Note:

$$
\begin{aligned}
g_1 &= +2.4 \text{ ft per station} \\
g_2 &= -1.4 \text{ ft per station} \\
2a_{100} &= -1.44 \text{ ft per station per station} \\
l &= 50 \text{ ft} \\
s_1 &= g_1 \frac{l}{100} = +1.2 \text{ ft per stage} \\
s_2 &= g_2 \frac{l}{100} = -0.7 \text{ ft per stage} \\
2a_{50} &= 2a_{100} \left(\frac{l}{100}\right)^2 = -0.36 \text{ ft per stage per stage} \\
R &= s_2 - s_1 = -1.9 \text{ ft per stage}
\end{aligned}
$$

3. From Eq. (13-27),

$$L = \frac{R}{2a} = \frac{-0.36}{-1.9} = 5.3 \text{ stages}$$

4. Choose the next larger full number of stages, i.e., six stages. From Eq. (13-25) there will be three stages on each side of the PI. Compute the value of $2a$ to be used. From Eq. (13-27),

$$
\begin{aligned}
2a &= \frac{R}{L} = \frac{-1.9}{6} = -0.31667 \\
a &= -0.15833
\end{aligned}
$$

5. Compute the elevations of the PVC and the PVT:

	Elev.		*Elev.*
PVI =	83.62	PVI =	83.62
$(-3)(1.2)$ =	$-$ 3.60	$(3)(-0.7)$ =	$-$ 2.10
PVC =	80.02	PVT =	81.52

6. Compute the plus of the PVC and the PVT:

	Plus		*Plus*
PVI	22 + 31.47	PVI	22 + 31.47
−(3)(50) −	1 + 50	+(3)(50)	1 + 50
PVC	20 + 81.47	PVT	23 + 81.47

7. For each stage point, compute the tangent correction and apply it to the tangent elevation, as shown in Table 13-1. From Eq. (13-23), $t = ax^2$.

TABLE 13-1

Stage points x	x^2	Tangent cor. ax^2	Tangent grade elev.	Curve grade elev.
0	0	0	80.02	80.02
1	1	−0.16	81.22	81.06
2	4	−0.63	82.42	81.79
3	9	−1.42	83.62	82.20
4	16	−2.53	84.82	82.29
5	25	−3.96	86.02	82.06
6	36	−5.70	87.22	81.52

The curve grade elevations can be computed from the second tangent as a check.

Note: Sometimes the length L of the curve is specified and no reference made to a minimum value for $2a$. In that case step 3 is omitted.

Some engineers prefer to find the value of a from the tangent correction at the PVT or the PVC. In the example, this would be computed as follows. From Eq. (13-23),

$$a = \frac{t}{x^2}$$

$$a = \frac{87.22 - 81.52}{36} = 0.15833 \qquad a = \frac{86.72 - 80.02}{36} = 0.15833$$

13-20. Street Pavement Crowns. Street crowns are often laid out on a parabolic curve. Figure 13-9 is useful in computing them.

13-21. Sight Distances. On long vertical curves over summits, the sight distance S may be the controlling element (see Fig. 13-10). The height of eye h_1 and the height of object to be observed h_2 are used with S to find $2a$. From Eq. (13-23),

$$h_1 = ap^2 \qquad h_2 = aq^2$$

$$S = \sqrt{\frac{h_1}{a}} + \sqrt{\frac{h_2}{a}}$$

$$a = \frac{h_1 + 2\sqrt{h_1h_2} + h_2}{S^2} \tag{13-29}$$

h_1 is usually taken as 4.5 ft and h_2 as 0.333 ft, so that

$$h_1 + 2\sqrt{h_1h_2} + h_2 = 7.283 \text{ ft}$$

and

$$aS^2 = 7.283 \text{ ft}$$

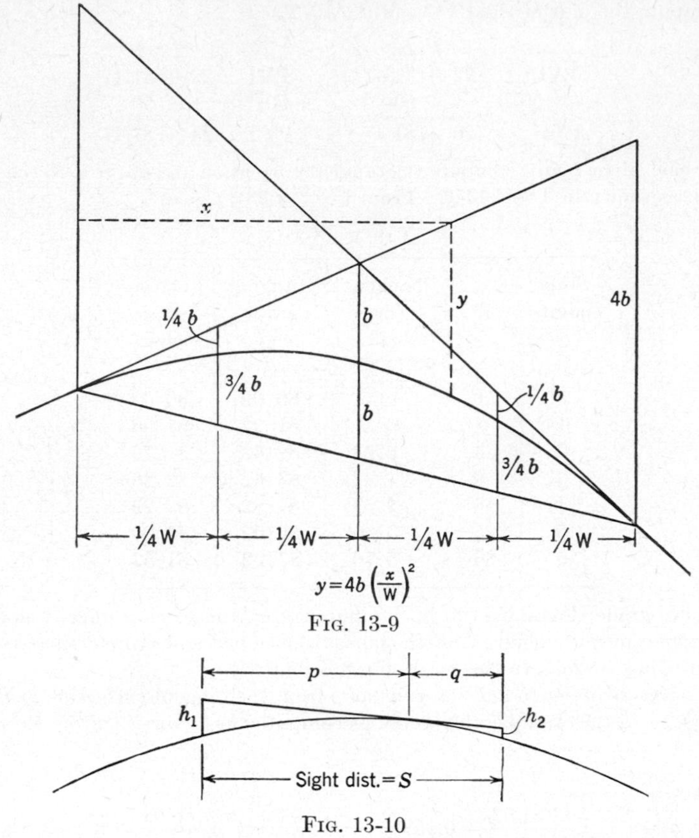

FIG. 13-9

FIG. 13-10

EARTHWORK QUANTITIES

When more earth must be moved to build the street than is necessary merely to shape the subgrade, the earth quantities for each cut and fill must be measured and computed.

13-22. Cross Sectioning. After the center line is staked out, short profiles are taken at right angles to the center line at uniform intervals, usually at every station and half station. Additional profiles are taken wherever there is a break in the grade of the ground along the center line. This process is called **cross sectioning.** Figure 13-11 shows two cross sections. Elevations are taken at the center line D, at uniform offsets on each side of the center line, A and H, and at any breaks, C and E. Figure 13-12 shows the form of notes for this work. The denominators of the fractions shown are the offsets, the position on the page gives the direction left or right, and the numerators are the rod readings. The elevations are written just above the rod readings.

13-23. Establishing the Grade. Profiles of the center line and often of the two offset lines are plotted, and a grade line is established with due consideration of the requirements of the street and of the cut and fill involved. The grade line is made up of uniform slopes of not less than 0.5 per cent (to provide drainage) connected by vertical curves. Grade

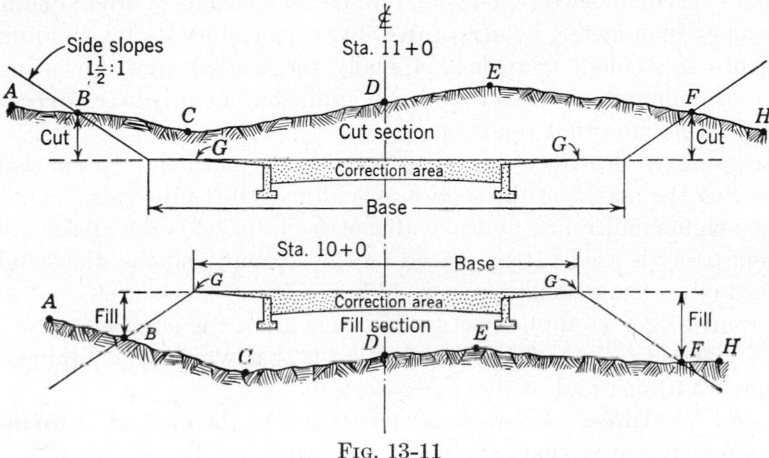

FIG. 13-11

elevations are computed for it. These refer to a certain point on the **typical cross sections,** or **templates,** and are called the **template grade.** A typical cross section shows the arrangement of the pavement, curbs, gutters, sidewalk areas, and other appurtenances and the **side slopes** of the cuts and fills. Without the ground profiles, Fig. 13-11

Sta	+	HI	−	Rod	Elev.	L			R	
B.M.	4.37	56.65			52.28					
TP #1	2.18	46.18	8.29		48.36					
10+0						45.5 0.7/50	40.0 6.2/25	41.6 4.6/0	42.4 3.8/13	41.0 5.2/50
	11.37	56.33	1.22		44.96	(A)	(C)	(D)	(E)	(H)
11+0						51.3 5.0/60	49.4 6.9/31	52.5 3.8/0	54.1 2.2/16	59.6 6.7/60

FIG. 13-12

would show typical cross sections. Usually the highest point in the typical section is chosen for the **template grade.** These are the points G in the figure.

13-24. Cross Sections. Each cross section is plotted on cross-section or profile paper. A convenient elevation is chosen for a heavy line on the paper where the section is to be plotted. The ground profile is plotted according to the field notes and the typical section is placed according to the template grade. Usually the horizontal scale adopted is 10 ft per in. and the vertical scale 5 or 10 ft per in.

13-25. Areas. The area of each section must be determined. This is usually accomplished by one of a number of methods of direct computation, or by planimeter, by **stripping,** by use of tables, or by dividing the area into convenient triangles. Usually the shaded area shown in Fig. 13-11 is computed once and for all and applied as a **template correction** to the simple area that remains.

Stripping is probably the best all-around method. It consists of measuring the height of the section, usually at 10-ft intervals. The sum of the heights multiplied by the width of the interval is the area. A long strip of paper is used. It is placed on the section, and the first height is marked off. The second height is added to it and marked off, and so on. The total height is applied to a scale that gives the square feet of area directly and also the number of cubic yards that would be contained in a volume 50 ft long and of that cross section.

13-26. Volumes. Volumes are computed by the **end-area formula.** This gives the area that would be generated by the section if it were moved from a point halfway from the previous section to a point halfway toward the next section. Accordingly, the area of each section is multiplied by this distance and divided by 27 to find the number of cubic yards.

13-27. Slope Stakes. If the quantities thus determined are found to give the greatest economy consistent with the requirements, the grade and location are accepted and **slope stakes** are set to guide the contractor. Often, more frequent and accurate sections are measured at this time. Slope stakes are placed at the points in the field that correspond to points *B* and *F* in Fig. 13-11 (see also Fig. 13-13). Each is marked with its station and offset, and the depth of the cut or the height of the fill from the ground at the stake, up or down to the template grade. Equivalent cut or fill values are placed on the center-line stakes.

13-28. To Set a Slope Stake. 1. Compute the grade rod for the station in question.

$$\text{Grade rod} = \text{H.I.} - \text{template grade} \tag{13-30}$$

2. Have the rodman hold the rod at the offset where it is estimated the slope stake should be set. Read the rod and compute the cut at this point.

$$\text{Cut} = \text{grade rod} - \text{rod}$$

(A minus value indicates a fill.)

3. Compute the offset on the typical section that would have this cut (or fill).

$$\text{Offset} = \tfrac{1}{2}\,\text{base} + \text{cut (or fill)} \times \text{slope}$$

4. If the offset computed does not agree with the offset of the stake, find the correct position by trial.

Some engineers prefer to scale the various offsets to the grade stakes from the cross sections before going into the field. These values are used as the first estimate.

13-29. Passing from Cut to Fill. When accurate quantities are desired, extra sections are taken at the boundaries between cuts and fills as shown in Fig. 13-13. The width of the base for cuts is usually greater than for fills to provide for drainage. There are, therefore, five lines to consider: the center line, the two edges of the fill base, and the two edges

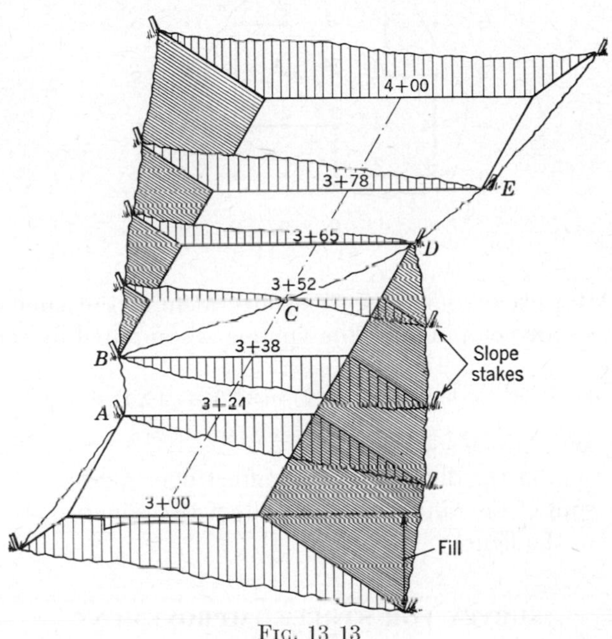

FIG. 13 13

of the cut base. The points where these lines are at ground level are found by trial in the field. The rod target is set at the value of the grade rod and the rod is moved along each of the lines until the target is at the H.I. If the rate of grade is steep, the target will have to be adjusted accordingly. These points are the following:

A	left edge of fill base
B	left edge of cut base
C	center line
D	right edge of fill base
E	right edge of cut base

These points are marked, the plus of each is recorded, a section is taken as shown, and the other slope stake is set. Often, sufficient accuracy is obtained when the points at A and D and their sections—the edges of the fill base—are omitted.

13-30. Borrow Pits. When extra earth is required, it is taken from selected areas called **borrow pits.** The quantity removed must be measured for payment. The area is laid out in squares 25 to 50 ft on a side as shown in Fig. 13-14, and elevations are taken at the corners of the

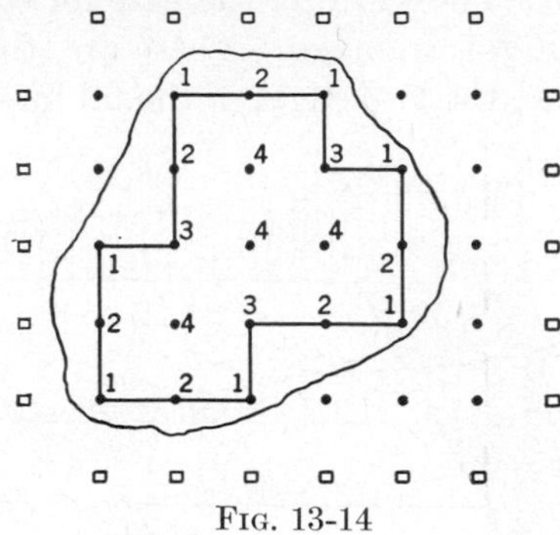

FIG. 13-14

squares. After excavation, elevations are taken at the same points and the differences are computed. The volume is computed by the following formula:

$$V = \tfrac{1}{4}(\Sigma 1 + 2\Sigma 2 + 3\Sigma 3 + 4\Sigma 4)A \qquad (13\text{-}31)$$

where A = area of one square
$\Sigma 1$ = sum of the differences that affect one square
$\Sigma 2$ = sum of the differences that affect two squares, etc., as shown in the figure

SURVEY FOR STREET IMPROVEMENT

13-31. Most street improvements consist of realigning the curbs and rebuilding the pavement. The preliminary survey map must show the existing locations of the boundary lines of the street (which are often the building lines), the curb lines, entrances to the street, details of street intersections, and all features near the curb or in the roadway that might be affected by changes in the curb location or in the curb and pavement grades.

13-32. Procedure for Horizontal Measurement. Figure 13-15 illustrates the minimum requirements for a plan for a street improvement. The scales used range from 1 in. = 30 ft to 1 in. = 60 ft.

Station 0 is set at the midpoint between the curbs and on the line AB by extending a tape from A to B. It should be permanently marked by a roofing nail, paint, or a chisel cut. A keel mark is placed at every station and half station as shown. Measurements are made with a steel tape; alignment is by eye by observing the curbs.

Where the direction changes, carry the measurement around the curved path of the center line as judged by eye. Set *F* and *G* halfway between the curbs by measurement. Measure the plus of *F* and of *G*. Carry the measurement forward and set *H* also halfway between the curbs by measurement. Mark each with a permanent mark. Set the *PI* by eye by sighting along *OF* and *GH*. Mark the *PI* permanently. Measure from station 3 to *PI* and from *PI* to station 4.

The angle at *PI* must be measured. According to the best practice, the *forward clockwise angle* from the backward course to the forward course should be measured as shown. Many engineers prefer to measure the deflection angle.

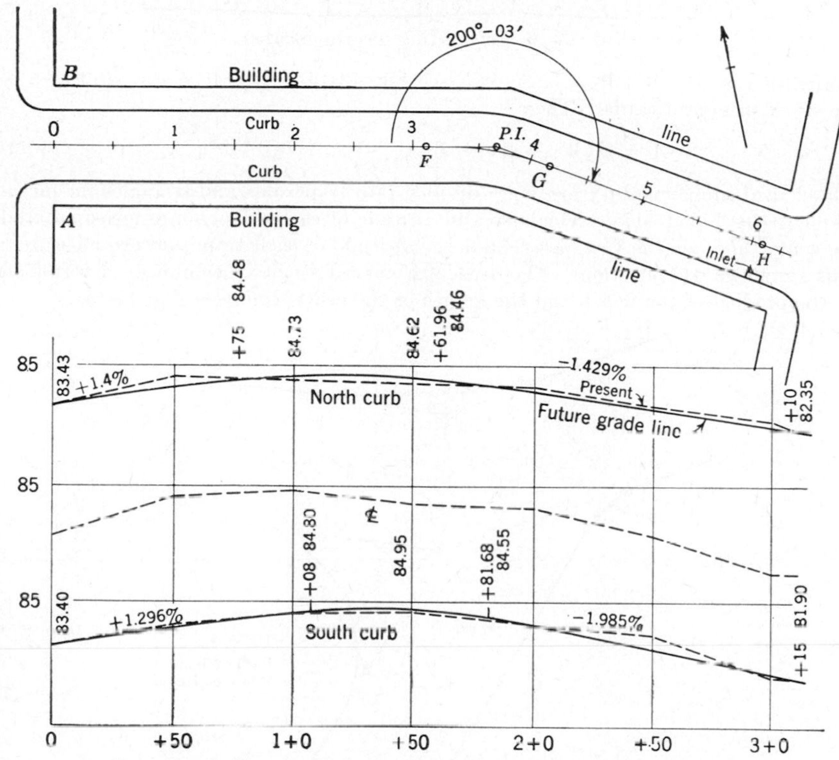

FIG. 13-15. Sketch of plan and profile for street improvement.

When the angle at the *PI* has been measured, the relative positions of the points *O*, *F*, *G*, and *H* are completely determined, and the length of the center line of the street is known. The necessary features can be located by polar coordinates from the points named, or by plus and offset.

When the traffic is heavy, it is frequently necessary to run a supplementary traverse along each side of the street. These are usually placed on the line of the face of each curb. They should be tied to the main traverse at frequent intervals.

13-33. Estimating Quantities for City Streets. The chief cost of street improvements are the installation of new pavement and new curbs, and resetting old curbs. The square yards of pavement surface must therefore be determined accurately. If excavation or fill is required, the quantities are determined as described in Secs. 13-22 to 13-29.

13-34. Determining Pavement Areas. The width of the roadway from face-of-curb to face-of-curb is measured at each 50-ft point. Let these measurements be

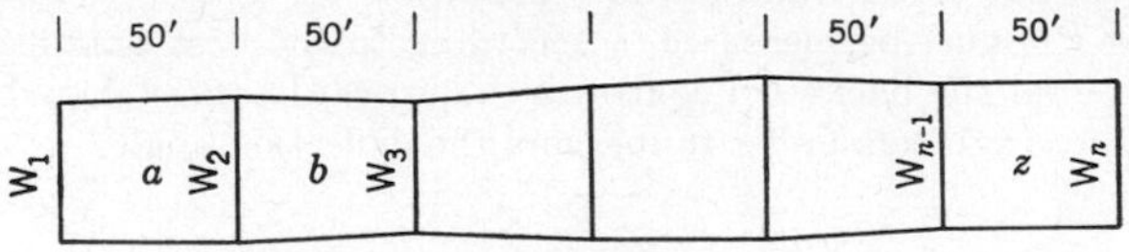

FIG. 13-16. Estimating pavement area.

represented by $W_1, W_2, W_3, \ldots, W_n$ (see Fig. 13-16). Let A be the total area of the street in square yards. Then

$$A = (\tfrac{1}{2}W_1 + W_2 + W_3 + \cdots + \tfrac{1}{2}W_n)\tfrac{50}{9} \tag{13-32}$$

Odd-shaped supplementary areas are divided into trapezoids and triangles for measurement (see Fig. 13-17). The base and altitude of each triangle are measured and the center line—or the two bases—and the altitude of each trapezoid are measured. This simplifies computation. The area of a curved street is accurately determined by the product of the width and the length of the center line (see Fig. 13-18).

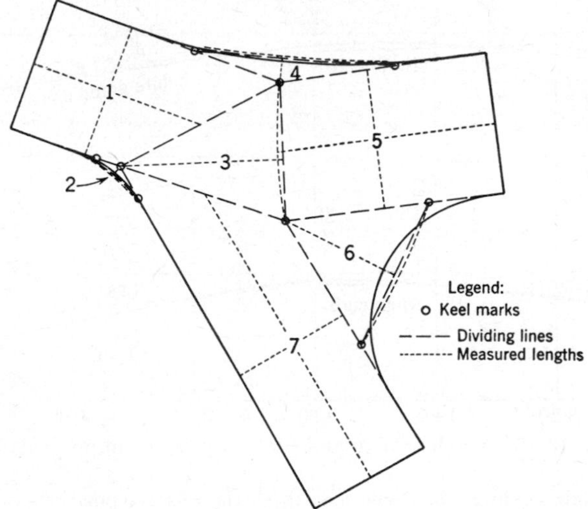

FIG. 13-17. Typical treatment of odd-shaped area. The circles represent keel marks which are lined by eye. The various measurements are shown by dotted lines. Note that that side of the triangles 2, 4, and 6 which crosses the curved curb is placed by estimation so that the included and excluded areas balance.

13-35. Using the Tape for Summing Lengths. To determine the length of new curbing required, each curbstone must first be examined and accepted or rejected. Then the total footage that must be replaced is measured. This requires the addition of many short measurements. It is best accomplished as follows. The rear tapeman holds the zero end of the tape at the beginning of a rejected stone. The head tapeman unreels enough tape to reach the other end of the stone and takes hold of the tape with his left hand at that point. The rear tapeman releases the zero end, comes forward, and takes hold of the tape with his right hand touching the left hand of the head tapeman. The head tapeman releases his left hand and they move forward to the next rejected stone, letting the zero end of the tape drag behind them. When the next rejected stone is reached, the rear tapeman places the point on the tape held in his right hand at the beginning of the stone, and the process is repeated. They keep a tally of the number of tape lengths. This process provides a quick method of handling the tape and gives the required sum without addition.

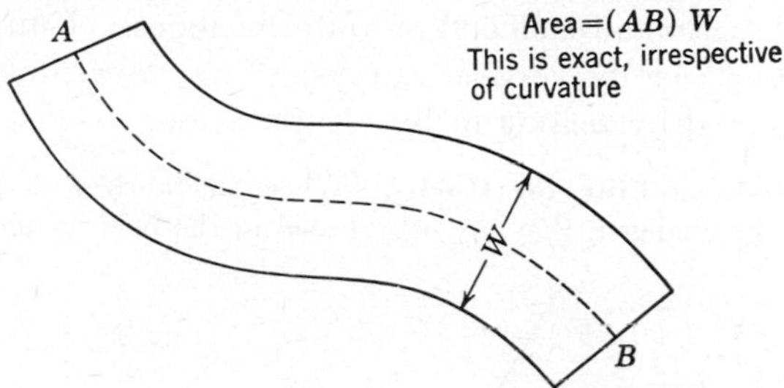

FIG. 13-18. Area of curved streets.

Usually other items must be measured for the estimate, but one of the two principles outlined can be applied.

13-36. Profiles Required for a Street. In order to establish the best grade for the finished street it is usually necessary to measure three to seven profiles as follows (see Fig. 13-19): the center line, the top of each curb, both gutters, and a line along each sidewalk about 18 to 24 in. from the face of the curb. These sidewalk lines are placed at the limit of any changes in sidewalk grade necessitated by resetting the curb. Except on very steep slopes, the rodman can estimate the proper position along the

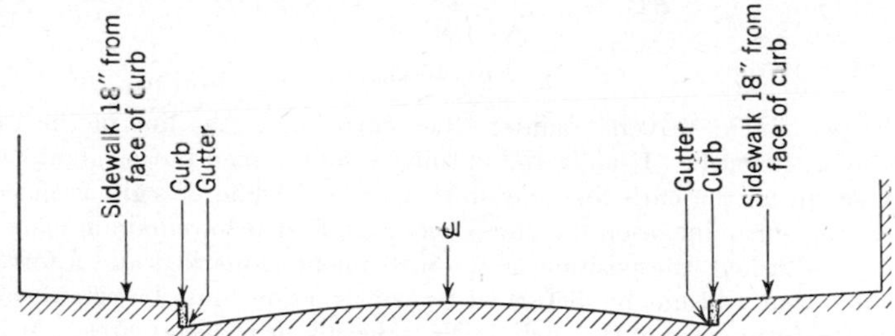

FIG. 13-19. Seven profiles usually determined for street improvement.

line to hold the rod from the keel station marks in the center of the street. The profiles are usually plotted directly above or below the street plan (see dashed lines in Fig. 13-15). Only three out of the seven profiles are shown in the figure. Note that three datum lines are used. The gutter and sidewalk lines would be plotted on the same datum as their respective curbs. The horizontal scale is usually the same as that for the plan. The vertical scale is ten to twenty times as large.

13-37. Plans for Street Improvement. On the map are indicated any changes to be made in the curb locations. Full lines are placed on the profiles to indicate the future curb and gutter profiles. They are placed where they appear nearly to average the existing grades. Frequently they must meet the grade of an existing curb which is not to be changed. Often they are restricted by the existing sidewalk elevation.

The exact positions of these projected profiles are defined by noting the plus and elevation at each end of uniform slopes. Uniform slopes are frequently connected by vertical curves. The elevations of each 50-ft point is determined by scaling or by computation.

13-38. Establishing Line for Curbs. When the horizontal positions of the curb lines are to be changed, they are established in the field by measurement from

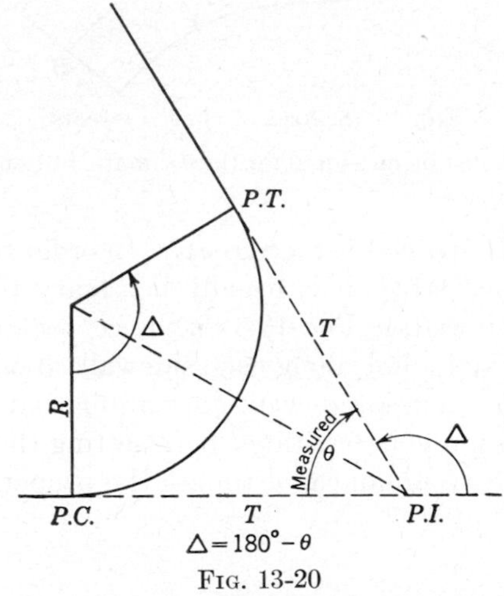

FIG. 13-20

any of the permanent traverse points. The curb line is the line of the top of the outside face of the curb. Usually the old line is maintained but straightened. The transit is set up on the curb line and shifted until the line of sight is placed on an average existing curb line or on the line of any curb that is to remain in place. When this line is established, the position of the instrument is marked and a foresight set. An offset line that will not be disturbed by construction must be established, 18 to 36 in. from the future face of the curb, in the sidewalk or parkway area. It is usually impossible to use the transit on such a line because of shade trees, lamp posts, or other obstructions. The pluses are measured along the face of the curb. At each 50-ft point an offset mark is set by measuring from the line of sight. The transitman lines in the end of a 6-ft rule and a mark is placed at some 6-in. graduation. Cross marks are then chiseled in the sidewalk where stakes and tacks cannot be used.

13-39. Lines for Curves of Short Radii. The curb lines of most city streets are straight lines connected by circular curves of very short radius at street intersections (see Fig. 13-20). To give line for short radii curves, proceed as follows. By eye, mark the *PI* of the two straight curb lines to be joined. With the transit, measure the angle between these lines. Subtract this angle from 180° to find the angle subtended by the curve, that is, Δ.

Compute T by Eq. (13-1) and set the PC and the PT. Find the center of the curve by swinging intersecting arcs from the PC and the PT or by intersecting perpendiculars from these points. Mark the offset points by measuring several radii from the center.

13-40. Setting Grades for Curbs. Curb grades are established by any method desired, but usually the elevations of the line marks are determined and the "cut or fill" is computed at each mark.

13-41. Marking Gutter Grades. When the actual curbs have been set, the gutter grades are marked with keel on the curbs at each 50-ft point. These are established by any method desired. Wherever the grades are on a uniform slope, the method of "shooting in" grades will save time.

13-42. Grades for Roadway Pavements. For a number of reasons, the cross section of a street pavement is usually made in the shape of a parabola, whose principle axis is vertical.

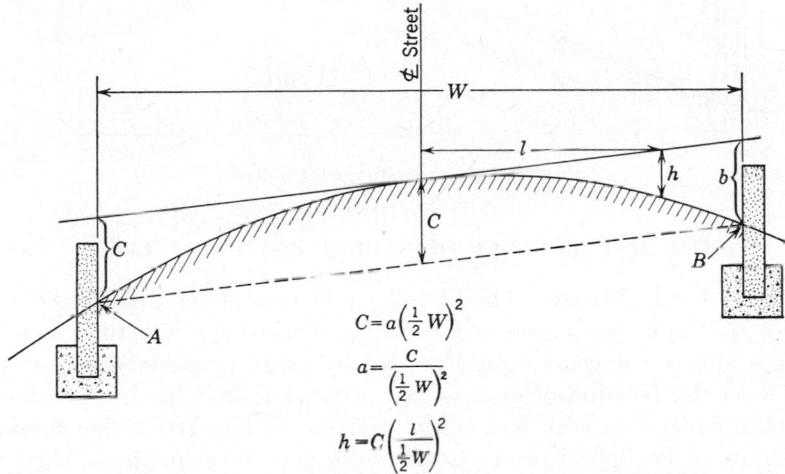

FIG. 13-21. Elevation of finished surface of street at any point between curbs.

Figure 13-21 represents the cross section of a street. The height of the **crown** C is based on the width of the roadway and the type of surface used. C can be reduced when the surface material drains well. It is often taken as $\frac{1}{30}W$.

When the curbs have been set, the gutter grades at A and B are marked with keel on the curbs at each 50-ft point, as previously described. A row of uniformly spaced steel pins are driven into the subgrade extending across the street at each 50-ft point. They are usually spaced at convenient fractions of W. Two 1- × 2-in. wooden blocks are cut to a length equal to the crown and held against the curbs with their lower ends on the gutter grade marks (Fig. 13-22). A string is tightly stretched between their tops. Keel marks are placed on the pins at the height of the string. The h (tangent offset) distances can be computed by the formula shown in Fig. 13-21, and marks are placed on the pins for the finished crown. Frequently, tables are made up for h distances for various crowns for various fractions of W from the center line. The grades for manholes and other features in the street are set in this way. The pins are removed as the paving progresses. Sometimes wooden templates are made to be used instead of the pins.

13-43. Surveys for New Streets. When new streets are planned, more engineering skill is required, but the survey details are simplified.

Usually the street intersections are monumented and further alignment is based on the monuments. Profiles are run on the center line and on the side lines of the right-of-way. Thereafter the procedure is much the same as for street improvement.

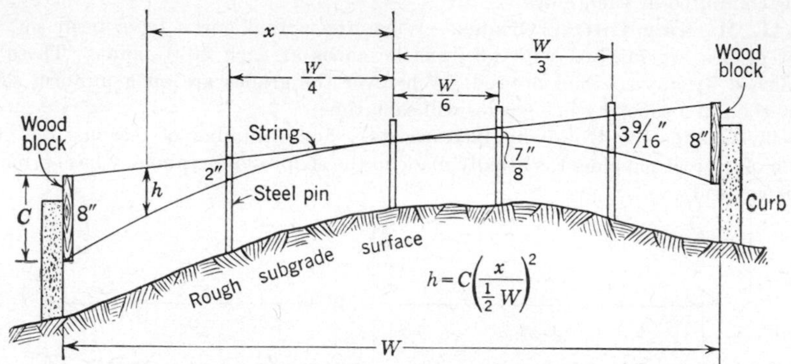

FIG. 13-22. Grade stakes for a street.

THE MAINTENANCE OF STREET BOUNDARY LINES

13-44. Importance. The physical positions of the lines that bound street rights of way are of signal importance to the abutting property owners and to the general public. Nearly every property description is based on the position of at least one street line and the intersection of another street line with it. If the position of any street line becomes indefinite or is allowed to move by even a very small distance, the positions of many nearby parcels can no longer be precisely determined. This creates confusion, improper location of buildings, lawsuits, and serious damage to title. Moreover, city ordinances regulating structures like porches and steps that are built within the rights-of-way cannot be properly enforced nor can zoning laws that regulate the positions of buildings near property lines be properly followed. It is difficult to plan accurately land acquisition or construction for street and other municipal improvements.

It follows that street lines should be monumented at the earliest possible moment and the established monuments should be carefully protected and maintained.

Monuments are usually placed at the intersections of street boundaries or at the intersections of lines within the rights of way a few feet offset from the boundaries. Sometimes they are placed at the intersections of the street center lines. Each system has its advantages and disadvantages.

Monuments placed at the intersection of the street boundaries are the least likely to be disturbed, but they cannot be used when the building line coincides with the street line and in general they seldom can be used directly. Monuments on offset lines are the easiest to use but they can be mistaken for corner monuments. Monuments on the center lines provide the simplest system but they are difficult to use where traffic is heavy.

Every city should have ordinances that cover the following points:

1. Subdivision plans should not be approved if they are not properly monumented.

2. The maintenance of streets should not be taken over unless they are properly monumented.

3. Fines should be levied for the disturbance of monuments.

4. When the lines of a street are lost the monuments should be reset under the direction of a land surveyor who is very familiar with the property lines in the vicinity.

5. The position of monuments should be carefully witnessed by short measurements to marks on durable structures when possible, and by measurements to adjacent monuments.

It will often be found that, in the course of time, the street lines that surround a block have become established by use in such a way that they are not quite parallel with the lines on the opposite sides of the streets. This condition is not serious and there is seldom any justification for correcting it.

CONTROL SURVEYS

13-45. Vertical Control. So much construction carried out by the city and so much private construction depends on the official grades for streets, sewers, and other facilities that it is essential for every city to establish and maintain a system of accurate benchmarks. These can be set on permanent buildings, fire hydrants, firmly established monuments, etc. They should be accurately surveyed and the net adjusted. A graphical index should be maintained of these benchmarks by plotting their positions on a city map together with a reference number. A card index should be maintained by reference number that gives the description and the official elevation of each benchmark.

13-46. Horizontal Control. Every city should have a tax map, a map of underground facilities, and a topographic map. Without these, real estate taxes cannot be properly levied, underground facilities are difficult to design and to maintain, and city improvements and expansion cannot be properly planned. Many cities have thought it wise to establish an over-all horizontal control system to serve these surveys and to hold the positions of street lines.

Usually a triangulation system is established on the roofs of the buildings. The positions of the triangulation stations are brought down to street level by small local triangles, and traverses are run from these points over the street monuments. Usually auxiliary stations are required to complete these traverses. Nearly all auxiliary stations should be monumented, as it will be found that they will be of great value in the future.

Engineers differ on the accuracy that should be required. It can be argued that, since city property is so valuable, great accuracy is necessary. On the other hand, if each block is considered as a separate unit—as previously described—traverses of second-order accuracy will be satisfactory. In this case a first-order triangulation net will give adequate control.

13-47. Coordinate Systems. If possible, the coordinates should be based on the state coordinate system so that other surveys can be easily tied to the city net. Some engineers believe that if the scale correction or the sea-level correction is too large at the locality of the city, state coordinates should not be used because they will not quite check with the actual measurements and will thus cause confusion. The proper choice may depend on the accuracy desired.

If state coordinates are to be used, the triangulation net should be arranged so that strong ties are established with triangulation stations of the U.S. Coast and Geodetic Survey.

When a city has already established a coordinate system, it can be converted to state coordinates by the formulas for translation and rotation given in the next section. Often it is possible to set up special tables for rapid conversion for a particular case.

13-48. Formulas for Translation and Rotation of Coordinates. In Fig. 13-23, let O be the origin of an old coordinate system with axes N and E, and let S be the origin of a new system with axes Y and X. The coordinates of S on the old system are Δn and Δe and the clockwise angle from the old to the new axes is θ.

Let n and e be the old coordinates of any point P and y and x be the new coordinates of P. Then

$$y = a + b = (n - \Delta n) \cos \theta + (e - \Delta e) \sin \theta$$

$$x = \overline{SF} - c = (e - \Delta e) \cos \theta - (n - \Delta n) \sin \theta$$

or

$$y = n \cos \theta + e \sin \theta - (\Delta n \cos \theta + \Delta e \sin \theta) \tag{13-33}$$

$$x = e \cos \theta - n \sin \theta - (\Delta e \cos \theta - \Delta n \sin \theta) \tag{13-34}$$

Let

$$\Delta y = \Delta n \cos \theta + \Delta e \sin \theta \tag{13-35}$$

$$\Delta x = \Delta e \cos \theta - \Delta n \sin \theta \tag{13-36}$$

Then

$$y = n \cos \theta + e \sin \theta - \Delta y \tag{13-37}$$

$$x = e \cos \theta - n \sin \theta - \Delta x \tag{13-38}$$

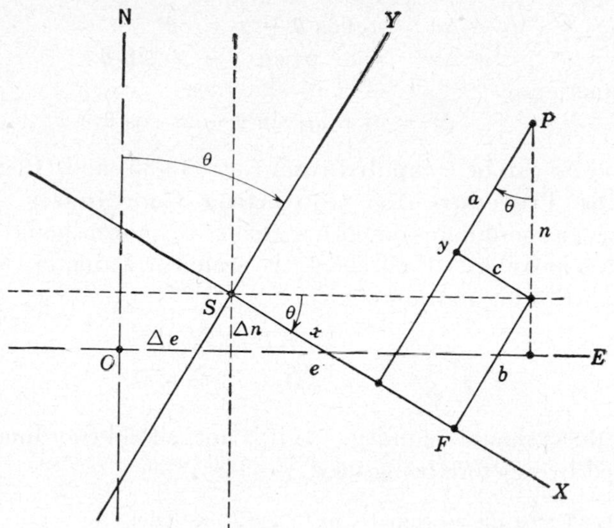

FIG. 13-23

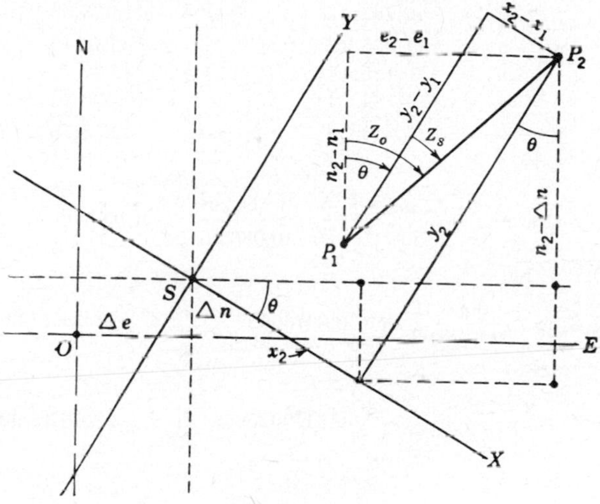

FIG. 13-24

In these formulas Δy, Δx, and θ are all constants that can be determined once and for all for the transformation.

In order to determine the constants θ, Δy, and Δx, the coordinates of at least two points must be determined on both coordinate systems. In Fig. 13-24,

$$\tan Z_0 = \frac{e_2 - e_1}{n_2 - n_1} \qquad \tan Z_s = \frac{x_2 - x_1}{y_2 - y_1} \tag{13-39}$$

$$\theta = Z_0 - Z_s \tag{13-40}$$

Also

$$n_2 - \Delta n = y_2 \cos \theta - x_2 \sin \theta$$

$$\Delta n = n_2 - y_2 \cos \theta + x_2 \sin \theta \qquad (13\text{-}41)$$

In like manner

$$\Delta e = e_2 - y_2 \sin \theta - x_2 \cos \theta \qquad (13\text{-}42)$$

Then Δy and Δx can be computed from Eqs. (13-35) and (13-36).

13-49. The Procedure for Converting Coordinates. The two points for which coordinates on both systems are known should be located as far apart as possible. First check the scale correction of the two systems by the following formula:

$$s = \frac{\sqrt{(x_2 - x_1)^2 + (y_2 - y_1)^2}}{\sqrt{(n_2 - n_1)^2 + (e_2 - e_1)^2}} \qquad (13\text{-}43)$$

The scale value s should be unity. If it is not, all old coordinates should be multiplied by s before being used.

Example 1. Given the coordinates of P_1 and P_2 as follows:

$$\begin{aligned} n_2 &= 13{,}968.51 & e_2 &= 1{,}542.24 \\ n_1 &= 2{,}732.54 & e_1 &= 11{,}628.39 \\ n_2 - n_1 &= +11{,}235.97 & e_2 - e_1 &= -10{,}086.15 \\ y_2 &= 16{,}425.47 & x_2 &= 3{,}342.73 \\ y_1 &= 3{,}184.18 & x_1 &= 10{,}598.46 \\ y_2 - y_1 &= +13{,}241.29 & x_2 - x_1 &= -\ 7{,}255.73 \end{aligned}$$

By Eq. (13-43),

$$s = \frac{\sqrt{(-7255.73)^2 + (13{,}241.29)^2}}{\sqrt{(11{,}235.97)^2 + (-10{,}086.15)^2}} = 1.000000$$

By Eq. (13-39),

$$\tan Z_0 = \frac{-10086.15}{+11235.97} = -0.8976662 \qquad Z_0 = -41°54'47.69''$$

By Eq. (13-39),

$$\tan Z_s = \frac{-7255.73}{+13{,}241.29} = -0.5479625 \qquad Z_s = -28°43'15.92''$$

By Eq. (13-40),

$$\theta = -41°54'47.69'' - (-28°43'15.92'') = -13°11'31.77''$$

$$\sin \theta = -0.22821762 \qquad \cos \theta = +0.97361015$$

By Eq. (13-41), solving for coordinates of both P_1 and P_2,

$$\Delta n = 2{,}732.54 - 3{,}100.15 - 2418.76 = -2786.37$$

$$\Delta n = 13{,}968.51 - 15{,}992.00 - 762.87 = -2786.36$$

By Eq. (13-42), similarly,

$$\Delta e = 11{,}628.39 + 726.69 - 10{,}318.77 = +2036.31$$

$$\Delta e = 1{,}542.24 + 3748.58 - 3{,}254.52 = +2036.30$$

By Eq. (13-35),

$$\Delta y = -2712.83 - 464.72 = -3177.55$$
$$\Delta x = +1982.56 - 635.90 = +1346.66$$

Thus we have the necessary constants sin θ, cos θ, Δy, and Δx. With these, any pair of coordinates can be converted.

Example 2. The new coordinates of P_1 and P_2 are computed from the old values by Eqs. (13-37) and (13-38):

$$P_1\text{: } y_1 = 2{,}660.43 - 2653.80 + 3177.55 = 3{,}184.18$$
$$x_1 = 11{,}321.52 + 623.61 - 1346.66 = 10{,}598.47$$
$$P_2\text{: } y_2 = 13{,}599.88 - 351.97 + 3177.55 = 16{,}425.46$$
$$x_2 = 1{,}501.54 + 3187.86 - 1346.66 = 3{,}342.74$$

Second Method.[1] The transformation equations, Eqs. (13-37) and (13-38), can be written, and the necessary constants found, in such a way that any scale change can be taken care of automatically. From Fig. 13-24,

$$\cos\theta = \cos(Z_0 - Z_s) = \cos Z_0 \cos Z_s + \sin Z_0 \sin Z_s$$

Let $\delta n = n_2 - n_1$, etc. Then

$$\cos Z_0 = \frac{\delta n}{\sqrt{\delta n^2 + \delta e^2}} \qquad \cos Z_s = \frac{\delta y}{\sqrt{\delta n^2 + \delta e^2}}$$

$$\sin Z_0 = \frac{\delta e}{\sqrt{\delta n^2 + \delta e^2}} \qquad \sin Z_s = \frac{\delta x}{\sqrt{\delta n^2 + \delta e^2}}$$

Substituting,

$$\cos\theta = \frac{\delta n\,\delta y + \delta e\,\delta x}{\delta n^2 + \delta e^2}$$

Let S equal the scale change. Then, on the new system, each length from the old system must be multiplied by S:

$$\cos\theta = \frac{S\,\delta n\,\delta y + S\,\delta e\,\delta x}{S^2\,\delta n^2 + S^2\,\delta e^2}$$

$$S\cos\theta = \frac{\delta n\,\delta y + \delta e + \delta x}{\delta n^2 + \delta e^2} \qquad (13\text{-}44)$$

Similarly,

$$S\sin\theta = \frac{\delta e\,\delta y - \delta x\,\delta n}{\delta n^2 + \delta e^2} \qquad (13\text{-}45)$$

Also, correcting Eqs. (13-37) and (13-38) for scale change,

$$\Delta y = nS\cos\theta + eS\sin\theta - y \qquad (13\text{-}46)$$
$$\Delta x = eS\cos\theta - nS\sin\theta - x \qquad (13\text{-}47)$$

which are the transformation equations. The constants that must be found are S cos θ, S sin θ, Δy, and Δx. The first two can be found by Eqs. (13-44) and (13-45) and the other two as follows: From Eqs. (13-46) and (13-47),

$$\Delta y = n_2S\cos\theta + e_2S\sin\theta - y_2 \qquad (13\text{-}48)$$
$$\Delta x = e_2S\cos\theta - n_2S\sin\theta - x_2 \qquad (13\text{-}49)$$

Example. Given the coordinates of P_1 and P_2 as follows:

$n_2 =$	13,968.51	$e_2 =$	1,542.24	$y_2 =$	16,427.11	$x_2 =$	3,343.06
$n_1 =$	2,732.54	$e_1 =$	11,628.39	$y_1 =$	3,184.50	$x_1 =$	10,599.52
$\delta n =$	+11,235.97	$\delta e =$	−10,086.15	$\delta y =$	+13,242.61	$\delta x =$	−7,256.46

[1] Suggested by Robert Singleton.

By Eqs. (13-44) and (13-45),

$$S \cos \theta = \frac{11{,}235.97(13{,}242.61) + (-10{,}086.15)(-7256.46)}{(11{,}235.97)^2 + (-10{,}086.15)^2}$$
$$S \cos \theta = 0.97370735$$
$$S \sin \theta = \frac{-10.086.15(13{,}242.61) - (-7256.46)(11{,}235.97)}{(11{,}235.97)^2 + (-10{,}086.15)^2}$$
$$S \sin \theta = -0.22824004$$

By Eqs. (13-48) and (13-49),

$$\Delta y = 13{,}968.51(0.97370735) + 1542.24(-0.22824004) - 16{,}427.11$$
$$\Delta y = -3177.87$$
$$\Delta y = 2731.54(0.97370735) + 11{,}628.39(-0.22824004) - 3184.50$$
$$\Delta y = -3177.87$$
$$\Delta x = 1542.24(0.97370735) - 13{,}968.51(-0.22824004) - 3343.06$$
$$\Delta x = 1346.80$$
$$\Delta x = 11{,}628.39(0.97370735) - 2732.54(-0.22824004) - 10{,}599.52$$
$$\Delta x = 1346.81$$

PROBLEMS

13-1. What are the degrees of curvature, by the two definitions, of a curve with a radius of 500 ft?

Compute the deflection angles for the half stations (50-ft points) for the following curves:

13-2. $R = 600$; $\Delta = 36°28'$; plus of PI = 20 + 31.92.

13-3. Compute the subchords required for Prob. 13-2.

13-4. $D = 15°$ arc definition; $\Delta = 86°54'$; plus of PI = 10 + 42.64.

13-5. Compute the subchords required for Prob. 13-4.

13-6. $D = 10°$ chord definition; $\Delta = 46°18'$; plus of PI = 47 + 32.29.

13-7. Compute the subchords required for Prob. 13-6.

13-8. For a compound curve, given $\Delta = 42°10'$; $\Delta_2 = 16°$; $R_1 = 200$ ft; $T_1 = 160$ ft. Compute Δ_1, T_2, and R_2.

13-9. Given diverging tangents for which $\theta = 22°18'$; $T_2 = 1000$ ft; common R = 300 ft; to find Δ_1, Δ_2, and T_1. The figure showing the constructed lines measuring for this curve is not given in the book.

13-10. Compute the vertical curve from the following data. Required, elevations at every 25-ft point. Plus of PVI = 20 + 31.68; elevation 62.40; $g_1 = -6.30$; $g_2 = +1.20$; rate of change of grade 4 per cent per 100-ft station. Make the check computations.

13-11. Given the coordinates for two points on an old and a new coordinate system as follows:

	P_1	P_2
North..........	12,468.12	3,628.72
East...........	1,064.27	16,477.08
y.............	9,065.47	129.58
x.............	269.32	15,626.39

find the values of S, θ, Δy and Δx.

CHAPTER 14

MINE SURVEYS

14-1. Introduction. Mine surveys are made to determine the relative positions and elevations of underground workings, geological formations, and surface features; to measure quantities; and to establish line and grade for future operations or construction. They are used to plan operations, to show the progress of work, to determine the quantities of ore or coal removed, to estimate the location and quantity of ore or coal available, to locate the positions of property boundaries underground, to plan the direction and slope for underground connections, and to control the direction, grade, and extent of new work.

14-2. Survey Control. A mine survey should be based on a system of well-monumented horizontal and vertical control stations established on the surface. These are connected by triangulation, traverse, or leveling. Control should be brought into the workings wherever possible and marked as permanently as conditions permit. Underground, both horizontal and vertical control are usually carried by traverse, although leveling is often used. The traverse courses must follow the workings. These are often steeply inclined and seldom straight for more than short distances. Vertical angles as well as horizontal angles are measured at each station, and nearly all lengths are measured on a slope usually along, or parallel to, the line of sight so that the vertical angle recorded gives the slope of the tape. Very steep slopes that approximate 90° are often used.

14-3. Mining Transits. A mining transit (see Fig. 14-1) should have the complete equipment of an engineer's transit—a telescope level, a full vertical circle, stadia wires, and a compass. It should be fully protected from the drip and damp usually present in mines. The tripod should have adjustable legs for cramped setups, and a trivet and a bracket should be available for setups in especially difficult positions (see Fig. 14-2).

Some engineers prefer to use a striding level to establish the direction of the horizontal axis on important, steeply inclined sights. Collars must then be provided on the horizontal axle to take the striding level.

14-4. Auxiliary Telescopes. A mining transit has an auxiliary telescope for steep sights. It is attached either at the top of the main tele-

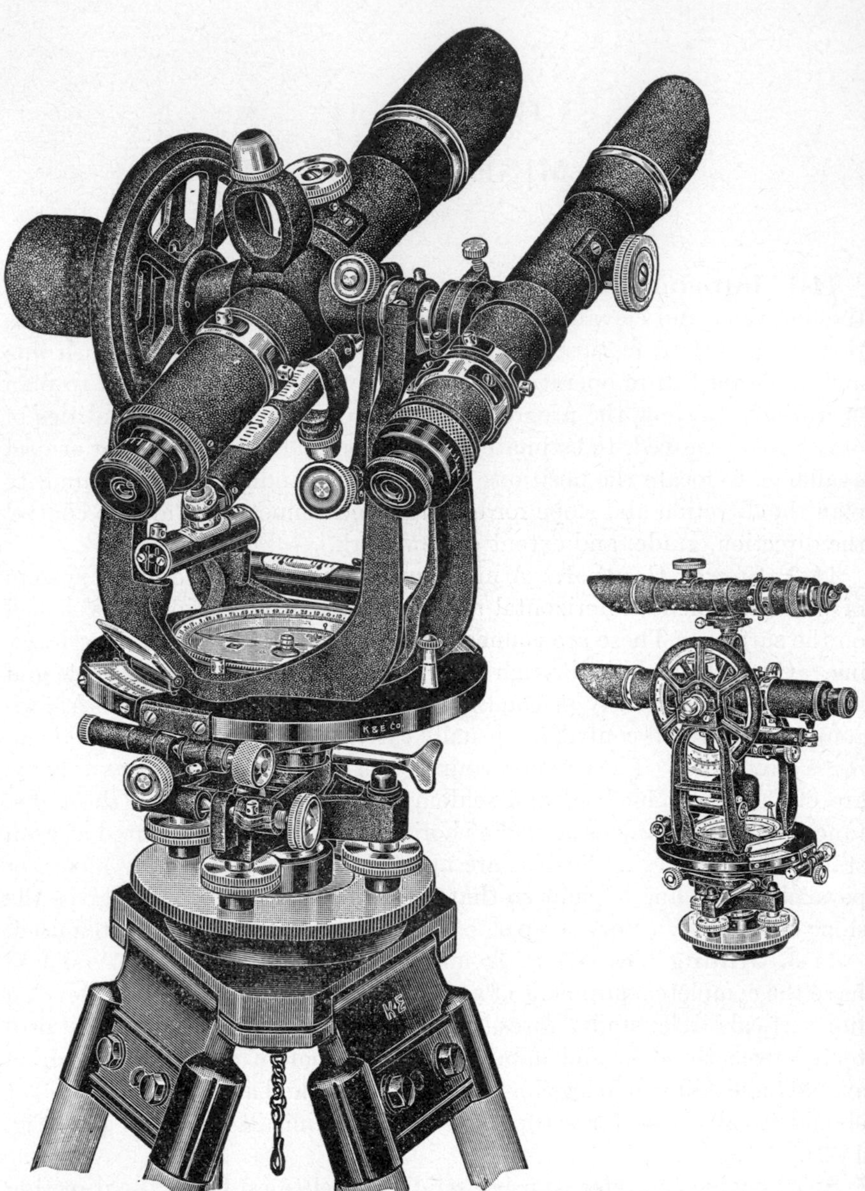

FIG. 14-1. A mining transit showing the auxiliary telescope in both positions. (*Keuffel & Esser Co.*)

scope or at the end of the horizontal axis. It is often made so that it can be placed in either position. Its line of sight is adjusted so that it is parallel to the line of sight of the main telescope. It should have a detachable, prismatic eyepiece. Both telescopes should have a short minimum focus, and lower power—about 18 to 20 diameters.

14-5. Adjustments of the Auxiliary Telescope. Before the auxiliary telescope is adjusted, the transit adjustments should be checked and rectified if necessary. The vertical cross hair of the auxiliary telescope should be placed in a plane that is perpendicular to the horizontal axis, by the usual method of rotating the reticule.

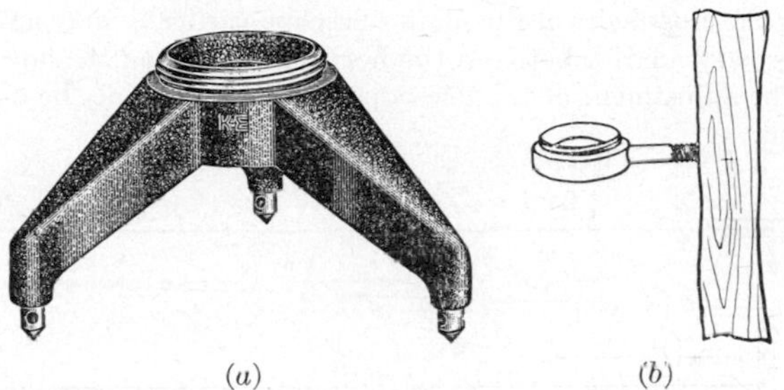

(a) (b)

FIG. 14-2. (a) A trivet (*Keuffel & Esser Co.*). (b) A bracket screwed into a mine timber.

To make the line of sight of an auxiliary telescope parallel to that of the main telescope, the procedure given below is followed:

1. *Distant-point Method.* Aim the main telescope at a well-defined point at least 2 miles distant. Bring the cross hairs of the auxiliary telescope on the point, either by the reticule adjustment screws or by external adjusting screws when these are provided.

2. *Parallel-lines-on-a-card Method.* In the process of carrying out this method, the distance between the line of sight of the main telescope and that of the auxiliary telescope is accurately determined. This distance is the eccentricity. It must also be used to compute effect of eccentricity on angular measurement, as explained later. The procedure is described for a side telescope, but it can be applied to either.

a. Measure, as accurately as possible from their external dimensions, the distance between the centers of the two telescopes.

b. Construct a card with a single horizontal line and two vertical lines separated by this distance.

c. Place the card about 200 ft from the instrument and rotate the card until, when the instrument is turned slightly in azimuth, the cross hairs remain on the horizontal line.

d. Point the main telescope at one intersection.

e. Adjust the line of sight of the auxiliary telescope until it falls on the other intersection. Usually an external adjusting device is provided. If not, the reticule must be moved.

f. Place the card as near as the telescopes will focus and align the horizontal line as before.

g. If, when the main telescope is properly aimed, the auxiliary-telescope cross hairs do not fall on the other intersection, mark where they fall. The eccentricity can then be computed as shown in Fig. 14-3. The value of m will be minus when the lines diverge. Make up a new card with the lines separated by the correct eccentricity.

h. Repeat the process with the new card. Check the separation at both distances.

To adjust an auxiliary telescope that can be mounted both at the side and on top, first adjust the auxiliary telescope in the side position by either method described. The vertical movement should be made with an external adjusting device, if provided, rather than with the cross hairs.

Place the auxiliary telescope in the top position and adjust as before, but make the vertical adjustment with the cross hairs of the main telescope and make the horizontal adjustment with the external adjusting device of the auxiliary telescope. If no external adjusting device is provided, the reticule must be adjusted. The adjustment at the side position must then be checked.

Since moving the cross hairs of the main telescope vertically may affect their lateral position, the perpendicularity between the horizontal axis and the line of sight should be checked. The adjustment of the telescope level will certainly be disturbed. This must be rectified.

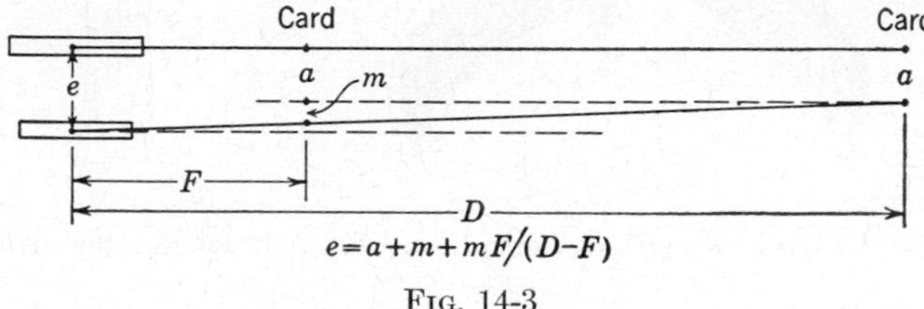

FIG. 14-3

14-6. Measurement of Angles with Auxiliary Telescopes. Since a top telescope prevents reversed sights, the use of the side telescope is considered better practice. No correction is necessary for a vertical angle measured with a side telescope and the corrections to a horizontal angle can be eliminated by measuring the angle an equal number of times direct and reversed.

With a top telescope, horizontal angles can be measured directly but all vertical angles must be corrected for eccentricity.

14-8. Horizontal-angle Corrections for a Side Telescope. In Fig. 14-4, 0 is the instrument station where the desired angle V is to be measured from A to B. When the side telescope is used, the instrument will measure the angle v_d when the instrument is direct and v_r when it is reversed. H_a and H_b are the horizontal components of OA and OB, respectively, and e is the eccentricity of the side telescope. From the figure,

$$\sin c_a = \frac{e}{H_a} \qquad \sin c_b = \frac{e}{H_b} \tag{14-1}$$

$$\text{Direct } V = v_d - c_a + c_b \tag{14-2a}$$

$$\text{Reversed } V = v_r + c_a - c_b \tag{14-2b}$$

Adding,

$$1 \text{ D.R.} = 2V = v_d + v_r$$

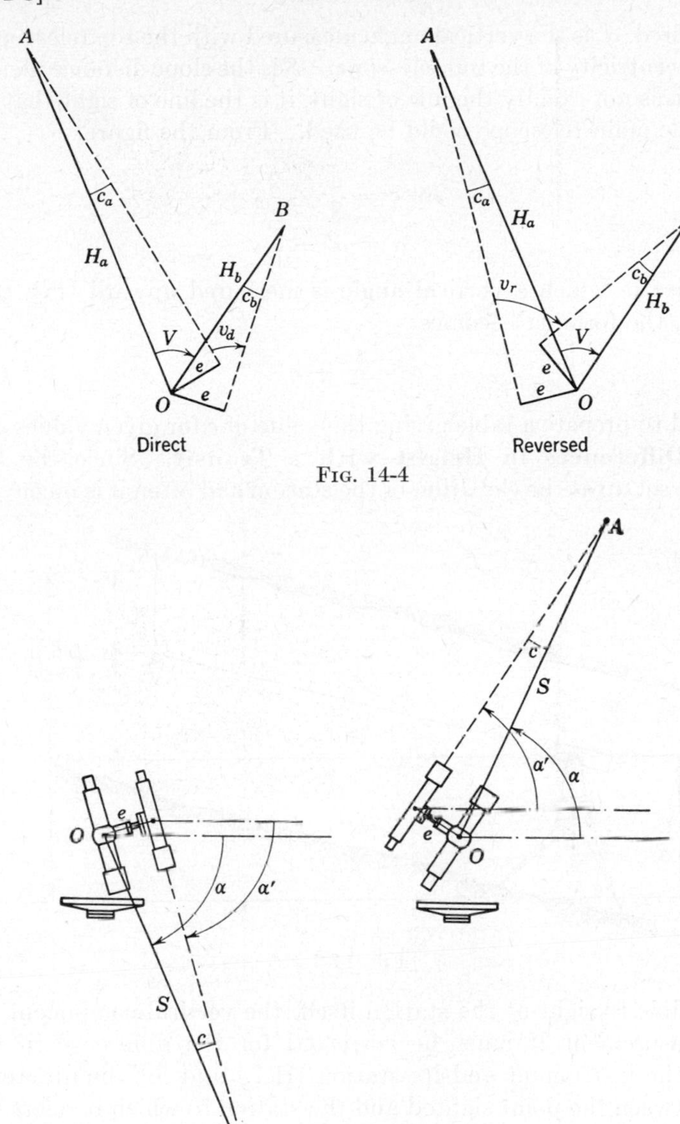

FIG. 14-4

FIG. 14-5

It is usual to prepare a table giving the values of c for given values of H so that the measured angles can be corrected easily, if they are not doubled in the field.

14-8. Vertical-angle Corrections for a Top Telescope. In Fig. 14-5, 0 is the instrument center, A is the station sighted, α is the vertical angle desired, α' is the vertical angle measured with the top telescope, and e is the eccentricity of the top telescope. S is the slope distance measured. While this is not exactly the line of sight, it is the line of sight that would exist if the main telescope could be used. From the figure,

$$\sin c = \frac{e}{S} \tag{14-3}$$

$$\alpha = \alpha' - c \tag{14-4}$$

In the case in which a vertical angle is measured upward with the top telescope, the formula becomes

$$\alpha = \alpha' + c \tag{14-4b}$$

It is usual to prepare a table giving the value of c for given values of S.

14-9. Differences in Height with a Transit. Since the transit cannot be set up at the elevation of the station and often it is inconvenient

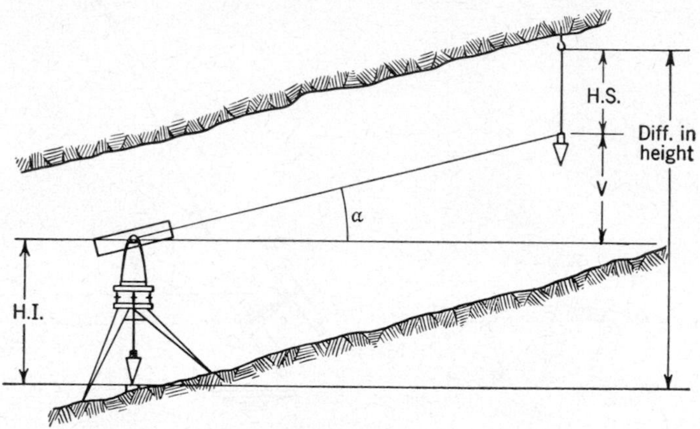

FIG. 14-6

or impossible to sight at the station itself, the vertical component of the slope measurement V must be corrected for the difference in height between the instrument and its station (H.I.) and for the difference in height between the point sighted and the station to which it refers (H.S.) (see Fig. 14-6). The H.I. and the H.S. are given the signs that must be used to compute the total difference in height by the following formula:

$$\text{Diff. in ht.} = \text{H.I.} + V + \text{H.S.} \tag{14-5}$$

Figure 14-6 shows the arrangement for which plus values are given for both H.I. and H.S. When the instrument is under the station, the H.I. is recorded as a minus value and when the station is below the point sighted, the H.S. is given a minus value. Finally, when the vertical angle is minus, V has a minus value. Thus all three can always be added algebraically to find the difference in height with its proper sign.

14-10. Underground Traverses. Traverse stations are usually placed in the roof to prevent disturbance. They consist of a hook (spad) driven into a timber or into a wooden plug in a drill hole. They are usually marked by tags of noncorrosive metal stamped with the station number. A plumb bob is hung from the hook either for a sight or to give position to the instrument. Often the top of the bob is sighted for elevation. Sometimes a pin stuck through the string or a special target is used. The target is usually illuminated by shining a cap lamp or a hand lamp on the notebook or other white surface held behind the target. The silhouette provides a good target and the transit cross hairs can be seen against the white surface.

14-11. Use of the Transit. Mining transits usually have a center mark on the top of the telescope that is directly over the instrument center when the telescope is level. To set up under a plumb bob is somewhat a trial-and-error procedure, because the instrument center changes position as the instrument is leveled. The telescope is first set so that the vertical angle reads zero, the instrument is then leveled, and finally it is shifted to the plumb bob. It is then releveled and its position checked.

If the vertical circle is graduated all the way to 90°, the telescope can be set so that the vertical angle reads 90° with the eyepiece up. Then the eyepiece can be centered under the bob.

The transitman should carry a hand lamp to read the instrument and to shine into the objective to illuminate the cross hairs when necessary.

14-12. Tapes. Since nearly all tape measurements are made on a slope, the tape should be as long as the total slope distance between stations. Because of the irregularities of the workings, this can seldom be greater than 200 ft so that a 200-ft tape is usually adequate. If a longer distance is required, an intermediate station can be used. A short, light tape about 10 ft long should be carried for the many necessary short measurements.

14-13. Tape Corrections. On the surface and in some of the workings, measurements with the tape are made in the usual manner. For this purpose the tape is standardized and the corrections are applied as described in Chap. 3.

14-14. Slope Measurements. In most workings, the length measurements are made practically along, or parallel to, the line of sight so that the vertical angle measured by the instrument gives the value of the slope of the tape. Usually an estimated 20-lb tension is applied. The tape is supported only at the instrument and at the station, if the length is less than 100 ft. Otherwise, a third tapeman supports the tape at the 100-ft point by raising it until it is in line with the line of sight.

The rear tapeman holds the zero mark over the center mark on the telescope or at the end of the horizontal axis, as close to the instrument as possible without touching the instrument with the tape. The head tapeman holds the tape close to the point sighted, applies the tension, and reads the graduation opposite the station.

Usually no corrections are made except to reduce the slope measurement to its horizontal and vertical components by multiplying by the cosine and the sine, respectively, of the vertical angle.

14-15. Accurate Slope Measurements. When accurate slope measurements are necessary, special methods are required. The measurements may be corrected for as many conditions as are necessary. In addition to the usual corrections, allowance must be made for the reduction of the effect of sag due to the slope of the tape and for the differences in the stretch throughout the tape due to its own effective weight in the vertical component of the slope.

14-16. Standardization of Tape. In the following paragraphs are given all possible corrections. To use them the tape should be standardized when fully supported.

Select any reasonably smooth surface that does not slope more than 2 per cent and that has not been exposed to the sun for several hours. Mark off the length with the standard tape. This is best accomplished by scratching a piece of copper with a needle. Place the tape to be standardized in position under a tension of 20 lb and record the actual length. If both tapes are steel, no temperature correction is necessary so long as the length of the standard tape at 68°F is known (see also Chap. 3). Then the standard correction to be applied to measurements, C_{st}, is found as follows:

$$C_{st} = \text{actual length at } 68°\text{F} - \text{nominal length at } 68°\text{F}$$

14-17. Measurement Procedure. The tape may be supported as desired. It may hang free or it may be supported anywhere in the span, so long as these supports are on a straight line joining the end supports. The more intermediate supports used, the less will be the error introduced by errors in observing the tension. A tension handle should be used at the head end and a tape thermometer should be in place near one end of the tape. The following data must be recorded:

L, the slope distance indicated by the tape
l_1, l_2, etc., the length of each span
F, the tape temperature, °F
α, the vertical angle of the tape
p or q, the tension, lb; the letter p indicates that the tension is measured at the upper end of the tape; q indicates that the tension is measured at the lower end

Note: It is evident that, when the tape slopes, p would include some or all of the weight of the tape, depending on the angle of slope.

14-18. Corrections. The slope distance is computed by adding the following corrections algebraically to the slope distance as indicated by the tape (all lengths are in feet):

$$C_t = C_{st}\frac{L}{T} \tag{14-6}$$

$$C_f = 0.00000645L(F - 68°) \tag{14-7}$$

$$C_p = \frac{L(t_1 - t_0)}{28{,}000{,}000S} \tag{14-8}$$

where $$t_1 = p - \frac{wL}{2}\sin\alpha$$

or $$= q + \frac{wL}{2}\sin\alpha$$

$$C_s = -kl \tag{14-9}$$

where C_t = correction to L for error in tape length
C_{st} = actual tape length minus nominal tape length
L = measured slope length
T = nominal tape length
C_f = correction for temperature
F = temperature in degrees Fahrenheit
C_p = correction for tension
t_0 = tension, lb, used when tape was standardized
p = tension, lb, at upper end of tape
q = tension, lb, at lower end of tape
S = cross-section area of tape, sq in.
w = weight of tape, lb, per linear foot
α = vertical angle of slope
C_s = sag correction, which must be computed and applied for each span
k = constant from Chart 14-1 or 14-2
l = length of a span

Example. To find the slope distance, given:

Measured slope distance, L	182.13
Tape supported at 0, 100, 182.13 ft:	
l_1	100
l_2	82.13
Tape temperature, F	58°F
Vertical angle of tape, α	45°15′
Tension at upper end, p	30 lb
Tape standardized when supported throughout, 20 lb; tension 68°F; standardized length 199.984, C_{st}	−0.016
Area of tape cross section, S	0.0038 sq in.
Weight of tape per linear foot, w	0.013 lb

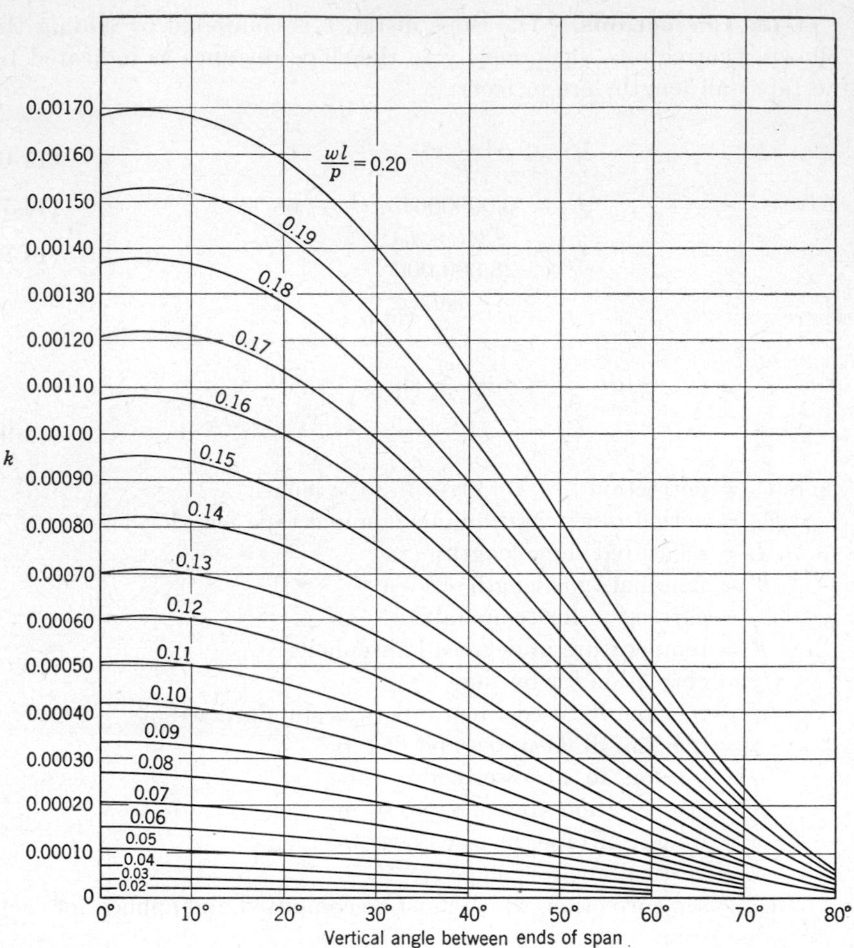

CHART 14-1. Values of k for correcting steep slope measurements when the tension is applied at upper end.

$$C_t = -\frac{0.016(182)}{200} = -0.015$$

$$C_f = 0.00000645(182)(58 - 68) = -0.012$$

$$t_1 = 30 - \frac{0.013(182)(0.701)}{2} = 29.16$$

$$C_p = \frac{182(29.16 - 20)}{28{,}000{,}000(0.0038)} = +0.016$$

$$\frac{wl_1}{p} = \frac{0.013(100)}{30} = 0.043 \qquad k_1 = 0.00004$$

$$\frac{wl_2}{p} = \frac{0.013(82)}{30} = 0.036 \qquad k_2 = 0.00003$$

$$C_s = -100(0.00004) - 82(0.00003) = -0.006$$

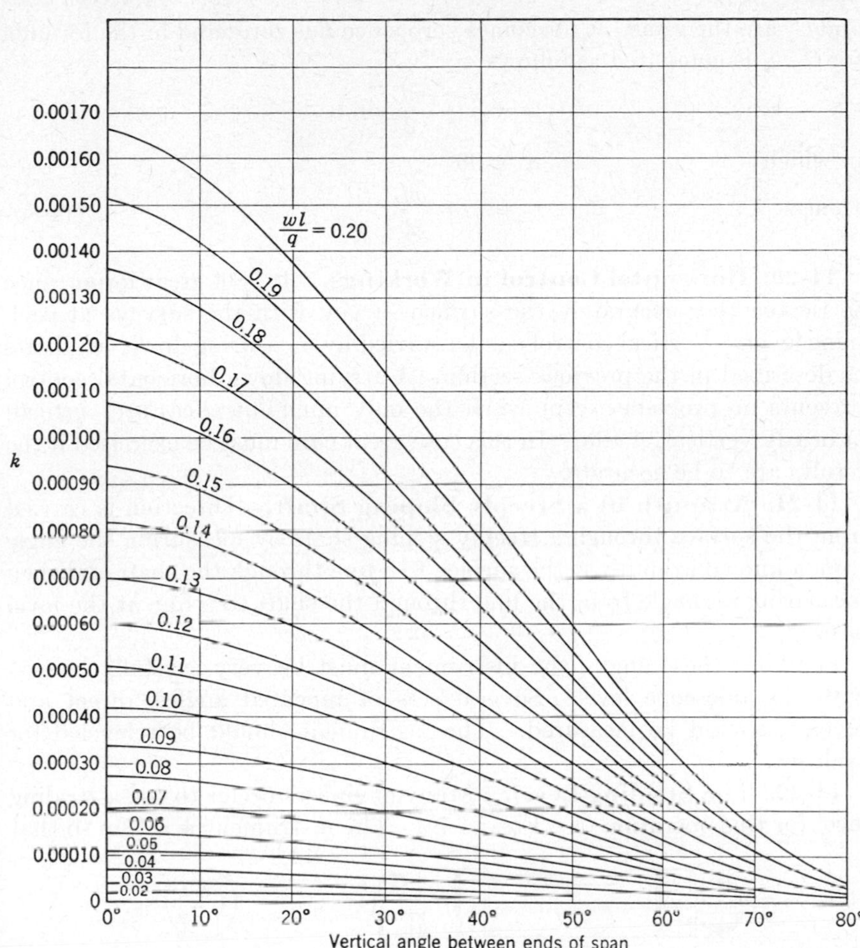

CHART 14-2. Values of k for correcting steep slope measurements when the tension is applied at lower end.

Adding,

$$
\begin{aligned}
L &= 182.13 \\
C_t &= - \quad 0.015 \\
C_f &= - \quad 0.012 \\
C_p &= + \quad 0.016 \\
C_s &= - \quad 0.006 \\
\text{Corrected slope distance} &= 182.10
\end{aligned}
$$

14-19. Leveling down a Vertical Shaft. Levels are usually carried down a vertical shaft by tape measurements. A weight is usually attached at the lower end in place of the spring balance. The corrections used for a slope distance given in the previous section may be used. C_t

and C_f are the same. C_s becomes zero since k is zero; and in the formula for C_p, t_1 is computed as follows:

$$t_1 = q + \frac{wL}{2} \sin \alpha$$

in which $\alpha = 90°$

Hence $$t_1 = q + \frac{wL}{2} \qquad (14\text{-}8a)$$

14-20. Horizontal Control in Workings. It is of great importance to tie together accurately the surface survey with the surveys at each mine level. Vertical control can be carried by measuring down the shafts as described in the previous section. Carrying down horizontal control presents no problem except when the only mine entrances are vertical, or nearly vertical, shafts. In this case great care must be exercised if the results are to be accurate.

14-21. Azimuth in a Steeply Sloping Shaft. Direction is carried from the surface through a steeply sloping shaft by measuring the angle from a known azimuth at the surface to a line through the shaft and then measuring an angle from the line through the shaft to a line at the level desired.

For both these angles the instrument must be very carefully leveled with the telescope level. Several sets of repeated angles, direct and reverse, should be measured. The instrument should be releveled for each set.

14-22. The Striding Level. Many engineers prefer to use a striding level for this operation (see Fig. 14-7). The instrument is set up so that

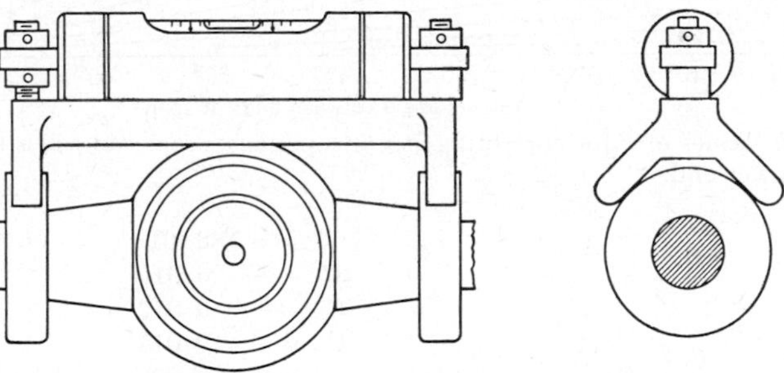

FIG. 14-7. A striding level. It is shown in place, resting on collars on the telescope axle. It can be turned end for end.

one pair of opposite leveling screws is at right angles to the direction of the shaft. After the instrument is leveled with the plate levels, the telescope is aimed at the point at the far end of the shaft. The striding level is then carefully centered with the leveling screws, and the pointing is perfected. The angle is then turned to the station that is approximately level with the instrument. The angle is measured in units of four repetitions, each as follows:

1. Telescope direct, level normal
2. Telescope direct, level reversed
3. Telescope reversed, level normal
4. Telescope reversed, level reversed

This process eliminates any error of adjustment of the striding level or of the transit.

14-23. Single Vertical Shaft. When only one shaft is available, two weighted wires are hung in the shaft as far apart as possible. The azimuth of the two wires is measured at the surface and then used as a basis for the azimuths of the underground traverse. The method is often called **shaft plumbing.**

When the workings extend a considerable distance from the shaft, this operation must be carried out with great care. If, for example, the wires were 4 ft apart and the combined error of the surface and underground operations was 0.01 ft in azimuth, the resulting error at 3,000 ft from the shaft would be

$$3000' \times \frac{0.01}{4} = 7.5'$$

14-24. Equipment. The maximum amount of weight necessary is 50 lb, although often 20 lb will be found to be adequate. Any shape will do. It should be hung from the wire by a ring so that its center of gravity will hang in line with the wire without bending the wire.

Music spring wire (piano wire) is usually used. The breaking strength of this wire is 110 lb for a diameter of 0.02 in. and 226 lb for 0.03 in. The larger is recommended for a 50-lb weight. Often, key rings on a few links of a chain are inserted at instrument height so that the wire farthest from the instrument can be seen. The wire should be mounted on a ratchet reel so that it can be easily handled.

The weight should be immersed in water or oil to steady the wire. A glass vessel is the most convenient, because the vessel's position must be adjusted until neither the weight nor the wire touches it. A cover should be arranged to protect the liquid from falling water.

14-25. Mounting the Wires. It is convenient to hang the wires over a sheave mounted on a cross beam. The wires must then be led over a guide to steady them (see Fig. 14-8). Great care must be exercised to prevent the wires from touching any object. A light should be passed around each end and observed from the other end to make sure that no object interferes. The distance between the wires should be carefully measured at the top and bottom as a further check. Some engineers move each wire laterally a short distance at the top and make sure that the same movement occurs at the bottom.

Air currents often give great difficulty. If the height plumbed is not greater than 300 ft, this difficulty is not often encountered.

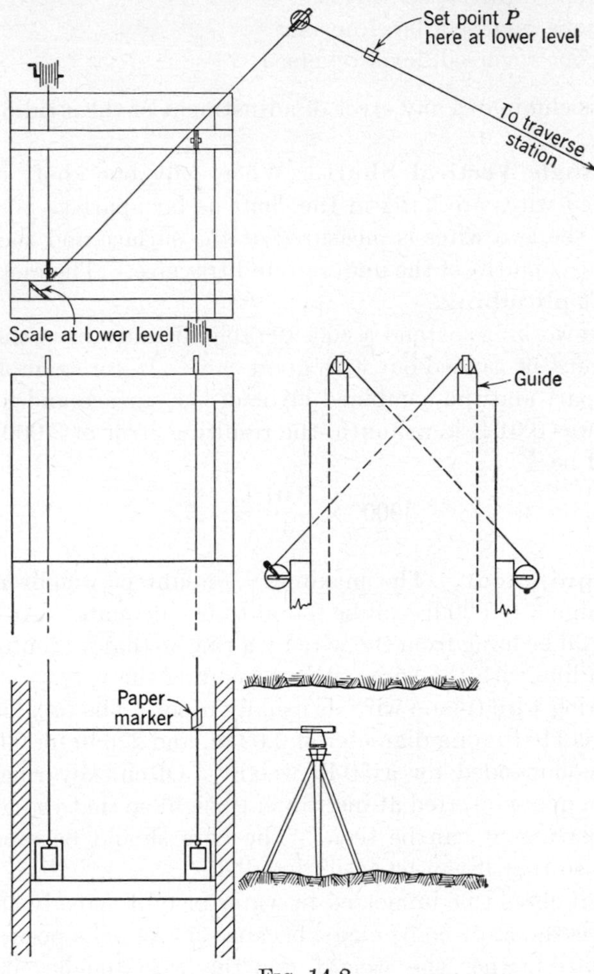

FIG. 14-8

14-26. Transit Operation. On the surface, the transit should be placed as near the wires as possible and bucked into alignment with them. If the transit is near enough, the near wire can be focused out entirely, and no chain or ring is needed. A piece of paper should be placed on the near wire to identify it. The telescope should be kept direct throughout this procedure, and the angle to a distant control station should be measured by repetition, with the telescope direct throughout. The transit should then be bucked into line with the telescope reversed and the angle measured by repetition with the telescope reversed. The average of the two angle determinations is used. Since the focus is so short, the focusing lens must move through a considerable distance to focus on the two wires. This introduces the angular error of the focusing draw. Also the line of sight may be slightly eccentric to the vertical axis. If any error exists, the telescope will be in slightly different positions when it is bucked in direct and reverse. By taking the average the error is eliminated.

A plumb bob is attached to the instrument and its position carefully marked on a stake. The average of the two bucked-in positions is used.

In the mine, the same procedure is used if the transit can be placed in line with the wires. If the sight to the nearest traverse station is very short, the exact position of the transit must be especially carefully marked after it is bucked in. Sometimes a piece of flat lead is placed under the plumb bob to receive the mark. It is better to hang a bob from the roof in such a way that it can be adjusted over the telescope center mark. Some engineers prefer to align a plumb bob with the line of sight when the telescope is pointed at the distant traverse station (see point P in Fig. 14-8). The bob should be as near the instrument as focus will permit. It is used as a backsight when the next angle in the workings is measured. This method is more accurate if the line of sight does not pass exactly under the mark on the telescope. In every case the average of the bucked-in positions is used. Usually the wires continue to swing throughout the observations. A small scale may be mounted behind the wires to assist in determining their mean positions.

14-27. Triangulation. When the transit cannot be set up in line with the wires, a small triangulation net must be established. To obtain the strongest triangles, the transit should be placed as near the wires as focus will permit and as nearly on line with them as possible (see Fig. 14-9). Two transit positions should be used if possible. All the sides of the triangles are measured (including the distance between the wires) and the angle between the wires is measured with the transit. If a second transit position cannot be found, the figure will be strengthened by establishing an auxiliary point at D and measuring the lengths of the dotted lines. The angles at B are computed by triangles CBD and DBA. This gives an independent determination of the angle CBA. The work should be refined until the following three methods of finding angle B check closely:

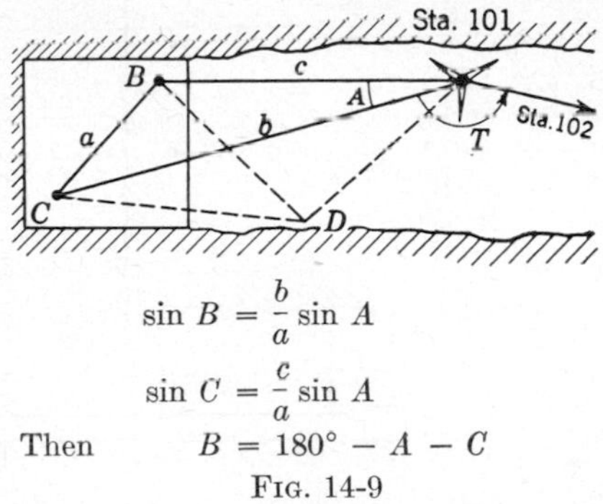

$$\sin B = \frac{b}{a} \sin A$$

$$\sin C = \frac{c}{a} \sin A$$

Then $$B = 180° - A - C$$

FIG. 14-9

$$\sin B = \frac{b}{a} \sin A$$

$$B = 180 - A - C$$

where $$\sin C = \frac{a}{c} \sin A$$

$$B = \text{angle } CBD + \text{angle } DBA$$

where $$\cos \angle DBA = \frac{(\overline{AB}^2 + \overline{BD}^2 - \overline{AD}^2)}{2\,\overline{AB} \times \overline{BD}} \tag{14-10a}$$

$$\cos \angle CBD = \frac{(\overline{BC}^2 + \overline{BD}^2 - \overline{CD}^2)}{2\,\overline{BC} \times \overline{BD}} \tag{14-10b}$$

Other methods of triangulation will occur to the reader.

Sometimes three or four wires are used. A number of results are thus obtained, with a resulting increase in the accuracy of the average.

14-28. Alternative Method. Sometimes it is possible to establish a long horizontal wire within the workings as shown in Fig. 14-10. A transit is bucked in so that, as it sweeps the wire, the wire remains on the cross hair. A line is then established on the surface with the transit. The transit should also be bucked in reversed, and the average of the two surface lines used. The transit should be leveled with great care, either with the telescope level or with a striding level, as previously described.

A position along the wire can be established by turning the transit 90°.

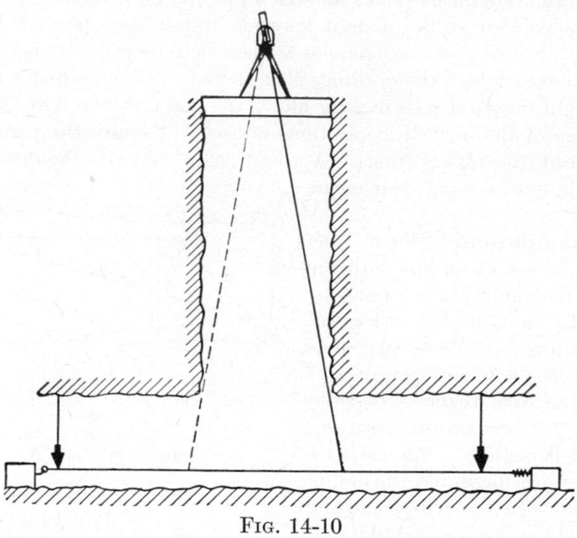

FIG. 14-10

14-29. Two Vertical Shafts. When two vertical shafts are available, a single wire is hung in each at A and D and the coordinates of the wires are determined by connections with the surface control (see Fig. 14-11). An underground traverse is run between the two wires, the azimuth of one underground course is assumed, and the coordinates are computed from an origin at one wire. Thus the true azimuth of AD can be computed from the surface coordinates, and the underground azimuth of AD can be computed from the underground coordinates. The difference between them is the error of the assumed azimuth and is therefore equal to the error of the bearings of each course. They are corrected accordingly. The length of each course is adjusted in proportion to the error of the length of AD as determined underground.

An example of this computation is shown in Table 14-1. It is based on the operation shown in Fig. 14-11. The survey data shown have all been reduced to the horizontal.

TABLE 14-1. EXAMPLE OF COMPUTATION OF TRAVERSE BETWEEN TWO SHAFTS

Surface Control

Sta.	*Bearing* / *Length*	*cos* / *sin*	*N*	*E*
1–*A*	S72°18′45″W	0.303825	98.17	388.23
	132.64	0.952728	−40.30	−126.37
		A	57.87	261.86
2–*D*	N82°29′59″W	0.130531	429.21	214.62
	144.54	0.991444	+18.87	−143.30
		D	448.08	71.32
Diff. *A* to *D*:			+390.21	−190.54

Tan br. *A* to *D*: −190.54/390.21 = −0.488301

Br. *A* to *D*: N26°01′35″W

Length *A* to *D*: 390.21/cos 26°01′35″ = 434.25

Underground Traverse

Sta.	*Bearing* / *Length*	*cos* / *sin*	*N*	*E*
A–*B*	N60°00′00″W	0.500000		
	152.89	0.866025	+ 76.44	−132.41
B–*C*	N0°14′48″E	0.999991		
	163.18	0.004305	+163.18	+ 0.70
C–*D*	N35°29′56″W	0.814127		
	161.23	0.580687	+131.26	− 93.62
Diff. *A* to *D*:			+370.88	−225.33

Tan br. *A* to *D*: −225.33/370.88 = −0.607555

Br. *A* to *D*: N31°16′52″W

Length *A* to *D*: 370.88/cos 31°16′52″ = 433.97

Adjustment of Underground Traverse

A to *D* control (fixed):	N26°01′35″W	434.25
A to *D* underground:	N31°16′52″W	433.97
Diff.	5°15′17″	0.28

Correction to bearings: 5°15′17″ clockwise

Correction to lengths: $\frac{0.28}{433.97}$ × length of course

Corrected Underground Traverse

Sta.	*Bearing* / *Length*	*cos* / *sin*	*N*	*E*
A–*B*	N54°44′43″W	0.577212	57.87	261.86
	152.99	0.816594	+ 88.31	−124.93
		B	146.18	136.93
B–*C*	N5°30′05″E	0.995394		
	163.28	0.095870	+162.53	+ 15.65
		C	308.71	152.58
C–*D*	N30°14′39″W	0.863887		
	161.33	0.503686	+139.37	− 81.26
		D	448.08	71.32

14-30. Side Measurements from Traverse. The positions of points on the roof, walls, and floor of workings are measured by azimuths and distances, distances and offset, and any of the other methods used in surface surveys (see Fig. 14-12).

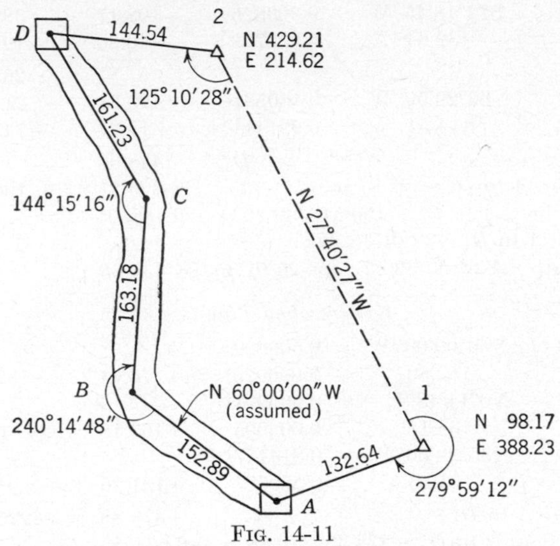

Fig. 14-11

14-31. Stopes. Most of the ore is excavated from rooms off the passageways. Such a room is usually called a **stope.** Its connection with the passageway is a **mill hole,** or if it slopes down from the stope, a **chute.**

Frequent measurements of each stope are made to determine the quantity of ore or waste material removed and the quantity of loose material that remains on the floor. A traverse connection from the main traverse is run through the mill hole and measurements are made from it. Elevations on the floor and on the roof are taken at the intersection of a grid system, and the area is carefully measured by plus and offset.

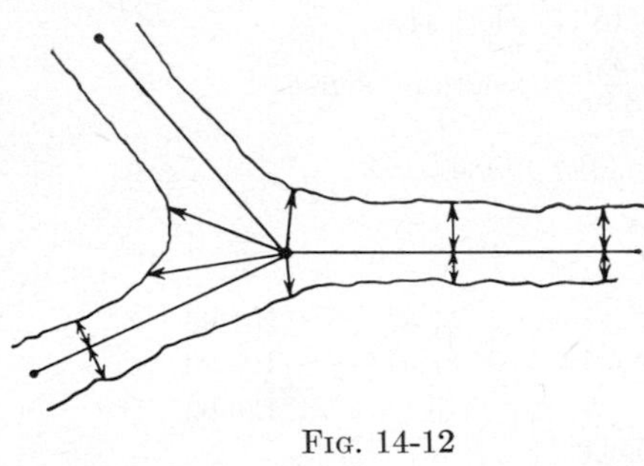
Fig. 14-12

14-32. Enumeration of Traverse Stations. Stations are usually numbered in three digits. Stations in the first level (that nearest the surface) are numbered from 100 to 199; in the second level, 200 to 299; etc.

14-33. Example of a Typical Traverse. Figure 14-13 illustrates a typical traverse. The notes for this traverse are given in Fig. 14-14, and

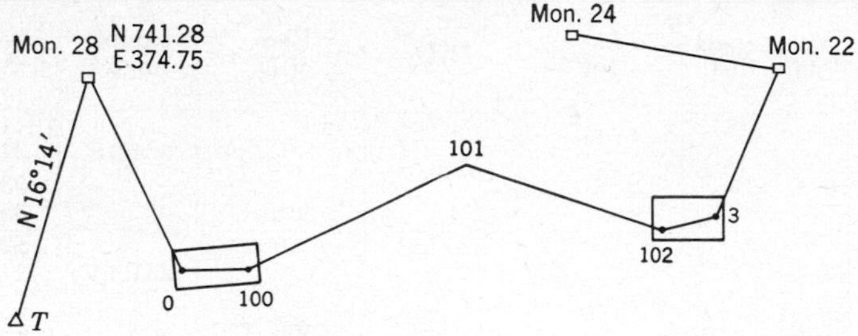

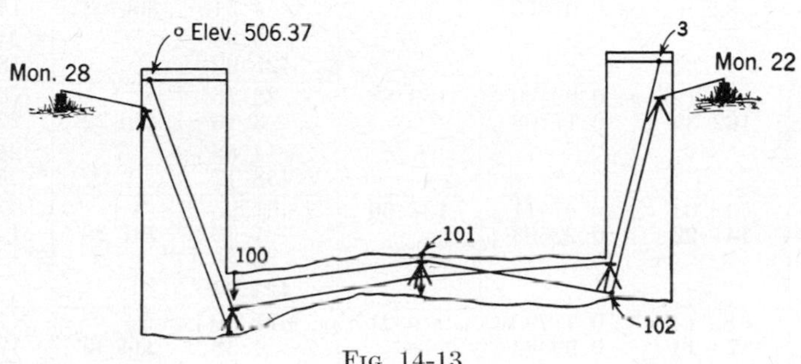

FIG. 14-13

HI	Sight	Vernier ° '	Sl. Dist	Vert ∡ ° '	H.S	Remarks
Mon 28	T	0 00				
	0	317 09	137.52			Spad in Beam Elev. 207.81
	T	" "	Hor.			
	0	274 18				
0	Mon 28	0 00				
-4.31	100	114 32	82.16	-78 10	+6.34	Spad in shaft timber
	Mon 28	" "				at [illegible] Side Telescope
	100	231 14				
100	0	0 00				
-3.16	101	154 07	162.39	+10 12	+4.58	Spad. Side telescope
	0	" "				
	101	306 02				
101	100	0 00				
-1.32	102	225 10	141.22	-13 04	0	Spike in floor
	100	" "				at East shaft
	102	90 22				
103	101	0 00				
+4.38	3	147 36	78.14	+83 14	0	Spad in beam. Side
	101	" "				telescope
	3	297 28				
3	102	0 00				
	Mon 22	123 24	108.24			Side telescope
	102	" "	Hor.			
	Mon 22	244 36				

FIG. 14-14

Sta.	Vert. angle slope dist.	cos sin	H.D.	Elev. V.D.	Double angles	Bearings Angles
						N16°14′E
					634°18′	317 09
						333°23′
						359 60
Mon 28			137.52	0		S26°37′E
					231°14′	115 37
0				506.37		
0	−78°10′	0.20507	16.85	−80.41		N89°00′E
	82.16	0.97875		− 4.31	306°02′	153 01
				+ 6.34		242°01′
100				427.99		180 00
100	+10°12′	0.98420	159.82	28.76		N62°01′E
	162.39	0.17708		3.16	450°22′	225 11
				4.58		287°12′
101				458.17		359 60
101	−13°04′	0.97411	137.56	−31.93		S72°48′E
	141.22	0.22608		− 1.32	297°28′	148 44
				0		
102				424.92		
102	+83°14′	0.11783	9.21	+77.60		N75°56′E
	78.14	0.99303		4.38	244°36′	122 18
				0		198°14′
3				506.90		180 00
3			108.24			N18°00′E
Mon 22						

Sta.	Bearing Length	cos sin	N	E
Mon. 28	S26°37′E	0.89402	741.28	374.75
0	137.52	0.44802	−122.95	+ 61.61
0	N89°00′E	0.01745	618.33	436.36
100	16.85	0.99985	+ 0.29	16.85
100	N62°01′E	0.46921	618.62	453.21
101	159.82	0.88308	+ 74.99	+141.13
101	S72°48′E	0.29571	693.61	594.34
102	137.56	0.95528	− 40.68	+131.41
102	N75°56′E	0.24305	652.93	725.75
3	9.21	0.97001	+ 2.24	+ 8.93
3	N18°14′E	0.94979	655.17	734.68
Mon. 22	108.24	0.31289	+102.81	+ 33.87
Mon. 22			757.98	768.55

FIG. 14-15. Traverse computation.

the computations are shown in Fig. 14-15. The operations are evident from the foregoing sections.

14-34. Mine Maps. Figure 14-16 shows a map of the passageways in a mine. In addition to the plan, a longitudinal section and a transverse section are shown. They coincide with the axes of the coordinate system. It is evident that the drawings of the passageways overlap so much that the map is confusing.

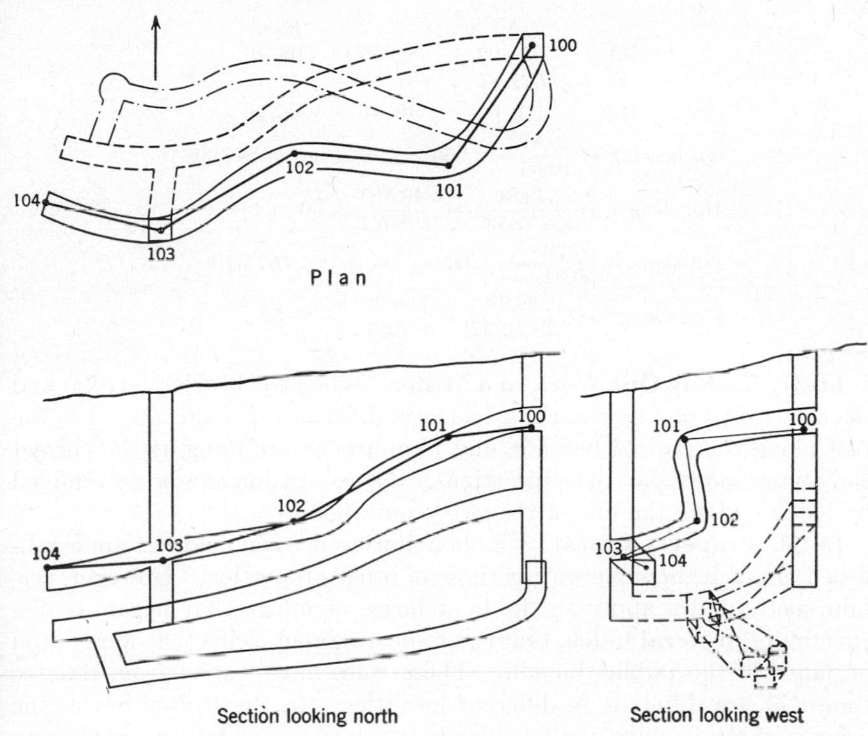

FIG. 14-16

To avoid this, each level is usually plotted separately. Sometimes the coordinate system is oriented so that the least confused views will be obtained, and different colors are used for different levels. Usually each stope is plotted on a special map to a large scale. The stope sheets are kept up to date as the work progresses.

14-35. Mine Models. Sometimes each level is mapped on a sheet of glass and the whole mounted in the proper relative positions, or rather intricate models are made of reinforced plastic material. The excavations are represented as solids. All these devices help to show the relative positions of the parts of the mine so that future work can be planned more easily.

14-36. Computational Problems. Many problems are encountered in determining the direction, slope, and length to be followed in laying out workings. These are usually most easily solved by coordinates.

Example. Assume that it is required to build a connection in the mine from the point *A* at one level to the point *B* on another level. The coordinates and elevation of each point have been determined by the mine survey. Find the bearing, slope, and length of the line *AB*.

	N	*E*	*Elev.*
A	572.92	981.65	208.46
B	562.87	931.84	142.17
Diff.	10.05	49.81	66.29

$$\text{tan bearing} = \frac{49.81}{10.05} = 4.9562 \qquad \text{Bearing} = \text{S}78°36'\text{W}$$

$$\text{Hor. length} = \frac{49.81}{\sin 78°36'} = \frac{49.81}{0.98027} = 50.81$$

$$\text{tan slope} = \frac{66.29}{50.81} = 1.3047 \qquad \text{Slope} = -52°32'$$

$$\text{Slope length} = \frac{66.29}{\sin 52°32'} = \frac{66.29}{0.79371} = 83.52$$

14-37. To Lay Out Work in a Mine. When the bearing (strike) and the slope (dip) of a new working has been determined, spads are set in the roof along the desired bearing, and plumb bobs are hung at the correct elevations and slope. The direction of the excavation can be determined by looking along the line of the two plumb bobs.

14-38. Property Lines. The boundaries of mine holdings are established in much the same way as those of other properties. However, certain special rules apply to the boundaries of mines that began as discoveries of **mineral lodes,** that is, ore-bearing rock veins that were found on lands of the public domain. These rules have varied from time to time and are different in different localities. In the United States the process of establishing the boundaries is somewhat as follows, subject to local variations:

When a prospector discovers a lode, he may stake out a claim whose size and shape is specified by law. The center line of the claim can be established in any direction desired, it may consist of one or more courses, its total length must not exceed 1500 ft, and it must usually pass through the point of discovery. The side lines must not be more than 300 ft each side of the center line. The end lines must pass through the ends of the center line; they can take any desired direction but they must be parallel to each other. The maxima of 1500 and 300 ft may be reduced by local laws. Special rules govern the markers to be used to show the boundaries. Sometimes a surveyor is employed to assist in staking out the claim. The process of staking out a claim as described constitutes the **location.** The location may be officially recorded by filing a **certificate of location** which includes a description of the claim which must define a legally acceptable size and shape and give a general reference to a nearby landmark.

When the owner has completed $500 worth of improvements on the claim and he has filed a certificate of location, he may have the claim patented. A survey for patent must be made by a **mineral surveyor** who is appointed by the federal government and is bonded. He measures the actual boundaries as they are marked on the ground and ties them to the nearest public land corner. If no corner exists within two miles, he establishes a location monument in a prominent position and runs a tie from that. He makes ties to patent corners of adjoining claims and locates the intersections of any overlapping claims. He must place permanent marks at the corners of the claim and at the ends of the center line. His work must conform with a high standard of accuracy. If the description in the original certificate agrees closely with his measurements and if there is no encroachment on property of others, the certificate is allowed to stand; otherwise a new certificate must be filed. If desired, and if no rights of others are affected, an entirely new location and a new certificate can be established in the field at this time.

When the location is surveyed and is proved to be acceptable, the mineral surveyor files a **return.** The return consists of a description in the form of special field notes and a plat.

When the patent is granted, the owner has exclusive rights to develop the lode. While in most countries the limits of rights are vertical planes through the boundaries, in the United States the **rule of apex** applies. The apex is the highest point of the lode. Usually a surface discovery is at the apex, although an apex may be discovered underground from other diggings. The owner of the apex may follow the lode beyond his side boundaries but he is limited by vertical planes through his end boundaries or his end boundaries prolonged. The rights to follow the particular lode beyond the side boundaries are often called **extralateral rights.**

It is evident that claims may overlap without the knowledge of the original locators and later cause confusion. Extralateral rights are often difficult to establish when there is a question whether or not the same lode or two different lodes were discovered on two adjacent claims. Frequently, adjacent claims of unknown value are purchased by the owner of a productive claim to protect his extralateral rights.

It is evident from the foregoing that the prospector should make every effort to locate his claim so that the greatest part of the lode is contained between the two parallel end planes of the claim. The side lines are not so important because of the extralateral rights.

In addition to the claim, a **mill site** may be patented. This should be a rectangle 330 by 660 ft. It is located off the claim but adjacent to it, and it does not carry extralateral rights.

It is evident that, to conduct a survey for a lode mine properly, the surveyor must be familiar with a large special body of law, have a precise knowledge of the regulations that govern his work, and be familiar with

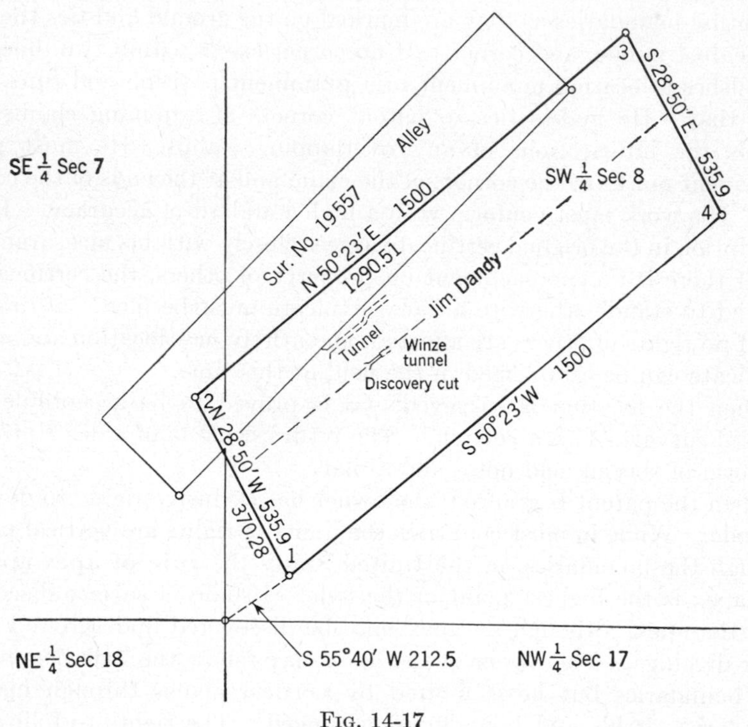

FIG. 14-17

the survey marks and customs in the locality. Accordingly, it is seldom wise for an engineer in charge of the surveys of the mine workings to attempt to redetermine a lost boundary without consultation with a surveyor who is experienced in this work.

Figure 14-17 shows a typical plat for a patent survey. Note that the side 4-1 is parallel to the center line, although not exactly 300 ft from it. For further details of the requirements for a mineral patent survey, the reader is referred to "Manual of Instructions for the Survey of the Public Lands of the United States," prepared and published by the Bureau of Land Management, 1947, from which Fig. 14-17 has been adapted.

PROBLEMS

14-1. Compute the precise slope distance from the following data. Use all corrections.

$$L = 191.39 \qquad l_1 = 100.00 \qquad l_2 = 91.38 \qquad F = 75°\text{F}$$
$$\alpha = 52°18' \qquad p = 40 \text{ lb} \qquad C_{st} = +0.009$$
$$S = 0.0040 \text{ sq in.} \qquad w = 0.014 \text{ lb}$$

14-2. From the following data, which apply to Fig. 14-9, compute the azimuth station 101 to station 102.

$$\text{Azim. } CB = 38°12' \qquad A = 16°28' \qquad T = 172°39'$$
$$a = 7.362 \qquad b = 20.374 \qquad c = 14.972$$
$$CD = 12.177 \qquad BD = 10.279 \qquad AD = 10.182$$

14-3. The following are data from an underground traverse. Find the coordinates of stations *B* and *C*.

Hor. dist.: *AB* 123.36; *BC* 197.42; *CD* 364.07
Clockwise angles: at *B*, *A* to *C*, 127°49′
at *C*, *B* to *D*, 261°56′
Coordinates *A*, North 261.38; East 461.89
D, North 346.43; East 1016.61

14-4. From field notes in Fig. 14-18, compute the coordinates and the elevation of station 203.

Azim Δ 10-Mon 20° 168°22′ Coor Mon 20 N 10000 E 452 86
⊼

HI	Sight	Vernier	Sl. Dist	Vert ∡	H.S.	Remarks
		° ′		° ′		
Mon 20	Δ #10	0 00				
	A	98 17	176.48			Spad in beam Elev. 731.48
	Δ #10	" "	Hor.			
	A	196 35				
A	Mon 20	0 00				
−5.61	201	126 10	92.61	−81 03	0.00	Spike in floor N. Shaft
	Mon 20	" "				Level #2 Side telescope
	201	254 08				
201	A	0 00				
+3.28	202	192 16	68.18	+3 02	+2.68	Spad in roof timber
	A	" "				
	202	24 32				
202	201	0 00				
−2.01	203	167 58	75.31	+2 05	−0.96	Chisel X in Floor East Side
	201	" "				
	203	335 57				

Fig. 14-18

CHAPTER 15

HYDROGRAPHIC SURVEYS AND STREAM GAGING

HYDROGRAPHIC SURVEYS PROPER

15-1. Definition. Hydrographic surveys proper can be defined as surveys made to measure the position of the bottom of bodies of water. They are chiefly used to make nautical charts for navigation but they are often essential for planning and controlling engineering projects like bridges, tunnels, dams, reservoirs, docks, and other river and harbor improvements.

15-2. Method. Usually the depth of the water is measured by taking soundings at selected points. If the height of the water surface is changing, the measured depth must be reduced to a common datum. The final data are usually given in terms of depths from a selected water surface level or **stage.** For engineering projects, they are usually given in elevations. Under certain conditions, elevations can be measured directly.

Horizontal survey control must be established on shore from which to indicate the positions where the soundings are to be taken and also to determine the actual positions of the soundings when they are made. Usually tide or water-stage gauges must be kept in operation to establish the common datum and to give the height of the water when each sounding is taken. Vertical control must be established to connect these gauges with shore elevations and with each other.

15-3. Wire Drag. Isolated rocks, wrecks, or other obstructions can be missed by soundings. Where these might exist, a wire drag is used to make sure of their absence at any desired depth. The wire drag is a wire, held level by buoys, which is towed over the area.

Sounding Equipment

15-4. Measuring Devices. For most engineering work, soundings are taken from small boats with a sounding pole or a leadline.

A **sounding pole** is usually a pole 15 ft in length made of $1\frac{1}{2}$-in.-round lumber, capped with a metal shoe at each end and sometimes weighted. It is marked symmetrically from each end. Between soundings it is turned end for end without removing it from the water. A 15-ft pole can be used to depths up to 12 ft.

A **leadline** is usually a length of sash cord, or tiller rope of Indian hemp, or braided flax, with a sounding lead attached to the end. Sometimes the line is made with a phosphor-bronze wire center. Some engineers use a light chain. The type of chain used for sash weights is best, because it does not kink and it is not so hard on the leadsman's hands. The lead weight should weigh at least 8 lb for depths up to 40 ft; 12- to 14-lb weights are satisfactory for greater depths. The weight should be somewhat streamlined and have an eye at the top for attaching the cord. It often has a cup-shaped cavity in the bottom so that it may be armed with lard or tallow to pick up samples from the bottom (see Fig. 15-1). A lead-filled pipe with a board at the top is used where the bottom is very soft. The weight penetrates the mud and stops where the board strikes the mud surface.

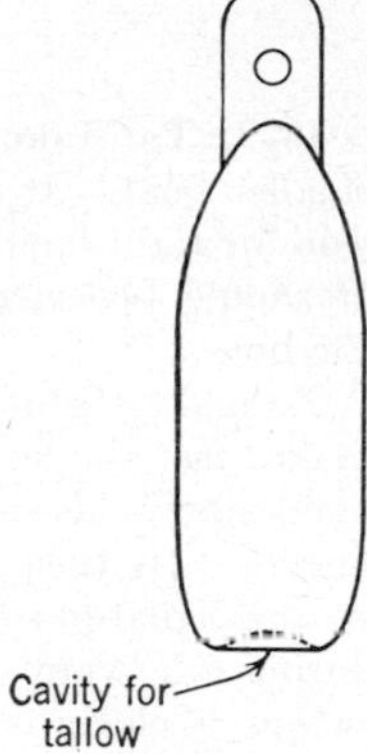

FIG. 15-1. A sounding lead.

15-5. The Toggle. For soundings from a rapidly moving boat, the leadsman swings the lead to heave it forward. When the lead is heavy he must have a toggle on the line to hold it. This is a small, round piece of wood seized to the line or held in a clove hitch so that it is perpendicular to the line. It is placed up to 18 ft from the lead but not far enough to allow the lead to strike the tops of the waves. When it is held by a hitch in the line it must be attached before the line is marked.

15-6. Stretching the Line. A line shrinks in length when wet and stretches out again when dry, but a new line will stretch while it is in use. This stretch can be removed by winding it tightly around a post and wetting it. When it dries, the slack is taken up and it is wet again. This is repeated until it has very little slack when dry.

15-7. Marking the Line. The toggle and the lead should be attached, and the line soaked for 24 hr, before it is marked. It is then stretched with a tension equal to the lead in water and temporarily marked. It can be dried and marked permanently later. It should be soaked for 24 hr before it is used for sounding. After use its length should be tested.

15-8. Suggested Marking System. The U.S. Coast and Geodetic Survey recommends the following systems of marking the poles and the leadlines:

Poles. Make a small permanent notch at each half-foot. Paint the entire pole white and the spaces between the 2- and 3-, the 7- and 8-, and the 12- and 13-ft marks black. Paint a ½-in. red band at the 5- and 10-ft marks, a ½-in. black band at each of the other foot marks, and ¼-in. bands at the half-foot marks. These ¼-in. bands are black where the pole is white, and vice versa.

Leadlines. A leadline is marked in feet as follows:

Feet	*Marks*
2, 12, 22, etc	Red bunting
4, 14, 24, etc	White bunting
6, 16, 26, etc	Blue bunting
8, 18, 28, etc	Yellow bunting
10, 60, 110	One strip of leather
20, 70, 120	Two strips of leather
30, 80, 130	Leather with two holes
40, 90, 140	Leather with one hole
50	Star-shaped leather
100	Star-shaped leather with one hole

The intermediate odd feet (1, 3, 5, 7, 9, etc.) are marked by white seizings.

Soundings

15-9. To Make a Sounding. A sounding platform should be built for use in the smaller boat. It should be placed well forward of amidships and on the starboard side for right-handed leadsmen. It consists of a platform with a strong 3-ft railing. It should be extended far enough over the side to prevent the line from striking the boat.

To make a sounding from the starboard side of a moving boat, the leadsman first makes fast the bitter end of the line forward of his position, coils the line in his left hand, and holds the toggle in his right hand with the line between his first and second fingers. He then stands on the platform facing forward with lead hanging outboard. At the signal to sound, he swings the lead back and forth and then heaves it forward during a forward swing. When the boat is moving rapidly, or the water is deep, he swings it clockwise in a complete vertical circle and releases it while it is moving upward and forward. He pays out the line freely as the lead pulls it but hauls in any slack caused by the forward movement of the boat. As the boat moves over the lead, he "feels" the bottom by raising and dropping the lead and pulls the line taut at the moment it is vertical. He calls the mark at the surface of the water, attempting to take the mean of the waves. He then hauls in the line for the next cast. As he brings in the line he sometimes coils it in his left hand or, more often, an assistant handles it.

15-10. Echo Soundings. Devices usually called **fathometers** are available that make a continuous, accurate record of the depth of water below the boat or ship in which they are installed. Essentially, they create a sound wave in the water near the water surface and record the time interval from the moment of the original sound until the return of the echo from the bottom. They are adjusted to read depth in accordance with the velocity of sound in the type of water in which they are being used. The operation of these instruments is thoroughly described by K. T. Adams in U.S. Department of Commerce Coast and Geodetic Survey Hydrographic Manual Special Publication 143, revised edition, 1942.

15-11. Sounding Craft. Rowboats, motor launches, and small ships can be used for making soundings. When very small boats must be used, it is well to fasten two together to make a catamaran to give a firm support for the sounding platform.

Sometimes when motor launches are used, a log drag is rigged and towed astern to regulate the speed. Lines are attached to each end so that the angle of the log and hence its resistance can be regulated.

15-12. Ranges. Ranges are usually used to indicate positions where soundings are to be taken. Each range is usually marked at one end by

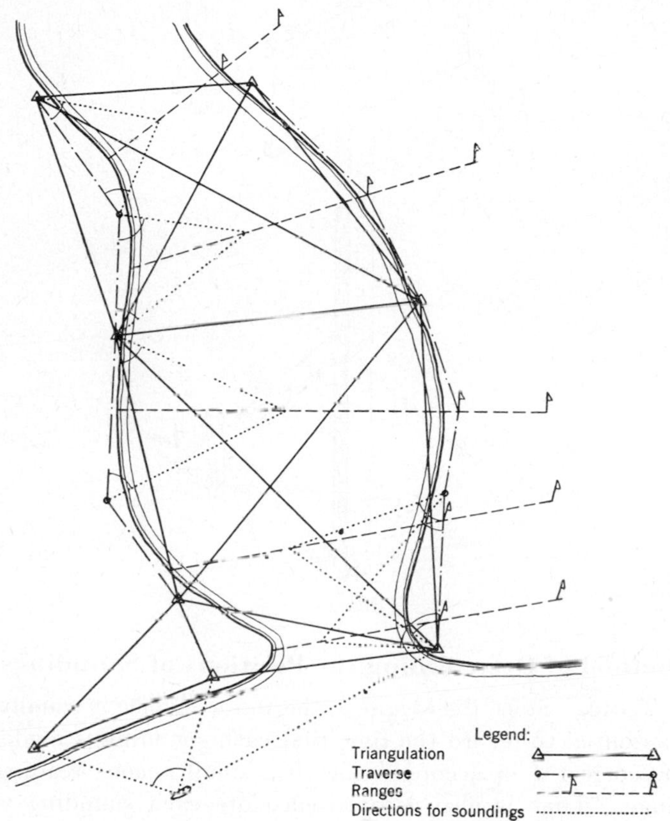

FIG. 15-2. Survey scheme for a hydrographic survey.

two signals on shore or, in shallow water, by two buoys, or by a combination of both. Cross ranges should be used where soundings are to be taken several times at the same points. The positions for the signals are established from the horizontal control (see Fig. 15-2).

Sometimes a range is marked by a single signal or a buoy. The boat then follows a compass course from the mark.

15-13. Signals. Signals can be constructed in a variety of ways. They should be readily seen and easily distinguished from each other.

Usually a vertical post is mounted at a convenient height and colored; square pieces of cloth are then stretched on crossarms nailed to the post (see Fig. 15-3). When signals are to be observed by sextant (see later paragraphs), they should be established preferably at nearly the same height as the eyes of the observers in the sounding launch.

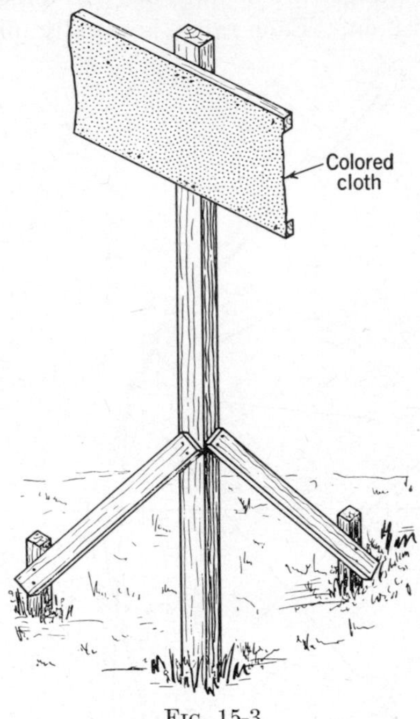

FIG. 15-3

Method of Determining the Positions of Soundings

15-14. Time. Since the height of the water surface is usually changing, it is essential to record the time that each sounding is made so that it may be corrected in accordance with a simultaneous reading of the water gauge. Time is also used to correlate each sounding with the readings of the instruments trained on the boat to locate the soundings.

15-15. Survey Methods. Soundings can be located by the following methods. Often these methods are combined. The reader will find that a special combination is usually best for any particular project.

1. Intersecting ranges
2. Graduated line or wire
3. One range line and one angle from shore
4. Two angles from shore
5. One angle and a stadia distance from shore

6. Two angles (measured with sextants) from the boat to three (or four) signals or buoys
7. One range and one sextant angle from the boat to two signals or buoys
8. Uniform speed and time intervals

15-16. Intersecting Ranges. Figure 15-4 illustrates a method used to control soundings during dredging operations for driving piles for

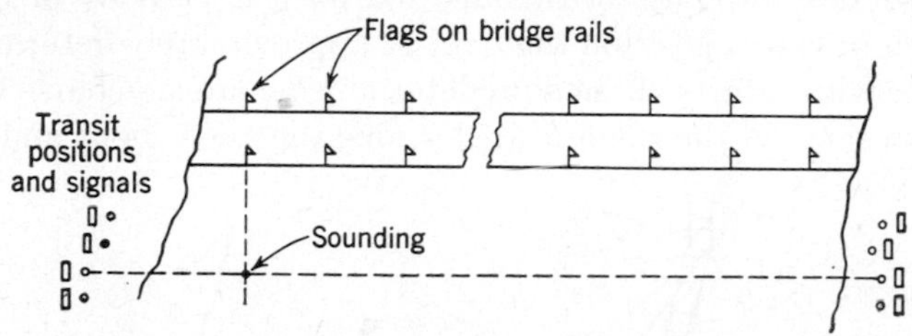

FIG. 15-4. Intersecting ranges.

bridge pier foundations. Permanent transit positions and signals are set on shore and flags are attached to the bridge rails. The transit can be placed on either shore and oriented with the opposite signal. The boat is placed on line with a pair of flags on the bridge, and the transitman places the boat on his line by signals with a flag.

15-17. Graduated Line. Figure 15-5 shows a method for sounding a small stream. Stakes are set on shore where profiles of the bottom are

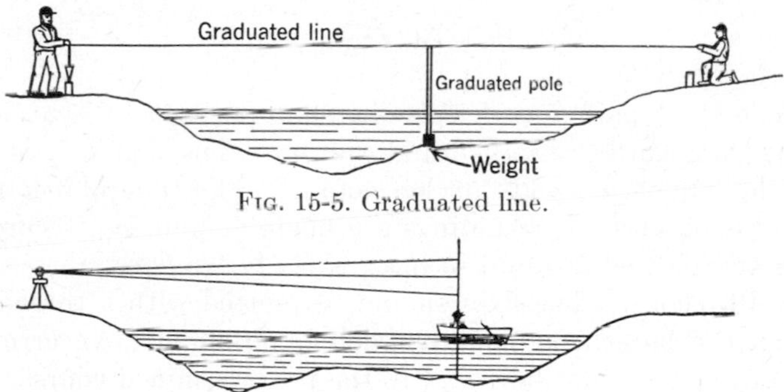

FIG. 15-5. Graduated line.

FIG. 15-6. Stadia method.

required. The depth of water can be read on the weighted pole. The position of the pole is given by the graduated line. The two men can hold the pole erect where desired. The elevation of the water surface is taken with a level and a rod from time to time.

15-18. Stadia Method. Figure 15-6 shows a convenient method for a small body of water. Level stadia shots are taken on a stadia rod held on the bottom. The stadia intercept and the azimuth locate the sounding. Either the center wire or the water surface can be read for depth of water. To avoid confusing the notes, the transitman rather than the boatman takes the rod reading at the water surface; he can see it easily with his telescope.

15-19. Two Sextant Angles. As explained in later paragraphs, an angle can be measured with a sextant from a moving platform. The three-point method is employed when two sextant angles are measured. Figure 15-7 shows the use of this method for a large body of water.

A launch runs to a position that can be approximately determined from shore objects and starts off on a predetermined compass course and speed. Two observers read the sextant angles, one the angle right and the other

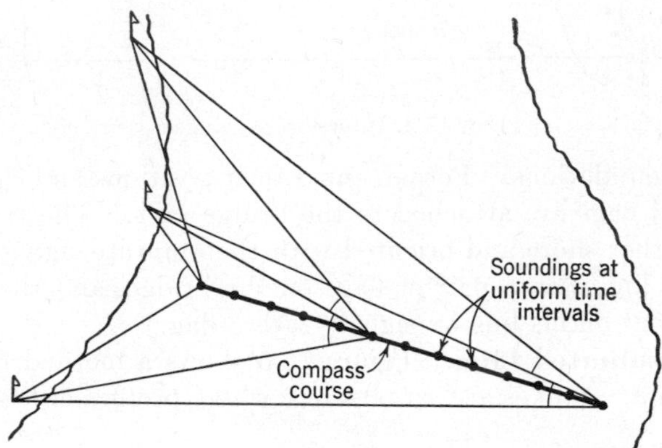

FIG. 15-7. Three-point method.

the angle left. A pair of angles gives a fix (a position). Fixes are taken as fast as they can be observed and plotted. The first pair is taken as soon as the launch is steady on her course. The time is recorded each time a fix is observed. Soundings are made at uniform, recorded time intervals and plotted at uniform intervals between fixes.

15-20. Plotting. The fixes should be plotted with a three-arm protractor on the launch, as soon as each is observed. An error can be noted if any fix does not fall close to the predetermined course. Soundings are plotted temporarily along the compass course and corrected later for interval and depth.

Example of a Hydrographic Survey

15-21. An example of a well-organized hydrographic survey is described here. It combines several methods of locating soundings (see Fig. 15-2).

15-22. Water Gauge. A recording water gauge is established as soon as possible (see later paragraphs).

15-23. Reconnaissance. Any existing maps or air photographs are obtained. A triangulation system is laid out that will connect both shores and serve as horizontal control.

15-24. Preliminary Survey. The triangulation net is surveyed. A traverse is run along each shore and the shore line, and any desirable shore objects are mapped, particularly those that can be used as range markers. Locations for any necessary auxiliary water gauges are selected. At least three benchmarks are established at each water gauge location and connected by levels with a standard datum.

15-25. Plan. Range lines are laid out on the map. They should run as nearly perpendicular as possible to the channels. Often existing structures like church spires can be used to mark one end of a range.

15-26. Location Survey. The locations for the range marks are established in the field by scaled measurements from the horizontal control. They should be built as soon as possible, tied to the survey net, and their final positions plotted on the map. Transit positions are staked out along the shore in such a way that the entire length of each range can be observed from at least two transit positions. A range signal or some other point on the horizontal control system must be visible from each transit position so that the instrument can be oriented. Each transit position is tied to the horizontal control.

15-27. Soundings. Both transitmen and the recorder in the boat synchronize their watches. The transits are set up at the proper points to cover the range to be sounded. The transitmen orient their transits to the azimuth of the coordinate

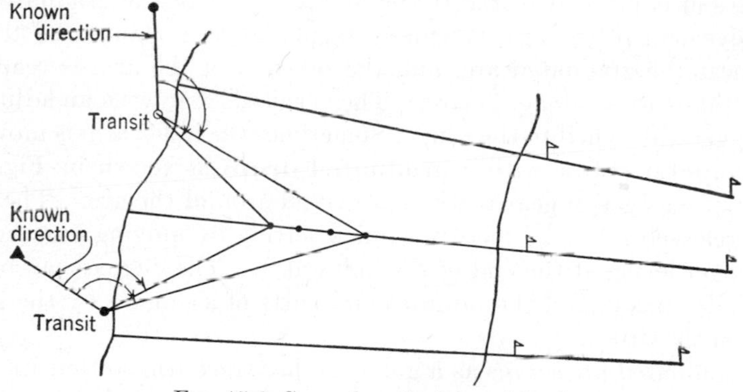

FIG. 15-8. Several methods combined.

system and pick up the boat in their instruments. Each man keeps his cross hair on the boat continuously. Starting at one shore, the boat proceeds along the range at a constant speed. The first sounding is taken at an exact minute of time. At the moment the sounding is taken, a red or a white flag is displayed on the boat. The transitmen point their instruments at the flag and let this pointing stand. They record the minute of time, the color of the flag, and the azimuth, to the nearest minute. They again place their cross hairs on the boat (see Fig. 15-8).

The recorder records the time, the color of the flag, and the value of the sounding. Soundings are taken every 20 sec but a flag is displayed only every minute. The color of the flag is chosen more or less by chance but a single color should not be displayed more than four times consecutively. The recorder calls the time and records the depth for each sounding, but he records the color of the flag and the time only when the flag is displayed.

The chief difficulty in making any hydrographic survey is the proper correlation of the data. Although in the example given the data are recorded at three different localities, the method is reasonably foolproof. Corresponding data are first collected according to the times indicated. They are then checked by the colors of the flags. Finally, when the angles are plotted the intersections should fall very nearly on the range and at fairly uniform intervals.

The recorded depths are corrected according to the water-gauge readings before being placed on the chart.

The Sextant

15-28. Construction of a Sextant. Figure 15-9 is an illustration of one form of a sextant. It is a hand instrument originally designed to measure the altitudes of heavenly bodies from the deck of a ship. The important feature of a sextant is that it can be used to measure angles from a moving platform. In hydrographic surveys it is used on board the sounding boat to measure horizontal angles between control points.

The instrument consists essentially of two parts, the **frame** and the **index arm.** A handle is attached to the back of the frame, and on the front of the frame are mounted the **graduated arc,** the **horizon mirror,** and a **holder** for low-power telescopes. The index arm carries the **index mirror** and is pivoted to the frame at the center of the graduated arc. The movement of the arm is usually regulated by a clamp and tangent screw near the graduated arc, and the position of the arm is read by a vernier that travels along the arc. The vernier is read with an adjustable reading glass attached to the arm. Sometimes the index arm is moved by a **micrometer screw** with a **graduated drum** as shown in Fig. 15-9. The screw engages in gear teeth that extend around the arc. The screw can be released from the teeth for quick setting by moving together two small finger levers at the end of the index arm. The degrees can be read by an index mark, and the minutes and parts of a minute by the graduations on the drum.

The graduated arc serves as a guide to the index arm so that its movement is in the plane of the arc. The two mirrors are adjusted perpendicular to the plane of the arc so that the angle measured is the angle between the projection of the two sight lines on the plane of the arc.

Usually a peep sight and one or two telescopes of different low power are provided. The instrument can be used without any of these. The peep sight improves the focus of the objects and helps keep the line of sight parallel to the plane of the arc. The greater the power of the telescope the more accurate the results will be, but the higher the power the smaller the field of view and the more unsteady the objects viewed appear.

In addition, the instrument has two sets of colored **filters** that may be swung into position. With these, the brightness and color of the objects viewed may be regulated. A colored eyepiece filter is also often provided. The filters are chiefly used for sun observation in navigation. Under unusual conditions they are used in hydrographic surveys.

FIG. 15-9. A sextant. 1, frame; 2, index arm; 3, handle; 4, graduated arc; 5, horizon mirror; 6, opening over mirror; 7, telescope holder; 8, telescope; 9, index mirror; 10, micrometer screw; 11, graduated drum; 12, finger levers; 13, gear teeth; 14, filters. (*U.S. Navy.*)

15-29. To Measure a Horizontal Angle. To measure a horizontal angle, the observer holds the sextant in his right hand with the arc horizontal and the mirrors uppermost. He looks through the telescope, parallel to the circle, over the top edge of the horizon mirror at the left-hand mark. He moves the index arm until the image of the right-hand mark reflected by the index mirror appears in the horizon mirror. It is convenient to hold the index arm by the clamp screw (or the finger lever). He then clamps the index arm (or allows the screw to engage with the gear teeth) and makes a fine adjustment with the tangent screw (or the drum screw) until the image of the right-hand mark apparently coincides with the left-hand mark. The angle between the marks can then be read on the graduated arc.

It must be remembered that the angle measured is in the plane of the arc. The arc must therefore be held very nearly horizontal in order to obtain a true horizontal angle. Signals should therefore be built at as nearly eye height above the water as possible. If this is impossible, the observer must superimpose points on the projections of the marks that are at eye height. Fortunately, the shore line can usually be seen so that it is not difficult to project mentally the marks perpendicular to this line.

When the instrument is in proper adjustment, the two points will not appear to be superimposed unless they are very nearly on the plane of the arc. When they appear to be above or below each other, the instrument is tilted left or right. If one of the marks fades out, the instrument is tilted forward or back, or the eye is not in the center of the eyepiece.

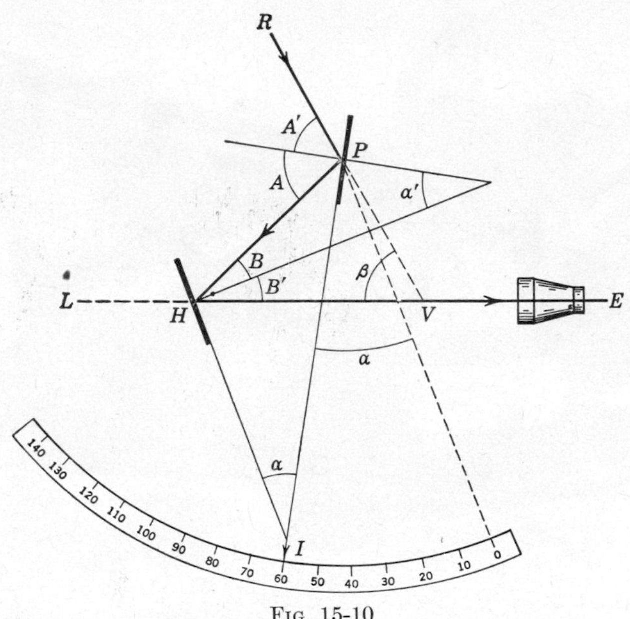

FIG. 15-10

Since the mirrors must be slightly offset to each other, a certain parallax is introduced between them so that angles to nearby points are not exactly accurate. The vertex of the angle measured moves backward as the angle grows smaller (see Fig. 15-10). In hydrographic works this error is negligible.

15-30. Optical Principles. It is an optical principle that when a ray of light is reflected by two mirrors, it changes direction by twice the angle between the mirror surfaces. The change of direction is measured by the trace of the ray in a plane mutually perpendicular to the mirrors. This is illustrated in the case of the sextant in Fig. 15-10, in which the plane of the paper is perpendicular to both mirrors. Since the change in the direction of the ray is affected only by the angle between the mirrors, as long as this angle remains fixed, **simultaneous rotation of both mirrors** in the plane mutually perpendicular to them will not affect the change in direction of the ray.

15-31. Principles of the Sextant. Figure 15-10 shows the principles of the sextant. The observer's eye is at E. He sees the left-hand point over mirror H at L. In the mirror H he sees the right-hand point R after two reflections. He adjusts the index arm P.I. (pivoted at P) until the index mirror at P is at the proper angle to make R and L apparently coincide.

The actual angle between R and L is β, V is the moving vertex, and α is the angle between the mirrors. Since the angle of reflection equals the angle of incidence,

$$A = A' \qquad B = B'$$

$$\alpha = \alpha' \qquad \text{(sides perpendicular)}$$
$$\alpha' = A - B \qquad \text{(exterior angle)}$$
$$\beta = A + A' - (B + B') \qquad \text{(exterior angle)}$$
$$\beta = 2A - 2B$$
$$\alpha = \frac{\beta}{2}$$

Therefore the angle between the mirrors is always equal to $\frac{1}{2}\beta$. The mirror P is adjusted so that it is parallel to mirror H when the index arm is at zero. As the index arm is moved around the scale, it turns P and thus introduces an angle between the mirrors equal to its movement along the scale. The scale is numbered in values equal to twice the actual angle on the scale so that the value of β is read directly.

15-32. Geometric Requirements of the Sextant. It is evident from the foregoing that in addition to an accurately graduated and centered arc, the geometric requirements of the sextant are the following:

1. The two mirrors shall be perpendicular to the plane of the graduated arc so that the change in the direction of the line of sight will be in the plane of the graduated circle by which it is measured.

2. The index should read zero or its reading should be known, when the two mirrors are parallel.

3. The lines from the observer's eye to the objects viewed should be roughly parallel to the arc of the graduated circle. Since the observer views the coincidence at the top of the horizon mirror, the peep sight should be at the same distance above the arc as the top of the mirror. When a telescope is used, the observer automatically places his eye nearly on the optical axis. The optical axis should therefore be parallel to the graduated arc and pass through the top of the horizon mirror.

15-33. Sextant Adjustments. The adjustments should be made in the order given.

1. *To make the index mirror perpendicular to the graduated arc.* Stand the sextant on a table and place the eye near the index mirror so that the mirror is between the eye and the vernier (or drum). Compare the reflection of the graduated arc (that is

seen in the index mirror) with the actual arc (that is seen to the observer's right of the index arm). Adjust the mirror by tilting it forward or backward until the reflection of the arc appears to be continuous with the arc itself. Check this with the index arm in different positions.

2. *To make the horizon mirror perpendicular to the graduated arc.* Set the vernier at approximately zero. Aim at some well-defined distant point, like a star, with the arc vertical. Move the index arm back and forth slightly. The image of the star will move up and down. Adjust the horizon mirror by tilting it forward or backward until, when the index arm is moved, the image of the star, in passing, will coincide with the star itself.

3. *To make the horizon mirror parallel to the index mirror when the vernier is set at zero (i.e., to eliminate any index correction).* Set the vernier at zero. Aim at some well-defined distant point like a star. Adjust the horizon mirror by turning it in the plane of the graduated arc until the image of the star coincides with the star itself. Recheck the previous adjustment of the horizon mirror.

Note: Few sextants will hold their adjustment even temporarily. It is therefore customary to determine the index correction before using the sextant. In this case, instead of adjusting the mirror, obtain the coincidence of a distant point by moving the index arm and then read the vernier. If the reading is "on the scale," that is, greater than zero, subtract it from all observed values, and vice versa.

4. *To make the optical axis parallel to the plane of the graduated arc and at the height of the top of the horizon mirror above the arc.*

a. Set the telescope holder so that the top of the mirror appears in the center of the field of view, in a low-power telescope.

b. Place the instrument on a table. Sight along the plane of the graduated arc and place a mark in line on a wall or other vertical surface 20 ft or so away. Measure the height of the axis of the telescope above the graduated arc and set another mark on the wall above the first by this amount. Adjust the telescope holder so that the center of field of view strikes the second mark.

Some telescopes have a pair of lines equally spaced from the center of the optical axis to guide this operation. Rotate these lines until they are parallel to the arc and adjust the telescope holder until the mark falls halfway between them. The accuracy of this adjustment does not have to be very great.

Water Gauges

15-34. Purpose. Water gauges are usually established for hydrographic surveys for the following purposes:

1. To determine the elevation of the surface of the water when each depth measurement is made. Depth measurements can then be used to determine the elevations of the bottom or they may be corrected to a common height of water.

2. In tidal water, to establish a datum relative to a certain stage of the tide.

3. In rivers, to determine the height of the water at a certain point so that the river flow may be estimated.

4. In reservoirs, to estimate the impounded volume.

Tide gauges are also used to establish mean sea level as a basis for geodetic leveling. This work is not part of a hydrographic survey.

15-35. Types of Water Gauges. Water gauges are of three general types:

1. *Staff Gauges.* A staff gauge (Fig. 15-11) usually consists of a painted board marked in feet and tenths and numbered upward. It is firmly attached in a vertical position in the water. The U.S. Coast and Geodetic Survey has developed a scale made of sections with vitrified coatings that may be attached to the board. These sections are easily cleaned and are

FIG. 15-11. Tide staff and well float in operation. (*U.S. Coast and Geodetic Survey.*)

highly resistant to the destructive effect of the weather and impurities in the water. A staff gauge is read by noting the readings of the crests and the troughs of several waves. The two values are recorded and the average used.

In rough water readings may be facilitated by attaching half-inch glass tubing to the face of the staff. The tube is open at the top, and the bottom is partly closed by a cork with a notch cut in it. Sometimes a cork chip is floated on the water surface inside the tube, to show the water surface. Since the water in the tube must move through the notch in the cork, small waves create little movement. However, larger waves affect the gauge so that crests and troughs must be recorded.

2. *Well Floats.* A float gauge consists of a float suspended in a **stilling well** (Fig. 15-12). The well is often made of pipe or boards. An orifice is provided at the bottom large enough to ensure a proper response. Such an orifice has a tendency to clog and so must be carefully maintained. The float may carry a high, graduated rod that extends through a hole at the top of the stilling well. It is read at an index mark. The rod must be numbered downward. The float is more often connected with a lightweight bronze tape or chain that runs over a pulley to a counterweight. The tape may be graduated and read at an index mark.

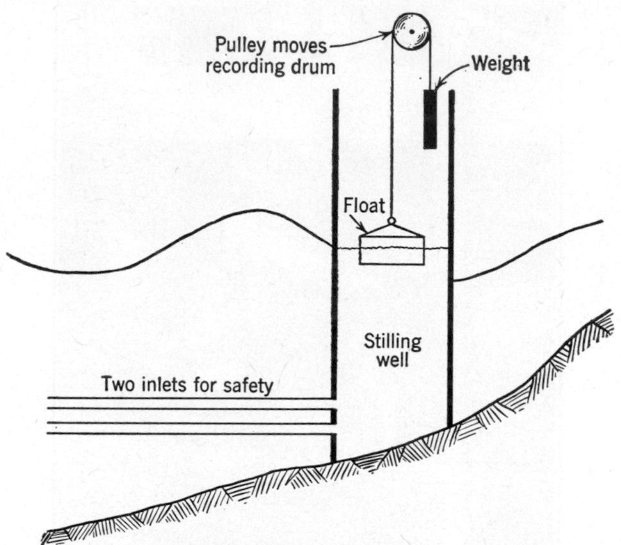

FIG. 15-12. Principles of a well float.

3. *Automatic Recorders.* Many automatic recorders are on the market (Fig. 15-13). These have well floats as described in the previous paragraph, and they are operated by clockwork driven by weights or a synchronous motor. Most of them record graphically on coordinate paper; some of them print the heights at uniform time intervals. Most will run for a week and some for a year without attention. Most of them will operate over a wide range of water heights, but some special recorders can be set to give a record to 0.001 ft over a comparatively short range.

15-36. Leveling Required. It is essential that the elevations of the water surface should be tied to the shore level net. At least one (preferably three) benchmarks of this net should be established on firm ground near the water gauge. Usually a staff gauge is established in connection with any of the well-type gauges. The elevation of the zero of the staff gauge should be determined from these benchmarks at frequent intervals.

15-37. Reading the Gauges. When the well gauge is visited, the staff reading and the well-gauge reading are compared. The well gauge may be set to give the same readings as the staff but it is better practice merely to note the staff reading and the well-gauge reading on the record. At all events, the well-gauge record is always reduced to corresponding staff readings by means of the various comparisons and thereafter designated as staff readings.

FIG. 15-13. Recording gauge. (*W. & L. E. Gurley.*)

The importance of frequently checking the well-gauge record with the staff readings cannot be overemphasized. Too many things can happen to a well gauge to rely on it alone.

15-38. To Compute Bottom Elevations and Depth of Water. When the elevations of the bottom are desired, the computation is as follows:

Sounding depth	10.3
Staff reading at time of sounding (subtract)	−4.8
Depth below staff zero	5.5
Elevation staff zero	192.7
Depth below staff zero (subtract)	−5.5
Elevation bottom	187.2

To compute depth of water,

Staff reading at the water stage chosen for datum (usually mean low water)	8.3
Staff reading at moment of sounding (subtract)	−14.6
	− 6.3
Sounding	10.9
Correction	− 6.3
Water depth used	4.6

15-39. Location of Water Gauges. The gauge should be placed as near the sounding operations as possible. Where river or tidal currents are present, the water level will be different at different points. When a large area is covered where currents are present, a principal gauge should be established and one or two secondary gauges should be placed above and below the work. Staff readings between these points can then be interpolated. Gauges should be located where the water is open—not near a constricted passage. The best location is the end of a wharf. This gives easy access for the recorder and for leveling and permits a fairly unobstructed water flow.

15-40. To Establish a Tidal Datum. In tidal waters, if the project affects navigation, the depths of the water must be referred to a certain tidal stage of the water surface. The height of this surface will depend on local conditions and therefore it cannot be transferred from some other locality by levels or computed from geodetic mean sea level. Before the procedure for establishing a local tidal datum is described, the general theory of tides must be outlined.

General Theory of Tides

15-41. Theory. The moon and the earth revolve around each other about once every lunar month in ellipses along which the curvature of the paths and the velocities of the two bodies balance the effect of gravitation between them. These two forces are balanced only at the center of mass. At points on the earth nearer the moon than the earth's center of mass, the moon's gravitational pull is slightly greater than the centrifugal force. At points farther away than the center of mass, the moon's pull is slightly less than the centrifugal force. These unbalances tend to alter the shape of the ocean surface that would be formed by terrestrial forces alone. They tend to form the surface of sea into a prolate ellipsoid whose axis passes through the moon. That is to say, there is one bulge toward the moon and one away from it (see Fig. 15-14). As the earth rotates under the moon, this ellipsoid passes around the earth. At any point on the earth's surface a high-water and a low-water stage occur twice in a lunar day as the two bulges in ellipsoid pass by.

Approximately every month, the moon moves from a point between

$18\frac{1}{2}$ and $28\frac{1}{2}°$ south of the equator to a point at an equal angle north of the equator, and back. The axis of the sea ellipsoid follows the moon, so that when the moon is north of the equator there is a bulge in the sea centered under the moon that travels in northern latitudes and another bulge centered opposite to the moon that travels in southern latitudes. In northern latitudes, under these conditions, the high tide under the moon is higher than the high tide opposite the moon. In southern latitudes at the same time, the reverse is true (see Fig. 15-14).

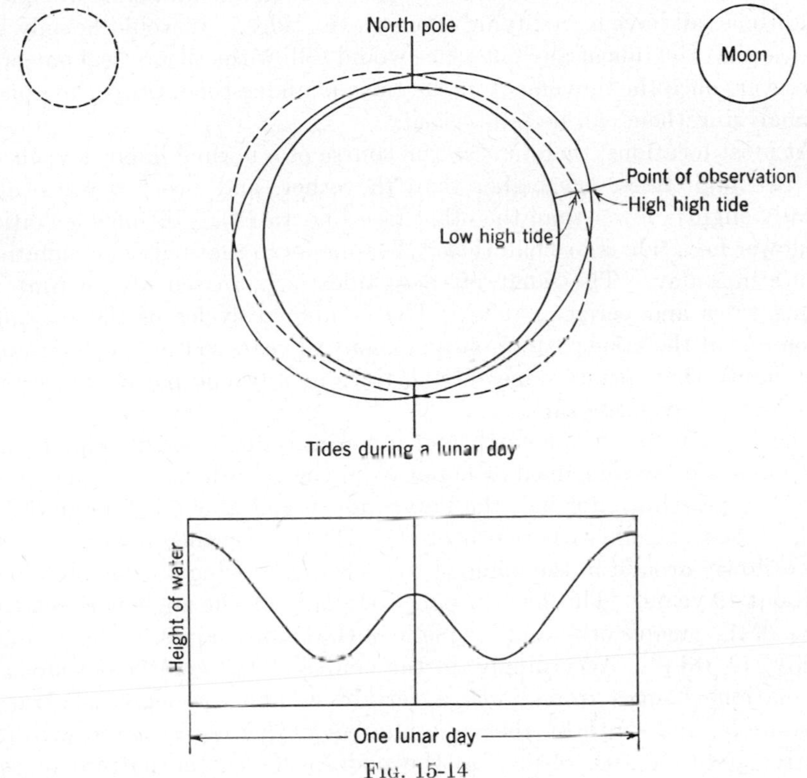

FIG. 15-14

Tides that occur when the moon is north or south of the equator are called **tropic tides.** When the moon is at the equator and the two high tides are equal they are called **equatorial tides.**

When the moon comes nearer the earth in its elliptical path, the range of the tides is greater. These are called the **perigean tides.** When the moon is farthest away the tides are called **apogean.**

The sun has an effect exactly like that of the moon but slightly less than half as great. Since the plane of the moon's orbit is always close to the plane of the earth's orbit, that is, to the ecliptic (the two planes make a constant angle with each other of slightly over 5°), the sun's effect has very nearly a simple phase relationship with that of the moon. The sun increases the effect of the moon when the sun, moon, and earth are in line (at the new moon or the full moon) or decreases it when the sun and moon are at right angles when viewed from the earth (at the moon's first or third quarter). The tides of increased range are called **spring tides,** those of decreased range, **neap tides.**

The effect of the friction of the water, the depth of the sea, the shape of the bottom and the adjacent land, the ocean currents, and meteorological conditions, all have a modifying effect on the tides. It would be only by chance that the tide at any one place would follow the theoretical pattern. However, once the movement of the tides is understood, proper methods of analyzing them can be established.

At most locations, therefore, in the course of a normal lunar day, there are two high tides, one higher than the other, and two low tides, one usually slightly lower than the other (see Fig. 15-14). In some localities the lower high tide is so small that it disappears so that only one high tide occurs in a day. The range of these tides is increased at the time of spring tides and perigean tides. The combined cycles of the sun and moon repeat the same pattern once in about 19 years so that, to determine thoroughly the various stages of the tides at any one place, the record must extend over 19 years.

The movements of the earth and moon are illustrated in Fig. 15-15. The moon's orbit is inclined at about 5° to the ecliptic, so that the moon is below the ecliptic for half the lunar month and above it for the other half. The two points where it is on the ecliptic are called nodes. These move slowly around in the plane of the ecliptic, making a complete circle in about 19 years. This has the effect of changing the angle between the plane of the moon's orbit and the plane of the earth's equator over a range of 28½ to 18½°. Accordingly, in the course of the month, the moon's declination changes from north a certain number of degrees between these limits and south by the same number of degrees in phase with its relative position with respect to the nodes. Thus, the pattern of the moon's declination and the position of the sun repeat themselves each time the same node returns to the same position with respect to the sun, that is, once in about 19 years.

15-42. Tidal Datum. The usual datum levels established from the tides are the following:

Mean sea level is the height at any locality for which $\int y\,dt = 0$, in which y is the height of the water above the datum and t is the elapsed time. For practical purposes it is the average of the heights of the water taken every hour. Mean sea level is used as a datum for geodetic levels but plays little part in hydrographic surveying.

Mean low water (MLW) is the average of all the heights of each low water. It is used as a datum to which depths are referred. It is used on the Atlantic and Gulf Coasts of the United States, in the Canal Zone, and Puerto Rico.

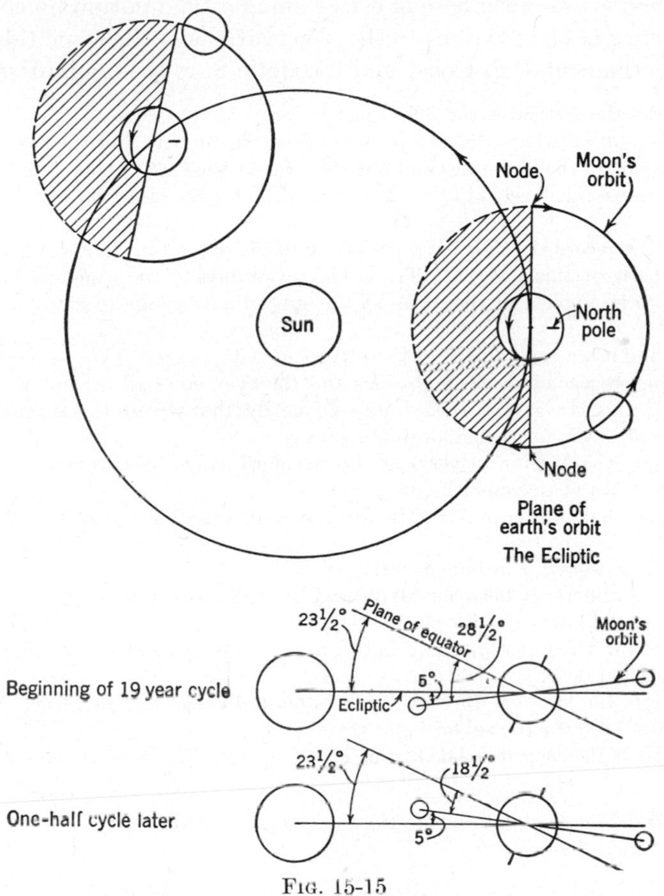

FIG. 15-15

Mean lower low water (MLLW) is the average of all heights of the lower of the two low waters that occur in each lunar day. It is used as a reference for depths on the Pacific Coast of the United States, in Alaska, Hawaii, and the Philippines, with the exception of Wrangell Narrows, Alaska, where depths refer to a height 3 ft below MLLW.

To Establish a Tidal Datum

15-43. Methods. Since it would be necessary to observe the height of the tide for 19 years to establish correctly a tidal datum independently at any locality, various approximations must be used. The U.S. Coast and Geodetic Survey has available certain constants that can be applied to a series of observations of the order of one year in length that will approximate the results of a 19-year series. These are based on the basic harmonic movements of the tide heights.

A method is described here of establishing a tidal datum by comparing a short series of observations to the observations of the same tides at the nearest permanent U.S. Coast and Geodetic Survey standard station.

Essentially the method is the following:

1. At the station where the datum is required (station *A*) find the average of the staff heights of all the high tides and low tides for at least seven complete lunar days. This is the mean tide level (MTL). Find the MTL for the same tides at the standard gauge *B*.

2. Find the recorded MTL for a period of 19 years for the standard gauge. The difference between this and the MTL for the seven days at the standard gauge is the correction to be applied to the MTL at the subordinate gauge to give the corrected MTL.

3. Find the mean range (Mn) of these tides at both gauges. This is the difference between the average of all the high tides and the average of all the low tides.

4. Divide the Mn at the subordinate gauge by the Mn at the standard gauge. This is the Mn ratio for the subordinate gauge.

5. Multiply the Mn for 19 years at the standard gauge by this ratio. This gives the corrected Mn at the subordinate gauge.

6. Subtract half the corrected Mn from the corrected MTL to find MLW at *A* (corrected).

If MLLW is desired, continue as follows:

7. Find the difference between MLW and MLLW at both gauges. This is called the mean diurnal low water inequality (DLQ).

8. Divide the DLQ at the subordinate gauge by the DLQ at the standard station. This is the DLQ ratio.

9. Multiply the DLQ for 19 years at the standard gauge by this ratio. This gives the corrected DLQ for the subordinate station.

10. Subtract the corrected DLQ from MLW to find MLLW at *A* (corrected). Thus

A = subordinate station where tidal datum is desired
B = standard gauge

MTL at *B* 7 days	14.09
MTL at *B* 19 years	14.06
Correction	0.03
MTL at *A*	20.14
Correction	−0.03
MTL at *A* corrected	20.11
Mn at *A*	4.67
Mn at *B* 7 days	7.49
Mn at *B* 19 years	7.64

$$\frac{4.67}{7.49} = 0.623 \text{ Mn ratio}$$

$$0.623 \times 7.64 = 4.76 \text{ Mn at } A\text{, corrected}$$

$$20.11 - \tfrac{1}{2}(4.76) = 17.73 \text{ MLW at } A\text{, corrected}$$

17.80	MLW at A 7 days		10.35	MLW at B 7 days
15.57	MLLW at A 7 days		7.86	MLLW at B 7 days
2.23	DLQ at A 7 days		2.49	DLQ at B 7 days

$$\frac{2.23}{2.49} = 0.896 \quad \text{DLQ ratio}$$

2.83 DLQ at B 19 years

$0.896 \times 2.83 = 2.54$ DLQ at A corrected

17.73	MLW at A
−2.54	DLQ at A corrected
15.19	MLLW at A corrected

Figure 15-16 shows the U.S. Coast and Geodetic Survey form for making this computation. It also shows the high-water interval HWI and the low-water interval LWI. The first is the time lag between the passage of the moon and the resulting high tide and the second is the similar lag for low tide.

The other abbreviations in this form are:

HHW	higher high water
LHW	lower high water
DHQ	mean diurnal high water inequality

STREAM GAGING

15-44. Definition. Stream gaging may be defined as the process of measuring the discharge of a stream. A continuous record of the rate of discharge gives all the data necessary for any volumetric computations desired. Such a record can be approximated by spot measurements at certain stages of the stream.

15-45. Purpose. The record of the rate of discharge is used to estimate the hourly character of the discharge and the quantity of discharge that may be expected on each day, week, month, or year. These data are required for many engineering projects, including: water supply, irrigation, flood control, erosion control, estimates of runoff and rainfall, and the design of culverts and bridges, sewage disposal plants, hydroelectric power projects, and all other projects that require impounding reservoirs.

15-46. Available Data. The U.S. Geological Survey carries on a program of gaging streams throughout the country. This bureau maintains permanent stream-gaging stations in nearly every drainage area of importance. Public and private agencies maintain permanent gaging stations for special engineering works under their control.

Since stream discharge is a function of rainfall, the surveyor can usually obtain good estimates of the required quantities by comparing the results of a comparatively short record of a stream discharge with the continuous records of nearby permanent stations with due allowance for the local conditions of rainfall and runoff.

DEPARTMENT OF COMMERCE
U. S. COAST AND GEODETIC SURVEY
Form 248
Ed. May, 1928

TIDES: Comparison of Simultaneous Observations

(A) Subordinate station Anacortes, Wash. Lat. 48°31′N. Long. 122°36′W.
(B) Standard station Seattle, Wash. Lat. 47°37′N. Long. 122°20′W.
Chief of party ______ Time Meridian: (A) 120° W. (B) ______

DATE.	(A) STATION. Time of—		(B) STATION. Time of—		(A)–(B) Time difference.		(A) STATION. Height of—		(B) STATION. Height of—		(A)–(B) Height difference.	
Year. 1922	HW.	LW.	HW.	LW.	HW.	LW.	HW.	LW.	HW.	LW.	HW.	LW.
Mo. D.	Hours.	Hours.	Hours.	Hours.	Hours.	Hours.	Feet.	Feet.	Feet.	Feet.	Feet.	Feet.
Sept. 24	7.4	0.1	7.2	0.6	0.2	−0.5	22.7	15.4	18.3	7.3	4.4	8.1
	18.5	12.3	18.8	13.0	−0.3	−0.7	23.2	18.7	19.2	11.0	4.0	7.7
25	8.6	1.0	8.0	1.5	0.6	−0.5	22.6	15.0	18.2	7.0	4.4	8.0
	18.9	13.0	19.4	13.8	−0.5	−0.8	22.9	19.8	18.7	12.3	4.2	7.5
26	9.9	1.9	9.4	2.3	0.5	−0.4	22.9	15.2	18.1	7.3	4.8	7.9
	20.0	14.4	20.5	15.0	−0.5	−0.6	22.9	20.7	18.4	13.5	4.5	7.2
27	11.0	2.9	10.4	3.2	0.6	−0.3	23.0	15.7	18.1	8.0	4.9	7.7
	20.5	16.0	21.5	16.4	−1.0	−0.4	22.1	20.9	17.1	13.8	5.0	7.1
28	12.5	4.0	12.0	4.3	0.5	−0.3	22.7	15.6	18.0	8.2	4.7	7.4
	21.6	18.0	22.9	17.8	−1.3	−0.2	21.0	20.4	16.1	13.6	4.9	6.8
29	13.5	5.0	—	5.5	0.6	−0.5	22.7	15.8	—	8.3	4.7	7.5
	23.2	19.0	12.9	19.0	−0.9	0.0	20.7	20.0	18.0	13.0	4.9	7.0
30	—	6.0	0.1	6.6	—	−0.6	—	16.3	15.8	8.9	—	7.4
	14.3	20.4	14.0	20.0	0.3	0.4	23.2	19.8	18.4	12.7	4.8	7.1
							HHW. 7	HLW. 7	HHW. 7	HLW. 7	HHW. 7	HLW. 7
Sums							160.6	140.3	128.8	89.9	31.8	50.4
Means							22.94	20.04	18.40	12.84	4.54	7.20
							LHW.	LLW.	LHW.	LLW.	LHW.	LLW.
Sums							132.0	109.0	103.6	55.0	28.4	54.0
Means							22.00	15.57	17.27	7.86	4.73	7.71

	HW.	LW.	
(1)=	−0.09	−0.36	=Mean difference in time of high and low water respectively.
(2)=	−0.02	−0.02	=Correction for difference in longitude. (Table on back of form.)
(3)=	−0.11	−0.38	=(1)+(2)=Mean difference in high and low water intervals, respectively.

Feet. / Feet.

(4)= 22.94 =Mean HHW height at (A).
(5) = 20.04 =Mean HLW height at (A).
(6)= 22.00 =Mean LHW height at (A).
(7) = 15.57 =Mean LLW height at (A).
(8)= 0.94 =(4)−(6)=2DHQ at (A).
(9) = 4.47 =(5)−(7)=2DLQ at (A).
(10)= 22.47 =½[(4)+(6)]=Mean HW height at (A).
(11)= 17.80 =½[(5)+(7)]=Mean LW height at (A).
(12)= 4.67 =(10)−(11)=Mn at (A).
(13)= 20.14 =½[(10)+(11)]=MTL at (A).

(14)= 4.54 =Mean HHW difference.
(15)= 7.20 =Mean HLW difference.
(16)= 4.73 =Mean LHW difference.
(17)= 7.71 =Mean LLW difference.
(18)= −0.19 =(14)−(16)=2DHQ difference.
(19)= −0.51 =(15)−(17)=2DLQ difference.
(20)= 4.64 =½[(14)+(16)]=Mean HW difference.
(21)= 7.46 =½[(15)+(17)]=Mean LW difference.
(22)= −2.82 =(20)−(21)=Mn difference.
(23)= 6.05 =½[(20)+(21)]=MTL difference.

(24)= 0.623 =(12)÷[(12)−(22)]=Mn ratio.
(25)= 0.832 =(8)÷[(8)−(18)]=DHQ ratio.
(26)= 0.898 =(9)÷[(9)−(19)]=DLQ ratio.

Results from comparison of Stations A and B.	HWI.	LWI.	MTL.	Mn.	DHQ.	DLQ.
Length of Series.	Hours.	Hours.	Feet.	Feet.	Feet.	Feet.
Accepted values for standard station, from ______	4.48	10.70	14.06	7.64	0.86	2.83
Differences and ratios: (3), (23), (24), (25), (26) ______	−0.11	−0.38	6.05	×0.623	×0.832	×0.898
Corrected values for subordinate station ______	4.37	10.32	20.11	4.76	0.72	2.54

Mean LW on staff at subordinate station=MTL−½Mn = ______ feet.
Mean LLW on staff at subordinate station=MTL−½Mn−DLQ= 15.19 feet.

Computed by ______, ______ (Date.) Verified by ______, ______ (Date.)
11—6944

FIG. 15-16. Form 248, Ed. May 1928. (*U.S. Coast and Geodetic Survey.*)

Continuous Records

15-47. Continuous Record of Discharge. An accurate, continuous record of the rate of discharge is obtained essentially by keeping a continuous record of the height of the water at a given point with a water-level gauge and calibrating or rating these heights by measuring the rate of discharge at various stages of the stream. The rate of discharge is determined by measuring the velocity at a measured cross section.

15-48. The Three Sections. To establish such a system, three cross sections of the stream must be selected: a control section, a gaging section, and a measuring section. As they are all interdependent, they usually must be chosen at the same time (see Fig. 15-17).

FIG. 15-17. Complete stream-gaging installation. The dam provides the *control section*. The pool it creates is the *gauging section*. The stilling well and the automatic water-stage recorder are in the towerlike gauge house. The well is connected with the pool. The current meter is positioned in the *measuring section* from the car on the cableway.

15-49. The Control Section. A control section is the cross section of the stream that controls the height of the water in the pool where the gauge is located. The control section should be permanent in shape and high enough above any downstream section to prevent the downstream section from causing water to back up over it at times of flood. It should be narrow and V-shaped so that it will increase the height of the water surface a maximum amount for a minimum increase in discharge, and the banks should be high enough to contain flood water. Accordingly, the section should be followed by rapids, should not contain loose rock or other easily displaced material, and should have steep rock sides. Often existing dams are used or small dams are built for the purpose.

Since the character of the discharge is affected by the impounding effects of pools or lakes, the control section must be at or below the outlet of any important body of water that feeds the length of the stream to be gaged.

15-50. The Gaging Section. The gaging section is at the pool created by the control section. The staff gauge and the stilling well of the float gauge are connected with the pool, usually by two pipes for safety. The gauge house must be above the well, high enough to clear flood waters, and it is made accessible by a short bridge from the bank. An automatic recorder is usually installed.

15-51. The Measuring Section. The measuring section is the cross section of the stream where the discharge is measured at different stages, usually with current meters. It can be located at any place along the stream where, at all times, the discharge is the same as at the control section; this may be either above or below the control section. Usually the measuring section is within 1000 ft of the control section.

The stream should be reasonably straight above and below the measuring section and the bottom of the stream at the section should be reasonably smooth. On wide streams it is convenient to choose a section at a bridge (preferably without piers) from which the velocity-measuring device can be manipulated.

15-52. Current Meters. A current meter consists of a wheel with cups, vanes, or blades so arranged that when it is submerged, the water current rotates the wheel; the wheel moves at a rate that is nearly in proportion to its velocity. It makes an electric contact at each revolution or at every fifth revolution as desired. The observer has a bell or a receiver activated by the contact so that he can count the revolutions in a given period of time. The readings therefore are in feet of flow measured for a certain length of time. The meter is rated by towing it through still water at given speeds.

Figure 15-18 shows two types of current meters. They may be mounted on rods or lines. Streamlined weights are used to keep the lines vertical. The rods or lines are marked so that depths can be measured. Vanes are used to keep the meters headed into the current. The wheel should be as friction-free as possible to give accurate results, especially in slow currents. The meters should be held stationary while they are being read. Any sidewise movement introduces a vector that increases the speed of rotation, and vertical movement has a small similar effect.

15-53. Use of the Current Meter. The velocity of the water in a stream is not uniform over a cross section. Measurements must therefore be made according to a prearranged pattern. The most accurate method is described first.

The cross section of the stream is determined by soundings. The area is divided into vertical strips equal in width, here called segments (see Fig. 15-19). The meter is held at various depths along the vertical center line of each segment. The depths are arranged so that the meter positions are at the centers of uniform portions of the area of the segment (see Fig. 15-20). The number of depths at which readings are taken in each segment is proportional to the depth of the segment.

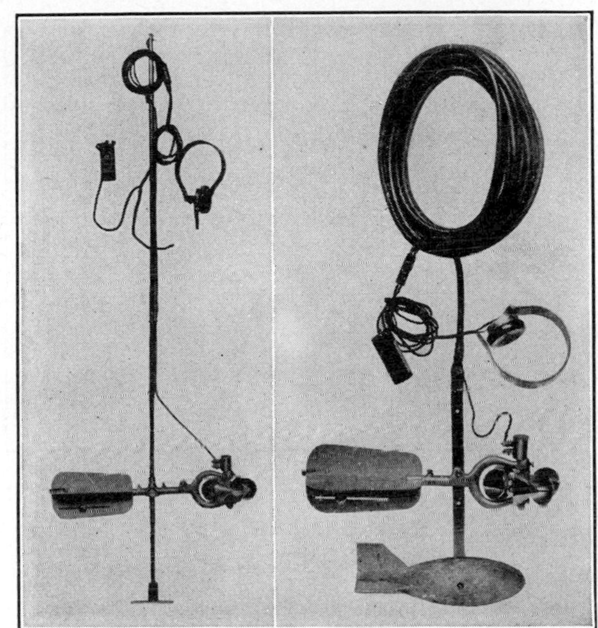
FIG. 15-18. Current meters. (*W. & L. E. Gurley.*)

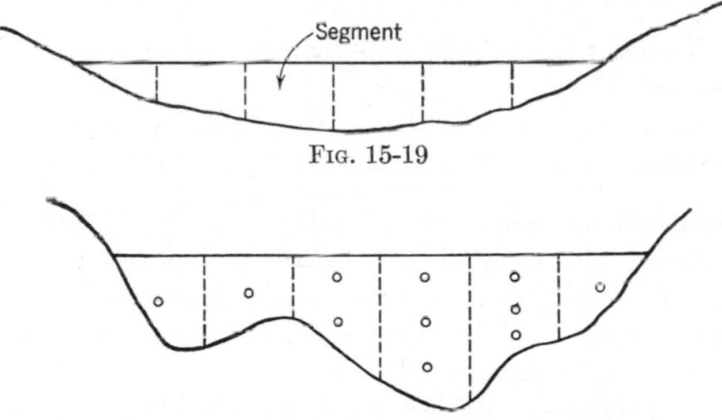

FIG. 15-19

FIG. 15-20. Meter positions.

The average of the velocities for each segment is computed separately from these determinations and multiplied by the area of the segment to obtain the discharge rate of the segment. If a very accurate average velocity in the segment is desired, the depths of the meter positions are plotted as ordinates against the velocities as abscissas (see Fig. 15-21). The points are joined by a smooth curve and the total area is measured by a planimeter. The result is multiplied by the width of the segment to determine the discharge rate for the segment. The sum of the discharge rates of the segments is, of course, the total discharge rate of the stream.

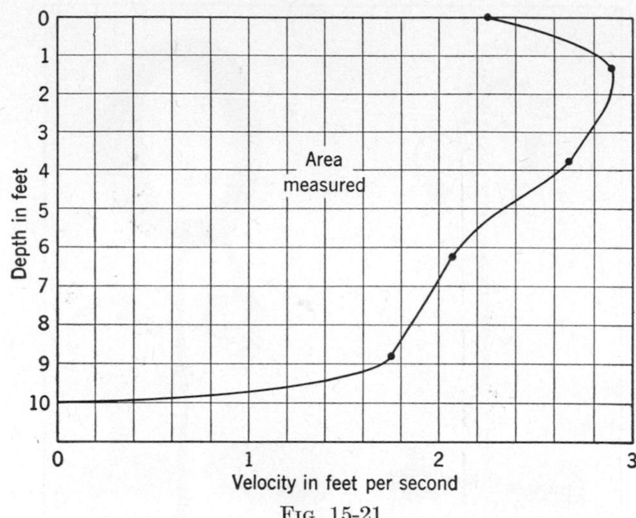

FIG. 15-21

15-54. Nature of Stream Flow. It has been found that curves representing velocities at different depths along a vertical line (as described in the previous paragraph) are parabolic in shape and the average velocity is nearly always equal to:

1. The average of the velocities at two-tenths and eight-tenths of the depth (very nearly)
2. The velocity at six-tenths the depth (approximately)
3. Nine-tenths the velocity at the surface (approximately)

15-55. Approximations. Relationship 1 (above) is so nearly correct that it is almost always used. Relationship 2 is used to save time. Relationship 3 is used in swift water where the meter would become unmanageable at depth. The meter is held just under the surface. This method is excellent for flood conditions.

15-56. Bridge Piers. Bridge piers disturb the flow, particularly downstream. Measurements should be made as far as the measuring apparatus can be extended beyond the upstream rail of the bridge. Meter positions should be selected so that none comes in front of a pier.

15-57. Velocities by Floats. When a current meter is not available, floats can be used to measure velocity. They are liberated at the different segments and timed over a course of 50 to 200 ft. Surface, rod, and "kite" floats are used (see Fig. 15-22).

Surface floats give the surface velocity. This, multiplied by 0.9 or 0.85, is taken as the average velocity.

Rod floats may be rods or tubes weighted so that they float in a vertical position. They are assumed to give the average velocity.

Kite floats are connected by a cord to an underwater device, here called a kite. The kite has metal vanes or other devices that offer resistance. They are assumed to move at the velocity of the water at the depth of the kite. The kite is set at six-tenths of the depth and assumed to move at the average velocity.

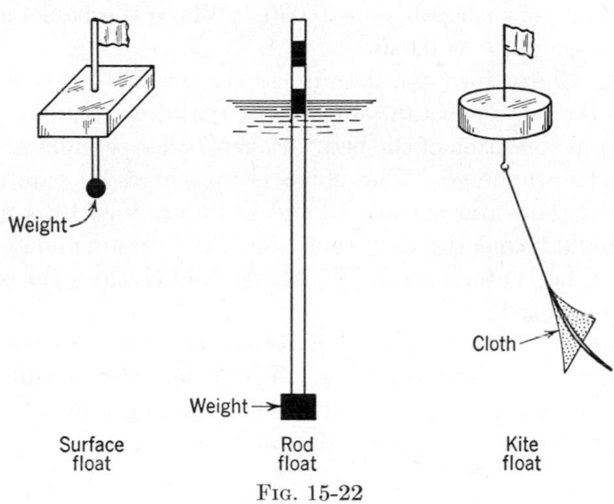

FIG. 15-22

15-58. Open-channel Formula. When neither current meters nor floats can be used, a rough approximation of stream velocity can be found by applying the Chézy formula for open channels with Kutter's coefficient. The Chézy formula is expressed as follows:

$$V = C\sqrt{rs}$$

where V = mean velocity, ft per sec. When V is multiplied by the areas of the cross section in square feet, the flow in cubic feet per second results.

r = hydraulic mean depth, ft. This is the number of square feet in the average cross section of the water, divided by the length of the wetted perimeter. The wetted perimeter is the actual length of the line in the water cross section where the water and the ground are in contact.

s = longitudinal slope of the water surface. This is the fall divided by the horizontal length.

C = coefficient. Kutter's coefficient is usually used.

Kutter's coefficient is expressed as follows:

$$C = \frac{41.65 + \dfrac{1.811}{n} + \dfrac{0.00281}{s}}{1 + \left(41.65 + \dfrac{0.00281}{s}\right)\dfrac{n}{\sqrt{r}}}$$

where r and s are as in the Chézy formula and n is a value taken from the list below.

Values of n. For the average stream, $n = 0.035$. Where the banks and bottom are very smooth, $n = 0.030$. Where the banks and bottom are especially rough, $n = 0.040$.

To use the Chézy formula, a uniform stretch of the stream must be found where there is a constant slope, little variation in cross section, and a fairly uniform condition of the bed. Several cross sections are measured to obtain a true average. The slope is measured by careful leveling. The heights of the water surface should be determined by a hook gauge. The slope should be measured on each side of the stream and at the center, if possible. The center is usually slightly higher than the sides. The average slope is used.

15-59. Weirs. When desirable, a weir can be constructed in the stream itself or on an existing dam. The usual weir formulas can then be applied. This is an excellent method of gaging a small stream or a stream with a dam that is easily adaptable.

PROBLEMS

15-1. Draw a line illustrating the usual heights of water plotted vertically against a lunar day plotted horizontally, similar to the graph in Fig. 15-14, for the following tidal conditions. Use four diagrams:

a. Neap, equatorial, apogean
b. Spring, equatorial, apogean
c. Spring, tropic, apogean
d. Spring, tropic, perigean

15-2. From the following data, compute the staff reading for the MLW and the MLLW data at *A*. *B* is a standard station. Values are in staff readings.

	A 7 days	*B* 7 days	*B* 19 years
MTL	10.29	8.36	8.40
Mn	6.42	4.21	4.78
MLW	7.08	6.26	6.01
MLLW	6.50	5.87	5.77

15-3. Compute the discharge in cubic feet per second of a stream where the current is measured in four segments with a current meter. The segments are each 4 ft wide and each reading was taken at the depths stated and recorded in feet of flow for 20 sec.

Segment no.	1	2	3	4
Depth of meter, ft	$D^* = 3'$	$D = 5'$	$D = 3'$	$D = 2'$
0.5	25	28	27	20
1.5	29	42	49	36
2.5	17	14	15	
3.5	...	17		

* D indicates the depth of each segment.

15-4. Compute the discharge in cubic feet per second by the Chézy formula from the following data:

Average cross section............ 75 sq ft
Wetted perimeter................ 32 ft
Slope.......................... 0.007
Banks and bottom very rough

CHAPTER 16

BRIDGE AND TUNNEL SURVEYS

16-1. Types of Surveys. Bridge and tunnel surveys vary widely in complexity in accordance with the terrain, the size of the structure, and the extent of curvature in the alignment. In addition to the reconnaisance, four types of surveys are required; these may be combined if the project is simple. They are the **preliminary survey,** the **survey for design,** the **control surveys for location,** and the **location survey.**

16-2. Preliminary Survey. This consists of a topographic survey of the general area that will include the structure. It may involve only a few cross sections at the ends of a proposed bridge or tunnel or it may require an aerial survey covering several square miles. It must show the precise location of rail, highway, and drainage facilities, a survey of ground-water conditions, and it may require the location and elevation data for a system of borings. Often a complete hydrographic survey is necessary. This may include a study of the drainage area, currents, and the effect of tides or floods.

16-3. Survey for Design. Once the position of the structure is definitely chosen, accurate survey data must be provided for the detailed design. Frequently the exact sizes of the structural members themselves depend on the position of existing structures. For example, the piers of a bridge crossing railroad tracks must be placed so that proper track clearances and alignment can be maintained. Highway tunnels often must be fitted to existing city property lines and frequently the new structure must join old structures. Surveys for detailed design must be rigidly accurate and so thoroughly checked that there can be not the slightest possibility of a blunder.

16-4. Control Survey for Location. While the preliminary survey and the survey for design must be based on a control system, the control system for the location of a bridge or a tunnel is such an important part of the work that it is often handled separately, and planned and executed with special care without regard to the other control.

For example, frequently a long bridge is required over a navigable waterway with a special span over the channel. The exact position of the whole structure is of little importance. This position can be selected well enough from the data on the preliminary survey map. But the relative positions of the structural parts that are erected at the two ends must be exactly correct so that the center span will fit when it is erected.

In this and similar cases a somewhat rudimentary triangulation system will be sufficient for the preliminary survey map but a very precise triangulation survey is essential for the control of the location.

The control for location requires a considerable volume of computations to reduce the triangulation and traverse data so that survey measurements for marking the basic location position can be determined.

16-5. Location Survey. Once the location computations have been completed, the basic location positions must be established and from them points must be set that give position for building the structure. This survey is often of considerable difficulty and in the case of an underwater structure it is a very exacting procedure.

16-6. Combined Surveys. It is evident that both vertical and horizontal control systems must be established as soon as possible. These may involve only a line of levels and a straight traverse but, for water crossings, they usually involve triangulation and often reciprocal leveling. Sometimes these systems can be laid out at the time of the preliminary survey. They can be measured at first with a low degree of accuracy for mapping, and later the accuracy can be improved by more precise measurements for the more accurate data required in design, location computations, and location.

16-7. Permanence. As a rule the control systems will be used again and again throughout the work. For this reason stations and benchmarks should be carefully built for permanence. The preliminary survey control may be marked by stakes but if the same system is to be used later, these should be replaced by concrete monuments with copper or bronze center marks. Reference marks and measurements should be carefully determined.

16-8. Triangulation. The triangulation stations should be placed to comply, as far as possible, with the following requirements:

1. A primary station should be placed near each end of the structure. Tunnels require additional stations near each shaft, and bridges require them at points from which the bridge piers can be set by intersection.

2. The system should consist of one or two well-shaped quadrangles and both diagonals should be observed.

3. At least two bases should be measured, each consisting of the entire side of a quadrangle. The stations should be placed where the bases can be easily measured. Often short connections must be made from an end of the base to a triangulation station located on a nearby roof or other high point. Bases should be measured by invar tapes standardized by the National Bureau of Standards. If invar tapes are not available,

all measurements must be made at night or on cloudy days. The tape should be fully supported if possible. This reduces the errors due to wind or to variations in tension. If this is impossible, stakes must be set at proper grades to support the tape at frequent intervals. It is often advisable to use a weight supported over a frictionless pulley like a bicycle wheel, rather than a spring balance, to ensure proper tension.

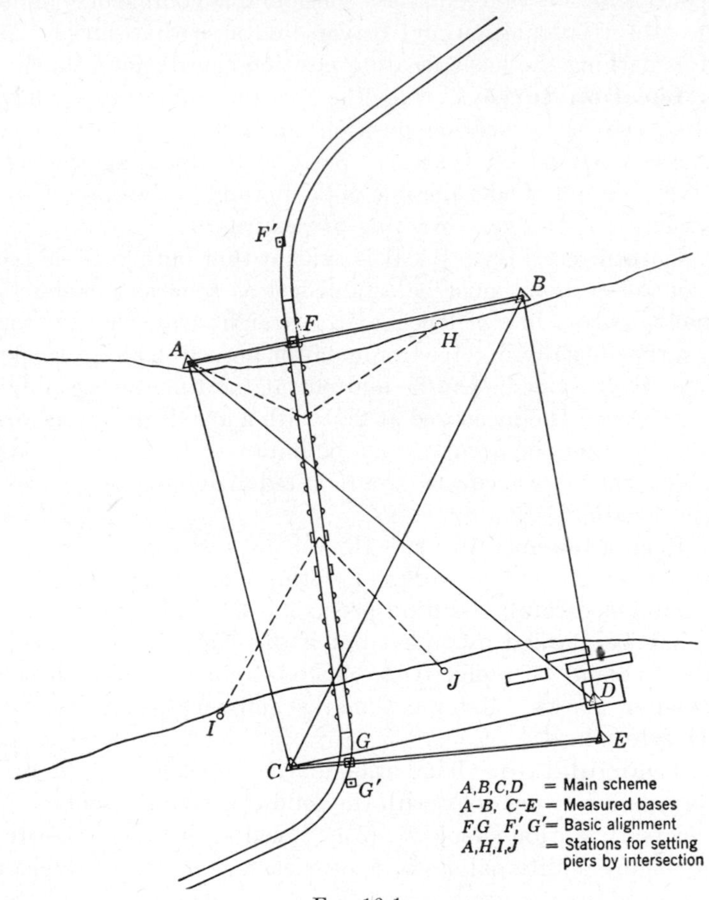

FIG. 16-1

16-9. Typical Triangulation Systems. Figure 16-1 shows a typical triangulation system for a bridge. The stations *A*, *B*, *C*, and *D* form the main scheme quadrilateral. The measured bases are *AB* and *CE*. The triangle *CDE* carries the base to *CD*. *G* and *F* are set on the center line and *F′* and *G′* are established from them. *F′* and *G′* are very permanently marked and targets are placed over them. *H*, *I*, and *J* are placed where strong intersections can be obtained for the centers of the piers. They are tied in by traverse or triangulation. When precise positions on the center line at the bridge piers are required, intersections from A and H are checked with the center line and similarly from I and J. The nearest pair are used. The transit is oriented by sighting the most convenient main scheme point. Permanent targets should be erected for this purpose.

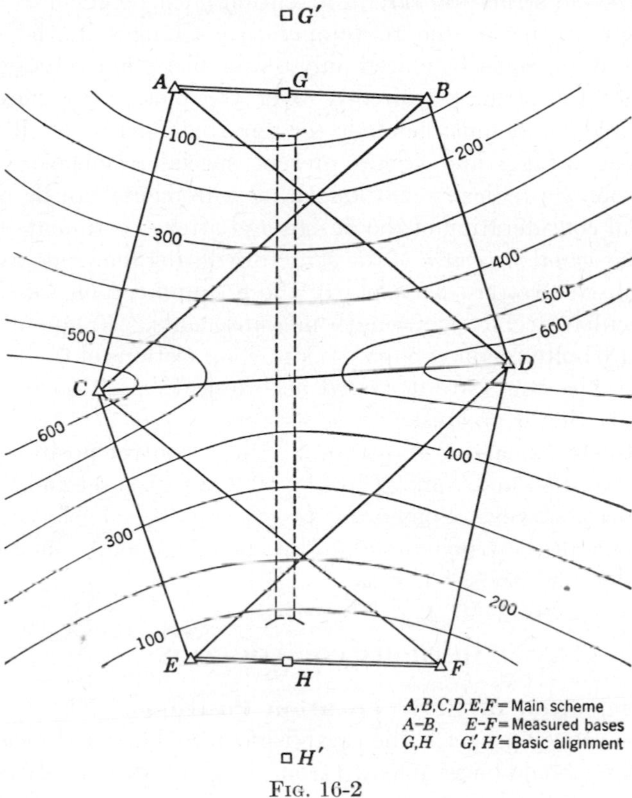

FIG. 16-2

Figure 16-2 shows a triangulation system for a tunnel that is to be built through a ridge where surveying is difficult. It is impossible to establish a satisfactory traverse across the ridge over the tunnel center line or to measure a base from C to D.

Two quadrilaterals are used and AB and EF are measured for bases. H and G are set on the center line, G' and H' are set very permanently, and targets are built over them.

16-10. Instruments. It is usually advisable to use optical-reading direction theodolites for the triangulation and for establishing the basic location control. They are especially useful for establishing bridge piers by intersection, as angles can be set off with them quite accurately. It will then be necessary to move the first points set by only a small amount when the established angle is checked by careful measurement.

Primary control levels should be run with first-order instruments and precise rods.

16-11. Control on Existing Bridges. Often an existing bridge parallels the new structure. When a traverse or a line of levels is established on the old structure, careful attention must be given to the movement of survey marks due to temperature changes in the structure. Permanent marks must be placed only on the piers themselves or on each fixed end of the structure directly over the pier. Any measurement between these points must be made (or repeated) just before it is used.

Suspension bridges and arches present special problems. On these, elevations change radically and length measurements can be made only after careful consideration of the design and after experiments have been made on the *actual movement* of the structure during temperature changes.

Long railroad trestles, particularly those supported on wooden bents, present peculiar problems of length measurements. Frequently the rail joint bars are bolted tight enough to cause long sections of track to expand as a unit. The bents are deflected according to position of the joints where the friction is greatest.

16-12. Basic Location Positions. Basic control positions such as F and G in Fig. 16-1 and G and H in Fig. 16-2 are often set and tied in during the control survey. Sometimes they are not established until later. The positions for these points and all basic control points must be chosen with considerable care.

METHODS FOR LOCATION

16-13. Selection of Basic Location Positions. The first step in the location survey proper is the establishment of the basic location positions. These should be established from the location control system with the greatest care and thereafter used as fundamental positions. These points should be placed as near the structure as possible but in such a position that they can be protected from destruction by the construction operations. They should be arranged so that they can be used with a minimum of computation.

If possible, a point should be placed at each end of the structure on the center line. Failing this, the points should be placed on an offset line at an even number of tens of feet offset. If intervisible points are impossible, a pair of points at each end should be set in the direction of the center line. If curved alignment is used, this direction will be different at each end.

When shafts are used in tunneling, a surface point and often a distant alignment point should be marked at each shaft. Sometimes it is possible to establish an accurate surface traverse along the entire center line of a tunnel. Sometimes only a series of accurate directions can be established on the surface.

FIG. 16-3. A Berger vertical collimator. It establishes a vertical line of sight. The eyepiece end of the telescope shown rotates on its axis to direct the line of sight up or down by turning a prism mounted on it at the instrument center. Two objective lenses are used. (*C. L. Berger & Sons, Inc.*)

When bridge piers must be located in water, basic location positions must be established from which the piers are located by intersection. Shore marks must be established by which the transits can be oriented.

When the basic location points have been placed, permanent targets should be built over each marker that is to be frequently sighted. Sometimes these are designed so that a transit can be set up under the target.

16-14. Alignment in Tunnels with Shafts. Alignment can be carried into tunnels by any of the methods described for mine surveying in Chap. 14. The most accurate device for carrying positions down a shaft is a vertical collimator (see Fig. 16-3). This device establishes a vertical telescopic line of sight either up or down. It is usually arranged so that a plumb bob can be used instead of a very short, downward sight.

The level vial can be adjusted so that it centers when the vertical axis is vertical by leveling the instrument and turning it in azimuth. The line of sight can be placed in line with the vertical axis by aiming at a

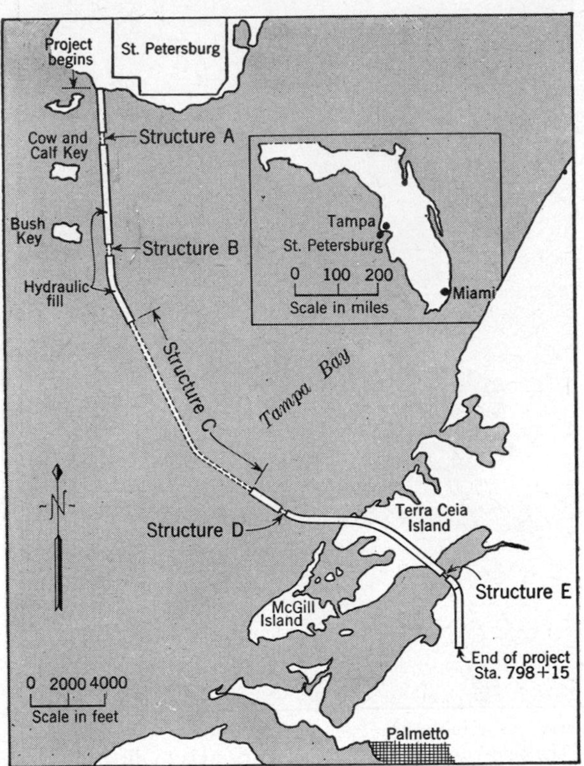

Fig. 16-4. Providing horizontal control for construction of Tampa Bay Bridge meant carrying line across 11 miles of open water. Most difficult section was control for 4½-mile-long structure *C*. (*Peter A. Hakman in Civil Engineering.*)

well-defined point, turning the instrument in azimuth and adjusting the cross hairs until they remain on a point. The instrument may be set up either in the tunnel or on a special stand built out over the shaft.

Alignment is often carried on the roof, as in mining. Usually a scale is mounted at each instrument station so that the line can be run into the tunnel repeatedly to check its position and to note any movement, particularly in tube construction (see Fig. 16-9). A record is kept of the scale values that fall on line and are therefore used as instrument positions. Frequently, a device that carries a plumb bob is arranged so that the position of the plumb bob can be read on the scale. In large tunnels a special bracket that hangs from the roof is arranged for the instrument, and a small platform is arranged for the observer.

16-15. Curved Alignment. When the alignment of a tunnel is curved, the entire alignment data is, of course, computed beforehand.

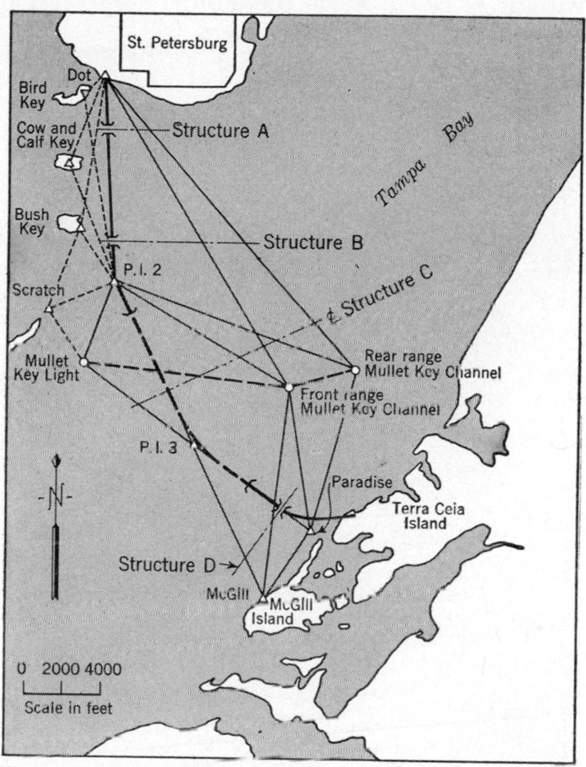

FIG. 16-5. Preliminary triangulation system was established from three existing aids to navigation—Mullet Key Light and Front range and Rear range, Mullet Key Channel. These points had been previously triangulated by Corps of Engineers to third-order accuracy and proved satisfactory for preliminary alignment and location of borings and index piles. (*Peter A. Hakman in Civil Engineering.*)

Chords or secants should be chosen so that they utilize the longest sight distances available. The basic line is carried over these lines and the intermediate points are filled in.

16-16. Levels. Levels are handled much as they are in ordinary surveys. In tube construction through mud-deposited material, the tube will usually move up and down to a small extent with changing depths of water above it. The water has a tendency to compress the mud and thus depress the tube. It rises again when the depth of water decreases. Careful attention must be given to the stage of the tide or the river.

OPERATIONS

16-17. Precautions. In general all survey operations should be carried out with great care. Since the desired accuracies are high and

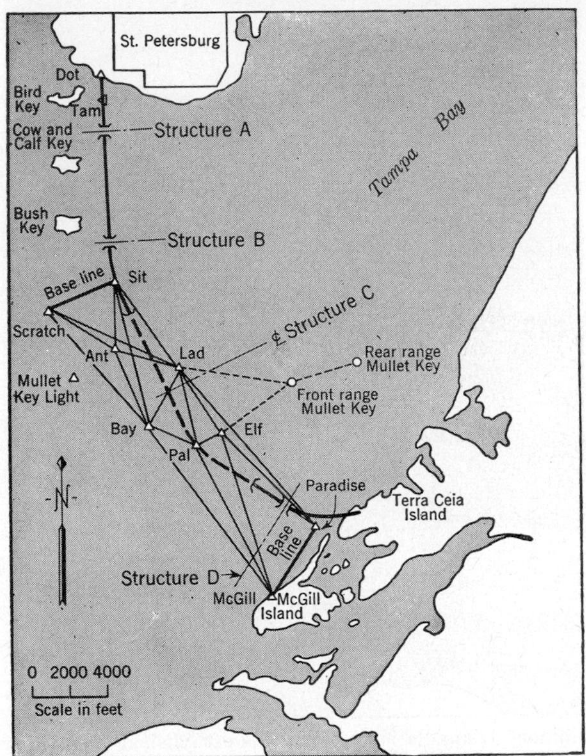

FIG. 16-6. Final triangulation system for construction surveys established base lines as shown and control points at strategic positions such as PI's and other points so located as to give strong intersections. Sight distances were limited to about 4000 ft. Using a Wild T3 reading directly to tenths of a second, angle measurement proceeded rapidly and average closure of angles before adjustment was 0.28 second. (*Peter A. Hakman in Civil Engineering.*)

the triangles are often comparatively small, especial attention must be given to centering both instruments and targets. Since a blunder usually results in almost prohibitive expense, measurements must be made repeatedly. Permanent marks should be used as much as possible even for minor positions. Care and planning are more important for surveys of this kind than for almost any other types of surveying.

EXAMPLES

16-18. Bridge Survey. The surveys conducted for the Tampa Bay Bridge are an outstanding example of the methods used to solve a very difficult bridge survey problem. Peter A. Hakman, A.M. ASCE, Resident Engineer, Parsons, Brinckerhoff,

FIG. 16-7. Survey tower is erected in two independent sections. Central steel-pipe H-pile section supports trivet on which instrument rests. Timber pile clusters support timber deck on which surveyors stand. Pills for seasickness were a standard item of surveying gear on this part of the job. (*Peter A. Hakman in Civil Engineering.*)

Hall & MacDonald, New York, who was in charge of these surveys, published a short article[1] in *Civil Engineering*, July, 1953, which described this work.

Figure 16-4 shows the general arrangement for the bridge, developed from the reconnaissance on which the plan was based.

Figure 16-5 shows the triangulation scheme for the preliminary survey. No base-line measurement was necessary. *Mullet Key Light*, *Front range Mullet Key Channel*, and *Rear range Mullet Key Channel* were lights whose positions had been

[1] "First-order Survey for Construction across Tampa Bay," p. 43.

Fig. 16-8. General view of the main span. (*Peter A. Hakman in Civil Engineering.*)

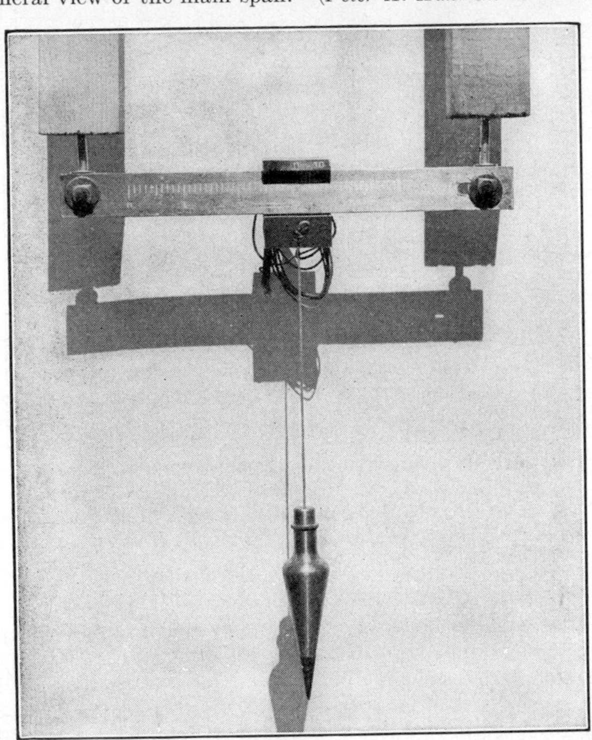

Fig. 16-9. A roof scale and vernier. It can be set to the nearest 0.002 ft. (*Massachusetts Metropolitan District Commission and C. L. Berger & Sons, Inc.*)

determined to third-order accuracy by the U.S. Corps of Engineers. The lines between these lights were used as a base. The angle measurements were held to third-order accuracy. The triangulation system was used to determine positions for borings, index piles, and stations on which were based the computations for the alignment plan. It therefore served as a basis for the preliminary survey and the survey for design.

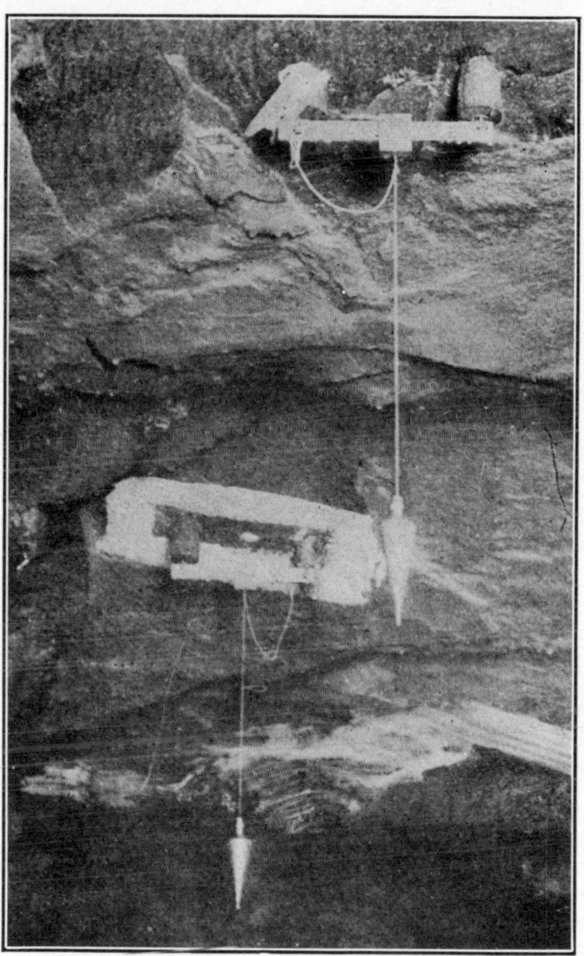

FIG. 16-10. Alignment stations with scales and verniers. (*Massachusetts Metropolitan District Commission and C. L. Berger & Sons, Inc.*)

Figure 16-6 shows the triangulation system used to control the location. The two base lines were measured over the water, which, at some points, reached depths as great as 12 ft. First, permanent marks were installed that consisted of 4-in. pipes driven into the bay bottom. In some cases these were stiffened with tremie concrete poured around the pipe. They were placed at intervals of about 2000 ft to serve as the end marks for sections. Intermediate points were marked by 2½-in. pipes driven at 150-ft intervals. The tape was supported only at these markers. A tension of 25 to 30 lb was applied. An invar tape was used which had to be calibrated at frequent intervals because of the corrosive effect of the salt water. Each 2000-ft section was measured ten to twenty times to reach the desired accuracy.

The final results showed a probable error for the north base of 1:625,000, for the south base 1:1,290,000; the calculated value of one base carried through the triangulation from the other checked with its measured value to an accuracy of 1:600,000.

FIG. 16-11. Using a light box to show the plumb bob. The transit is a Berger 6¼ in. (*Massachusetts Metropolitan District Commission and C. L. Berger & Sons, Inc.*)

The angles were measured by first-order methods and the least-squares adjustment gave angular corrections that ranged from 0.00 to 0.54 second.

Figure 16-7 shows station *ANT* on the final triangulation net and Fig. 16-8 shows the main bridge span under construction.

16-19. Tunnel Survey. A number of excellent procedures were utilized in the survey for the Chelsea Tunnel built under the direction of the Construction Division, Commonwealth of Massachusetts Metropolitan District Commission, Charles W. Greenough, Commissioner, Frank W. Gow, Engineer of Construction Division. Mr. Gow and Mr. Louis Berger, Vice President, C. L. Berger & Sons, Inc., collected the illustrations used here.

The alignment was carried through the tunnel on stations in the roof established by scales with verniers from which plumb bobs could be suspended (see Figs. 16-9 and 16-10). Figure 16-11 shows how a plumb bob was illuminated. A box with a side made of tracing cloth is illuminated from behind and held so that the plumb bob and the instrument cross hair are silhouetted against it.

FIG. 16-12 Bucking in a Berger transit on a line marked overhead. A lateral adjuster is being used. (*Massachusetts Metropolitan District Commission and C. L. Berger & Sons, Inc.*)

FIG. 16-13. A Berger lateral adjuster. It is used to facilitate bucking in. (*C. L. Berger & Sons, Inc.*)

The transit positions were bucked in on line with the roof stations as shown in Fig. 16-12. The transit was mounted on a lateral adjuster (Fig. 16-13). This device slides the transit a short distance left and right when the screw is turned.

Figure 16-14 shows the operation of measuring the cross section of the tunnel for clearances and for determining quantities. The device used is called a **sunflower.** It is shown in detail in Fig. 16-15. It gives the polar coordinates of the position of the end of the rod.

Fig. 16-14. Taking cross sections with a sunflower. (*Massachusetts Metropolitan District Commission and C. L. Berger & Sons, Inc.*)

Fig. 16-15. A sunflower for cross sectioning in a tunnel. (*Massachusetts Metropolitan District Commission and C. L. Berger & Sons, Inc.*)

Part III. Procedures for Precise Control

CHAPTER 17

PRECISE LEVEL NETS

17-1. Introduction. In any area where construction is concentrated, a fundamental requirement is a precise level net. Such a net can be built up piecemeal but it is more satisfactory to establish at least a framework of very accurate levels that is adjusted simultaneously.

The Board of Surveys and Maps of the Federal government, in its "Specifications for Horizontal and Vertical Control Surveys," approved by the Board on May 9, 1933, and issued in mimeographed form for the guidance of all mapping agencies of the United States government, classified leveling in four different grades as to its accuracy and specified the criteria to be used in classifying the work. The following is quoted from these specifications:

First-order leveling.—First-order leveling should be used in developing the main level net of the United States. The lines should be so placed that eventually no point in the country will be more than about 50 miles from a bench mark established by leveling of this order. All the lines should be divided into sections 1 to 2 kilometers in length, and each section should be run forward and backward, the two runnings of a section not to differ more than 4 mm. $\sqrt{K}$ or 0.017 foot $\sqrt{M}$, where K is the length of the section in kilometers and M its length in miles.

Second-order leveling.—Second-order leveling should be used in subdividing loops of first-order leveling until no point within the area is much more than 12½ miles from a first- or second-order bench mark. Second-order leveling will include lines run by first-order methods, but in only one direction, between bench marks previously established by first-order leveling, and all double-run lines of leveling whose sections, run in a backward and forward direction, check within the limits of 8.4 mm. $\sqrt{K}$ or 0.035 foot $\sqrt{M}$, where K is the length of the section in kilometers and M its length in miles.

Third-order leveling.—Third-order leveling may be used in subdividing loops of first- or second-order leveling, where additional control may be required. Third-order lines should not be extended more than 30 miles from lines of the first- or second-order; they may be single-run lines but must always be loops or circuits closed upon lines of equal or higher order. Closing checks are not to exceed 12 mm. $\sqrt{\text{kilometers in circuit}}$ or 0.05 foot $\sqrt{\text{miles in circuit}}$.

Leveling of lower order.—Leveling that allows closures greater than the limit stated for third-order work, such as trigonometric leveling, barometric leveling, or "flying" levels, shall be considered as belonging to the lower order of work.

No bench marks established by leveling that is less accurate than that of the third order, as described above, shall be marked by standard bench-mark tablets, except that in mountainous regions inaccessible to ordinary spirit-level lines, standard marks may be used on the mountain summits to mark elevations determined by trigonometric leveling; such marks should be stamped in a distinguishing manner. Elevations inferior to the third order in accuracy shall not be published in such a way as to be confused with standard work of the third or higher orders.[1]

Note that in the above specifications for first and second order, the distances entered in the formulas are one-half the distances actually leveled.

17-2. Instruments. Only the best levels and rods that can be purchased should be used for precise leveling. They require fewer setups, give the desired accuracy without overcareful observations, and cut down the quantity of reruns required. Even on a small net they save their cost many times over. Instruments suitable for this work are described in Chap. 2.

17-3. Method of Leveling. Many methods have been successfully employed for precise leveling, as described in Chap. 2. The optical micrometer method and the three-wire method give the best results. In the United States, the three-wire method is considered best except when the net is confined to a small area. It is used as an example in this chapter and is described in detail in later paragraphs. The method covers all the requirements of precise leveling so that other methods can be derived from it.

17-4. Net Adjustment. The level net can be adjusted by one of the methods described in Chap. 6, but the results will be more accurate if it is adjusted by the method of least squares as described in Chap. 22 or by the **Braaten circuit-reduction method** described in this chapter. The Braaten method cannot always be applied, but it gives the same results as the method of least squares and it is usually much easier to use and more easily understood.

THE THREE-WIRE METHOD

17-5. General Procedure. For this method, the three wires necessary in the instrument are the central horizontal wire and two stadia wires placed equidistant from the central wire. The stadia wires should be set close to the central wire to bring them into the best part of the optics and to keep the intercept on the rod short so that greater differences in elevation between the rod and the instrument can be used. A setting

[1] Quoted from Howard S. Rappleye, "Manual of Leveling Computation and Adjustment," U.S. Coast and Geodetic Survey Special Publication 240.

that gives a stadia multiplier constant of about 300 is excellent. When a rod graduated in hundredths of a yard (the **yard rod**) is used, the constant should be 100/0.3. If more than one instrument is employed, the stadia intercepts should be checked and the results reduced to a common value.

All three wires are read and the average is used. The blister must be independently centered for each reading. The upper and the lower intervals (half stadia intercepts) are compared to avoid blunders. The total stadia intercept is used to measure the excess or deficiency in the length of the plus sights with respect to the minus sights and to determine the total length of the level lines. The total lengths are used to compute the accuracy of the work and to determine the lengths of the links in the net for weighting them in the net adjustment.

The error of the instrument adjustment is determined at regular periods and the necessary correction is applied to each run in accordance with the length of the unbalance between the total horizontal lengths of the plus and minus sights.

DETAILED PROCEDURE

17-6. Instruments. It is assumed in this description that the stadia multiplier constant of the level is 100/0.3. Two rods should be used, and it is preferable that they be graduated in hundredths of a yard. With this arrangement the sum of the three-wire readings gives the rod reading directly in feet, and the stadia intercept multiplied by 1000 gives the length of the sight in feet. Conventional rods can be used without difficulty, however.

Before leveling is begun, the instrument should be exposed to air temperature (not in the sun) for 30 min to 1 hr to eliminate temperature differentials that might affect the liquid in the vial.

17-7. Rod Schedule. The rods are arbitrarily marked *A* and *B*. The *A* rod is called the B.M. rod. The *A* rod is held on the benchmark and the *B* rod on the first turning point. When the instrument is moved, the *B* rod is left at the first turning point and the *A* rod is moved to the second turning point. When the instrument is moved again, the *A* rod is held where it is and the *B* rod is moved. The rods are "leapfrogged" in this manner throughout the section (see Fig. 17-1). If it should happen that the *B* rod normally comes to the B.M. at the end of a section of levels, it is not used. Instead, the *A* rod is moved to the B.M. Thus both sights at this instrument position are taken on the *A* rod. This procedure eliminates any difference in index correction of the rods.

A	*B*	*A*	*B*	*A*	
□	•	•	•	□	
BM	*TP*	*TP*	*TR*	*BM*	

A	*B*	*A*	*B*	*A*	*A*
□	•	•	•	•	□
BM	*TP*	*TP*	*TP*	*TP*	*BM*

FIG. 17-1

17-8. Order of Sighting. The *A* rod is always sighted first. Thus, the first sight at each consecutive instrument position is alternately a plus sight and a minus sight. If the *A* rod is substituted for the *B* rod at a benchmark, the plus sight is taken first, as would be normally the case. This procedure neutralizes the effect of changing conditions, like sinking of the level or changing refraction.

17-9. Rod Temperatures. With invar ribbon rods, it is necessary to know only the average of the rod temperatures, throughout the section. If the work proceeds without interruption and no sudden change in temperature occurs, it is sufficient to record the two rod temperatures at the beginning and end of the sections. The average of the four temperature readings is used.

17-10. Section. A section is the leveling between temporary or permanent benchmarks. It should not be greater than 4000 ft in length. It should be leveled forward and backward, preferably under different weather conditions like morning and afternoon. The difference in elevation obtained by these two runs should check within the limits of accuracy desired. If this condition is met, the average difference is used. If the limit of error is exceeded, other runs must be made until there exists a forward and a backward run that check within the limit. Then the value of the elevation difference is computed as follows:

1. Reject any run that obviously contains a blunder.
2. Take the average of all the runs. Reject any run that differs from this average by more than 150 per cent of the error limit.
3. Find the separate average of the remaining forward runs and that of the remaining backward runs. Take the mean of these averages as the final value.

Two principles govern these rules. The first is that the best average will result when all blunder-free values are used no matter how much they differ. The second is that since forward and backward runs are leveled under different conditions, each should equally affect the average.

Example 1. Assume that a section of 3600 ft is being leveled for first-order accuracy. The closure required is computed as follows: From Table 1-3 for first-order accuracy, the limit of error L in feet is expressed as follows:

$$L = 0.012\sqrt{M}$$

where M is distance leveled, in miles.

Since the section is closed by leveling twice the length of the section, this may be written

$$L = 0.012\sqrt{2M} \qquad \text{or} \qquad L = 0.017\sqrt{M}$$

where M is the length of the section in miles. Reducing this to feet,

$$L = 0.000234\sqrt{F_s}$$

where F_s is the length of the section in feet. Then

$$\begin{aligned} L &= 0.000234\sqrt{3600} = 0.014 \text{ ft} \\ 150\%\, L &= 1.5 \times 0.014 \qquad = 0.021 \text{ ft} \end{aligned}$$

Assume that the results of the leveling of the section were the following, in the order given. The signs of the results of the backward runs are reversed according to practice.

	Rejection limits
$F = -10.354$	$-10.369 - 0.021 = -10.390$
$B = -10.370$	
$F = -10.352$	$-10.369 + 0.021 = -10.348$
$B = -10.396$	
$F = -10.375$	
Sum $= -51.847$	
Av. $= -10.369$	

The runs had to be continued until the fifth run to obtain a backward and forward pair that checked within ± 0.014 ft. The rejection limits are computed from the average of all the runs as shown. The fourth run is then rejected.

To compute the final differences, the separate average of all the retained forward runs and all the retained backward runs are averaged, thus:

	F	B
	-10.354	-10.370
	-10.352	
	-10.375	
Sum =	-31.081	
Av. =	-10.360	-10.370

These are each corrected for unbalance in length of sights and then entered in the computation sheets. There they are corrected for rod temperatures and for the orthometric correction.

Usually orthometric corrections are not required. The principle on which these are based is described in Chap. 12. The values of the corrections can be computed from the nomograms in Charts 17-1 and 17-2. To use the nomograms the following data are required:

1. Mean elevation of the section (h)
2. Mean latitude ϕ

The diagonal line gives the correction to be subtracted from the observed elevation for each minute of increase in latitude ($\Delta\phi$) from the first B.M. to the last B.M. The correction is added for decreasing latitude.

Example 2.

Mean elevation, 2000 ft
Lat. B.M. No. 1, 60°00′; Lat. B.M. No. 2, 60°02′
Observed elevation B.M. No. 2, 1500.000′
From nomogram, 0.00267
Total correction $C_0 = 2(-0.00267)$ ft
Elevation B.M. No. 2 = 1499.995

17-11. Field Notes. The form of field notes for use with the yard rod is given in Fig. 17-2. The station numbers in Column 1 refer to instrument positions rather than to turning points. The plus sights are on the left-hand side of the notes and the minus sights on the right-hand side. The first three numbers in Columns 2 and 6 are the wire readings at station 8. In Columns 3 and 7 are the upper and lower intervals, respectively, between wires. Immediately below is the difference between the intervals. A difference of three or more would probably indicate a blunder. In Columns 5 and 9 are the sums of the intervals; these are the stadia intercepts.

Between the heavy lines in Columns 2 and 6 are the running totals of the wire readings. Between the heavy lines in Columns 3 and 7 are three times the middle wire reading for each sight. This value plus the interval difference gives the value

From: B.M. 29 to B.M. 30 Forward Date

1	2	3	4	5	6	7	8	9	10
Sta	Wire Readings	Wire Interval	+ Sight	Sum of Interv.	Wire Readings	Wire Interval	− Sight	Sum of Interv.	Remarks
	2749	102			2648	99			Rod A 72°F
8	2647	101			2549	100			Rod B 72°F
	2546	+1		203	2449	−1		199	
	7942	7941	7942	203	7646	7647	7646	199	
	1429	99			1564	98			
9	1330	101			1466	100			
	1229	−2		200	1366	−2		198	
	11930	3990	3988	403	12042	4398	4396	397	
	1148	87			1074	89			
10	1061	88			0985	88			
	0973	−1		175	0897	+1		177	
	15112	3183	3182	578	14998	2955	2956	574	
	0643	103			0521	101			
11	0540	103			0420	102			
	0437	0		206	0318	−1		203	
	16732	1620	1620	784	16257	1260	1259	777	
	2041	101			1606	99			
12	1940	102			1507	99			
	1838	−1		203	1408	0		198	
	22551	5820	5819	987	20778	4521	4521	975	
	−20778		22551	−975			20778		
	+1.773	ft.		+12	ft.				Rod A 74°F
	+1								Rod B 74°F
	+1.774	ft.							
		+12	(+0.057) =		+.684	nearest	= +1		

FIG. 17-2

Error of Instrument Adjustment Date
Determination of C

Sta.	Wire Readings	Wire Interval	+ Sight	Sum of Intervals	Wire Readings	Wire Interval	− Sight	Sum of Intervals	Remarks
	1765	90			1872	10			
	1675	91			1862	9			
	1584	−1		181	1853	+1		19	
	5024	5025	5024	181	5587	5586	5587	19	
	1947	89			1688	12			
	1858	90			1676	12			
	1768	−1		179	1664	0		24	
	10597	5574	5573	360	10615	5028	5028	43	
	−10615		10597	43			10615		
	−018			317					
	$C = +\frac{18}{317}$		$= +.057$						

FIG. 17-3

of the sight in feet. The value of the sight is the same as the sum of the three wire readings. It should be checked accordingly. The sights are recorded in Columns 4 and 8. Their sum at the bottom of the column should agree with the totals of the wire readings. Between the heavy lines in Columns 5 and 8 are the running totals of the stadia intercepts.

The difference in elevation in feet for the section is found by subtracting the total minus-wire readings from the total of the plus-wire readings and inserting the decimal point. This difference is later corrected for instrument adjustment.

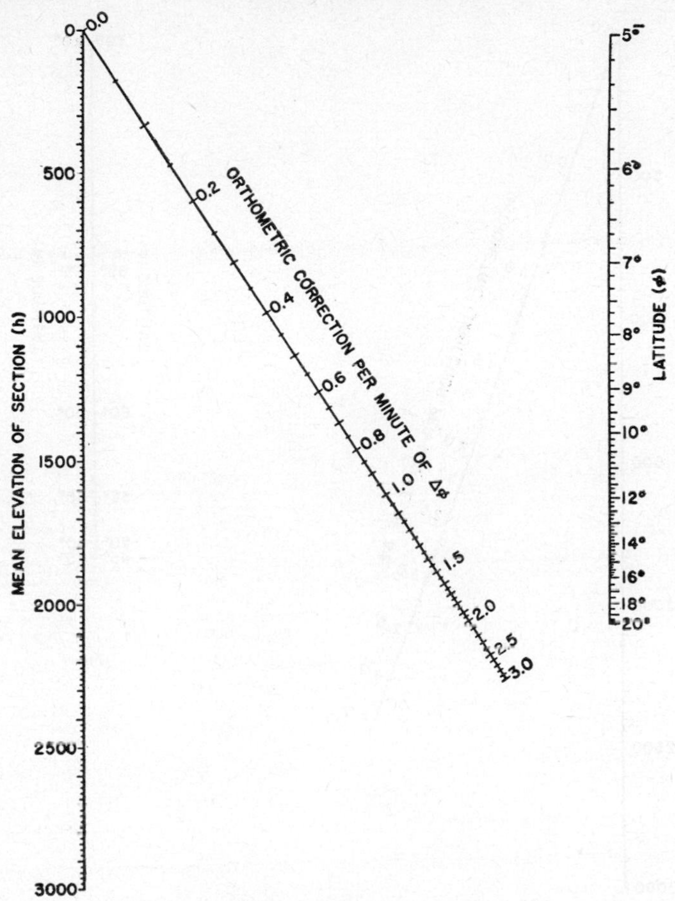

$$\text{Correction} = -Ch\,\Delta\phi = -f(\phi)h\,\Delta\phi$$

A straight line joining the given h and ϕ intersects the correction scale at the required orthometric correction per minute of difference of latitude ($\Delta\phi$). Since the sign of the correction is opposite to that of $\Delta\phi$, it is minus on sections of increasing latitude and plus on sections of decreasing latitude.

With h in meters, yards, or feet, the unit of the correction will be millimeters, milliyards, or millifeet, respectively.

CHART 17-1. Nomogram for the computation of the orthometric correction for use between latitudes 5 and 20°. (*U.S. Coast and Geodetic Survey.*)

The number of feet by which the plus sights are in excess of the minus sights is the total in Column 5 less the total in Column 9. This value, multiplied by the correction factor C for the error in the instrument adjustment, is the correction to be added. In this case it is +0.001 ft, which when added gives the final difference +1.774 ft.

17-12. Correction for Instrument Adjustment. The correction factor C for instrument adjustment is the correction to be added algebraically to the elevation difference per ft of excess in length of plus sights.

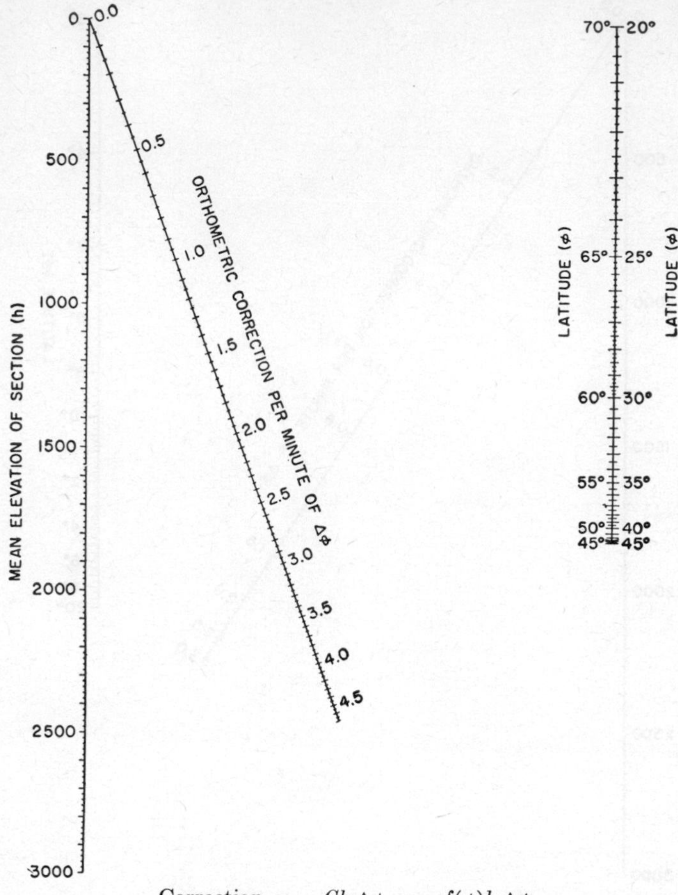

$$\text{Correction} = -Ch\,\Delta\phi = -f(\phi)h\,\Delta\phi$$

A straight line joining the given h and ϕ intersects the correction scale at the required orthometric correction per minute of difference of latitude ($\Delta\phi$). Since the sign of the correction is opposite to that of $\Delta\phi$, it is minus on sections of increasing latitude and plus on sections of decreasing latitude.

With h in meters, yards, or feet, the unit of the correction will be millimeters, milliyards, or millifeet, respectively.

CHART 17-2. Nomogram for the computation of the orthometric correction for use between latitudes 20 and 70°. (*U.S. Coast and Geodetic Survey.*)

To find C select two turning points A and B about 200 to 300 ft apart (see Fig. 2-27). Set up the instrument near B, take the long plus sight to A and the short minus sight to B. Move to a point near A, and take a long plus sight to B and a short minus sight to A. Arrange the field notes exactly as though a loop had been run from A through B and back to A again (see Fig. 17-3).

The difference in elevation and therefore the error for the loop is -0.018 ft. The plus sights exceed the minus sights in length by 317 ft. The correction therefore is $+18/317 = +0.057$ ft per ft of excess plus sights.

In Fig. 17-2 the excess of the length of the plus sights is 12 ft. Therefore the correction is $12(+0.057) = 0.001$ ft. If the sum of the lengths of the minus sights were greater, the excess of plus sights would have a minus sign and the correction, $+0.057$, would be multiplied by a minus quantity.

COMPILATION SHEET

Date	Designation of section	F or B	Rod temp.	Length of section	Elevation differences (B run, sign reversed)	Divergence of B run	B.M.	Elevations and mean diff.	Remarks
7-5	24-25	F	72	2500	+20.781	+ 9	24	183.461	
7-5		B	73		+20.790			+ 20.786	
							25	204.247	
7 6	25-AR2	F	02	1600	− 5.236				
7-6		B	63		− 5.250	+ 6		− 5.253	
7-6		F	65		− 5.256	+15	AR2	198.994	
7-6	AR2-31	F	60	3200	−13.428	− 6		− 13.431	
7-7		B	62		−13.434	+ 9	31	185.563	
7-9	31-32	F	68	3700	− 0.008	+12		− 0.002	
		B	70	3700	+ 0.004	+21	32	185.561	

FIG. 17-4

17-13. Compilation Sheet. A compilation sheet is made up containing the data from all the sections in each link. It will be remembered that a link is the leveling between junction points. Figure 17-4 shows a form for the compilation sheet. The results of all forward and backward runs for each section are entered from the field notes. The elevation differences are treated as previously described to obtain the accepted forward and backward averages. This can usually be accomplished by inspection. The divergence of the *B*-run average from the *A*-run average is computed and recorded as shown. In the same column a running total is kept of these values from section to section. The running total should remain nearly within the limits of closure.

+ Rod	Stad	Check	− Rod	Sta.	Check	Elev.	Sta.	Stad
8.266[1]	161[4]		3.491	171		98.461	B.M. 29	
8.105[2]	165[5]	−0.0013[6]	3.320	168	+0.0010	+8.104		+326
7.940[3]		8.1037[7]	3.152		3.3210	106.565	H.I.	+326
24.311[8]		8.1037[10]	9.963		3.3210	−3.321		−339
0.326[9]	326[11]		0.339	339		103.244	T.P. 1	− 13[12]
6.574	216		4.623	204				
6.358	217	−0.0003	4.419	200	+0.0013	+6.358		+433
6.141		6.3577	4.219		4.4203	109.602	H.I.	+420
19.073		6.3577	13.261		4.4203	−4.420		−404
0.433	433		0.404	404		105.182	T.P. 2	+ 16
6.203	182		2.819	188				
6.021	178	+0.0013	2.631	185	+0.0010	+6.022		+360
5.843		6.0223	2.446		2.6320	111.204	H.I.	+376
18.067		6.0223	7.896		2.6320	−2.632		−373
0.360	360		0.373	373		108.572	T.P. 3	+ 3
4.671	158		8.743	170				
4.513	153	+0.0017	8.573	173	−0.0010	+4.515		+311
4.360		4.5147	8.400		8.5720	113.087	H.I.	+314
13.544		4.5147	25.716		8.5720	−8.572		−343
0.311	311		0.343	343		104.515	T.P. 4	− 29
5.281	235		7.812	239				
5.046	235	0	7.573	236	+0.0010	+5.046		+470
4.811		5.0460	7.337		7.5740	109.561	H.I.	+441
15.138		5.0460	22.722		7.5740	−7.574		−475
0.470	470		0.475	475		101.987	B.M. 30	− 34
Sums	1.900	30.0444		1.934	26.5193			

FIG. 17-5. Form of field notes for conventional Philadelphia rod. Three-wire leveling.

Finally the F and the B runs for each section are averaged to obtain the difference in elevation. These are applied successively to the elevations to obtain the elevation of each benchmark. The final difference for the link is then corrected for rod temperature and for the orthometric corrections. The resulting elevations are called the observed elevations. These are used as a basis for the net adjustment.

17-14. Use with a Conventional Rod. Figure 17-5 shows a form of notes that can be used with the conventional Philadelphia rod marked in feet, tenths, and hundredths of a foot. This form of notes is arranged to fit a standard field notebook. There are six columns for the left-hand page and three columns for the right-hand page. The standard 25 lines are used, 5 lines for each instrument position. The first three columns apply to the backsights, the next three apply to the foresights, and the right-hand page is used to carry the elevations and the cumulative unbalance of the stadia intercepts.

1, 2, 3 are wire readings.

$$4 = 1 - 2,\ 5 = 2 - 3,\ 6 = \frac{4 - 5}{3},\ 7 = 2 + 6.$$

$$8 = 1 + 2 + 3,\ 9 = 1 - 3,\ 10 = \frac{8}{3},\ 11 = 4 + 5.$$

12 is the cumulative unbalance of the stadia intercepts.

Checks. 7 must equal 10, and 9 must equal 11.

Sums. The sums at the bottom of the page are, respectively, the sum of the lengths of the foresights (in stadia intercepts), the sum of the foresights, and the corresponding values for the backsights.

THE BRAATEN CIRCUIT-REDUCTION METHOD OF ADJUSTMENT[1]

17-15. Introduction. The Braaten method at its present stage of development can be applied to level nets of certain types. As these are the types that are usually encountered in small level nets, the method is extremely useful. It gives the same results as the method of least squares but is more easily understood, requires less computation, and can be computed with a slide rule. The reader is referred to Chap. 6 for methods of preparing a net for adjustment and for the terms used.

17-16. Principle. The entire procedure for this method is based on successive applications of the single principle stated in the next paragraphs.

[1] Norman F. Braaten, mathematician, Section of Levels, U.S. Coast and Geodetic Survey, originated and developed this method. For a description of the method and demonstration of the theory by Braaten, see Howard S. Rappleye, "Manual of Level Computation and Adjustment," U.S. Coast and Geodetic Survey Special Publication 240.

When two links of known length, such as l_1 and l_2 in Fig. 17-6, that extend from known elevations meet at a common junction L with a third link that extends from the rest of the level net, it is possible to compute the length of an "equivalent link," which will have the same effect on the adjustment of the net as the two links l_1 and l_2.

If such an equivalent link is substituted for l_1 and l_2, the circuit closure A_0 can also be handled so that the rest of the net is not affected.

Accordingly, when the equivalent link is introduced, the circuit closures (as A_1 and A_3) adjacent to the two links that are eliminated must be changed so that when the two links are finally adjusted separately, A_1 and A_3 will be zero.

17-17. Procedure. Thus the first step in the adjustment is to modify the net by substituting an equivalent link for l_1 and l_2 and to transfer the error A_0 to the adjacent circuits A_1 and A_3 in proportion to the lengths of the links l_1 and l_2.

The resulting new net can then be treated in the same way and thus a second pair of links can be combined, one of which is the equivalent link. This process can be continued until there exist only two links in the net, the last equivalent link and one original link. These links will be in series and lie between two known elevations with equal closure errors (but of opposite sign) on each side. This error can then be eliminated by adjusting the links in proportion to their lengths.

Working backwards, the last equivalent link can be separated into the two links which it represents and its adjustment can be distributed between them according to their size. This process can be continued backward until the adjustment for every link has been computed.

17-18. Forms. The method requires a series of small computations. To prevent errors, Braaten has devised a number of forms, some of which are shown in Figs. 17-7 and 17-10 to 17-13. At the top of each form is a stylized level net. The actual level nets are fitted into these stylized forms. The computations can then be carried out mechanically by successively solving the formulas given.

17-19. Definitions. A link can be considered a length and an observed difference elevation between two junctions, two fixed elevations, or between a junction and a fixed elevation. It is represented by a line with an arrow. The arrow indicates the direction in which the difference in elevation is figured. Although the usual method of computing differences of elevations can be used, throughout this explanation of the Braaten method the U.S. Coast and Geodetic system of B.M.-minus-B.M. method of computing elevations is used. This method is as follows:

1. *Difference in Elevation.* When the arrow flies from A to B, the difference is computed from the formula

$$\text{Diff. in elev.} = \text{elev. } A - \text{elev. } B$$

so that, if the elevation of A is 10 ft and that of B is 7 ft, the difference would be +3 ft. Under the usual procedure as given in Chaps. 2 and 22 it would be −3 ft. Differences are therefore given the opposite sign from the usual practice.

2. *Error of Closure* (usually shortened to *Closure*) is computed **counter-clockwise** around the circuit. This is contrary to the usual procedure as given in Chaps. 2 and 22.

17-20. Symbols. Subscripts are used for all symbols to indicate to which circuit or link they refer.

A = error of closure of a circuit
C = error of closure of certain subsidiary circuits
v = computed correction to the observed difference in elevation in a link; the algebraic sum of the v's for a circuit must be equal to but of opposite sign to A or C
l = observed length of a link
K = change in the value of A when an equivalent link is substituted
D = sum of several lengths; it is used when the sum is used in several formulas
E = length of an equivalent link
l_x = same as E

DEVELOPMENT OF THE FORMS

Form *A*

17-21. Example. Given the level net shown in Fig. 17-6. This is placed in Form A, Fig. 17-7, as shown in Fig. 17-8. In placing it in Form A the signs of the elevation differences are made to correspond with the direction of the arrows in the form.

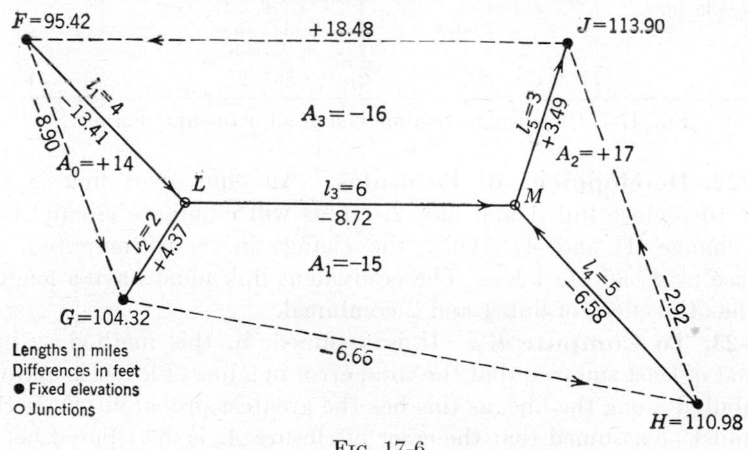

FIG. 17-6

The corrections can now be computed by first entering the known values, then following the form down the left-hand side to the last link required (in this case link 5), and then, starting with v_5, following the form upward to v_1. The last two columns are checks. They will be explained later. Figure 17-9 shows the usual accessory computations.

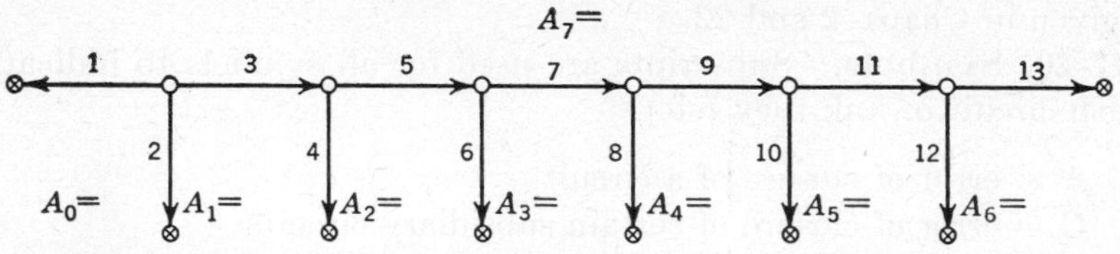

			v/l	Check
$l_1 =$				
$l_2 =$	$A_0 =$			
$E_1 = \frac{l_2 l_1}{l_1 + l_2} =$	$K_1 = \frac{l_2 A_0}{l_1 + l_2} =$	$v_1 = -A_0 + v_2 =$		
$l_3 =$		$v_2 = -A_1 + v_3 + v_4 =$		
$l_4 =$	$A_1 =$			
$E_2 = \frac{l_4(E_1 + l_3)}{E_1 + l_3 + l_4} =$	$K_2 = \frac{l_4(K_1 + A_1)}{E_1 + l_3 + l_4} =$	$v_3 = \frac{l_3(K_1 + A_1 - v_4)}{E_1 + l_3} =$		
$l_5 =$				
$l_6 =$	$A_2 =$	$v_4 = -A_2 + v_5 + v_6 =$		
$E_3 = \frac{l_6(E_2 + l_5)}{E_2 + l_5 + l_6} =$	$K_3 = \frac{l_6(K_2 + A_2)}{E_2 + l_5 + l_6} =$	$v_5 = \frac{l_5(K_2 + A_2 - v_6)}{E_2 + l_5} =$		
$l_7 =$				
$l_8 =$	$A_3 =$	$v_6 = A_3 + v_7 + v_8 =$		
$E_4 = \frac{l_8(E_3 + l_7)}{E_3 + l_7 + l_8} =$	$K_4 = \frac{l_8(K_3 + A_3)}{E_3 + l_7 + l_8} =$	$v_7 = \frac{l_7(K_3 + A_3 - v_8)}{E_3 + l_7} =$		
$l_9 =$				
$l_{10} =$	$A_4 =$	$v_8 = -A_4 + v_9 + v_{10} =$		
$E_5 = \frac{l_{10}(E_4 + l_9)}{E_4 + l_9 + l_{10}} =$	$K_5 = \frac{l_{10}(K_4 + A_4)}{E_4 + l_9 + l_{10}} =$	$v_9 = \frac{l_9(K_4 + A_4 - v_{10})}{E_4 + l_9} =$		
$l_{11} =$				
$l_{12} =$	$A_5 =$	$v_{10} = -A_5 + v_{11} + v_{12} =$		
$E_6 = \frac{l_{12}(E_5 + l_{11})}{E_5 + l_{11} + l_{12}} =$	$K_6 = \frac{l_{12}(K_5 + A_5)}{E_5 + l_{11} + l_{12}} =$	$v_{11} = \frac{l_{11}(K_5 + A_5 - v_{12})}{E_5 + l_{11}} =$		
$l_{13} =$	$A_6 =$	$v_{12} = -A_6 + v_{13} =$		
		$v_{13} = \frac{l_{13}(K_6 + A_6)}{E_6 + l_{13}} =$		

FIG. 17-7. Circuit-reduction method of adjustment. Form A.

17-22. Development of Formulas. An equivalent link is to be found to replace link 1 and link 2. This will eliminate circuit A_0 and thus change A_1 and A_3. Only the change in A_1 is computed. The increase in A_1 is called K_1. The equivalent link must have a length E_1 that has the effect of link 1 and 2 combined.

17-23. To Compute K_1. It is assumed, in this method as in the method of least squares, that the total error in a line of levels is uniformly distributed along the line, as this has the greatest probability. It therefore must be assumed that the error of closure A_0 is distributed between

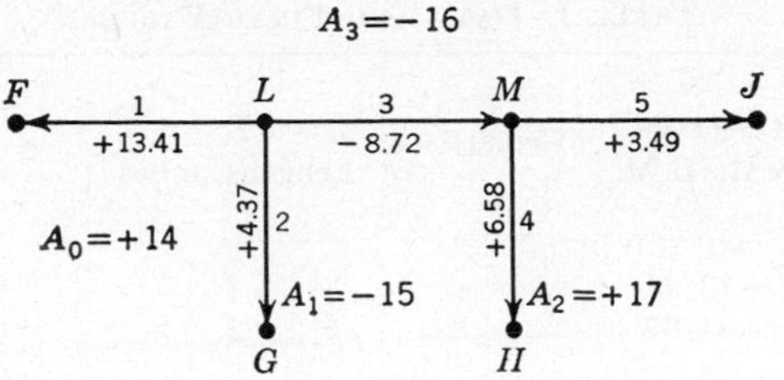

						v/l	Check
$l_1 =$	4						
		$A_0 =$	+14				
$l_2 =$	2						
$E_1 = \frac{l_2 l_1}{l_1 + l_2} =$	1.33	$K_1 = \frac{l_2 A_0}{l_1 + l_2} =$	+ 4.67	$v_1 = -A_0 + v_2 =$	− 9.37	−2.34	
$l_3 =$	6						
		$A_1 =$	−15	$v_2 = -A_1 + v_3 + v_4 =$	+ 4.63	+2.31	
$l_4 =$	5						1, 2, 3 / 0.00
$E_2 = \frac{l_4(E_1 + l_3)}{E_1 + l_3 + l_4} =$	2.97	$K_2 = \frac{l_4(K_1 + A_1)}{E_1 + l_3 + l_4} =$	− 1.19	$v_3 = \frac{l_3(K_1 + A_1 - v_4)}{E_1 + l_3} =$	+ 0.19	+0.03	
$l_5 =$	3						
		$A_2 =$	+17	$v_4 = -A_2 + v_5 =$	−10.56	−2.11	
				$v_5 = \frac{l_5(K_2 + A_2)}{E_2 + l_5} =$	+ 6.44	+2.15	3, 4, 5 / 0.01

FIG. 17-8

link 1 and link 2 according to their lengths. Therefore

$$\text{Error}_1 = \frac{l_1}{l_1 + l_2} A_0 \tag{17-1}$$

$$\text{Error}_2 = \frac{l_2}{l_1 + l_2} A_0 \tag{17-2}$$

The elimination of A_0 will make it necessary to place that part of the error A_0 that is assigned to l_2 into the closure A_1 (see Chap. 6). In effect, error$_2$ is subtracted from A_0 and added to A_1. Thus K_1, the change in A_1, is equal to error$_2$, or

$$K_1 = \frac{l_2 A_0}{l_1 + l_2} \tag{17-3}$$

as shown in Form A.

17-24. To Compute E_1. Since errors must be assumed to be uniformly distributed along any line of levels, the total error in a link must be allowed to affect the adjustment only in accordance with the reciprocal of the length of the link. The effect of two links then would be the sum of the reciprocals of their lengths. Accordingly, the reciprocal of the length of the equivalent link must be equal to the sum of the reciprocals

TABLE I. DATA AND FINAL VALUES

Links: Name	Links: No.	Observed B.M.–B.M.	Corrections	L Lengths, miles	Final values: v	Final values: Links
L–F	1	+13.41	v_1	4	−0.09	+13.32
L–G	2	+ 4.37	v_2	2	+0.05	+ 4.42
L–M	3	− 8.72	v_3	6	+0.01	− 8.71
M–H	4	+ 6.58	v_4	5	−0.11	+ 6.47
M–J	5	+ 3.49	v_5	3	+0.06	+ 3.55
J–F		+18.48				

FINAL ELEVATIONS

Computation: B.M.'s and links	Computation: Elevations and differences	Checks: B.M.'s and links	Checks: Elevations and differences
F	95.42	*L*	108.74
+1	+ 13.32	−2	− 4.42
L	108.74	*G*	104.32
−3	+ 8.71		
M	117.45	*M*	117.45
−5	− 3.55	−4	− 6.47
J	113.90	*H*	110.98

TABLE II. SUBSTITUTION OF v'S IN CIRCUITS

	v	0: Sign	0: Value	1: Sign	1: Value	2: Sign	2: Value	3: Sign	3: Value
1	−0.09	+	−0.09					−	+0.09
2	+0.05	−	−0.05	+	+0.05				
					−0.01				+0.01
3	0			−	ø			+	ø
4	−0.11			−	+0.11	+	−0.11		
5	+0.06					−	−0.06	+	+0.06
A			+0.14		−0.15		+0.17		−0.16
	Sums		0		+0.01		0		−0.01

FIG. 17-9

of the two links eliminated. Thus

$$\frac{1}{E_1} = \frac{1}{l_1} + \frac{1}{l_2}$$
$$E_1 = \frac{l_1 l_2}{l_1 + l_2} \qquad (17\text{-}4)$$

as shown in form A.

We now have, in effect, an equivalent link whose length is E_1 running from a fixed elevation F to junction L and the actual link 3 running from L to junction M. Since they are in series, they can be handled together.

17-25. To Compute K_2. It is evident that the formula for K_2 can be found by making proper substitutions in the formula for K_1. These are the following: for K_1 substitute K_2, for l_2 substitute l_4, for A_0 substitute $K_1 + A_1$, and for l_1 substitute $E_1 + l_3$. This gives

$$K_2 = \frac{l_4(K_1 + A_1)}{E_1 + l_3 + l_4}$$

as shown in Form A.

17-26. To Compute E_2. In the formula for E_1 make the same substitutions as in the previous paragraphs. This gives

$$E_2 = \frac{l_4(E_1 + l_3)}{E_1 + l_3 + l_4}$$

17-27. Continuation. It is evident that this process can be continued indefinitely. In the form, it is carried out to K_6, E_6, and l_{13}. But actual computations are carried out only as far as necessary. In Fig. 17-8 only the actual part of the form used in the example is shown.

17-28. To Compute v_5. To demonstrate the development of the formulas for the corrections (the v's), only that part of the form used in the example is covered. The rest of the form is similar.

In the example, the net has been reduced to two links in series, the equivalent link extending to M and the actual link from M to J. Since the new error of closure for them has been computed, they, in effect, join two points of known elevation. As they are in series, the error of each can be computed in proportion to its length. Thus, using Eq. (17-1) as a basis,

$$\text{Error}_5 = \frac{l_5(K_2 + A_2)}{E_2 + l_5}$$

Since a correction is equal to the error it eliminates, but with the opposite sign, $v_5 = -\text{error}_5$.

In the case of link 5, the arrow is opposite to counterclockwise; therefore the sign of v_5 is changed, and

$$v_5 = \frac{l_5(K_2 + A_2)}{E_2 + l_5}$$

To compute v_4,

$$A_2 = \text{error}_4 + \text{error}_5$$
$$A_2 = -v_4 - v_5$$

Changing signs for the direction of the arrows

$$A_2 = -v_4 + v_5$$

and

$$v_4 = -A_2 + v_5$$

as shown in Form *A*.

17-29. To Compute v_3. When the value of v_4 is added to the closure A_2, it is subtracted from A_1 on the other side of link 4. Thus, as for v_5,

$$v_3 = \frac{l_3(K_1 + A_1 - v_4)}{E_1 + l_3}$$

17-30. To Compute v_2.

$$A_1 = \text{error}_2 + \text{error}_3 + \text{error}_4$$
$$A_1 = -v_2 - v_3 - v_4$$

Changing signs for the direction of the arrows,

$$A_1 = -v_2 + v_3 + v_4$$

and

$$v_2 = -A_1 + v_3 + v_4$$

17-31. To Compute v_1.

$$A_0 = \text{error}_1 + \text{error}_2$$
$$A_0 = -v_1 - v_2$$

Changing signs for the direction of the arrows,

$$A_0 = -v_1 + v_2$$

or

$$v_1 = -A_0 + v_2$$

17-32. To Compute v/l. In the column v/l are placed the values v_1/l_1, v_2/l_2, etc. The sum of such values around a point should vary from zero only by the effect of rounding off in the last places in computation. The sign of the v/l is changed when the arrow points toward the point in question. The checks for certain points are shown in the check column.

It is evident that, since the total correction to a pair of links that meet at a junction point is apportioned according to their lengths, the effect of the total correction is to impose a single *rate of correction* along their combined lengths. The rate of correction measured *from the junction* is equal on both links but of opposite sign. Since the final rates applied to the links is the result of successive application of rates of correction to pairs of links, the sum of the rates is zero.[1] In every case v/l is a rate and therefore the sums of the v/l's is zero.

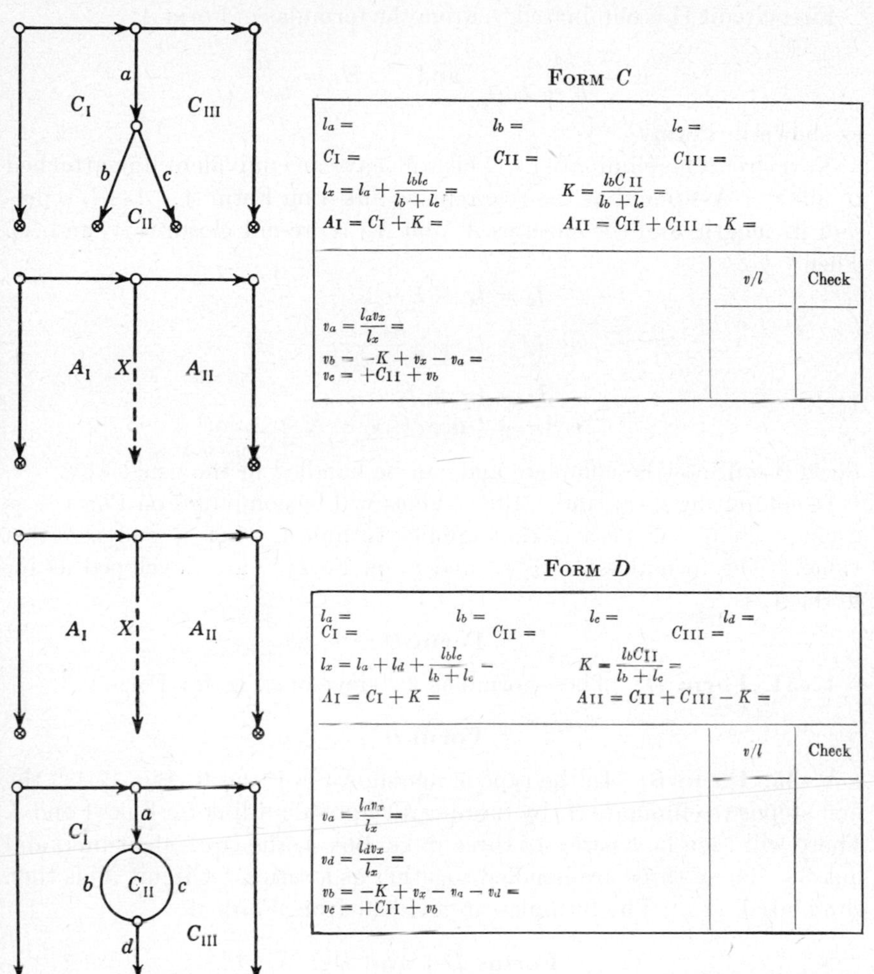

FORM C

$l_a =$ $l_b =$ $l_c =$

$C_I =$ $C_{II} =$ $C_{III} =$

$l_x = l_a + \frac{l_b l_c}{l_b + l_c} =$ $K = \frac{l_b C_{II}}{l_b + l_c} =$

$A_I = C_I + K =$ $A_{II} = C_{II} + C_{III} - K =$

	v/l	Check
$v_a = \frac{l_a v_x}{l_x} =$ $v_b = -K + v_x - v_a =$ $v_c = +C_{II} + v_b$		

FORM D

$l_a =$ $l_b =$ $l_c =$ $l_d =$

$C_I =$ $C_{II} =$ $C_{III} =$

$l_x = l_a + l_d + \frac{l_b l_c}{l_b + l_c} =$ $K = \frac{l_b C_{II}}{l_b + l_c} =$

$A_I = C_I + K =$ $A_{II} = C_{II} + C_{III} - K =$

	v/l	Check
$v_a = \frac{l_a v_x}{l_x} =$ $v_d = \frac{l_d v_x}{l_x} =$ $v_b = -K + v_x - v_a - v_d =$ $v_c = +C_{II} + v_b$		

FIG. 17-10. Circuit-reduction method of adjustment. Forms C and D.

Form C

17-33. Form C (Top). Form C, shown in Fig. 17-10, is used to obtain an equivalent link when a connection between a fixed elevation and the

[1] For example, around junction M, a rate of correction is computed for links 3 and 4 and another rate of correction for the links 4 and 5. They are then combined.

net is not a single link but two links that join before reaching the net (as shown on the form). For example, link 4 in Form A might have been an inverted Y as shown on Form C. The problem is to find an equivalent link for the inverted Y.

First circuit II is eliminated. From the formulas in Form A,

$$K = \frac{l_b C_{\text{II}}}{l_b + l_c} \qquad \text{and} \qquad E_1 = \frac{l_b l_c}{l_b + l_c}$$

as shown in Form C.

Next circuit I is eliminated. This will leave an equivalent link attached to link a. Assume that the two replace link 4 on Form A. Let l_x represent its length, and the closures A_{I} and A_{II} represent closures A_1 and A_2. Then

$$l_x = l_a + E_1$$
$$l_x = l_a + \frac{l_b l_c}{l_b + l_c}$$
$$A_{\text{I}} = C_{\text{I}} + K$$
$$A_{\text{II}} = C_{\text{II}} + C_{\text{III}} - K$$

Form A will now be complete and can be handled in the usual way.

In computing v_a, v_b, and v_c the v values will be computed on Form A as usual. There will be a v_4 that applies to link 4. Let v_x represent this value. The formulas for v_a, v_b, and v_c on Form C are developed as for Form A.

Form *D*

17-34. Form *D*. These formulas are developed as for Form C.

Form *B*

17-35. Form *B*. In the type of net shown in Form B (Fig. 17-11) the first step is to eliminate A_0 by forming an equivalent link for links 1 and 2. There will then be a series of three links, link 4, the equivalent link, and link 3. These three are handled together as a series. Circuit A_1 is then eliminated, etc. The formulas are evident from Form A.

Forms *D*-1 and *D*-2

17-36. Forms *D*-2 and *D*-1. Forms D-2 and D-1, shown in Figs. 17-12 and 17-13, are used when there is an interior circuit more complicated than those shown in Form D or Form B. An interior circuit is a circuit that does not pass through a fixed elevation. Form D-2 is an elaboration of Form D-1 and is developed by the same methods. Only Form D-1 will be discussed.

17-37. Form *D*-1. Consider the level net shown in Fig. 17-14, 1. The adjustment of this net would be the same as that shown in Fig.

17-14, 2 in which the two fixed points A' and A'' have been combined into one. Only links can be adjusted; therefore, the elevations of fixed points have no effect on adjustment except to establish circuit closure errors. By combining A' and A'' we merely indicate the existing fact that no adjustment can take place between them.

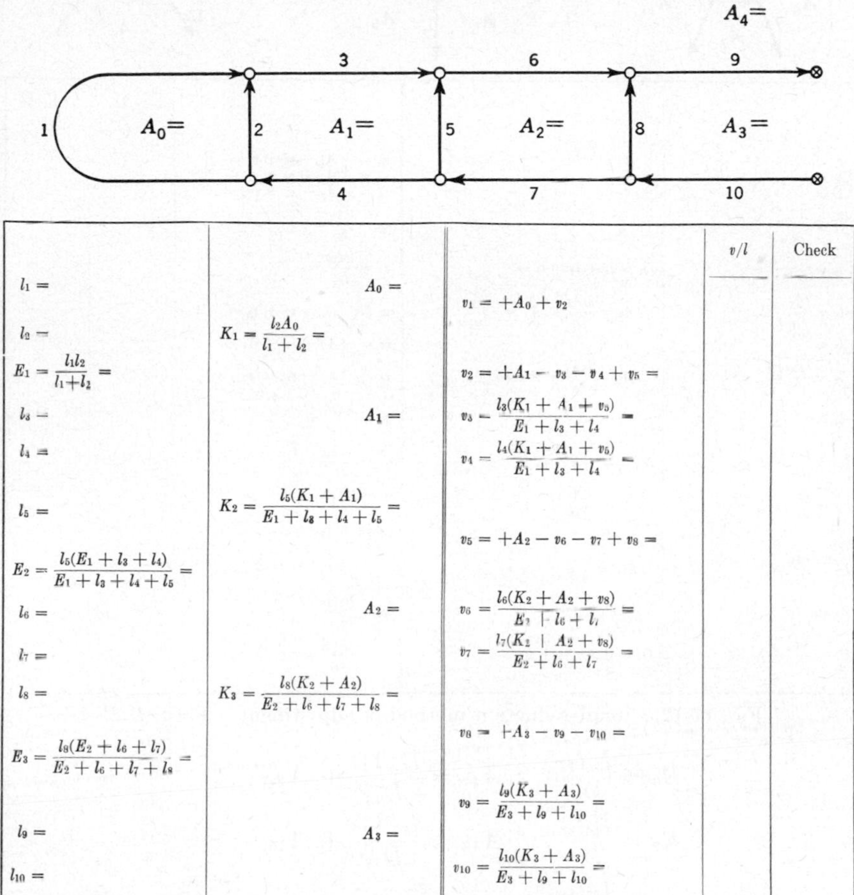

			v/l	Check
$l_1 =$	$A_0 =$	$v_1 = +A_0 + v_2$		
$l_2 =$	$K_1 = \dfrac{l_2 A_0}{l_1 + l_2} =$			
$E_1 = \dfrac{l_1 l_2}{l_1 + l_2} =$		$v_2 = +A_1 - v_3 - v_4 + v_5 =$		
$l_3 =$	$A_1 =$	$v_3 = \dfrac{l_3(K_1 + A_1 + v_5)}{E_1 + l_3 + l_4} =$		
$l_4 =$		$v_4 = \dfrac{l_4(K_1 + A_1 + v_5)}{E_1 + l_3 + l_4} =$		
$l_5 =$	$K_2 = \dfrac{l_5(K_1 + A_1)}{E_1 + l_3 + l_4 + l_5} =$			
		$v_5 = +A_2 - v_6 - v_7 + v_8 =$		
$E_2 = \dfrac{l_5(E_1 + l_3 + l_4)}{E_1 + l_3 + l_4 + l_5} =$				
$l_6 =$	$A_2 =$	$v_6 = \dfrac{l_6(K_2 + A_2 + v_8)}{E_2 + l_6 + l_7} =$		
$l_7 =$		$v_7 = \dfrac{l_7(K_2 + A_2 + v_8)}{E_2 + l_6 + l_7} =$		
$l_8 =$	$K_3 = \dfrac{l_8(K_2 + A_2)}{E_2 + l_6 + l_7 + l_8} =$			
		$v_8 = +A_3 - v_9 - v_{10} =$		
$E_3 = \dfrac{l_8(E_2 + l_6 + l_7)}{E_2 + l_6 + l_7 + l_8} =$				
		$v_9 = \dfrac{l_9(K_3 + A_3)}{E_3 + l_9 + l_{10}} =$		
$l_9 =$	$A_3 =$			
$l_{10} =$		$v_{10} = \dfrac{l_{10}(K_3 + A_3)}{E_3 + l_9 + l_{10}} =$		

FIG. 17-11. Circuit-reduction method of adjustment. Form B.

For later convenience let $l_{16} + l_{17} + l_{18}$ be represented by D_{13}.

Now consider the net in Fig. 17-14, 3. From the explanation of Form A it is evident that, for adjustment purposes, this net is equivalent to that in Fig. 17-14, 2, if E_0 is the equivalent of links 16 and 17, E_2 that of links 16 and 18, and E_3 that of links 17 and 18 and the proper transfers of closure errors are made. Based on this plan, we have, from Eqs. (17-4) and (17-3),

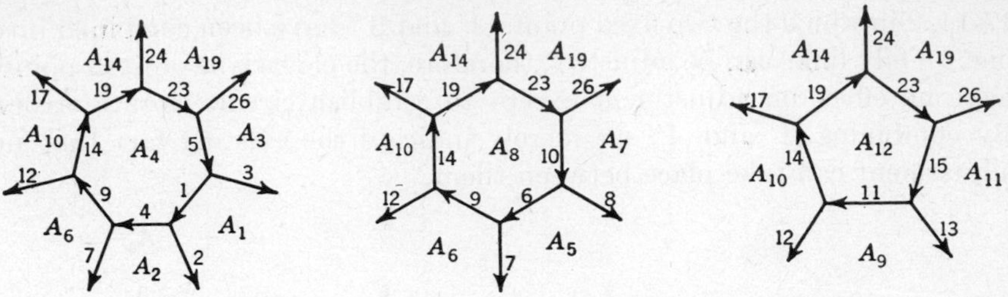

			v/l	Check
$l_1 =$	$A_1 =$	$v_1 = -A_1 - v_2 + v_3 =$		
$l_2 =$	$A_2 =$	$v_2 = +A_2 + v_4 + v_7 =$		
$l_3 =$	$A_3 =$	$v_3 = -A_3 - v_5 + v_{26} =$		
$l_4 =$		$v_4 = \frac{l_4 v_6}{l_6} =$		
$l_5 =$	$D_1 = l_1 + l_2 + l_3 =$	$v_5 = \frac{l_5 v_{10}}{l_{10}} =$		
$l_6 = \frac{l_1 l_2}{D_1} + l_4 =$	$A_5 = \frac{l_2 A_1}{D_1} + A_2 =$	$v_6 = -A_5 - v_7 + v_8 =$		
$l_7 =$	$A_6 =$	$v_7 = +A_6 + v_9 + v_{12} =$		
$l_8 = \frac{l_2 l_3}{D_1} =$	$A_7 = \frac{l_3 A_1}{D_1} + A_3 =$	$v_8 = -A_7 - v_{10} + v_{26} =$		
$l_9 =$		$v_9 = \frac{l_9 v_{11}}{l_{11}} =$		
$l_{10} = \frac{l_1 l_3}{D_1} + l_5 =$	$D_5 = l_6 + l_7 + l_8 =$	$v_{10} = \frac{l_{10} v_{15}}{l_{15}} =$		
$l_{11} = \frac{l_6 l_7}{D_5} + l_9 =$	$A_9 = \frac{l_7 A_5}{D_5} + A_6 =$	$v_{11} = -A_9 - v_{12} - v_{13} =$		
$l_{12} =$	$A_{10} =$	$v_{12} = +A_{10} + v_{14} + v_{17} =$		
$l_{13} = \frac{l_7 l_8}{D_5} =$	$A_{11} = \frac{l_8 A_5}{D_5} + A_7 =$	$v_{13} = -A_{11} - v_{15} + v_{26} =$		
$l_{14} =$		$v_{14} = \frac{l_{14} v_{16}}{l_{16}} =$		
$l_{15} = \frac{l_6 l_8}{D_5} + l_{10} =$	$D_9 = l_{11} + l_{12} + l_{13} =$	$v_{15} = \frac{l_{15} v_{20}}{l_{20}} =$		

FIG. 17-12. Circuit-reduction method of adjustment. Form D-2.

$$E_0 = \frac{l_{16} l_{17}}{D_{13}} \qquad A_{17} = \frac{l_{17} A_{13}}{D_{13}} + A_{14}$$

$$E_2 = \frac{l_{16} l_{18}}{D_{13}} \qquad A_{18} = \frac{l_{18} A_{13}}{D_{13}} + A_{15}$$

$$E_3 = \frac{l_{17} l_{18}}{D_{13}} \qquad A_{20} = -A_{17} - A_{18} - A_{19}$$

The net in Fig. 17-14, 3, can be drawn as shown in Fig. 17-13, 4.

Let $l_{21} + l_{22} + l_{23}$ be represented by D_{20}. Since each pair is in series, let

$$l_{21} = E_0 + l_{19} \qquad l_{22} = E_2 + l_{20} \qquad l_{25} = E_3$$

Substituting,

$$l_{21} = \frac{l_{16} l_{17}}{D_{13}} + l_{19} \qquad l_{22} = \frac{l_{16} l_{18}}{D_{13}} + l_{20} \qquad l_{25} = \frac{l_{17} l_{18}}{D_{13}}$$

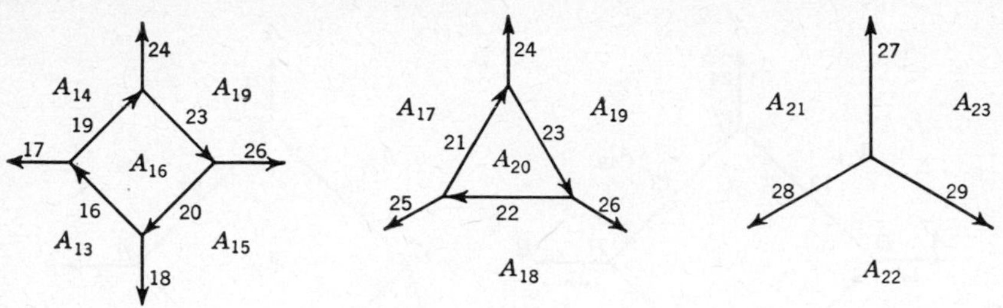

			v/l	Check
$l_{16} = \dfrac{l_{11}l_{12}}{D_9} + l_{14} =$	$A_{13} = \dfrac{l_{12}A_9}{D_9} + A_{10} =$	$v_{16} = -A_{13} - v_{17} + v_{18} =$		
$l_{17} =$	$A_{14} =$	$v_{17} = +A_{14} + v_{19} + v_{24} =$		
$l_{18} = \dfrac{l_{12}l_{13}}{D_9} =$	$A_{15} = \dfrac{l_{13}A_9}{D_9} + A_{11} =$	$v_{18} = -A_{15} - v_{20} + v_{26} =$		
$l_{19} =$		$v_{19} = \dfrac{l_{19}v_{21}}{l_{21}} =$		
$l_{20} = \dfrac{l_{11}l_{13}}{D_9} + l_{15} =$	$D_{13} = l_{16} + l_{17} + l_{18} =$	$v_{20} = \dfrac{l_{20}v_{22}}{l_{22}} =$		
$l_{21} = \dfrac{l_{16}l_{17}}{D_{13}} + l_{19} =$	$A_{17} = \dfrac{l_{17}A_{13}}{D_{13}} + A_{14} =$	$v_{21} = -A_{17} - v_{24} + v_{25} =$		
$l_{22} = \dfrac{l_{16}l_{18}}{D_{13}} + l_{20} =$	$A_{18} = \dfrac{l_{18}A_{13}}{D_{13}} + A_{15} =$	$v_{22} = -A_{18} - v_{25} + v_{26} =$		
$l_{23} =$	$A_{19} =$	$v_{23} = -A_{19} + v_{24} - v_{26} =$		
$l_{24} =$	$A_{20} = -A_{17} - A_{18} - A_{19} =$	$v_{24} = \dfrac{l_{24}v_{27}}{l_{27}} =$		
$l_{25} = \dfrac{l_{17}l_{18}}{D_{13}} =$		$v_{25} = \dfrac{l_{25}v_{28}}{l_{28}} =$		
$l_{26} =$	$D_{20} = l_{21} + l_{22} + l_{23} =$	$v_{26} = \dfrac{l_{26}v_{29}}{l_{29}} =$		
$l_{27} = \dfrac{l_{21}l_{23}}{D_{20}} + l_{24} =$	$A_{21} = \dfrac{l_{21}A_{20}}{D_{20}} + A_{17} =$	$v_{27} = -A_{21} + v_{20} =$		
$l_{28} = \dfrac{l_{21}l_{22}}{D_{20}} + l_{25} =$	$A_{22} = \dfrac{l_{22}A_{20}}{D_{20}} + A_{18} =$	$v_{28} = -A_{22} + v_{29} =$		
$l_{29} = \dfrac{l_{22}l_{23}}{D_{20}} + l_{26} =$		$v_{29} = \dfrac{l_{29}(A_{22} + K_1)}{l_{29} + E_1} =$		
$E_1 = \dfrac{l_{27}l_{28}}{l_{27} + l_{28}} =$	$K_1 = \dfrac{l_{28}A_{21}}{l_{27} + l_{28}} =$			

FIG. 17-13. Circuit reduction method of adjustment. Form D-1.

The net in Fig. 17-14, 5, is equivalent, for adjustment purposes, to that shown in Fig. 17-14, 4, by proper substitution of equivalent links and transfer of circuit closures, as was carried out for the change from Fig. 17-14, 2, to Fig. 17-14, 3. Thus

$$E_4 = \frac{l_{21}l_{23}}{D_{20}} \qquad A_{21} = \frac{l_{21}A_{20}}{D_{20}} + A_{17}$$

$$E_5 = \frac{l_{21}l_{22}}{D_{20}} \qquad A_{22} = \frac{l_{22}A_{20}}{D_{20}} + A_{18}$$

$$E_6 = \frac{l_{22}l_{23}}{D_{20}}$$

The net can now be drawn as shown in Fig. 17-14, 6.

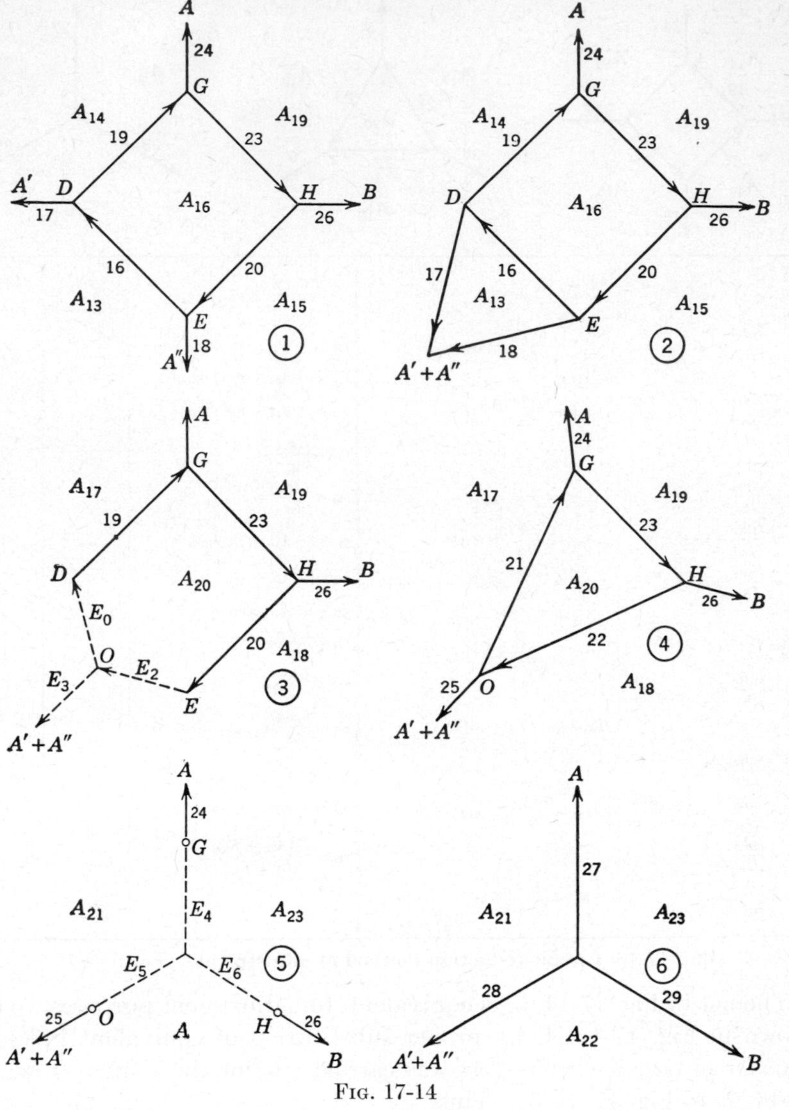

FIG. 17-14

Let $l_{27} = E_4 + l_{24}$, $l_{28} = E_5 + l_{25}$, and $l_{29} = E_6 + l_{26}$. Substituting,

$$l_{27} = \frac{l_{21}l_{23}}{D_{20}} + l_{24} \qquad l_{28} = \frac{l_{21}l_{22}}{D_{20}} + l_{25} \qquad l_{29} = \frac{l_{22}l_{23}}{D_{20}} + l_{26}$$

Eliminating circuit A_{21} as in Form A,

$$E_1 = \frac{l_{27}l_{28}}{l_{27} + l_{28}} \qquad K_1 = \frac{l_{28}A_{21}}{l_{27} + l_{28}}$$

The *formulas for the v's* are developed as shown for Form A.

The *formulas in Form D*-1 are those which have been developed in the previous paragraphs with the exception of several at the beginning, i.e., formulas for l_{16}, l_{18}, l_{20}, A_{13}, and A_{15}. These are necessary only when an

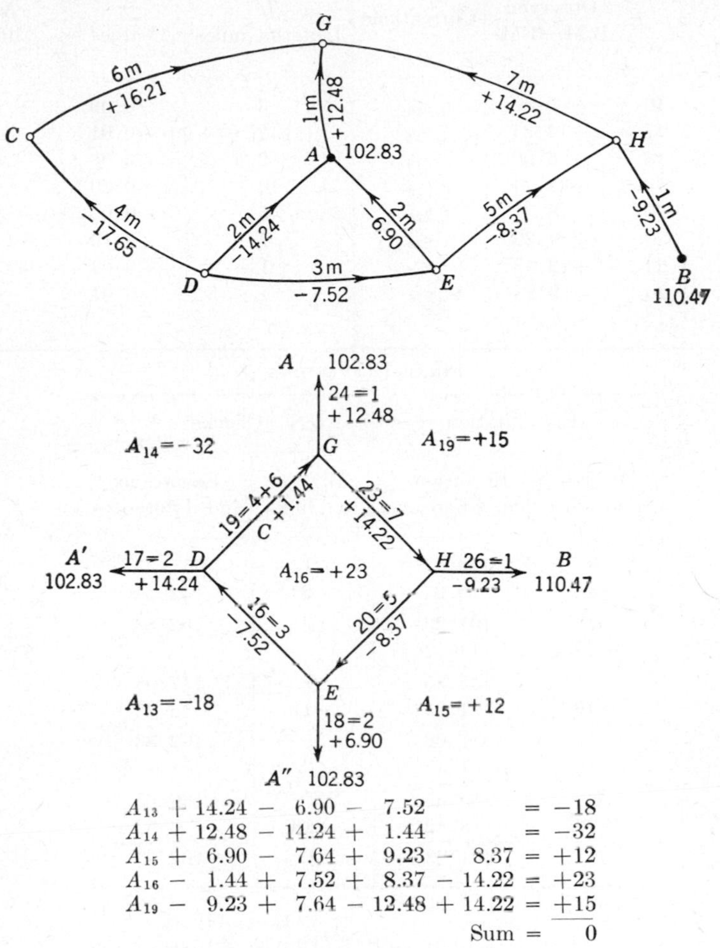

FIG. 17-15

interior circuit of more than four sides is to be adjusted. In that case Form D-2 is first used to reduce the loop to four sides. When the original internal circuit has only four sides (as in the example), these values can be set down at once.

Example. Figures 17-15 to 17-18 show an example of the adjustment of a net by Form D-1. This is the same net that is adjusted by least squares in Chap. 22. Note that fixed point A has been given three positions, A, A', and A''. This does not affect the adjustment in any way, as no adjustment can take place between fixed points.

TABLE I. DATA AND FINAL VALUES

Links		Observed B.M.–B.M.	Corrections	L Lengths, miles	v Values	Adj. links
Name	No.					
E–D	16	− 7.52	v_{16}	3	+0.09	− 7.43
D–A	17	+14.24	v_{17}	2	+0.01	+14.25
E–A	18	+ 6.90	v_{18}	2	−0.08	+ 6.82
D–G	19	+ 1.44	v_{19}	10	+0.29	+ 1.73
H–E	20	− 8.37	v_{20}	5	−0.05	− 8.42
G–H	23	+14.22	v_{23}	7	−0.10	+14.12
G–A	24	+12.48	v_{24}	1	+0.04	+12.52
H–B	26	− 9.23	v_{26}	1	−0.01	− 9.24
A–B		− 7.64				

FINAL ELEVATIONS

Computation		Checks	
B.M.'s and links	Elevations and differences	B.M.'s and links	Elevations and differences
B	110.47	G	115.35
+26	− 9.24	−24	−12.52
H	101.23	A	102.83
+23	+14.12		
G	115.35	D	117.08
+19	+ 1.73	−17	−14.25
D	117.08	A	102.83
+16	− 7.43		
E	109.65	E	109.65
+20	− 8.42	−18	− 6.82
H	101.23	A	102.83

$$G = 115.35$$
$$G\text{–}C = -16.21 + (6/10)(0.29) = -16.04$$
$$C = 99.31$$

FIG. 17-16

APPLICATION OF METHOD

17-38. Fitting Actual Nets to the Forms. The process of adapting an existing level net to the available forms can be handled as follows:

1. Draw the usual sketch of the net according to its actual geographical arrangement.

2*a*. Choose the form or forms that apply.

2*b*. Remember that any fixed point can be split up into several points or several can be combined into one.

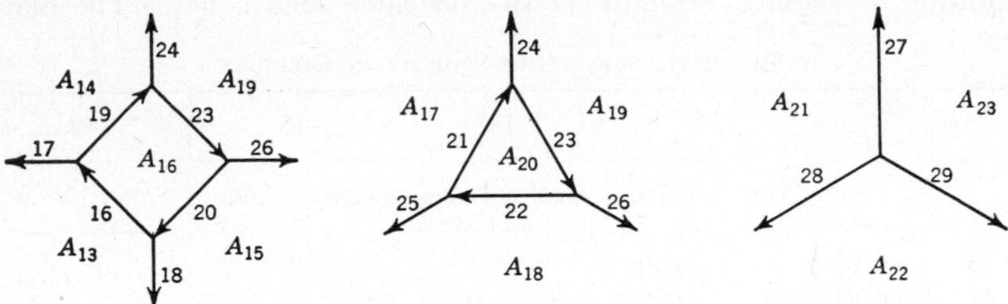

			v/l	Check
$l_{16}=\frac{l_{11}l_{12}}{D_9}+l_{14}=$ 3	$A_{13}=\frac{l_{12}A_9}{D_9}+A_{10}=$ −18	$v_{16}=-A_{13}-v_{17}+v_{18}=+$ 9.37	+3.12	
$l_{17}=$ 2	$A_{14}=$ −32	$v_{17}=+A_{14}+v_{19}+v_{24}=+$ 0.60	+0.30	+ + **16, 18, 20** 0.00
$l_{18}=\frac{l_{12}l_{13}}{D_9}=$ 2	$A_{15}=\frac{l_{13}A_9}{D_9}+A_{11}=$ +12	$v_{18}=-A_{15}-v_{20}+v_{26}=-$ 8.04	−4.02	− + + **16, 17, 19** +0.01
$l_{19}=$ 10		$v_{19}=\frac{l_{19}v_{21}}{l_{21}}=$ +28.30	+2.83	− + + **19, 23, 24** +0.01
$l_{20}=\frac{l_{11}l_{13}}{D_9}+l_{15}=$ 5	$D_{13}=l_{16}+l_{17}+l_{18}=$ + 7	$v_{20}=\frac{l_{20}v_{22}}{l_{22}}=$ − 4.51	−0.90	+ − + **20, 23, 26** 0.00
$l_{21}=\frac{l_{16}l_{17}}{D_{13}}+l_{19}=$ +10.86	$A_{17}=\frac{l_{17}A_{13}}{D_{13}}+A_{14}=$ −37.14	$v_{21}=-A_{17}-v_{24}+v_{25}=$ +30.73	+2.83	
$l_{22}=\frac{l_{16}l_{18}}{D_{13}}+l_{20}=$ + 5.86	$A_{18}=\frac{l_{18}A_{13}}{D_{13}}+A_{15}=$ + 6.86	$v_{22}=-A_{18}-v_{25}+v_{26}=-$ 5.29	−0.00	+ − + **21, 22, 25** +0.01
$l_{23}=$ 7	$A_{19}=$ +15	$v_{23}=-A_{19}+v_{24}-v_{26}=$ −10.16	−1.45	
$l_{24}=$ 1	$A_{20}=-A_{17}-A_{18}-A_{19}=$ +15.28	$v_{24}=\frac{l_{24}v_{27}}{l_{27}}=$ + 4.29	+4.29	
$l_{25}=\frac{l_{17}l_{18}}{D_{13}}=$ + 0.57		$v_{25}=\frac{l_{25}v_{28}}{l_{25}}=$ − 2.12	−3.72	
$l_{26}=$ 1	$D_{20}=l_{21}+l_{22}+l_{23}=$ +23.72	$v_{26}=\frac{l_{26}v_{29}}{l_{29}}=$ − 0.55	−0.55	
$l_{27}=\frac{l_{21}l_{23}}{D_{20}}+l_{24}=$ + 4.20	$A_{21}=\frac{l_{21}A_{20}}{D_{20}}+A_{17}=$ −30.14	$v_{27}=-A_{21}+v_{28}=$ +18.00	+4.29	
$l_{28}=\frac{l_{21}l_{22}}{D_{20}}+l_{25}=$ + 3.25	$A_{22}=\frac{l_{22}A_{20}}{D_{20}}+A_{18}=$ +10.63	$v_{28}=-A_{22}+v_{29}=$ −12.14	−3.74	+ + + **27, 28, 29** 0.00
$l_{29}=\frac{l_{22}l_{23}}{D_{20}}+l_{26}=$ + 2.73		$v_{29}=\frac{l_{29}(A_{22}+K_1)}{l_{29}+E_1}=$ − 1.51	−0.55	
$E_1=\frac{l_{27}l_{28}}{l_{27}+l_{28}}=$ + 1.83	$K_1=\frac{l_{28}A_{21}}{l_{27}+l_{28}}=$ −13.15			

FIG. 17-17. Circuit-reduction method of adjustment. Form *D*-1.

3. Place the designations of the junctions and fixed elevations on the form.

4. Place the observed differences in elevations on the proper links on the form with signs that apply to the direction of the arrows on the forms.

5. Compute the circuit closures, the v's, and the checks.

The most important operation is the choice of the proper forms. There are often several correct ways in which the forms can be used.

If there is no interior circuit, Form *A* can be used with the occasional addition of Form *C*. Figure 17-19 illustrates such a net. The black

TABLE II. SUBSTITUTION OF *v*'S IN CIRCUITS

	v	13		14		15		19	
		Sign	Value	Sign	Value	Sign	Value	Sign	Value
16	+0.09	+	+0.09						
17	+0.01	+	+0.01	−	−0.01				
18	−0.08	−	+0.08			+	−0.08		
					+0.29				
19	+0.28			+	~~+0.28~~				
20	−0.05					+	−0.05		
23	−0.10							+	−0.10
24	+0.04			+	+0.04			−	−0.04
26	−0.01					−	+0.01	+	−0.01
A			−0.18		−0.32		+0.12		+0.15
	Sums		0		−0.01		0		0

FIG. 17-18

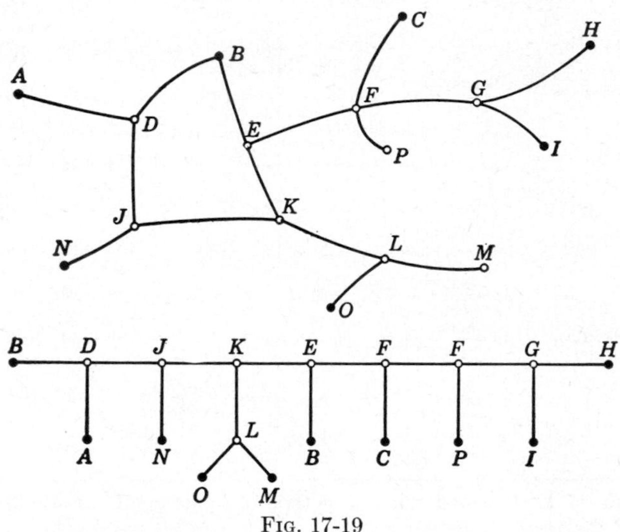

FIG. 17-19

circles indicate fixed elevations. Since *BDJKE* passes through the fixed elevation *B*, it is not an interior loop. Note that on the form, fixed elevation *B* appears twice. At point *F* two links join the main line. Note that on the form, *F* appears twice connected by a link of zero length.

If there is an interior circuit not covered by Forms C or B, the D forms are used as demonstrated in the original example.

17-39. To Attach One Form to Another. It must be remembered that any link that leads from a fixed elevation to a junction can be replaced by, or placed in series with, the equivalent link to which another circuit has been reduced. This was illustrated in the use of Form C.

Figure 17-20 shows a net that can be resolved into Form B attached to Form A. In the ordinary use of Form B the net is reduced to a series of links between two fixed points. The fixed elevations would be D and A

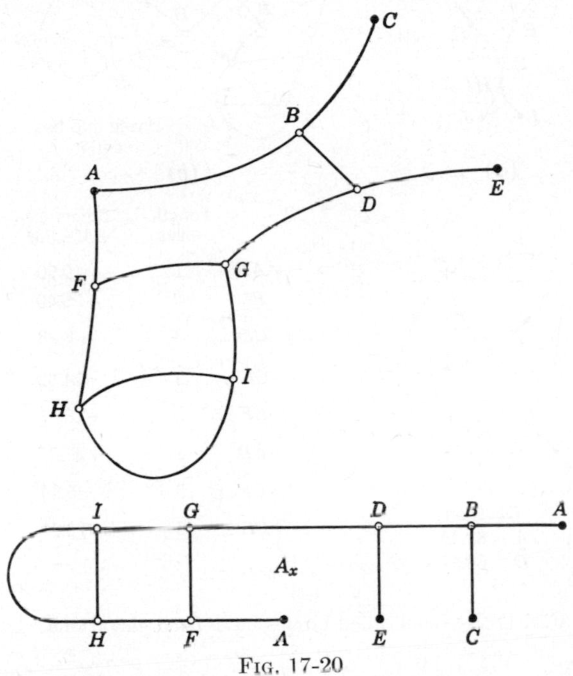

FIG. 17-20

and the links would be DG, the equivalent link, and FA. Since the three can be handled together, they are made into link 1 of Form A. A_1 of Form A is made equal to A_x shown. Evidently A_x can be computed.

17-40. Limits of the Use of the Method. It is evident that the process of fitting actual nets to the form can be extended indefinitely. In Special Publication 240,[1] Rappleye says, after mentioning simple uses of Forms B and D,

> However, this method can be applied to even more complicated networks, though the practical limit seems to have been reached when a network requires the reduction of about 15 circuits or includes an inner circuit without fixed points and having from 5 to 7 links.

[1] *Ibid.*

PROBLEMS

17-1. Place the level net shown in Fig. 17-21*a* on standard forms.

17-2. Separate the level net shown in Fig. 17-21*b* into two parts and place on standard forms.

17-3. Adjust the level net shown in Fig. 17-21*c*. The elevation differences are given in accordance with the B.M.-minus-B.M. system.

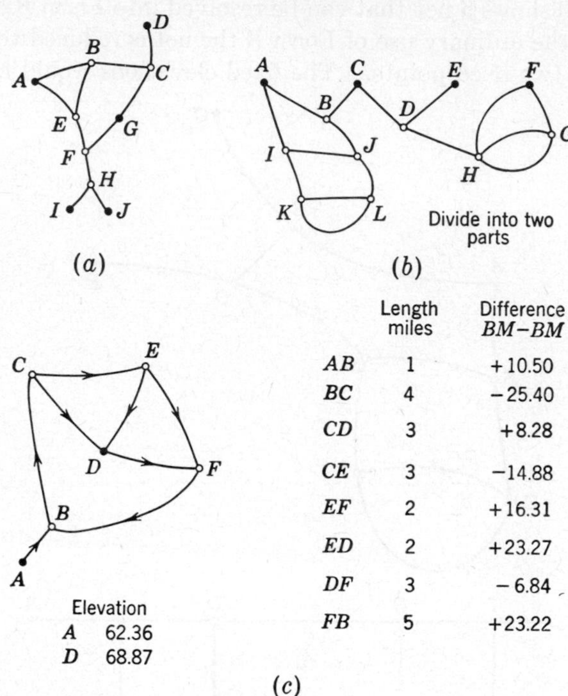

	Length miles	Difference *BM–BM*
AB	1	+10.50
BC	4	−25.40
CD	3	+8.28
CE	3	−14.88
EF	2	+16.31
ED	2	+23.27
DF	3	−6.84
FB	5	+23.22

FIG. 17-21. Black-filled circles have fixed elevations.

CHAPTER 18

PRECISE TRIANGULATION

18-1. The Requirements of Precise Triangulation. The methods of handling triangulation given in Chap. 6 are adequate for most requirements. When the triangulation scheme is of considerable importance, factors that increase the accuracy but were not mentioned in Chap. 6 must be considered. These are covered in this chapter. If further accuracy is required, the scheme should be adjusted by the method of least squares given in Chap. 22.

The following subjects are covered in this chapter:

1. The analysis of the **strength of figure**
2. Trigonometric leveling
3. Accounting for spherical excess
4. Precise base measurement
5. Other precautions

18-2. Orders of Accuracy. In addition to the specifications for orders of accuracy given in Table 1-3, the U.S. Coast and Geodetic Survey has developed criteria for the various elements of triangulation which, when adhered to, ensure more certain attainment of the desired order of accuracy. In addition, they eliminate the possibility that inaccurate methods might combine by chance in such a way as to indicate a higher degree of accuracy than is actually reached throughout the work. These criteria are listed in Table 18-1.

STRENGTH OF FIGURE

18-3. Introduction. The U.S. Coast and Geodetic Survey has developed a very rapid and convenient method of evaluating the strength of a triangulation net. Before the arrangement of an important net is adopted, its strength should be evaluated by this method. The method gives, for any triangulation net, a close estimate of the probable error of the length determinations caused by any assumed probable error of direction measurement. If the error of a net is found to be too large, three remedies are available:

TABLE 18-1. TRIANGULATION*

Criterion for	First-order	Second-order	Third-order
Strength of figure:			
Desirable limit, ΣR_1 between bases..	80	100	125
Maximum limit, ΣR_1 between bases.	110	130	175
Desirable limit, R_1 single figure....	15 (R_2 = 50)	25 (R_2 = 80)	25 (R_2 = 120)
Maximum limit, R_1 single figure....	25 (R_2 = 80)	40 (R_2 = 120)	50 (R_2 = 150)
Discrepancy between computed length and measured length of base or adjusted length of check line, not to exceed..........	1 in 25,000	1 in 10,000	1 in 5,000
Triangle closure:			
Average, not to exceed...........	1 second	3 seconds	5 seconds
Maximum, not to exceed..........	3 seconds	5 seconds	10 seconds
Side checks:			
Regular quadrilaterals; maximum difference of sides, in units of diff. of 1 sec in sixth place of log sine of smallest angle involved..........	2 times diff.	4 times diff.	
Side equation test:			
Approximate average correction to a direction, not to exceed........	0.4 second	0.8 second	
Usual number of observations:			
Positions with 1-second direction theodolite......................	16	8 (16 mod.†)	4
Positions with 2-second direction theodolite......................	24	12	4
Sets of 6 direct and reverse with 10-sec repeating theodolite.......	5–6	2–3	1–2
Base measurement:			
Actual error of base not to exceed‡..	1 in 300,000	1 in 150,000	1 in 75,000
Probable error of base not to exceed.	1 in 1,000,000	1 in 500,000	1 in 250,000
Discrepancy between two measures of a section not to exceed........	10 mm $\sqrt{k}$	20 mm $\sqrt{k}$	25 mm $\sqrt{k}$
Astronomical azimuth, probable error of result........................	0.3 second	0.5 second	2.0 seconds

* From "Manual of Geodetic Triangulation," U.S. Coast and Geodetic Survey Special Publication 247.

† Sixteen positions with 5-second abstract rejection limit, for second-order triangulation executed by a first-order party, in a scheme that is supplemental to first-order triangulation.

‡ The estimated value of all uncorrected errors. These consist of the sum of the estimated uncorrected residuals of the systematic errors plus the probable effect of the combination of the accidental errors inherent in the methods, estimated from experience. These estimates are independent of the results attained.

1. A measured base may be added.
2. Extra lines or directions may be observed.
3. The arrangement of the net may be improved.

18-4. Principle. The method is based on an expression for the square of the probable error (L^2) that would occur in the sixth place of the logarithm of any side, if the computations were carried from a known side through a single chain of triangles after the net had been adjusted for the side and angle conditions. The expression is based on the probable error of angular measurement and assumes no error in the known side. It can be demonstrated that

$$L^2 = \tfrac{4}{3}d^2 \frac{D - C}{D} \sum \delta_A^2 + \delta_A\delta_B + \delta_B^2 \tag{18-1}$$

in which d = probable error of an observed direction, sec

D = number of directions observed in the net from a given line to the side in question; the directions at the ends of the known line are not counted so that D = total directions observed less 2

C = number of angle and side conditions to be satisfied in the net from the known line to the side in question

δ_A = difference per second in the sixth place of logarithms of the sine of the distance angle A* of each triangle in the chain used

δ_B = same as δ_A but for the distance angle B

The method of determining these quantities is covered in later paragraphs.

For convenience, let R represent the terms in the equation affected by the shape of the figure; then

$$R = \frac{D - C}{D} \sum \delta_A^2 + \delta_A\delta_B + \delta_B^2 \tag{18-2}$$

and

$$L^2 = \tfrac{4}{3}d^2R \tag{18-3}$$

The value of R computed for the strongest chain of triangles is called R_1 and that for the second strongest chain R_2.

Since *the strength of a figure is almost exactly equal to the strength of the strongest chain as expressed in Eq.* (18-1), R_1 is a measure of the strength of figure. A maximum value to be allowed for R_1 is often established.

Example. Assume that the maximum probable error desired is 1 part in 25,000 and the probable error of direction measurement d is 1.25 seconds. Since L is the probable error of a logarithm, it represents the logarithm of the ratio of the true

* The distance angle A of a triangle is the angle opposite the side to be computed, that is, the side in common with the next triangle of the chain. Distance angle B is that opposite the known or previously computed side.

value and a value containing the probable error. In this case

$$L = \text{the sixth place in log } (1 \pm 1/25{,}000)$$
$$= \pm 17$$
$$L^2 = 289$$

Solving Eq. (18-3) and substituting,

$$R_{max} = \tfrac{3}{4} \times \frac{L^2}{d^2}$$

$$R_{max} = \tfrac{3}{4} \times \frac{289}{1.56} = 139 \text{ (say 140)}$$

The U.S. Coast and Geodetic Survey recommends maximum values of R as shown in the following table:

	1st order		2d order		3d order	
	R_1	R_2	R_1	R_2	R_1	R_2
Single independent figure:						
Desirable	15	50	25	80	25	120
Maximum	25	80	40	120	50	150
Net between bases:						
Desirable	80	. .	100	. . .	125	
Maximum	110	. .	130	. . .	175	

The value of R_1 can also be used to determine a choice between alternate proposed nets. The net with the smaller R_1 is used. R_2 is usually computed as well. When the two values of R_1 are very nearly the same and the values of R_2 differ greatly, the net with the smaller R_2 is chosen.

TO FIND THE VALUE OF *R*

18-5. Table 18-2. Evidently the value of the expression $\delta_A^2 + \delta_A\delta_B + \delta_B$ must be computed for each triangle in the chain used. Table 18-2 is arranged to give these values at once. To use this table the approximate values of the angles of the net planned must be measured during the reconnaissance, either by direct measurement or by plotting the net on a map and scaling them with a protractor. Values to the nearest degree are usually more than accurate enough.

The first part of Fig. 18-1 shows how Table 18-2 is used. The A and B angles of the triangles are selected according to the chain to be examined. In the first three columns are shown the triangles and the values of the corresponding A and B angles for the strongest chain. The smaller of the two angles is read along the top of Table 18-2 and the larger down the side. Interpolation is by estimation. The resulting values are shown in Column Σ in Fig. 18-1. The sum of these is used to compute R_1.

The second strongest chain is shown in the next set of columns in Fig. 18-1. The sum is used to compute R_2.

TABLE 18-2. FACTORS FOR DETERMINING STRENGTH OF FIGURE*

	10°	12°	14°	16°	18°	20°	22°	24°	26°	28°	30°	35°	40°	45°	50°	55°	60°	65°	70°	75°	80°	85°	90°
°																							
10	428	359																					
12	359	295	253																				
14	315	253	214	187																			
16	284	225	187	162	143																		
18	262	204	168	143	126	113																	
20	245	189	153	130	113	100	91																
22	232	177	142	119	103	91	81	74															
24	221	167	134	111	95	83	74	67	61														
26	213	160	126	104	89	77	68	61	56	51													
28	206	153	120	99	83	72	63	57	51	47	43												
30	199	148	115	94	79	68	59	53	48	43	40	33											
35	188	137	106	85	71	60	52	46	41	37	33	27	23										
40	179	129	99	79	65	54	47	41	36	32	29	23	19	16									
45	172	124	93	74	60	50	43	37	32	28	25	20	16	13	11								
50	167	119	89	70	57	47	39	34	29	26	23	18	14	11	9	8							
55	162	115	86	67	54	44	37	32	27	24	21	16	12	10	8	7	5						
60	159	112	83	64	51	42	35	30	25	22	19	14	11	9	7	5	4	4					
65	155	109	80	62	49	40	33	28	24	21	18	13	10	7	6	5	4	3	2				
70	152	106	78	60	48	38	32	27	23	19	17	12	9	7	5	4	3	2	2	1			
75	150	104	76	58	46	37	30	25	21	18	16	11	8	6	4	3	2	2	1	1	1		
80	147	102	74	57	45	36	29	24	20	17	15	10	7	5	4	3	2	1	1	1	0	0	
85	145	100	73	55	43	34	28	23	19	16	14	10	7	5	3	2	2	1	1	0	0	0	0
90	143	98	71	54	42	33	27	22	19	16	13	9	6	4	3	2	1	1	1	0	0	0	0
95	140	96	70	53	41	32	26	22	18	15	13	9	6	4	3	2	1	1	0	0	0	0	
100	138	95	68	51	40	31	25	21	17	14	12	8	6	4	3	2	1	1	0	0	0		
105	136	93	67	50	39	30	25	20	17	14	12	8	6	4	3	2	1	1	0	0			
110	134	91	65	49	38	30	24	19	16	13	11	7	5	3	2	2	1	1	1				
115	132	89	64	48	37	29	23	19	15	13	11	7	5	3	2	2	1	1					
120	129	88	62	46	36	28	22	18	15	12	10	7	5	3	2	2	1						
125	127	86	61	45	35	27	22	18	14	12	10	7	5	4	3	2							
130	125	84	59	44	34	26	21	17	14	12	10	7	5	4	3								
135	122	82	58	43	33	26	21	17	14	12	10	7	5	4									
140	119	80	56	42	32	25	20	17	14	12	10	8	6										
145	116	77	55	41	32	25	21	17	15	13	11	9											
150	112	75	54	40	32	26	21	18	16	15	13												
152	111	75	53	40	32	26	22	19	17	16													
154	110	74	53	41	33	27	23	21	19														
156	108	74	54	42	34	28	25	22															
158	107	74	54	43	35	30	27																
160	107	74	56	45	38	33																	
162	107	76	59	48	42																		
164	109	79	63	54																			
166	113	86	71																				
168	122	98																					
170	143																						

* From U.S. Coast and Geodetic Survey Special Publication 247.

The triangles that form the strongest and the second strongest chain can usually be chosen by estimation. However, when there is doubt, the values in Table 18-2 can be used as a check. For example, there are four possible routes through quadrangle *ABCD*. Three are shown in Fig. 18-1. They have values of Σ of 20, 30,

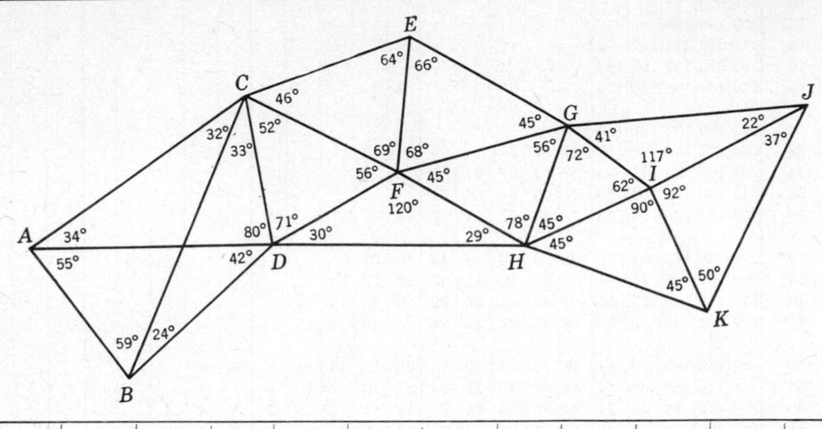

Δ	A	B	Σ	Δ	A	B	Σ	Δ	A	B	Σ
ABD	83°	42°	6	*ABC*	89°	32°	12	*ABC*	59°	32°	17
ACD	34	65	14	*BCD*	24	122	18	*ACD*	34	80	11
CDF	71	56	4					*CDF*	52	56	7
CEF	46	64	7					*DFH*	30	29	41
EFG	66	45	7					*FGH*	45	56	10
FGH	45	78	5								
GHI	72	62	2					*GHI*	45	62	8
HIK	45	45	13					*GIJ*	41	22	46
IJK	92	37	8					*IJK*	92	50	3
Sum			66					Sum			143

$$n = 22 \qquad n' = 22 \qquad s = 11 \qquad s' = 11 \qquad D = 42$$

$$C = (22 - 11 + 1) + (22 - 22 + 3) = 15$$

$$\frac{D - C}{D} = \frac{42 - 15}{42} = 0.64 \qquad R_1 = (0.64)(66) = 42 \qquad R_2 = (0.64)(143) = 92$$

$$L^2 = \frac{4}{3} d^2 \frac{D - C}{D} \sum (\delta_A{}^2 + \delta_A \delta_B + \delta_B{}^2) = \frac{4}{3} (1.25)^2 42$$

$$L = \pm 9.35$$

Fig. 18-1

and 28, respectively. The fourth route, by triangles *ABD* and *BCD*, is obviously not as strong as the others (Σ when computed for it is 60). The estimated strongest route turned out to be the strongest route in fact. But the estimated second strongest route through the quadrangle had to be discarded for a stronger one, as shown.

In the other figures the routes were properly estimated as shown.

18-6. To Compute C. C is computed by the following formula:

$$C = (n' - s' + 1) + (n - 2s + 3) \tag{18-4}$$

where n = total number of lines
n' = number of lines observed in both directions
s = total number of stations
s' = number of occupied stations

The first parenthesis in the formula is the number of angle conditions and the second is the number of side conditions. These formulas are derived in Chap. 22.

Figure 18-2 shows the computation of C and $(D - C)/D$ for several types of nets.

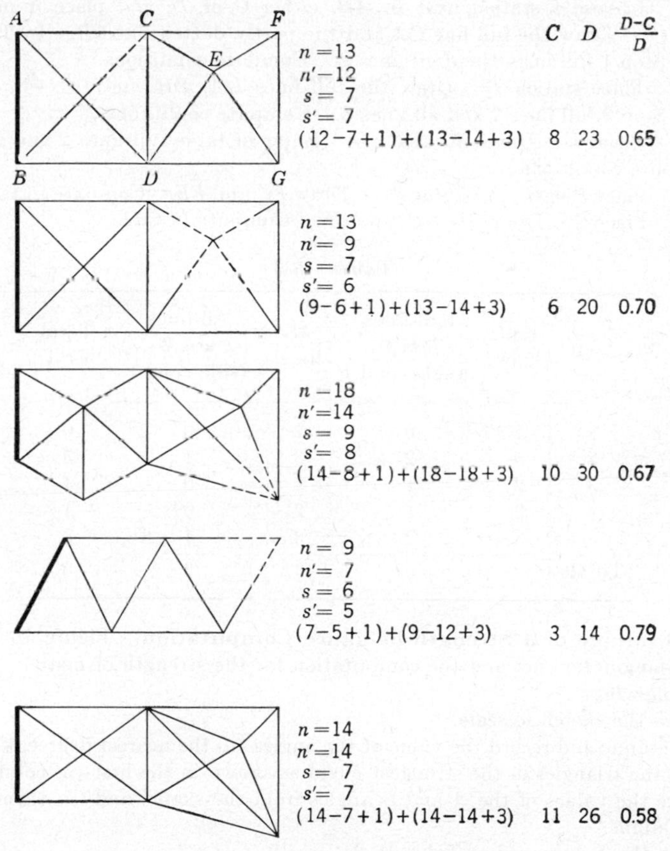

FIG. 18-2

Instead of using Eq. (18-4), the number of conditions can be determined by a sketch. To do this proceed as follows:

First draw a sketch of the triangulation net as shown in Fig. 18-1. Use a heavy line for the base or known side and use dotted lines where the directions are not observed. Then build up the figure by copying the sketch as follows. After the base and its two end stations are drawn, follow the rule stated below. Place a station next to any line drawn previously. Draw all the lines to that station from stations already drawn. Record an angle condition for each line without dots less one. Record a side condition for each line (full or dotted) less two. When the net is completed, the sums will be the total number of angle and side conditions.

Example. Referring to Fig. 18-1, follow the steps listed below. Place stations *A* and *B* and draw the base *AB*.

Step 1. Choose a station next to *AB*, either *C* or *D*, and place it in position. (Assume *C*.) Draw the full line *CA* and the partly dotted line *CB*. In Table 18-1, enter for Step 1 full lines 1 and all lines 2. Compute conditions.

Step 2. Place station *D*. Draw the full lines *DA*, *DB*, and *DC*. In the table, enter for Step 2 full lines 3 and all lines 3. Compute conditions.

Step 3. Place *E*. Draw *EC* and *ED*. Enter in table full lines 2 and all lines 2, and compute conditions.

Step 4. Place *F* or *G*. (Assume *F*.) Draw *FC* and *FE*. Compute in table.

Step 5. Place *G*. Draw *GD*, *GE*, and *GF*. Compute in table.

TABLE 18-3

Step	Full lines	Full lines less 1 (angle cond.)	All lines	All lines less 2 (side cond.)	Total cond.
1	1	0	2	0	0
2	3	2	3	1	3
3	2	1	2	0	1
4	2	1	2	0	1
5	3	2	3	1	3
Totals	. . .	6	. . .	2	8

18-7. Example of a Strength-of-figure Computation. Figure 18-1 shows a typical triangulation net and the computation for the strength of figure. The steps are the following:

1. Draw the sketch to scale.
2. Determine and record the value of the angles (to the nearest degree, if possible).
3. List the triangles in the strongest chain as shown in the first (Δ) column.
4. Take the values of the *A* and *B* angles from the sketch and list them in the *A* and *B* columns.
5. Take the Σ values from Table 18-2.
6. Repeat the same process for the second strongest chain.
7. When there is any doubt as to which route is stronger, check any route through each independent figure by finding the sum of the Σ values for each. The smaller

the Σ values are, the stronger the route. Note that, for the first quadrangle, the route in the third set of columns is stronger than the route used in the second chain It must be substituted.

8. Add the values in the Σ columns. These are the values of $\Sigma(\delta_A^2 + \delta_A\delta_B + \delta_B^2)$. In the example, values are 66 for the first chain and 143 for the second.

9. Compute $D = 42$.

10. Compute $C = 15$.

11. Find $R_1 = 42$ and $R_2 = 92$.

12. Check the value of R_1 against 140 or whatever value was chosen for the maximum value of R_1.

13. If desired, find the probable error of the side JK. If the probable error of angular measurement is assumed to be ±1.25 seconds, the substitution will be as shown at the bottom of the figure, $L = \pm 9.35$. This is the probable error of the sixth place of logarithms so that the probable error of the logarithm of JK is ±0.00000935. The antilog is 1.000022, which is the ratio between the true value of JK and JK plus or minus its probable error. Hence the probable error of JK due to errors in direction measurement is

$$\frac{22}{1{,}000{,}000} = \frac{1}{45{,}000}$$

It must be remembered that this is a determination of probable error. It is not safe to assume that the accuracy will be 1 part in 45,000. The actual result has an equal chance of having a larger or a smaller error.

TRIGONOMETRIC LEVELING

18-8. Purpose. When the bases of a triangulation net are at different elevations, their measured lengths must be reduced to a common elevation datum. Lengths between stations on federal control systems are given in terms of sea-level distances. When a net is tied to such a system the measured length of each base must be reduced to sea-level value. When the elevation of a base line cannot be determined by spirit leveling, elevations are carried through the triangulation system by trigonometric leveling. Vertical angles are recorded at the time the horizontal angles are measured, and the differences in elevation are computed from the lengths determined from the triangulation computations and the corrections for refraction and curvature. Since the effects of refraction are somewhat irregular, accurate results cannot be expected.

18-9. Formula. The differences in elevation between station marks can be computed by the approximate formula given below. It is sufficiently accurate for all but very extensive triangulation. When the difference in elevation is computed by this formula for observations in both directions on the same line, the inaccuracies are nearly eliminated in the average. The *increase* in elevation, that is,

$$\text{Elev.}_2 - \text{elev.}_1 = h_1 - h_2 + s(\tan\alpha + 2.509s \times 10^{-8})$$

where the units are in feet throughout

subscript 1 refers to the station where the observation was made

subscript 2 refers to the station observed

h = height of instrument or target above the actual station mark

s = the horizontal distance between stations

α = the vertical angle observed at the station

18-10. Procedure. When possible, observations should be made in both directions, preferably simultaneously so that refraction conditions will be more nearly the same. As many lines should be observed as possible.

Daytime observations should be made between noon and 4 P.M. when refraction is least variable. Night observations are more accurate.

Due to the shape of triangulation nets, when determinations are made along a number of lines a net of trigonometric leveling results. If this is adjusted properly the irregularities of refraction are somewhat eliminated.

Since the results are not, and need not be, very accurate, a rigidly mathematical adjustment is unnecessary. The method of simultaneous adjustment by estimation is recommended (see Chap. 6). Errors in trigonometric leveling can be assumed to be proportional to the square root of the length of line observed. Therefore, when the sketch is drawn for the estimate, the triangulation figure will be distorted so that each correction is proportional to the square of its length. If a least squares solution is used, the lengths should be squared before beginning the work.

SPHERICAL EXCESS

18-11. Principles. Since the horizontal angles of each triangle are measured in a plane perpendicular to the direction of gravity at the position of the station, and not in parallel planes, their sum should be larger than 180°. This excess is usually so small that, except in very large nets, it is usually neglected. As shown later, there are two reasons for taking it into consideration in smaller nets and there are various approximate formulas for computing it. Since it is always computed as though the section of the spheroid where the triangle is located were spherical, it is called *spherical excess.* The following approximate formula is sufficiently accurate for all but geodetic triangulation:

Approximate spherical excess (in seconds) = 0.0066 × base × altitude of triangle (in statute miles).

The required values can be scaled from an accurate sketch. The values obtained have two uses:

1. To test the closure of a triangle in the field, the spherical excess should be subtracted from the sum of the angles before comparing it to 180°.

2. In making a very careful least squares adjustment as in Chap. 22, slightly more accurate results will be secured if, before the observed angles are used, one-third the spherical excess of each triangle is sub-

tracted from the value of each angle in the triangle. This removes a systematic error that would otherwise be treated like an accidental error.

PRECISE BASE MEASUREMENT[1]

18-12. Introduction. The methods of taping described in Chap. 3 can be used successfully for precise base-line measurement by including certain refinements described in the following paragraphs.

18-13. Measurement. The best results can probably be obtained with flat-wire, band-steel tapes 300 ft long used at night. Lengths from 100 to 500 ft have been found satisfactory. Steel tapes should not be used except at night or on heavily overcast days. Invar tapes can be used in the sun. The measurement should be made with two different tapes, both carefully standardized.

The base should be divided into approximately equal sections 1000 to 4000 ft long and arranged so that each section (except the last) terminates at a tape end.

The different measurements of each section should check within the following limits for first-order accuracy:

$$0.057 \text{ ft } \sqrt{S}$$

where S is the number of **stages.** A stage is here defined as 10,000 ft. For example, if the length of a section were 3900 ft, the maximum allowable difference between any two measurements would be

$$0.057 \sqrt{0.3900} \text{ ft} = 0.036 \text{ ft}$$

Temperature. Two thermometers should be used, one attached at each end.

Tension. With most tapes 30 lb should be used. A spring balance is satisfactory but it must be checked frequently in a horizontal position against a weight hung over a frictionless pulley as described in Chap. 3.

Support. The tape should not be supported throughout, as this introduces too much friction. Supports should be placed at intervals of 100 ft. Greater intervals should not be used and smaller intervals are unnecessary. The supports should hold the tape at a uniform grade from one end to the other. The slope should not exceed 10 per cent.

18-14. Measurement over Posts. Over any ground that is not paved the tape should be supported on posts (see Fig. 18-3). At each tape end 4 × 4 posts with beveled tops are used, and 1 × 6 boards are used for intermediary supports. The boards are placed beside the tape and

[1] Adopted from U.S. Coast and Geodetic Survey methods. See Manual of Geodetic Triangulation Special Publication 247.

the tape is supported by short pieces of wire hanging from nails in the sides of the boards.

Once the alignment between the end monuments has been staked out, the posts and boards are placed in position with a transit and a 100-ft steel tape. The tops of posts must be set with sufficient accuracy so that the tape ends will come reasonably near the middle, and the side of the boards should be set between ¼ and ½ in. of the line.

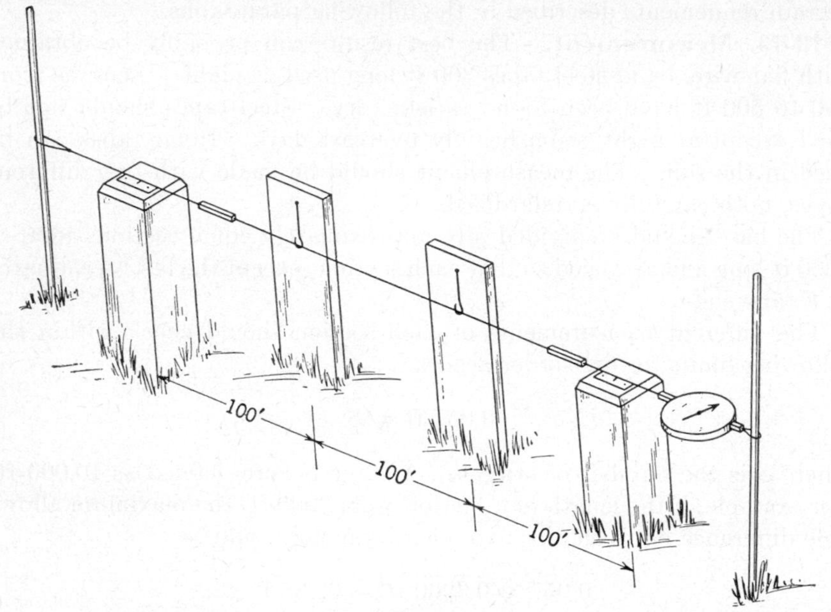

FIG. 18-3

When a post is set, a pencil line is ruled on the top of the post on the exact line. Then the nails for the stirrups are driven into the sides of the boards so that the tape will be held at an elevation that is on a straight grade between the tops of each pair of posts. The height can be best determined by sighting from the top of one post to the next by eye or with low-power binoculars.

A small strip of copper of the same thickness as the tape is then nailed to the top of each post so that one side of the strip is on the pencil line.

When a post is to be used for the end of a section, it is firmly braced by diagonal boards nailed to the sides.

A firm bench should be built over the monuments marking the ends of the base line and arranged so that the position can be plumbed up to it.

Measurement Procedure

The tape is placed in position on the posts and in the stirrups. A dry rag should be passed over it to make sure no dirt or other material adheres to it.

When all is ready, the head stretcher man (tension man) applies the tension gradually. The rear stretcher man keeps the tape at the mark as nearly as possible. The rear contactman flexes the tape behind the post to bring the tape exactly to the mark. When correct, he calls "mark." The tension man checks the tension and the head contactman scratches a mark with an awl, made of a needle set in the end of a round piece of wood, on the copper strip opposite the tape graduation. He should be careful to keep the needle away from the tape so that the graduation will not be marred.

The tension should be relieved and then applied again for a check. If satisfactory, the thermometers are read and the tape is carried forward without allowing it to touch the ground. This is best accomplished by several men distributed along the tape who hold it high enough to clear.

If the marks begin to come too near the end of the copper strips due to the differences in measurement between those used to set the posts and the base measurement, a second mark is placed on the strip and the distance between the marks is measured with a boxwood scale. This is recorded as a "setup" if the distance should be added to the measurement and a "setback" if it should be subtracted.

The boxwood scale should be graduated in units of 0.002 ft. If such a scale is not available, the usual 50 scale (50 ft per in.) can be used. The reading must then be divided by 600 to obtain values in feet.

When the end of a tape length does not come at the terminal monument, a section post should be established at the last tape end and special arrangements made to make the final measurement.

It is wise to measure the base in both directions with both tapes. The numerous marks on the copper strips can be identified by little signs. Levels are run over the tops of the posts and the benches built over the end points.

18-15. Measurement on Pavements. It has been found that when tapes are supported throughout, too much friction exists for high accuracy. When measurements are made over pavements, small supports should be used to support the tape at proper intervals. They should be just high enough to keep the tape off the ground. They must support the tape without friction. The U.S. Coast and Geodetic Survey uses supports 6 in. high that have ball-bearing rollers to carry the tape.

The supports should be placed so that the two ends of the tape are at the ground (see Fig. 18-4). In the figure a 300-ft tape is used with supports at the 50-, 150-, and 250-ft points.

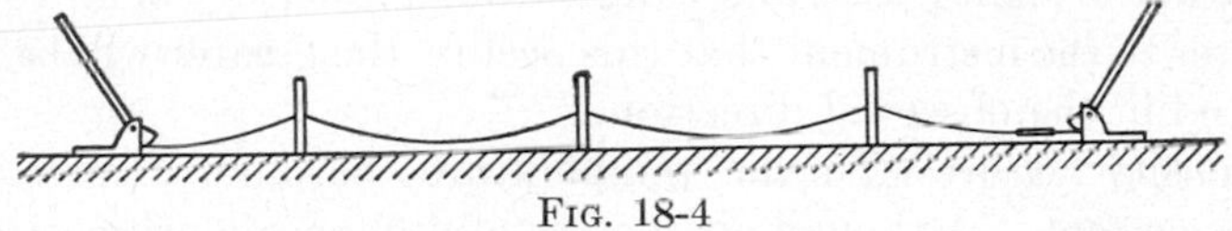

FIG. 18-4

It might be noted that the effect of sag in this case is the same as it would be if the tape were supported at the 0-, 100-, and 200-ft points.

Special stretchers must be constructed to apply the pull near the ground. Figure 18-4 shows a possible design. The stretcher man presses the base against the ground with his foot. It is well to attach a strip of rubber under the base to provide friction. Some engineers prefer to apply the tension by a weight hung over a bicycle wheel.

Tape ends can be marked on small strips of Bristol board of the same thickness as the tape. They are held in place by adhesive. They must be aligned along a pencil mark placed on line with a transit. The point is pricked with the needle awl.

Section ends should be marked by brass bolts or screws grouted in drill holes in the pavement. A scratched line or a hole made with a center punch may be used to give the exact point.

Levels are run over the positions of the tape ends and points of support.

In other details the method is the same as taping over posts.

18-16. Measurements on a Railroad Rail. Measurements can be made very successfully on the top of a railroad rail where not too many trains pass. The base is usually laid out so that a long railroad tangent intervenes between the end stations. The measurement is then carried off the rail where the track curves out of line. If this arrangement is impossible, the end monument must be set at a short offset from the rail and the distance transferred by a small triangle.

The measurement is made very much in the same way as over a pavement. Section marks are made with a center punch on the rail or scratched with a glass cutter. The stretcher should have a grooved base that fits the rail to keep it in position. Keel marks should be used liberally to prevent loss of the points.

A rail moves longitudinally by a small amount whenever a train passes over it. The movement of each rail on which a section mark is placed must be measured very carefully to determine the correction to be applied to the measured length of the sections. About 10 twentypenny nails are driven into different cross ties on alternate sides of the rail in the vicinity of the section marks. They are driven so that their heads are almost in contact with the base of the rail and flush with the top of the base. A sharp line is cut in the rail and the top of the nail at right angles to the rail. Whenever the measurement is brought to the section point, the average movement is computed and applied as a setup or setback.

If a train passes during the actual measurement of a section and if there is sufficient warning, the same process is attempted for the rail being marked last. Otherwise the section measurement must be repeated.

Levels are run over the positions of the tape ends.

In other respects the method is much the same as over pavements.

OTHER PRECAUTIONS

18-17. Lateral Refraction. Atmospheric refraction that causes the line of sight to bend in azimuth is called lateral refraction. It occurs when the line of sight passes at an angle from one density of air to another. The nearer to the instrument that this occurs, the greater will be the error introduced in the observed direction.

Air density varies with the temperature, barometric pressure, and moisture content. Although air masses of different densities are usually stratified in horizontal layers, layers or columns of different densities that are sloping or vertical are often encountered. Conditions that produce these effects should be understood.

Air is heated chiefly by contact with the ground. When the sun is shining, bare and dry ground becomes warmer and heats the air more rapidly. The rate is less for ground covered by vegetation and very much less over wooded areas. Large water surfaces hardly change temperature at all.

On calm sunny days, vertical columns of heated air of low density rise from the warmer surfaces. The effect increases as the day advances. After dark, the warmer surfaces cool off faster and tend to reach an equilibrium with other surfaces. On clear nights, however, the barer areas radiate heat faster and late at night become cooler than others so that the convection currents are reversed.

Accordingly, when sight lines cross different types of ground cover at angles, they should be observed in the early evening or on cloudy days.

Sloping or vertical surfaces are usually covered by air of different densities. Currents of air move up or down sloping ground and surround poles and buildings wherever their temperatures differ from those of their surroundings.

Therefore no sight should be laid out too close to any object.

Convection currents are especially prevalent over cities due to the great variety of surfaces and the effect of chimneys and heated buildings. When lines must be observed over cities, they should be laid out to avoid obvious sources of warm air and observed when the sky has been overcast for 12 hr or more. A slight breeze is helpful. Often many observations must be made under different weather conditions before the triangles can be properly closed.

18-18. Targets. Targets for triangulation observations are difficult to design as they must be viewed from many directions. They must be free from **phase,** that is, unequal illumination over the surface presented to the observer, as he will have a tendency to point to the brighter side. Usually flat boards can be arranged to face each observing position. Sometimes they must be removable to prevent their shadows from creating phase on other boards.

Heliotropes are excellent in haze and smoke. A mirror not more than 2 in. in diameter should be used with various coverings to cut down its diameter to a minimum of about $\frac{3}{8}$ in. Heliotropes require constant attention in use to keep the sun's reflection directed at the observer (see Fig. 1-6).

Lights can be used successfully in the daytime when a powerful bulb is used. Almost any type of light can be used for both day or night observations provided that it is bright enough to be recognized and not so bright that it creates glare. Very low illumination is often necessary at night. A light should always be observed with an instrument having two vertical cross hairs so that the light can be centered between them. A single line cannot be centered on a small light.

Usually lights must be powered by batteries so that small bulbs set in parabolic reflectors must be used. Old type automobile headlights are excellent. The lens in front must be removed and it is well to substitute plain glass to protect the reflector. However, when a light with a parabolic reflector is not aimed exactly at the observer, the brightest point will appear to be at one side of the reflector. Great care must be used in pointing these lights.

Battery lights are usually powered by dry cells. The number of cells used can be changed to regulate their brightness. Often it is convenient to take the power from a special jack or from the cigarette lighter of an automobile. A rheostat must then be used to control their brightness.

18-19. Accurate Centering. When triangulation sights are relatively short, the importance of accurately centering the targets and the instruments cannot be overemphasized. One-sixteenth inch introduces an error of 1 second at 1074 ft.

When a net is built up of small triangles, unless the target and the instrument position for any station exactly replace each other in horizontal position, errors accumulate very rapidly. If possible, carefully centered permanent target and signal positions should be installed at each station.

18-20. Communications. Communication between the observer and the light or heliotrope keeper is frequently necessary. Portable radio sets have been used with considerable success. At night the international code can be signaled by flashing the target light. The observer must then be equipped with a portable light that can be aimed at the station with which he wishes to communicate. Lights with shutters for signaling are available.

PROBLEMS

18-1 to 18-3. Compute the values of R_1 and R_2 for the figures shown:

18-1. Solve Fig. 18-5.
18-2. Solve Fig. 18-6.
18-3. Solve Fig. 18-7.

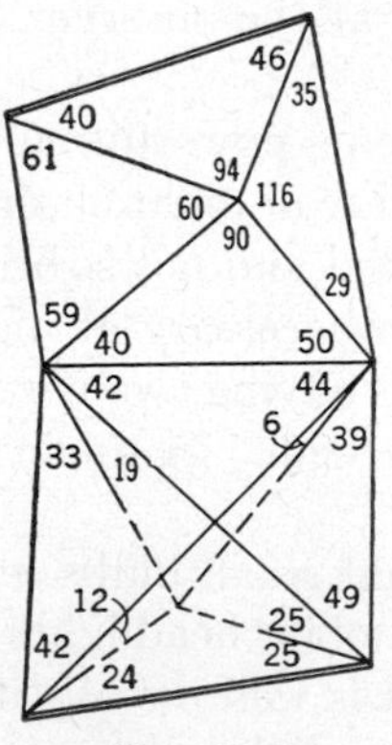

Fig. 18-5

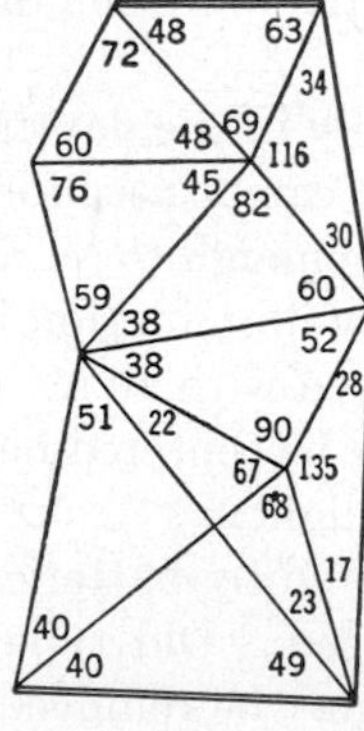

Fig. 18-6

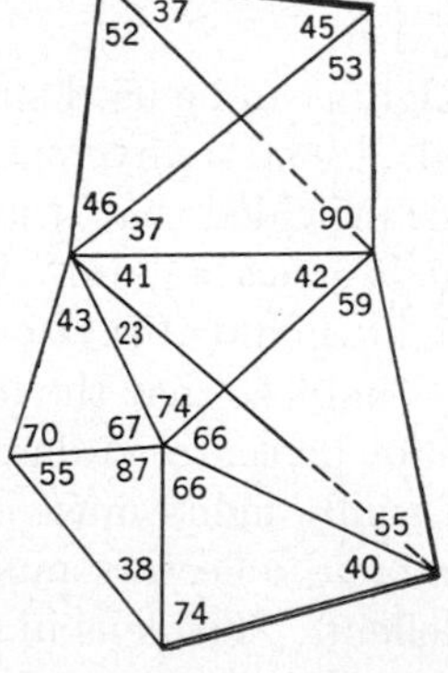

Fig. 18-7

CHAPTER 19

PRECISE TRAVERSE NETS

19-1. Net Adjustment. The methods of precise measurement and the proper adjustment of a single traverse have been covered in previous chapters. Approximate methods of adjustment of a traverse net are given in Chap. 6. It remains to describe a precise method of adjusting a traverse net.

19-2. Theory. It is impossible to reach a mathematically perfect theory for the adjustment of a traverse without a more definite relationship between angle and distance measurement than exists in practice. Accordingly, a number of methods are used. The method suggested here is excellent as long as the control to which the traverse is adjusted is well designed and of at least as high an actual accuracy as the traverse.

19-3. Method. In accordance with general practice, the angles of the net are adjusted first. The latitudes and departures are then computed from bearings that result from the adjusted angles. These latitudes and departures are then adjusted in proportion to the lengths of the courses to obtain an exact closure throughout the net.

Each of these three adjustments is made separately but each is a simultaneous adjustment of the whole net. A least-squares adjustment of each can be made by conventional methods or by the Braaten method. The Braaten method is used in the example.

RULES FOR SIGNS

19-4. Introduction. Since the Braaten method which is to be used was designed for an unusual set of sign conventions, it is well to review the rules for applying signs.

19-5. Frame of Reference. The results of a traverse survey are usually referred to a frame of reference called a grid or a coordinate system. In this frame of reference the sign conventions are the usual ones whatever method is used to adjust the survey. Plus indicates clockwise azimuth, north and east. The Y axis runs north, the X axis runs east, and the coordinates take their signs accordingly.

19-6. Survey. The survey data can be considered to be measurements of **changes** in azimuth, y coordinate, or x coordinate. The sign of a **change** can be chosen as desired. Usually *plus* indicates an *increase* but in the Braaten method it indicates a *decrease*.

The *arrow* on a link indicates the *direction* in which the sign stated for the link applies. Whenever desired, the arrow and the sign can be reversed simultaneously.

19-7. The Error of Closure. The error of closure must be computed in the same direction for all loops in the net. It is usually computed clockwise. In the Braaten method it is computed counterclockwise.

In computing the error of closure in the direction chosen, when the arrow of any link is contrary, the sign stated for the link must be reversed in the computation. A plus error of closure indicates that the total plus values of the links (thus computed for the closure) are too large.

It is evident that, since in the Braaten method the direction of the computation of the closures and the sign of the links are both reversed from the usual method, the final sign of the error of closure will be the same as for the usual method.

19-8. The Sign of the v's. In all methods the v's are corrections to the links, and the formulas for computing them are arranged so that the

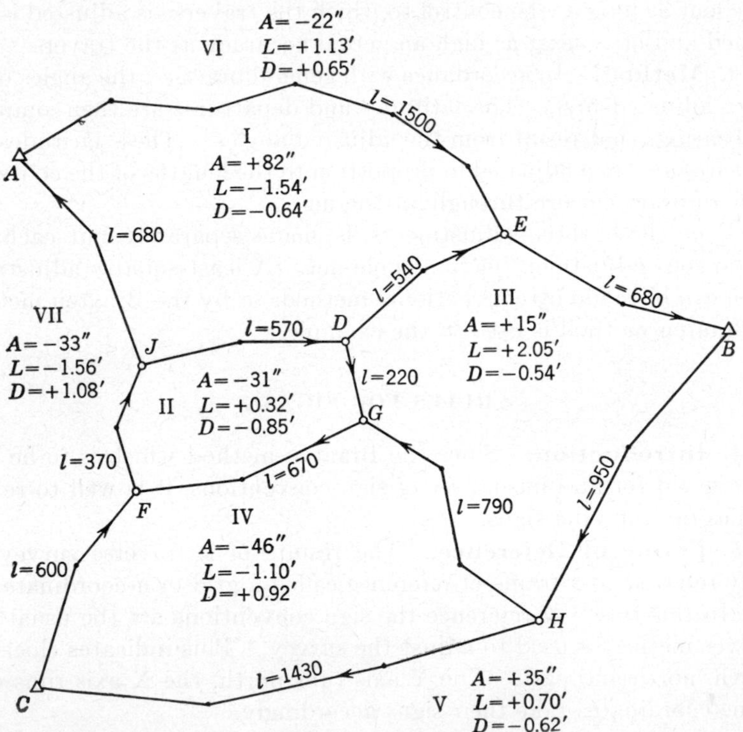

FIG. 19-1. Closures computed clockwise.

v's must be added (algebraically) to the changes, provided each change (link) is given the sign that applies when the link is computed in the direction of the arrow.

Since in the Braaten method the arrows are placed on the links always in the same way, the sign of the v's for each link must be correct for the given arrow.

METHOD OF SOLUTION

19-9. Example. The traverse net in Chap. 6 (Fig. 6-20) was adjusted by the simultaneous-inspection method. Here it will be adjusted by the Braaten method.

A **sketch** is drawn up to scale as before (see Fig. 19-1). It can be assumed that the signs of the links (not shown) indicate *increases* and the errors of closure were computed clockwise—both the usual practice. They must be handled properly when placed on the Braaten forms.

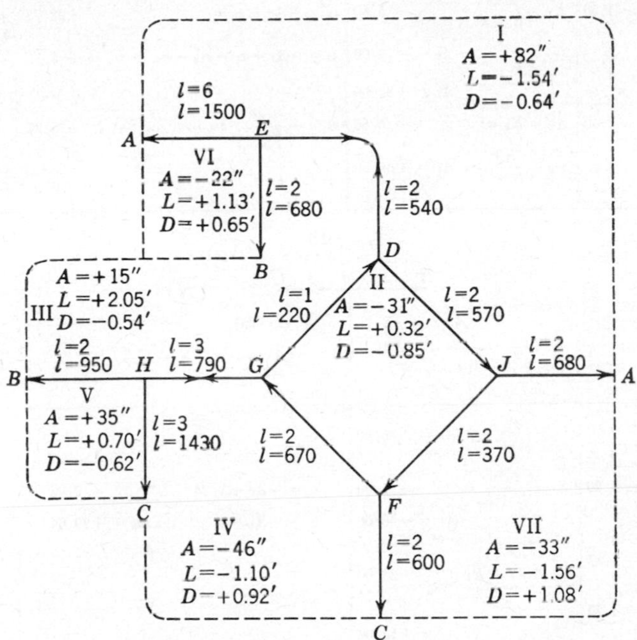

FIG. 19-2. Closures computed counterclockwise.

The net is now drawn up for the Braaten method as shown in Fig. 19-2. It will be found that this figure represents the same net if the letters for the junctions and fixed points are examined. The figure consists of Braaten Form D-1 with two Forms A attached, one to link 24 and the other to link 17 (see Fig. 19-4). It will be noted that for these two links there is a conflict in the direction of the arrows. This affects nothing but the v's. It will be discussed when the v's are computed.

The length l of each link is stated. The number of courses is the value of l for the angle adjustment; the length in feet is the l for the latitude and the departure adjustments.

The closures have been given the same sign as in the original net but the net has been turned over right for left. Thus the sign of the links and the v's are reversed so that the Braaten signs for the v's can be used without change, when applied to the original net.

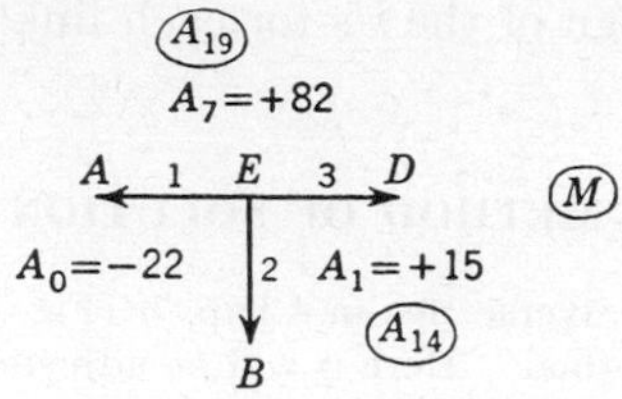

A_{19}

$A_7 = +82$

A 1 E 3 D (M)

$A_0 = -22$ 2 $A_1 = +15$

A_{14}

B

			v/l	Check
$l_1 = 6$	$A_0 = -22.00$			
$l_2 = 2$				
$E_1 = \frac{l_2 l_1}{l_1 + l_2} = 1.50$	$K_1 = \frac{l_2 A_0}{l_1 + l_2} = -5.50$	$v_1 = -A_0 + v_2 = +22.78$	$+3.80$	
$l_3 = 2$	$A_1 = +15.00$	$v_2 = -A_1 + v_3 - v_{19} - v_{3n} + v_{1n} = +0.78$		1, 2, 3 $+0.01$
	$A_7 = +82.00$		$+0.39$	
$l_{24} = E_1 + l_3 = 3.50$	$A_x = K_1 = -5.50$	$v_3 = -\frac{l_3}{E_1 + l_3} v_{24} = -8.37$	-4.18	
	$A_{19} = \frac{l_1 A_0}{l_1 + l_2} + A_7 = +65.50$			

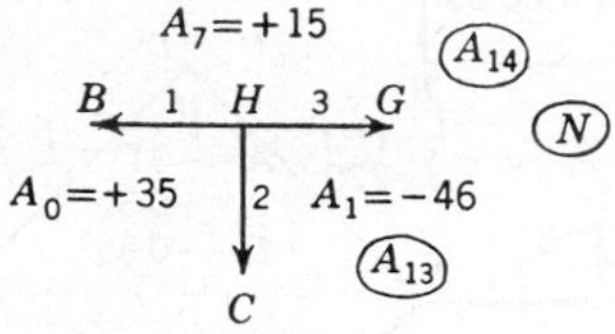

$A_7 = +15$ A_{14}

B 1 H 3 G (N)

$A_0 = +35$ 2 $A_1 = -46$

A_{13}

C

			v/l	Check
$l_1 = 2$	$A_0 = +35.00$			
$l_2 = 3$				
$E_1 = \frac{l_2 l_1}{l_1 + l_2} = 1.20$	$K_1 = \frac{l_2 A_0}{l_1 + l_2} = +21.00$	$v_1 = -A_0 + v_2 = -5.36$	-2.68	
	$A_1 = -46.00$	$v_2 = -A_1 + v_3 - v_{16} + v_{18} = +29.64$		
$l_3 = 3$				
	$A_{13} = A_1 + K_1 = -25.00$		$+9.88$	1, 2, 3 0.01
$l_{17} = E_1 + l_3 = 4.20$				
	$A_7 = +15.00$			
	$A_y = \frac{l_1 A_0}{l_1 + l_2} + A_7 = +29.00$	$v_3 = -\frac{l_3}{E_1 + l_3} v_{17} = -21.60$	-7.20	
	$A_{14} = A_x + A_y = +23.50$			

Fig. 19-3

19-10. Computations on Forms A. Figure 19-3 shows the computations for Forms A. These are slightly modified when they join Form D-1. Note that the junction of the first Form A is marked M and the second, N.

The first step is to find an equivalent length E_1 for links 1 and 2 in the usual way (see Fig. 19-3). The error of closure A_0 is thrown proportionally into A_1 and A_7.

But A_{7m}[1] is A_{19} on Form D-1. The proportional part of A_{0m} is therefore added to A_{19}. A_{1n} is A_{13} on Form D-1. This is handled likewise. A_{1m} and A_{7n} are both the same closure. They are A_{14} on Form D-1. Proportional parts of both A_{0m} and A_{0n} are added to the original value of A_{14} (i.e. +15). Formulas have been introduced to indicate these operations. The right-hand half of Fig. 19-3 cannot be computed until Fig. 19-4 is completed.

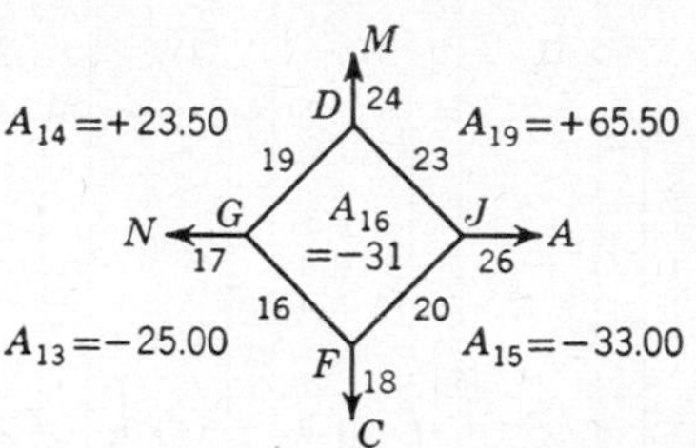

$l_{10} = 2$	$A_{13} = -25.00$	$V_{16} = -A_{13} - V_{17} + V_{18} = -1.37$	v/l	Check
$l_{17} = 4.20$	$A_{14} = +23.50$	$V_{17} = +A_{14} + V_{19} + V_{24} = +30.24$		
$l_{18} = 2$	$A_{15} = -33.00$	$V_{18} = -A_{15} - V_{20} + V_{26} = +3.87$		19, 23, 24 +0.02
$l_{19} = 1$		$V_{19} = \frac{l_{19}V_{21}}{l_{21}} = -7.91$	-7.91	
$l_{20} = 2$	$D_{10} = l_{10} + l_{17} + l_{18} = 8.20$	$V_{20} = \frac{l_{20}V_{22}}{l_{22}} = +2.43$		
$l_{21} = \frac{l_{16}l_{17}}{D_{13}} + l_{19} = 2.02$	$A_{17} = \frac{l_{17}A_{13}}{D_{13}} + A_{14} = +10.70$	$V_{21} = -A_{17} - V_{24} + V_{25} = -15.98$	-7.91	
$l_{22} = \frac{l_{16}l_{18}}{D_{13}} + l_{20} = 2.49$	$A_{18} = \frac{l_{18}A_{13}}{D_{13}} + A_{15} = -39.10$	$V_{22} = -A_{18} - V_{25} + V_{26} = +3.03$	$+1.22$	21, 22, 25 +0.06
$l_{23} = 2$	$A_{19} = +65.50$	$V_{23} = -A_{19} + V_{24} - V_{26} = -24.15$	-12.08	
$l_{24} = 3.50$	$A_{20} = -A_{17} - A_{18} - A_{19} = -37.10$	$V_{24} = \frac{l_{24}V_{27}}{l_{27}} = +14.65$	$+4.19$	
$l_{25} = \frac{l_{17}l_{18}}{D_{13}} = 1.02$		$V_{25} = \frac{l_{25}V_{28}}{l_{28}} = +9.37$	$+9.19$	
$l_{26} = 2$	$D_{20} = l_{21} + l_{22} + l_{23} = +6.51$	$V_{26} = \frac{l_{26}V_{29}}{l_{29}} = -26.70$	-13.35	
$l_{27} = \frac{l_{21}l_{23}}{D_{20}} = 4.12$	$A_{21} = \frac{l_{21}A_{20}}{D_{20}} + A_{17} = -0.81$	$V_{27} = -A_{21} + V_{28} = +17.25$	$+4.19$	
$l_{28} = \frac{l_{21}l_{22}}{D_{20}} + l_{25} = 1.79$	$A_{22} = \frac{l_{22}A_{20}}{D_{20}} + A_{16} = -53.29$	$V_{28} = -A_{22} + V_{29} = +16.44$	$+9.18$	27, 28, 29 +0.02
$l_{29} = \frac{l_{22}l_{23}}{D_{20}} + l_{26} = 2.76$		$V_{29} = \frac{l_{29}(A_{22} + K_1)}{l_{29} + E_1} = -36.85$	-13.35	
$E_1 = \frac{l_{27}l_{28}}{l_{27} + l_{28}} = 1.25$	$K_1 = \frac{l_{28}A_{21}}{l_{27} + l_{28}} = -0.25$			

FIG. 19-4

19-11. Form D-1. Form D-1 (see Fig. 19-4) can now be entered. A_{13}, A_{14}, and A_{19} have just been computed. l_{24} and l_{17} are $E_{1m} + l_{3m}$ and $E_{1n} + l_{3n}$, respectively. Form D-1 is now computed in the usual way (see Fig. 19-4). All of the v's computed can be used as they are computed except v_{24} and v_{17}, which are composite links from Forms A.

19-12. Computation of v's in Forms A. The problem in Form A is to separate the v that applies to link 3 and the v that applies to E_1 (the equivalent link). This is accomplished by multiplying the value of v_{24} or v_{17} by the ratio that l_3 bears to the

[1] Subscripts indicate which of the two Forms A is used.

whole composite link $E_3 + l_1$. The result would have the sign of v_{24} or v_{17} which applies when the arrow is outward on Form *D*-1. Actually the arrow for link 3 is inward toward Form *A* contrary to Form *A*, and therefore the ratio is given a minus sign as shown in Fig. 19-3. The computation is continued in the usual way.

SUBSTITUTION OF *v*'S IN CIRCUITS

		v	I		II		III		IV		V		VI		VII	
EA	1	+22.8	−	−22.8									+	+22.8		
EB	2	+ 0.8					+	+ 0.8					−	− 0.8		
ED	3	− 8.4	+	− 8.4			−	+ 8.4								
HB	1*n*	− 5.4					−	+ 5.4			+	− 5.4				
HC	2*n*	+29.6							+	+29.6	−	−29.6				
HG	3*n*	−21.6 [7]					+	−21.6 [7]	−	+21.6 [7]						
FG	16	− 1.4			−	+ 1.4			+	− 1.4						
FC	18	+ 3.9							−	− 3.9					+	+ 3.9
GD	19	− 7.9			−	+ 7.9	+	− 7.9								
JF	20	+ 2.4			−	− 2.4									+	+ 2.4
DJ	23	−24.2 [1]	+	−24.2 [1]	−	+24.2 [1]										
JA	26	−26.7	+	−26.7											−	+26.7
A				+82.0		−31.0		+15.0		−46.0		+35.0		−22.0		−33.0
Sums				− 0.1		+ 0.1		+ 0.1		− 0.1		0		0		0

Links	Braaten-method cor.*	Simultaneous-inspection method	Error of simultaneous inspection
EA	−22.8	−15	+7.8
EB	+0.8	−7	−7.8
ED	+8.4	+5	−3.4
HB	+5.4	0	−5.4
HC	+29.6	+35	+5.4
HG	−21.7	−10	+11.7
FG	+1.4	0	−1.4
FC	−3.9	+1	+4.9
GD	+7.9	−3	−10.9
JF	−2.4	0	+2.4
DJ	+24.1	+28	+3.9
JA	−26.7	−34	−7.3

* Signs changed where Braaten forms have arrows opposite to original layout.

FIG. 19-5. Substitution of *v*'s and comparison with simultaneous-inspection method.

19-13. Substitution of *v*'s in Circuits. The *v*'s are substituted into the circuits as used in the Braaten method (see Fig. 19-5). They are corrected for errors due to rounding off in the usual way. If the arrows on the original circuits are made to agree with the arrows of the Braaten forms by changing the signs of the links, the *v*'s can be applied as they are listed. If the Braaten form had not been a turned-over replica of the original net, the signs of all the *v*'s would have to be reversed.

The resulting v's have been compared to those found for the same net in Chap. 6 with the simultaneous method by inspection (see Fig. 19-5). The errors with the inspection method are as high as 12 seconds.

The angles are now corrected as in Chap. 6.

19-14. Coordinates. The latitudes and departures can now be computed from the resulting bearings. The resulting latitudes and departures are adjusted separately by exactly the same formulas as those used for the angles. This operation is not shown.

PROBLEMS

19-1. Adjust the traverse net shown in Fig. 19-6 by the Braaten method.

19-2. Complete the adjustment of the latitudes and the departures in Fig. 19-1 by the Braaten method.

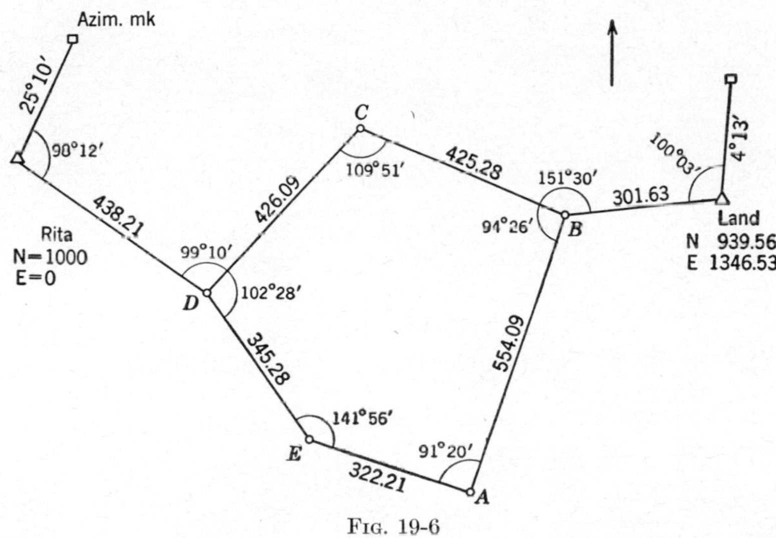

FIG. 19-6

Part IV

CHAPTER 20
AERIAL MAPPING

20-1. Introduction. Mapping from aerial photographs is the best mapping procedure yet developed for most large projects. It is faster and cheaper than any other method, it provides more complete and more accurate topographic detail, and it has very few limitations. It has been used successfully for maps varying in scale from 1:1,000,000 to 1 in. = 50 ft, and contour lines can be mapped accurately and economically to intervals as small as 1 ft.

The photographs necessary for the maps provide information difficult to acquire by other means. They aid in geological investigations, soil surveys, the location of property lines and other boundaries, and the construction of tax maps. They also provide inventories of land use, sources of construction material, and the location of trees for highway landscaping. Often they can be used for reconnaissance and they are invaluable for military intelligence.

Aerial mapping is not economical for surveys of small areas. The break-even size is probably somewhere between 30 and 100 acres. This varies widely with circumstance. The process cannot be used successfully over certain kinds of terrain, chiefly the following:

1. Certain desert or plains areas that photograph as uniform shades without texture. Photographs of these areas actually do not show the ground surface.

2. Deep canyons or high buildings that conceal the ground surface in the photographs.

3. Areas covered with dense conifer or tropical rain forests.

The major users of aerial mapping methods are the civilian and military mapping agencies of the government. Some of these agencies carry out almost continuous mapping programs. In addition, the method is used extensively for highway, railroad, pipe-line, and other route surveys and it is being used more and more for maps for large plant installations, reservoirs, and other projects involving considerable acreage.

Many unusual uses for aerial mapping have been discovered; for example, monthly estimates of quantities in coal piles, spot photographs for the control of radar altimetry, studies of shore erosion, etc. The advantages and possibilities of aerial mapping are so extensive that the method should be given careful consideration whenever any mapping project is planned.

GENERAL CONSIDERATIONS

20-2. Ground Control. It is impossible to make a map from aerial photographs without survey control on the ground. On the other hand it is possible to extend the map beyond the control or to bridge from one control area to another by means of the images on the photographs themselves. The accuracy of the map is always increased by the quantity of ground control available but the necessity for ground control is continually being reduced by improved methods and by an increase in the accuracy of the cameras, plotters, and other equipment employed; by the arrangement of the coverage of the photographs; and, to some extent, by the determination of camera position by radar.

20-3. Plotters. Except for very minor considerations, a map is an orthogonal projection of the topographic details, while a photograph is a perspective projection of them. Accordingly, the relative positions of points on a photograph are at least slightly different from their relative positions on a map. Moreover, if contour lines are required, the data on two photographs of the same ground must be used simultaneously. It follows that the data from the ground control and those from the photographs must be combined by some means so that a map results. This could be accomplished by direct measurements on the photographs and by mathematical computations, but the process would be hopelessly complicated. Many practical methods have been developed to accomplish this purpose; they depend on various types of equipment, from very simple to very complex.

Most of the successful methods depend chiefly on one of several types of instruments called **plotters.** These instruments make it possible for an operator to place the photographs in their proper geometric relationships with respect to the plotted positions of the ground control (to orient them in space) and to find and plot the contour lines and the planimetry while viewing the photographs. Plotters vary widely in accuracy and complexity. The art of determining results by measurements on photographs by plotters, or otherwise, is called photogrammetry. The greatest use of photogrammetry is in aerial mapping.

Every plotter or procedure devised for determining contour lines from aerial photographs is based on stereovision. This subject is covered in the next section.

THE PRINCIPLES OF STEREOVISION

20-4. Depth perception is the mental process of determining relative distance of objects from the observer from the impressions received through the eyes. Numerous impressions are received that serve as clues to depth. Those clues that concern photogrammetry are the following:

1. Head parallax
2. Accommodation
3. Convergence
4. Retinal disparity

Head parallax is the apparent relative movement of objects at different distances from the observer when the observer moves.

Accommodation. The lens of the eye can be flattened or made more convex in accordance with the requirements placed on it. The process is called accommodation. It is flattened to focus distant points on the retina and made convex for nearby points. The brain is aware of the condition of the lens and thus receives an approximate clue to distance. The ability to focus the eyes in this way begins to lessen in the forties and is usually completely lost in the sixties.

Convergence. The desire to view something clearly causes the two eyes to turn so that the image of the desired object is placed on the most sensitive part of each retina (the fovea). Since the eyes converge more for nearby points and the brain is aware of their relative positions, their convergence is a clue to distance. Usually the eyes accommodate automatically in accordance with the convergence required.

Retinal Disparity. Since the two eyes are at different positions, the pictures they receive are slightly different. For example, when a small object stands between an observer and a large object, the part of the larger object observed is different for each eye (see Fig. 20-1). The difference between the images on the retinas is called retinal disparity. Since it is a function of the relative distance of the objects viewed, it provides a distance clue. In this case it makes it possible to judge how far in front of the book the cylinder stands. Actually it is a very strong clue and heavily relied on by the individual. Moreover, it is the only clue actually used in photogrammetry. The importance of the other clues is that they must be disregarded, as will be explained later.

FIG. 20-1

20-5. Stereopairs. Since a photograph is a perspective projection, it represents, geometrically, the type of view seen by one eye. When two photographs are made of the same object from different positions and then arranged by some means so that the right-hand photograph is seen by the right eye and the left-hand photograph is seen by the left eye, retinal disparity is established and the observer can distinguish depth. Two such photographs are called a stereopair. The clue of head parallax does not exist nor does the clue of accommodation, since the images on each photograph are all at the same distance from the eye. However, if the observer does not move his head, he does not expect the clue of head parallax, and if the objects shown in the picture do not vary greatly in distance, the accommodation clue is not expected either. By certain arrangements, the convergence clue that fits the accommodation necessary to focus the photographs can be provided. But it has been found

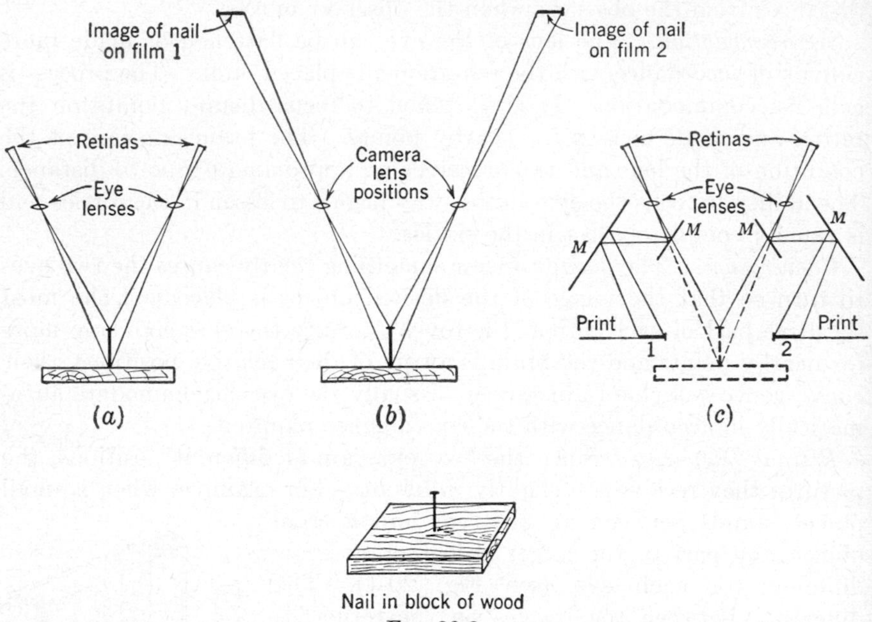

FIG. 20-2

that the ability to reject a convergence clue that does not agree with the accommodation can be quickly learned so that this is a somewhat unnecessary refinement.

Two devices are used for viewing stereopairs, the **stereoscope** and the **anaglyph.**

20-6. The Stereoscope. There are two kinds of stereoscopes, the **mirror stereoscope** and the **lens stereoscope.** Figure 20-2 shows the principle of the mirror stereoscope arranged to give the proper convergence. It is drawn somewhat out of scale for better illustration.

Drawing *a* shows the eyes observing a nail in a block of wood. Note that the head of the nail is to the left in the left retina and to the right in the right retina. The three clues, retinal disparity, convergence, and accommodation are present.

Drawing *b* shows the process of photographing the nail from two positions. The camera lens is placed first in the position of the left eye and then in the position of the right eye. Separate photographs are taken at each position. Images are formed on the film as shown. Contact prints from these negatives are placed in the mirror stereoscope as shown in drawing *c*. (Only the images of the nail are drawn.) The four mirrors (marked *M*) transfer the light to the eyes exactly (except for accommodation) as if it had come from the nail in a block as shown by the dotted lines. The accommodation clue may give an impression of how far dis-

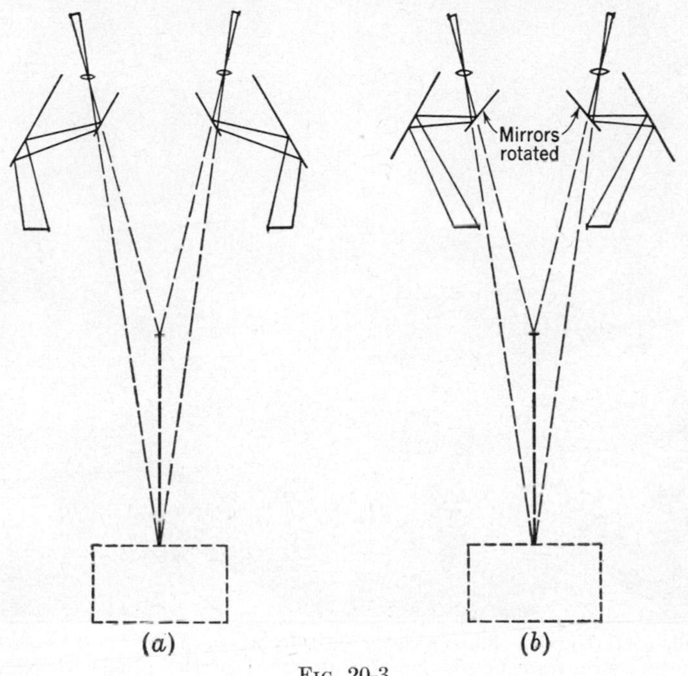

FIG. 20-3

tant the photographs are from the eye but it does not indicate the height of the nail. The convergence and retinal disparity are exactly as before and are sufficient for the observer to see the nail in three dimensions. Most people with normal eyesight will see this at once.

20-7. Rejection of the Convergence Clue. If the photographs with the images of the nail are moved outward as shown in Fig. 20-3*a*, or if the same impression is created by rotating the mirrors as shown in *b*, the observer is presented with a geometric pattern and therefore the convergence clue that would result if all points in the nail had been moved proportional amounts away from the observer, without changing the horizontal distances. Since the observer has a definite idea of how far away the pictures are and since he knows that the photographs are at the original distance, he recognizes that something is wrong. He then does one of two things: he either rejects all depth clues and no longer sees any depth, or he rejects the convergence clue and sees no change in the picture. In this case, if the movement is made smoothly he will be unconscious of any change.

When a trained observer sees an arrangement of this kind without its having been positioned for proper convergence, he will be able to change

FIG. 20-4*a*. A stereopair. Place a card vertically between the two photographs and let each eye concentrate on the prominent object in the respective photographs. Depth should appear.

the convergence of his eyes until the images are matched on the retinas, that is, **fused.** He rejects the convergence clue and sees depth immediately. The images can be separated until the eyes begin to diverge. Since this can never occur with real objects, even trained observers will usually be unable to fuse the images when there are one or two degrees of divergence.

Figure 20-4*a* shows a stereopair where convergence and accommodation do not match. Depth is perceived by retinal disparity when the left photograph is viewed with the left eye and the right photograph with the right eye. A card held vertically between the photographs helps the observer accomplish this. Figure 20-4*b* shows another stereopair.

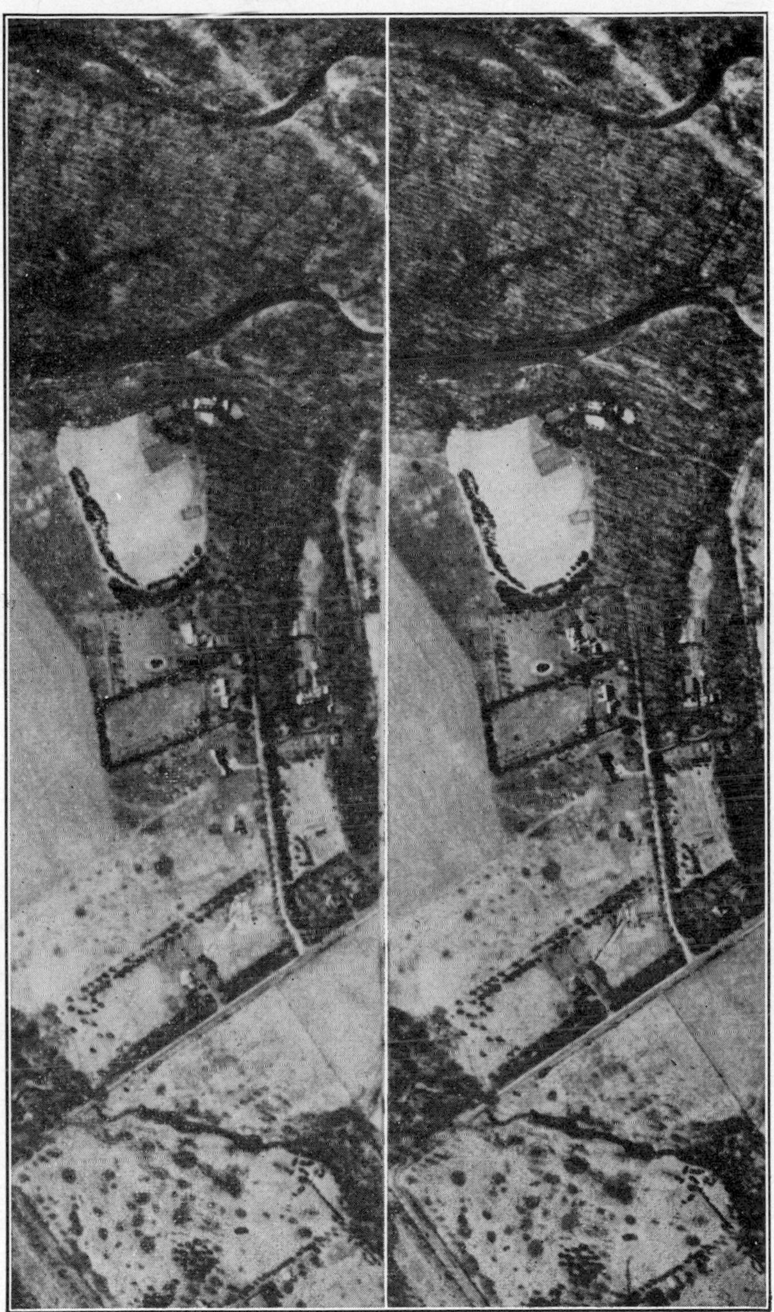

FIG. 20-4*b*. A stereopair.

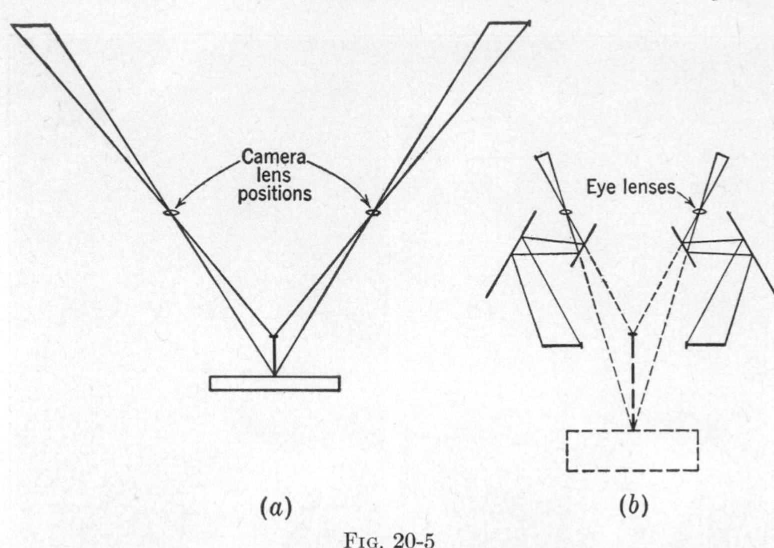

FIG. 20-5

FIG. 20-6. A Fairchild mirror stereoscope. Accommodation lenses are shown. A binocular eyepiece is turned back out of the way. In use it magnifies the photographs viewed. (*Fairchild Camera and Instrument Corp.*)

20-8. High Convergence. Most aerial pictures are taken so that the required convergence is greater than possible for any individual, as shown in Fig. 20-5*a*. In this case the observer will not see them in stereo (i.e., see depth) until they are arranged as shown in Fig. 20-5*b* in which they have been moved apart until a more normal convergence has occurred. As shown, the geometry presented, and thus the retinal disparity, is that

FIG. 20-7. Prismatic binoculars make it possible to extend the range of a stereoscope. (*Harrison C. Ryker, Inc.*)

which would occur if distances from the observer had been increased without changing horizontal distances. Accordingly, when the photographs are taken in this manner, the observer will see the object exaggerated in height according to the retinal disparity clue. This has the fortunate effect of increasing his ability to determine depth.

20-9. The Binocular Stereoscope. Mirror stereoscopes are sometimes equipped with binoculars which act like telescopes as in Figs. 20-6 and 20-7. They enlarge the photographs so that detail is more easily seen, and they make it possible to place the photographs at greater distances from the eyes.

20-10. The Lens Stereoscope. A lens stereoscope has no mirrors; it consists of a single lens for each eye. This makes it possible to focus the eyes on photographs only about 4 in. away from the eyes. The two photographs can then be brought so close to the eyes that proper convergence can be maintained without causing the photographs to interfere with each other as shown in Fig. 20-8. Since the photographs are very close to the eyes, the images occupy larger angular dimensions and therefore appear enlarged. Moving the images apart has the same effect as with the mirror stereoscope.

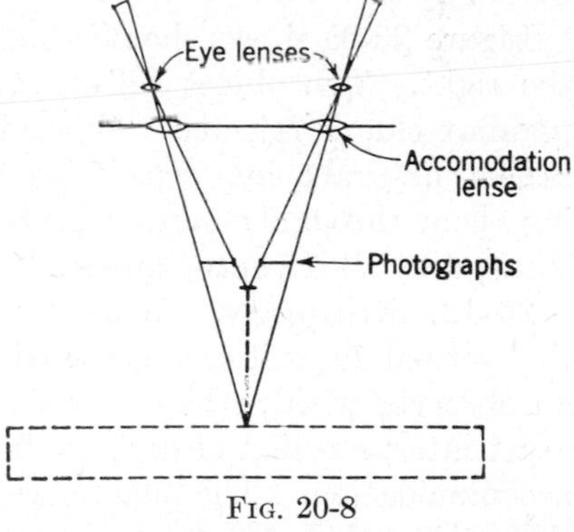

FIG. 20-8

20-11. The Anaglyph. The so-called anaglyph provides another means of arranging the photographs so that each is seen by the proper eye only. An anaglyph can be created by printing half tones, by color photographic printing, or by projection. The photographs are superimposed so that the various objects nearly match and are printed or projected in complementary colors, usually green and red. They are viewed with a green filter over one eye and a red filter over the other eye so that each eye sees only one photograph. Figure 20-9*a* shows this arrangement. The convergence and the accommodation are matched so that the observer sees depth at once. When the images are projected, the two photographs can be moved apart with exactly the same results as with the mirror stereoscope.

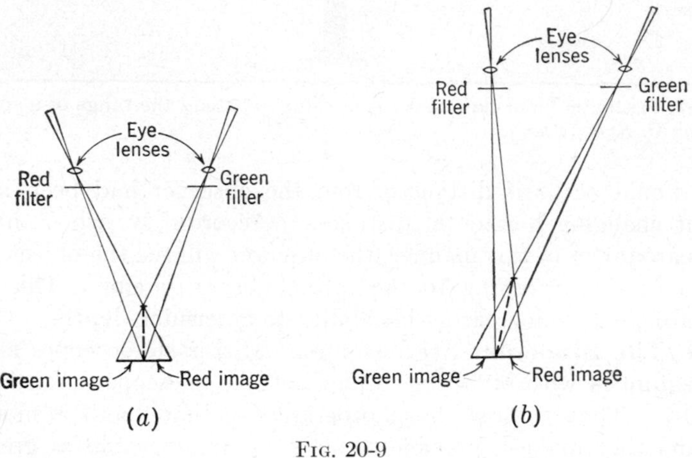

FIG. 20-9

Figure 20-9*b* shows the effect of raising the eyes and moving them to the right. The object will apparently elongate and sway. The head parallax clue is rejected. When it is desired to view color photographs in their natural colors, separation is attained by polarizing them and viewing them through polarized glasses, as is done in 3-*D* motion pictures. These are called **vectographs.**

20-12. Summary. When photographs are taken and arranged so that retinal disparities are created that are within the range of experience, an observer with perhaps a little training will see them in stereo. His exact interpretation of depth will depend somewhat on convergence and accommodation. The images are best placed so that, to match them on the retinas, the convergence is not much greater than would be expected at normal reading distance. The convergence can be decreased until the eyes are parallel, but if it is carried much beyond this point even a trained observer will fail to see depth.

Unless the photographs are arranged so that the convergence and the retinal disparity are exactly as would occur in actual objects, the observer will not interpret depth in its proper proportion to horizontal distance. However, whenever the observer can see several objects simultaneously, he can judge their *proportional* heights.

THE AERIAL CAMERA

20-13. Characteristics of Aerial Cameras. Although there are many types of air mapping cameras, their basic characteristics are very much the same. Cameras used in the United States have the following features:

1. The usual size of the negative is 9 by 9 in.
2. The lens is fixed at infinity focus.
3. Wide-angle lenses are almost universally used. The most generally used lens is the metrogon shown in Fig. 20-10. It has an angular field of 93° and a nominal focal length of 6 in. A large field requires less flying, fewer photographs, less ground control, and gives a higher mapping accuracy. However, steep ground slopes photographed near the edge of the field of a wide-angle camera often conceal topographical detail.

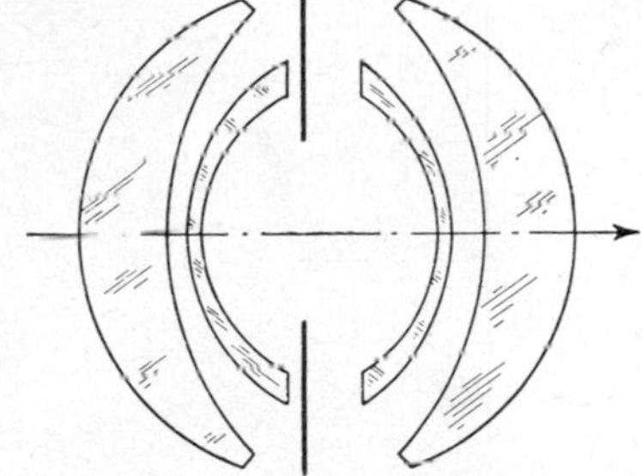

FIG. 20-10. Cross section of the Bausch and Lomb metrogon lens.

4. A focal length of 6 in. is becoming standard, but 4 to 8 in. is also used.
5. Since shutter speeds must be quite fast to avoid movement of the images on the negative, large apertures are important. These are difficult to attain in a wide angle lens. About the largest in use is $f/6$.*
6. The lenses are achromatic; i.e., they are made up of elements that together focus light of all wave lengths within the photographic range on the same point on the negative.
7. Although glass plates are used successfully, most cameras are designed for film. Film provides many more exposures for one loading, it can be handled and developed with greater facility, and it takes up less space. However, film shrinks when it is processed, and sometimes it shrinks proportionally more in width than in length, although in modern films permanent shrinkage is very slight. Shrinkage introduces errors that are difficult to eliminate entirely and, accordingly, glass gives the greatest accuracy.

* This indicates that the focal length divided by the diameter of the lens aperture is 6.

8. Most mapping cameras are equipped with devices that operate the shutter and move the film automatically. A device called an **intervalometer** can be set to make exposures at desired time intervals. Some provide a moving mark that can be regulated to follow the apparent movement of objects on the ground. It can be set to make exposures when the ground images have moved over any desired fraction of the negative area. These are independent of altitude.

9. Cameras are mounted so that they may be turned in azimuth to keep them aligned with the flight path when the airplane must crab toward the wind. They are also supported in gimbals so that they may

FIG. 20-11. A modern precision Fairchild aerial camera. (*Fairchild Camera and Instrument Corp.*)

be kept pointing directly downward despite movements of the airplane. Dash pots are provided to damp any tendency to swing, and levels are provided to aid in judging the vertical. Since the levels are affected by any accelerations of the airplane as well as by gravity, expert flying is necessary to aid in keeping the camera properly aimed. Automatic stabilization by gyros is sometimes used. With present techniques cameras can be held at an average of 1 degree of the vertical, and the error rarely exceeds 3°. Figure 20-11 shows a modern precision aerial camera made by Fairchild Camera and Instrument Corp.

THE PERSPECTIVE CENTER IN LENSES

20-14. Lens Nodes. Figure 20-12 shows a lens forming an image at point I of a point O on the object. All of the rays from the point O that

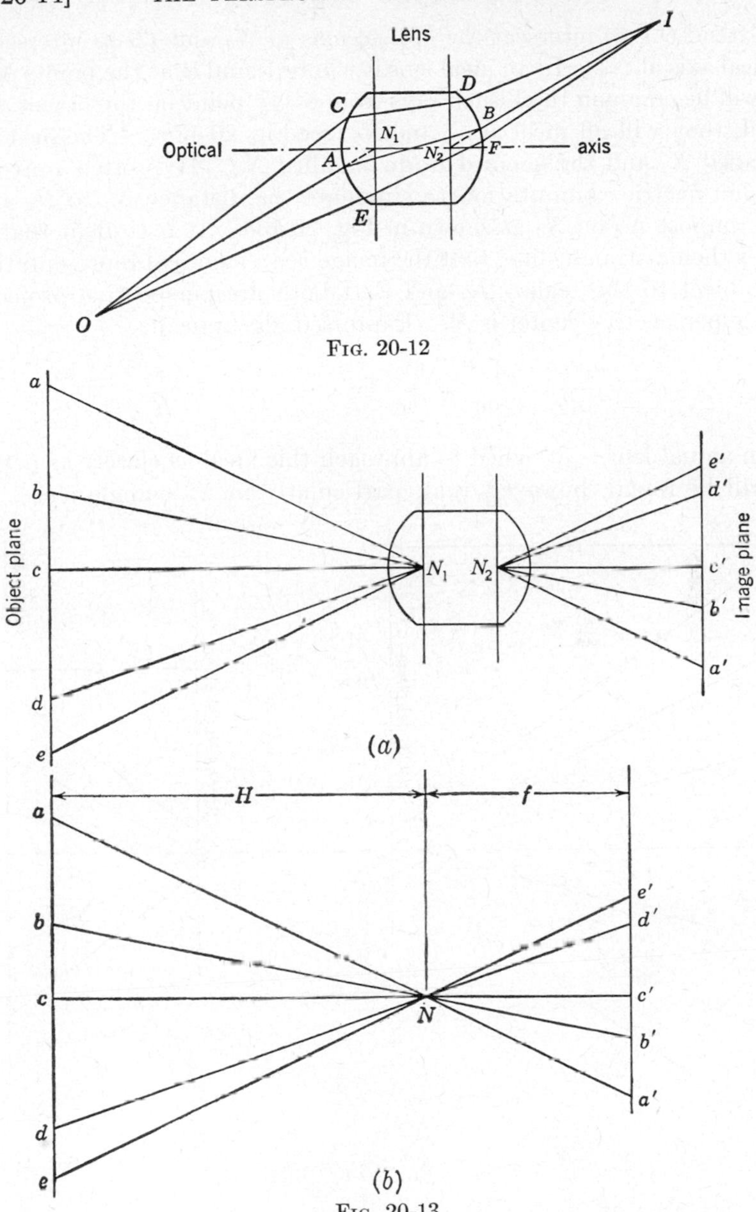

FIG. 20-12

(a)

(b)

FIG. 20-13

pass through the lens are collected at I. The rays that strike the lens near its top emerge with their directions changed slightly downward, like the ray $OCDI$. Those that strike near the bottom are changed slightly upward, like the ray $OEFI$. Evidently between these extremes there must be a ray that emerges parallel to its original direction, like the ray $OABI$.

Extend OA to intersect the optical axis at N_1 and IB to intersect the optical axis at N_2. In an ideal lens it will be found that the points N_1 and N_2 will be common to all such rays from every point on the object. In a word, they will all meet at N_1 and N_2 (see Fig. 20-13*a*). The **first node** is called N_1 and the **second node** is called N_2. It is often convenient in photometric computation to eliminate the distance N_1 to N_2 and to superimpose N_2 on N_1 as shown in Fig. 20-13*b*. It is evident that each ray is then a straight line, that the image is an identical representation of the object to the scale f/H, and that both are perspective projections whose perspective center is N. Expressed algebraically,

$$\frac{a'b'}{ab} = \frac{a'c'}{ac} = \frac{b'c'}{bc} = \cdots, \text{etc.} = \frac{f}{H}$$

An actual lens is designed to approach this ideal as closely as possible. It will be found, however, that, particularly in wide-angle lenses, it is

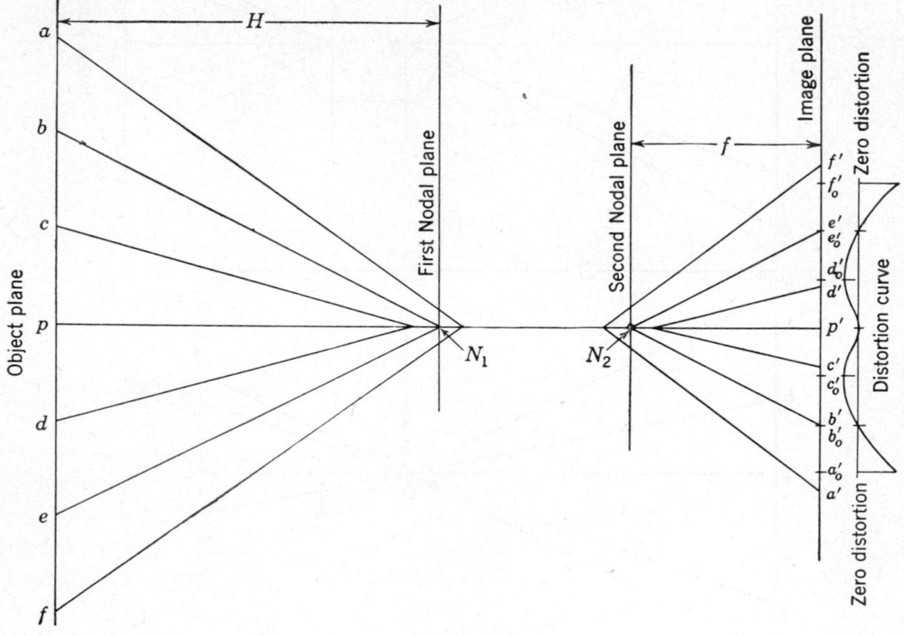

FIG. 20-14

impossible to design or to manufacture them so that all rays that emerge from the lens parallel to their original directions radiate exactly from a single point. Figure 20-14 shows one of the many possibilities that may exist. The resulting errors are determined by calibrating the cameras.

20-15. Camera Calibration. First the lens is permanently set so that the best focus (**definition**) is obtained over the whole area of the negative. Then a photograph is taken of a set of uniformly spaced targets shown a, b, c, p, d, e, and f placed on a line perpendicular to the camera axis. The resulting images should be uniformly spaced. If they are not, some condition like that shown in Fig. 20-14 must exist. Usually, collimators[1] are used instead of distant targets, and measurements are made on the image plane by means of a micrometer microscope.

20-16. Distortion. Lens distortion is defined as the variation in the positions of the images from a true perspective. The term is often incorrectly applied to the variations of a photograph from an orthographic projection, that is, variations from a map. This has nothing to do with lens distortions. The most important distortions of a lens are the type shown in Fig. 20-14. These are radial distortions in the image plane radiating from the center of the negative or, more exactly, from the principal point which is the foot of the perpendicular from the perspective center.

To calibrate a lens it is necessary to determine the position of the perspective center N_2 that creates the least distortion and to evaluate the distortions from a true perspective from that point.

First, any pair of symmetrically placed images are assumed to be properly positioned (in this case b' and e'). Then

$$f = H \frac{b'e'}{be}$$

With this value, the true positions of the images f'_o, d'_o, etc., can be computed thus:

$$p'd' = \frac{f}{H} pd \qquad \text{etc.}$$

The error in position of each image, f'_o to f', etc., is measured and plotted. The result is the **distortion curve** shown. Errors outward from the center are called plus. The assumed value of f is then adjusted by computation to the best advantage; usually a value is chosen that will make the greatest plus and the greatest minus distortions equal. The final value of f is used as the principal distance in computations; it is called the calibrated focal length CFL. It usually differs slightly from the focal length that was computed from the design values of curvature, index of refraction, and the positions of the lens elements. The focal length computed from the design values is known as the **equivalent focal length** EFL. In photogrammetric computation, the letter f indicates the principal distance.

[1] A collimator is like a transit telescope focused at infinity with the reticule illuminated. When observed from the objective end, it places the cross hairs at infinity in the direction of its line of sight. Thus, collimators can simulate distant points.

20-17. Pillow and Barrel Distortions. Figure 20-15 shows the appearance of a rectangular grid photographed with the lens used in Fig. 20-14. Positive errors form a pillow-shaped pattern and are called **pillow distortions.** These appear in the outside of the diagram. Negative errors (near the center in this example) form a pattern something like a barrel and are called **barrel distortions.**

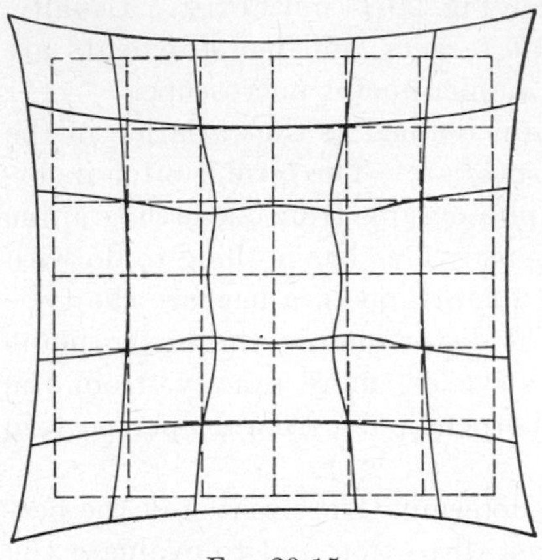

Fig. 20-15

20-18. Corrections. For accurate photogrammetry, the photographs must be corrected for lens distortions. This can be accomplished in a number of ways. Sometimes a special correction plate is used in the camera. In some systems, the lens used to print the transparent positive (the diapositive) used in the plotter is designed to correct the distortions; sometimes the plotter has a device that eliminates the distortions mechanically.

20-19. Tangential Distortions. Errors in mounting the lens elements sometimes cause distortions perpendicular to radial lines. These can only be corrected by changing the mountings.

20-20. The Principal Point. The foot of the perpendicular to the image plane dropped from the perspective point is known as the **principal point.** Aerial mapping cameras usually have some device for marking this point on each exposure. One device consists of small metal shapes of various kinds at the middle of each of the four sides of the negative plane; they photograph as silhouettes. They are placed so that lines connecting opposite pairs intersect at the principal point. They are often set at given distances apart across the negative so that changes in the size of the negative can be measured (see Fig. 20-16).

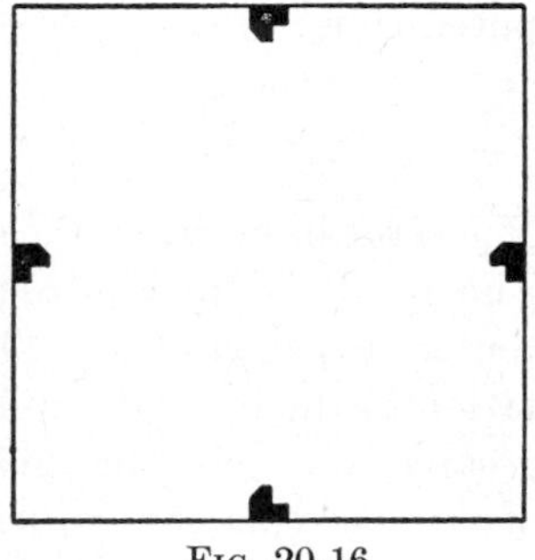

Fig. 20-16

PROCEDURE FOR AERIAL MAPPING

20-21. Methods. Many successful methods have been devised for mapping by aerial photography. The choice of a method depends on the field conditions imposed and the results desired. Military mapping, particularly in the theater of operations, imposes the most difficult field conditions and therefore includes the greatest variety of procedures. Methods for accurate contour mapping under normal conditions vary chiefly in camera and plotting equipment. Vertical photographs are used almost exclusively; these are photographs taken with the camera pointing as nearly vertically downward as is possible in an aircraft. Accuracy is increased by lower-altitude flying, more dense control, and more accurate plotters. The procedure for this type of mapping is outlined in the steps given below. They are given in the usual chronological order, but several can be accomplished simultaneously.

1. *Specifications.* These are set up for the map desired.

2. *Survey Control.* Vertical and horizontal survey control **systems** are established in the area to be mapped.

3. *Flight Plan.* This is worked out.

4. *Photography.* The airplane is flown along a series of straight, parallel flight paths as shown in Fig. 20-17. Photographs are taken at regular intervals in such a way that each successive photograph overlaps about 60 per cent of the previous photograph measured along the direction of flight. A series of such photographs is called a **flight strip.** The overlap described is called the **forward overlap.**

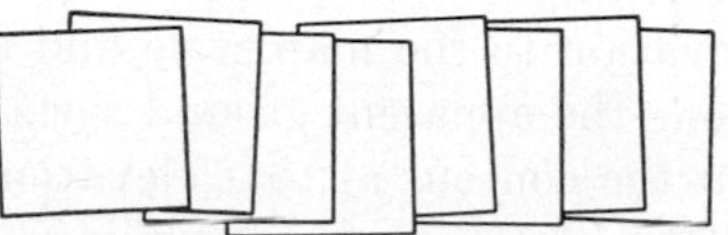

Coverage of photographs on a flight strip. Each overlaps the next by 60 per cent (forward overlap).

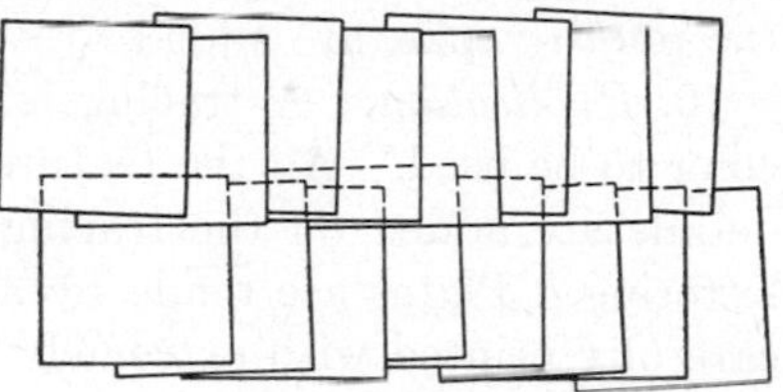

Flight strips overlap by 10 to 20 per cent (side overlap).

FIG. 20-17. The area to be photographed is covered by a series of flight strips.

It is evident that every ground point appears on at least two photographs and therefore the flight strip is completely covered by stereopairs. The appearance of a stereopair viewed stereoscopically is called a stereomodel.

The flight paths are spaced so that no point on the ground will be missed. Usually this requires a **side overlap** of 10 to 20 per cent. Government specifications usually require 30 per cent.

5. *Photocontrol Points.* These are selected on the photographs. They are well-defined point images that can be identified on the ground with certainty. At least four photocontrol points should appear in every stereomodel, preferably located near the corners.

6. *Survey Ties.* These are made for elevations on as many photocontrol points as possible, and horizontal ties are made to at least one point (two are much better) on the first and last stereomodel in each flight strip and one on every third or fourth stereomodel throughout the strip.

7. *Manuscript Map Layout.* The original map, called the manuscript map, should be drawn to at least twice the scale of the finished map.

Sheets are prepared with those photocontrol points whose horizontal positions have been surveyed plotted to the scale chosen.

8. *Plotting.* The contours and other topographic details are placed on the map with one of the many types of plotters. By optical means these devices provide a stereomodel from any desired stereopair. They also provide, either physically or optically, a movable mark called a **floating mark** that can be seen with the model; the height and horizontal position of this mark with respect to the manuscript map can be set. With the aid of this device the stereomodel can be adjusted so that the photocontrol points fit their proper positions with respect to the map in three dimensions. This recreates to scale the geometry that existed at the time the photographs were taken. The floating mark can then be set at a given elevation to the map scale and datum. Once it is set, it can be moved along the apparent ground surface in the stereomodel and thus can trace out the contour at that elevation.

The plotter and its auxiliary equipment are the heart of photogrammetry. Various types will be described later.

9. *Field Editing.* Photographs of the manuscript map must be taken into the field where names, boundary lines, and topography, not shown on the photographs, are added by ordinary survey procedures.

10. *Publication.* A tracing is made of the manuscript map for each color to be used. All the features of that color are traced. Names and details are added on the tracing prepared for the color desired for the lettering. Plates are made by a lithographic process and the maps are carefully printed with extremely accurate registry.

SPECIFICATIONS

20-22. Chief Elements. Scale and accuracy are the most important elements in a map. They determine the usefulness of the map and its cost. When a map is planned, the specifications for these elements should be given careful consideration in the light of the purpose for which the map is to be used.

20-23. Scale. The scale should be as small as possible so that the maximum information is available within a minimum area; but the scale should be large enough to make it possible to take off information quickly to the desired accuracy and to delineate new construction without difficulty.

20-24. Accuracy. The accuracy of horizontal position may be specified by a ratio or by a limit of error on a single sheet. A ratio specification should be used for special purpose maps when the use to which they will be put can be forecast. The ratio specification always reduces the cost of the survey control system and usually reduces costs throughout.

General purpose maps necessarily must be specified by a limit of error for the sheet.

The accuracy of vertical position is established by the contour interval chosen. There is no reason to hold the accuracy to better than half a contour interval. The form of the ground cannot be shown to better than the nearest 10 ft by 20-ft contours no matter how accurately they are located. However, when only a part of the map requires great vertical accuracy or when relative accuracy between nearby points is all that is necessary, economies will result by including these requirements specifically in the specifications.

20-25. General-purpose Maps. When a general-purpose map like a U.S. Geological Survey quadrangle is made up, since it may be used for any purpose the scale and the contour interval must be set arbitrarily. The following specifications are generally accepted:

1. *Maximum Error in Horizontal Position on Map.* This is $\frac{1}{50}$ or $\frac{1}{40}$ in. Thus, for a map with a maximum dimension of 20 in., the error could not exceed

$$\frac{0.02}{20} = 1:1000 \qquad \text{or} \qquad \frac{0.025}{20} = 1:800$$

2. *Maximum Error in Elevation.* A profile taken anywhere on the map shall not be in error by more than one-half the contour interval throughout 90 per cent of its length. The true profile shall be determined by spirit leveling on the ground.

20-26. Special-purpose Maps. A good example of a special-purpose map is the preliminary survey for a highway. Such a map must be used for proper locating, cost estimates of cut and fill, and often as a basis for the final plans. An average scale for such a map is 40 ft to the in. The most useful area of the map, where careful measurements will be made, will be a strip about 800 ft wide near the proposed highway center line. Each sheet may cover a length of 4000 ft. Details near the center line should be plotted horizontally to the nearest foot and a 5-ft contour interval should be used.

If the specifications for a general-purpose map were used, the plotting error would have to be held to 1:4000, which would necessitate a second-order triangulation arc at least 2000 ft wide. At least three lines of spirit leveling would have to be run, one along the center line and one on each side.

If, on the other hand, the position error of 1 ft were held only 400 ft each side of the line, and if the accuracy of the contours beyond 400 ft were not held to half the contour interval, the only control required would be an ordinary transit traverse and a line of benchmarks along the center line. Photocontrol elevations on the sides could be taken by altimetry or by stadia.

Both specifications would provide equally useful maps. The second would reduce the cost 33 to 50 per cent. In general, substantial savings can be made if careful study is given to the actual requirements of any special-purpose map.

FLIGHT PLAN[1]

20-27. Features. A flight plan must state at what altitude to fly, give a layout of the flight lines, and state at what interval to make exposures.

20-28. Altitude. The height above the ground is dictated by the accuracy of the process to be used and the contour interval desired. A process is rated by its C-factor,[2] a value that must be obtained by experience with the process. The C-factor is the number by which the contour interval is multiplied to obtain the maximum height above the ground at which the photographs can be taken to ensure compliance with the half-contour-interval specifications. C-factors for various processes vary from 600 to 1500.

Since costs vary nearly inversely as the square of the flying height, the greatest height above the ground should be used that will be sure to give the accuracy desired. In many types of plotters the optics establish a relationship, within certain limits, between the size of the photograph and the size to which it is magnified to reach the map scale. Thus, once the flying height above the ground is established, the scale of the manuscript map follows. The altitude of flight must often be changed in mountainous country in order to retain the same flying height above the ground and thus the same scale for the manuscript map.

For example, assume that the multiplex plotter and a camera with metrogon 93° and a 6-in. nominal focal length lens is used. The C-factor of this process can be taken as 600 with proper survey control. Assume that the desired contour interval is 20 ft. The height above the ground is computed thus:

$$600 \times 20 = 12{,}000'$$

The range of height above the model of the perspective center of the multiplex projector is, at map scale,

Minimum	0.95′ (290 mm)
Optimum	1.18′ (360 mm)
Maximum	1.51′ (460 mm)

[1] Most of the principles contained in the paragraphs on flight planning are taken from the prize article by Morris M. Thompson, A New Approach to Flight Planning, *Photogrammetric Engineering*, vol. 16, pp. 49–62, March, 1950.

[2] Originally suggested by Russell K. Bean, U.S. Geological Survey.

The optimum height divided by the actual height above ground gives the optimum manuscript map scale:

$$\frac{1.18}{12{,}000} = 1{:}10{,}169 \qquad \text{assume } 1{:}10{,}000$$

The range of height above ground is then

$$10{,}000 \times 0.95 = 9{,}500'$$
$$10{,}000 \times 1.51 = 15{,}100'$$

If the elevation of the area to be mapped ranges from 1000 to 3000 ft, the flying altitude might be specified at 13,800 ft. This would give a range in height above ground of 12,800 to 10,800 ft, well within the limits of the plotter when the manuscript map has the scale 1:10,000. If in some areas the ground rose to 5000 ft, a higher altitude would have to be specified for those areas.

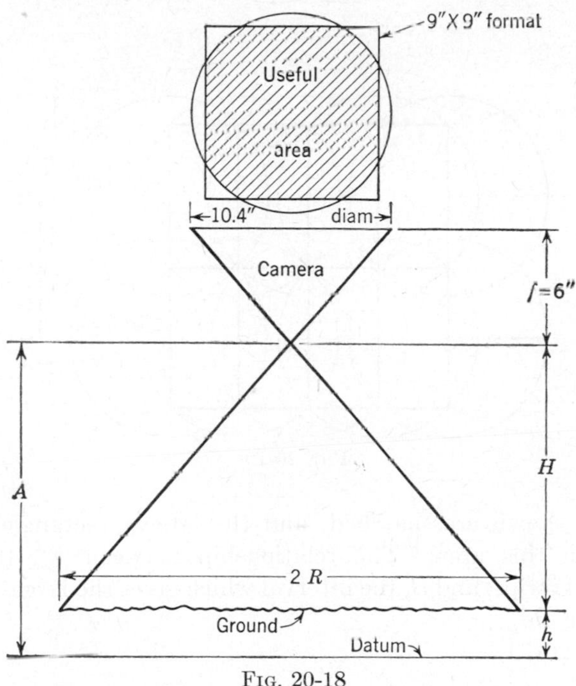

FIG. 20-18

20-29. Flight Lines and Intervals. The aim of spacing the flight strips and of setting the intervals between photographs is to cover completely the area with a minimum of stereomodels. The relation between the separation of the flight lines and the separation between photographs must therefore be arranged to give the greatest area to each stereopair.

The 6-in. metrogon lens gives a useful circle at the negative plane 11 to 12 in. in diameter. Due to errors in directing the camera and in following the flight lines, this should be reduced to at least 10.4 in., as in Fig. 20-18. The 9- by 9-in. format reduces this circle, but it is evident in Fig. 20-19 that it does not control the spacing. Since the stereomodels must fit each other, the useful stereoareas must be assumed to be rectangles having a width equal to the interval B between exposures (see Fig. 20-19). The

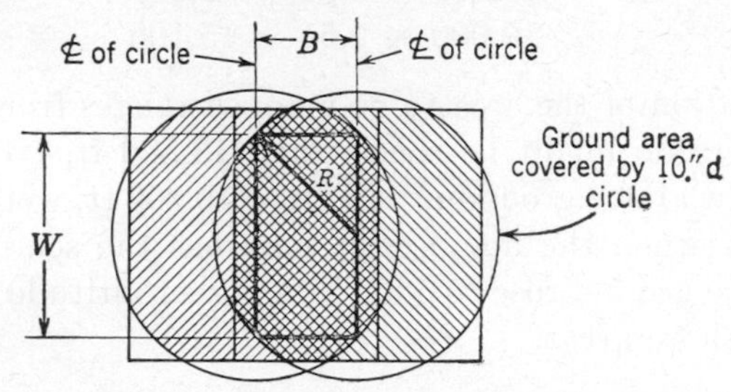

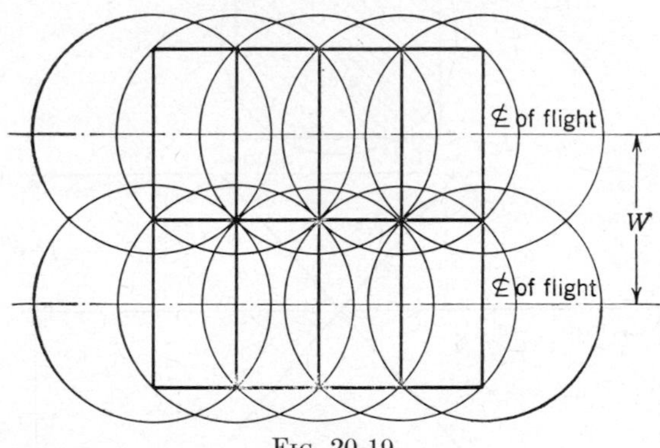

FIG. 20-19

stereoarea is shown crosshatched, and the largest rectangle possible is drawn within this area. The relationship between W, the distance between flight strips, and B, the interval which gives the largest stereoarea A, must be found.

From Fig. 20-19

$$A_s = BW$$

$$\tfrac{1}{2}W = \sqrt{R^2 - B^2} \qquad (20\text{-}1)$$

Substituting,

$$A_s = 2B\sqrt{R^2 - B^2}$$

Then

$$\frac{dA_s}{dB} = 0 = R^2 - 2B^2$$

$$R = B\sqrt{2} \qquad (20\text{-}2)$$

Substituting in Eq. (20-1),

$$\tfrac{1}{2}W = \sqrt{2B^2 - B^2}$$
$$W = 2B \qquad \text{for max. area} \tag{20-3}$$

But from Fig. 20-18

$$\frac{2R}{H} = \frac{10.4}{6}$$
$$R = 0.867H$$

Substituting in Eq. (20-2),

$$B = 0.61H$$

Substituting in Eq. (20-3),

$$W = 1.22H$$

where H = height above ground.

For the example used

$$B = 0.61 \times 10{,}800' = 6600'$$
$$W = 1.22 \times 10{,}800' = 13{,}200'$$

Thus, an exposure should be taken every 6600 ft and the separation of the flight strips should be 13,200 ft.

It is often more convenient to express B in the form of a ratio or to express it in terms of a distance on the negative plane, thus:

$$\frac{B}{H} = 0.61 = \frac{b}{f} \tag{20-4}$$
$$b = 0.61 \times 6'' = 3.66''$$

where b is the distance on the negative plane between exposures.

Flight paths should be laid out parallel to the longest dimension of the area; this reduces flying time. There is some advantage in running them parallel to the direction of mountain ranges to prevent concealment. Often they are run due north and south or due east and west.

Unless the area to be mapped is exactly covered by a certain number of flight paths spaced at the computed value of W, W should be reduced to introduce one more flight path to utilize the excess photography for increasing the side overlap.

The flight-path center lines are laid out on any existing map. Where no map exists, a series of very high altitude photographs are taken and pieced together to make a mosaic. The flight strips are then laid out on the mosaic.

PLOTTING

20-30. The Multiplex. The theory of fitting the stereomodels to control, and of plotting the contours and planimetry, is best explained by using the multiplex plotter as an example. Figure 20-20 shows a photograph of this plotter.

The instrument consists essentially of a set of small projectors mounted on a bar and directed downward. Each projector can be moved by slow-motion screws on the three major axes of position. They can be translated parallel to each axis or rotated around each axis, giving six motions in all.

FIG. 20-20. The Multiplex projector ready for use. (*U.S. Geological Survey.*)

The X axis is taken parallel to the bar, the Y axis is horizontal and perpendicular to the bar, and the Z axis is vertical. The x tilt is rotation around the X axis, y tilt around the Y axis, and swing or z tilt is rotation around the Z axis. The projectors are mounted so that they rotate around the exterior node of the projection lens. Below the projectors is a flat surface, held flat to very close tolerances. On this is placed the manuscript map with the horizontal control points plotted on it.

The operator works with a **tracing table,** shown in Fig. 20-21. It provides a screen (or platen) which intercepts the projections of the photographs so that the operator can see them. At the center of the screen is a white light which serves as the floating mark. This is very small in size (0.004 in. in diameter) and arranged so that its rays radiate from a point exactly at the surface of the platen. Its brightness can be controlled.

The platen can be raised or lowered exactly perpendicular to the surface of the map, and its height with respect to an established datum parallel to the map surface can be read precisely on a vernier or a counter. Directly under the floating mark is a pencil that can be raised to clear the

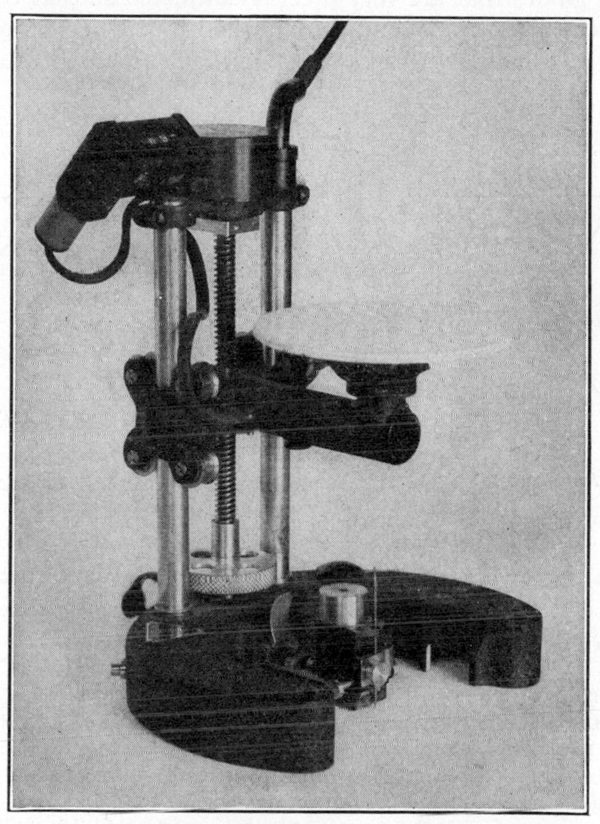

FIG. 20-21. Tracing table for the Multiplex. (*U.S. Geological Survey.*)

map or lowered to rest on its surface. Two small lamps illuminate a small area of the map around the pencil.

The base of the tracing table is supported on three agate feet which slide easily and determine the plane to which the instrument is adjusted.

20-31. The Projectors. The lens of each projector is usually the wide-angle type to cover the same field as the metrogon lens. It has a small aperture which enables it to be made essentially distortion free and have a considerable depth of focus. The range is 290 to 460 mm with an optimum at 360 mm.

20-32. Diapositives. A small glass diapositive about 2 in. square is made from each photograph in a special projector printer that is set to reduce the size of the original negative by the ratio of the principal distance of the projector lens to the principal distance of the camera used (corrected for film shrinkage). The printer lens is designed to remove the radial distortions of the camera lens. The diapositive is held in the projector in guides that are very precisely adjusted so that the principal point is on the optical axis, and the emulsion is exactly perpendicular to the optical axis and exactly at the established principal distance of the

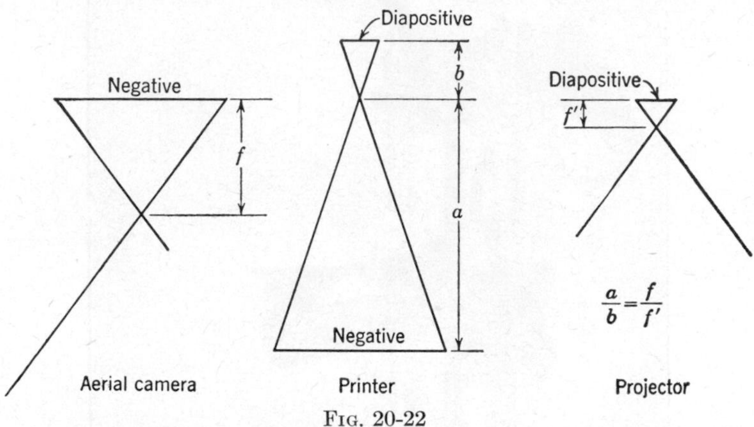

FIG. 20-22

lens. With this arrangement (Fig. 20-22), the relative directions of the projections of the various images are exactly the same as their relative directions when they were received by the aerial camera.

20-33. Setting Up the Multiplex. The diapositives are placed in the series of projectors in the order in which the photographs were taken in the flight strip. It is then possible to adjust the projectors so that, to the map scale, they take the exact positions and orientations occupied by the aerial camera when the negatives were exposed. This exactly recreates the geometry that existed in the field when the photographs were taken. It follows that the image of each object in the field is projected toward its correct position with relation to all the others. Since it is projected from two projectors, the intersection of the two rays of projection places it at its correct position in space above the map (Fig. 20-23). All of the intersections taken together form a true model.

The adjustment of the projectors, in present-day practice, is carried out stereoscopically in two steps. First two projections are brought into relative orientation by manipulating the two projectors until the two rays for every object intersect and thus form a model. The model is then brought into **absolute orientation** by manipulating the projectors as a unit by moving the supporting bar until the model is adjusted to fit the space positions of the photocontrol points. The space position of each photocontrol point is given by placing the tracing table over its plotted position and raising the platen to its correct elevation.

20-34. Stereoviewing. Before the exact procedure for these two operations is covered, the stereoscopic conditions must be analyzed. Figure 20-24*a* illustrates the conditions that exist after a model has been placed in both relative and absolute orientation. The true model is formed along the line representing the ground, A_t, B_t, and C_t by the intersections of the corresponding pairs of rays from the two projectors. This model cannot be seen, although it represents the true position of the ground with respect to the map.

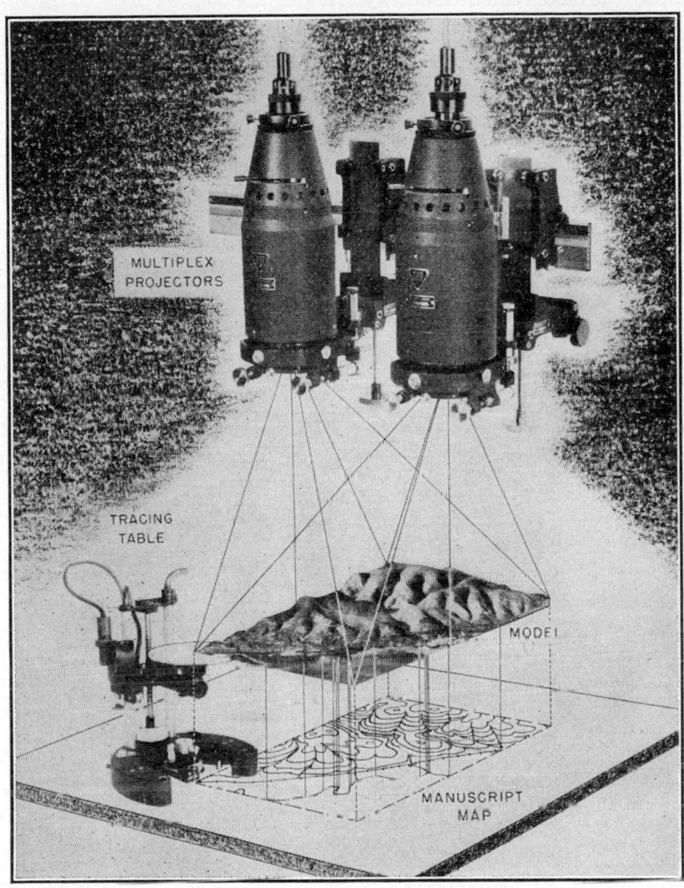

FIG. 20-23. The true model is formed as shown. It cannot be seen. The stereomodel, with heights exaggerated, is seen instead. By moving the platen, on which the stereomodel is seen, points on the true model are found. Their horizontal positions are transferred to the map by the pencil. (*U.S. Geological Survey.*)

A blue-green filter is placed over one projector and a red filter over the other. When the platen is placed in the model area its surface becomes an anaglyph. Viewed with a blue-green glass over one eye and a red glass over the other eye in the proper order, a stereomodel (not the true model) appears as shown in Fig. 20-25.

In Fig. 20-24, three positions are shown for the platen. In position 1 the platen is above the true model. The point A_t on the true model is seen on the platen by the left eye at a and by the right eye at a'. The point will appear to be at A_s on the stereomodel. The floating mark, seen by both eyes, will appear in its actual position and much higher than A_s. When the platen is lowered as in position 2, the stereomodel moves up unnoticed by the operator so that when the platen is on the true model, the floating mark appears to be on the stereomodel. Thus, the positions of points on the true model can be found by bringing the floating mark to the stereomodel.

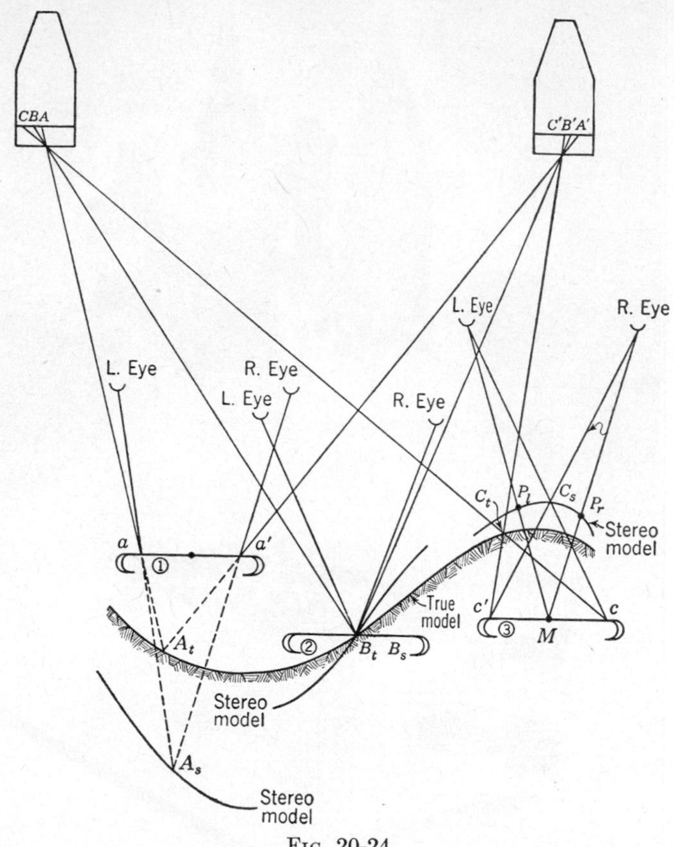

FIG. 20-24

When the platen is below the true model, an interesting psychological phenomenon occurs. The novice observer usually rejects the impression that the floating mark is underground at M. Instead, he sees it split into two marks at P_l and P_r. Later he learns to see it underground.

20-35. Relative Orientation. The first two projections on the flight strip are brought into relative orientation first. The projectors are placed in approximate positions and the platen is placed in the model area. By raising and lowering the platen and manipulating the two projectors, the stereomodel can be made to appear in one part of the model even though the adjustment is far from perfect. The adjustment is then perfected by continuing this process until perfect coincidence occurs, that is, until each image is clearly defined at a single point on the platen surface at each corner of the model and at the middles of the long sides. By following a certain order, each successive adjustment has a minimum effect on the preceding adjustment. At best it is a process of successive approximations.

The process is known as **clearing the y parallax,** because in the proper order of procedure the operator moves the two images of any point together in the direction of the y-axis and thus removes any difference or parallax between the y positions of the point. Actually the x parallax is also properly set. The whole process can be accomplished by moving only one projector.

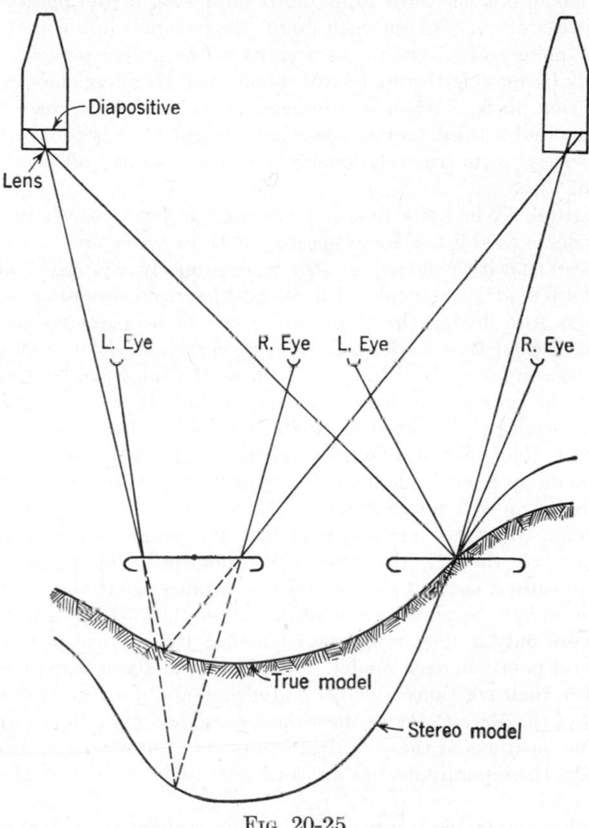

FIG. 20-25

When five of the six points agree, the sixth point, and all the other points on the model, will agree. The result is a true model of the ground. It may not be at the map scale and, with respect to the map, it may be swung on a line connecting the two projection points (the exterior nodes of the two projectors), and it may not be in proper orientation, vertically or horizontally, with the map.

20-36. Absolute Orientation. The true model is adjusted to control by two steps. First it is scaled and then leveled. The scale of the true model is equal to the ratio of the distance between the exterior nodes of the projector lenses divided by the distance in the field between the exterior nodes of the camera lenses (the **air base**), when the photographs were taken.

To orient and scale the model, at least two photocontrol points of known horizontal and vertical positions and a third point of known elevation are required within the area of the model. The platen is set at the elevation of one of these points and placed so that the pencil is exactly over the plotted position. The model is moved so that the corresponding image is brought to the floating mark. The floating mark is then placed at the elevation of, and over, the second control point and the scale is shrunk or expanded chiefly by x-movement of the one of the projectors. It is then turned chiefly by swinging both. The swing can be accomplished by rotating the map if desirable. The process is repeated until both points are satisfied.

Finally the model is leveled by adjusting it until each of the photocontrol points is at its correct elevation. To set each point, the control table is set at the known elevation of the point and moved to the vicinity of the proper image. The image is moved vertically by moving the model as a whole until the image has the same elevation as the floating mark. When two images on the model are correct for position and for elevation and a third (not in the same straight line) is correct for elevation, the model is exactly in its true relationship with the map and all points are in their correct position.

20-37. Bridging. When the first true model has been placed in position, the second true model is established by adjusting the third projector.

It can be shown that if the images of only three properly arranged objects from the third projector are made to coincide with the positions and elevations established for them in the first true model, the third projector will be perfectly positioned and, accordingly, the second true model will be in its correct position. Obviously, since the second model can be established from the first, the third can be established from the second and the process can be carried on indefinitely without control. It has been found that accidental errors accumulate so rapidly, however, that it is not safe to establish more than one or two stereomodels without vertical control. When control is available on a subsequent model, adjustments must often be made of the previous models set up without control.

When horizontal control is reached, it is often necessary to eliminate scale errors proportionately back through the intervening models. This is accomplished by moving each projector a computed proportional distance along the x-direction.

Once this is complete the plotting can start. Often, however, it is more economical to utilize the work only to determine the horizontal and vertical positions of a large number of control points in each model. Usually about six or eight well-distributed points are chosen, their positions scaled from the map and their elevations determined from the height of the platen. With these data, each model can be set up and plotted separately. The positions of these control points are also plotted on the manuscript map so that when the separate models are complete, they can be fitted in their proper position.

Frequently when control is insufficient, the photographs are run through another type of more accurate plotter to obtain sufficient control points, much as described for the multiplex.

20-38. Plotting. Finally, when the models are in position, the topography can be plotted. The tracing table (the platen) is set at the elevation of a desired contour. The floating mark is moved against the ground as shown in Fig. 20-25, the tracing pencil is lowered, and the table is moved so that the floating mark remains in contact with the ground. Other contours are plotted by changing the height of the tracing table. It must be remembered that, since only the stereo model can be seen, when **planimetry** is plotted, the floating mark must be kept at the ground surface in order to find correct horizontal positions. For example, when a road is plotted, the platen must be moved up and down as the road runs over summits and sags. While the plotting is in progress, exceptionally well-defined points are chosen and marked along the edges of the flight strip so that adjoining strips can be accurately tied together. This is of assistance particularly when sufficient control is lacking.

TYPES OF PLOTTERS

20-39. Similarity. The multiplex, just described, illustrates the principles of all plotters that can handle relief. All of them create a model to scale by recreating the geometry, directly or indirectly. All of them depend on stereovision to find the surface of the model. Each has special features and certain advantages and disadvantages.

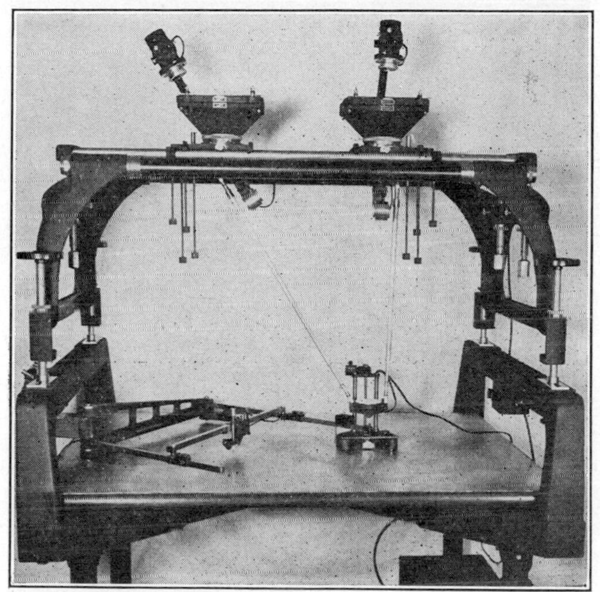

FIG. 20-26. A Kelsh plotter. (*Harry T. Kelsh.*)

20-40. Classification. Plotters can be classified according to the degree to which they restitute the original geometry:

1. Projectors (complete restitution)
2. Direct viewers (partial restitution)
3. Scanners (minor restitution)

20-41. Projectors. In addition to the multiplex, the most representative projector plotters are the Kelsh plotter and the Zeiss stereoplanigraph.

The Kelsh Plotter. The Kelsh plotter (Fig. 20-26) is like two projectors of the multiplex. The diapositives are contact prints. The projectors are each illuminated by a small light focused on the lens aperture and aimed at the platen by rods attached to the tracing table. The small lights require no cooling. A distortion-free (Hypergon) lens is used. The distortions of the camera lens are eliminated by cams that, by raising and lowering the projector lenses, keep the perspective centers at the same heights as the tracing table is moved toward the edge of the field.

When the two projectors have been relatively oriented, the whole model can be moved to fit the control without disturbing the relative orientation.

The Zeiss Stereoplanigraph. In principle, the Zeiss stereoplanigraph is like the Kelsh plotter (see Figs. 20-27 and 20-28). The projectors, how-

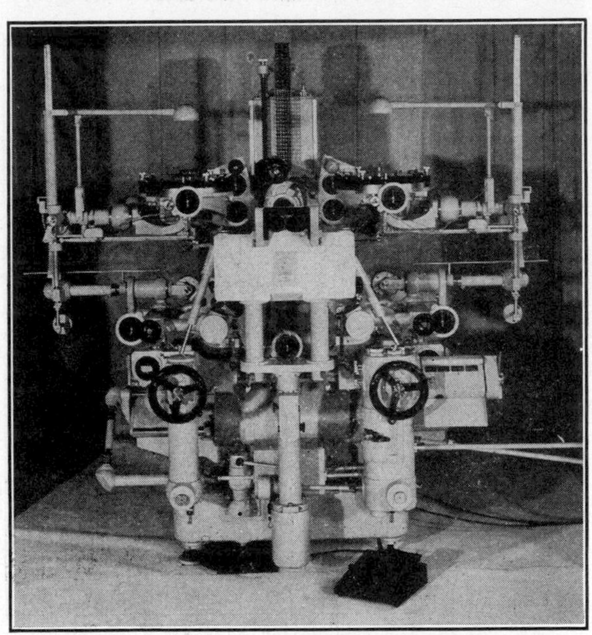

FIG. 20-27. Zeiss stereoplanigraph Model C-8. (*Corps of Engineers, U.S. Army.*)

ever, are separated so that the images do not overlap and the floating mark consists of two marks at the appropriate separation for the projectors. Each image is viewed by a microscope focused on the image plane at the mark, as in Fig. 20-29. A long optical train carries the image to the eye. The two microscopes can be moved together with control wheels in the X, Y, and Z axes. The x and y motions are transferred to a pencil on a separate plotting table. The scale can be changed by changing gears. Various devices are used to maintain precise focus throughout.

The projected image viewed by each eye can be interchanged by an optical shift so that when work on a model is complete, the right-hand projector can be made the left-hand projector for the next model and thus its exact orientation is maintained.

The instrument is extremely accurate. Tests have shown that it can bridge eight models successfully.

20-42. Direct Viewers. The direct viewers, usually called orthographic plotters, are exemplified by the KEK plotter, not shown, and the

FIG. 20-28. Zeiss stereoplanigraph showing plotting table. (*Corps of Engineers, U.S. Army.*)

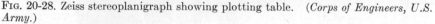

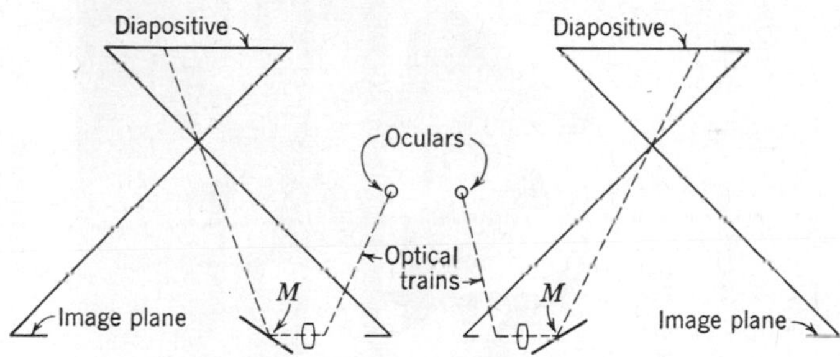

FIG. 20-29. Principles of projector plotter with images viewed through microscopes.

Wernstedt-Mahon plotter, Fig. 20-30. These are based on the principle shown in Fig. 20-31.

In Fig. 20-31 let L and L' be the perspective centers and P and P' be the picture planes of an aerial camera when two exposures were made of the ground AB. If the centers of rotation of the two eyeballs were placed at L and L' (these points are actually close together) and the pictures were placed at P_1 and P'_1, respectively, so that L and L' were then perspective points, the observer would see the pictures in the same geometric relation-

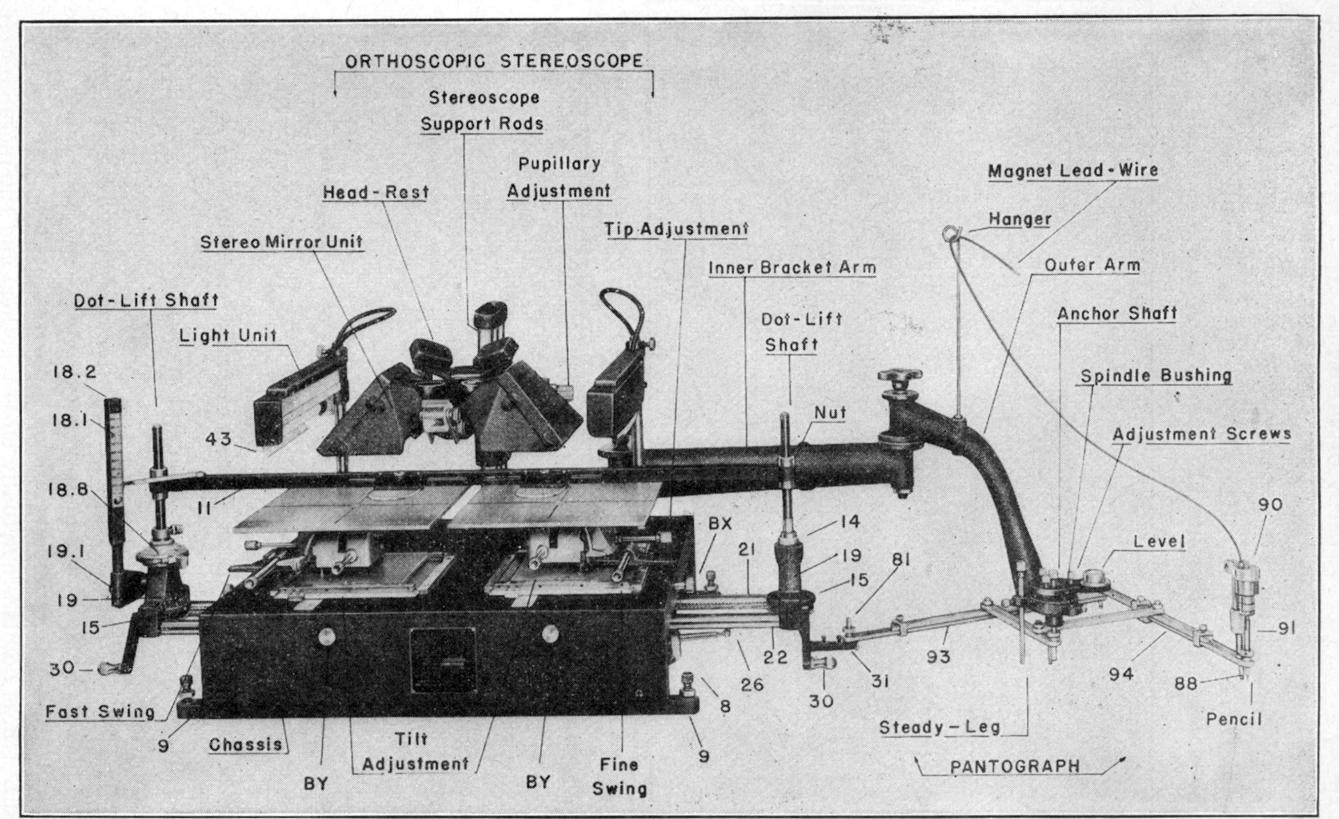

FIG. 20-30. New type of Wernstedt-Mahon plotter developed and built by Harrison C. Ryker, Inc. (*Harrison C. Ryker, Inc.*)

ship as occurred when the pictures were taken. If the bar f carried two properly spaced dots over the pictures, the observer would see them as a single floating mark at F.

The complete geometry would be recreated to scale. The floating mark could be raised or lowered by raising or lowering the bar f and the contours could be traced by the bar as with a projection plotter.

Unfortunately, it is difficult to design a plotter that can place the eyeballs at the perspective center of the photographs. Usually the plotters place the centers of the eyeballs too far away from the image plane. As a

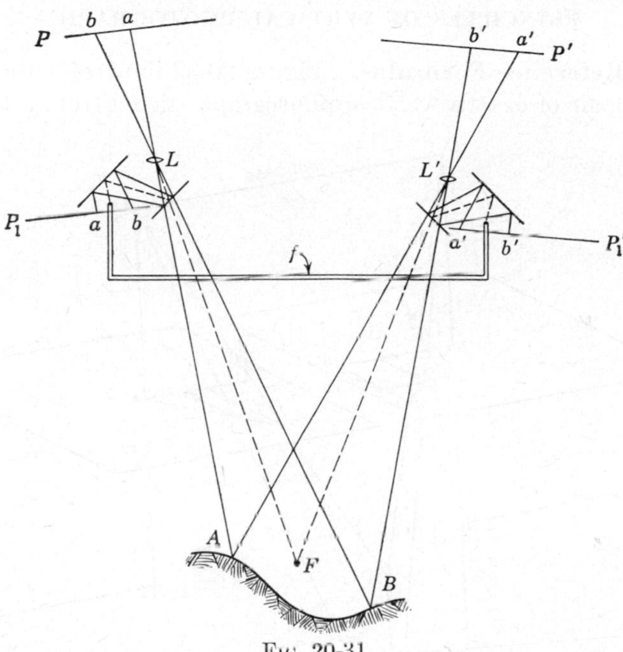

FIG. 20-31

result, the perspective distance is too great and heigths are exaggerated parallel to the principal distance. This difficulty can be overcome by regulating the vertical scale so long as the photographs are nearly vertical. When they contain a tilt, the principal distance is not parallel with the increase in height and therefore small errors are introduced. It is then impossible entirely to eliminate y parallax.

Errors in x parallax are not noticed, as they are translated into slightly erroneous heights, but errors in y parallax will cause the floating mark to split into two marks, one further along the Y axis than the other.

Various methods have been worked on to eliminate these difficulties.

20-43. Scanners. There is a class of plotters that use a floating mark consisting of two dots placed or projected on the surface of the photographs themselves; they thus scan the photographs directly. They are usually known as stereocomparators. The map drawn by one of these instruments is a perspective view of one of the photographs. This must be transformed to an orthographic projection. Unless exactly vertical photographs or **rectified** photographs (see Sec. 20-79) are used, special corrections must be applied.

Before discussing the scanning plotters, the basic elements of vertical photography should be covered.

PRINCIPLES OF VERTICAL PHOTOGRAPHY

20-44. Reference Formulas. Figure 20-32 illustrates the geometry of a stereopair of exactly vertical photographs taken from a flight strip.

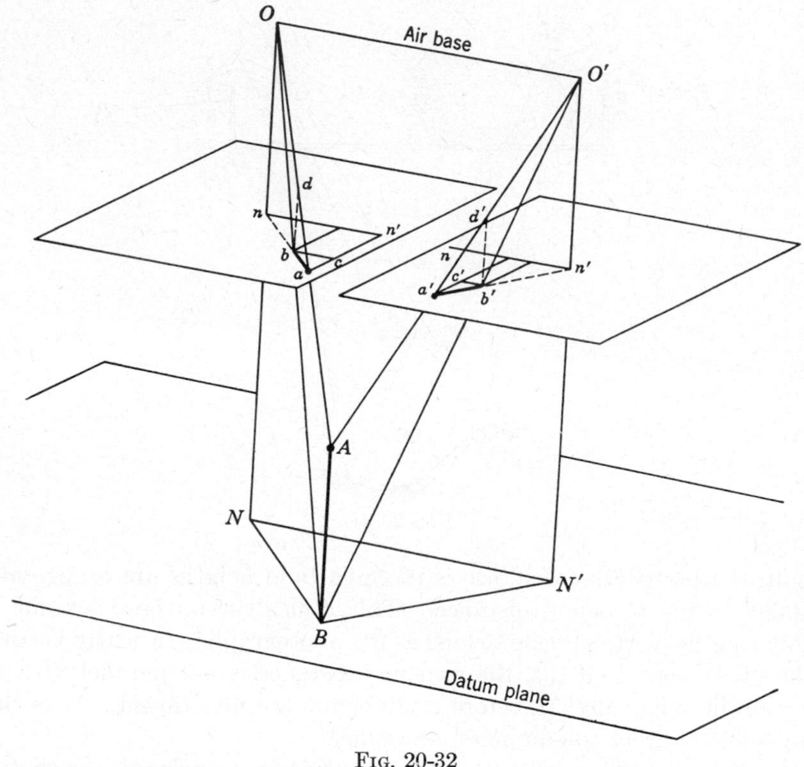

Fig. 20-32

The camera image plane has been placed at the perspective distance below the lens instead of above it. This does not change the geometry but it simplifies the explanation.

A represents a point on the ground and *B* is its orthogonal projection to the datum plane. The perspective centers are *O* and *O'*, and *ON* and

$O'N'$ are perpendicular to the datum plane and to the image planes. N and N' are at the nadirs and n and n' are at the principal points.

If A represents a point on the ground surface, the line AB would be a projection into the ground. It is convenient to think of it, however, as a physical object that can be photographed, like a flagpole. Its height represents the elevation of the point A above the datum. The images of the flagpole would be ab and $a'b'$. These lines represent the radial shift called **parallax difference** or merely **parallax** caused by the elevation of A above the datum. The **air base** is $OO' = NN'$. The lines nn' are the images of NN' on each photograph. These are exactly in line and represent the X axis for each photograph. They are not necessarily aligned with the fiducial marks on the photograph, as the pictures might have been crabbed. The x parallaxes are bc and $b'c'$ and the y parallaxes are ac and $a'c'$. The y parallaxes are always equal. The height of the flagpole on the model is bd or $b'd'$.

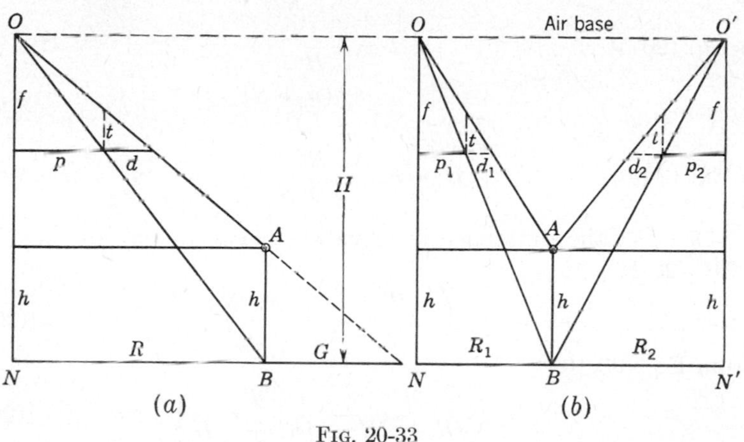

FIG. 20-33

In Fig. 20-33a is shown the radial plane $ONAB$ from Fig. 20-32. This may be allowed to represent the radial plane to any object. In Fig. 20-33b are shown the projections of the two radial planes in Fig. 20-32 onto the plane $ONO'N'$. In these projections vertical heights are projected at their true size but horizontal lengths are foreshortened proportionally.

From Fig. 20-33a the following formulas are evident:

$$\frac{p}{R} = \frac{f}{H} = \frac{t}{h} \tag{20-5}$$

$$\frac{d}{t} = \frac{R}{H - h} = \frac{p + d}{f} \tag{20-6}$$

$$\frac{p}{p + d} = \frac{R}{R + G} = \frac{H - h}{H} \tag{20-7}$$

From these, the following may be derived:

From Eq. (20-6) $$d = t\frac{R}{H - h}$$

From Eq. (20-5) $$t = h\frac{p}{R}$$

Substituting in Eq. (20-6), $$d = \frac{hp}{H - h} \tag{20-8}$$

From Eq. (20-5) $$p = \frac{fR}{H}$$

Substituting in Eq. (20-8), $$d = \frac{fRh}{H(H - h)} \tag{20-9}$$

From Eq. (20-9) $$h = \frac{dH^2}{fR + dH} \tag{20-10}$$

From Eq. (20-6) $$d = t\frac{p + d}{f}$$

From Eq. (20-5) $$t = h\frac{f}{H}$$

Substituting, $$d = \frac{h(p + d)}{H} \tag{20-11}$$

From Eq. (20-6) $$H - h = \frac{Rf}{p + d} \tag{20-12}$$

Let $v = h - h'$, the difference in elevation between two points. Then from Eq. (20-12)

$$v = f\left(\frac{R'}{p' + d'} - \frac{R}{p + d}\right) \tag{20-13}$$

and from Eq. (20-10)

$$v = H^2\left(\frac{d}{fR + dH} - \frac{d'}{fR' + d'H}\right) \tag{20-14}$$

It will be noted that these formulas also apply to each half of Fig. 20-33*b*. Hence from Eq. (20-5)

$$p_1 = R_1\frac{f}{H}$$

and $$p_2 = R_2\frac{f}{H}$$

Adding, $$p_1 + p_2 = (R_1 + R_2)\frac{f}{H}$$

From Eq. (20-6) $$d_1 = R_1\frac{t}{H - h}$$

and $$d_2 = R_2\frac{t}{H - h}$$

Adding, $$d_1 + d_2 = (R_1 + R_2)\frac{t}{H - h}$$

When the point A is moved without changing h, all vertical heights and the value of $R_1 + R_2$ remain unchanged. Since the right-hand values of the equations above remain constant, the left-hand values also remain constant, that is, $p_1 + p_2$ and $d_1 + d_2$ are the same wherever A is placed horizontally.

When A is moved over to the line $O'N'$ in Fig. 20-33*b*, the figure is identical to Fig. 20-33*a*, except that p becomes $p_1 + p_2$, d becomes $d_1 + d_2$, and R becomes $R_1 + R_2$. Accordingly, in all the formulas from Eq. (20-5) to Eq. (20-14), the sums of these values can be substituted for their single value. The letters usually employed to designate these sums are the following:

$p_1 + p_2 = b$	x parallax of datum[1]
$d_1 + d_2 = \Delta p$	parallax difference
$R_1 + R_2 = B$	air base

20-45. The Parallax Bar. Equation (20-8) can be written

$$\Delta p = b\frac{h}{H - h} \qquad \text{or} \qquad h = \frac{\Delta p H}{\Delta p + b} \tag{20-15}$$

This expresses a useful relationship between Δp and h, a difference in elevation. A parallax bar is a device for measuring Δp and thus is a means of determining differences in elevation. Figure 20-34 illustrates the principle involved.

C is a point of known elevation. H is the airplane altitude (usually taken from the altimeter reading) minus the elevation of C. A more accurate value of H can be found by measuring a length L on the ground and the length of the photographic image l. Equation (20-5) can be written

$$H_a = f\frac{L}{l} \tag{20-16}$$

where H_a is the height of the airplane above the average elevation of the two ends of the length measured.

The value b in Eq. (20-15) is the distance $p_1 + p_2$ which is $nc + cn'$ (on the lower part of the figure); nc and cn' can be measured on the photographs.

Thus, with H and b known, the height above C of any point A can be determined by measuring Δp.

20-46. Procedure. The principal point is found on each photograph from the fiducial marks. It is then transferred to the other photograph by finding the corresponding ground images. This is usually accomplished under a stereoscope. Thus n and n' are marked on both photographs.

[1] Also $b = nn'$.

The photographs are then mounted side by side and oriented so that the line nn' on one photograph is in line with the line nn' on the other. The distance between the center of the photographs nn_2' (upper part of Fig. 20-34) and the distance cc between the images of C is measured. Then

$$b = nn_2' - cc$$

The parallax bar has two glass disks called lenses mounted on it. Each has a small dot at the center for an index mark. One disk is stationary

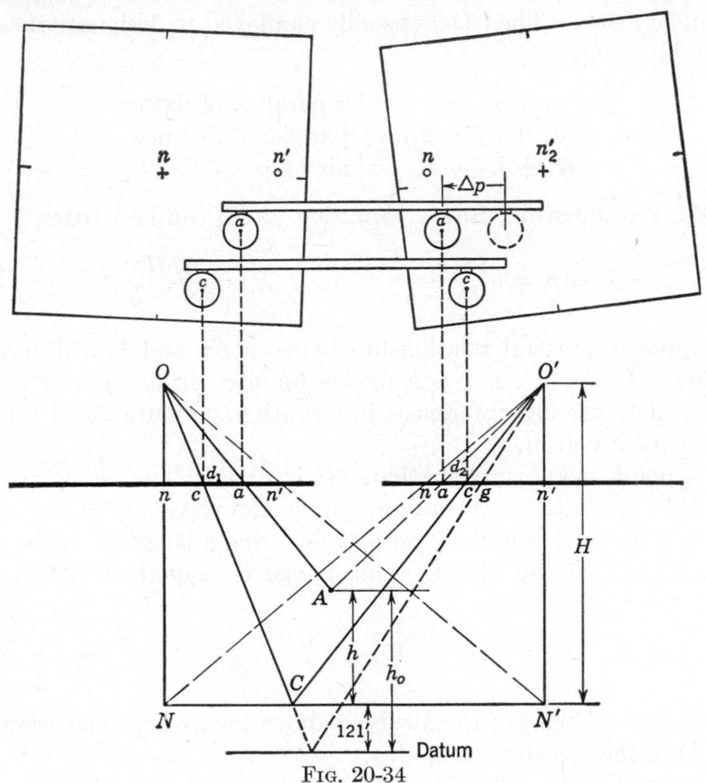

FIG. 20-34

on the bar but the other can be moved along the bar in the direction of the x parallax by a micrometer screw. The movement can be read on the micrometer indicator. The two dots are first placed on the images of the point of known elevation C and then moved to the images of the point whose elevation is required. The movable dot must be changed in position along the bar. This movement is the value of Δp. The difference in height h can then be computed by Eq. (20-15).

20-47. The Stereocomparator. A stereocomparator plotter is a plotter based on the parallax bar. The Abrams contour finder, the Fairchild stereocomparagraph, and the Brock-and-Weymouth stereometer are examples of this type of plotter. Figure 20-35 is a photograph of the Abrams contour finder. Figure 20-36 illustrates how it operates.

The parallax bar is attached to a drafting machine to keep it parallel to the line nn', a stereoscope is mounted over it, and a plotting pencil is attached to the bar. The observer sees the dots as a floating mark that can be moved over the stereomodel to trace the contours.

One of the lenses can be moved in the usual way along the length of the bar. This lens is called the x-parallax index. The other lens, the y-parallax index, can be moved in the y direction to compensate for small tilt.

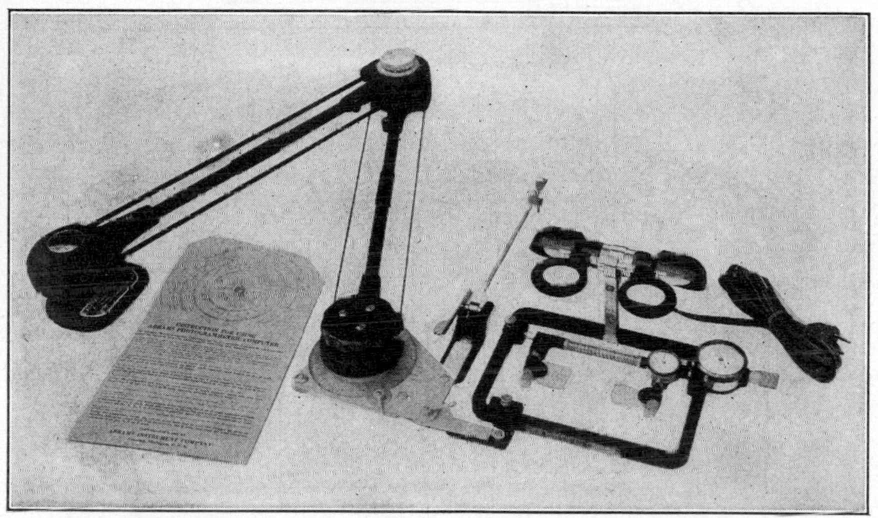

FIG. 20-35. Abrams contour finder. (*Abrams Aerial Survey Corporation.*)

It is evident that, when the photographs are viewed in the stereoscope, if the dots are placed in contact with the two images of the same point, the floating mark will appear to have the same elevation as that point. When the dots are moved closer together, the floating mark will appear to rise above the ground. Thus, by changing the separation of the marks, the same effect is attained as by raising or lowering the floating mark in the projector plotters or the direct viewers; and contours can be plotted by setting the marks at the proper separation.

There is an important difference, however. It will be noted that the pencil exactly follows the images of the photograph under the stationary y-parallax dot. If the contours had been painted on the ground when that photograph was taken, their forms and relative positions on the photograph would be identical to the contours traced as described above. Since the photograph is a perspective projection, the plotted contours would be a perspective projection of the real contours. It is therefore necessary to correct their positions to orthographic projection.

20-48. Near-vertical Photographs. Since the dots on the lenses on the parallax bar are in contact with the photographs, the photographs must be mounted on a plane surface. There is no way to tilt them to correct for the small tilts that occur in actual photography. Accordingly,

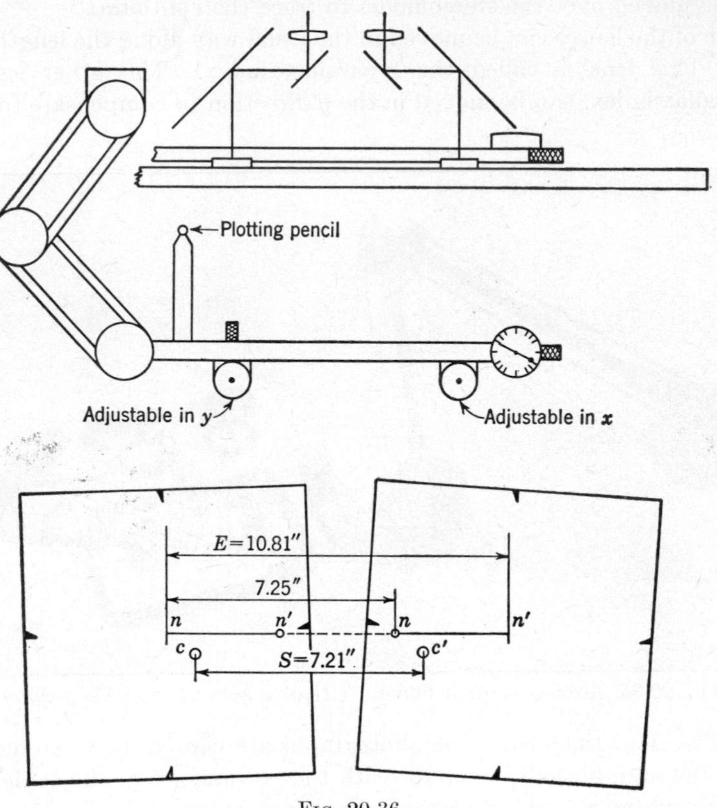

Fig. 20-36

to give correct results either the photographs must be rectified or a correction template must be constructed and applied. A photograph is rectified by photographing it at a certain angle so that the result will be the same as would have been obtained if it had been photographed exactly vertically at first. The process is described later.

20-49. Operational Procedure. Adjust the stereoscope and the parallax bar so that the index marks fuse comfortably under the stereoscope when the dial is near the center of its run (Fig. 20-36). Measure the distance between the dots (assume that this is 7.25 in.).

Align the photographs and adjust their separation so that the distance n between the two points is 7.25 in. as shown in Fig. 20-36. Measure E (10.81 in.) and S (7.21 in.).

S is the distance between the images of a single survey control point C of known elevation (121 ft).

Referring to Fig. 20-34,

$$b_c = p_1 + p_2 = E - S$$

or

$$b_c = 10.81'' - 7.21'' = 3.60''$$

The value of H_c is best determined by a measured base on the ground, laid out approximately at the perpendicular bisector of NN'.

Assume (1) ground length $L = 1342$ ft, (2) average image length on the two photographs = 4.31 in., (3) average elevation of the two ends of the base = 110 ft, and (4) $f = 8.25$.

Then by Eq. (20-16) the height of camera above base is

$$H_b = 8.25 \frac{1342}{4.31} = 2569'$$

Then

$$\begin{aligned} H_b &= 2569 \\ \text{Add} \quad l &= \ \ 110 \\ \text{Camera altitude} \quad H_0 &= 2679 \\ \text{Subtract elevation } C &= \ \ 121 \\ H_c &= 2558 \end{aligned}$$

It simplifies the work if a formula is found for Δp based on zero elevation rather than the elevation of C. This is developed below.

Compute the value of Δp_c for the difference in elevation from C to the zero contour. By Eq. (20-15)

$$\Delta p_c = b_c \frac{h_c}{H_c - h_c} = 3.60 \frac{-121}{2558 - (-121)} = -0.163''$$

Compute b_0, the b value for heights measured from the zero contour h_0. As shown in Fig. 20-34,

$$h_0 = no + cn' = cg$$

$$b_0 = b_c + \Delta p_c = 3.60'' - 0.16'' = 3.44''$$

The value of the parallax difference Δp_0 from the zero contour up to any required elevation h_0 can now be computed by Eq. (20-15):

$$\Delta p_0 = b_0 \frac{h_0}{H_0 - h_0} = 3.44 \frac{h_0}{2679 - h_0}$$

Compute the Δp_0 for the elevation of each survey control point and for each contour line required.

20-50. To Set the Parallax Bar. Set the y parallax dial at zero and adjust the drafting machine so that the two dots lie on the line joining the principal points. The parallax bar will now remain parallel to the X axis and therefore will correctly measure x parallax differences (the Δp_0 values).

20-51. To Set the Photographs. The alignment of the photographs according to the lines joining the principal points is not accurate enough for measuring parallaxes. The right-hand photograph must be adjusted so that the correct parallaxes will be read both at the top and the bottom of the pictures. Since a correction chart will still be necessary due to uncorrected errors in tilt, shrinkage, lens errors, etc., the photographs are usually placed so that all the corrections will have the same sign.

Choose two widely separated survey control points, one near the top, and the other near the bottom of the photographs. Set the x-parallax index at the Δp_0 value for one of the points plus 0.03 in. Place the left-hand dot over the image of this control point on the left-hand photograph, and slide the right-hand photograph parallel to the line of the principal points until the floating mark appears exactly on the ground on this control point. Pass a pin through the image of this point on the right-hand photograph and thus pin it down.

Set the x-parallax index at the Δp_0 value for the second control point plus 0.03 in. Place the left-hand dot on the proper image on the left-hand photograph and rotate the right-hand photograph until the floating mark appears on the ground at the control point. Permanently fasten the right-hand photograph in this position. With this arrangement the parallaxes will be very nearly the computed value plus 0.03 in.

20-52. To Construct the Correction Template. Find the dial reading for each survey control point. Subtract the Δp_0 for that point from this reading. This is the dial correction. Write it on the control point on the left photograph. Draw the contours of correction on this photograph as though the dial corrections were elevations.

20-53. To Plot the Map. Mark the control points. Draw the drainage pattern. Set the dial at the value of the Δp_0 for the contour to be drawn plus the correction in the vicinity. Draw the contour. Change the dial setting whenever the floating mark crosses a correction contour. Plot the planimetry.

20-54. Orthogonal Projection. As stated previously, the map is a perspective projection equivalent to (in this case) the left-hand photograph. Thus, each contour is drawn at a scale that depends on the height of the camera lens above it. If, for example, it is decided to plot all contours at the scale used for the zero contour, the size of the figure of each contour must be changed by the ratio

$$R = \frac{H_0 - h_0}{H_0}$$

For example, in the case described, the 100-ft contour would be changed in size by the ratio

$$R_{100} = {}^{2579}\!/\!_{2679} = 0.963$$

Similarly, the 200-ft contour would be changed in size by the ratio

$$R_{200} = {}^{2479}\!/\!_{2679} = 0.925$$

These are not large ratios and they might be disregarded. If they are to be followed, each contour should be redrawn at the calculated reduced scale, keeping the position of the principle point unchanged. This can be accomplished by a pantagraph, by projection, or by other graphical means. The planimetry must also be replotted according to the contour on which it lies. Often, only rudimentary planimetry is plotted with the contour lines. When the final details are traced, the tracing is then slightly shifted as required to agree with the planimetry already plotted.

TILTED PHOTOGRAPHS

20-55. Definition. Tilted photographs can be classified under three heads: near verticals, obliques, and high obliques. Near verticals are taken as nearly vertically as possible, and they usually have a tilt of less than 5 degrees. Obliques have large angle of tilt. High obliques include the horizon in the photograph.

20-56. Nomenclature and Geometry. Figure 20-37 represents a tilted photograph. The reference surface is a horizontal plane at any convenient height.

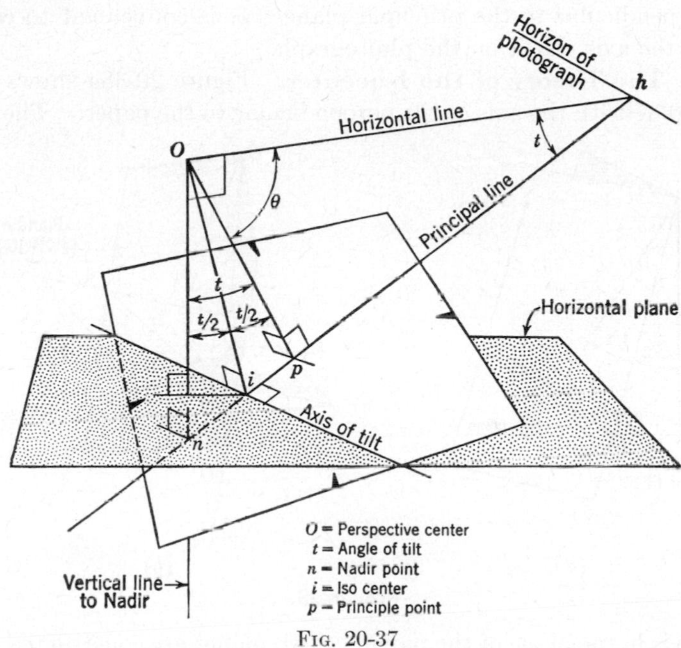

FIG. 20-37

On is a vertical line and therefore perpendicular to the reference surface. The point *n* is at the intersection of this line and the photograph plane. *Op* is perpendicular to the photograph and therefore passes through *p*, the principal point, that is established by the camera calibration and shown by the fiducial marks. The plane *Onp* is the tilt plane. It is called the **principal plane.** Since it contains perpendiculars to both the photograph plane and the horizontal it is perpendicular to both. The angle *nOp* is evidently the angle of tilt *t*. The **principal line** *nph* is the intersection of the principal plane and the photograph.

The horizon of the photograph is the perspective projection, onto the plane of the photograph, of the line of intersection (at infinity) of all horizontal planes. The horizon is therefore the intersection of a horizontal plane through O, with the plane of the photograph.

20-57. Construction. The line Oi is constructed so that it is the bisector of the angle t. The intersection of line Oi with the photograph is the isocenter i. Then

$$pi = f \tan \frac{t}{2} \tag{20-17}$$

A horizontal plane (shaded in the figure) is passed through the point i. The line of intersection of this plane with the photograph plane is perpendicular to the principal plane, since it is the intersection of two planes both perpendicular to the principal plane. It is convenient to consider this line the axis of tilt on the photograph.

20-58. The Theory of the Isocenter. Figure 20-38a shows a view of Fig. 20-37 with the axis of tilt perpendicular to the paper. The princi-

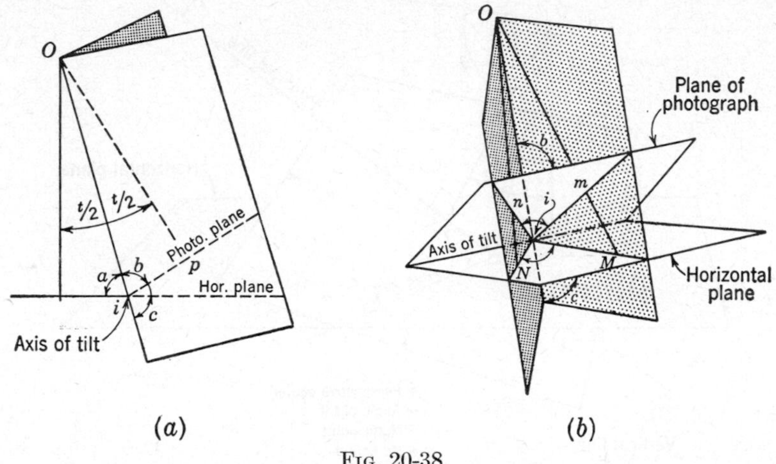

FIG. 20-38

pal plane is in the plane of the paper. Two planes are constructed so that they form a dihedral angle whose edge coincides with the line Oi. The planes can be in any desired direction. From similar triangles, angles a, b, and c are equal. Therefore the photograph plane and the horizontal plane make equal angles with the edge of the dihedral angle.

Figure 20-38b shows another view of Fig. 20-38a. M and N represent object points on the horizontal plane, contained in the sides of the dihedral angle, and m and n are their projections on the photograph and therefore contained in the sides of the dihedral. Since the photograph plane and the horizontal plane make equal angles with the edge of the dihedral, the angle nim equals the angle MiN by symmetry.

Thus lines drawn to images from the isocenter form angles that are equal to the corresponding angles on a horizontal plane, independent of tilt.

THE RADIAL LINE PLOT

20-59. Purpose. The radial line plot, often called **phototriangulation** is an extremely useful means of determining the true horizontal positions of objects shown on near-vertical photographs. The method can be applied to almost any photogrammetric process. Its use greatly reduces the required density of horizontal control.

20-60. Theory. It has been shown that, on exactly vertical photographs, a change of height of an object will cause its image to shift along a line that radiates from the principal point. Therefore, angles measured on the photograph at the principal point are true horizontal angles independent of the elevations of the ground objects or the altitude of the camera. It has also been shown that, on tilted photographs, angles measured at the isocenter are true horizontal angles independent of tilt, provided that all objects photographed have the same elevation.

It follows that, in near-vertical photography, in which the isocenter is near the principal point, angles measured in the vicinity of these points are very nearly equal to the true horizontal angles, independent of tilt or elevation. This principle is the basis of the radial line plot.

Figure 20-39 shows two near-vertical photographs on which images have been chosen that appear on both. The points n and n' chosen are often not exactly at the principal points but are well-defined images near the center of the photographs. The photographs are placed so that the lines nn' are in line with each other. Radial lines are drawn through the selected images as shown. Since the angles between the radial lines are true, the intersections of pairs of corresponding lines give the true positions of the selected points with respect to each other, independent of image shifts due to tilt or elevation.

Figure 20-39*b* shows that, when the separation of the photographs is changed, the points remain at their true relative positions but the scale is changed. It is evident that if two of the points chosen were horizontal control points, the photographs could be shifted until the proper intersections coincided with the plotted positions of the control points. This would establish scale and azimuth. Other photographs can be added and the method extended indefinitely. By adjusting the photographs slightly, all intersections representing horizontal control points can be placed in their proper positions. Once the radial lines have been made to fit the control, the true horizontal position can be found for *any point* by drawing intersecting radial lines. A number of practical methods of utilizing this system have been devised.

20-61. Points Selected on Photographs. In common with other photogrammetric methods, nine points are usually chosen for control on each photograph. These consist of a point near the principal point and

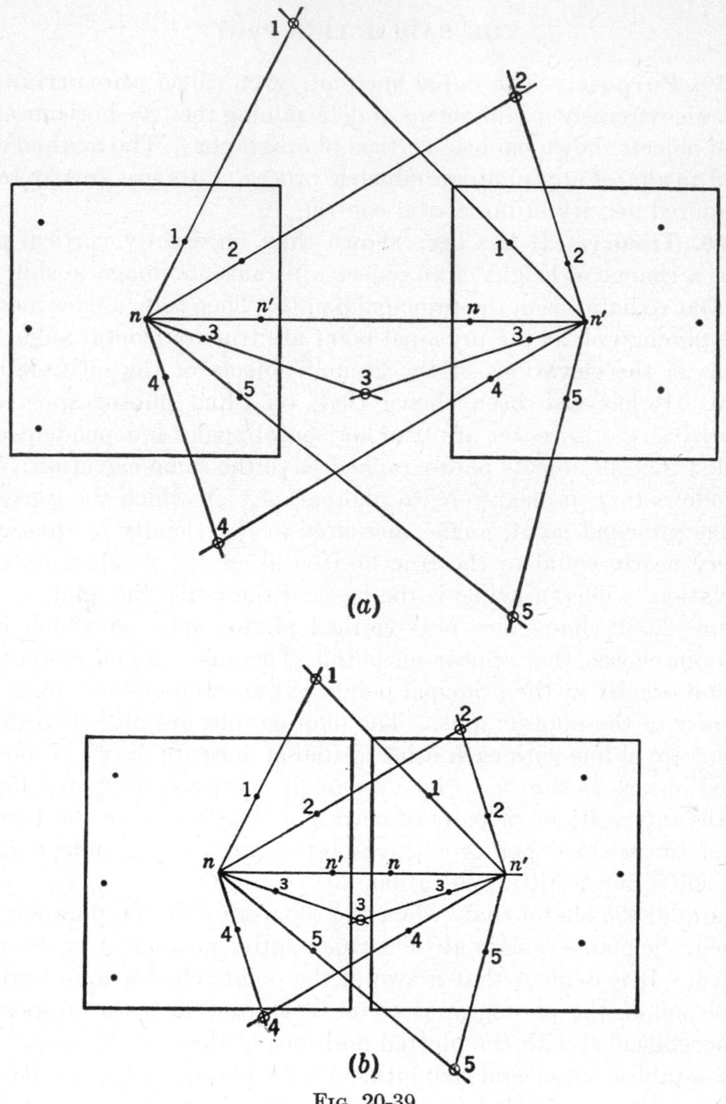

FIG. 20-39

a point near the center of each edge that forms the side of the flight strip. These three, with the corresponding images from the adjacent photographs, make up the nine points.

20-62. The Hand-template Method. Radial lines are drawn through each point selected and a tracing is made of these lines for each photograph. After the horizontal control is plotted, the first two tracings at the beginning of the flight strip are adjusted to the control. The strip is then built up by fitting successive tracings until the next horizontal control points are reached. There, any necessary corrections to scale and azimuth are made by slight adjustments of all the tracings. Adjacent flight strips are, of course, tied together by the points common to both. The final adjustment, when all of the flight strips are in place, gives a very accurate position to all the points. Any of these points can be used, therefore, as horizontal control points.

20-63. The Slotted-template Method. The hand-template method can be greatly facilitated by matching and adjusting the templates

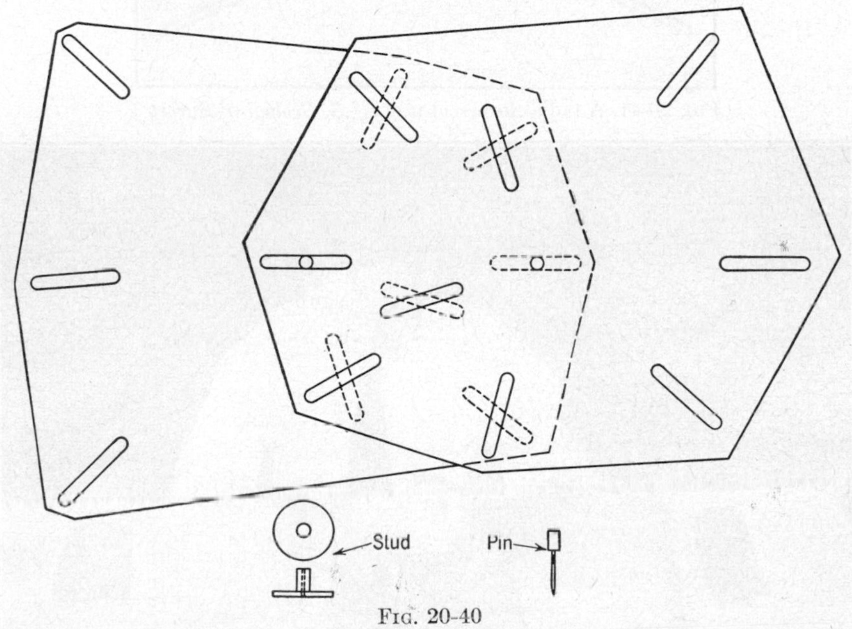

FIG. 20-40

mechanically rather than visually. The slotted-template method is designed for this purpose. Instead of tracings, heavy cardboard sheets are cut with slots that are aligned with the radial lines by a special cutter (see Fig. 20-40). Studs, like the one shown in the figure, are placed in the slots. The slots keep the studs in their proper relative positions when the scale is changed as the templates are moved together or apart. A complete set of slotted templates can be made up for the whole area. When these are assembled, all the studs will be in their true position. Were it not for friction, the whole assembly (or laydown) could be changed in scale. The actual procedure for assembling the templates is described in the next paragraphs.

First, a stud is glued over each horizontal control position. The pin shown facilitates this operation. A single flight strip is then built up

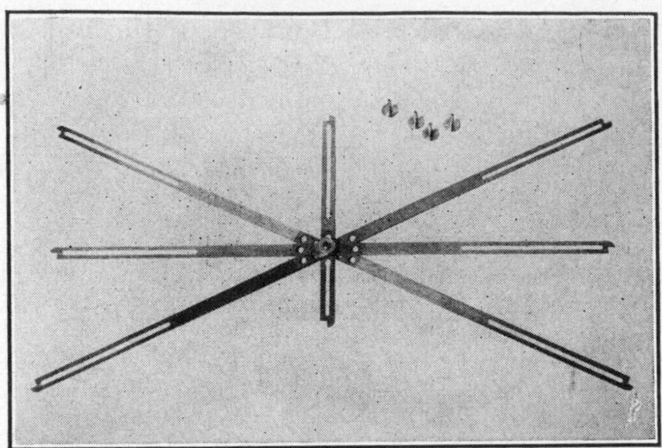

FIG. 20-41. A radial intersector. (*U.S. Geological Survey.*)

FIG. 20-42. Adjusting a radial intersector to fit the selected points on a photograph. (*U.S. Geological Survey.*)

beginning at one end. When the next horizontal control studs are reached, the whole strip is adjusted to fit. Usually alternate flight strips are first compiled and then the intermediate strips are added. When the templates are in place, the horizontal positions of the studs are pricked with the pins.

Figure 20-41 shows a device called a **radial intersector.** It is built of slotted metal arms that can be adjusted to any arrangement of angles as shown in Fig. 20-42. Radial intersectors are often used instead of the cardboard templates. Figure 20-43 shows a radial intersector laydown fitted to horizontal control. Control points are shown by small triangles.

FIG. 20-43. A laydown of radial intersectors. Horizontal control positions, to which it is fitted, are shown by small triangles. (*U.S. Geological Survey.*)

20-64. Application of the Radial Line Plot. The radial line plot can be used by itself or to improve or simplify almost any photogrammetric procedure. Its application to one of the most accurate of all processes, the Brock-Weymouth process, is an excellent example. In this process it is used to determine horizontal positions and also to obtain the first approximations to the altitude of the camera.

20-65. The Brock-Weymouth Process. A radial plot is established as described. Once the laydown is complete, the radial distances for each photograph are scaled to each photocontrol point of known elevation. The altitude is sometimes computed for each point by Eq. (20-12).

$$H = \frac{Rf}{p + d} + h$$

The average for all the points is determined. This practically eliminates the effect of tilt in the altitude determination. However, an approximate altitude is usually used.

The Δp_c values are then measured on the stereopairs in a very accurate stereocomparator. By comparing these with the values computed by Eq. (20-15),

$$\Delta_p = \frac{bp}{H - h}$$

the first approximation to the tilts and scales can be determined on precomputed graphs.

The two photographs (in the form of diapositives) are placed in **rectifiers** (later described) where they are projected on a ground-glass screen

FIG. 20-44. The Brock rectifiers. They are among the earliest rectifiers ever built and they give extremely high accuracy. (*Aero Service Corp.*)

(see Fig. 20-44). The tilts and scales are set according to the first approximation and adjusted until the x parallaxes (the Δp values in the X axis) agree with the values computed by Eq. 20-15 and the y parallaxes are eliminated. The ground glass is replaced by an emulsified plate and an exposure is made. This produces a negative that has the same geometric characteristics as a photograph taken of the ground at an established altitude with the camera exactly vertical. The contours and planimetry can then be drawn in a stereocomparator without approximations. Figures 20-45 and 20-46 show this operation. The resulting

FIG. 20-45. Brock stereometer. (*Aero Service Corp.*)

FIG. 20-46. Drawing contours on the Brock stereometer. (*Aero Service Corp.*)

perspective projection is reduced to orthographic projection by tracing each contour at its proper scale in a variable scale projector.

Note that the radial line plot gives the actual fundamental data for the computation of horizontal position and sometimes for the tilt.

MOSAICS

20-66. Simple Mosaics. Since vertical photographs look so much like the ground, a set can be fitted together to form a maplike photograph

FIG. 20-47. Making a mosaic. (*Aero Service Corp.*)

of the ground. As the photographs usually overlap, only the best part of each photograph need be used. Since they are taken at slightly varying altitudes and they contain tilts, they often do not fit each other very well and it is best to rephotograph them before they are used to bring them to a desired scale and to eliminate some of the tilt. Approximate applications of some of the radial-line-plot principles can be used to accomplish this.

20-67. Controlled Mosaics. When the photographs are carefully placed so that the horizontal control points agree with their previously plotted positions, a controlled mosaic is obtained. When this is photographed to form a single picture, a nearly perfect planimetric map results that shows a wealth of detail that could not possibly be included in a conventional map. While horizontal distances are not exactly correct in every case, on the whole they are quite accurate and the mosaic can be used for many purposes more effectively than a map.

Making controlled mosaics is an art. Figure 20-47 shows the process. The part of each photograph that is to be used is torn out so that the thin edges will lie flat on the previous photograph that has been placed. Figure 20-48 shows a finished product.

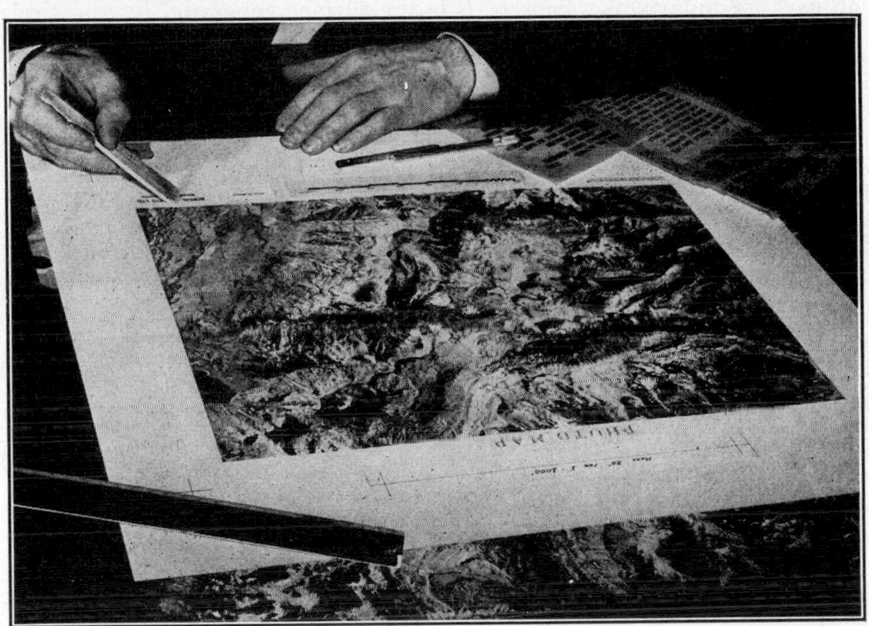

FIG. 20-48. A completed mosaic. (*Aero Service Corp.*)

RESECTION AND ORIENTATION IN SPACE

20-68. Basic Theory. It can be assumed that for any photograph the perspective distance f and the position of the principal point p are known from the calibration of the camera. When these data are available, the position and orientation of the photograph in space with respect to ground coordinates can be computed, when the elevations and horizontal positions are known of at least three properly arranged ground points that can be identified on the photograph. Other combinations of these nine coordinates are sufficient.

From the camera data and the measured positions of the images of the three ground points, it is possible to compute the three plane angles measured at O between each pair of rays to the ground points. In Fig. 20-49, two of these angles, A and B, are shown. The length f is known and the distances pd, pe, and pg can be measured. Then Od and Oe can be computed, *de* can be measured, and angle *A* can be computed, as the sides of the triangle *Ode* are all known. Similarly angles *B* and *C* (not shown) can be computed.

Angles *A*, *B*, and *C* completely define the position and orientation of the photograph. This is shown in Fig. 20-50. The locus of *A* is the surface generated by the perimeter of the circle *DEO* rotated about *DE* as an axis. Similarly the locus of *B* is the surface generated by the circle *OEG* rotated about *EG*. These two surfaces intersect in a curved line. When the locus of *C* is generated, the three surfaces must meet above

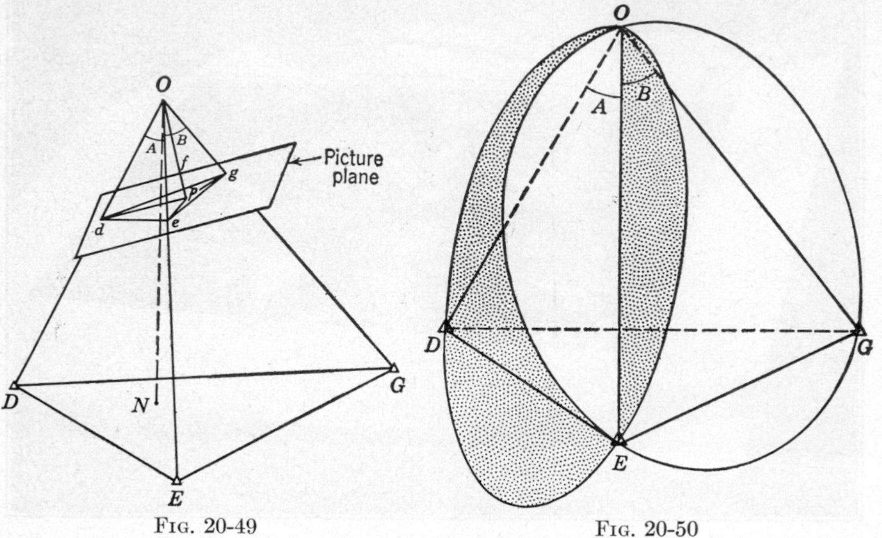

FIG. 20-49

FIG. 20-50

the ground at a single point *O*, unless, by chance the arrangement is indeterminate.

20-69. Mathematical Solutions. There are many ways of computing the position of this point and of the tilt and the direction of tilt of the photograph. In photogrammetric practice, space resection is nearly always accomplished by some method of physically recreating the actual geometric relationships and fitting these to the control positions, as described in previous sections of this chapter. However, one method is outlined here to demonstrate how computational methods can be applied.

20-70. Example of a Computational Method. Lay out the angles *A*, *B*, and *C* as shown in Fig. 20-51. At any convenient scale, mark off *OH* equal to *OD* (Fig. 20-49), by estimation. At the same scale, mark off *HI* equal to *DE*, *IJ* equal to *EG*, and *JK* equal to *GD*. *KO* should then be equal to *OH*. If it is not, different estimates for the length of *OH* should be made and the process repeated, as shown, until the condition is satisfied.

When a good graphical solution is found, the same process can be carried out by computation, since in each triangle two sides and the angle opposite one of them are known. By trial and error, final values can be determined for the lengths in Fig. 20-49 of OD, OE, and OG, and hence the angles at D, E, and G. The coordinates of N (the nadir on the plane DEG) and the length ON can then be computed directly by plane and spherical trigonometry. The tilt and the direction of tilt are then easily determined.

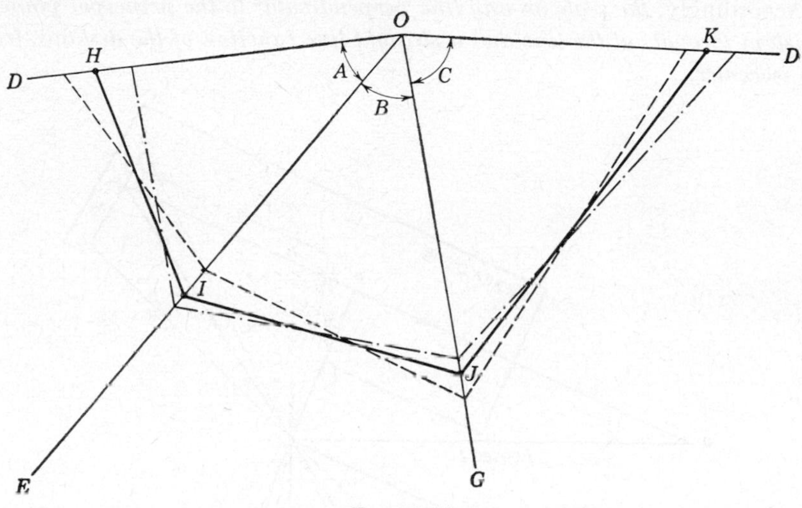

FIG. 20-51

20-71. Scale in Tilted Photographs. In certain photogrammetric processes, the scale relationships inherent in tilted photographs must be understood. These relationships are developed in the following paragraphs.

In Fig. 20-52, triangles VPO and vmO are in the principal plane; lines POp and Non are perpendicular to the picture plane and the ground plane, respectively; line IOi passes through the isocenter i; vO and VO are parallel to the ground plane and the picture plane, respectively; and t is the angle of tilt.

Let S_m be the scale of the photograph on a line perpendicular to the principal plane at any point m. Then

$$S_m = \frac{ab}{AB} = \frac{Om}{OM} \tag{20-18}$$

In the similar triangles vmO and VOM,

$$\frac{Om}{OM} = \frac{vm}{VO} = \frac{vi + im}{VO}$$

Triangle vOi is isosceles, since its base angles are equal to $90° - t/2$. Therefore

$$vi = f \csc t$$

Substituting, $$S_m = \frac{f \csc t + y_m}{H \csc t} = \frac{f + y_m \sin t}{H} \tag{20-19}$$

when y_m is measured *away from* the horizon v.

Accordingly, *the scale on any line perpendicular to the principal plane is constant throughout the line and a straight line function of the distance from the isocenter.*

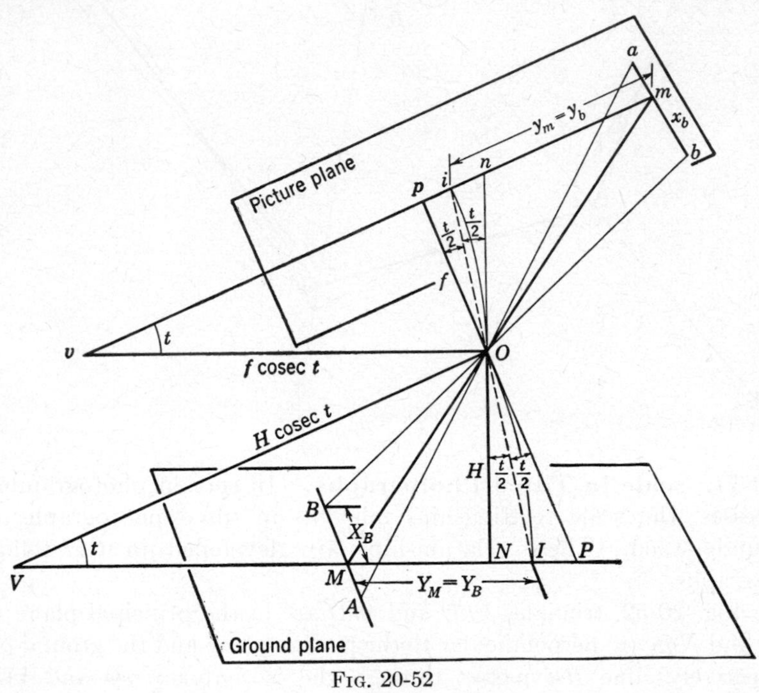

FIG. 20-52

Let σ_m represent a special kind of scale defined in the following way (see Fig. 20-52):

$$\sigma_m = \frac{y_m}{Y_M}$$

Let $$f' = f \csc t$$
$$H' = H \csc t$$

and let $$Y = Y_M$$
$$y = y_m$$

In the similar triangles vmO and VOM

$$\frac{H' - Y}{H'} = \frac{f'}{f' + y}$$

$$Y = H' - \frac{H'f'}{f' + y}$$

$$\frac{y}{Y} = \frac{f' + y}{H'}$$

By substitution
$$\frac{y_m}{Y_M} = \frac{f \csc t + y_m}{H \csc t}$$

$$\sigma_m = \frac{f + y_m \sin t}{H} \tag{20-20}$$

From Eq. (20-19)
$$\sigma_m = S_m$$

Accordingly, the scale of a length measured parallel to the principle plane from a point opposite the isocenter, in reference to the length of the corresponding line on the ground plane, is equal to the scale of the perpendicular to the principal plane at the terminus of the length. Thus

$$S_m = \sigma_m = \frac{f + y_m \sin t}{H} \tag{20-21}$$

Let ΔS represent the rate of change of S or σ along the principle line with respect to y:

$$\Delta S = \frac{dS}{dy} = \frac{\sin t}{H} \tag{20-22}$$

From Eq. (20-21), the scale S_i at the isocenter is expressed

$$S_i = \frac{f}{H} \qquad \text{and} \qquad H = \frac{f}{S_i} \tag{20-23a}$$

Substituting,
$$\sin t = \frac{f\,\Delta S}{S_i} \tag{20-23b}$$

20-72. Relation between Coordinates on the Ground and Coordinates on the Photograph. Let rectangular coordinate systems be arranged on the photograph and the ground plane as shown in Fig. 20-52. For any point B and its image b, the coordinates are the following: on the ground plane, X_B and Y_B, and on the photograph x_b and y_b. From Eq. (20-20)

$$\frac{y_b}{Y_B} = \frac{f \csc t + y_b}{H \csc t} \tag{20-24}$$

Let
$$Y'_B = \frac{f}{H} Y_B \qquad \text{or} \qquad Y_B = Y'_B \frac{H}{f}$$

Substituting,
$$\frac{y_b}{Y'_B}\frac{f}{H} = \frac{f \csc t + y_b}{H \csc t}$$

Solving for y_b,

$$y_b = \frac{f(\csc t)Y'_B}{f \csc t - Y'_B}$$

or

$$y_b = \frac{f(\csc t)Y_B \dfrac{f}{H}}{f \csc t - Y_B \dfrac{f}{H}} \tag{20-25}$$

Let

$$X'_B = X_B \frac{f}{H}$$

From Eq. (20-21)

$$\frac{x_b}{X_B} = \frac{y_b}{Y_B} \qquad \text{and} \qquad \frac{x_b}{X'_B} = \frac{y_b}{Y'_B}$$

From Eq. (20-24), as modified above,

$$\frac{x_b}{X'_B} = \frac{y_b}{Y'_B} = \frac{f \csc t}{f \csc t - Y'_B}$$

$$x_b = \frac{f(\csc t)X'_B}{f \csc t - Y'_B} \tag{20-26}$$

or

$$x_b = \frac{f(\csc t)X_B \dfrac{f}{H}}{f \csc t - Y_B \dfrac{f}{H}} \tag{20-27}$$

ANDERSON-RIHN[1] METHOD

20-73. Principles Involved. In some photogrammetric processes, it is occasionally necessary to solve a space-resection problem. In the United States, the Anderson-Rihn method is usually employed. The method is approximate and it is used chiefly to determine the value and the direction of tilt of near-vertical photographs. The method depends on the graphical determination of the **scale point** of certain lines. The concept of the scale point was first introduced by Anderson and is explained below.

In Fig. 20-53, *ab* represents the image of *AB*. The scale of *ab* (S_{ab}) is defined thus:

$$S_{ab} = \frac{ab}{AB} = \frac{l}{L} \tag{20-28}$$

The point S_p is the scale point, the point at which the scale perpendicular to the principal plane is equal to S_{ab}. The length y which locates the scale point is found as follows.

[1] R. O. Anderson and Jack Rihn.

In Fig. 20-53 construct $a'b' = d$ parallel to the ground plane so that $d = l$. Construct r through a and parallel to AB. Then, by similar

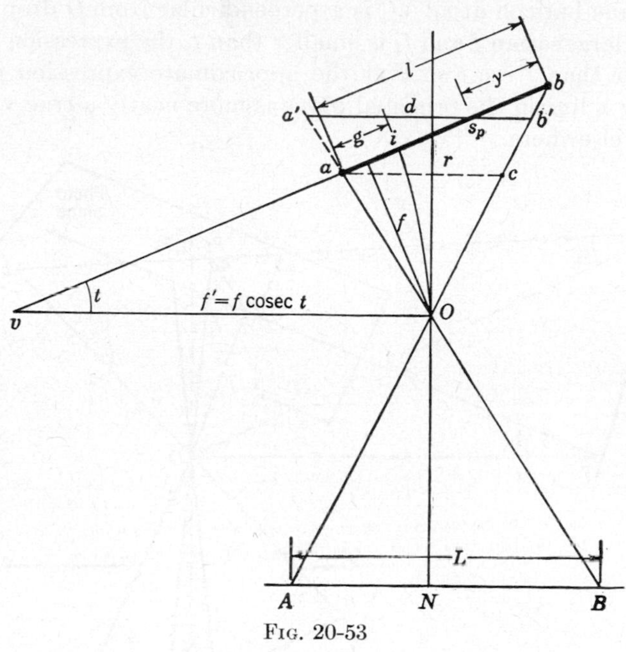

FIG. 20-53

triangles,

$$\frac{\overline{ac}}{\overline{a'b'}} = \frac{\overline{aO}}{\overline{a'O}} = \frac{\overline{av}}{\overline{vS_p}} \qquad \text{and} \qquad \frac{\overline{ac}}{\overline{ab}} = \frac{\overline{vO}}{\overline{vb}}$$

Hence

$$\frac{r}{d} = \frac{f' - g}{f' - g + l - y} \qquad \text{and} \qquad \frac{r}{l} = \frac{f'}{f' - g + l}$$

Since $d = l$ by construction, these may be equated:

$$\frac{f' - g}{f' - g + l - y} = \frac{f'}{f' - g + l}$$

Solving for y,

$$y = \frac{g(f' - g + l)}{f'} \tag{20-29}$$

When the tilt is small, f' has a very large value so that the coefficient $f' - g + l$ is very nearly equal to f', and hence

$$y = g \qquad \text{approximately} \tag{20-30}$$

Thus the scale point S_p is at the same distance from one end of the line ab as the isocenter is from the other end.

When a line is not in the principal plane, the same geometry exists, as shown in Fig. 20-54. ab represents any line. Extended, it strikes the picture plane horizon at v_1. f_1 is a perpendicular from O dropped to ab. Since f_1 is larger than f and t_1 is smaller than t, the expression $f_1 \csc t_1$ is even larger than $f \csc t$ so that the approximate expression previously derived for a line in the principal plane is more nearly a true value when the line is elsewhere.

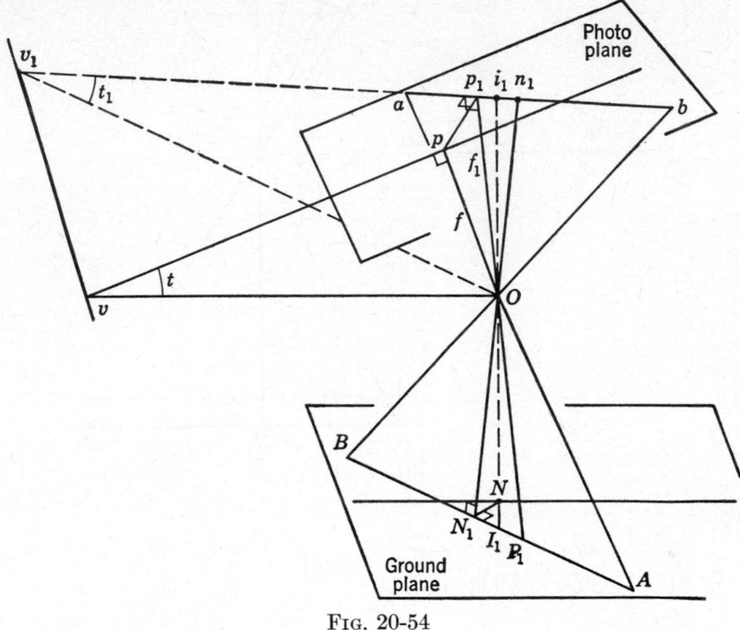

FIG. 20-54

The point p_1 (the foot of a perpendicular from O) is at the foot of a perpendicular dropped from p. This can be demonstrated as follows:

The line f_1 is perpendicular to ab by construction. It must therefore lie in a plane perpendicular to ab. Such a plane is parallel to, or contains, every line perpendicular to the photograph plane. Therefore, since it contains O, it must contain the line f. Its trace in the photograph plane pp_1 must be perpendicular to ab.

When the tilt is small, i_1 can be very nearly found by dropping a perpendicular from i (not shown), as iO will not be far from perpendicular to the photograph plane.

Thus, to find the scale point S_p of any line on a photograph, drop a perpendicular to the line from the isocenter. This will fall approximately at i_1. Then find the point on the line that is as far from one end of the line as the foot of the perpendicular is from the other end of the line. Note that if the line does not extend far enough to reach the perpendicular, the line should be extended. In this case the scale point is beyond the other end of the line, as can be shown by a demonstration similar to that shown in Fig. 20-53. In either case, the scale point is the same distance from the center of the line as the foot of the perpendicular but on the other side of the center.

Robert Singleton, who reviewed this book, suggests an excellent proof and analysis of the Anderson-Rihn method. It is given in the next paragraphs.

Let ab be any line on the photograph, as in Fig. 20-54. Let a have the coordinate y_a and b the coordinate y_b (see Sec. 20-72). Then, by similar triangles,

$$\frac{ab}{AB} = \frac{y_b - y_a}{Y_b - Y_a} \tag{20-31}$$

It is to be remembered that the y coordinates are measured from the isocenter in the direction that is away from the horizon. From Eq. (20-20)

$$y_a = Y_a \frac{f + y_a \sin t}{H} \tag{20-32}$$

$$y_b = Y_b \frac{f + y_b \sin t}{H} \tag{20-33}$$

Substituting these values of y_a and y_b in Eq. (20-31),

$$\frac{ab}{AB} = \frac{f}{H} + \frac{\sin t}{H} \frac{Y_b y_b - Y_a y_a}{Y_b - Y_a} \tag{20-34}$$

Let p be the scale point. Then from Eq. (20-21)

$$\frac{ab}{AB} = \frac{f}{H} + \frac{\sin t}{H} y_p \tag{20-35}$$

Equating Eq. (20-35) and Eq. (20-34),

$$y_p = \frac{Y_b y_b - Y_a y_a}{Y_b - Y_u}$$

Subtracting from y_b this value of y_p,

$$y_b - y_p = -Y_a \frac{y_b - y_a}{Y_b - Y_a}$$

From Eq. (20-31)

$$y_b - y_p = -Y_a \frac{ab}{AB}$$

If there were no tilt, $-Y_a\, ab/AB$ would be equal to $-y_a$, which is the value used. Since the tilt is small, very little error is introduced by this approximation.

Singleton has developed the following approximate expression for the relative scale error:

$$\frac{dS_p}{S_p} = \frac{\sin^2 t}{f^2} y_a y_b$$

Let $y_a = 6.36$ in. $y_b = -6.36$ in. (so that the line is the diagonal of a 9×9 photograph), $f = 6$ in. and $t = 5°$. Then the maximum relative scale error for vertical 6-in. pictures is 0.0085, or less than 1 per cent.

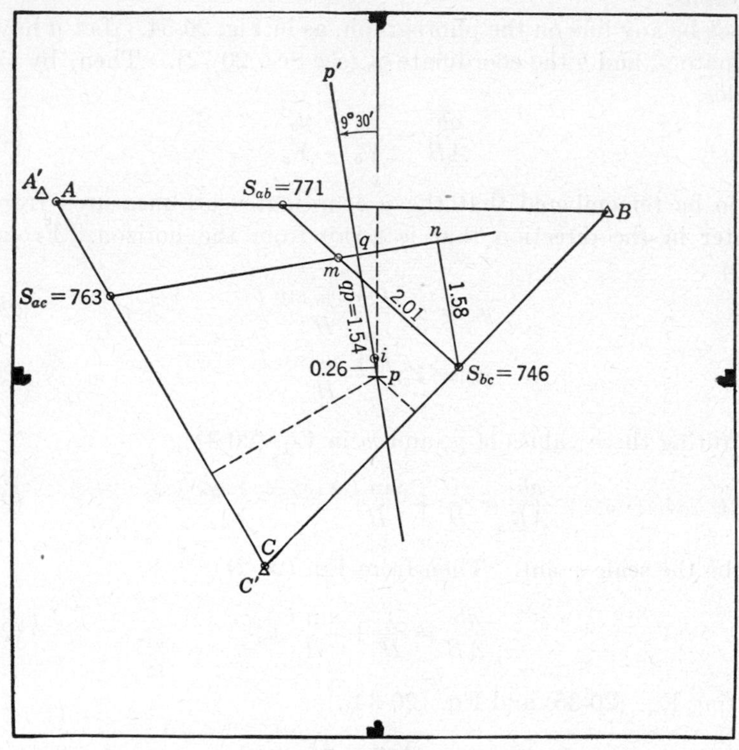

FIG. 20-55

20-74. Procedure. Figure 20-55 shows a photograph in which A', B, and C' are control points of known elevation and position. The computation is shown in tabulated form in Table 20-1. The elevation of the lowest point B is used for datum and the values in Column a are computed accordingly. Column b gives the radial distances $p + d$ measured on the photograph from p to each point.

Proceed with the following steps:

1. Compute approximate height of camera station above B by Eq. (20-12). Use for R the survey length of the line nearest the principle point, in this case BC', and for $p + d$ its measured length on the photograph before correction (6.17). Use the average height of the ends of this line above B for h.

2. Compute the elevation displacement d for A and C by Eq. (20-11). Height above B is used for h, $p + d$ is the radial distance to the point measured on the photograph, and H is the height of the air station above B. The displacements are 0.17 in. for A' and 0.03 in. for C'.

TABLE 20-1. COMPUTATION
f = 6.000 in.

		a	*b*	*c*	*d*	*e*	*f*
Point	Elev., ft	Ht. above B, ft	Radial on photo, in.	Displacement ab/H, in.	Survey length, ft	Cor. photo length, in.	Scale, $10^6\ e/d$
A	500	300	4.56	0.17			
					8975	6.92	771
B	200	0	....	0			
					8233	6.14	746
C	300	100	2.76	0.03			
					6872	5.24	763
A							

1. $H = \dfrac{Rf}{p+d} + h = \dfrac{8233(6)}{6.17} + 50 = 8056$

2. $d = \dfrac{h(p+d)}{H} = \dfrac{300(4.56)}{8056} = 0.17$

$= \dfrac{100(2.76)}{8056} = 0.03$

3. S_{bc} to $m = \dfrac{763 - 746}{771 - 746}(S_{bc} \text{ to } S_{ab}) = \dfrac{17}{25} 2.96 = 2.01$

4. $\Delta S = \dfrac{763 - 746}{1.58} = 10.8$

5. $S_i = 763 - 1.54(10.8) = 746$

6. $\sin t = \dfrac{f\Delta S}{S_i} = \dfrac{6(10.8)}{746} = 0.0869$

7. $t = \text{arc} \sin 0.0869 = 4°59'$

8. $pi = f \tan \dfrac{t}{2} = 0.262$

9. $S_i = S_{ac}(qi)\Delta S = 763(1.54 - 0.26)10.8 - 749$

10. $H = \dfrac{f}{S_i} = \dfrac{6}{749(10^{-6})} = 8011$

Set off the computed displacements toward the principal point and mark the new positions A and C. These are the positions that would have been photographed if A and C had had the elevation of B. Draw AB, AC, and BC. Drop perpendiculars from the principal points to each. (The principal point must be used instead of the isocenter.) Mark the scale points on each line.

Compute the scales. Column d gives the survey lengths of the lines (usually computed from the ground coordinates of the points) and Column e gives the lengths of the corrected lines as measured on the photograph. Column f gives the scales of the lines to three significant figures. They are multiplied by some value (in this case 10^6) to eliminate decimals. The scales are therefore given in inches per 1,000,000 ft.

Draw a line connecting the scale points for the largest and the smallest scales S_{ab} and S_{bc}. Equations (20-18) and (20-22) show that the rate of change of scale is constant along any line in the picture plane; therefore, the point m on this line where the scale is equal to S_{ac} can be found by proportion.

3. The length S_{bc} to S_{ab} measured on the photograph is 2.96 in. Compute S_{bc} to m as shown. It is 2.01 in. Place m accordingly.

A line drawn from S_{ac} to m will, then, have a constant scale and must therefore be perpendicular to the principal plane and parallel to the axis of tilt. Draw the line pp' perpendicular to $S_{ac}m$ to give the position of the principal plane.

4. The rate of change of scale ΔS along pp' is the same as on any line parallel to it. Draw the perpendicular $S_{bc}n$ to $S_{ac}n$. Measure it (1.58) and compute ΔS (10.8). This is equivalent to the rate of change of scale per inch along the line pp'.

5. Compute the scale S_i at the principal point. Measure qp (1.54). Then $S_i = S_{ac} - \overline{qp}\Delta S$.

6. Compute t by Eq. (20-23a). $t = 4°59'$.

7. Compute pi by Eq. (20-17) giving 0.262 in. Plot the isocenter i. The isocenter is always on the principal plane and in the direction of increasing scale from the principal point (see Fig. 20-53).

8. Compute a better value for the altitude. First compute a better value for S_i using the distance from q to the isocenter now determined. This gives 749. Scale the angular direction of the principal plane.

20-75. Errors. This example was set up artificially from coordinates computed by Eqs. (20-25) and (20-27). The tilt, direction of tilt, and the altitude chosen for this computation are the correct values in this case. A comparison between these values and the values determined is given below.

	True values	*Values determined*
Tilt.	4°35′	4°59′
Direction of tilt.	14°40′	9°30′
Altitude, ft.	8000	8011

The results can be improved by repeating the procedure, using the better value for the altitude ($H = 8011$) to compute the elevation displacements (Computation 1) and the position determined for the isocenter for controlling the scale points.

THE PERSPECTIVE GRID

20-76. Horizon on Photograph. When only planimetric detail is required and the topography is relatively flat, large areas can be mapped very rapidly by utilizing high-oblique photography. Since the actual horizon appears in a high-oblique photograph, the theoretical perspective horizon can be placed on the photograph by correcting the image of the actual horizon; the corrections are for the height of the airplane given by the altimeter reading, the effect of earth curvature, and the refraction caused by the atmosphere. When the perspective horizon has been placed on the photograph, this, with the altitude, gives complete data for mapping.

This principle has been utilized in a number of mapping systems and various methods have been developed to transfer topographic details to the map. One of these methods utilizes the perspective grid. The perspective grid method was developed by the Canada Topographical Survey[1] and has been used with considerable success. An understanding of the perspective grid is of importance in the study of photogrammetry.

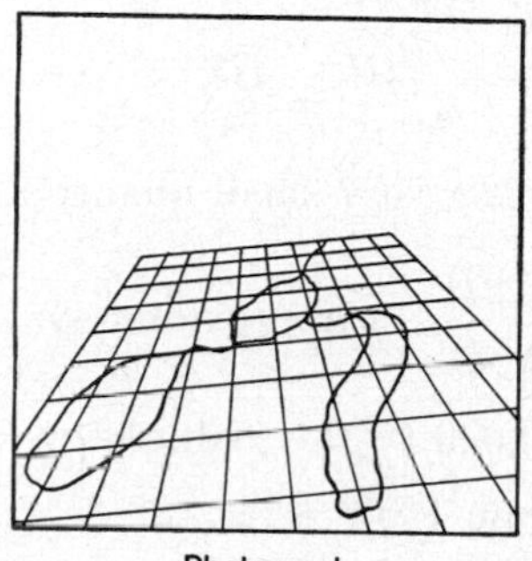

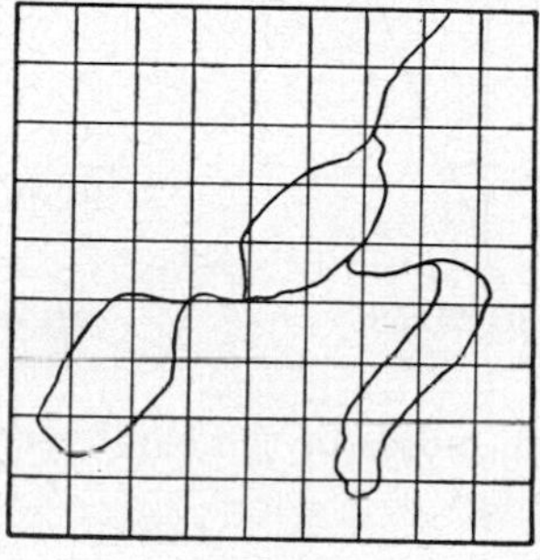

FIG. 20-56. Once the grid is placed on photograph, the map details can be drawn by eye.

20-77. Principle. A perspective grid (as the term is used in photogrammetry) is a representation, on the plane of the photograph, of a rectangular grid on the ground. When such a grid is placed over the photograph, the details of the photograph can be transferred by eye to their proper positions with respect to the corresponding rectangular grid laid out at map scale, as seen in Fig. 20-56.

20-78. To Construct the Grid. Figure 20-57 illustrates a high-oblique photograph with the image of the actual horizon near the top. The line *ab* is drawn tangent to the curved horizon line and equally spaced from it at its two ends. Such a line is perpendicular to the direction of gravity and therefore is perpendicular to the principal plane. The principal line $h'p$ is drawn perpendicular to *ab* through the principal point *p*. The perspective horizon is drawn parallel to *ab* at the proper distance above it. This distance is found from the considerations stated in the next paragraph.

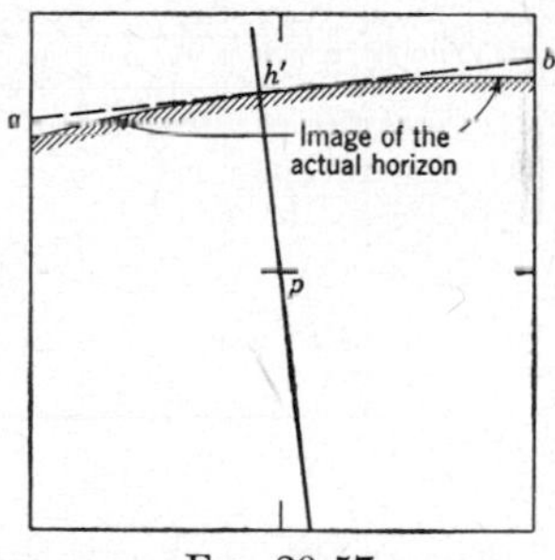

FIG. 20-57

[1] Topographic Survey Bulletin 62, 1932.

Figure 20-58 shows the geometry involved without allowance for refraction. The air station is at S at an altitude of H above the earth's surface. α' is the angle of depression of the horizon and r is the radius of the earth. From the figure

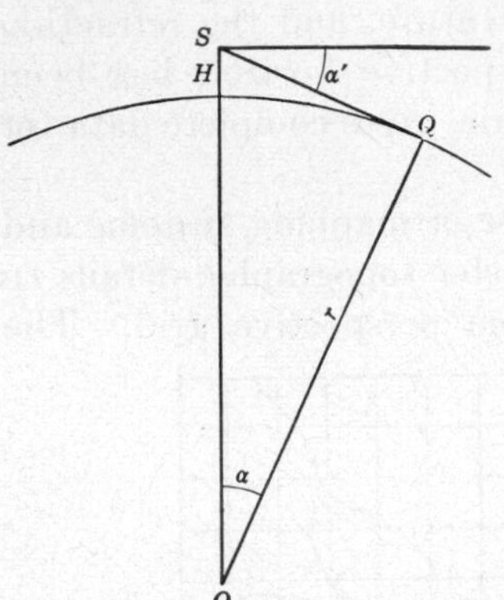

FIG. 20-58

$$SO = r + H$$

$$\tan \alpha' = \frac{SQ}{r} = \frac{\sqrt{(r + H)^2 - r^2}}{r}$$

$$= \sqrt{\frac{2H}{r} + \frac{H^2}{r^2}}$$

Since H^2/r^2 is a very small quantity,

$$\tan \alpha' = \sqrt{\frac{2H}{r}} \quad \text{approximately} \tag{20-36}$$

Taking the radius of the earth as 21,000,000 ft, this reduces to

$$\tan \alpha' = 0.000309 \sqrt{H}$$

$$\alpha' = 64'' \sqrt{H}$$

When reduced for refraction,

$$\alpha = 60'' \sqrt{H}$$

where α is the angle between the perspective horizon and the true horizon.

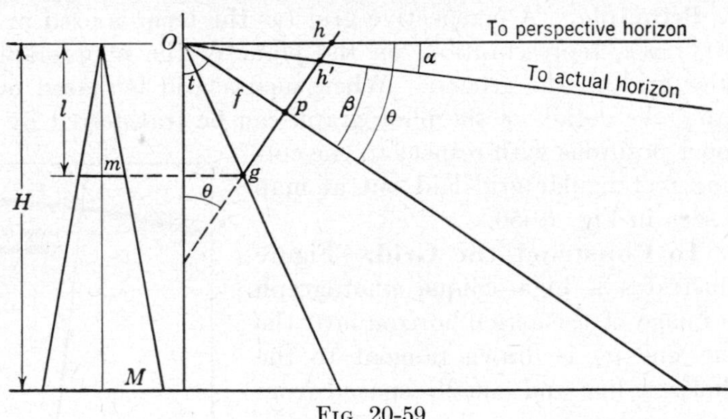

FIG. 20-59

Figure 20-59 shows the conditions that exist when the photograph is taken. The principal plane is in the plane of the paper. The perspective point is O. The photograph has been moved to the positive position gh.

$$\tan \beta = \frac{\overline{h'p}}{f}$$

$$\theta = \alpha + \beta$$

When θ is the angle of depression of the photograph ($\theta = 90° - t$),

$$\overline{ph} = f \tan \theta$$

Example. Given

$$H = 6000' \qquad ph' = 3.03''$$
$$\alpha = 60'' \sqrt{6000} = 77.5' = 1°17.5'$$
$$\tan \beta = \frac{3.03}{6} = 0.505$$
$$\beta = 26°48'$$
$$\theta = 1°17.5' + 26°48' = 28°06'$$

Then $$ph = f \tan \theta = 3.204''$$

The distance $ph = 3.204$ in. is laid off to establish h, and the perspective horizon is drawn parallel to ab (see Fig. 20-60). A line eg is laid off at the bottom of the photograph or below the photograph perpendicular to the principal line. Its exact position is established in such a way that the grid lines will be spaced along it at intervals that are easy to lay out and will represent a ground distance expressed in round numbers. The grid lines should not be too infrequent nor so closely spaced that they interfere with the photographic detail.

The geometry involved is shown in Fig. 20-59. g is a point near the bottom of the photograph. At the left is shown the projection, on a vertical plane perpendicular to the principal plane, of the rays from the ground grid. From the figure

$$l = \overline{gh} \cos \theta \qquad \text{and} \qquad \frac{m}{M} = \frac{l}{H}$$

Substituting,

$$\overline{gh} = \frac{mH}{M \cos \theta} \qquad \text{or} \qquad m = \frac{\overline{gh}\, M \cos \theta}{H} \tag{20-37}$$

Assume that it is desired to space the grid lines so that they represent 1000-ft intervals on the ground; $M = 1000$ ft. Assume gh is approximately 9 in. By Eq. (20-37)

$$m = \frac{(9)(1000)(0.882)}{6000} = 1.323$$

Choose 1.25 in. for m. Then

$$\overline{gh} = \frac{(1.25)(6000)}{(1000)(0.882)} = 8.50$$

Lay off hg equal to 8.50 in. (Fig. 20-60), mark off $m = 1.25$-in. intervals, and draw the perspective grid lines through h as shown.

The position of the horizontal lines may be computed by Eq. (20-25) or they may be placed graphically. To use the graphical method, find the vanishing point v on the horizon where lines at 45° to the principal plane meet. Since, at the isocenter, angles have their true size and the tangent of 45° is unity, *hv* must equal *ih*, the distance from the isocenter to the horizon. It has been shown that this distance is equal to $f \csc t$; therefore

$$\overline{hv} = f \csc t = \frac{f}{\cos \theta}$$

$$\overline{hv} = \frac{6}{0.882} = 6.80$$

Then any line drawn through *v* from the corner of a grid square will be a diagonal of every grid square that it touches.

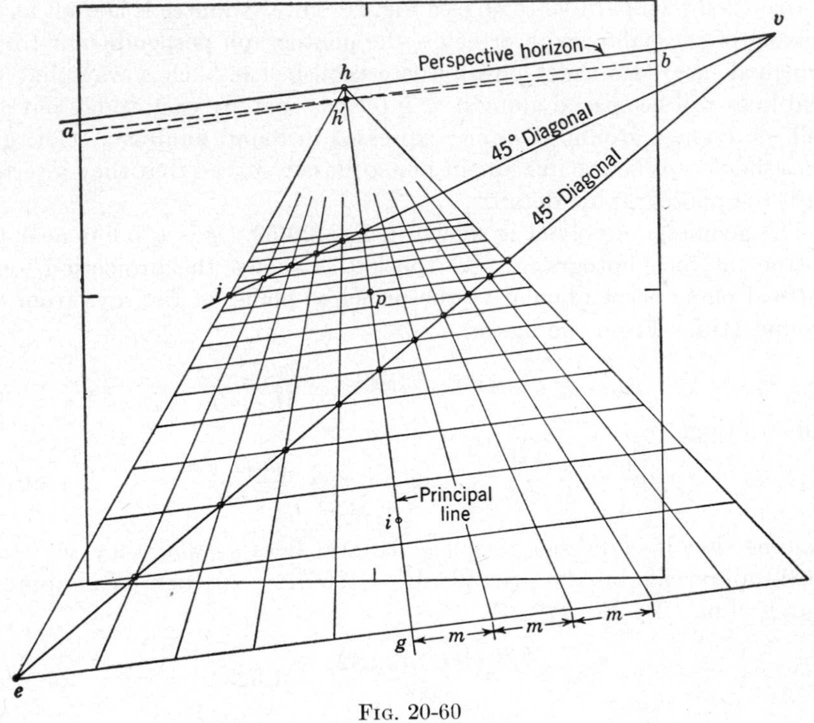

FIG. 20-60

Draw *ev*. At each intersection of a line, draw a line parallel to *eg*. A second line *jv* can be drawn to complete the grid.

RECTIFICATION

20-79. Definition. Rectification is the process of rephotographing an aerial photograph so that the effects of tilt are eliminated. The term is also applied when the effects of tilt are only modified. Often the photograph is also enlarged or reduced as part of the process.

Figure 20-61 shows that rectification is entirely possible, provided the lens can be focused as required. This can be easily accomplished as explained in later paragraphs. *A* represents the plane on which the photograph was taken and *B* is the horizontal plane on which the photograph would have been taken if the camera had been exactly vertical. It is evident that if the tilted photograph is rephotographed at plane *C* where $f_0 = f$, the resulting picture will have the same arrangement of images as would have occurred on plane *B*. Also, if the original picture is

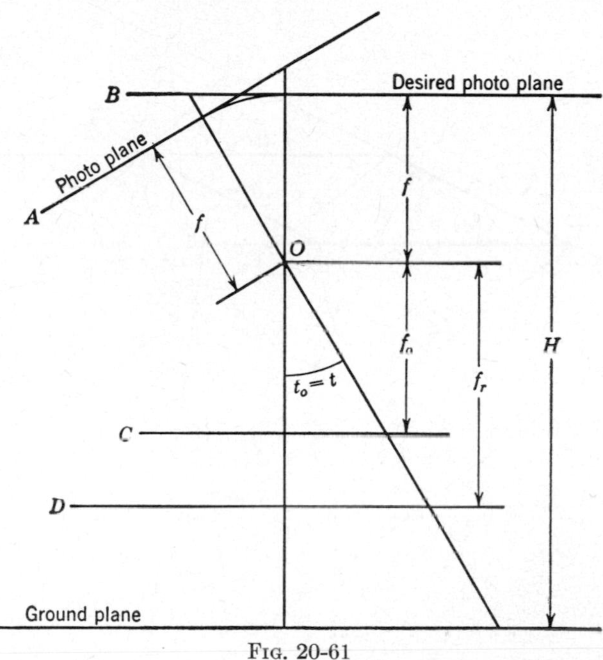

FIG. 20-61

rephotographed on plane *D*, the arrangement of images would still be the same except that the picture would have been enlarged by the ratio f_r/f.

A rectifier can therefore be designed that is based on this geometry. It would have an easel to support the photograph corresponding to plane *A*, a properly oriented lens of the proper focal length placed at *O*, and the plate at the plane *C* or *D*.

However, not all of the geometric requirements shown in Fig. 20-61 need be maintained. It will be shown that the angle of the easel t_0 need not be equal to the angle of tilt *t* if certain other requirements are maintained. It may even be held fixed for any value of *t*, if desired.

Figure 20-62 shows the geometric principles of a rectifier. *VO* is the plane of the photograph (the object plane) and *VR* is the plane of the rectified picture. The rectifier is constructed so that f_0 is equal to f in the photograph, the angle t_0 is equal to t, and f_r is such that f_r/f_0 gives the magnification (in this case a slight reduction) desired. As these are exactly the conditions shown in Fig. 20-61, a true rectification will result.

It can be shown that if certain changes are made, as illustrated, the rectified print will be identical to a print made previous to the changes. The changes are the following:

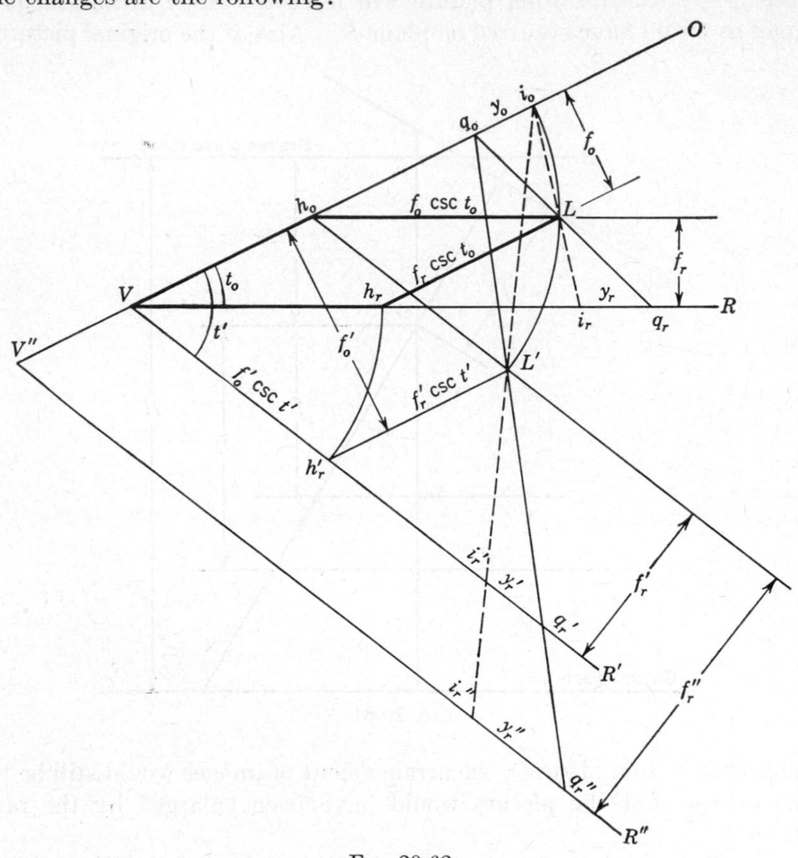

FIG. 20-62

1. The plane VR is rotated to any desired position VR'.

2. The perspective point L is moved to L' so that the parallelogram $h_0L\,Vh_r$ has the same length sides as before.

Let q_r represent any image point whose distance from the isocenter is y_r. The length y_r can be computed by Eq. (20-20), which can be written

$$y_m = Y_M \frac{f \csc t + y_m}{H \csc t}$$

In this equation $y_m = y_r$, $Y_M = y_0$, $f = f_r$, $H = f_0$, and $t = t_0$. Substituting,

$$y_r = y_0 \frac{f_r \csc t_0 + y_r}{f_0 \csc t_0} \qquad \text{or} \qquad y_r = \frac{y_0 f_r \csc t_0}{f_0 \csc t_0 - y_0}$$

In like manner

$$y'_r = y_0 \frac{f'_r \csc t' + y'_r}{f'_0 \csc t'} \qquad \text{or} \qquad y'_r = \frac{y_0 f'_r \csc t'}{f'_0 \csc t' - y_0}$$

But $\quad f_r \csc t_0 = f'_r \csc t'$

and $\quad f_0 \csc t_0 = f'_0 \csc t' \quad$ by construction

Therefore $\quad y'_r = y_r$

Equation (20-21) shows that, if the y values are satisfied, the x values (distances perpendicular to the principal plane) are also satisfied. Hence the projection on VR' is the same as the projection on VR.

If any other magnification factor is required, the plane $V''R''$ may be used. From Eq. (20-5)

$$\frac{p}{R} = \frac{f}{H}$$

Hence

$$\frac{y'_r}{y''_r} = \frac{f'_r}{f''_r}$$

20-80. Fixed-angle Rectifier. In the geometric requirement discussed so far, the value $f'_0 \csc t'$ was held fixed and equal to the value of $f \csc t$ in the photograph. This makes it necessary to change the positions of the parts of the rectifier for each photograph. Instead, the rectifier can be built with fixed parts. That is to say, t' and L' can be fixed. For such a rectifier $f'_0 \csc t'$ cannot be changed.

To use such a rectifier, the photograph must first be enlarged or reduced until $f \csc t$ is equal to the $f'_0 \csc t'$ of the rectifier. The isocenter and the perspective horizon can be marked on the photograph, if desired, and their distance made equal to the known value for the rectifier by enlargement or reduction. The picture must be placed on the rectifier so that the isocenter comes at the position of the isocenter on the rectifier easel.

20-81. Orientation and Focus of Lens. The preceding paragraphs give the position of the lens. It remains to find its proper orientation and focal length so that all points in the object plane will be focused on the image plane.

It can be shown (Fig. 20-63) that the lens plane must intersect the intersection of the object and the image planes and that the focal length must be the length of a perpendicular dropped to the lens plane from the horizon point on either the object plane or the image plane.

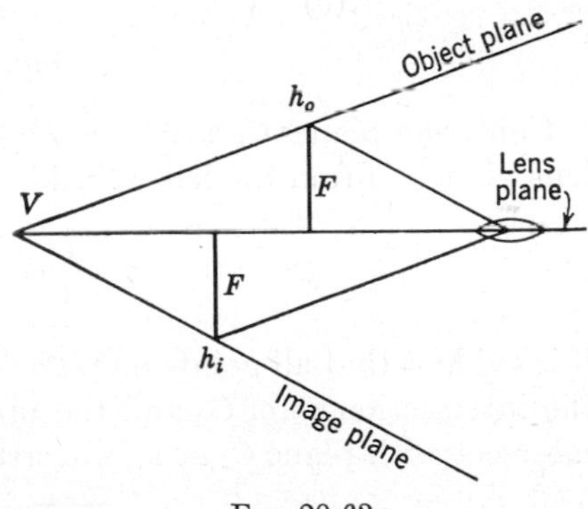

FIG. 20-63

In Fig. 20-64a, a_0M is the object plane and h_0 is the horizon point. The point h_0 must be focused at infinity so that it must lie in the focal plane of the lens, that is, at a distance F from the plane of the lens. Construct the lens plane LM so that $h_0g = F$.

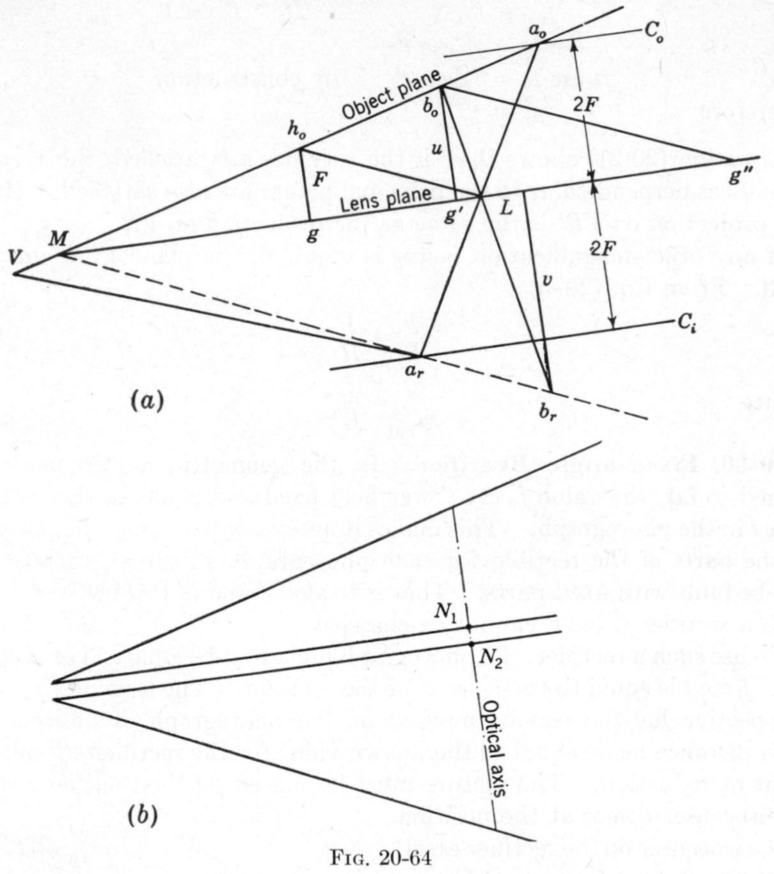

FIG. 20-64

Construct planes C_0 and C_i, as shown, each at a distance of $2F$ from the lens plane. From the lens relationship

$$\frac{1}{u} + \frac{1}{v} = \frac{1}{F}$$

it is evident that all points on plane C_0 will be imaged on plane C_i. From the intersection a_0 of C_0 and the object plane construct a_0L extended to intersect with plane C_i at a_r, construct a_rM. By similar triangles

$$\overline{a_0M} = 2\,\overline{a_0h_0}$$
$$\overline{a_0a_r} = 2\,\overline{a_0L}$$

Therefore $\overline{h_oL}$ is parallel to $\overline{Ma_r}$. But the object plane V_{a_r} must be parallel to h_0L due to the requirements of rectification. Therefore, V must coincide with M.

This can be further demonstrated by proving that any point on the object plane will be focused on the image plane. Let b_0 be any object

FIG. 20-65. Rectifying camera. Removes tilt and changes scale of photo. (*U.S. Coast and Geodetic Survey.*)

point at the object distance u from the plane of the lens and b_r be its image at the image distance v. Construct b_0g'' parallel to h_0L. Then by similar triangles

$$\frac{u}{v} = \frac{\overline{g''L}}{\overline{LM}} \qquad \frac{u}{F} = \frac{\overline{g''L + LM}}{\overline{LM}} = \frac{\overline{g''L}}{\overline{LM}} + 1$$

Substituting, $$\frac{u}{F} = \frac{u}{v} + 1 \qquad \text{or} \qquad \frac{1}{F} = \frac{1}{v} + \frac{1}{u}$$

Therefore b_0 and b_r are at conjugate.

Thus the plane of the lens must intersect the intersection of the object and image planes, and the focal length of the lens must be equal to the perpendicular dropped to the lens plane from the horizon points of either.

From Fig. 20-62, $Vh_0 = f_r \csc t$, $h_0L = f_0 \csc t$, and the angle at h_0 is equal to $180° - t$. Thus triangle Vh_0L can be solved and F computed.

20-82. Node Separation. As explained in previous sections of this chapter and shown in Fig. 20-13, the actual computation for a rectifier must include the separation of the nodes of the lens N_1N_2. The computations remain the same but at the end the two planes are separated as shown in Fig. 20-64*b*.

PROBLEMS

20-1. Aerial photographs are to be taken on 9- by 9-in. negatives with a camera that has a principal distance of 8 in. If the ground distance parallel to the line of flight to be covered by each exposure is about 9000 ft, at what height should the photographs be taken?

20-2. What will be the approximate scale of the negatives in Prob. 20-1?

20-3. The principal distance of the camera is 8.25 in., the original negatives are 9 in. square, and the desired scale of the contact photographs is approximately 1:12,000. If the end overlap is to be 60 per cent and the side overlap 25 per cent, what is:

a. The desired altitude of flight?

b. The distance along the ground between successive exposures in each flight?

c. The distance along the ground between adjacent flights?

20-4. On a negative that was made with the airplane at an altitude of 10,000 ft above the datum plane, the distances from the principal point to the images of two objects A and B are 3.47 and 2.94 in., respectively. If it is known that object A is 100 ft above the datum plane and object B is 80 ft below the datum plane, what are the amounts and directions of the relief displacements of the objects?

20-5. A photograph of a ground surface which is composed of two planes at different elevations connected by a steep cliff is taken at an altitude of 3600 ft above the datum plane with a camera that has a principal distance of 6 in. The average elevation of the high part of the ground is 400 ft above the datum plane, and the average elevation of the low part is 400 ft below the datum plane. Determine the scale of the photograph for each of the following ground elevations:

a. The datum plane

b. The high ground

c. The low ground

20-6. The altitude of the camera lens above the datum plane is 3000 ft; the base of the stack ST is 40 ft below the datum plane; and the distances ps and pt from the principal point of the actual photograph to the images of the bottom and the top of the stack on the same side of p are, respectively, 3.88 and 4.00 in. Determine the approximate height of the stack.

20-7. On a photograph that is taken from an approximate altitude of 12,000 ft above the datum plane are shown an image of a scale-line terminal A, which is 800 ft below the datum plane, and an image of a scale-line terminal B, which is 300 ft above the plane. On the photograph, the distance AB scales 4.94 in. and the distance from

p to B scales 3.62 in.; also, angle pAB is 45° and angle pBA is 30°. If the scale, or horizontal, distance between the scale-line terminals is 10,000 ft, what is the approximate scale of the photograph?

20-8. It is desired to make prints with a scale ratio of 1.282. If the principal distance of the projecting camera is 6 in., what are the approximate distances:

a. From the negative to the lens?

b. From the lens to the copyboard?

20-9. The horizontal distance between the exposure stations for two successive photographs of a strip, as scaled from the plotted positions of the principal points on the base map, is found to be 1840 ft. The height of the airplane above the datum plane when the exposures were made was 10,000 ft, and the principal distance of the camera is 12 in. The distances on the photographs from the principal points to the images of a certain point A to be plotted are as follows: On the first photograph, the distance measured in the direction of the line of flight is 3.52 in. and the distance perpendicular to that line is 1.63 in. On the second photograph, the distance in the direction of the line of flight is 1.27 in. Determine the approximate horizontal ground distances, parallel and perpendicular to the line of flight, from the ground nadir point for the first photograph to the point A; and determine the approximate height of the point A above the datum plane.

20-10. Measuring in the direction of an air base, the parallax of point A is 68 mm in photograph 1 and -13 mm in photograph 2. The principal distance is 210 mm, the elevation of point A is 450 ft, and that of the air base is 9000 ft.

a. What is the altitude?

b. In the same photographs the parallaxes of point B are 103 and $+16$ mm. What is the elevation of B?

20-11. Flying altitude 12,000 ft, parallax 4.10 in., parallax difference 0.05 in.

a. What is the elevation?

b. Can you determine the focal length?

20-12

	Point A	*Point B*
Elevation................	170 ft	770 ft
Parallax..................	150 mm	50 mm
Distance from P.P.........	3400 ft	1000 ft

Find the altitude and focal length.

20-13. Air base 5000 ft, altitude 12,000 ft, principal distance 8.25 in. What is the parallax difference between two points having elevations of 100 and 800 ft, respectively?

20-14. Compute the following, assuming no tilt.

Given	*Find*
Principal distance f = 8.250 in.	Relief displacement d
Radial distance R = 4000 ft	
Elevation h = 820 ft	
Altitude A = 12,000 ft	

20-15. Compute the following, assuming no tilt.

Given	*Find*
Total photographic parallax $p + d$ = 3.796 in.	Elevation h
Altitude A = 12,000 ft	
Parallax difference d = 0.176 in.	

20-16. A camera of 6-in. focal length is flown at an altitude of 10,000 ft. Determine the scale, in terms of feet per inch, on the general land surface and on top of a mesa 1100 ft high.

20-17. A camera of 8.25-in. focal length is flown at altitude such as to produce photos on a scale of 1:20,000. What variations in scale from print to print would result from departures of 400 ft from the correct flight altitude?

20-18. On a 9- by 9-in. photo taken with a camera of 12-in. focal length on a scale of 1:20,000, a mountain peak 2500 ft higher than the point directly under the camera is found to lie at a measured distance of 4 in. from the center of the photo.

a. Calculate its true distance from the center point on a planimetric map of the same scale.

b. What would the amount of displacement have been if the true position of the peak had been only 1.5 in. from the center?

c. If a camera of 8.25-in. focal length had been used to take a photo on the same scale, what would the radial displacement have been?

20-19. In compiling a planimetric map by the radial-line method, it is found that a point which appears to be 3.7 in. from the center of a photo (scale 1:20,000, focal length of camera 8.25 in.) is actually 4.1 in. from the center point. Calculate its elevation relative to that at the center point.

20-20. On a 10- by 10-in. photo taken with a camera of 5.37-in. focal length on a scale of 2 in. to the mile, point *A* (elevation 1300 ft below the center point) lies 4 in. S50°E from the center point, and point *B* (elevation 1900 ft above the center point) lies 7 in. N10°E from point *A*. Determine the true direction and distance from point *A* to point *B*.

20-21. A symmetrical, conical peak 4200 ft high and with sides sloping 35° is photographed from a point 9800 ft directly above the apex. Determine graphically the distortion of contours due to relief, and make two maps on a scale of 1 in. = 2000 ft, one showing the true position of 500-ft contours on a map and the other showing the apparent position of the contour lines on the photo. Would distortion of this type be appreciable on hills not more than 500 ft high?

20-22. A symmetrical, conical hill 1500 ft high and with sides sloping 30° is photographed from a height of 14,000 ft, at a point 3000 ft to one side of the apex. Prepare maps showing true and apparent positions of 200-ft contour lines as in Prob. 20-21, above, and on the same scale.

20-23. The coordinate system of a radial line plot was laid out at a scale of 1 in. = 1,000 ft. The intersection of the center lines of a photocontrol point was found to be located at a radial distance of 7.00 in. from the principal point of one of the templates upon which it appeared. The elevation of the photocontrol point was 400 ft. The radial distance from the principal point on the photograph to the image photocontrol point was 3.00 in. The principal distance of the camera was 10 in. What was the altitude of the lens?

20-24. A stereopair of plates were aligned in a Brock stereocomparator. The plates were adjusted so that the fixed cross hairs appeared to be on the ground at the principal point of the left-hand plate. This point had an elevation of 200 ft. The plates were moved parallel to the air base and then adjusted so that the cross hairs appeared on the ground at a photocontrol point having an elevation of 700 ft. The true length of the air base was 8000 ft. The altitude was 12,000 ft and the principal distance was 14 in. Find the relative movement between the plates parallel to the air base. Were the plates moved toward each other or away from each other?

20-25. A photograph was made at an altitude of 12,000 ft. The principal distance of the camera was 12 in. It is desired to project the contours obtained by this photograph on a reducing-enlarging projector, so that the sea-level scale is 1:20,000. What is the ratio between the distance u this photograph is held from the projecting lens and the distance v from the lens to the screen where the tracing is made (i.e., find v/u), when the 1000-ft contour is being traced.

20-26. An enlarging lens having a focal length of 7 in. is used to make an enlargement of 1.5 diameters (i.e., 1.5 times original size). Find u and v.

20-27. Assume an altitude of 12,000 ft and a principal distance of 210 mm. The distance from the principal point to the image of an object was found to be 82.0 mm on the photograph. The object was located 5000 ft from the principal point on the ground. Find the elevation of the object.

20-28. Principal distance 10 in., altitude 12,000 ft. Two points appeared on each of two photographs. In each case, measuring toward the other conjugate center parallel to the air base from the principal points to the points in question, the following distances were determined:

Point	*Photo* 1	*Photo* 2
A	3.48″	0.52″
B	2.25″	1.05″

The elevation of point *A* was 800 ft. What was the elevation of point *B*?

20-29. Find the angle of tilt and the direction of tilt on a photograph by the Anderson-Rihn method. Assume that the actual images of the ground points were located on the photograph according to the following coordinates. The coordinate axes are aligned with the fiducial marks and the origin is at the principal point. The coordinates are given in inches. The principal distance f was 6.000 in.

Location of Images

	A	*B*	*C*
x	−3.00	+2.60	+2.50
y	+1.00	+3.80	−2.00

Ground Data

Elevation *A* = 100″	*B* = 400′	*C* = 500′
Lengths *A-B* = 7748′	*B-C* = 8725′	*C-A* = 8348′

Part V. Appendix

CHAPTER 21
PROBABILITY

THE THEORY OF ACCIDENTAL ERRORS

The Scope of This Chapter. As stated in other chapters, the sources of errors in measurement can be classified under three mutually exclusive heads: (1) blunders, (2) systematic errors, and (3) accidental errors. This chapter is confined to the discussion of accidental errors only.

21-1. Accidental Errors. The accidental error of a determination of a quantity is the difference between (1) the true value of the quantity and (2) a determination of this value that is free from blunders or systematic errors.

21-2. Limits of Precision. No matter how precise the measuring equipment or how skillful the observer, the determination of the value of a quantity can never be perfect. It must consist of the selection of a range of quantities within which the measured quantity lies. The **limits of precision** can be defined as the smallest range within which the value of a quantity can be correctly placed.

For example, if an angle is measured once with a 20-second instrument by an observer able to estimate correctly the vernier reading to the nearest 10 seconds, the accidental error in this determination will be within the range ±5 seconds. The limits of precision of this operation are, then, ±5 seconds (see Fig. 21-1).

21-3. Accidental Errors and the Laws of Chance. The error may fall anywhere between ±5 seconds, depending on the exact size of the angle. *Since chance governs the exact size of the angle, the size and sign of the accidental error depend on the laws of chance.*

The probability of the occurrence of any *exact* value is $1/\infty$, but the probability of the occurrence of an error of some value within a certain range or group is finite. In the discussion of the example stated, we will study the probability of the measured value falling within each of a series of groups, each group 1 second in extent. This group will be called the unit of error of measurement.

If a series of groups is established as shown in Fig. 21-2 in such a way that any error between ±½ second will be given the value 0 and any error between +½ second and +1½ seconds will be given the value +1, etc., there will be an equal chance that the error will fall in any one of the eleven groups depending on the true sign of the angle. It may be argued that for the groups called −5 seconds and +5 seconds the chance is only one-half that of the others. But this introduces a refinement unnecessary to the discussion.

Accordingly, there are eleven possibilities for values of errors: −5, −4, −3, −2, −1, 0, +1, +2, +3, +4, and +5 seconds. The probability of any one of them is $1/11$ or 0.0909 and the sum of their probabilities is, of course, unity. This may be illustrated graphically in a frequency diagram by plotting the value of each group as an abscissa and the probability of each group as an ordinate (see Fig. 21-2).

21-4. The Sum of Two Accidental Errors. If two adjacent angles are measured as described above (assuming for the moment that the vernier can be set exactly at zero at the beginning of each measurement) and their sum computed, the same possibilities for accidental errors as described above would exist in each of the angles measured. *In the sum, any pair of these eleven possibilities would be equally probable.*

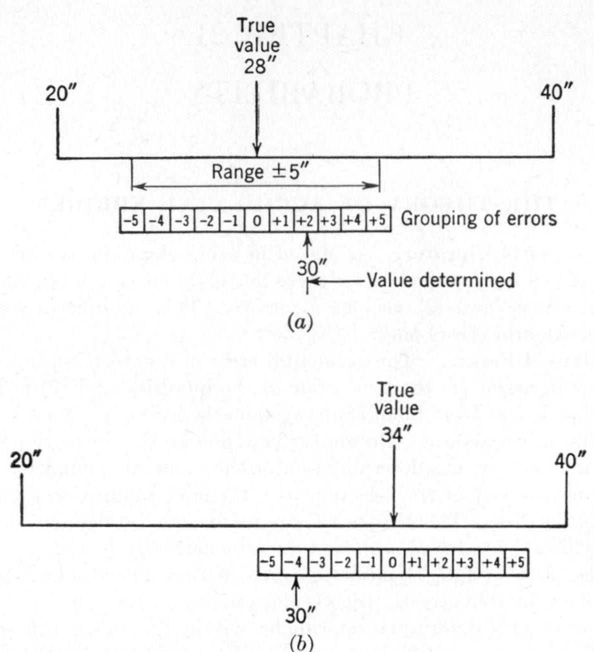

FIG. 21-1. The possible range of errors of one determination. (*a*) The true value is 28 in. The observer will estimate that it is nearly halfway between 20 and 40 in. and will call it 30 in. The error is then +2 in. (*b*) The true value is 34 in. The observer will again call it 30 in. The error is then −4 in.

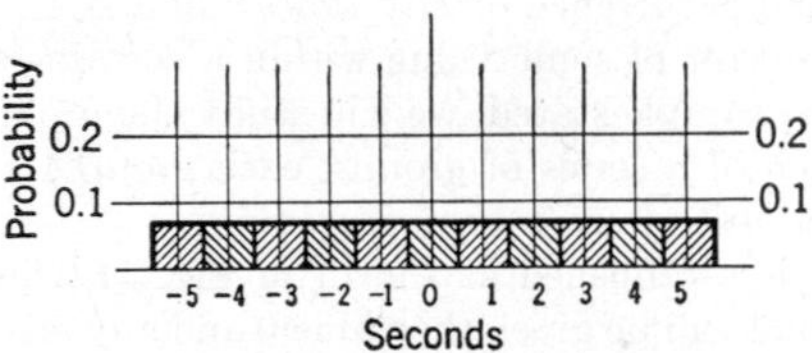

FIG. 21-2. The probability of errors of one determination.

There would be $(11)^2$, or 121, possibilities, ranging from −10 to +10 seconds, giving a total of 21 groups. Only one pair of errors would produce an error in the sum equal to −10 seconds, i.e., −5 seconds in the first angle and −5 seconds in the second angle. But −9 could occur in two different ways, viz., −5 in the first angle and −4 in the second, or −4 in the first and −5 in the second. Similarly, −8 could occur in three ways, etc. Dividing the number of possibilities in each group by the total number of possibilities (121), the probability of each group is obtained thus:

Value of error group	Possibilities	Probability	Probability in decimals
±10	1	1/121	0.0083
±9	2	2/121	0.0165
±8	3	3/121	0.0248
±7	4	4/121	0.0331
±6	5	5/121	0.0413
±5	6	6/121	0.0496
±4	7	7/121	0.0579
±3	8	8/121	0.0661
±2	9	9/121	0.0744
±1	10	10/121	0.0826
0	11	11/121	0.0909

The values are plotted in Fig. 21-3.

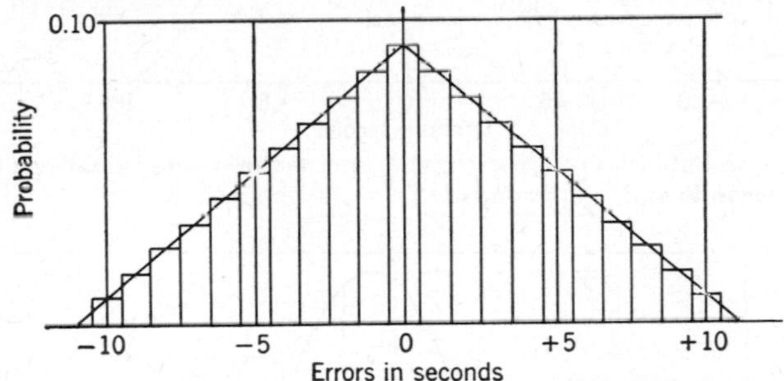

FIG. 21-3. The probabilities of errors in the sum of two determinations.

21-5. The Sum of Three Accidental Errors. If three adjacent angles were measured in the same way, the range of error in the sum would be ±15 seconds and there would be $(11)^3$, or 1331, possibilities and 31 groups. The possibilities of an error of −15 in the sum would be 1; of −14, 3; of −13, 6; of −12, 10; of −11, 15, etc. They are listed below and plotted in Fig. 21-4.

Value of error group	Possibilities	Probability	Value of error group	Possibilities	Probability
±15	1	0.0008	±7	45	0.0338
±14	3	0.0023	±6	55	0.0413
±13	6	0.0045	±5	66	0.0496
±12	10	0.0075	±4	75	0.0563
±11	15	0.0113	±3	82	0.0616
±10	21	0.0158	±2	87	0.0654
±9	28	0.0210	±1	90	0.0676
±8	36	0.0270	0	91	0.0684

Figure 21-4 also shows the curve for the sum of three determinations, but for a limit of precision of ±3 seconds instead of ±5 seconds.

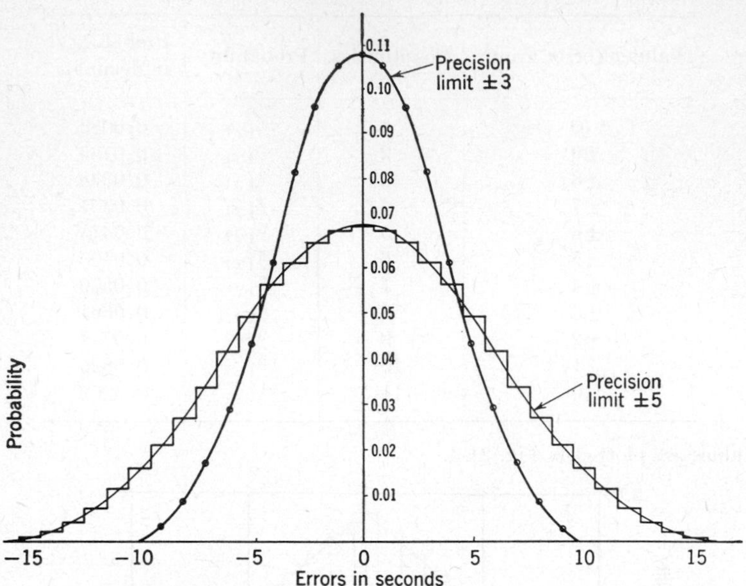

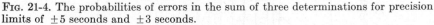

FIG. 21-4. The probabilities of errors in the sum of three determinations for precision limits of ±5 seconds and ±3 seconds.

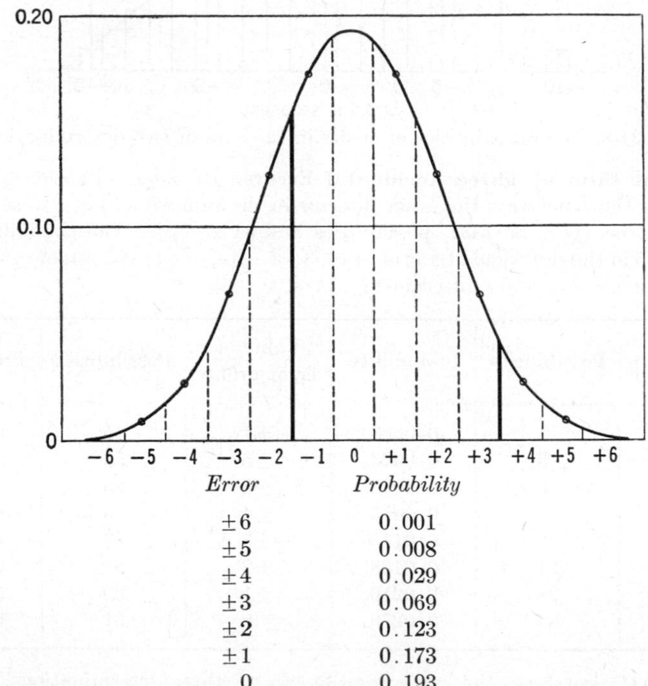

Error	*Probability*
±6	0.001
±5	0.008
±4	0.029
±3	0.069
±2	0.123
±1	0.173
0	0.193

FIG. 21-5. Probabilities of errors in the sum of six determinations.

21-6. Meaning of Curves. A smooth curve can be drawn connecting the ordinates in each of these figures. Such a curve can be called a **frequency distribution curve** or a **probability curve.** A similar curve for the sum of six determinations is shown in Fig. 21-5. The limit of precision is here taken as ± 1, the unit of error measurement, 1; there are $(3)^6$, or 729, possibilities, and they fall within the limits of ± 6.

It is evident that there is a certain curve for each number of determinations which are added together to make the sum. Of course, if the limit of precision is changed, the curve is affected. But the effect is only to change the scales of the abscissas and the ordinates. The characteristics of the curve remain the same. For example,

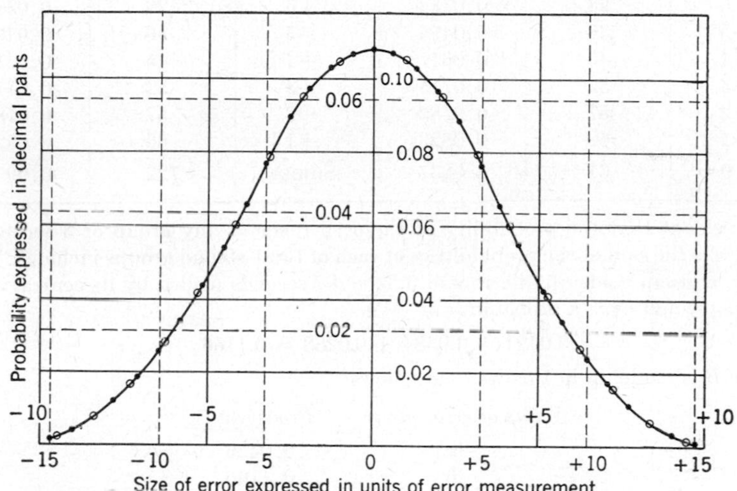

FIG. 21-6. This curve shows the probability of accidental errors when a value is obtained by the sum of three determinations. The dots and the dotted coordinate lines refer to a limit of precision of ± 5 units. The circles and full lines refer to a limit of precision of ± 3 units. The scales to which the dots were plotted differ from the scales to which the circles were plotted as shown. For example, the probability of making an error of 5 units is about 0.050 for the dots and about 0.044 for the circles.

in Fig. 21-6 the two curves from Fig. 21-4, which are each based on the sum of three determinations but have different precision limits, are plotted at different scales so that they can be superimposed as shown.

21-7. Effect of Grouping. Consider the curves in Fig. 21-7. These are based on the errors that would occur if the limit of precision were 4 seconds. At the upper left is the curve for one determination grouped by single seconds. There are nine possibilities; each has a probability of $\frac{1}{9}$ or 0.1111. The sum of all probabilities is $9 \times \frac{1}{9}$ = unity.

At the upper right is a curve based on the same conditions, but the group unit used is 3 seconds. There are three possibilities; each has a probability of $\frac{1}{3}$ or 0.3333.

In the main diagram at the lower center, Curve *A* shows the probabilities for three determinations computed for group units of 1 second, as follows:

Total range	$3 \times \pm 4 = \pm 12$
Total possibilities	$(9)^3 = 729$

Value of error group	Possibility	Probability	Value of error group	Possibility	Probability
−12	1	0.0014	+12	1	0.0014
−11	3	0.0041	+11	3	0.0041
−10	6	0.0082	+10	6	0.0082
−9	10	0.0137	+9	10	0.0137
−8	15	0.0206	+8	15	0.0206
−7	21	0.0288	+7	21	0.0288
−6	28	0.0384	+6	28	0.0384
−5	36	0.0494	+5	36	0.0494
−4	45	0.0617	+4	45	0.0617
−3	52	0.0713	+3	52	0.0713
−2	57	0.0782	+2	57	0.0782
−1	60	0.0823	+1	60	0.0823
0	61	0.0837	Sums	729	0.9999

It is evident that the probability of the occurrence of any group of 3 seconds in this curve is the sum of the probabilities of each of the 1-second groups included. For example, a group made up of errors of 5, 6, and 7 seconds (called by its central value, 6 seconds) would have a probability of

$$0.0494 + 0.0384 + 0.0288 = 0.1166$$

Curve *B* is made up in this way, as follows:

Value of error group	*Probability*
−9	0.0425
−6	0.1166
−3	0.2112
0	0.2483
+3	0.2112
+6	0.1166
+9	0.0425
Sum	0.9889

Very nearly the same curve will result if the original grouping shown at the upper right is used as a basis of computation. The points on such a curve are shown by circles. The computation is as follows:

Total range $3 \times \pm 3 = \pm 9$
Total possibilities $(3)^3 = 27$

Value of error group	*Possibility*	*Probability*
−9	1	0.037
−6	3	0.111
−3	6	0.222
0	7	0.259
+3	6	0.222
+6	3	0.111
+9	1	0.037
Sums	27	0.999

This curve is less accurate, as not so many possibilities are taken into account.

It is evident from the foregoing analysis that

1. The probability of the occurrence of an error within any range is equal to the sum of the probabilities of all of the groups within that range.

2. If the size of a single group is used as a unit of abscissa and the probability of each of such groups is used as ordinate, the area under any section of any one of these curves is equal to the probability of an error falling within that range.

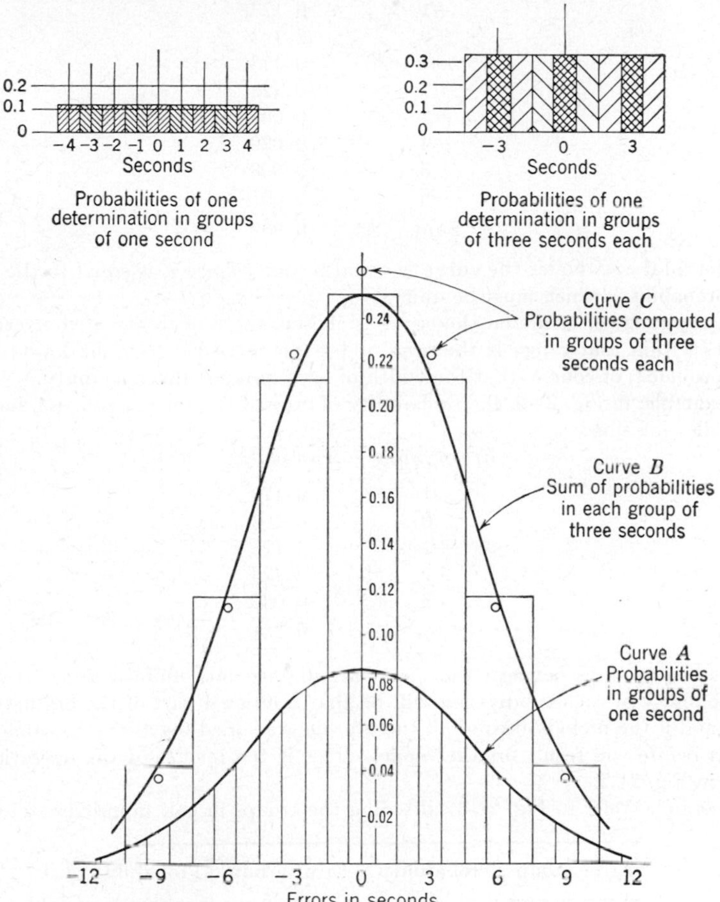

FIG. 21-7. The probabilities of errors in the sum of three determinations for precision limits of ±4 seconds in groups of 1 second and 3 seconds.

3. Since the probability of the occurrence of an error falling within the range of the whole curve is unity, the area under the whole of any curve is unity.

21-8. Characteristics Common to All Curves. All of these curves, accordingly, have certain characteristics in common:

1. Each ordinate gives, in decimal parts, the probability of the group of errors to which it refers, i.e., the group of errors of which it is the center.

2. The sum of the probabilities is unity. For example, in Fig. 21-5, the sum of all the probabilities can be computed thus:

Error group	*Probability*
+6	0.001
+5	0.008
+4	0.029
+3	0.069
+2	0.123
+1	0.173
0	0.193
−1	0.173
−2	0.123
−3	0.069
−4	0.029
−5	0.008
−6	0.001
Sum	0.999

3. The total area under the curve is equal to unity since it is equal to the sum of every probability (which must be unity).

4. If any range of errors on the curve is chosen, the probability of the occurrence of errors within that range is the area under the curve between the limits of that range, provided, of course, that the width of each group is taken as unity.

For example, in Fig. 21-5, the probability of errors between −1 and +3, inclusive, is the following sum:

Error group	*Probability*
−1	0.173
0	0.193
+1	0.173
+2	0.123
+3	0.069
Sum	0.731

If the error groups between the limits are divided into uniform fractional parts, the probability of each subdivision will be that fractional part of the ordinate at its center; and if the probabilities of all these parts are added together, the same probability as before will result for the range. This is the reverse of the operation performed in Fig. 21-7.

For example (refer to Fig. 21-5), dividing the groups in half (approximately),

Error group	Probability	Error group	Probability
−1¼	0.163	+1¼	0.163
−¾	0.183	+1¾	0.139
−¼	0.191	+2¼	0.110
+¼	0.191	+2¾	0.080
+¾	0.183	+3¼	0.056
			2⟌1.459
			0.730

If the error groups were divided into infinitesimals, the operation is merely an extension of the above.

5. According to paragraph 4, the probability of all the errors is obviously all the area under the curve, which is unity.

6. As the number of determinations is increased, the curve rapidly approaches a characteristic shape.

21-9. The Need for a Probability Curve. Since a different curve results when the number of determinations is changed, precise computations involving accidental errors would be too arduous to be practical. Accordingly, an approximate curve is used that is usually called **the probability curve.** It is the curve which would result if the number of determinations were infinite and the unit measure of the errors were infinitesimal. The shape of the curve, of course, must vary with the precision, but its characteristics remain the same.

There is very little approximation in using such a curve. The curve for the sum of four or five determinations is practically the same as the probability curve, and an actual measurement even when measured only once nearly always really includes several determinations. For example, in the illustration used at the beginning of this chapter, setting the vernier at zero is subject to the same errors as reading it. Thus, in determining the sum of two angles, four determinations of this nature are made.

Human error must also be made up of a number of accidental errors. An error-frequency curve based on human error always approximates the probability curve.

THE PROBABILITY CURVE

The probability curve can be defined as a curve that shows the relationship between the size of an error and the probability of its occurrence, provided the precision of the measurement is known. It is based on the laws of chance, and it is absolutely true only when the number of measurements is infinite and the unit of error of measurement is infinitesimal. It is usually called the normal probability curve to distinguish it from a frequency diagram made up of errors that are not accidental and therefore not "normal." From it may be determined, to a very close approximation, the most probable behavior of accidental errors. An example of the curve is shown in Fig. 21-8. The equation of the curve, which will be derived, may be written

$$y = \frac{h}{\sqrt{\pi}} e^{-h^2x^2}$$

where e = base of natural logarithms, 2.718
x = size of the error
y = probability, usually expressed in decimal parts, of the occurrence of an error of the size x
h = number that expresses the precision of the observations; large values of h indicate high precision and cause the peak of the curve to be high and narrow

21-11. Derivation of the Probability Curve.[1] Consider some type of observation in which the error e could be either $+a$ or $-a$ and nothing else and in which $+a$ and $-a$ have equal probabilities of occurrence. A system of errors of this nature is called a **binominal distribution** of errors. The sum m_n of the errors of n observa-

[1] This form of proof was suggested by Robert Singleton. It is not meant to be a rigid mathematical treatise but a demonstration of how the probability curve follows from the foregoing considerations.

tions can be expressed as

$$m_n = e_1 + e_2 + \cdots + e_n$$

If n is odd, m_n may be equal to any one of the following: $\pm a$, $\pm 3a$, $\pm 5a$, . . . , $\pm na$. If n is even, the possible values are 0, $\pm 2a$, $\pm 4a$, . . . , $\pm na$.

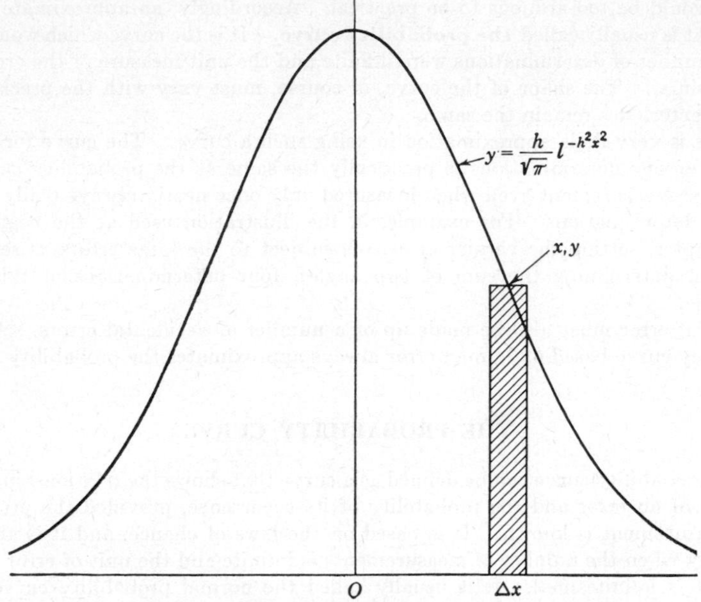

FIG. 21-8. Typical probability curve.

For any particular set of such observations, let p equal the number of errors of the value $+a$ and q equal the number of errors of the value $-a$. Let $u = p - q$, the number of errors of $+a$ in excess of the number of errors of $-a$. Then

$$m_n = ua \tag{21-1}$$

This particular value $m_n = ua$ will occur in n observations whenever $+a$ occurs p times. But, from the theory of combinations, the number of combinations of n things taken p at a time is expressed as

$$C_p{}^n = \frac{n!}{p!(n-p)!}$$

By definition, $p + q = n$, and $p - q = u$. Adding, $p = (n + u)/2$. Substituting for p its equal, $(n + u)/2$,

$$C_p{}^n = \frac{n!}{\left(\frac{n+u}{2}\right)!\left(\frac{n-u}{2}\right)!} \tag{21-2}$$

From the theory of combinations, the total number of possible arrangements of $+a$ and $-a$ (two errors) in the sum of n observations is 2^n. Thus, the probability y_n that the particular value of the sum m_n will be ua is expressed as $C_p{}^n/2^n$, or

$$y_n u = \frac{n!}{\left(\frac{n+u}{2}\right)!\left(\frac{n-u}{2}\right)!2^n}$$

In like manner, the probability that m_n has the value $(u+2)a$ is expressed as

$$y_n(u+2) = \frac{n!}{\left(\frac{n+u+2}{2}\right)!\left(\frac{n-u-2}{2}\right)!2^n}$$

Let $\Delta y_n/\Delta u$ represent the change in the probability for the smallest change in u. Note that, from the hypothesis, the smallest change in u is 2. Then

$$\begin{aligned}\frac{\Delta y_n}{\Delta u} &= \frac{y_n(u+2) - y_n(u)}{2}\\ &= \frac{n!}{2\left(\frac{n+u}{2}\right)!\left(\frac{n-u}{2}\right)!2^n}\left[\frac{(n-u)/2}{(n+u+2)/2} - 1\right]\end{aligned}$$

This expression can be checked by inserting values for n and u. It reduces to

$$\frac{\Delta y}{\Delta u} = -y_n \frac{u+1}{n+u+2} \tag{21-3}$$

If n increases indefinitely, this will approach the differential equation for the normal probability curve.

It has been shown in Sec. 21-7 that the grouping of the errors can be changed without changing the characteristics of the probability curve. It will be found convenient to use, instead of a group equal to a, a group equal to some function of n. Let such a function be $1/h\sqrt{2n}$.

The same argument can be applied throughout, with $1/h\sqrt{2n}$ substituted for a. In this case

$$m_n = u\frac{1}{h\sqrt{2n}}$$

Let the expression $u/h\sqrt{2n} = x$; then

$$u = hx\sqrt{2n} \tag{21-4}$$

Replacing u by its equal $hx\sqrt{2n}$ in Eq. (21-3) above,

$$\frac{\Delta y_n}{h\sqrt{2n}\,\Delta x} = -y_n\frac{hx\sqrt{2n}+1}{n+hx\sqrt{2n}+2}$$

or

$$\frac{\Delta y_n}{\Delta x} = -y_n\frac{2nh^2x + h\sqrt{2n}}{n+hx\sqrt{2n}+2} \tag{21-5}$$

and now, if n increases indefinitely, this approaches

$$\frac{dy}{dx} = -2h^2xy \tag{21-6}$$

whose solution is

$$y = ce^{-h^2x^2} \tag{21-7}$$

This function expresses the probability of the occurrence of a total error of the size $m_n = x$ in n observations as n increases indefinitely, when each observation can have only one of two fundamental errors ($+a$ and $-a$). It remains to show that the expression is true when each observation can have any one of a number of sizes of fundamental errors provided that they are accidental errors. Section 21-6 and Fig. 21-6 show that the characteristics of a curve based on three fundamental errors, $+1$, 0, and -1, and in which $n = 6$, has the same characteristics as a curve based on eleven fundamental errors in which $n = 3$. A similar set of computations made for other systems of errors will show that they all have the same characteristics as a curve based on two fundamental errors.

It remains to evaluate the constant c.

Since the value of y at any point on the curve represents the probability of the occurrence of the infinitesimal dx at error x, the probability of the occurrence of a range of errors from x_1 to x_2 will be the sum of the probabilities of the infinitesimals ax from x_1 to x_2. Such a sum is the integral of the curve from x_1 to x_2. Thus, *the area under the curve between two values of error equals the probability of errors occurring within that range.* Since the probability that an error will occur somewhere under the whole curve is unity, the area under the whole curve (i.e., between $x_1 = -\infty$ and $x_2 = +\infty$) is unity; hence

$$\int_{-\infty}^{+\infty} ce^{-h^2x^2}\,dx = 2c\int_0^{\infty} e^{-h^2x^2}\,dx = 1 \qquad (21\text{-}8)$$

Note the different limits.

Representing the second integral by I,

$$I = \int_0^{\infty} e^{-h^2x^2}\,dx$$

The quantity I would have the same value if the integration were performed in the direction of the Y axis, or

$$I = \int_0^{\infty} e^{-h^2y^2}\,dy$$

Hence

$$I^2 = \int_0^{\infty}\int_0^{\infty} e^{-h^2(x^2+y^2)}\,dy\,dx$$

The above can be expressed in polar coordinates as well as in rectangular coordinates. This evaluates I, as shown below:

$$I^2 = \int_0^{\pi/2}\int_0^{\infty} e^{-h^2\rho^2}\rho d\,\rho\,d\theta$$

$$= \frac{1}{2h^2}\int_0^{\pi/2} e^{-h^2\rho^2}\Big]_0^{\infty}\,d\theta = \frac{\pi}{4h^2}$$

Since

$$e^{-h^2\rho^2}\Big]_0^{\infty} = -1$$

$$I = \frac{\sqrt{\pi}}{2h} \qquad (21\text{-}9)$$

Substituting in Eq. (21-8),

$$2c\,\frac{\sqrt{\pi}}{2h} = 1$$

$$c = \frac{h}{\sqrt{\pi}}$$

Substituting in Eq. (21-7),

$$y = \frac{h}{\sqrt{\pi}} e^{-h^2x^2} \tag{21-10}$$

which is the equation of the probability curve.

21-12. Pierce's Tables.[1] To use the foregoing function conveniently, h must be separated out.

Denote the probability of an error falling between x_1 and x_2 by $P_{x_1x_2}$. Then

$$P_{-a,+a} = \frac{h}{\sqrt{\pi}} \int_{-a}^{+a} e^{-h^2x^2}\, dx = \frac{2}{\sqrt{\pi}} \int_0^a e^{-h^2x^2} h\, dx$$

Let $t = hx$; then $dt = h\, dx$, and when $x = a$, $t = ha$.

Substituting these in the integral will not change its value; therefore it will still equal $P_{-a,+a}$. Therefore

$$P_{-a,+a} = \frac{2}{\sqrt{\pi}} \int_0^{t=ha} e^{-t^2}\, dt \tag{21-11}$$

B. O. Pierce has computed and published, in a book entitled "A Short Table of Integrals," a table entitled "The Probability Integral" giving the values of this integral for limits of t from 0 to 2.00. When h is known the table can be used for three purposes:

1. To find the probability of errors between any given absolute size, for example, from $-a$ to $+a$. Compute $t = ha$, and enter the table with this value of t.
2. To find the probability of errors between two absolute sizes, for example, $-b$ to $-a$ and $+a$ to $+b$. Compute $t = ha$ and $t = hb$ and subtract the tabular values.
3. To find the range of errors that would have a given probability. From the table find the value of $t = ha$ for the given probability. The range would then be

$$\pm a = \frac{t}{h}$$

21-13. Measures of Precision. It is clear that the probability will always be the same for any given product of ha. For any given probability, if the error is decreased and thus the precision increased, h will be increased proportionally. Obviously, h is inversely proportional to the error and therefore directly proportional to the precision. Accordingly, *h is a measure of precision.*

Another measure of precision is the **probable error.** Actually it is proportional to the size of the errors and inversely proportional to the precision. The probable error can be defined as the error of such magnitude that half of the errors are smaller in size and half of them are larger. Accordingly, the probable error is the error with a probability of 0.5. More accurately stated, if ϵ represents the probable error, the probability of an error falling between $-\epsilon$ and $+\epsilon$ is 0.5 (see Fig. 21-9). By interpolation in Pierce's "Tables" it will be found that the value of t which corresponds to a probability of 0.5 is 0.4769. Then

$$t = h\epsilon = 0.4769$$

$$\epsilon = \frac{0.4769}{h} \tag{21-12}$$

[1] B. O. Pierce, "A Short Table of Integrals," Ginn & Company, Boston, 1929.

A third measure of precision often used is the **standard error** or **standard deviation,** usually expressed by σ. It is defined as follows:

$$\sigma = \sqrt{\frac{\Sigma x^2}{n}} \tag{21-13a}$$

where Σx^2 is the sum of the squares of the errors and n is the number of measurements and hence the number of errors.

When the actual errors of several determinations are known, σ can be computed by Eq. (21-13*a*). It remains to relate h and σ so that h and ϵ can also be computed from actual measurements.

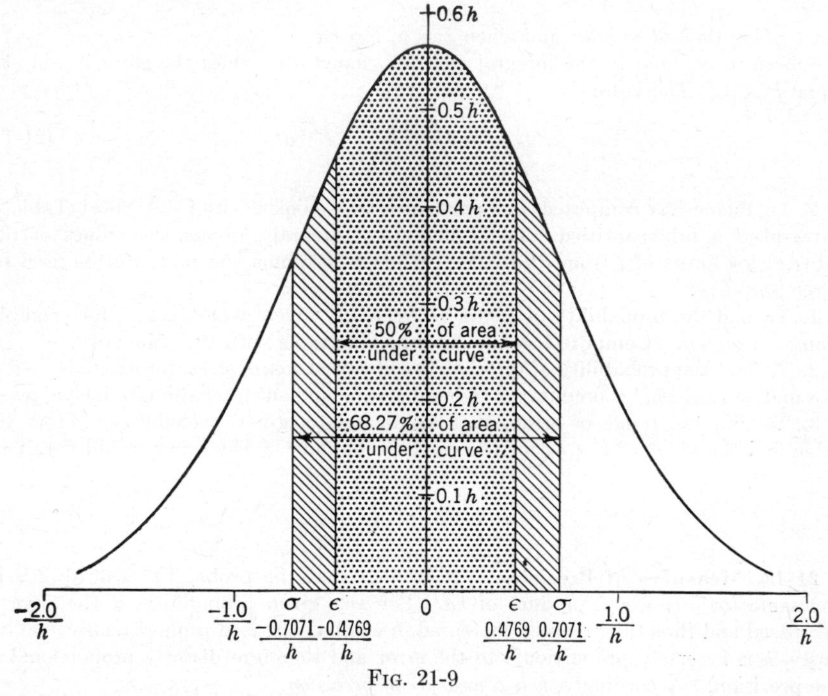

FIG. 21-9

21-14. Determination of Relationship between Measures of Precision. The probability of any error x is the number of errors having the value x divided by the total number of errors:

$$P_x = \frac{n_x}{n} \quad \text{or} \quad n_x = nP_x \tag{21-13b}$$

If the value of each error is squared and then multiplied by the number of errors of that size, the sum of such values is Σx^2; hence

$$\sum x^2 = \int_{-\infty}^{+\infty} n_x x^2 \, dx$$

By Eq. (21-13*b*),

$$\sum x^2 = n \int_{-\infty}^{+\infty} P_x x^2 \, dx$$

By Eq. (21-10), $P_x = y = \frac{h}{\sqrt{\pi}} e^{-h^2x^2}$

Substituting, $\frac{\Sigma x^2}{n} = \frac{h}{\sqrt{\pi}} \int_{-\infty}^{+\infty} e^{-h^2x^2} x^2 \, dx$

or $\sigma^2 = \frac{2h}{\sqrt{\pi}} \int_0^{\infty} e^{-h^2x^2} x^2 \, dx$

Rewriting, $\sigma^2 = -\frac{1}{h\sqrt{\pi}} \int_0^{\infty} xe^{-h^2x^2}(-2h^2x \, dx)$

The formula for integration by parts is

$$\int_0^{\infty} u \, dv = \Big[uv \Big]_0^{\infty} - \int_0^{\infty} v \, du$$

Taking $u = x$ and $du = dx$,

$$dv = e^{-h^2x^2}(-2h^2x \, dx)$$
$$v = e^{-h^2x^2}$$

Then $\sigma^2 = -\frac{1}{h\sqrt{\pi}} \left(\Big[xe^{-h^2x^2} \Big]_0^{\infty} - \int_0^{\infty} e^{-h^2x^2} \, dx \right)$

The value of the bracket is zero, and the value of the integral is, by Eq. (21-9), $\sqrt{\pi}/2h$.

Substituting, $\sigma^2 = -\frac{1}{h\sqrt{\pi}} \left(-\frac{\sqrt{\pi}}{2h} \right)$

$$\sigma = \sqrt{\frac{\Sigma x^2}{n}} = \frac{1}{\sqrt{2}\, h} \qquad (21\text{-}14)$$

Also, by Eq. (21-12), $\epsilon = \frac{0.4769}{h} = 0.4769 \sqrt{2} \sqrt{\frac{\Sigma x^2}{n}}$

$$\epsilon = 0.6745 \sqrt{\frac{\Sigma x^2}{n}} \qquad (21\text{-}15)$$

Collecting the expressions of precision,

$$h = \sqrt{\frac{n}{2\,\Sigma x^2}} \qquad \sigma = \sqrt{\frac{\Sigma x^2}{n}} \qquad \epsilon = 0.6745 \sqrt{\frac{\Sigma x^2}{n}}$$

It should be noted that the above values are based on the theoretical probability curve in which the errors x are computed from the true value of the quantity, i.e., the average of an infinite number of measurements.

21-15. Rules for Computation Involving Measures of Precision. From the above, the following laws can be derived:

1. The expressions of precision always bear a constant relationship to each other. For example, from Eqs. (21-12) and (21-14)

$$\epsilon = 0.6745\sigma = \frac{0.4769}{h} \qquad (21\text{-}16)$$

2. When a measurement consists of the algebraic sum of several part measurements, each having a different standard error, the standard error of the measurement is equal to the square root of the sum of the squares of the standard errors of the various part measurements.

Let m be a measurement which is the sum of two determinations a and b, that is,

$$m = a + b$$

Let $x_1, x_2, x_3, \ldots, x_n$ be the errors in n determinations of a and $y_1, y_2, y_3, \ldots, y_n$ be the errors in n determination of b.

The various determinations of m can then be assumed to be the following:

$$\begin{aligned} m_1 &= a_1 + x_1 + b_1 + y_1 \\ m_2 &= a_2 + x_2 + b_2 + y_2 \qquad \text{etc.} \end{aligned}$$

The errors can then be expressed thus:

$$\begin{aligned} &\text{For } m_1 \qquad x_1 + y_1 \\ &\text{For } m_2 \qquad x_2 + y_2 \qquad \text{etc.} \end{aligned}$$

By definition, the standard error for m, σ_s, is expressed as follows:

$$\begin{aligned} n\sigma_s^2 &= (x_1 + y_1)^2 + (x_2 + y_2)^2 + \cdots \\ n\sigma_s^2 &= \Sigma x^2 + 2\Sigma xy + \Sigma y^2 \end{aligned}$$

Since x and y are each as likely to be plus as minus, Σxy approaches zero as n becomes large. Dropping this term,

$$\sigma_s = \sqrt{\frac{\Sigma x^2}{n} + \frac{\Sigma y^2}{n}}$$

Again, from the definition of standard error,

$$\sigma_s = \sqrt{\sigma_a^2 + \sigma_b^2}$$

In like manner it may be proved that if

$$\begin{aligned} m &= a - b \\ \sigma_s &= \sqrt{\sigma_a^2 + \sigma_b^2} \end{aligned}$$

When m consists of the algebraic sum of a number of measurements, they may be grouped to make pairs and the standard error will be again as stated previously. For example, if

$$\begin{aligned} m &= a + b - c + e - e \cdots \\ \sigma_s &= \sqrt{\sigma_{a+b}^2 + \sigma_c^2 + \sigma_{d+e}^2 + \cdots} \\ \sigma_s &= \sqrt{\sigma_a^2 + \sigma_b^2 + \sigma_c^2 + \sigma_d^2 + \sigma_e^2 + \cdots} \end{aligned} \tag{21-17}$$

3. When a measurement consists of the algebraic sum of n part measurements, each having the same standard error σ_a, the standard error of the measurement σ_q is equal to $\sqrt{n}$ times the standard error of each part measurement. By Eq. (21-17)

$$\sigma_q = \sqrt{n\sigma_a^2} = \sigma_a \sqrt{n} \tag{21-18}$$

4. When a measurement consists of the average of several measurements of the same quantity, each having the same standard error σ_a the standard error σ_0 of the average measurement is equal to $1/\sqrt{n}$ times the standard error of the several measurements. By Eq. (21-17)

$$\sigma_0 = \frac{\sigma_a \sqrt{n}}{n} = \frac{\sigma_a}{\sqrt{n}} \tag{21-19}$$

5. When a measurement m consists of the product of several measurements a, b, $c, \ldots$, having different standard errors $\sigma_a, \sigma_b, \sigma_c, \ldots$, the standard error of the measurement σ_p is found as follows:

Let Δa, Δb, Δc, . . . , represent increments in the measurements a, b, c, . . . , and let Δm represent the resulting increment in m. Thus

$$m + \Delta m = (a + \Delta a)(b + \Delta b)(c + \Delta c) \cdots$$

Multiplying together the factors on the right and discarding the products of increments, there results

$$m + \Delta m = abc \cdots + (ab \cdots)\,\Delta c + (ac \cdots)\,\Delta b + (bc \cdots)\,\Delta a + \cdots$$

but $$m = abc$$

Subtracting, $$\Delta m = (ab \cdots)\,\Delta c + (ac \cdots)\,\Delta b + (bc \cdots)\,\Delta a + \cdots$$

If the increments are caused by accidental errors, there results, by Eq. (21-17),

$$\sigma_p = \sqrt{(ab \cdots)^2\sigma_c^2 + (ac \cdots)^2\sigma_b^2 + (bc \cdots)^2\sigma_a^2 + \cdots}$$

or, if desired,

$$\sigma_p = \sqrt{(abc \cdots)^2\left[\left(\frac{\sigma_a}{a}\right)^2 + \left(\frac{\sigma_b}{b}\right)^2 + \left(\frac{\sigma_c}{c}\right)^2 + \cdots\right]}$$

6. In the general case m is any function whatever of the measured quantities a_1, a_2, a_3, . . . , and is expressed by

$$m = f(a_1, a_2, a_3, \ldots)$$

An increment in m caused by various increments in a_1, a_2, a_3 expressed by x_1, x_2, x_3, . . . , can be computed as follows:

$$\Delta m = \frac{\partial m}{\partial a_1}x_1 + \frac{\partial m}{\partial a_2}x_2 + \frac{\partial m}{\partial a_3}x_3 + \cdots$$

If the increments are caused by accidental errors there results, by Eq. (21-17),

$$\partial_m{}^2 = \left(\frac{\partial m}{\partial a_1}\sigma_1\right)^2 + \left(\frac{\partial m}{\partial a_2}\sigma_2\right)^2 + \left(\frac{\partial m}{\partial a_3}\sigma_3\right)^2 + \cdots \qquad (21\text{-}20)$$

7. It is evident from Eq. (21-16) that these laws apply to the probable error ϵ and to the reciprocal of the precision, $1/h$, exactly as they apply to the standard error σ.

21-16. Application to Actual Observations. It has been stated that the expressions determined above are based on the assumption that the value of the quantity from which the errors were computed was the average of an infinite number of measurements and therefore the true value of the quantity. Sometimes in actual practice the true value of a quantity is known and sometimes it must be inferred from the average of the measurements. For example, assume that an expert rifleman fires at a target under such conditions that no systematic errors exist. The actual distance from the center of the bull's eye would represent the true error of each shot. If he continued to shoot, the average of the positions of the shots would continue to approach the bull's eye but theoretically would never quite reach it. If the errors were based on the distances to the center of the bull's eye, the above formulas apply. If the errors were based on the distance to the average of the shots, other formulas are necessary.

Each error based on the average of a finite number of measurements should be increased by the error of the average in order to obtain the true error.

Let v represent the values of errors based on the average, x represent the true errors, and δ the error of the average.

$$x = v + \delta$$

$$\sigma = \sqrt{\frac{\Sigma x^2}{n}} = \sqrt{\frac{\Sigma(v + \delta)^2}{n}}$$

But $$\Sigma(v + \delta)^2 = \Sigma v^2 + 2\Sigma v\delta + n\delta^2$$

and since $\Sigma v = 0$, the middle term equals zero.

Therefore $$\sigma = \sqrt{\frac{\Sigma v^2 + n\delta^2}{n}}$$

The approximation usually used for δ is taken from Eq. (21-19):

$$\delta = \sigma_0 = \frac{\sigma}{\sqrt{n}}$$

Substituting for δ and solving for σ, the standard error of a single measurement is

$$\sigma = \sqrt{\frac{\Sigma v^2}{n - 1}} \tag{21-21}$$

Since all measures of precision are proportional, the following table can be written:

TABLE 21-1. MEASURES OF PRECISION

	Standard error σ		Probable error ϵ		Precision h	
One of n determinations	$\sqrt{\frac{\Sigma x^2}{n}}$	$\sqrt{\frac{\Sigma v^2}{n-1}}$	$0.6745\sqrt{\frac{\Sigma x^2}{n}}$	$0.6745\sqrt{\frac{\Sigma v^2}{n-1}}$	$\sqrt{\frac{n}{2\Sigma x^2}}$	$\sqrt{\frac{n-1}{2\Sigma v^2}}$
Average of n determinations	$\sqrt{\frac{\Sigma x^2}{n^2}}$	$\sqrt{\frac{\Sigma v^2}{n(n-1)}}$	$0.6745\sqrt{\frac{\Sigma x^2}{n^2}}$	$0.6745\sqrt{\frac{\Sigma v^2}{n(n-1)}}$	$\sqrt{\frac{n^2}{2\Sigma x^2}}$	$\sqrt{\frac{n(n-1)}{2\Sigma v^2}}$

where Σx^2 or Σv^2 is the sum of the squares of the errors by which each determination differs from the true value of the quantity or from the average of the determinations, respectively.

21-17. Collected Formulas for Computation. Table 21-2 gives formulas for combining measures of precision or accuracy. In these formulas ϵ or $1/h$ or any related measures of precision can be substituted for σ.

21-18. Other Measures of Accuracy. Frequently it is advantageous to measure the accuracy by other standards than ϵ, σ, or $1/h$. For example, in evaluating the accuracy of an instrument, it would be well to know the size of the error that would never be exceeded. One hundred per cent of the errors would be equal to or smaller than such an error. Theoretically such an error is always infinitely large. From a practical standpoint, therefore, some other percentage must be used. Ninety per cent is an excellent value. When 90 per cent of the errors are smaller than an error of a certain size, that error might be called the 90 per cent error. By this definition, the probable error, ϵ, would be called the 50 per cent error. Other percentages may be used.

Equations (21-14) and (21-15) demonstrate that the formulas for computing the size of an error that is appropriate to a certain per cent are the same except for a

TABLE 21-2

$$s = a + b - c + d - e + \cdots$$
$$\sigma_s = \sqrt{\sigma_a^2 + \sigma_b^2 + \sigma_c^2 + \sigma_d^2 + \sigma_e^2 + \cdots} \qquad (21\text{-}22)$$

$$q = a_1 + a_2 - a_3 + a_4 - a_5 + \cdots + a_n$$
$$\sigma_q = \sigma_a \sqrt{n} \qquad (21\text{-}23)$$

$$o = \frac{a_1 + a_2 + a_3 + \cdots + c_n}{n}$$
$$\sigma_o = \frac{\sigma_a}{\sqrt{n}} \qquad (21\text{-}24)$$

$$p = abc \ldots$$
$$\sigma_p = \sqrt{(abc \ldots)^2 \left[\left(\frac{\sigma_a}{a}\right)^2 + \left(\frac{\sigma_b}{b}\right)^2 + \left(\frac{\sigma_c}{c}\right)^2 + \cdots\right]} \qquad (21\text{-}25)$$

$$m = f(a_1, a_2, a_3 \ldots)$$
$$\sigma_m = \sqrt{\left(\frac{\partial m}{\partial a_1}\sigma_1\right)^2 + \left(\frac{\partial m}{\partial a_2}\sigma_2\right)^2 + \left(\frac{\partial m}{\partial a_3}\sigma_3\right)^2} \qquad (21\text{-}26)$$

single constant which can be computed from Pierce's "Tables." It is evident that the following equation is true:

$$\text{Size of } x\% \text{ error} = C_x \sqrt{\frac{\Sigma v^2}{n - 1}}$$

where C_x is the constant appropriate to the $x\%$ error. The appropriate values of C are given in Table 21-3 and in Chart I (page 706).

TABLE 21-3

% Error	C	% Error	C	% Error	C
5	0.0627	40	0.5244	70	1.0364
10	0.1257	45	0.5978	75	1.1503
15	0.1891	50 e	0.6745	80	1.2816
20	0.2534	55	0.7554	85	1.4395
25	0.3186	60	0.8416	90	1.6449
30	0.3853	65	0.9346	95	1.9599
35	0.4538	68.27 σ	1.0000	100	

CHAPTER 22

LEAST SQUARES

PRINCIPLES

22-1. Redundant Measurements. No important survey is complete without its full complement of check measurements. They must be arranged so that there will be available a total of at least two routes through the survey that connect each station or benchmark with every other corresponding mark in the system. They provide the only sure check against blunders and are the means of gauging the accuracy of the work. They must be made with the same care and precision employed in other measurements and accordingly they may be equally relied upon to determine results. Of necessity, check measurements are redundant, and it would be only by chance that they would give exact agreement with the other determinations. By combining the results of the check measurements with the others, final values are obtained that are usually more accurate than those that would have been obtained by either alone. The process of combining the results is, of course, known as the survey adjustment.

If blunders and systematic errors have been eliminated from a survey, the only cause for lack of agreement among redundant determinations are the accidental errors in the actual measurements themselves. It has been shown in Chap. 21 that accidental errors obey the laws of chance and that, accordingly, when only accidental errors exist, the more nearly a measurement agrees with its true value, the higher is its probability of occurrence.

It follows that a set of corrections should be applied to the measurements which will result in making the measurements geometrically consistent and yet be that set of corrections which will eliminate the set of errors that has the greatest probability of occurrence.

The probability of the occurrence of errors is stated by the probability equation, Eq. (21-10),

$$y = \frac{h}{\sqrt{\pi}} e^{-h^2x^2} \qquad (21\text{-}10)$$

where h = measure of precision and therefore a constant for a set of measurements each made in the same way

x = size of the error considered

y = probability of occurrence of that error

The probability of any particular set of measurements occurring at the same time is the product of their individual probabilities. If $x_1, x_2, x_3, \ldots, x_n$ is a set of errors and $y_1, y_2, y_3, \ldots, y_n$ are their respective probabilities, the probability y_s of the occurrence of the set is the following:

$$y_s = y_1 y_2 y_3 \cdots y_n$$

Substituting for $y_1y_2y_3 \cdots y_n$ their values from Eq. (21-10) above,

$$y_s = \left(\frac{h}{\sqrt{\pi}}\right)^n e^{-h^2(x_1^2+x_2^2+x_3^2+\cdots x_n^2)}$$

The value of this function is greatest when the expression $x_1^2 + x_2^2 + x_3^2 + \cdots x_n^2$ is a minimum. (Note that it is part of a negative exponent.) Accordingly, the set of errors that has the greatest probability of occurrence is that in which the sum of squares of the errors is a minimum.

Since corrections equal errors (with their signs changed), the set of corrections desired is that in which the sum of the squares of the corrections is minimum. The method of determining these values is called the **method of least squares** and an adjustment of survey data by this method is often called a **least-squares adjustment.**

22-2. The Arithmetic Mean. A least-squares adjustment of several measurements of the same quantity is the arithmetic mean of the measurements. A least-squares adjustment of this kind can be made as follows: Let 1000.29, 1000.35, and 1000.20 be three determinations of a certain distance. Let v_1, v_2, and v_3 be the corrections sought, and let M represent the adjusted value:

$$\begin{aligned} M &= 1000.29 + v_1 \qquad & v_1 &= M - 1000.29 \\ M &= 1000.35 + v_2 \qquad & v_2 &= M - 1000.35 \qquad \text{condition equations} \\ M &= 1000.20 + v_3 \qquad & v_3 &= M - 1000.20 \end{aligned}$$

Let U represent the function whose minimum is required:

$$\begin{aligned} U &= v_1^2 + v_2^2 + v_3^2 \\ U &= (M - 1000.29)^2 + (M - 1000.35)^2 + (M - 1000.20)^2 \\ \frac{dU}{dM} &= 2M - 2(1000.29) + 2M - 2(1000.35) + 2M - 2(1000.20) \end{aligned}$$

Equating to zero to find the minimum and solving,

$$\begin{aligned} 3M &= 3000.84 \\ M &= 1000.28 \qquad \text{normal equation} \end{aligned}$$

Note that this is equal to the arithmetic mean of three determinations.

22-3. Use of Average for Adjustment. Whenever possible, the principle of the arithmetic mean is used to adjust survey data. It is used in length measurement, the adjustment of the three angles of a triangle, etc. When the most probable set of corrections are desired, all corrections that affect each other must be computed simultaneously. Accordingly, the arithmetic mean can be used for this purpose only when the corrections are related by very simple conditions.

22-4. The Need for the Least-squares Adjustment. Consider the level net in Fig. 22-1. Each link is the same length and the elevation differences noted were computed in the direction of the arrows. The corrections must be such that the elevation difference between any pair of benchmarks shall be the same by whatever route it is computed. Using v's to denote corrections, this can be stated thus:

$$\begin{aligned} (a + v_1) + (b + v_2) + (c + v_3) + (d + v_4) &= 0 \\ (a + v_1) + (e + v_5) + (d + v_4) &= 0 \\ (b + v_2) + (c + v_2) - (e + v_5) &= 0 \end{aligned}$$

It will be noted that if the last two equations are added, the first equation is obtained. Since the first equation is therefore not independent, it can be dropped.

Substituting the observed elevation differences in the last two equations,

$$(+21.14 + v_1) + (-5.20 + v_5) + (-15.96 + v_4) = 0$$
$$(+5.08 + v_2) + (-10.06 + v_3) - (-5.20 + v_5) = 0$$

Combining numbers,

$$\left.\begin{aligned} v_1 + v_5 + v_4 - 0.02 &= 0 \\ v_2 + v_3 - v_5 + 0.22 &= 0 \end{aligned}\right\} \quad \text{condition equations}$$

These are the conditions that must be satisfied and they are therefore called the condition equations. Since v_5 appears in both, the v's cannot be determined separately for each equation and therefore the usual method (i.e., making the v's equal) cannot be utilized to obtain the most probable values of the v's. Accordingly, some other method like the method of least squares must be applied.

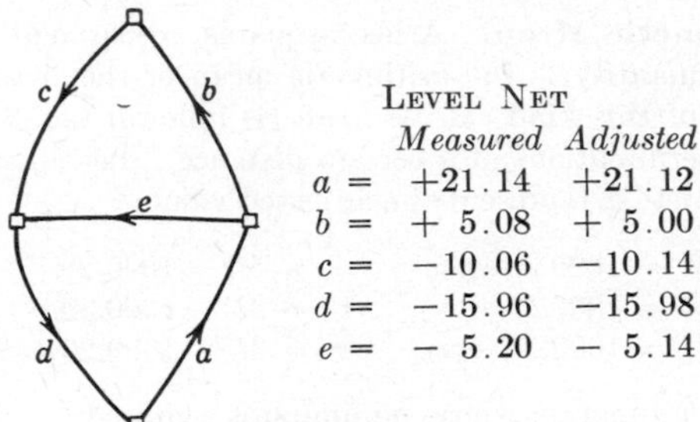

	Measured	*Adjusted*
a =	+21.14	+21.12
b =	+ 5.08	+ 5.00
c =	−10.06	−10.14
d =	−15.96	−15.98
e =	− 5.20	− 5.14

Fig. 22-1. The simplest problem requiring least squares.

There are many ways of proceeding with a least-squares adjustment. Let us solve for v_1 and v_2:

$$v_1 = -v_5 - v_4 + 0.02$$
$$v_2 = -v_3 + v_5 - 0.22$$
$$U = v_1^2 + v_2^2 + v_3^2 + v_4^2 + v_5^2$$

Substituting for v_1 and v_2,

$$U = (-v_5 - v_4 + 0.02)^2 + (-v_3 + v_5 - 0.22)^2 + v_3^2 + v_4^2 + v_5^2$$

Taking partial derivatives,

$$\left.\begin{aligned} \tfrac{1}{2}\frac{\partial U}{\partial v_3} &= 2v_3 - v_5 + 0.22 = 0 \\ \tfrac{1}{2}\frac{\partial U}{\partial v_4} &= 2v_4 + v_5 - 0.02 = 0 \\ \tfrac{1}{2}\frac{\partial U}{\partial v_5} &= -v_3 + v_4 + 3v_5 - 0.24 = 0 \end{aligned}\right\} \quad \text{normal equations}$$

Solving simultaneously and correcting the measurements,

$$v_5 = +0.06 \qquad e' = -\ 5.14$$
$$v_4 = -0.02 \qquad d' = -15.98$$
$$v_3 = -0.08 \qquad c' = -10.14$$

Substituting in the condition equations,

$$v_2 = -0.08 \qquad b' = +\ 5.00$$
$$v_1 = -0.02 \qquad a' = +21.12$$

These corrected measurements or adjusted differences may now be checked by substitution in the level net:

$$a' + e' + d' = 21.12 - 5.14 - 15.98 = 0$$
$$b' + c' - e' = 5.00 - 10.14 + 5.14 = 0$$

22-5. Corrections Minimized Must Be Corrections to Actual Observations. It must be carefully noted that the least-squares method is based on the behavior of accidental errors. The rules which govern this behavior are true only when the errors are errors that occur by pure chance. Observational errors (free from blunders and systematic errors) are of this nature. Of course linear functions of such errors are also of the same nature, but other functions are not. For example, in ordinary triangulation, when angles are measured, the errors in the lengths of the sides depend on the errors in the measured angles through the sine relationship. The errors in the sides therefore do not obey the laws of chance alone. They depend on the accidental angular errors modified by the sines of the angles.

If the corrections to the lengths of the sides are minimized (corrections found whose sum of squares is minimum), their most probable value will not be found. Instead the corrections to the angles must be minimized.

22-6. Weights. When the measurements of a survey consist of the sums of different numbers of equally precise observations, the corrections to each measurement must be weighted according to the square root of the number of observations.

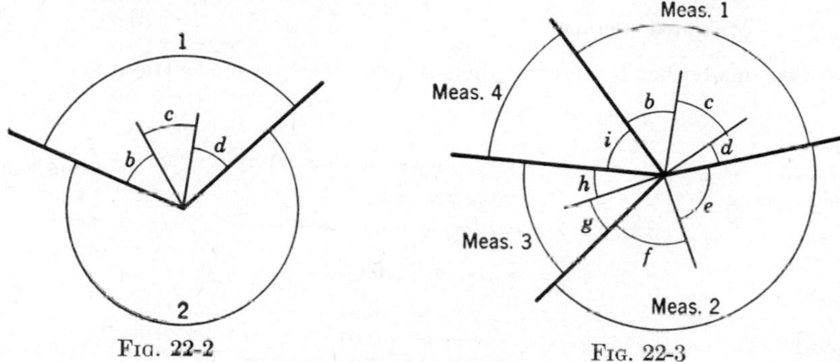

FIG. 22-2 FIG. 22-3

For example, in Fig. 22-2, angle 1 was determined by measuring b, e, and d—three equally precise observations, while the angle 2 was measured in one observation. The correction to 1 must be equal to $\sqrt{3}$ times the correction to 2. Then

$$v_1 = \sqrt{3}\, v_2$$
$$v_1^2 = 3v^2$$
$$\tfrac{1}{3}v_1^2 = v_2^2$$

This may be approached in a different way. Note Fig. 22-3.

Let a_1 = correction to each observation in measurement No. 1

n_1 = number of observations in measurement No. 1

v_1 = total correction to measurement No. 1

and likewise for all the measurements by changing the subscripts. Then

$$v_1 = n_1 a_1 \qquad v_2 = n_2 a_2 \qquad \text{etc.}$$

But from the theory of least squares, the function to be minimized is the sum of the squares of the corrections to all the observations; thus

$$U = n_1 a_1^2 + n_2 a_2^2 + \cdots$$

Substituting for the a's their values from above,

$$U = \frac{1}{n_1} v_1^2 + \frac{1}{n_2} v_2^2 + \cdots$$

But since the derivatives will be equated to zero, the coefficients of the v^2's may be multiplied by any constant:

$$U = \frac{K}{n_1} v_1^2 + \frac{K}{n_2} v_2^2 + \cdots$$

It is evident, then, that the v^2's must be weighted in proportion to the reciprocals of the number of observations that they represent.

The weights used in this way are the same weights used to determine a weighted arithmetic mean. For example, assume that two level lines were run between the same two points using the same precision but different length routes. By weighted arithmetic mean,

Line	*Length*	*Weight*		*Elev. diff.*	*Weighted value*
1	8 miles	⅛	or 1 ×	10.03	10.03
2	2 miles	½	or 4 ×	10.08	40.32
Sums			5		50.35
Weighted average					10.07

By least squares, let D represent the adjusted difference in elevation:

$$10.03 + v_1 = D$$
$$10.08 + v_2 = D$$

Subtracting,

$$v_1 = v_2 + 0.05 \qquad \text{condition equation}$$
$$U = \tfrac{1}{8}v_1^2 + \tfrac{1}{2}v_2^2$$
$$U = \tfrac{1}{8}(v_2 + 0.05)^2 + \tfrac{1}{2}v_2^2$$
$$\tfrac{1}{2}\frac{\partial U}{\partial v_2} = \tfrac{1}{8}v_2 + \tfrac{1}{8}(0.05) + \tfrac{1}{2}v_2 = 0$$
$$v_2 = -0.01$$
$$v_1 = +0.04$$

Line 1: $10.03 + 0.04 = 10.07$
Line 2: $10.08 - 0.01 = 10.07$

which are the same as the result by the weighted mean.

PROCEDURE FOR APPLYING LEAST SQUARES

22-7. Solutions. The examples in the preceding paragraphs illustrate the principles of the method of least squares and also serve to demonstrate, in an elementary way, the method of solving a least-squares problem. There are so many kinds of problems and so many ways of handling even the same problem that it is difficult to give rules for solution. The steps common to most solutions are listed in the next paragraph. Following them are examples of least-squares solutions of the types of problems the surveyor is likely to encounter. There are many other methods and variations in the solutions that could have been used to solve these problems. However, by following the solutions given, the surveyor should be able to handle the problems most likely to occur in practice and the student who wishes to make a thorough study of the method of least squares should be able to embrace the principles involved with a minimum expenditure of time.

1. Express each desired adjusted result in the form of the observed value plus a symbol (usually a v with a subscript) to represent the correction.

2. Express the interrelationships between the adjusted results imposed by the geometric conditions in the form of equations.

3. Eliminate identical equations and simplify. The equations that remain are the condition equations.

4. Write the function expressing the sum of the squares of the corrections (the v's). Multiply each v^2 term by its weight. The weight is a constant which is proportional to the reciprocal of the number of observations (or the number of quantities affected by purely accidental errors) of equal precision which together require the correction v. When the observations are of different precisions, each v^2 term should be multiplied by a weight proportional to the reciprocal of the probable error of the observation to which it applies.

5. Make substitutions in this function from the condition equations.

6. Take a partial derivative of this function with respect to each v necessary. Equate each derivative to zero. This gives the normal equations. There should be one for each necessary value.

When correlates are used, the process requires more explanation. It is explained in the section describing the application of least squares to a level net.

7. Solve the normal equations simultaneously.

8. The results of the solution of normal equations are used as required.

ORDINATES AT ESTABLISHED ABSCISSAS

22-8. Example 1. Suppose that the descriptions of the four lots shown in Fig. 22-4 gave no side line lengths, but indicated that each side line ran to a back line that was

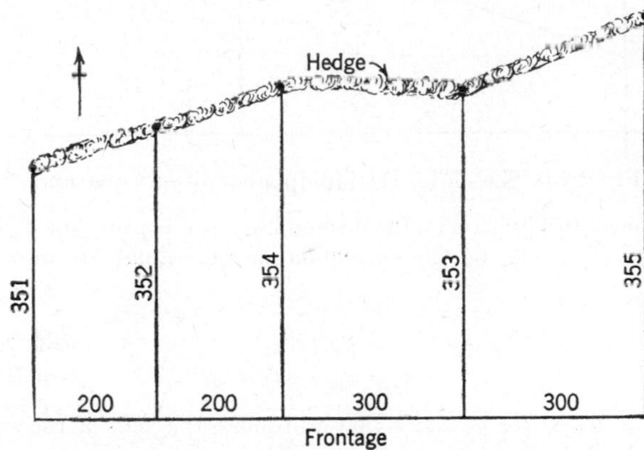

FIG. 22-4. Line to be straightened by least squares.

known to be originally straight over-all. Actual measurements from a well-established front line to an existing hedge gave the distances shown on each side line. It was known that the hedge was originally laid out along stakes, since lost, that marked the corners of the lots. It is desired to determine the lengths of the side lines that would most nearly reestablish the straight back line.

If it is assumed that the movements of hedge from the stakes occurred entirely by chance, the problem should be solved by least squares. An approximation can be reached graphically but this requires judgment and is not so satisfactory to the owners.

The numbers in this problem can be simplified by choosing a unit of 100 ft to express distances along the frontage and by subtracting 350 ft from each side line. The true values can be restored when the computation is complete.

The problem is one of fitting a straight line to a series of ordinates measured at various established abscissas (see Figs. 22-5 and 22-6).

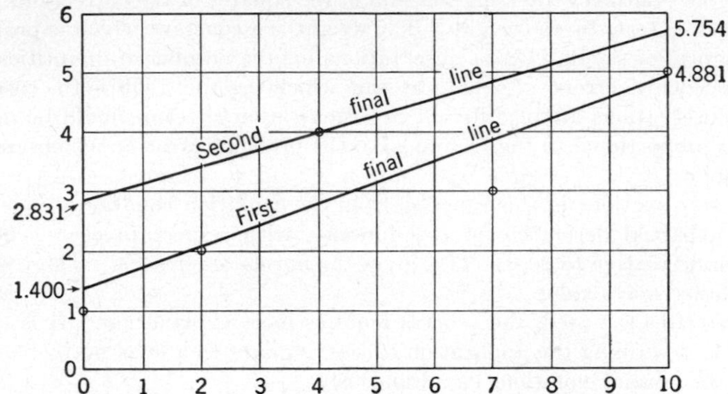

FIG. 22-5. Least-squares adjustments of a survey line.

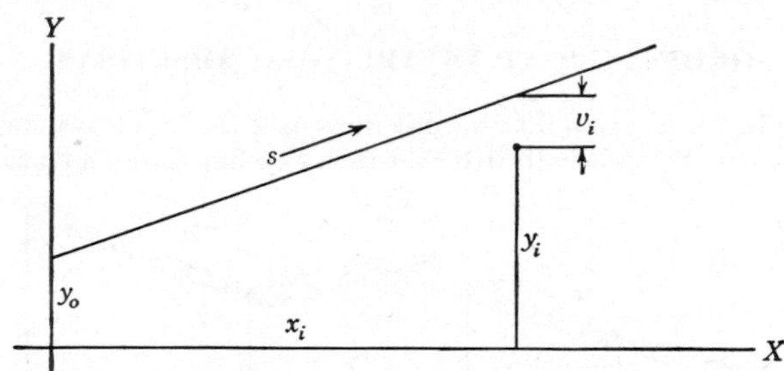

FIG. 22-6. Sketch for setting up a condition equation.

Let the line shown in Fig. 22-6 be the desired line, intercepting the Y axis at y_0 and having the slope s. Let v_i be the correction to an ordinate y_i measured at any abscissa x_i.

Theory

$$y_i + v_i = y_0 + sx_i$$

or

$$v_i = y_0 + sx_i - y_i \qquad \text{condition equation}$$

Values must be found for y_0 and s that will make the sum of the squares of the various v's a minimum.

$$U = \sum_{i=1}^{i=n} v_i^2$$

$$U = \sum_{i=1}^{i=n} (y_0 + sx_i - y_i)^2$$

Taking partial derivatives and equating to zero,

$$\tfrac{1}{2}\frac{\partial U}{\partial y_0} = 0 = ny_0 + s\sum_{i=1}^{i=n} x_i - \sum_{i=1}^{i=n} y_i$$

$$\tfrac{1}{2}\frac{\partial U}{\partial s} = 0 = s\sum_{i=1}^{i=n} x_i^2 + y_0\sum_{i=1}^{i=n} x_i - \sum_{i=1}^{i=n} x_iy_i$$

Denoting

$$a = \sum_{i=1}^{i=n} x_i \qquad b = \sum_{i=1}^{i=n} y_i \qquad c = \sum_{i=1}^{i=n} x_i^2 \qquad d = \sum_{i=1}^{i=n} x_iy_i$$

$$0 = ny_0 + as - b \qquad \text{normal equation (1)}$$
$$0 = ay_0 + cs - d \qquad \text{normal equation (2)}$$

Application. Divide the distances of the side lines from the west corner by 100 to obtain the x values. Subtract 350 ft from the measured lengths of the side lines to obtain the y values. List them in a table:

Line	x	y	x^2	xy
1	0	1	0	0
2	2	2	4	4
3	4	4	16	16
4	7	3	49	21
5	10	5	100	50
Σ	23	15	169	91
	a	b	c	d

Compute the remaining two columns and find the sum of each.

Substitute in the normal equations (n = number of measurements):

$$0 = 5y_0 + 23s - 15$$
$$0 = 23y_0 + 169s - 91$$

Solve simultaneously with the following results:

$$y_0 = +1.400 \qquad s = +0.3481$$

The lengths of the side lines can be computed thus:

Line	Cor.	y_0	s	x	Length
1	350	+1.400			= 351.400
2	350	+1.400	+0.3481 ×	2	= 352.096
3	350	+1.400	+0.3481 ×	4	= 352.792
4	350	+1.400	+0.3481 ×	7	= 353.837
5	350	+1.400	+0.3481 ×	10	= 354.881

22-9. Example 2. Suppose that in Example 1 a monument existed on the line between the two central lots (the line 400 ft from the westerly corner) 354 ft from the front line. It was definitely established that it marked the back line.

This resolves itself into the problem of fitting a straight line to a series of ordinates measured at various established abscissas but with one ordinate fixed.

It is convenient to place the origin at the fixed position, i.e., 400 ft from the westerly line and 354 ft from the front.

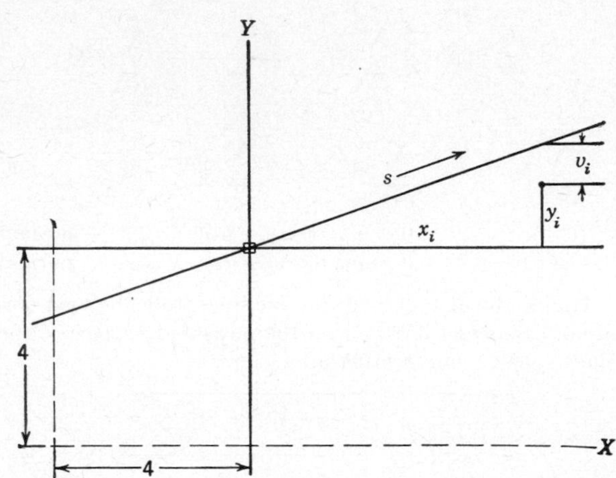

FIG. 22-7. Sketch for setting up a condition equation.

Theory. From Fig. 22-7

$$y_i + v_i = sx_i$$

$$v_i = sx_i - y_i \qquad \text{condition equation}$$

Then

$$U = \sum_{i=1}^{i=n} v_i^2 = \sum_{i=1}^{i=n} (sx_i - y_i)^2$$

$$\tfrac{1}{2}\frac{\partial U}{\partial s} = 0 = s\sum_{i=1}^{i=n} x_i^2 - \sum_{i=1}^{i=n} x_i y_i$$

Denoting

$$c = \sum_{i=1}^{i=n} x_i^2 \qquad d = \sum_{i=1}^{i=n} x_i y_i$$

$$0 = cs - d \qquad \text{normal equation}$$

Application. Divide the distances of the side lines from the west corner by 100 and subtract 4 for the x values. Subtract 4 for the y values. List them in the table.

Line	x	y	x^2	xy
1	−4	−3	16	+12
2	−2	−2	4	+ 4
3	0	0	0	0
4	+3	−1	9	− 3
5	+6	+1	36	+ 6
			65	19
			c	d

Complete the table and substitute in the normal equation:

$$0 = 65s - 19$$
$$s = 0.2923$$

Compute the lengths of the side lines:

Line	Cor.	s	x	Length
1	354	$+(0.2923)$	$\times\ (-4)$	$= 352.831$
2	354	$+(0.2923)$	$\times\ (-2)$	$= 353.415$
3	354	$+(0.2923)$	$\times\ 0$	$= 354.000$
4	354	$+(0.2923)$	$\times\ (+3)$	$= 354.877$
5	354	$+(0.2923)$	$\times\ (+6)$	$= 355.754$

RANDOM POSITIONS

22-10. Example. Usually a straight line is established by setting monuments along it. When it is found later that the monuments do not mark a straight line, it can be assumed that the errors are perpendicular to the line that was meant to be

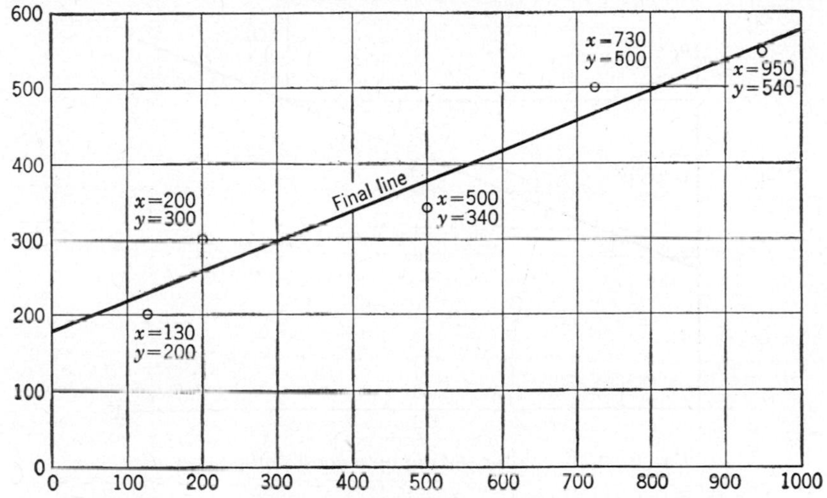

FIG. 22-8. Survey line to be "averaged" among several monuments.

laid out originally. To determine the most likely line, a survey should be made of the existing monuments and their coordinates computed. Figure 22-8 shows the arrangement of the monuments. Round numbers are used for simplicity.

Theory. Fig. 22-9. Let the desired line be represented by the general equation

$$y = a + mx \qquad \text{condition equation}$$

where

$$m = \tan \psi$$

Let P represent any marker having the coordinates $x_i y_i$. Let O_i represent the offset (perpendicular distance) from the final line:

$$O_i = b_i \cos \psi = b_i \frac{1}{\sqrt{1 + \tan^2 \psi}}$$

but

$$b_i = y_i - a - x_i \tan \psi$$

Therefore

$$O_i = \frac{y_i - a - x_i \tan \psi}{\sqrt{1 + \tan^2 \psi}}$$

Let

$$U = \sum_{i=1}^{i=n} O_i^2 = \sum_{i=1}^{i=n} \left(\frac{y_i - a - mx_i}{\sqrt{1 + m^2}} \right)^2$$

$$\frac{dU}{da} = 0 = \sum_{i=1}^{i=n} y_i - na - m \sum_{i=1}^{i=n} x_i$$

$$\frac{1}{n} \sum_{i=1}^{i=n} y_i = a + m \frac{1}{n} \sum_{i=1}^{i=n} x_i \qquad \text{normal equation (1)}$$

This shows that whatever may be the values of a and m, the line passes through the point which has for its x coordinate the average of the x's and for its y coordinate the average of the y's.

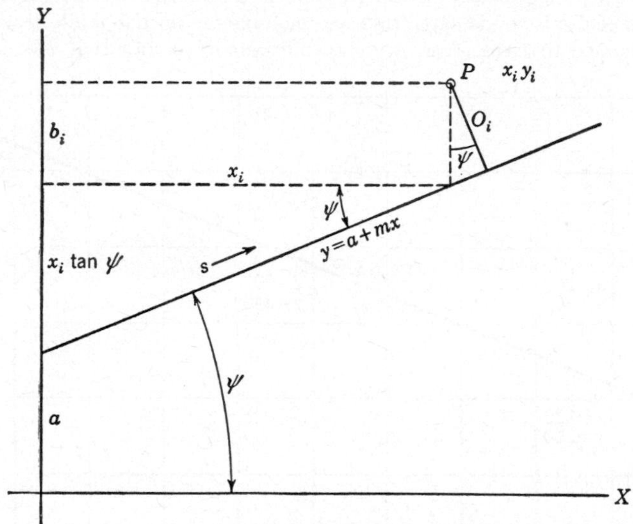

FIG. 22-9. Sketch for setting up a condition equation.

The coordinates of this point are computed in the table: $x_0 = 502$, $y_0 = 376$. To compute the slope m, the origin of coordinates can be shifted to this point as follows:

$$x' = x - x_0 \qquad x' = x - 502$$
$$y' = y - y_0 \qquad y' = y - 376$$

Then

$$U = \sum_{i=1}^{i=+n} O_i^2 = \sum_{i=1}^{i=+n} \left(\frac{y_i' - mx_i'}{\sqrt{1 + m^2}} \right)^2$$

$$\frac{dU}{dm} = 0 = m^2 \sum_{i=1}^{i=+n} y_i'x_i' + m \sum_{i=1}^{i=+n} x_i'^2 - m \sum_{i=1}^{i=+n} y_i'^2 - \sum_{i=1}^{i=+n} x_i'y_i'$$

Denoting $p = \sum_{i=-n}^{i=+n} x_i'^2 \quad q = \sum_{i=-n}^{i=+n} y_i'^2 \quad r = \sum_{i=-n}^{i=+n} x_i'y_i'$

$$0 = m^2 + m\frac{p-q}{r} - 1 \qquad \text{normal equation (2)}$$

No.	x	y	x'	y'	x'^2	y'^2	$x'y'$
1	130	200	−372	−176	138,384	30,976	+65,472
2	200	300	−302	− 76	91,204	5,776	+22,952
3	500	340	− 2	− 36	4	1,296	+ 72
4	730	500	+228	+124	51,984	15,376	+28,272
5	950	540	+448	+164	200,704	26,896	+73,472
	2510	1880			482,280	80,320	190,240
Av.	502	376			p	q	r

Application. Complete the table and substitute values in normal equation (2) to obtain m:

$$m^2 + \frac{482{,}280 - 80{,}320}{190{,}240}m - 1 = 0$$
$$m = +0.39822$$

Substitute in normal equation (1) to obtain a:

$$376 = a + (0.39822)502$$
$$a = 176.094$$

The final equation for the line is therefore

$$y = 176.094 + 0.39822x$$
$$\tan\psi = m$$
$$\psi = 21°42'48.4''$$

The bearing of the line is evidently N68°17′11.6″E.

The values of the offsets can be computed by substitution in the expression for O_i.

$$O_i = (y_i - a - mx_i)\cos\psi$$
$$O_i = (y_i - 176.094 - 0.39822x_i)0.92905$$

The bearing O_i of the offset line is, of course, N21°42′48.4″W. The substitutions are shown in the following table:

No.	1	2	3	4	5
x_i	130.000	200.000	500.000	730.000	950.000
mx_i	51.769	79.644	199.110	290.701	378.309
a	176.094	176.094	176.094	176.094	176.094
$a + mx_i$	227.863	255.738	375.204	466.795	554.403
y_i	200.000	300.000	340.000	500.000	540.000
$y_i - a - mx_i$	−27.863	+44.262	−35.204	+33.205	−14.403
O_i	−25.886	+41.122	−32.706	+30.849	−13.381

Check: offsets should add to zero.

The offsets may be laid out in the field from the monuments. The back bearing is used for plus values and the forward bearing for minus values. The reversal of sign occurs because the measurement in the field is made from the monuments to the line, whereas the computations are based on offsets from the line to the monuments.

DISTANCES AT VARIOUS AZIMUTHS

22-11. Example. Captain F. B. T. Siems, U.S. Coast and Geodetic Survey, in an article entitled "Latitude and Longitude by Altitude Observation," *Surveying and Mapping*, pages 12–15, April, 1945, published a method that he developed to determine a mean by least squares in the method of finding astronomical position by altitudes of celestial bodies at various azimuths (see Chap. 9).

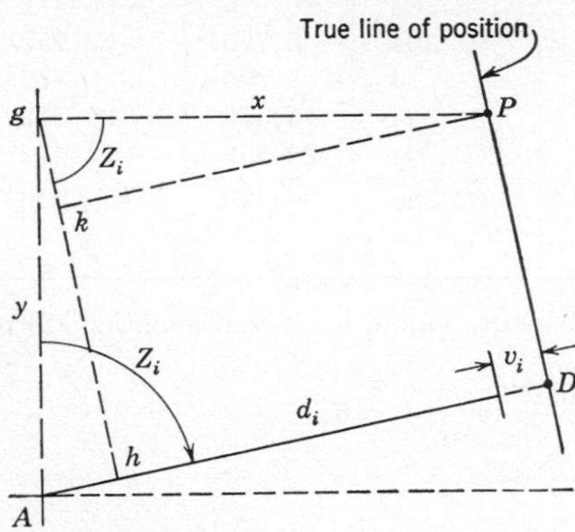

FIG. 22-10. Sketch for setting up a condition equation.

Theory. In Fig. 22-10, A represents the assumed position and P the true position. The differences in latitude and longitude of P with respect to A are y and x. If there were no error in altitude, all of the lines of position would pass through P.

Let d_i be the observed intercept on a star at azimuth Z_i and let v_i be the correction to the intercept to reach the true line of position PD. From g drop the perpendicular gh to AD and from P drop the perpendicular Pk to gh. Then

$$d_i + v_i = Pk + hA$$

$$v_i = x \sin Z_i + y \cos Z_i - d_i$$

$$U = \sum_{i=1}^{i=n} v_i^2 = \sum_{i=1}^{i=n} (x \sin Z_i + y \cos Z_i - d_i)^2$$

$$\frac{1}{2}\frac{\partial U}{\partial x} = 0 = x \sum_{i=1}^{i=n} \sin^2 Z_i + y \sum_{i=1}^{i=n} \sin Z_1 \cos Z_i - \sum_{i=1}^{i=n} d_i \sin Z_i$$

$$\frac{1}{2}\frac{\partial U}{\partial y} = 0 = y \sum_{i=1}^{i=n} \cos^2 Z_i + x \sum_{i=1}^{i=n} \sin Z_i \cos Z_i - \sum_{i=1}^{i=n} d_i \cos Z_i$$

Let $\qquad a_i = \sin Z_i \qquad b_i = \cos Z_i \qquad d = d_i$

Then

$$0 = x \sum_{i=1}^{i=n} a_i^2 + y \sum_{i=1}^{i=n} a_ib_i - \sum_{i=1}^{i=n} a_id_i \qquad \text{normal equation (1)}$$

$$0 = x \sum_{i=1}^{i=n} a_ib_i + y \sum_{i=1}^{i=n} b_i^2 - \sum_{i=1}^{i=n} b_id_i \qquad \text{normal equation (2)}$$

Application. Assume that the azimuths and intercepts of five altitudes were those shown in the appropriate columns in the table. The table is computed as shown (see Fig. 22-11):

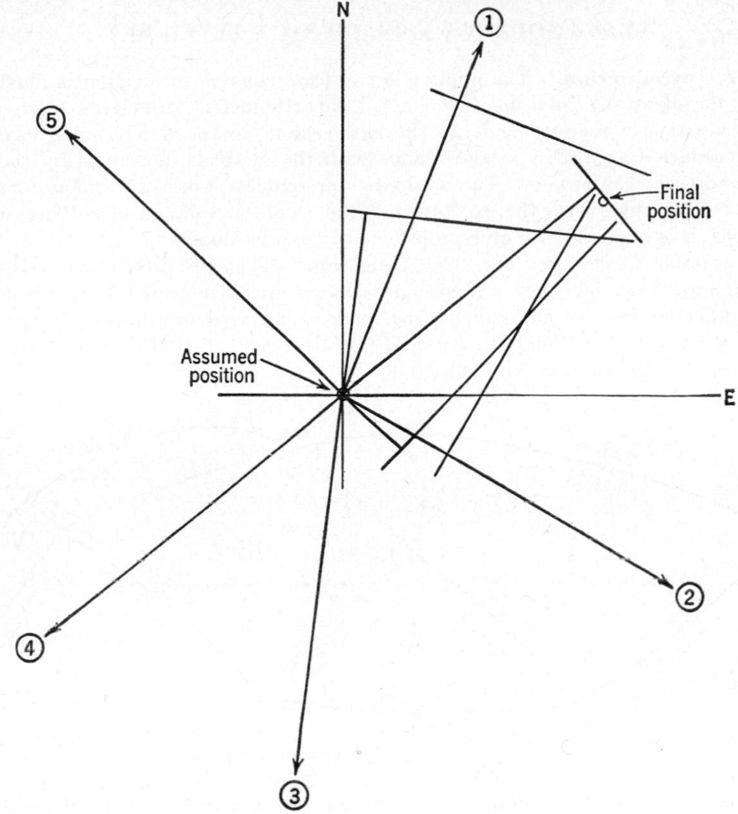

FIG. 22 11. Solution of an altitude fix.

No.	Azim. Z	a sin Z	b cos z	d	a^2	ab	ad	b^2	bd
1	20°50′	+0.3556	+0.9346	+3.3	+0.1265	+0.3323	+1.1735	+0.8735	+3.0842
2	119 40	+0.8689	−0.4950	+1.2	+0.7550	−0.4301	+1.0427	+0.2450	−0.5940
3	186 50	−0.1190	−0.9929	−1.9	+0.0142	+0.1182	+0.2261	+0.9859	+1.8865
4	230 00	−0.7660	−0.6428	−3.4	+0.5868	+0.4924	+2.6044	+0.4132	+2.1855
5	310 20	−0.7623	+0.6472	−0.8	+0.5811	−0.4934	+0.6098	+0.4189	−0.5178
					+2.0636	+0.0194	+5.6565	+2.9365	+6.0444

Substituting in the normal equations,

$$0 = 2.0636x + 0.0194y - 5.6565$$
$$0 = 0.0194x + 2.9365y - 6.0444$$

Solving simultaneously,

$$y = +2.04$$
$$x = +2.72$$

LEAST SQUARES APPLIED TO A LEVEL NET

22-12. Introduction. The application of least squares to leveling is illustrated here by the adjustment of a small level net. No orthometric corrections[1] are required and no mention of them is made, as they are seldom necessary in a small level net. The procedure described is in accordance with the methods developed by the U.S. Coast and Geodetic Survey. The methods represent the work of so many members of this bureau that, with the exception of the Doolittle method of solving normal equations, it is impossible to give proper credit to individuals.

The example chosen (see Fig. 22-12) illustrates all of the problems that usually occur in adjusting a level net. If special problems are encountered, or if it is desired to use other methods of adjustment, the reader is referred to Howard S. Rappleye, "Manual of Leveling Computation and Adjustment," U.S. Coast and Geodetic Survey Special Publication No. 240, 1948.

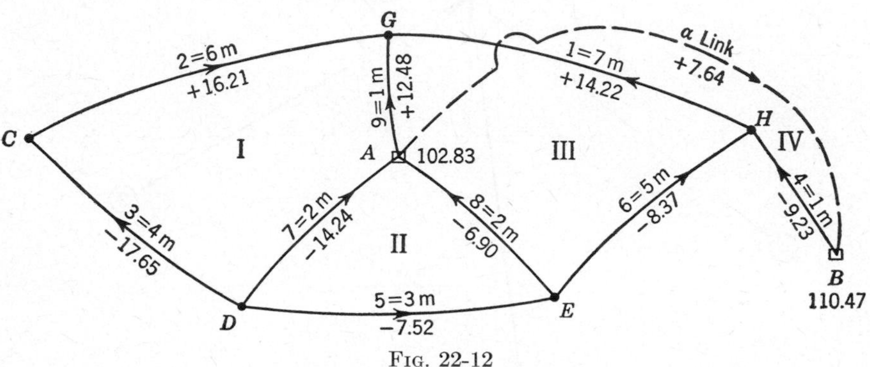

Fig. 22-12

It is thought that the example given here will serve not only as an illustration of level-net adjustment, but also as an introduction to the use of correlates and to the Doolittle method, both of which are invaluable in the adjustment of most surveys by least squares.

Principles

22-13. Definitions. A level net consists of benchmarks and lines of levels joining them. One or more benchmarks must have a fixed elevation. Benchmarks where level lines join are called junctions. That part of a level line which extends between adjacent junctions is called a link. A benchmark of fixed elevation is a junction between the net to be adjusted and previous leveling. Fixed elevations must be assumed to be connected by a single route of nonadjustable links, here called α **links.**

A circuit consists of the minimum number of links that provide two routes between

[1] Orthometric corrections are applied to observed differences in elevation determined by spirit leveling. Spirit leveling follows level surfaces. But a level surface is slightly nearer sea level at the poles than at the equator and the higher the surface, the greater the effect. Therefore the rate of change of elevation of a level surface with respect to sea level is a function of its elevation and latitude. The correction is a function of the rate of change and of the latitude difference of the ends of the spirit level line. The total orthometric correction for the example given is of the order of 0.0001 ft. It is neglected.

adjacent junctions. When a level net has been plotted approximately to scale, the circuits can be picked out very readily by inspection. It is slightly more difficult to do this in the unusual case when two lines of levels cross each other without being connected. The circuits are numbered in roman numerals in Fig. 22-12.

22-14. Imposed Conditions. The elevation differences represented by the links that compose each circuit must be adjusted so that each circuit closes without error. When this has been completed, all routes through the net will give the same difference in elevation between any pair of junctions.

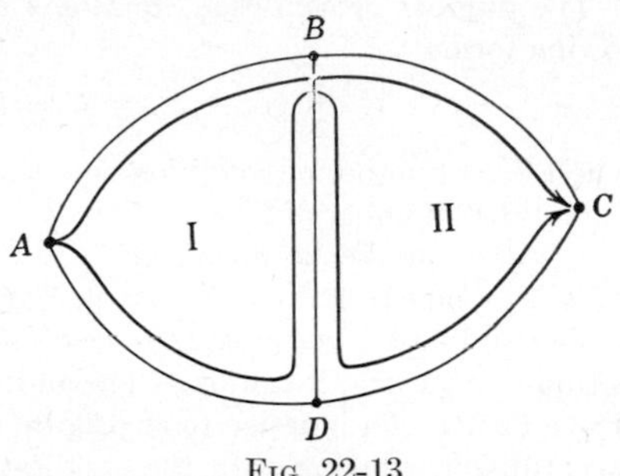

FIG. 22-13

The latter statement can be demonstrated as follows. Figure 22-13 shows two adjusted circuits I and II. Since they are adjusted, the routes from A to C shown by the two arrows must give the same elevation difference. But the route by the long arrow must give the same difference as the route ADC; for the only difference is link BD, and this comes in with both signs so that its effect is cancelled. Thus, since circuits ABD and BCD were adjusted, circuit $ABCD$ must have been automatically adjusted.

This principle may be applied indefinitely. In Fig. 22-14 all of the circuits have been adjusted so that all three routes shown from A to J must give the same difference in elevation. This is the case because, where they differ from the route shown by the full line, they merely take another route around an adjusted circuit.

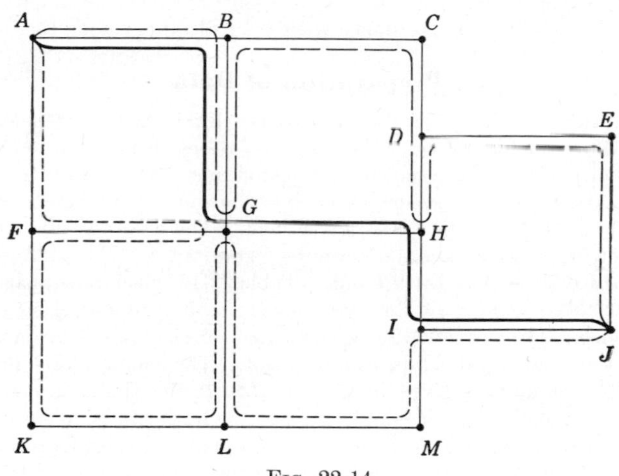

FIG. 22-14

From these, any combination of routes can be shown to be the same; for example,

Full line remains the same
Dashed line changes to $ABCDEJ$
Dotted line changed to $AFKLMIJ$

Any other routes can be treated in the same way.

22-15. Required Condition Equations. Since the conditions imposed are only that all the circuits shall close, a condition equation for each circuit is all that is required.

When condition equations are formed for other routes that are not circuits (for example, in Fig. 22-12, *GADEHG*) they will be found to be identical with the sum of certain circuit equations and therefore need not be considered.

The number of condition equations required may be further checked by the following formula:

$$C = L - J + 1$$

where C = number of condition equations (circuits)
L = number of links
J = number of junctions

This is demonstrated in the next paragraphs.

The simplest level circuit consists of two junctions connected by two links. It introduces but one condition. The addition of a link to any figure adds one condition. The addition of a junction to any figure together with one link adds no new condition. Accordingly, starting with the simplest case, the addition of each excess link over junction adds one condition. Expressing this as an equation,

$$C = 1 + [L - 2 - (J - 2)]$$
$$C = L - J + 1$$

Note: An α link must be counted as a link.

22-16. Completion of Adjustment. When the corrections to the elevation differences represented by the links have been computed by the least-squares adjustment, the total correction for each link is apportioned to the differences between intermediate benchmarks in accordance with their distances along the link.

Preparation of Data

22-17. Sketch. A sketch of the plan of the level net is drawn approximately to scale, as shown in Fig. 22-12. The fixed elevations A and B are given a characteristic symbol. The α link is shown by a dotted line. The circuits are selected and numbered. Arrows are placed on each link showing the direction in which the observed difference in elevation was determined. The observed difference in elevation is written in, each link is given a number, and its length is put down.

22-18. Table 22-1. The Data Used. Table 22-1 is filled in (see page 659). The name of each link should show the direction of the link by the order of the names of the junctions. The correction to be computed for each link is given the arbitrary name v with the number of the link as a subscript. The length of each link, expressed in any convenient unit, is given in Column L. If the stadia intercept has been determined for each observation, the lengths of the links can be taken from the level notes. Otherwise the net should be plotted on a map and the values scaled.

The reciprocal of the length of each link is computed, to be used later to weight the v's. As weights need not be absolute values, but need be only in correct proportion to each other, they all can be multiplied by any convenient constant K. This constant is usually chosen to keep the weights in the vicinity of unity to facilitate later computation. Here K is given the value 10 miles. Accordingly, the last column consists of the values of K/L, where both K and L are expressed in miles. The last two columns are filled in later when the values of the v's have been computed.

22-19. Table 22-2. Condition Equations. The condition equations are computed in Table 22-2. The circuits are taken in order, as shown in the first column. Moving around each circuit clockwise, the number of each link encountered is recorded in the second column. If the clockwise movement is in the direction of the arrow on the link in Fig. 22-12, a plus sign is placed in front of the link number. Otherwise, a minus sign is inserted. When the resulting sign of the link number is plus, the difference in elevation is inserted under its own sign in one of the next two columns. If the link number is minus, the difference in elevation is inserted under the sign contrary to its own. The algebraic sum of the differences, with the signs found as described, is the error of the closure of the circuit. It is placed in the error column.

The sum of the corrections to the links in a circuit when added to the error of the circuit must be equal to zero. The condition equations are made up accordingly. The signs that are given to the corrections (the v's) in the condition equations are the same as the signs of the link numbers. This rule must be followed because, to systematize the computation, the corrections are all assumed to be values that are added to the observed differences with their original signs.

The condition equations are shown in the last column. They need not be written in. They are given here for demonstration only.

A check loop is formed of the links surrounding circuits I, II, and III. As shown in Chap. 6, the error of the loop is equal to the sum of the errors of the circuits enclosed. Such a check loop is usually used for a check on the arithmetic.

Theory

22-20. Theory. Tables 22-3 and 22-4 can now be computed by a mechanical operation. To demonstrate the mathematics involved, the computations will be carried out here in full.

22-21. Weights. It was demonstrated earlier in the chapter that the v^2's must be weighted in inverse proportion to the number of equally precise observations they represent. It is safe to assume that the lengths of the sights are nearly the same and therefore that their precisions are equal and that the number of observations varies with the length of the distance leveled. The weights shown are therefore based on the lengths of the links.

22-22. Mathematics Involved. Substituting the values from Table 22-1, the function to be minimized is

$$U' = 1.4v_1^2 + 1.7v_2^2 + 2.5v_3^2 + 10v_4^2 + 3.3v_5^2 + 2v_6^2 + 5v_7^2 + 5v_8^2 + 10v_9^2$$

When the function to be minimized, and the condition equations that modify it, are set up, the solution can be carried forward in one of a number of ways. The most convenient method is known as the method of correlate equations, which utilizes the Lagrangian multiplier.

Quoting H. C. Mitchell, in "Definitions of Terms Used in Geodetic and Other Surveys,"[1] a correlate equation is "an equation derived from an observation or condition equation, employing undetermined multipliers, and expressing the condition that the sums of the squares of the residuals (or corrections) resulting from the application of these multipliers to the observation or condition equations shall be a minimum."

The expressions $-2C_1$, $-2C_2$, etc., are used for the Lagrangian multipliers. Multiplying each condition equation by $-2C$ and adding them all to the previous equation, a general expression is obtained:

$$\begin{aligned} U = {} & 1.4v_1^2 + 1.7v_2^2 + 2.5v_3^2 + 10v_4^2 + 3.3v_5^2 + 2v_6^2 + 5v_7^2 + 5v_8^2 + 10v_9^2 \\ & - 2C_1(+0.32 + v_2 - v_9 - v_7 + v_3) \\ & - 2C_2(+0.18 + v_7 - v_8 - v_5) \\ & - 2C_3(-0.27 + v_9 - v_1 - v_6 + v_8) \\ & - 2C_4(-0.12 + v_4 - v_6 + v_8) \end{aligned}$$

[1] U.S. Coast and Geodetic Survey Special Publication No. 242, p. 28, 1948.

By taking the partial derivative with respect to each v and equating to zero, the correlate equations are formed:

$\frac{1}{2}\frac{\partial U}{\partial v_1} = 1.4v_1 + C_3 = 0 \qquad v_1 = -0.7C_3$ correlate equation (1)

$\frac{1}{2}\frac{\partial U}{\partial v_2} = 1.7v_2 - C_1 = 0 \qquad v_2 = +0.6C_1$ correlate equation (2)

$\frac{1}{2}\frac{\partial U}{\partial v_3} = 2.5v_3 - C_1 = 0 \qquad v_3 = +0.4C_1$ correlate equation (3)

$\frac{1}{2}\frac{\partial U}{\partial v_4} = 10v_4 - C_4 = 0 \qquad v_4 = +0.1C_4$ correlate equation (4)

$\frac{1}{2}\frac{\partial U}{\partial v_5} = 3.3v_5 + C_2 = 0 \qquad v_5 = -0.3C_2$ correlate equation (5)

$\frac{1}{2}\frac{\partial U}{\partial v_6} = 2v_6 + C_3 + C_4 = 0 \qquad v_6 = -0.5C_3 - 0.5C_4$ correlate equation (6)

$\frac{1}{2}\frac{\partial U}{\partial v_7} = 5v_7 + C_1 - C_2 = 0 \qquad v_7 = -0.2C_1 + 0.2C_2$ correlate equation (7)

$\frac{1}{2}\frac{\partial U}{\partial v_8} = 5v_8 + C_2 - C_3 - C_4 = 0 \qquad v_8 = -0.2C_2 + 0.2C_3 + 0.2C_4$ correlate equation (8)

$\frac{1}{2}\frac{\partial U}{\partial v_9} = 10v_9 + C_1 - C_3 = 0 \qquad v_9 = -0.1C_1 + 0.1C_3$ correlate equation (9)

Substituting these values for the v's in the condition equations, the normal equations are formed:

$$0 = +0.32 + 0.6C_1 + 0.1C_1 - 0.1C_3 + 0.2C_1 - 0.2C_2 + 0.4C_1$$
$$0 = +0.18 - 0.2C_1 + 0.2C_2 + 0.2C_2 - 0.2C_3 - 0.2C_4 + 0.3C_2$$
$$0 = -0.27 - 0.1C_1 + 0.1C_3 + 0.7C_3 + 0.5C_3 + 0.5C_4 - 0.2C_2 + 0.2C_3 + 0.2C_4$$
$$0 = -0.12 + 0.1C_4 + 0.5C_3 + 0.5C_4 - 0.2C_2 + 0.2C_3 + 0.2C_4$$

Rearranging,

$0 = +1.3C_1 - 0.2C_2 - 0.1C_3 + 0.32$ normal equation (1)

$0 = -0.2C_1 + 0.7C_2 - 0.2C_3 - 0.2C_4 + 0.18$ normal equation (2)

$0 = -0.1C_1 - 0.2C_2 + 1.5C_3 + 0.7C_4 - 0.27$ normal equation (3)

$0 = - 0.2C_2 + 0.7C_3 + 0.8C_4 - 0.12$ normal equation (4)

The normal equations are now solved simultaneously to obtain the C's which can be substituted in the correlate equations to give the v's. The mathematics is obvious. The actual methods are formalized and are demonstrated by the method of computation given in the next section.

Method of Computation

22-23. Mechanical Method of Forming Equations. In practice, the correlate and the normal equations are formed from the condition equations by the mechanical method now described. It will be noted that the results are the same as those derived before.

22-24. Table 22-3. Correlate Equations. In Table 22-3 the nine correlate equations are made up from the four condition equations. In the first column are the numbers that apply to the v's and also to the correlate equations. In the second column are the reciprocals of K/L, i.e., L/K referring to the corresponding v's. The next four columns are numbered to correspond with the numbers of the C multipliers and hence with the numbers of the condition equations. In Column 1 are placed, in the proper lines, the coefficients of the v's that appear in condition equation (1) and hence in circuit I. Their signs and values, and the lines in which to place the coefficients, are taken from Table 22-2. The signs and the values used are those found in the condition equations (in this case the coefficients are unity). The line used for each v is given by its subscript. Columns 2, 3, and 4 are filled in for circuits II, III, and IV, respectively, in the same manner. In column Σ are the sums of the coefficients in each line. They are used to check the arithmetic later.

22-25. Table 22-4. Normal Equations. The normal equations are formed in Table 22-4. There is a normal equation for every condition and hence for every circuit. In this case there are four. Normal equations have a certain symmetry which is taken advantage of in the arrangement of the table. The rectangular form is numbered as shown both across and down. Both sets of numbers refer to the numbers of the normal equations. Two extra columns, N and Σ, are added as shown.

First the diagonal terms are inserted. These are placed in the spaces where the line and column numbers are the same. In line 1, Column 1, is placed the sum of the products obtained by squaring each value in Table 22-3, Column 1, and multiplying each square by the value in Column L/K in the same line.

In line 2, Column 2, is placed a corresponding sum from Table 22-3, Column 2, etc. Thus

$$0.6(+1)^2 + 0.4(+1)^2 + 0.2(-1)^2 + 0.1(-1)^2 = +1.3 \qquad \text{diagonal term 1}$$
$$0.3(-1)^2 + 0.2(+1)^2 + 0.2(-1)^2 = +0.7 \qquad \text{diagonal term 2}$$
$$0.7(-1)^2 + 0.5(-1)^2 + 0.2(+1)^2 + 0.1(+1)^2 = +1.5 \qquad \text{diagonal term 3}$$
$$0.1(+1)^2 + 0.5(-1)^2 + 0.2(+1)^2 = +0.8 \qquad \text{diagonal term 4}$$

To compute the other entries in the numbered columns, find, in each pair of numbered columns in Table 22-3, every line where there is an entry in both columns. Multiply the product of each pair by the value in the same line in Column L/K and add all such products.

For example, there is a pair in Columns 1 and 2 at line 7:

$$\text{Line 7: } (-1)(+1)(0.2) = -0.2$$

Enter this in Table 22-4 in line 1, Column 2, and in line 2, Column 1.

In Columns 3 and 4, pairs are found on lines 6 and 8:

$$\begin{aligned} \text{Line 6: } (-1)(-1)(0.5) &= +0.5 \\ \text{Line 8: } (+1)(+1)(0.2) &= \underline{+0.2} \\ \text{Sum} &\quad +0.7 \end{aligned}$$

Enter this in Table 22-4 in line 3, Column 4, and in line 4, Column 3.

In Columns 1 and 3, a pair is found on line 9:

$$\text{Line 9: } (-1)(+1)(0.1) = -0.1$$

Enter this in line 1, Column 3, and in line 3, Column 1.

In Column N are the constants in the equations. They are the errors shown in Table 22-2. When this column is complete, the four normal equations appear in lines 1, 2, 3, and 4, respectively. Note that they are the same as were found before by conventional mathematics.

Column Σ is used as a check. It is computed in exactly the same way as the numbered columns and then the value in Column N is added. For example, the value in line 1 is found as follows: Pairs in Columns 1 and Σ, Table 22-3, are found on lines 2, 3, 7, and 9:

$$
\begin{aligned}
\text{Line 2: } (+1)(+1)(0.6) &= +0.6 \\
\text{Line 3: } (+1)(+1)(0.4) &= +0.4 \\
\text{Line 7: } (-1)(0)(0.2) &= 0 \\
\text{Line 9: } (-1)(0)(0.1) &= 0 \\
\text{Sum } & \quad 1.00 \\
\text{Table 22-4, Column } N\text{, line 1} &= +0.32 \\
\text{Sum } & \quad +1.32
\end{aligned}
$$

The sum of all the values in the numbered columns and in Column N for each line in Table 22-4 should equal the value in Column Σ. For line 1,

$$+1.3 - 0.2 - 0.1 + 0.32 = +1.32$$

The numbers in parentheses in Table 22-4 below the heavy line are not actually written in as they have their counterparts where the column and the line numbers are interchanged. When they are not written in, a normal equation can be read by starting at the top of the column of the number of the normal equation desired and reading down to the diagonal term (which is included); thence read to the right, along the line through the value in Column N. This procedure is also used to obtain the sums to check with the values in Column N.

22-26. Table 22-5. Solution of Normal Equations. The normal equations are solved simultaneously to obtain the C values. Table 22-5 shows the method used. The first part is called the forward solution. Entries outside the heavy line are not necessary and therefore are not used in practice. They are included here only to show the algebra involved. The first 17 lines result in a value for C_4. This value is substituted in the previous equation to give the value for C_3, etc. The process of successive substitution is called the back solution.

22-27. Table 22-6. Doolittle Method. Table 22-6 shows the Doolittle method of solving normal equations. It is exactly like Table 22-5 except that unnecessary values are omitted, a Σ column (used for a check) is added, and the back solution is rearranged. It is explained in the second column in Table 22-6 and amplified in the following paragraphs. Note that values in the numbered columns are in each case the coefficients of the C indicated by the column number.

Line 1 is normal equation (1) copied from Table 22-4.

Line 2 is line 1 divided by the value found in line 1, Column 1, with its sign changed. At the beginning of line 2 is written $-C_1$ to aid in the back solution.

Line 3 is normal equation (2) copied from Table 22-4.

Line 4 is line 1 multiplied by the value found in line 2, Column 2. At the beginning of line 4 is written 1 to show that the multiplier comes from the line that begins with C_1.

Line 5 is the sum of lines 3 and 4.

Line 6 is line 5 divided by the value found in line 5, Column 2, with its sign changed. At the beginning of line 6 is written $-C_2$ to aid in the back solution.

Line 7 is normal equation (3) copied from Table 22-4.

Line 8 is line 1 multiplied by the value found in line 2, Column 3. At the beginning of line 8 is written 1 to show that the multiplier comes from the line that begins with C_1.

Line 9 is line 5 multiplied by the value found in line 6, Column 3. At the beginning of line 9 is written 2 to show that the multiplier comes from the line that begins with C_2.

Line 10 is the sum of lines 7 to 9.

Line 11 is line 10 divided by the value found in line 10, Column 3, with its sign changed. At the beginning of line 11, is written $-C_3$ to aid in the back solution.

The forward solution is completed in a similar manner.

The general rules expressing the operations between two horizontal lines are the following:

1. Write down the part of the normal equation that appears on the line in Table 22-4 having the number of the normal equation in question. Call this number n for this explanation.

2. Find the coefficient for C_n in each previous equation that is written just above a heavy horizontal line. Use each coefficient found to multiply the line directly above it. This will result in $n - 1$ multiplications. Write the result under Eq. (n) and add them all, including Eq. (n).

3. Divide the sum found in step 2 by the coefficient of C_n with its sign changed.

Column Σ is carried along exactly like the rest of the entries. Each value in Σ therefore should equal the sums of all the entries in the line in which it is found. But since the coefficients of all the various C's are recorded only on the C lines and the lines immediately above them (compare Tables 22-5 and 22-6), checks can be made only on such lines. It must be remembered that the coefficient of each C is -1 and that it must be included in the sum. For example, in line 11, adding,

$$\begin{array}{r} -1.000 \\ -0.447 \\ +0.120 \\ \hline -1.327 \end{array}$$

This is equal to the value -1.326 in the Σ column except for accumulated errors in computation.

In the back solution, only the C lines in the forward solution are used. Line 18 consists of the C line values in Column N. Line 19 consists of the C line values in Column 4, multiplied by the sum of the entries under C_4 in the back solution, i.e., by the value of C_4, -0.053. Line 20 consists of the C line values in Column 3 multiplied by the sum of the entries in C_3, i.e., $+0.144$. This process is continued. The sum of the entries in each column gives the C's.

Substitution of Values

22-28. Table 22-7. Substitution of C's. Once the values of the C's are found, they are substituted in the nine correlate equations given in Table 22-3. The substitutions are shown in Table 22-7. Each line gives the value of a v. For example, the substitution in line 8 is the following (from Table 22-3):

$$v_8 = 0.2(-C_2 + C_3 + C_4)$$
$$v_8 = 0.2(+0.312 + 0.144 - 0.053) = +0.08$$

The computed values are given in the column of first values. The adjusted values come from Table 22-8.

22-29. Table 22-8. Substitution of v's. The values of the v's from Table 22-7 are substituted in the condition equations as a check and to eliminate errors due to rounding off. The substitution is shown in Table 22-8. Table 22-8 is made up from Tables 22-7 and 22-2. The four columns represent the four condition equations. Note that a change had to be made in v_2 to cause Eq. (1) to check. v_2 was selected because the only equation in error was Eq. (1) and v_2 appears only in Eq. (1).

22-30. Table 22-1. The final values for the v's from Table 22-8 are entered in the proper column in Table 22-1. They are added to the observed differences in elevation to obtain the adjusted links in the last column.

22-31. Table 22-9. Finally the differences in elevation are computed in Table 22-9 using the adjusted links computed in Table 22-1.

LEAST SQUARES APPLIED TO TRIANGULATION FOR REDUCTION TO PLANE COORDINATES

22-32. Introduction. The application of least squares to triangulation is illustrated here by the adjustment of a small triangulation net to a plane coordinate system. The procedure described is in accordance with methods developed by the U.S. Coast and Geodetic Survey and used by that bureau to adjust first-order triangulation. The procedure is generally designated as the direction method and it represents the work of so many members of the bureau that, with the exception of the Doolittle method of solving normal equations, it is impossible to give proper credit to individuals.

The example chosen (Fig. 22-15) illustrates the general principles necessary to adjust triangulation for any conditions that may be imposed. When fewer conditions are required than those used in the example, the procedure may be curtailed by omitting the condition equations not applicable. It is thought that this example can be used as a model for the adjustment of any triangulation system that is to be reduced directly to plane coordinates. If special problems are encountered or if it is desired to use other methods, the reader is referred to the Walter F. Reynolds, "Manual of Triangulation Computation and Adjustment," U.S. Coast and Geodetic Survey Special Publication No. 138, 1934, which may be procured from the Superintendent of Documents, U.S. Government Printing Office, Washington, D.C.

The example consists of the triangulation net shown in Fig. 22-15. The positions of stations A, B, C, and D are fixed according to the coordinates given. All angles have been measured. The method of measurement is immaterial; it may be by repetition or by observed directions, so long as it is of the proper accuracy. It is assumed that all the measurements have been made by the same procedure and with uniform care so that they can be given the same weight.

At each station any required reductions to center have been carried out and the station adjustment has been completed. The list of directions for each station has been prepared (Table 22-10). It represents the results of these operations. The results for each station are expressed by clockwise angle measured from a single arbitrary direction to all directions observed. With this list, it is possible to compute the observed angle between any pair of lines by finding the difference between the stated directions of the two lines.

The problem is to find the most probable coordinates for stations E, F, and G. It has been shown in Chap. 21 that the most probable coordinates are those that require changes (corrections) in the observed directions such that the sum of the squares of the corrections is minimum. The corrections, of course, must be such that they create geometric consistency. The requirement of geometric consistency, therefore, imposes certain conditions on the corrections which can be expressed by condition equations.

22-33. Condition Equations. There are five types of condition equations that may occur in triangulation adjustment. The number of types that must be used depends on the geometric requirements involved. In the following paragraphs the five types are named and the geometric requirements which they represent are described.

1. *Angle Condition Equations.* In any net the direction of every line must be the same when computed through the angles by any route.

2. *Side Condition Equations.* When overlapping triangles are used, the length of any line must be the same when computed by any route.

3. *Length Condition Equations.* When more than one side is measured or its length otherwise fixed, the computed length of every fixed side must agree with its established length.

4. *Azimuth Condition Equations.* When the azimuth of more than one line is fixed, the computed azimuth of each fixed line must agree with its fixed azimuth.

5. *Position Condition Equations.* When the position of more than one point is fixed, the computed position of each fixed point must agree with its fixed position. This type of condition equation is not necessary when the two points are on the same side of a triangle (cf. a base), as their positions are not computed through the angles.

In the example chosen (Fig. 22-15), all five types of condition equations are required.[1] If station D were not fixed, the length condition equations and the azimuth condition equations could be omitted. If station C were also not fixed, the position condition equations could also be omitted. If the directions on the sides AF, FC, and FD were not observed, there would be no overlapping triangles and the side condition equations could be omitted.

When the required condition equations have been set up, the values of the corrections are computed by the method described for level nets. The process applied to the example chosen will now be described in detail.

Preparation of Data

22-34. Sketch. A sketch is drawn to approximate scale as shown in Fig. 22-15. Each direction observed at each station is given a number. It will be found convenient to number the directions clockwise around each station beginning at an outside line. The numbers are placed on the sketch.

22-35. Strongest Route. Beginning with a base, or a side of fixed length, the strongest route of single triangles through the net to the next fixed side is chosen. The angles of the chosen triangles are marked by arcs as shown. The strongest route may be found by computation of R's (Chap. 18) or by inspection. When two routes are strong, it makes little difference which route is chosen. When several fixed lengths exist, routes connecting them must be chosen.

22-36. Names for Angles of a Triangle. Each of the three angles of every triangle forming part of the strongest route is given a name and so marked. The angle opposite the known side from which the route starts is called B. The angle opposite the side through which the length is carried is called A. These two angles are called **length angles.** Angle A is called the **adjacent angle;** angle B is called the **opposite angle.** The third angle, C, is called the **azimuth angle.** In succeeding triangles the known side is that which was computed from the preceding triangle. These names may be omitted if neither length condition equations nor position condition equations are used. In this example, these letters should not be confused with station names.

[1] If two position equations are formed, one for C and one for D, length and azimuth position equations can be omitted. This procedure is not followed, as position equations are more difficult to compute.

22-37. Independent Figures. The independent figures in the net are noted. An independent figure is one that is connected to each adjacent figure by but one common side. In the example, there are two independent figures, *A, B, E, F* and *E, F, G, C, D*.

22-38. Table 22-11. Form for Final Computation. The final computation form (Table 22-11) is set up and partly filled out at this time. The angles are named quite differently for this form. The vertex opposite the known side is called 1 and the known side is called 2–3, the number 2 being the left-hand vertex when viewed from 1.

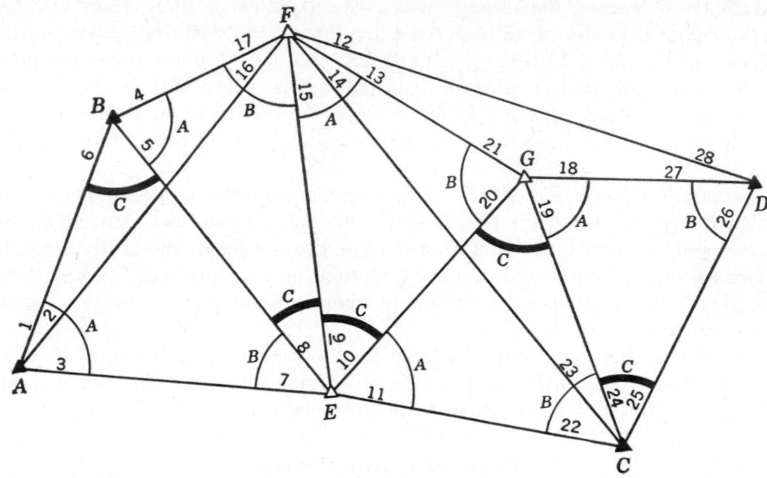

	y, ft	x, ft	Length	Azimuth
A	8,450.62	3,513.31	11,234.00	25°41′01.2″
B	18,574.71	8,382.15		
C	3,051.93	28,784.71	12,615.16	31°48′09.5″
D	13,773.16	35,432.84		

FIG. 22-15. Sketch for adjusting a triangulation system.

The column headed Stations is filled in first. In it are listed the angles in a systematic order. The net is divided into independent figures. Starting with the figure containing the first known side, each figure appears in the order of the direction used for computation. The triangles within each figure are set down according to the following rules (see Fig. 22-16):

1. The first angle, 1, is the angle opposite the known side in a triangle first used. Angles 2 and 3 are then chosen according to the previous definitions.

2. If there are any other angles opposite the known side, each is made angle 1 in order, taking them clockwise as viewed from the known side. Angles 2 and 3 are then numbered accordingly.

3. When all the angles opposite the known side have been used, the station where angle 1 was last located is used for angle 1 for every triangle in which that station appears. First, angle 2 is taken at the left end of the base and angle 3 is taken at

the end of each successive line clockwise from it as viewed from angle 1, without repeating any triangle. Then angle 2 is taken at the next clockwise line from its previous position and successive angles 3 are chosen clockwise from it as before. Thus the list is built up, always working clockwise, until all the triangles are recorded.

4. After the angles have been listed in order, the sides thus established are written in. The side 2–3 always precedes the angles, and the sides 1–3 and 1–2 follow the angles in that order as shown. Station names are written in as a guide.

5. In the column headed Sight Nos., write beside each angle the numbers of the observed directions used to compute each angle. Write the two numbers in numerical order but show the signs used to compute the angles. This is the standard method of denoting angles in the direction method, and it is used throughout this example.

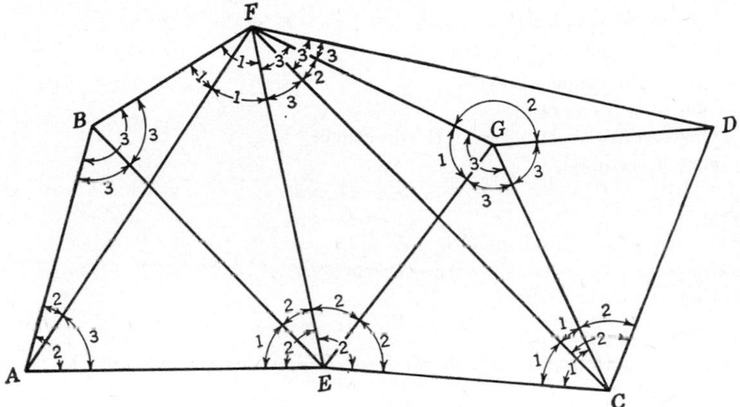

FIG. 22-16. Enumeration of angles.

6. In the angle column write the angles as computed from the observed directions. Find the sum of the three angles in each triangle. This value minus 180° is the error. It is recorded (to the right of the sum) with its sign changed as a required correction in the correction column and used in the angle condition equation for that triangle.

Formation of Condition Equations

22-39. Form Used for Condition Equations. Condition equations are uniformly written thus: zero equals the error plus the unknown corrections. As this example illustrates the direction method, corrections to directions are used. The unknown number of seconds correction to any direction is represented by, instead of x or v, the number of the direction to be corrected written in parentheses.

To avoid error, the corrections are computed throughout in terms of seconds to be added to the directions. It follows that corrections applied to the directions of the sides of an angle have the same signs and the same numbers as the direction by which the angle is computed.

The word "condition" is often omitted from the names of the condition equations.

22-40. Table 22-12. Angle Condition Equations. When the sum of the angles in every triangle is made equal to 180°, the direction of any line will be the same when computed by any route. If there are no overlapping triangles there must be an angle condition equation for each triangle to obtain this result. In a figure formed of overlapping triangles, it will be found that when certain of the triangles have been adjusted, all will have been automatically adjusted. To avoid wasting time by setting up identical angle equations, a means of computing the required number of equations is used.

In any figure, the direction of some one line must be, or be assumed to be, a reference direction. The direction of every subsequent line is completely established thereafter by the measurement of one angle for each line. Thus, for two lines one angle is required; for three lines, two angles, etc. Each angle that is measured in excess of this number is redundant and therefore requires a condition equation. Expressing this as an equation,

$$C_A = A - (L - 1) = A - L + 1$$

where C_A = no. of angle equations required
A = no. of angles[1]
L = no. of lines

Let D = no. of directions observed
s' = no. of stations occupied
n' = no. of lines observed in both directions

At any station occupied,

$$A = D - 1$$

In general, $A = D - s'$

But $L = D - n'$

Subtracting, $A - L = n' - s'$

Substituting in equation above,

$$C_A = n' - s' + 1$$

This formula is more useful than the first formula, as fewer items need be counted.

In the example,

Figure	n'	s'	C_A
ABFE	6	4	3
EFGCD	9	5	5
Total			8

As a check, the whole figure may be considered at once:

Figure	n'	s'	C_A
ABCDEFG	14	7	8

The triangles used in the angle equations are chosen arbitrarily within each independent figure. If side equations are to be used later, the triangles used in the angle equations should be chosen to avoid the small angles. This reserves the small angles for the side equations. Table 22-12 shows the angle equations. They are made up by using the error from Table 22-11 for the constant term and writing the names of the directions (and hence the names of the corrections) with their signs as used to compute the angles.

22-41. Side Condition Equations. In a figure having overlapping triangles, at least two independent routes are available by which length can be carried through the

[1] Since the station adjustments are complete, an angle closing the horizon is not redundant. In fact, it should not be considered an angle at all but only a means of measuring the others. At the central station in a central-point figure, it is difficult to remember that this angle does not establish a direction. It may be simpler just to remember that since the station adjustment is complete, this angle does not require a condition equation.

figure. If a condition equation is set up for every route possible, identical equations will result. The means of computing the required number of equations must be derived.

In a figure consisting of one triangle, only one route is available. For each pair of lines of known direction added, a new position can be computed. If the new position constitutes a new station, there is still but one route for computing lengths. If this procedure is continued, no condition equations will be required. Stated in other words, no condition equations are required as long as there are only twice as many lines as stations after the first triangle.

When a line of known direction is constructed between two existing stations, a new route is created for computing position. Thus, for every one of these extra lines a condition equation is required. Stating this as an equation,

$$C_S = n - 3 - 2(S - 3) = n - 2S + 3 \tag{22-2}$$

where C_S = no. of side condition equations
n = no. of sides
S = no. of stations

In the example we have

Figure	n	S	C_S
ABFE......	6	4	1
EFGCD.....	9	5	2
Total.................			3

The number of side condition equations can be determined graphically. Start by drawing a triangle to match a triangle at one end of the figure. Then mark a new station position. Draw two lines to it according to the location of lines on the actual net. Continue until the system of stations is built up. Now complete the figure by drawing the extra lines necessary between the existing points. Count the number of extra lines required. This is the number of excess lines and therefore the number of condition equations.

Pairs of routes through each independent figure are chosen arbitrarily for each condition equation. The only requirement is that the smallest angles must be used, as the lengths of sides opposite small angles are more sensitive to changes in the angles. The pairs of routes are treated as a loop. By beginning at one line, the length is carried out by one route and back by the other route to reach the original line. To avoid error, a formal system is adopted.

22-42. Method of Establishing the Side Equations. Table 22-13 shows the formation of the side equations. Any station called the pole is chosen so that the smallest angles are used. The lines that radiate from the pole are written down in a clockwise order. In the first figure in the example, B was chosen as the pole and the lines BF, BE, and BA result. By the sine formula

$$\begin{aligned}
\log BE &= \log BF - \log \sin(-8 + 9) + \log \sin(-15 + 17) \\
\log BA &= \log BE - \log \sin(-1 + 3) + \log \sin(-7 + 8) \\
\log BF &= \log BA - \log \sin(-16 + 17) + \log \sin(-1 + 2)
\end{aligned}$$

By adding these equations, rearranging, and changing signs we have

$$\begin{aligned}
0 = &+ \log \sin(-8 + 9) - \log \sin(-1 + 2) \\
&+ \log \sin(-1 + 3) - \log \sin(-15 + 17) \\
&+ \log \sin(-16 + 17) - \log \sin(-7 + 8)
\end{aligned}$$

Obviously the sum of the log sines added should equal the sum of the log sines subtracted. The difference is computed in Table 22-13.

22-43. Table 22-13. Side Equations. To set up Table 22-13, proceed as follows: List the sides in Column 1 according to the system described in the previous paragraph. For each side find the two triangles of which it forms a part. For side *BF* the triangles are *BFA* and *BFE*. Since we are proceeding from *BF* to the next line *BE*, we can say that the angle $-1 + 2$ is behind and that the angle $-8 + 9$ is ahead. Accordingly, list angle $-8 + 9$ in Column 2 under *angles ahead* and angle $-1 + 2$ in Column 6 under *angles behind*. Continue for the other sides.

Fill in the values of the observed angles from Table 22-11 and look up the log sines. Beside each log sine, record the difference per second of the log sines taken from the log sine table.

The error in logs is computed by subtracting the sum of the logs behind from the sum of the logs ahead.

22-44. Forming the Side Equations. The side equation for each pole is written underneath the table for that pole. The constant term used is the error of the logs. We must now express the effect on the error of the logs when each direction is corrected by the unknown number of seconds.

For example, if direction 1 is increased 1 second, it will change the log sine of angle $-1 + 3$ by 6 (in the 7th place) and the log sine of angle $-1 + 2$ by 64. This will cause a net change of 58 in the final log difference. This value is multiplied by the number of seconds in the unknown correction to the direction. It is expressed in the equation thus: 58(1). Note that the whole equation is divided by 100 as explained later.

The use of the signs can be explained in detail as follows. A correction of plus 1 second would decrease angle $-1 + 3$ by 1 second and thus reduce the log sin by 6 in the last place of decimals. The change therefore is -6. This correction in direction 1 would also decrease the log sin of angle $-1 + 2$ by 64, i.e., produce a change of -64. But a decrease in the logs behind has the same effect on the error as an increase in the logs ahead. Therefore the sign of the -64 is changed, giving $+64$. Combining -6 and $+64$, we obtain $+58$, which is the total effect on the error if direction 1 were increased 1 second.

Accordingly, the signs of the directions in Column 2 give the signs of the corrections, but the signs in Column 6 are reversed.

To prevent later computations from running to too many digits, the errors and coefficients must be kept about the same size in the various condition equations. Usually they are matched to the angle equations and hence depend on the estimated sizes of the errors as shown in Table 22-12. The errors in Table 22-12 are all not far from 1 second, so unity is chosen as a value to approach. The factor used could be 100,000 or 1,000,000, that is to say, the fifth or sixth decimal place in the logs could be taken as unity. The fifth place was chosen.

The computation of the values for the first side equation is shown in the table on page 653. The first side equation is condition equation (9). It is written out underneath the first tabular values.

22-45. Table 22-14. Length Condition Equation. Since the angle and side equations ensure geometric consistency, length equations need contain only corrections to angles forming the shortest route from one base (or fixed length) to another. The strongest route is chosen as shown by the arcs in Fig. 22-15. It is obvious from the sine formula that the sum of all the log sines of the *A* angles added to the log of the length *AB* should equal the sum of all the log sines of the *B* angles added to the log of the length *CD*.

Correction	Products		Final coefficient
	Ahead	Behind	
1	−0.06	+0.64	+0.58
2		−0.64	−0.64
3	+0.06		+0.06
7		+0.20	+0.20
8	−0.34	−0.20	−0.54
9	+0.34		+0.34
15		+0.07	+0.07
16	−0.45		−0.45
17	+0.45	−0.07	+0.38

Table 22-14 shows the computation. It is carried out in exactly the same manner as the computation for a side equation.

22-46. Table 22-15. Azimuth Condition Equation. Azimuth equations need contain only corrections to angles forming the shortest route from one fixed azimuth to another. The formation of the equation is shown in Table 22-15. It is self-explanatory.

22-47. Position Condition Equations. Since the length equations and azimuth equations ensure a perfect fit for length and azimuth, position equations need contain only corrections to angles that form part of the strongest route from one fixed position to another. Observed angles must be used, of course, as these are the angles that must be corrected. Two such angles determine a triangle so that only two observed angles in each triangle can be used. The third angle is concluded, that is, obtained by subtracting the sum of the two angles chosen from 180°.

First the position of the final point is computed through the strongest triangles, using any two observed angles and a concluded angle in each. Two position equations are set up, one for the error in y and the other for the error in x. In each equation, corrections only to the chosen observed directions are included. Each direction appears in one chosen angle and also in the concluded angle.

22-48. Table 22-16. Preliminary Coordinate Computation. Table 22-16 (first part) shows the form for computing the sides of the triangles to be used. In each triangle the C angle is concluded. Brackets surrounding the value of the angle indicate that it is concluded. The form is made up similar to Table 22-16 but unnecessary triangles are omitted.

The sum of the three angles of each triangle should be found for a check. It should be exactly 180°. The log sines of the observed angles are taken from Table 22-14. The log sines of the concluded angles must be looked up.

The final column is made up of six logs for each triangle. They are the following:

1. Log of known side
2. Colog sin angle 1
3. Log sin of angle 2
4. Log sin of angle 3
5. Sum of lines 1, 2, and 3 (log side 1–3)
6. Sum of lines 1, 2, and 4 (log side 1–2)

This arrangement is, of course, dictated by the sine formula.

Table 22-16 (second part) gives the form of coordinate computation. α stands for azimuth. The method is obvious. A check is obtained at each station. The check should be very close, as the triangles are perfect. Wherever a choice must be made of two values, that derived through the shortest route is used.

The resulting coordinates of station C are shown below. The errors are computed by subtracting the fixed values from the computed values.

	y	x
Computed.............	3052.04	28,784.67
Fixed..................	3051.93	23,784.71
Error..................	+0.11	−0.04

22-49. Theory of Position Condition Equations. It is now necessary to determine the effect on the errors stated above, of an increase of one second in each of the directions used to compute the coordinates. Consider any station N, in any triangulation net, and the final fixed station T (see Fig. 22-17). The part of the net from O to T is not shown. The side LN has been computed; the side NO is to be computed. The angles A, B, and C would be arranged as shown.

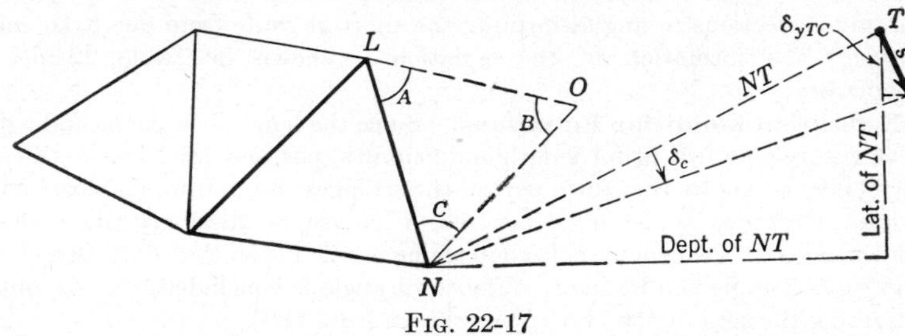

Fig. 22-17

If the A angle is increased, its sine will be increased, and the length NO and all subsequent lengths will be increased proportionally. Such an increase constitutes a change of scale, and the latitude of the line NT would be increased in the same proportion. Writing this in the form of an equation,

$$\frac{\delta_{yTA}}{\text{Lat. } NT} = \frac{\delta_{\sin A}}{\sin A}$$

or

$$\delta_{yTA} = \text{Lat. } NT \frac{\delta_{\sin A}}{\sin A}$$

where δ_{yTA} = increase in the y coordinate of station T for 1 second increase in A

$\delta_{\sin A}$ = increase in the sin A for 1 second increase in A

If the B angle is increased by a small amount its sine will be increased, and the length NO and all subsequent lengths will be decreased proportionally. Therefore, using the same notation,

$$\delta_{yTB} = -\text{Lat. } NT \frac{\delta_{\sin B}}{\sin B}$$

If the C angle is increased by 1 second, δ_C, the subsequent figure will be rotated clockwise around N. The coordinates of T will change by the vector s, in which

$$s = 0.00000485NT$$

By similar triangles

$$\frac{\delta_{yTC}}{s} = -\frac{\text{Dep. } NT}{NT}$$

or

$$\delta_{yTC} = -s\frac{\text{Dep. } NT}{NT}$$

Substituting for s,

$$\delta_{yTC} = -0.00000485 \text{ Dep. } NT$$

Let v_A, v_B, and v_C be the number of seconds in the corrections in angles A, B, and C, respectively. Then we may write

$$\Delta y_N = v_A \text{ Lat. } NT \frac{\delta_{\sin A}}{\sin A} - v_B \text{ Lat. } NT \frac{\delta_{\sin B}}{\sin B} - v_C(0.00000485) \text{ Dep. } NT$$

where Δy_N is the change in the value of the y coordinate of T. This may also be written as

$$\Delta y_N = v_A(y_T - y_n)\frac{\delta_{\sin A}}{\sin A} - v_B(y_T - y_n)\frac{\delta_{\sin B}}{\sin B} - v_C(0.00000485)(x_T - x_N)$$

It will be noted that the last term in this equation will change sign if the line NO is on the left side of C. When this is the case, an increase in C will change the sense of s. Accordingly, the signs of C must be carefully determined by this criterion. In the example, the following are the signs for the C angles:

C angle at	*Sign*
B	$-$
E	$+$
E	$+$
G	$-$

We now have a general expression for the Δy_N for any corrections in the three angles of any one triangle N. To use this expression, it is necessary to find the values of the 1-second differences in the sines of the A and B angles. Since it is more convenient to use similar differences for the log sines, the following substitution is made. In general, from the calculus

$$\frac{d(\log_e x)}{dx} = \frac{1}{x}$$

But

$$\log_{10} x = 0.4343 \log_e x$$

Hence

$$\frac{d(\log_{10} x)}{dx} = \frac{0.4343}{x}$$

and

$$\frac{\delta_A}{\delta_{\sin A}} = \frac{0.4343}{\sin A} \quad \text{very nearly}$$

where δ_A is the change in the log sin A for 1 second in A,

or

$$\frac{\delta_{\sin A}}{\sin A} = \frac{\delta_A}{0.4343}$$

Substituting this and a similar value in reference to B,

$$\Delta y_N = (v_A\delta_A - v_B\delta_B)\frac{y_T - y_n}{0.4343} - v_C(0.00000485)(x_T - x_n)$$

or

$$0.4343\Delta y_N = (v_A\delta_A - v_B\delta_B)(y_T - y_n) - v_C(0.00000211)(x_T - x_n)$$

To reduce the number of digits handled and to bring the coefficients approximately to unity, the equation is written as follows:

$$4.343\Delta y_n = 10^{-4}(y_T - y_n)(10^5\delta_A v_A - 10^5\delta_B v_B) + 10^{-4}(x_T - x_n)(-0.211)v_C$$

A similar derivation can be made for Δx_n. It will be noted that when a C angle is plus, an increase in C will increase x_T instead of decreasing the coordinate as it does y_T. Therefore the sign of the last term is changed:

$$4.343\Delta x_n = 10^{-4}(x_T - x_n)(10^5\delta_A v_A - 10^5\delta_B v_B) + 10^{-4}(y_T - y_n)(0.211)v_C$$

Obviously the condition equations are

$$0 = \text{error in } y + \Sigma\Delta y$$
$$0 = \text{error in } x + \Sigma\Delta x$$

or

$$0 = 0.4343 \text{ (error in } y) + 0.4343\,\Sigma\Delta y$$
$$0 = 0.4343 \text{ (error in } x) + 0.4343\,\Sigma\Delta x$$

22-50. Table 22-17. Forming the Position Condition Equations. Table 22-17 shows the form used for computing the position equations.

Column 1 contains the stations of the C angles and the signs of the C angles. Columns 2 and 3 contain the coordinates of the stations listed. Columns 4 and 5 contain 10^{-4} times each coordinate subtracted from the coordinate of the final station C. Columns 6 and 7 contain the tabular differences for 1-second change in the log sines of the A and B angles, respectively, each multiplied by 10^5. Columns 8, 10, 11, 13, 14, and 16 contain the products of the values in the columns stated on the form. *Note:* K is the constant 0.211 or 0.21. Columns 9, 12, and 15 contain the usual designations for the directions used in computing the A angles, the B angles, and the C angles, respectively. Instead of the actual concluded angles, their supplements are used, as the sine of a supplement is the same as the sine of an angle. The supplement is, of course, the sum of the other two angles in the triangle. The angles must be exactly the same as the angles used in the preliminary coordinate computation. Note that when the C angle is minus as given in Column 1, the sines of the directions are changed.

To make up the equations, the error found in the preliminary coordinate computation is first inserted. The coefficient for each correction is then found by adding all the coefficients applicable and inserting them as shown. The coefficients applicable to each angle are found in the two columns adjacent to that angle. The left-hand column gives the value for the y equations and the right-hand column gives the value for the x equations.

For example, to find the coefficients for the y equation for the directions 1 and 15, respectively, proceed as follows:

Direction 1: $-(-0.09) - (-0.43) = +0.52$
Direction 15: $-(+0.02) + (-0.26) + (-0.06) - (-0.26) = -0.08$

22-51. Summary of Condition Equations. The 15 condition equations obtained may be listed as follows:

Condition Equations

Numbers	*Name*	*Table*
1–8	Angle	22-12
9–11	Side	22-13
12	Azimuth	22-15
13	Length	22-14
14–15	Position	22-17

Computation

22-52. Least-squares Solution. The least-squares solution is made by the method of correlate equations, in accordance with the system demonstrated for a level net, Sec. 22-22.

The upper part of Table 22-18 shows the formation of the correlate equations from the condition equations and the lower part shows the formation of the normal equations from the correlates. There is no column for weights, as the angles were all measured with the same precision so that the weight of each is taken as unity. In contrast to the level net, there are many terms in the correlate equations having values other than unity. These represent the coefficients of the v's, which are not unity in any but the angle and azimuth condition equations.

The simultaneous solution of normal equations is shown in Table 22-19. Here the equations are not arranged numerically but in the order that will give the shortest solution. In a chain of quadrilaterals, the side equation for each quadrilateral should be introduced immediately after the three angle equations for the quadrilateral. In other types of triangulation each succeeding equation should be chosen to introduce as few new terms as possible.

Table 22-20 shows the back solution. The method was demonstrated for a level net. Note that in this table the coefficients of the C's are inserted before multiplication as well as the products.

Table 22-21 shows the computation of the values for the v's.

Table 22-22 shows the values of the v's taken from Table 22-21, rounded off, and substituted in the angle condition equations (from Table 22-12). At the bottom of each column in Table 22-22 is the sum of the v's for each triangle. Listed below is the error term taken from the angle condition equations. The sum of the v's plus the error term should be zero. Slight errors due to rounding off usually appear. In this case the first and second columns can be corrected by changing v_7 from -0.11 to -0.12. This could also be accomplished by changing v_3 from $+0.74$ to $+0.75$. But, by examining the exact values of the v's in Table 22-21, this change is found to be greater than the change made. A similar choice was made between v_{10} and v_{21} to correct Columns 6 and 8, etc.

Final Triangles and Coordinates

22-53. Completing the Computation. The correction column in Table 22-11 is now computed from the final v's. The two v's for each angle are added in accordance with the direction numbers listed in the first column. Each sum is entered beside the angle to which it refers. The sum of the corrections applied to the sum of the angles should equal 180°.

The corrections are applied to the seconds in the angles, and the logarithms are entered as described for Table 22-16.

The final coordinates are computed in Table 22-23 in the same manner as in Table 22-16. The final coordinates for C and D are compared with the fixed coordinates of these stations. The results are shown below:

	y	x
Sta. C fixed	3,051.93	28,784.71
final	3,051.92	28,784.71
Sta. D fixed	13,773.16	35,432.84
final	13,773.15	35,432.84

The list of the final coordinates for all the stations is now available:

Station	*y*	*x*
A	8,450.62	3,513.31
B	18,574.71	8,382.15
E	6,390.14	16,557.40
F	21,518.15	16,112.57
G	14,588.95	25,319.59
C	3,051.93	28,784.71
D	13,773.16	35,432.84

22-54. The Error of a Single Observation. In Chap. 21 it was shown that, when errors are computed from true values,

$$\sigma = \sqrt{\frac{\Sigma v^2}{n}}$$

and when the errors are computed from an average,

$$\sigma = \sqrt{\frac{\Sigma v^2}{n - 1}}$$

The change in the numerator under the radical is due to the error of the average from which the errors are computed. In other words, when the errors are computed from a deduced value rather than from a true value, the error of the deduced value must be recognized.

In the case of several measurements of the same quantity, the true value is deduced from one condition:

$$\Sigma v = 0$$

When two or more conditions are required to find the deduced values, the error of each condition must be recognized. By a derivation exactly like the derivation of Eq. (21-20) in Chap. 21, the general expression is found to be

$$\sigma = \sqrt{\frac{\Sigma v^2}{n - c}}$$

where c is the number of conditions required. In each case the numerator under the radical represents the number of redundant measurements.

In the case of the average, only one measurement is necessary. The rest, $n - 1$, are redundant. In Sec. 22-8, two conditions were applied so that the formula would be

$$\sigma = \sqrt{\frac{\Sigma v^2}{5 - 2}} = \sqrt{\frac{\Sigma v^2}{3}}$$

In this case two measurements would have established the line. Three measurements are redundant.

The rule is, therefore, that the numerator under the radical must be equal to the number of redundant measurements.

In the case of the triangulation net, there is a condition equation for each redundant measurement. Accordingly, the number under the radical is equal to the number of condition equations; thus

$$\epsilon = \pm 0.6745 \sqrt{\frac{\Sigma v^2}{C}}$$

where Σv^2 is the sum of the squares of the v's already computed and C is the number of condition equations used.

The value for Σv^2 can be computed best from Table 22-22, Column 2. In this particular adjustment, $\Sigma v^2 = 8.3159$ and $C = 15$. Substituting these values in the formula,

$$\epsilon = \pm 0.6745 \sqrt{\frac{8.3159}{15}} = 0.50''$$

TABLES SHOWING COMPUTATIONS

TABLE 22-1. LEVEL NET. DATA AND FINAL VALUES

Links		Observed diff. in elev.	Cor.	L Lengths in miles	Wt. of v's K/L	v	Adjusted links
Name	No.						
H–G	1	+14.22	v_1	7	1.4	−0.10	+14.12
C–G	2	+16.21	v_2	6	1.7	−0.18	+16.03
D–C	3	17.65	v_0	4	2.5	−0.11	−17.76
B–H	4	− 9.23	v_4	1	10.0	−0.01	− 9.24
D–E	5	− 7.52	v_5	3	3.3	+0.09	− 7.43
E–H	6	− 8.37	v_6	5	2.0	−0.05	− 8.42
D–A	7	−14.24	v_7	2	5.0	−0.01	−14.25
E–A	8	− 6.90	v_8	2	5.0	+0.08	− 6.82
A–G	9	+12.48	v_9	1	10.0	+0.04	+12.52
A–B	α	+ 7.64	0	0	—		

TABLE 22-2. LEVEL NET. COMPUTATION OF CONDITION EQUATIONS

Data				Error	Condition equations
Circuit	Link	+	−		
I	+2	16.21			
	−9		12.48		No. 1
	−7	14.24			$0 = +0.32 + v_2 - v_9 - v_7 + v_3$
	+3		17.65		
		30.45	30.13	+0.32	
II	+7		14.24		No. 2
	−8	6.90			$0 = +0.18 + v_7 - v_8 - v_5$
	−5	7.52			
		14.42	14.24	+0.18	
III	+9	12.48			
	−1		14.22		No. 3
	−6	8.37			$0 = -0.27 + v_9 - v_1 - v_6 + v_8$
	+8		6.90		
		20.85	21.12	−0.27	
IV	$+\alpha$	7.64			
	+4		9.23		No. 4
	−6	8.37			$0 = -0.12 + v_4 - v_6 + v_8$
	+8		6.90		
		16.01	16.13	−0.12	
Check	+2	16.21			
	−1		14.22		
	−6	8.37			
	−5	7.52			
	+3		17.65		
		32.10	31.87	+0.23	
Sum of circuits I, II, III				+0.23	

TABLE 22-3. LEVEL NET. CORRELATE EQUATIONS

Correlate equations	L/K	Condition equations (and C No.'s)				Σ
		1	2	3	4	
1	0.7			−1		−1
2	0.6	+1				+1
3	0.4	+1				+1
4	0.1				+1	+1
5	0.3		−1			−1
6	0.5			−1	−1	−2
7	0.2	−1	+1			0
8	0.2		−1	+1	+1	+1
9	0.1	−1		+1		0

TABLE 22-4. LEVEL NET. NORMAL EQUATIONS

No.	1	2	3	4	N	Σ
1	+1.3	−0.2	0.1		+0.32	+1.32
2	(−0.2)	+0.7	−0.2	−0.2	+0.18	+0.28
3	(−0.1)	(−0.2)	+1.5	+0.7	−0.27	+1.63
4		(−0.2)	(+0.7)	+0.8	−0.12	+1.18

TABLE 22-5. LEVEL NET. SOLUTION OF NORMAL EQUATIONS INCLUDING TERMS OMITTED IN DOOLITTLE METHOD

Forward Solution

		1	2	3	4	N
1 Normal eq. (1)	0 =	$+1.300C_1$	$-0.200C_2$	$-0.100C_3$		$+0.320$
2 Line 1 ÷ (−1.300)	0 =	$-1.000C_1$	$+0.154C_2$	$+0.077C_3$		-0.246
3 Normal eq. (2)	0 =	$-0.200C_1$	$+0.700C_2$	$-0.200C_3$	$-0.200C_4$	$+0.180$
4 Line 1 × (+0.154)	0 =	$+0.200C_1$	$-0.031C_2$	$-0.015C_3$		$+0.049$
5 Sum lines 3, 4	0 =	0	$+0.669C_2$	$-0.215C_3$	$-0.200C_4$	$+0.229$
6 Line 5 ÷ (−0.669)	0 =		$-1.000C_2$	$+0.321C_3$	$+0.299C_4$	-0.342
7 Normal eq. (3)	0 =	$-0.100C_1$	$-0.200C_2$	$+1.500C_3$	$+0.700C_4$	-0.270
8 Line 1 × (+0.077)	0 =	$+0.100C_1$	$-0.015C_2$	$-0.008C_3$		$+0.025$
9 Line 5 × (+0.321)	0 =	0	$+0.215C_2$	$-0.069C_3$	$-0.064C_4$	$+0.074$
10 Sum lines 7, 8, 9	0 =	0	0	$+1.423C_3$	$+0.636C_4$	-0.171
11 Line 10 ÷ (−1.423)	0 =			$-1.000C_3$	$-0.447C_4$	$+0.120$
12 Normal eq. (4)	0 =		$-0.200C_2$	$+0.700C_3$	$+0.800C_4$	-0.120
13 Line 1 × (0)	0 =	0	0	0		0
14 Line 5 × (+0.299)	0 =		$+0.200C_2$	$-0.064C_3$	$-0.060C_4$	$+0.068$
15 Line 10 × (−0.447)	0 =	0	0	$-0.636C_3$	$-0.284C_4$	$+0.076$
16 Sum lines 12, 13, 14, 15	0 =	0	0	0	$+0.456C_4$	$+0.024$
17 Line 16 ÷ (−0.456)	0 =	0	0	0	$-1.000C_4$	-0.053

Back Solution

		22	21	20	19	18
From line 17	0 =				$-1.000C_4$	-0.053
Hence $C_4 = -0.053$						
Substituting in line 11	0 =			$-1.000C_3$	$+0.024$	$+0.120$
Hence $C_3 = +0.144$						
Substituting in line 6	0 =		$-1.000C_2$	$+0.046$	-0.016	-0.342
Hence $C_2 = -0.312$						
Substituting in line 2	0 =	$-1.000C_1$	-0.048	$+0.011$		-0.246
Hence $C_1 = -0.283$						

TABLE 22-6. LEVEL NET. DOOLITTLE METHOD OF SOLVING NORMAL EQUATIONS

Forward Solution

	1	2	3	4	N	Σ
1 Normal eq. (1)	+1.300	−0.200	−0.100		+0.320	+1.320
2 Line 1 ÷ (−1.300)	$-C_1$	+0.154	+0.077		−0.246	−1.015*
3 Normal eq. (2)		+0.700	−0.200	−0.200	+0.180	+0.280
4 Line 1 × (+0.154)	1	−0.031	−0.015		+0.049	+0.203
5 Sum lines 3, 4		+0.669	−0.215	−0.200	+0.229	+0.483
6 Line 5 ÷ (−0.669)		$-C_2$	+0.321	+0.299	−0.342	−0.722*
7 Normal eq. (3)			+1.500	+0.700	−0.270	+1.630
8 Line 1 × (+0.077)		1	−0.008		+0.025	+0.102
9 Line 5 × (+0.321)		2	−0.069	−0.064	+0.074	+0.155
10 Sum lines 7, 8, 9			+1.423	+0.636	−0.171	+1.887
11 Line 10 ÷ (−1.423)			$-C_3$	−0.447	+0.120	−1.326*
12 Normal eq. (4)				+0.800	−0.120	+1.180
13 Line 1 × (0)			1	0	0	0
14 Line 5 × (+0.299)			2	−0.060	+0.068	+0.144
15 Line 10 × (−0.447)			3	−0.284	+0.076	−0.843
16 Sum lines 12, 13, 14, 15				+0.456	+0.024	+0.481
17 Line 16 ÷ (−0.456)				$-C_4$	−0.053	−1.055*

Back Solution†

	C_1	C_2	C_3	C_4
18 Col. N: lines 2, 6, 11, 17	−0.246	−0.342	+0.120	−0.053
19 Col. 4: lines 2, 6, 11 × (−0.053)		−0.016	+0.024	−0.053
20 Col. 3: lines 2, 6 × (+0.144)	+0.011	+0.046	+0.144	
21 Col. 2: lines 2 × (−0.312)	−0.048	−0.312		
22	−0.283			

* Should equal sum of values in line, for check.

† Below the heavy line are sums, therefore the values of the C's.

TABLE 22-7. LEVEL NET
Substitution of C's

v	Correlate equations with values for C's substituted	Values	
		First	Adjusted
1	$0.7(-0.144)$	-0.10	-0.10
2	$0.6(-0.283)$	-0.17	-0.18
3	$0.4(-0.283)$	-0.11	-0.11
4	$0.1(-0.053)$	-0.01	-0.01
5	$0.3(+0.312)$	$+0.09$	$+0.09$
6	$0.5(-0.144 + 0.053)$	-0.05	-0.05
7	$0.2(+0.283 - 0.312)$	-0.01	-0.01
8	$0.2(+0.312 + 0.144 - 0.053)$	$+0.08$	$+0.08$
9	$0.1(+0.283 + 0.144)$	$+0.04$	$+0.04$

TABLE 22-8. LEVEL NET. SUBSTITUTION OF v'S IN CONDITION EQUATIONS

		1		2		3		4	
		Sign	Value	Sign	Value	Sign	Value	Sign	Value
1	−0.10					−	+0.10		
2	−0.17	+	−0.17						
3	−0.11	+	−0.11						
4	−0.01							+	−0 01
5	+0.09			−	−0.09				
6	−0.05					−	+0.05	−	+0 05
7	−0.01	−	+0.01	+	−0.01				
8	+0.08			−	−0.08	+	+0.08	+	+0 08
9	+0.04	−	−0.04			+	+0.04		
N			+0.32		+0.18		0.27		−0.12
Sums			+0.01		0		0		0

TABLE 22-9. LEVEL NET. FINAL ELEVATIONS

Computation		Checks	
B.M.'s and links	Elevations and differences	*B.M.*'s and links	Elevations and differences
B	110.47	*G*	115.35
+4	− 9.24	−9	− 12.52
H	101.23	*A*	102.83
+1	+ 14.12		
G	115.35	*D*	117.08
−2	− 16.03	+7	− 14.25
C	99.32	*A*	102.83
−3	+ 17.76		
D	117.08	*E*	109.65
+5	− 7.43	+8	− 6.82
E	109.65	*A*	102.83
+6	− 8.42		
H	101.23		

TABLE 22-10. TRIANGULATION. LIST OF DIRECTIONS*

Sta.	No.	Directions			Sta.	No.	Directions		
A	1	00°	00′	00.0″	*E*	7	00°	00′	$\overline{59.8}$″
	2	18	16	16.7		8	47	09	50.1
	3	73	17	33.5		9	79	20	21.8
						10	127	55	33.4
B	4	00	00	00.3		11	186	17	37.1
	5	76	59	07.0					
	6	136	31	45.5	*F*	12	00	00	00.6
						13	15	07	15.4
C	22	00	00	$\overline{59.4}$		14	33	41	47.3
	23	40	16	13.8		15	66	28	16.6
	24	58	00	43.0		16	112	06	38.1
	25	106	31	55.9		17	137	18	39.3
D	26	00	00	00.0	*G*	18	00	00	01.1
	27	62	48	33.3		19	68	40	15.5
	28	80	02	30.4		20	132	17	27.9
						21	212	21	12.2

* The directions in this table are sets of clockwise angles. All the angles in each set are measured from the same assumed line at the station.

TABLE 22-11. COMPUTATION OF TRIANGLES

Sight Nos.	Stations		Angles			Cor.	Cor. angles	logs
	2–3	*A–B*						4.0505344
− 7 + 8	1	*E*	47°	09′	50.3″	−0.24″	50.06″	0.1347173
− 1 + 3	2	*A*	73	17	32.7	+1.18	33.88	9.9812685
− 5 + 6	3	*B*	59	32	38.5	−2.44	36.06	9.9355138
	1–3	*E–B*						4.1665202
	1–2	*E–A*						4.1207655
			179	58	121.5	−1.50	120.00	
	2–3	*E–A*						4.1207655
−15 + 16	1	*F*	45	38	21.5	−0.94	20.56	0.1457248
− 7 + 9	2	*E*	79	20	22.0	−0.40	21.60	9.9924386
− 2 + 3	3	*A*	55	01	16.8	+1.04	17.84	9.9134792
	1–3	*F–A*						4.2589289
	1–2	*F–E*						4.1799695 − 2
			179	59	60.3	−0.30	60.00	
	2–3	*E–B*						4.1665202
−15 + 17	1	*F*	70	50	22.7	−1.03	21.67	0.0247511
− 8 + 9	2	*E*	32	10	31.7	−0.16	31.54	9.7263305
− 4 + 5	3	*B*	76	59	06.7	+0.09	6.79	9.9886980
	1–3	*E–B*						3.9176018
	1–2	*E–E*						4.1799693
			178	119	61.1	−1.10	60.00	
	2–3	*A–B*						4.0505344
−16 + 17	1	*F*	25	12	01.2	−0.09	01.11	0.3708105
− 1 + 2	2	*A*	18	16	15.9	+0.14	16.04	9.4962568
− 4 + 6	3	*B*	136	31	45.2	−2.35	42.85	9.8375839
	1–3	*F–B*						3.9176017 + 1
	1–2	*F–A*						4.2589288 + 1
			179	59	62.3	−2.30	60.00	
	2–3	*E–F*						4.1799693
−20 + 21	1	*G*	80	03	44.3	+1.05	45.35	0.0065651
− 9 + 10	2	*E*	48	35	11.6	+0.66	12.26	9.8750369
−13 + 15	3	*F*	51	21	01.2	+1.19	02.39	9.8926415
	1–3	*G–F*						4.0615713
	1–2	*G–E*						4.0791759
			179	59	57.1	+2.90	60.00	
	2–3	*E–F*						4.1799693
−22 + 23	1	*C*	40	16	14.4	−0.93	13.47	0.1895015
− 9 + 11	2	*E*	106	57	15.3	+1.36	16.66	9.9807013
−14 + 15	3	*F*	32	46	29.3	+0.57	29.87	9.7334708
	1–3	*C–F*						4.3501721
	1–2	*C–E*						4.1029416 − 2
			178	119	59.0	+1.00	60.00	

TABLE 22-11. COMPUTATION OF TRIANGLES (*Continued*)

Sight Nos.	Stations		Angles			Cor.	Cor. angles	logs
	2–3	*E–G*						4.0791759
−22 + 24	1	*C*	58°	00′	43.6″	+0.06″	43.66″	0.0715221
−10 + 11	2	*E*	58	22	03.7	+0.70	04.40	9.9301505
−19 + 20	3	*G*	63	37	12.4	−0.46	11.94	9.9522434
	1–3	*C–G*						4.0808485
	1–2	*C–E*						4.1029414
			179	59	59.7	+0.30	60.00	
	2–3	*F–G*						4.0615713
−23 + 24	1	*C*	17	44	29.2	+0.99	30.19	0.5160896
−13 + 14	2	*F*	18	34	31.9	+0.62	32.52	9.5031876
−19 + 21	3	*G*	143	40	56.7	+0.59	57.29	9.7725110
	1–3	*C–G*						4.0808485
	1–2	*C–F*						4.3501719 + 2
			178	118	117.8	+2.20	120.00	
	2–3	*C–G*						4.0808485
−26 + 27	1	*D*	62	48	33.3	−0.24	33.06	0.0508591
−24 + 25	2	*C*	48	31	12.9	−0.39	12.51	9.8745912
−18 + 19	3	*G*	68	40	14.5	−0.07	14.43	9.9691853
	1–3	*D–G*						4.0062988
	1–2	*D–C*						4.1008929
			178	119	60.7	−0.70	60.00	
	2–3	*C–F*						4.3501721
−26 + 28	1	*D*	80	02	30.4	+0.48	30.88	0.0065927
−23 + 25	2	*C*	66	15	42.1	+0.60	42.70	9.9616085
−12 + 14	3	*F*	33	41	46.7	−0.28	46.42	9.7441284
	1–3	*D–F*						4.3183733
	1–2	*D–C*						4.1008932 − 3
			179	58	119.2	+0.80	120.00	
	2–3	*G–F*						4.0615713
−27 + 28	1	*D*	17	13	57.1	+0.72	57.82	0.5283364
−21 + 18	2	*G*	147	38	48.8	−0.52	48.28	9.7284656
−12 + 13	3	*F*	15	07	14.8	−0.90	13.90	9.4163915
	1–3	*D–F*						4.3183733
	1–2	*D–G*						4.0062992 − 4
			179	58	120.7	−0.70	120.00	

TABLE 22-12. ANGLE CONDITION EQUATIONS*

Triangle	Error	Eq. No.	Equation
ABE	+1.5	1	0 = +1.5 − (1) + (3) − (5) + (6) − (7) + (8)
AEF	+0.3	2	0 = +0.3 − (2) + (3) − (7) + (9) − (15) + (16)
BEF	+1.1	3	0 = +1.1 − (4) + (5) − (8) + (9) − (15) + (17)
EFG	−2.9	4	0 = −2.9 − (9) + (10) − (13) + (15) − (20) + (21)
EFC	−1.0	5	0 = −1.0 − (9) + (11) − (14) + (15) − (22) + (23)
EGC	−0.3	6	0 = −0.3 −(10) + (11) − (19) + (20) − (22) + (24)
FGD	+0.7	7	0 = +0.7 −(12) + (13) + (18) − (21) − (27) + (28)
GCD	+0.7	8	0 = +0.7 −(18) + (19) − (24) + (25) − (26) + (27)

*The correction to each direction is represented by the number of that direction placed in parentheses.

TABLE 22-13. SIDE CONDITION EQUATIONS
Use all the smallest angles

Pole at *B*:

1	2	3	4	5	6	7	8	9
Sides	Angles ahead				Angles behind			
BF	− 8 + 9	32°10′31.7″	9.7263310	+33.5	− 1 + 2	18°16′15.9″	9.4962559	+63.7
BE	− 1 + 3	73 17 32.7	9.9812678	+ 6.3	−15 + 17	70 50 22.7	9.9752497	+ 7.3
BA	−16 + 17	25 12 01.2	9.6291899	+44.8	− 7 + 8	47 09 50.3	9.8152832	+19.5
			9.3367887				9.3367888	
		−	9.3367888					
		−	1					

9. 0 = −0.01 + 0.58(1) − 0.64(2) + 0.06(3) + 0.20(7) − 0.54(8) + 0.34(9) + 0.07(15) − 0.45(16) + 0.38(17)

Pole at *G*:

Sides	Angles ahead				Angles behind			
GC	−10 + 11	58°22′03.7″	9.9301496	+12.9	−13 + 14	18°34′31.9″	9.5031837	+62.6
GE	−13 + 15	51 21 01.2	9.8926395	+16.9	−22 + 24	58 00 43.6	9.9284778	+13.1
GF	−23 + 24	17 44 29.2	9.4839039	+65.8	− 9 + 10	48 35 11.6	9.8750357	+18.6
			9.3066930				9.3066972	
		−	9.3066972					
		−	42					

10. 0 = −0.42 + 0.19(9) − 0.32(10) + 0.13(11) + 0.46(13) − 0.63(14) + 0.17(15) + 0.13(22) − 0.66(23) + 0.53(24)

TABLE 22-13. SIDE CONDITION EQUATIONS (*Continued*)

Pole at *D*:

Sides	Angles ahead				Angles behind			
DC	−18 + 19	68°40′14.5″	9.9691853	+ 8.3	−12 + 14	33°41′46.7″	9.7441293	+31.6
DG	−12 + 13	15 07 14.8	9.4163985	+77.9	−24 + 25	48 31 12.9	9.8745919	+18.6
DF	−23 + 25	66 15 42.1	9.9616079	+ 9.2	−21 + 18	147 38 48.8	9.7284639	−33.2
			9.3471917				9.3471851	
		−	9.3471851					
		+	66					

11. 0 = +0.66 − 0.46(12) + 0.78(13) − 0.32(14) + 0.25(18) + 0.08(19) − 0.33(21) − 0.09(23) + 0.19(24) − 0.10(25)

TABLE 22-14. LENGTH CONDITION EQUATION

1	2	3	4	5	6	7	8
A angles (adjacent)				*B* angles (opposite)			
AB		4.0505344		*CD*		4.1008928	
− 1 + 3	73°17′32.7″	9.9812678	+ 6.3	− 7 + 8	47°09′50.3″	9.8652832	+19.5
− 4 + 5	76 59 06.7	9.9886980	+ 4.9	−15 + 17	70 50 22.7	9.9752497	+ 7.3
−13 + 15	51 21 01.2	9.8926395	+16.9	−20 + 21	80 03 44.3	9.9934345	+ 3.7
−10 + 11	58 22 03.7	9.9301496	+12.9	−22 + 24	58 00 43.6	9.9284778	+13.1
−18 + 19	68 40 14.5	9.9691853	+ 8.3	−26 + 27	62 48 33.3	9.9491412	+10.8
		3.8124746				3.8124792	
		− 92					
		− 46					

13. 0 = − 0.46 − 0.06(1) + 0.06(3) − 0.05(4) + 0.05(5) − 0.17(13) + 17(15) − 0.13(10) + 0.13(11) − 0.08(18) + 0.08(19) + 0.20(7) − 0.20(8) + 0.07(15) − 0.07(17) + 0.04(20) − 0.04(21) + 0.13(22) − 13(24) + 0.11(26) − 0.11(27)

TABLE 22-15. AZIMUTH CONDITION EQUATION

BA (fixed azimuth)		205°41′01.2″
	−1 + 3	73 17 32.7
		278 58 33.9
		−180 00 00.0
AE		98 58 33.9
	−7 + 11	186 17 37.3
		285 16 11.2
		−180 00 0.00
EC		105 16 11.2
	−22 + 25	106 31 56.5
		211 48 07.7
		−180 00 00.0
CD		31 48 07.7
CD (fixed azimuth)		31 48 09.5
Error		−01.8

12. 0 = −1.8 − (1) + (3) − (7) + (11) − (22) + (25)

TABLE 22-16. PRELIMINARY COORDINATE COMPUTATION

Triangles

Sight Nos.	Stations		Angles			logs
	2–3	*A–B*				4.0505344
− 7 + 8	1	*E*	47°	09′	50.3″	0.1347168
− 1 + 3	2	*A*	73	17	32.7	9.9812678
	3	*B*	(59	32	37.0)	9.9355150
	1–3	*E–B*				4.1665190
	1–2	*E–A*				4.1207662
			179	58	120.0	
	2–3	*E–B*				4.1665190
−15 + 17	1	*F*	70	50	22.7	0.0247503
	2	*E*	(32	10	30.6)	9.7263273
− 4 + 5	3	*B*	76	59	06.7	9.9886980
	1–3	*F–B*				3.9176966
	1–2	*F–E*				4.1799673
			178	119	60.0	
	2–3	*E–F*				4.1799673
−20 + 21	1	*G*	80	03	44.3	0.0065655
	2	*E*	(48	35	14.5)	9.8750411
−13 + 15	3	*F*	51	21	01.2	9.8926395
	1–3	*G–F*				4.0615739
	1–2	*G–E*				4.0791723
			179	59	60.0	
	2–3	*E–G*				4.0791723
−22 + 24	1	*C*	58	00	43.6	0.0715222
−10 + 11	2	*E*	58	22	03.7	9.9301496
	3	*G*	(63	37	12.7)	9.9522442
	1–3	*C–G*				4.0808441
	1–2	*C–E*				4.1029387
			179	59	60.0	

Coordinates

α	*B–A*	205°41′01.2″	α	*A–B*	25°41′01.2″
Angle *B*		− 59 32 37.0	Angle *A*		+ 73 17 32.7
α	*B–E*	146 08 24.2	α	*A–E*	98 58 33.9
		179 59 60			179 59 60
		S33°51′35.8″E			S81°01′26.1″E
		18,574.71 *B*			8450.62 *A*
	4.0858074	−12,184.49		3.3139525	− 2060.40
	9.9192884	6,390.22 *E*		9.1931863	6390.22 *E*
BE	4.1665190		*AE*	4.1207662	
	9.7459836	8,382.15 *B*		9.9946487	3,513.31 *A*
	3.9125026	+ 8,175.28		4.1154149	+13,044.12
		16,557.43 *E*			16,557.43 *E*

TABLE 22-16. PRELIMINARY COORDINATE COMPUTATION (*Continued*)

α	*B–E*	146°08′24.2″	α	*E–B*	326°08′24.2″
Angle *B*		− 76 59 06.7	Angle *E*		+ 32 10 30.6
α	*B–F*	69 09 17.5	α	*E–F*	358 18 54.8
					359 59 60
		N69°09′17.5″E			N1°41′05.2″W

		18,574.71 *B*			6,390.22 *E*
	3.4688554	+ 2,943.44		4.1797795	+15,127.93
	9.5512588	21,518.15 *F*		9.9998122	21,518.15
BF	3.9175966		*EF*	4.1799673	
	9.9706005	8,382.15 *B*		8.4683574	16,557.43 *E*
	3.8881971	+ 7,730.31		2.6483247	− 444.97
		16,112.46 *F*			16,112.46 *F*

α	*E–F*	358°18′54.8″	α	*F–E*	178°18′54.8″
Angle *E*		+ 48 35 14.5	Angle *F*		− 51 21 01.2
α	*E–G*	406 54 09.3	α	*F–G*	126 57 53.6
		360 00 00			179 59 60
		N46°54′09.3″E			S53°02′06.4″E

		6,390.22 *E*			21,518.15 *F*
	3.9137462	+ 8,198.72		3.8406835	− 6,929.21
	9.8345739	14,588.94 *G*		9.7791096	14,588.94 *G*
EG	4.0791723		*FG*	4.0615739	
	9.8634377	16,557.43 *E*		9.9025490	16112.46 *F*
	3.9426100	+ 8,762.14		3.9641229	+ 9207.10
		25,319.57 *G*✓			25319.56 *G*

α	*E–G*	46°54′09.3″	α	*G–E*	226°54′09.3″
Angle *E*		+ 58 22 03.7	Angle *G*		− 63 37 12.7
α	*E–C*	105 16 13.0	α	*G–C*	163 16 56.6
		179 59 60			179 59 60
		S74°43′47.0″E			S16°43′03.4″E

		6390.22 *E*			14,588.94 *G*
	3.5235094	− 3338.18		4.0620891	−11,536.90
	9.4205707	3052.04 *C*		9.9812450	3,052.04 *C*
EC	4.1029387		*GC*	4.0808441	
	9.9843897	16,557.43 *E*		9.4588718	25,319.57 *G*
	4.0873284	+12,227.24		3.5397159	+ 3,465.10
		28,784.67 *C*			28,784.67 *C*

Table 22-17. Position Condition Equations

1	2	3	4	5	6	7	8	9	10	11	12	13	14	15	16
	y_n	x_n	10^{-4} $(y_T - y_n)$	10^{-4} $(x_T - x_n)$	10^5 δ_A	10^5 $-\delta_B$	y eq. Cols. 4 × 6	A	x eq. Cols. 5 × 6	y eq. Cols. 4 × 7	B	x eq. Cols. 5 × 7	y eq. Col. 5 × $(-K)$*	C	x eq. Col. 4 × $(+K)$*
$-B$	18,575	8,382	−1.55	+2.04	+.06	−.20	−0.09	− 1 + 3	+0.12	+0.31	− 7 + 8	−0.41	−0.43	{− 1 + 3; − 7 + 8}	−0.33
$+E$	6,390	16,558	−0.33	+1.22	+.05	−.07	−0.02	− 4 + 5	+0.06	+0.02	−15 + 17	−0.09	−0.26	{+ 4 − 5; +15 − 17}	−0.07
$+E$	6,390	16,558	−0.33	+1.22	+.17	−.04	−0.06	−13 + 15	+0.21	+0.01	−20 + 21	−0.05	−0.26	{+13 − 15; +20 − 21}	−0.07
$-G$	14,589	25,320	−1.15	+0.35	+.13	−.13	−0.15	−10 + 11	+0.05	+0.15	−22 + 24	−0.05	−0.07	{−10 + 11; −22 + 24}	−0.24

* K is always the number 0.211 or 0.21.

	y	x	
n	3052.04	28,784.67	Preliminary computation
n'	3051.93	28,784.71	Fixed
Error	+0.11	−0.04	
× 4.34	+0.48	−0.17	

y equation

14. $0 = +0.48 + 0.52(1) - 0.52(3) - 0.24(4) + 0.24(5) + 0.12(7) - 0.12(8) + 0.22(10) - 0.22(11) - 0.20(13) - 0.08(15) + 0.28(17) - 0.27(20) + 0.27(21) - 0.08(22) + 0.08(24)$

x equation

15. $0 = -0.17 + 0.21(1) - 0.21(3) - 0.13(4) + 0.13(5) + 0.74(7) - 0.74(8) + 0.19(10) - 0.19(11) - 0.28(13) + 0.30(15) - 0.02(17) - 0.02(20) + 0.02(21) + 0.29(22) - 0.29(24)$

TABLE 22-18. CORRELATE AND NORMAL EQUATIONS

	1	2	3	4	5	6	7	8	9	10	11	12	13	14	15	Σ
1	−1								+0.58			−1	−0.06	+0.52	+0.21	−0.75
2		−1							−0.64							−1.64
3	+1	+1							+0.06			+1	+0.06	−0.52	−0.21	+2.39
4			−1										−0.05	−0.24	−0.13	−1.42
5	−1		+1										+0.05	+0.24	+0.13	+0.42
6	+1															+1.00
7	−1	−1							+0.20			−1	+0.20	+0.12	+0.74	−1.74
8	+1		−1						−0.54				−0.20	−0.12	−0.74	−1.60
9		+1	+1	−1	−1				+0.34	+0.19						+0.53
10				+1		−1				+0.32			−0.13	+0.22	+0.19	−0.04
11					+1	+1				+0.13		+1	+0.13	−0.22	−0.19	+2.85
12							−1				−0.46					−1.46
13				−1			+1			+0.46	+0.78		−0.17	−0.20	−0.28	+0.59
14					−1					−0.63	−0.32					−1.95
15		−1	−1	+1	+1				+0.07	+0.17			+0.24	−0.08	+0.30	+0.70
16		+1							−0.45							+0.55
17			+1						+0.38				−0.07	+0.28	−0.02	+1.57
18							+1	−1			+0.25		−0.08			+0.17
19						−1		+1			+0.08		+0.08			+0.16
20				−1		+1							+0.04	−0.27	−0.02	−0.25
21				+1			−1				−0.33		−0.04	+0.27	+0.02	−0.08
22					−1	−1				+0.13		−1	+0.13	+0.08	+0.29	−2.53
23					+1					−0.66	−0.09					+0.25
24						+1		−1		+0.53	+0.19		−0.13	+0.08	−0.29	+0.38
25								+1			−0.10	+1				+1.90
26								−1					+0.11			−0.89
27							−1	+1					−0.11			−0.11
28							+1									+1.00

	1	2	3	4	5	6	7	8	9	10	11	12	13	14	15	N	Σ
1	+6	+2	−2						−1.26			+3.00	−0.33	−1.52	−2.03	+1.5	+ 5.36
2		+6	+2	−2	−2				+0.32	+0.02		+2.00	−0.38	−0.56	−1.25	+0.3	+ 6.45
3			+6	−2	−2				+1.19	+0.02			−0.01	+0.96	+0.68	+1.1	+ 5.94
4				+6	+2	−2	−2		−0.27	−0.80	−1.11		+0.20	+0.88	+0.81	−2.9	− 3.19
5					+6	+2			−0.27	−0.05	+0.23	+2.00	+0.24	−0.22	−0.18	−1.0	+ 6.75
6						+6		−2		+0.85	+0.11	+2.00	−0.04	−0.55	−0.98	−0.3	+ 5.00
7							+6	−2		+0.46	+1.82		−0.10	−0.47	−0.30	+0.7	+ 4.11
8								+6		−0.53	−0.46	+1.00	+0.07	−0.08	+0.29	+0.7	+ 2.99
9									+0.5486	+0.0765		−0.72	+0.1070	+0.4600	+0.6702	−0.01	+ 1.8423
10										+1.5262	+0.7205	0	−0.0309	−0.1726	−0.2793	−0.42	+ 1.3904
11											+1.1544	−0.10	−0.1577	−0.2299	−0.2801	+0.66	+ 2.3572
12												+6.00	−0.08	−1.30	−1.64	−0.18	+10.36
13													+0.2914	−0.0948	+0.4292	−0.46	− 0.3458
14														+1.0650	+0.5328	+0.48	− 0.8195
15															+1.6272	−0.17	− 2.0700

TABLE 22-19. SOLUTION

1	2	3	9	4	5	6	7	8
+6.00000	+2.00000	−2.00000	−1.26000					
C_1	−0.33333	+0.33333	+0.21000					
	+6.00000	+2.00000	+0.32000	−2.00000	−2.00000			
1	−0.66667	+0.66667	+0.42000					
	+5.33333	+2.66667	+0.74000	−2.00000	−2.00000			
	C_2	−0.50000	−0.13875	+0.37500	+0.37500			
		+6.00000	+1.19000	−2.00000	−2.00000			
	1	−0.66667	−0.42000					
	2	−1.33333	−0.37000	+1.00000	+1.00000			
		+4.00000	+0.40000	−1.00000	−1.00000			
		C_3	−0.10000	+0.25000	+0.25000			
			+1.54860	−0.27000	−0.27000			
		1	−0.26460					
		2	−0.10268	+0.27750	+0.27750			
		3	−0.04000	+0.10000	+0.10000			
			+1.14132	+0.10750	+0.10750			
			C_9	−0.09419	−0.09419			
				+6.00000	+2.00000	−2.00000	−2.00000	
			2	−0.75000	−0.75000			
			3	−0.25000	−0.25000			
			9	−0.01012	−0.01012			
				+4.98988	+0.98988	−2.00000	−2.00000	
				C_4	−0.19838	+0.40081	+0.40081	
					+6.00000	+2.00000		
				2	−0.75000			
				3	−0.25000			
				9	−0.01013			
				4	−0.19637	+0.39675	+0.39675	
					+4.79350	+2.39675	+0.39675	
					C_5	−0.50000	−0.08277	
						+6.00000		−2.00000
					4	−0.80162	−0.80162	
					5	−1.19838	−0.19838	
						+4.00000	−1.00000	−2.00000
						C_6	+0.25000	+0.50000
							+6.00000	−2.00000
						4	−0.80162	
						5	−0.03284	
						6	−0.25000	−0.50000
							+4.91554	−2.50000
							C_7	+0.50859
								+6.00000
							6	−1.00000
							7	−1.27148
								+3.72852
								C_8

OF NORMAL EQUATIONS

10	11	12	13	14	15	N	Σ_n
		+3.00000	−0.33000	−1.52000	−2.03000	+1.50000	+5.36000
		−0.50000	+0.05500	+0.25333	+0.33833	−0.25000	−0.89333
+0.02000		+2.00000	−0.38000	−0.56000	−1.25000	+0.30000	+6.45000
		−1.00000	+0.11000	+0.50667	+0.67667	−0.50000	−1.78667
+0.02000		+1.00000	−0.27000	−0.05333	−0.57333	−0.20000	+4.66333
−0.00375		−0.18750	+0.05062	+0.00999	+0.10749	+0.03750	−0.87437
+0.02000			−0.01000	+0.96000	+0.68000	+1.10000	+5.94000
		+1.00000	−0.11000	−0.50667	−0.67667	+0.50000	+1.78667
−0.01000		−0.50000	+0.13500	+0.02666	+0.28666	+0.10000	−2.33166
+0.01000		+0.50000	+0.01500	+0.47999	+0.28999	+1.70000	+5.39501
−0.00250		−0.12500	−0.00375	−0.12000	−0.07250	−0.42500	−1.34875
+0.07650		−0.72000	+0.10700	+0.46000	+0.67020	−0.01000	+1.84230
		+0.63000	−0.06930	−0.31920	−0.42630	+0.31500	+1.12560
−0.00278		−0.13875	+0.03746	+0.00740	+0.07954	+0.02775	−0.64704
−0.00100		−0.05000	−0.00150	−0.04800	−0.02900	−0.17000	−0.53950
+0.07272		−0.27875	+0.07366	+0.10020	+0.29444	+0.16275	+1.78136
−0.06372		+0.24423	−0.06451	0.08770	−0.25798	−0.14260	−1.56079
−0.80000	−1.11000		+0.20000	+0.88000	+0.81000	−2.90000	−3.19000
+0.00750		+0.37500	−0.10125	−0.02000	−0.21500	−0.07500	+1.74875
+0.00250		+0.12500	+0.00375	+0.12000	+0.07250	+0.42500	+1.34875
−0.00685		+0.02625	−0.00694	−0.00944	−0.02773	0.01533	−0.16779
−0.79685	−1.11000	+0.52625	+0.09556	+0.97056	+0.63977	−2.56533	−0.26029
+0.15969	+0.22245	−0.10546	−0.01915	−0.19451	−0.12822	+0.51411	+0.05216
−0.05000	+0.23000	+2.00000	+0.24000	−0.22000	−0.18000	−1.00000	+6.75000
+0.00750		+0.37500	−0.10125	−0.02000	−0.21500	−0.07500	+1.74875
+0.00250		+0.12500	+0.00375	+0.12000	+0.07250	+0.42500	+1.34875
−0.00685		+0.02625	−0.00694	−0.00944	−0.02773	−0.01533	−0.16779
+0.15807	+0.22020	−0.10439	−0.01896	−0.19254	−0.12692	+0.50891	+0.05164
+0.11122	+0.45020	+2.42186	+0.11660	−0.32198	−0.47715	−0.15642	+9.73135
−0.02320	−0.09392	−0.50524	−0.02433	+0.06717	+0.09954	+0.03263	−2.03011
+0.85000	+0.11000	+2.00000	−0.04000	−0.55000	−0.98000	−0.30000	+5.09000
−0.31938	−0.44490	+0.21092	+0.03830	+0.38901	+0.25643	−1.02822	−0.10433
−0.05560	−0.22510	−1.21093	−0.05830	+0.16099	+0.23858	+0.07821	−4.86568
+0.47502	−0.56000	+1.00000	−0.06000	0	−0.48500	−1.25000	+0.12000
−0.11875	+0.14000	−0.25000	+0.01500		+0.12125	+0.31250	−0.03000
+0.46000	+1.82000		−0.10000	−0.47000	−0.30000	+0.70000	+4.11000
−0.31938	−0.44490	+0.21092	+0.03830	+0.38001	+0.25643	−1.02822	−0.10434
−0.00920	−0.03726	−0.20045	−0.00965	+0.02665	+0.03949	+0.01295	−0.80545
+0.11875	−0.14000	+0.25000	−0.01500		−0.12125	−0.31250	+0.03000
+0.25017	+1.19784	+0.26047	−0.08635	−0.05434	−0.12533	−0.62777	+3.23021
−0.05089	−0.24368	−0.05299	+0.01757	+0.01105	+0.02549	+0.12771	−0.65714
−0.53000	−0.46000	+1.00000	+0.07000	−0.08000	+0.29000	+0.70000	+2.99000
+0.23750	−0.28000	+0.50000	−0.03000		−0.24250	−0.62500	+0.06000
+0.12722	+0.60920	+0.13248	−0.04392	−0.02762	−0.06372	−0.31928	+1.64285
−0.16528	−0.13080	+1.63248	−0.00392	−0.10762	−0.01622	−0.24428	+4.69285
+0.04432	+0.03508	−0.43784	+0.00105	+0.02886	+0.00435	+0.06552	−1.25864

TABLE 22-19. SOLUTION

8	10	11	12	13	14	15	N	Σ_n
	+1.52620	+0.72050	0	−0.03090	−0.17260	−0.27930	−0.42000	+ 1.39040
2	−0.00008		−0.00375	+0.00101	+0.00020	+0.00215	+0.00075	− 0.01749
3	−0.00002		−0.00125	−0.00004	−0.00120	−0.00072	−0.00425	− 0.01349
9	−0.00463		+0.01776	−0.00469	−0.00638	−0.01876	−0.01037	− 0.11351
4	−0.12725	−0.17726	+0.08404	+0.01526	+0.15499	+0.10216	−0.40967	− 0.04157
5	−0.00258	−0.01044	−0.05619	−0.00271	+0.00747	+0.01107	+0.00363	− 0.22577
6	−0.05641	+0.06650	−0.11876	+0.00713		+0.05760	+0.14844	− 0.01425
7	−0.01273	−0.06096	−0.01326	+0.00439	+0.00277	+0.00638	+0.03195	− 0.16439
8	−0.00733	−0.00580	+0.07237	−0.00017	−0.00477	−0.00072	−0.01083	+ 0.20803
	+1.31517	+0.53254	−0.01904	−0.01072	−0.01952	−0.12014	−0.67035	+ 1.00796
	C_{10}	−0.40492	+0.01448	+0.00815	+0.01484	+0.09135	+0.50971	− 0.76641
		+1.15440	−0.10000	−0.15770	−0.22990	−0.28010	+0.66000	+ 2.35720
	4	−0.24692	+0.11706	+0.02126	+0.21590	+0.14232	−0.57066	− 0.05790
	5	−0.04228	−0.22746	−0.01095	+0.03024	+0.04481	+0.01469	− 0.91397
	6	−0.07840	+0.14000	−0.00840		−0.06790	−0.17500	+ 0.01680
	7	−0.29189	−0.06347	+0.02105	+0.01324	+0.03054	+0.15298	− 0.78715
	8	−0.00459	+0.05727	−0.00014	−0.00377	−0.00057	−0.00857	+ 0.16463
	10	−0.21564	+0.00771	+0.00434	+0.00790	+0.04865	+0.27144	− 0.40814
		+0.27468	−0.06889	−0.13054	+0.03361	−0.08225	+0.34488	+ 0.37147
		C_{11}	+0.25080	+0.47524	−0.12236	+0.29944	−1.25557	− 1.35237
			+6.00000	−0.08000	−1.30000	−1.64000	−1.80000	+10.36000
		1	−1.50000	+0.16500	+0.76000	+1.01500	−0.75000	− 2.68000
		2	−0.18750	+0.05062	+0.01000	+0.10750	+0.03750	− 0.87437
		3	−0.06250	−0.00188	−0.06000	−0.03625	−0.21250	− 0.67438
		9	−0.06808	+0.01799	+0.02447	+0.07191	+0.03975	+ 0.43506
		4	−0.05550	−0.01008	−0.10236	−0.06747	+0.27055	+ 0.02745
		5	−1.22362	−0.05891	+0.16268	+0.24108	+0.07903	− 4.91667
		6	−0.25000	+0.01500		+0.12125	+0.31250	− 0.03000
		7	−0.01380	+0.00458	+0.00288	+0.00664	+0.03326	− 0.17117
		8	−0.71477	+0.00172	+0.04712	+0.00710	+0.10696	− 2.05472
		10	−0.00028	−0.00015	−0.00028	−0.00174	−0.00970	+ 0.01460
		11	−0.01728	−0.03274	+0.00843	−0.02063	+0.08650	+ 0.09316
			+1.90667	+0.07115	−0.44706	−0.19561	−1.80615	− 0.47104
			C_{12}	−0.03732	+0.23447	+0.10259	+0.94728	+ 0.24705

of Normal Equations (*Continued*)

12	13	14	15	N	Σ_n
	+0.29140	−0.09480	+0.42920	−0.46000	−0.34580
1	−0.01815	−0.08360	−0.11165	+0.08250	+0.29480
2	−0.01367	−0.00270	−0.02902	−0.01012	+0.23606
3	−0.00006	−0.00180	−0.00109	−0.00638	−0.02023
9	−0.00475	−0.00647	−0.01900	−0.01050	−0.11497
4	−0.00183	−0.01859	−0.01225	+0.04914	+0.00498
5	−0.00284	+0.00783	+0.01161	+0.00381	−0.23676
6	−0.00090		−0.00728	−0.01875	+0.00180
7	−0.00152	−0.00095	−0.00220	−0.01103	+0.05675
8	0	−0.00011	−0.00002	−0.00026	+0.00493
10	−0.00009	−0.00016	−0.00098	−0.00546	+0.00821
11	−0.06204	+0.01597	−0.03909	+0.16390	+0.17654
12	−0.00266	+0.01668	+0.00730	+0.06741	+0.01758
	+0.18289	−0.16870	+0.22553	−0.15574	+0.08389
	C_{13}	+0.92241	−1.23315	+0.85155	−0.45869
		+1.06500	+0.53280	+0.48000	−0.81950
	1	−0.38506	−0.51426	+0.38000	+1.35786
	2	−0.00053	−0.00573	−0.00200	+0.04659
	3	−0.05760	−0.03480	−0.20399	−0.64740
	9	−0.00880	−0.02585	−0.01429	−0.15639
	4	−0.18878	−0.12444	+0.49898	+0.05063
	5	−0.02163	−0.03205	−0.01051	+0.65365
	7	−0.00060	−0.00138	−0.00694	+0.03569
	8	−0.00311	−0.00047	−0.00705	+0.13544
	10	−0.00029	−0.00178	−0.00995	+0.01496
	11	−0.00411	+0.01006	−0.04220	−0.04545
	12	−0.10482	−0.04586	−0.42349	−0.11044
	13	−0.15561	+0.20803	−0.14366	+0.07738
		+0.13406	−0.03573	+0.49490	+0.59302
		C_{14}	+0.26652	−3.69163	−4.42354
			+1.62720	−0.17000	−2.07000
		1	−0.68681	+0.50750	+1.81346
		2	−0.06763	−0.02150	+0.50126
		3	−0.02102	0.10005	0.30111
		9	−0.07596	−0.04199	−0.45956
		4	−0.08203	+0.32892	+0.03337
		5	−0.04750	−0.01557	+0.96866
		6	−0.05881	−0.15156	+0.01455
		7	−0.00319	−0.01600	+0.08234
		8	−0.00007	−0.00106	+0.02041
		10	0.01007	−0.06124	+0.09208
		11	−0.02463	+0.10327	+0.11123
		12	−0.02007	−0.18529	−0.04832
		13	−0.27811	+0.19205	−0.10345
		14	−0.00952	+0.13190	+0.15805
			+0.24688	+0.47618	+0.72294
			C_{15}	−1.92879	−2.92831

TABLE 22-20. BACK

	1	2	3	9	4	5	6	7
N	−0.25000	+0.03750	−0.42500	−0.14260	+0.51411	+0.03263	+0.31250	+0.12771
15	+0.33833	+0.10749	−0.07250	−0.25798	−0.12822	+0.09954	+0.12125	+0.02549
	−0.65257	−0.20733	+0.13984	+0.49759	+0.24731	−0.19199	−0.23387	−0.04916
14	+0.25333	+0.00999	−0.12000	−0.08779	−0.19451	+0.06717	0	+0.01105
	−1.06543	−0.04201	+0.50468	+0.36922	+0.81805	−0.28250		−0.04647
13	+0.05500	+0.05062	−0.00375	−0.06454	−0.01915	−0.02433	+0.01500	+0.01757
	−0.03571	−0.03287	+0.00243	+0.04191	+0.01243	+0.01580	−0.00974	−0.01141
12	−0.50000	−0.18750	−0.12500	+0.24423	−0.10546	−0.50524	−0.25000	−0.05299
	+0.10624	+0.03984	+0.02656	−0.05189	+0.02241	+0.10735	+0.05312	+0.01126
11					+0.22245	−0.09392	+0.14000	−0.24368
					−0.37380	+0.15782	−0.23526	+0.40948
10		−0.00375	−0.00250	−0.06372	+0.15969	−0.02320	−0.11875	−0.05089
		−0.00354	−0.00236	−0.06010	+0.15061	−0.02188	−0.11200	−0.04800
8							+0.50000	+0.50859
							+0.00548	+0.00557
7					+0.40081	−0.08277	+0.25000	+0.39898
					+0.15992	−0.03302	+0.09974	C_7
6					+0.40081	−0.50000	−0.12003	
					−0.04811	+0.06002	C_6	
5		+0.37500	+0.25000	−0.09419	−0.19838	−0.15577		
		−0.05841	−0.03894	+0.01467	+0.03090	C_5		
4		+0.37500	+0.25000	−0.09419	+1.53383			
		+0.57519	+0.38346	−0.14447	C_4			
9	+0.21000	−0.13875	−0.10000	+0.52433				
	+0.11011	−0.07275	−0.05243	C_9				
3	+0.33333	−0.50000	+0.53824					
	+0.17941	−0.26912	C_2					
2	−0.33333	−0.03350						
	+0.01117	C_2						
1	−1.59678							
	C_1							

Solution

8	10	11	12	13	14	15
+0.06552	+0.50971	−1.25557	+0.94728	+0.85155	−3.69163	−1.92879
+0.00435	+0.09135	+0.29944	+0.10259	−1.23315	+0.26652	−1.92879
−0.00839	−0.17619	−0.57756	−0.19787	+2.37849	−0.51406	C_{15}
+0.02886	+0.01484	−0.12236	+0.23447	+0.92241	−4.20569	
−0.12138	−0.06241	+0.51461	−0.98611	−3.87937	C_{14}	
+0.00105	+0.00815	+0.47524	−0.03732	−0.64933		
−0.00068	−0.00529	−0.30859	+0.02423	C_{13}		
−0.43784	+0.01448	+0.25080	−0.21247			
+0.09303	−0.00308	−0.05329	C_{12}			
+0.03508	−0.40492	−1.68040				
−0.05895	+0.68043	C_{11}				
+0.04432	+0.94317					
+0.04180	C_{10}					
+0.01095						
C_8						

TABLE 22-21. VALUES OF v'S

v_1

1	−1.00	+1.59678
9	+0.58	+0.30411
12	−1.00	+0.21247
13	−0.06	+0.03896
14	+0.52	−2.18696
15	+0.21	−0.40505
		−0.43969

v_2

2	−1.00	+0.03350
9	−0.64	−0.33557
		−0.30207

v_3

1	+1.00	−1.59678
2	+1.00	−0.03350
9	+0.06	+0.03146
12	+1.00	−0.21247
13	+0.06	−0.03896
14	−0.52	+2.18696
15	−0.21	+0.40505
		+0.74176

v_4

3	−1.00	−0.53824
13	−0.05	+0.03247
14	−0.24	+1.00937
15	−0.13	+0.25074
		+0.75434

v_5

1	−1.00	+1.59678
3	+1.00	+0.53824
13	+0.05	−0.03247
14	+0.24	−1.00937
15	+0.13	−0.25074
		+0.84244

v_6

1	+1.00	−1.59678
		−1.59678

v_7

1	−1.00	+1.59678
2	−1.00	+0.03350
9	+0.20	+0.10487
12	−1.00	+0.21247
13	+0.20	−0.12987
14	+0.12	−0.50468
15	+0.74	−1.42730
		−0.11423

v_8

1	+1.00	−1.59678
3	−1.00	−0.53824
9	−0.54	−0.28314
13	−0.20	+0.12987
14	−0.12	+0.50468
15	−0.74	+1.42730
		−0.35631

v_9

2	+1.00	−0.03350
3	+1.00	+0.53824
4	−1.00	−1.53384
5	−1.00	+0.15577
9	+0.34	+0.17827
10	+0.19	+0.17920
		−0.51586

v_{10}

4	+1.00	+1.53383
6	−1.00	+0.12003
10	−0.32	−0.30181
13	−0.13	+0.08441
14	+0.22	−0.92525
15	+0.19	−0.36647
		+0.14474

TABLE 22-21. VALUES OF v'S (*Continued*)

v_{11}

5	+1.00	−0.15577
6	+1.00	−0.12003
10	+0.13	+0.12261
12	+1.00	−0.21247
13	+0.13	−0.08441
14	−0.22	+0.92525
15	−0.19	+0.36647
		+0.84165

v_{12}

7	−1.00	−0.39898
11	−0.46	+0.77298
		+0.37400

v_{13}

4	−1.00	−1.53383
7	+1.00	+0.39898
10	+0.46	+0.43386
11	+0.78	−1.31071
13	−0.17	+0.11039
14	−0.20	+0.84114
15	−0.28	+0.54006
		−0.52011

v_{14}

5	−1.00	+0.15577
10	−0.63	−0.59420
11	−0.32	+0.53773
		+0.09930

v_{15}

2	−1.00	+0.03350
3	−1.00	−0.53824
4	+1.00	+1.53383
5	+1.00	−0.15577
9	+0.07	+0.03670
10	+0.17	+0.16034
13	+0.24	−0.15584
14	−0.08	+0.33646
15	+0.30	−0.57864
		+0.67234

v_{16}

2	+1.00	−0.03350
9	−0.45	−0.23595
		−0.26945

v_{17}

3	+1.00	+0.53824
9	+0.38	+0.19925
13	−0.07	+0.04545
14	+0.28	−1.17759
15	−0.02	+0.03858
		−0.35607

v_{18}

7	+1.00	+0.39898
8	−1.00	−0.01095
11	+0.25	−0.42010
13	−0.08	+0.05195
		+0.01988

v_{19}

6	−1.00	+0.12003
8	+1.00	+0.01095
11	+0.08	−0.13443
13	+0.08	−0.05195
		−0.05540

v_{20}

4	−1.00	−1.53383
6	+1.00	−0.12003
13	+0.04	−0.02597
14	−0.27	+1.13554
15	−0.02	+0.03858
		−0.50571

v_{21}

4	+1.00	+1.53383
7	−1.00	−0.39898
11	−0.33	+0.55453
13	−0.04	+0.02597
14	+0.27	−1.13554
15	+0.02	−0.03858
		+0.54123

TABLE 22-21. VALUES OF v'S (*Continued*)

v_{22}

5	−1.00	+0.15577
6	−1.00	+0.12003
10	+0.13	+0.12261
12	−1.00	+0.21247
13	+0.13	−0.08441
14	−0.08	+0.33646
15	+0.29	−0.55935
		+0.30358

v_{23}

5	+1.00	−0.15577
10	−0.66	−0.62249
11	−0.09	+0.15123
		−0.62703

v_{24}

6	+1.00	−0.12003
8	−1.00	−0.01095
10	+0.53	+0.49988
11	+0.19	−0.31928
13	−0.13	+0.08441
14	+0.08	−0.33646
15	−0.29	+0.55935
		+0.35692

v_{25}

8	+1.00	+0.01095
11	−0.10	+0.16804
12	+1.00	−0.21247
		−0.03348

v_{26}

8	−1.00	−0.01095
13	+0.11	−0.07143
		−0.08238

v_{27}

7	−1.00	−0.39898
8	+1.00	+0.01095
13	−0.11	+0.07143
		−0.31660

v_{28}

7	+1.00	+0.39898
		+0.39898

Table 22-22. Substitutions, in the Angle Condition Equations, of the Corrections to Directions

	v	1	2	3	4	5	6	7	8
1	−0.44	+0 44							
2	−0.30		+0.30						
3	+0.74	+0 74	+0.74						
4	+0.75			−0.75					
5	+0 84	−0.84		+0.84					
6	−1.60	−1.60							
7	−0.11	+0.11	+0.11						
8	−0.36	−0.36		+0.36					
9	−0.52		−0.52	−0.52	+0.52	+0.52			
10	+0.14				+0.14		−0.14		
11	+0.84					+0.84	+0.84		
12	+0.37							−0.37	
13	−0.52				+0.52			−0.52	
14	+0.10					−0.10			
15	+0.67		−0.67	−0.67	+0.67	+0.67			
16	−0.27		−0.27						
17	−0.36			−0.36					
18	+0.02							+0.02	−0.02
19	−0.06						+0.06		−0.06
20	−0.51				+0.51		−0.51		
21	+0.54				+0.54			−0.54	
22	+0.30					−0.30	−0.30		
23	−0.63					−0.63			
24	+0.36						+0.36		−0.36
25	−0.03								−0.03
26	−0.08								+0.08
27	−0.32							+0.32	−0.32
28	+0.40							+0.40	
Sums		−1.51	−0.31	−1.10	+2.90	+1.00	+0.31	−0.69	−0.71
N		+1.52	+0.30	+1.10	−2.90	−1.00	−0.30	+0.70	+0.70

Table 22-23. Computation of Coordinates

α *B–A*		205°41′01.20″	α *A–B*		25°41′01.20″
Angle *B*		− 59 32 36.06	Angle *A*		+ 73 17 33.88
α *B–E*		146 08 25.14	α *A–E*		98 58 35.08
		179 59 60.00			179 59 60.00
		S33°51′34.86″E			S81°01′24.92″E
		18,574.71 *B*			8450.62 *A*
	4.0858100	−12,184.56		3.3139675	− 2060.48
	9.9192898	6,390.15 *E*		9.1932020	6390.14 *E*✓
BE	4.1665202		*AE*	4.1207655	
	9.7459807	8,382.15 *B*		9.9946482	3,513.31 *A*
	3.9125009	+ 8,175.25		4.1154137	+13,044.09
		16,557.40 *E*			16,557.40 *E*✓
α *B–E*		146°08′25.14″	α *E–B*		326°08′25.14″
Angle *B*		− 76 59 6.79	Angle *E*		+ 32 10 31.54
α *B–F*		69 09 18.35	α *E–F*		358 18 56.68
					359 59 60.00
		N69°09′18.35″E			N1°41′03.32″W
		18,574.71 *B*			6,390.14 *E*
	3.4688559	+ 2,943.44		4.1797816	+15,128.00
	9.5512541	21,518.15 *F*✓		9.9998123	21,518.14
BF	3.9176018		*EF*	4.1799693	
	9.9706012	8,382.15 *B*		8.4682228	16,557.40 *E*
	3.8882030	+ 7,730.42		2.6481921	− 444.83
		16,112.57 *F*✓			16,112.57 *F*
α *E–F*		358°18′56.68″	α *E–E*		178°18′56.68″
Angle *E*		48 35 12.26	Angle *F*		− 51 21 02.39
α *E–G*		406 54 08.94	α *F–G*		126 57 54.29
		360			179 59 60.00
		N46°54′08.94″E			S53°02′05.71″E
		6,390.14			21,518.15 *F*
	3.9137506	+ 8,198.81		3.8406829	− 6,929.20
	9.8345747	14,588.95 *G*✓		9.7791116	14,588.95 *G*
EG	4.0791759		*FG*	4.0615713	
	9.8634370	16,557.40		9.9025479	16,112.57 *F*
	3.9426129	+ 8,762.19		3.9641192	+ 9,207.02
		25,319.59 *G*✓			25,319.59 *G*
α *E–G*		46°54′08.94″	α *G–E*		226°54′08.94″
Angle *E*		58 22 04.40	Angle *G*		− 63 37 11.94
α *E–C*		105 16 13.34	α *G–C*		163 16 57.00
		179 59 60.00			179 59 60.00
		S74°43′46.66″E			S16°43′03.00″E

Table 22-23. Computation of Coordinates (*Continued*)

	6390.14 *E*		14,588.95 *G*
3.5235147	− 3338.22	4.0620937	−11,537.02
9.4205733	3051.92 *C*✓	9.9812452	3,051.93 *C*
EC 4.1029414		*GC* 4.0808485	
9.9843895	16,557.40 *E*	9.4588690	25,319.59 *G*
4.0873309	+12,227.31	3.5397175	+ 3,465.11
	28,784.71 *C*✓		28,784.70 *C*

α	*G–C*	163°16′57.00″	α	*C–G*	343°16′57.00″
Angle *G*		− 68 40 14.43	Angle *C*		+ 48 31 12.51
α	*G–D*	94 36 42.57	α	*C–D*	391 48 09.51
		179 59 60.00			360
		S85°23′17.43″E			N31°48′09.51″E

	14,588.95 *G*		3,051.92 *C*
2.9115799	− 815.79	4.0302445	+10,721.23
8.9052811	13,773.16 *D*	9.9293516	13,773.15 *D*
GD 4.0062988		*CD* 4.1008929	
9.9985916	25,319.59 *G*	9.7218064	28,784.71 *C*
4.0048904	+10,113.24	3.8226993	+ 6,648.13
	35,432.83 *D*		35,432.84 *D*✓

CHAPTER 23

THE THEORY OF STATE COORDINATE SYSTEMS

23-1. Introduction. The purpose of this chapter is to demonstrate how state coordinate systems have been established. The methods of actual computation of the systems are not included nor are any of the methods of geodetic reduction (i.e., the means of computing geodetic latitude and longitude from length and azimuth on the spheroid, or the inverse). The equations have therefore been carried only to the point where they illustrate the theory and not to the point where they can be easily used by the computer.[1]

23-2. Projections. Utilization of geodetic latitude and longitude requires geodetic reduction. If the fundamental horizontal control stations of the U.S. Coast and Geodetic Survey are to be used without geodetic reduction, some means must be found for expressing their positions in terms of plane rectangular coordinates. This necessitates establishing a relationship between positions on the surface of the spheroid and positions on a plane so that any given latitude and longitude position within the area chosen will have a corresponding definite plane coordinate position on the plane.

A system which relates each point on one surface to a definite corresponding point on another surface is called a *projection.* A projection which will serve the purpose indicated must satisfy three conditions, as follows:

1. It must be possible to compute from the latitude and longitude of a position on the spheroid the plane rectangular coordinates of the corresponding position on the plane. Such a computation can be made if the projection has a definite mathematical pattern or form. This form may have no geometric equivalent and therefore it may be impossible to express it graphically. It is only necessary that there exist an analytical relationship between positions on the two surfaces.

2. Angles computed by the resulting plane coordinates must be very nearly equal to the actual angles on the spheroid. This calls for a *conformal projection.* A conformal projection is one so designed that within infinitesimal areas, angles on the original surface are represented by equal angles on the projection. When this rela-

[1] The following publications of the U.S. Coast and Geodetic Survey are necessary for a complete understanding of the theory and of the methods of computation:

"Formulas and Tables for the Computation of Geodetic Positions," Special Publication No. 8, 1933.

Oscar S. Adams, "General Theory of the Lambert Conformal Conic Projections," Special Publication No. 53, 1918.

Adams, Oscar S. and Charles N. Claire, "Manual of Plane-Coordinate Computation," Special Publication No. 193, 1935.

Adams, Oscar S. and Charles N. Claire, "Manual of Traverse Computation on the Lambert Grid,"Special Publication No. 194, 1935.

Adams, Oscar S. and Charles N. Claire, "Manual of Traverse Computation on the Transverse Mercator Grid," Special Publication No. 195, 1935.

For further demonstration of the use of these systems see ASCE, "Horizontal Control Surveys to Supplement the Fundamental Net," Manuals of Engineering Practice—No. 20, 1940.

tionship is fulfilled, infinitesimal figures on both surfaces will be similar and their sides will be proportional. Thus, at any point the *scale of the projection* with respect to the spheroid will be the same in all directions.

3. The length of lines computed from the resulting plane coordinates must be very nearly equal to the actual lengths on the spheroid. Therefore, the scale of the plane-coordinate lengths with respect to the actual lengths represented by them must be close to unity. The scale relationship, however, is not and cannot be the same everywhere on the projection.

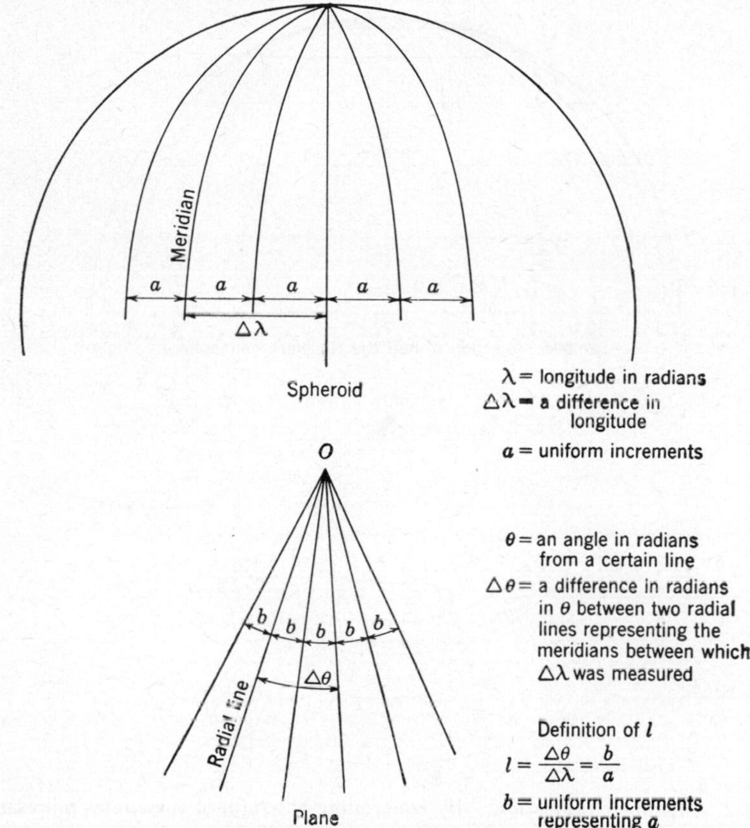

FIG. 23-1. Basic assumption for Lambert projection.

It is by controlling the scale over the projection plane that a spheroidal surface can be satisfactorily represented and the desired type of projection can be obtained.

23-3. Projections Used. Two types of projections are used for the various zones, the Lambert conformal conic two-parallel projection and the Transverse Mercator projection. The first of these best illustrates the principles of the design of projections.

23-4. The Lambert Conformal Conic Two-parallel Projection. The basic conception of this projection is that the meridians shall be represented on the projection by straight lines radiating from some point O (Fig. 23-1). The angles (b) between the lines are smaller and a definite fraction of the angles (a) which are the differences in longitude represented. Let the value of any angle $\Delta\theta$ between radial lines on the plane, divided by the value of the angle represented $\Delta\lambda$, be called l. Then

$$l = \frac{\Delta\theta}{\Delta\lambda}$$

where $\Delta\theta$ on the plane represents $\Delta\lambda$ on the spheroid.

Since, on the spheroid, the parallels of latitude intersect the meridians at right angles, they must intersect the radial lines on the projection plane at right angles to preserve conformality. Therefore they must be represented by circles centered at O (Fig. 23-1).

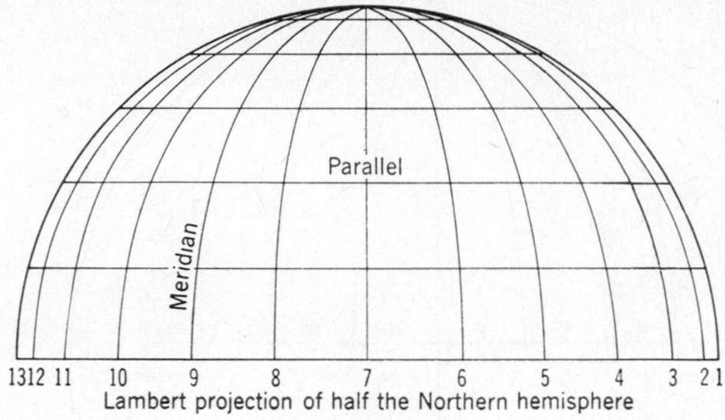

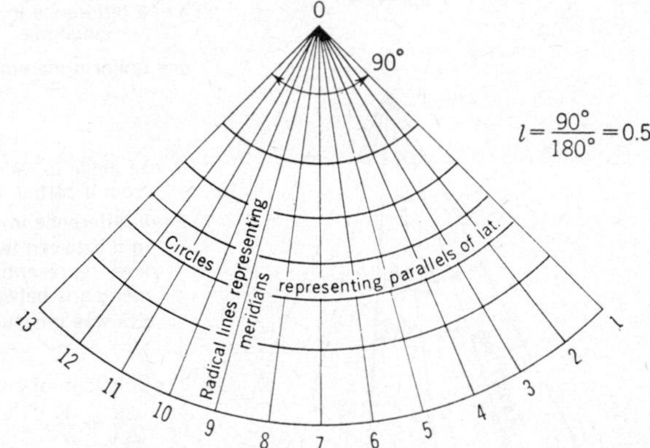

FIG. 23-2. A Lambert projection. By controlling the radii of the circles representing parallels of latitude, the spheroidal surface is represented so that any small figure on the spheroid will have the same shape as its representation on the projection.

Since such circles on the plane are intersected at equal intervals by the radial lines representing meridians and since on the spheroid the parallels of latitude are intersected by the meridians at equal intervals, the scale along each circle must be constant. Therefore the scale along the radial lines must be controlled so that the spheroidal surface may be represented conformally. This indicates that the circles must be spaced so that the requirements of the projection are fulfilled.

An expression for the radii of the circles can be found in terms of the latitude to be represented, a constant K, and l. Values for K and l can be determined which will cause the scale to be unity along any two desired circles. The scale will then vary as shown in Fig. 23-2.

23-5. Determination of Radii of Circles and Scale. The theory is developed for a *sphere* and later adapted to the spheroid (Fig. 23-3).

Let *abcd* be an infinitesimal area on a sphere bounded by meridians and parallels, and let *a′b′c′d′* be its representation on the projection plane.

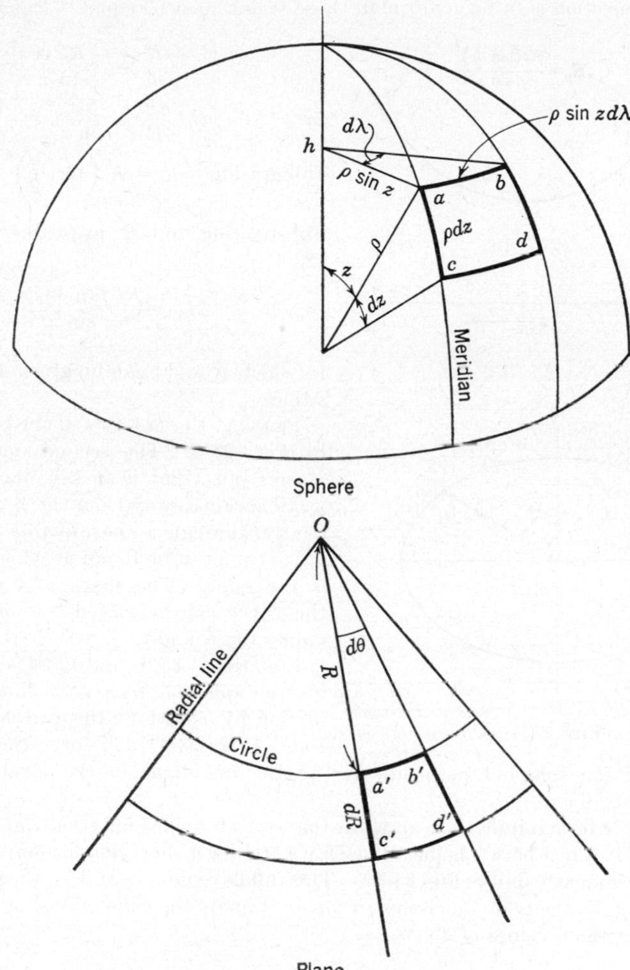

FIG. 23-3. Basic theory of Lambert projection.

Let λ = angle of longitude, radians
z = colatitude, radians ($\pi/2$ − latitude, radians)
ρ = radius of sphere
R = radius of circle

Then the scale S_ρ of the projection along the parallels is

$$S_\rho = \frac{a'b'}{ab} = \frac{R\ d\theta}{\rho \sin z\ d\lambda} = \frac{Rl}{\rho \sin z} \tag{23-1}$$

and the scale S_m of the projection along the meridian is

$$S_m = \frac{a'c'}{ac} = \frac{dR}{\rho\, dz} \tag{23-2}$$

Since the projection is to be conformal, these scales must be equal. Equating,

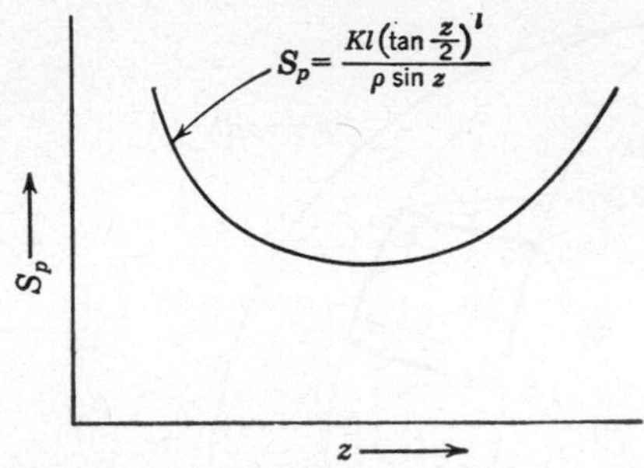

$$\frac{dR}{\rho\, dz} = \frac{Rl}{\rho \sin z}$$

or

$$\frac{dR}{R} = \frac{l dz}{\sin z}$$

Integrating, $$R = K\left(\tan \frac{z}{2}\right)^l \tag{23-3}$$

Substituting for R its value from Eq. (23-1),

$$S_\rho = \frac{Kl(\tan \frac{1}{2}z)^l}{\rho \sin z} \tag{23-4}$$

in which K and l can be given any desired values.

Plotting, Eq. (23-4) is of the form shown in Fig. 23-5. The second derivative is always plus, that is to say, the line is always curving away from the Z axis. Evidently, through a considerable range, two values of z can be found at which the scale is the same. The shape and position of the curve can be varied by changing the values for K and l.

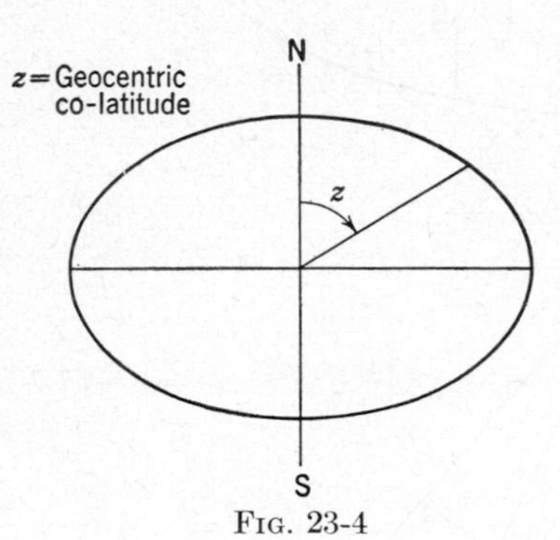

FIG. 23-4

Equations (23-3) and (23-4) can be used for the spheroid to a very close approximation by using for z the geocentric colatitude[1] (Fig. 23-4) and for ρ the radius of curvature of the spheroid perpendicular to the meridians in the locality of the projection.

It is possible to manipulate K and l so that at two values for z, not far apart, the scale is unity. Then between the values for z and for a short distance outside them, the scale will be nearly unity (Fig. 23-5). This fulfills requirement 3 for the projection.

Let z_1 and z_2 be the two values chosen for z. Equate the expressions for scale from Eq. (23-4) for these values of z:

$$\frac{Kl(\tan \frac{1}{2}z_1)^l}{\rho_1 \sin z_1} = \frac{Kl(\tan \frac{1}{2}z_2)^l}{\rho_2 \sin z_2} \tag{23-5}$$

Solving for l,

$$l = \frac{\log(\rho_1 \sin z_1) - \log(\rho_2 \sin z_2)}{\log \tan \frac{1}{2}z_1 - \log \tan \frac{1}{2}z_2} \tag{23-6}$$

By substituting desired values for z_1 and z_2 a value for l will be found which will cause the scale to be the same at z_1 and z_2.

[1] Isometric (conformal) latitude, which is very nearly the same as geocentric latitude, can be computed and used exactly.

A value for K can be found which will make the scale unity at z_1 and z_2 by substituting the value of l just found together with z_1 or z_2 in either half of Eq. (23-5), and

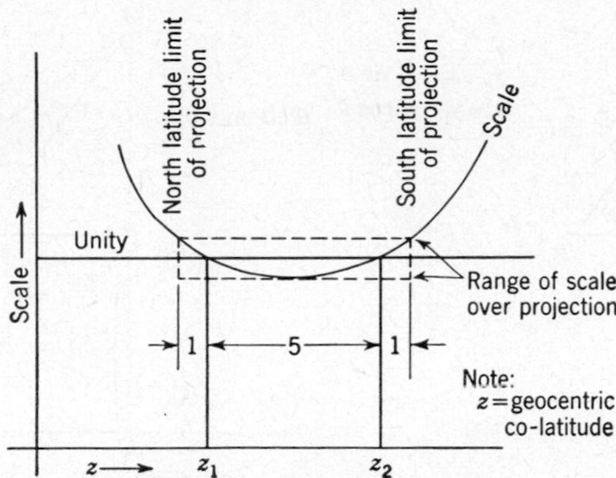

FIG. 23-5. Variation of scale over projection.

equating to 1. Thus

$$\text{Scale} = \frac{Kl(\tan \tfrac{1}{2}z_1)^l}{\rho_1 \sin z_1} = \frac{Kl(\tan \tfrac{1}{2}z_2)^l}{\rho_2 \sin z_2} = 1 \tag{23-7}$$

23-6. Plane Polar Coordinates. When the values of l and K have been determined, polar coordinates on the projection plane can be computed for any geodetic position on the spheroid.

Place origin at O (Fig. 23-6) and let the axis OM represent any given longitude. Let P be any point, and let

$\Delta\lambda$ = longitude value of OM minus longitude of P on spheroid

θ = central angle on plane measured counterclockwise from central meridian

Then

$$\theta = l\Delta\lambda \qquad \text{by definition}$$
$$R = K(\tan \tfrac{1}{2}z)^l$$

(The value of z is a function of latitude and can be found by geodetic computation.)

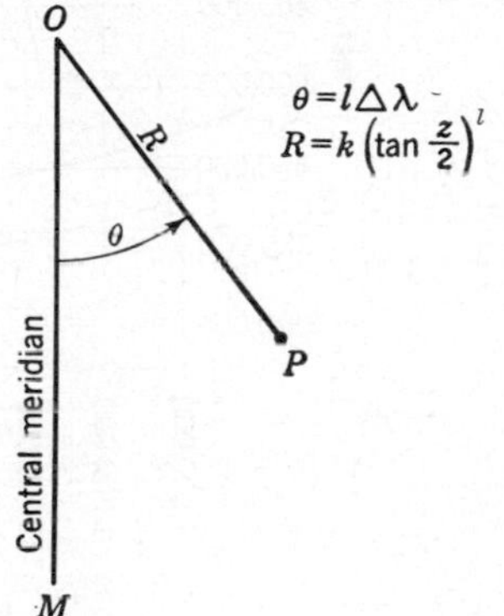

FIG. 23-6. Polar coordinates on projection plane.

23-7. Plane Rectangular Coordinates. With the polar coordinates known, rectangular coordinates can be easily computed (Fig. 23-7).

Let the Y axis be OM. Give the Y axis an x value, x_1, and give point O a y value, y_1, large enough in both cases to keep all coordinates positive. Let x_p and y_p be coordinates of the point P. Then

$$x_p = x_1 + R \sin \theta \tag{23-8}$$
$$y_p = y_1 - R \cos \theta \tag{23-9}$$

The plane coordinates of any geodetic position can be computed from these equations. The U.S. Coast and Geodetic Survey has published tables for each zone by which values for θ and R can be quickly determined. In actual computation another form of Eq. (23-9) is used for greater accuracy. It is derived as follows (Fig. 23-8):

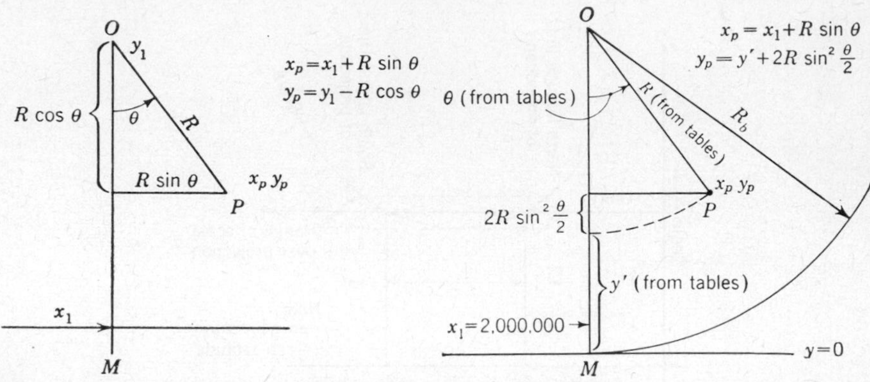

FIG. 23-7. Rectangular coordinates on the projection plane.

FIG. 23-8. Actual method of computing plane coordinates.

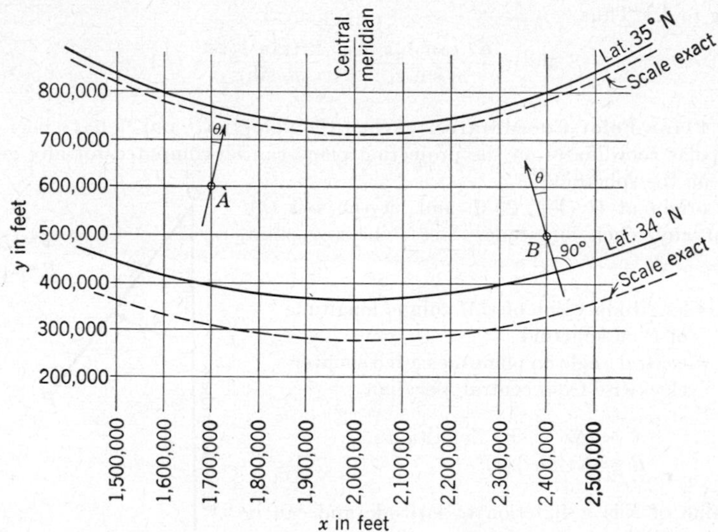

FIG. 23-9. The Lambert grid.

Let R_b = radius of the circle representing that latitude whose intersection with OM is given the value $y = 0$.

Let $$y' = R_b - R$$

Then $$y_p = y' + R \operatorname{vers} \theta$$

$$y_p = y' + 2R \sin^2 \frac{\theta}{2} \tag{23-10}$$

The value of the *scale* of the projection at any latitude can be computed from Eq. (23-4).

23-8. Appearance. Figure 23-9 shows the appearance of a Lambert projection.

23-9. The Transverse Mercator Projection. In this projection (Fig. 23-10) a certain meridian on the spheroid is chosen as a central meridian. It is represented on the plane by a straight line, here called a **base.** Straight lines on the spheroid, called **spheroid perpendiculars,** converge with increasing distance from the base and are perpendicular to the central meridian. Note that they are not parallels of latitude. On the projection they are represented by straight lines perpendicular to the base and are here called **projection perpendiculars.** This is, therefore, a special case of the Lambert projection in which $l = 0$. On the Lambert projection straight lines (meridians) were represented by straight lines radiating from a point. On the Transverse Mercator projection the scheme is rotated through 90° and straight

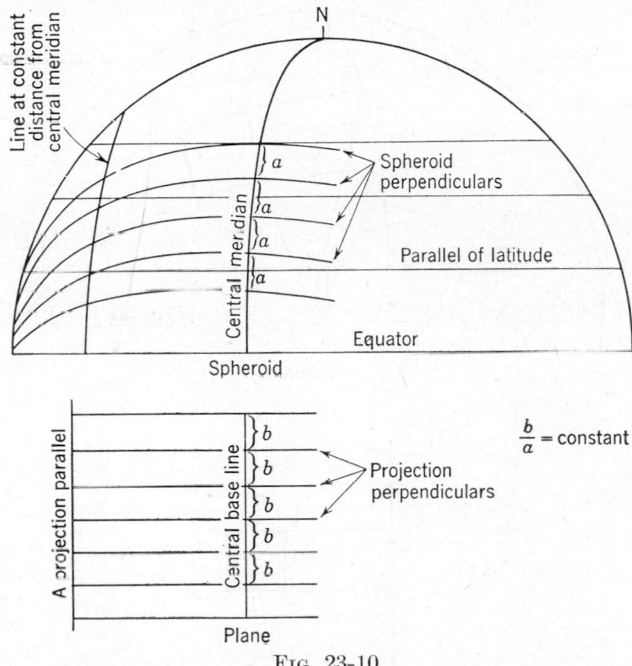

FIG. 23-10

lines (perpendiculars on the spheroid) are represented by straight lines (projection perpendiculars) which are parallel. Since the angle between them is zero (the lines being parallel), θ can be said to be 0, whence

$$l = \frac{\Delta\theta}{\Delta\lambda} = \frac{0}{\Delta\lambda} = 0$$

A line on the spheroid which is at a constant distance from the central meridian intersects the spheroid perpendiculars at right angles (very nearly), just as parallels of latitude intersect meridians at right angles. Such a line must be represented on the projection by a line which intersects the projection perpendiculars at right angles, to preserve conformality. All such lines are therefore parallel to the base and are here called **projection parallels** (Fig. 23-10).

Distances along the base (b, b, b, etc.) are made proportional to distances along the central meridian (a, a, a, etc.); that is, $b/a = K$. Thus the scale of the projection is uniform along the base.

On a *sphere* the perpendiculars to any meridian are great circles which converge to a point 90° from the meridian circle. If they are uniformly spaced along the central meridian they will likewise intersect equal arcs on every line which is at a constant distance from the meridian. Hence, the scale of the projection of a sphere must be constant along each projection parallel. Therefore since the scale is thus established on the projection parallels, the scale along the projection perpendiculars must be regulated so that the spheroidal surface may be represented conformally.

An expression for distances from the base along the projection perpendiculars can be found in terms of the distances along the spherical perpendiculars to be represented, the scale K at the central base, and the radius ρ of the surface of the spheroid

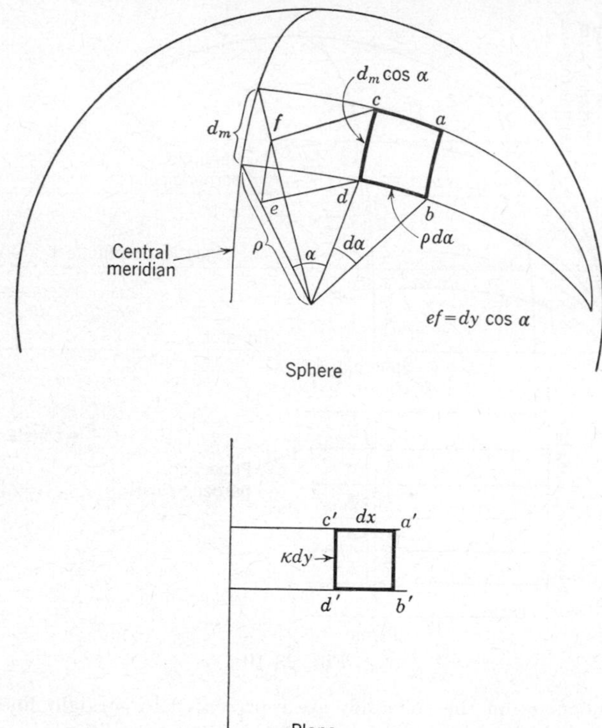

FIG. 23-11. Basic theory of Transverse Mercator projection.

along the spheroidal perpendiculars. Note that ρ is constant on a *sphere* and equal to its radius. Over the small area covered by a projection zone, ρ is nearly constant. Therefore the equations are developed for a sphere, which results in a very slight variation from the desired conformity.

A value for K can be established which will cause the scale to be unity along any two, but only two, projection parallels on each side of, and equidistant from, the central base. The scale will then vary as shown in Fig. 23-12.

23-10. Determination of Projection Distances and Scale along Perpendiculars. The theory is developed for a *sphere* and later adapted to the spheroid (Fig. 23-11).

Let *abcd* be an infinitesimal area on the sphere, bounded by perpendiculars and

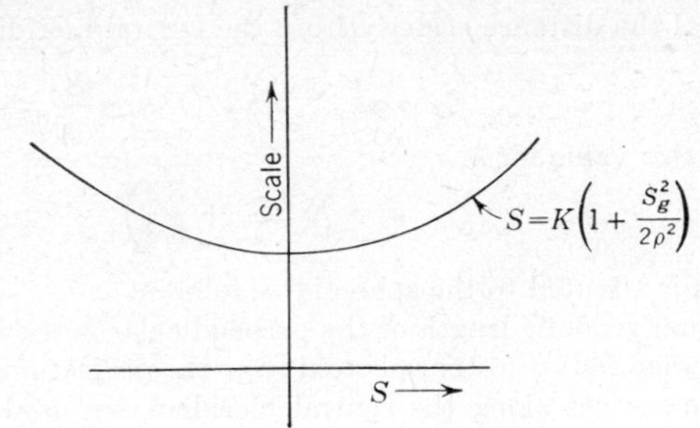

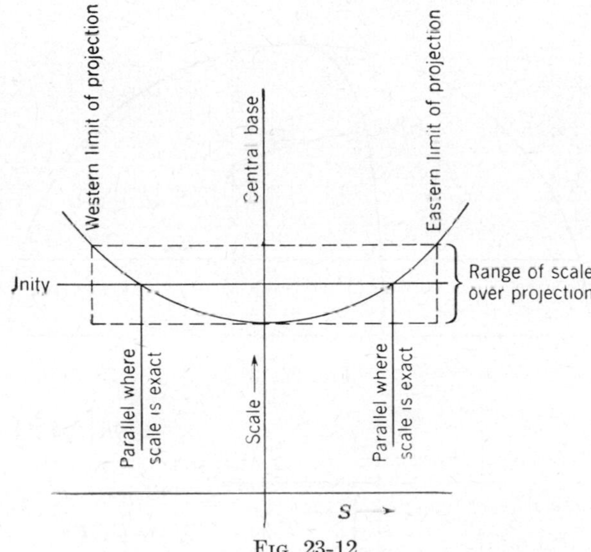

FIG. 23-12

parallels to the central meridian, and let $a'b'c'd'$ be its representation on the plane; then

$$\text{Projection scale along parallels} = \frac{c'd'}{cd} = \frac{K\,dm}{dm\cos\alpha}$$

$$\text{Projection scale along perpendiculars} = \frac{d'b'}{db} = \frac{dx}{\rho\,d\alpha}$$

But since the projection is to be conformal, these scales must be equal; hence, equating,

$$\frac{dx}{\rho\,d\alpha} = \frac{K\,dm}{dm\cos\alpha}$$

or

$$dx = K\rho\sec\alpha\,d\alpha$$

Approximating sec α by the first two terms of its series expansion,

$$dx = K\rho\left(1 + \frac{\alpha^2}{2} + \cdots\right)d\alpha$$

Integrating,

$$x = K\rho\left(\alpha + \frac{\alpha^3}{6}\right)$$

Let S_g equal the distance easterly from the central meridian; then

$$S_g = \rho\alpha \qquad \text{and} \qquad \alpha = \frac{S_g}{\rho}$$

Substituting this value for α,

$$x = K\left(S_g + \frac{S_g^3}{6\rho^2}\right) \tag{23-11}$$

The equation is adapted to the spheroid as follows:

1. The actual geodetic length of the perpendicular is used for S_g.
2. The average radius of the spheroid over the projection is used for ρ.
3. Actual distances along the central meridian are used as a basis for y values. (They are multiplied by a constant K, less than unity, to make the scale exact on two projection parallels.)

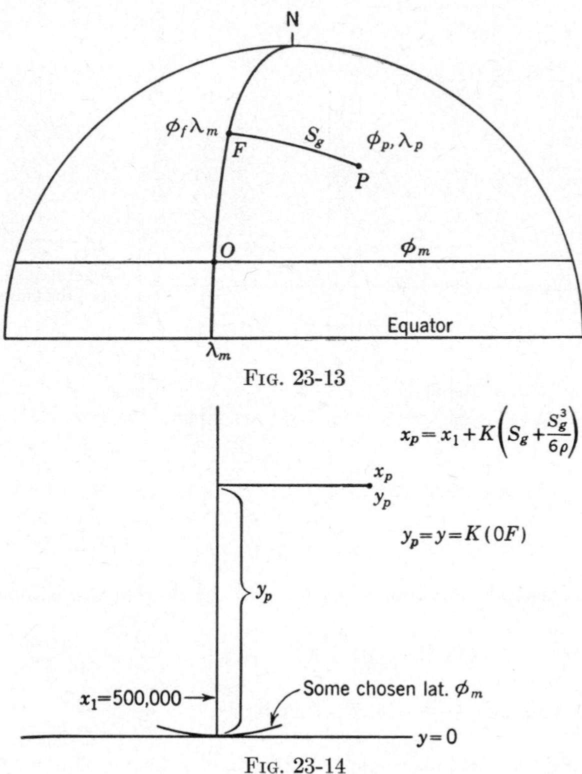

FIG. 23-13

FIG. 23-14

Taking the first derivative of Eq. (23-11),

$$\text{Scale} = \frac{dx}{ds} = K\left(1 + \frac{S_g^2}{2\rho^2}\right) \tag{23-12}$$

This is evidently a parabola and can be represented as shown in Fig. 23-12.

23-11. Plane Rectangular Coordinates. When a value for K has been established, plane rectangular coordinates can be computed for any geodetic position on the spheroid. Assume that the point P on the spheroid has the position latitude ϕ_p and longitude λ_p (see Fig. 23-13).

Choose a central meridian having the longitude λ_m and a point, O, upon it at latitude ϕ_m, from which distances along the central meridian are to be measured. By geodetic reduction it is possible to compute the following values on the spheroid:

S_g = length of the perpendicular from P to the central meridian
ϕ_f = latitude of the foot of the perpendicular
OF = distance along the central meridian from ϕ_m to ϕ_f

Then the coordinates of the point on the projection (Fig. 23-14) are

$$x_p = x_1 + K\left(S_g + \frac{S_g^3}{6\rho^2}\right)$$
$$y_p = K(OF)$$

As in any coordinate system, the *grid azimuth* of a line may be determined thus:

$$\tan A_s = \frac{x_1 - x_2}{y_1 - y_2} \tag{23-13}$$

23-12. Appearance. Figure 23-15 shows the appearance of the Transverse Mercator grid.

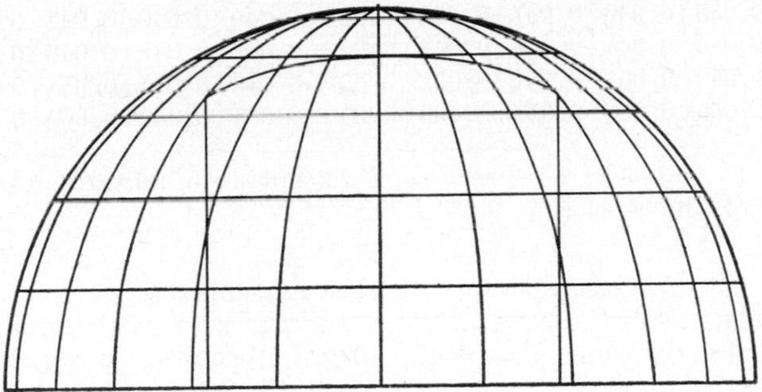

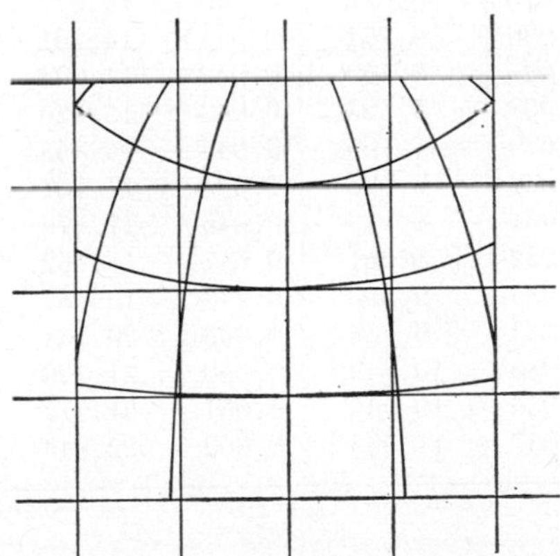

FIG. 23-15. Meridians and parallels plotted on a grid by a Transverse Mercator projection.

TABLES AND CHARTS

TABLE I. REQUIRED CLOSURES FOR SECOND-ORDER LEVELING

Closures in feet; distance leveled in miles

M	0	0.1	0.2	0.3	0.4	0.5	0.6	0.7	0.8	0.9
0	0	0.008	0.011	0.014	0.016	0.018	0.019	0.021	0.022	0.024
1	0.025	0.026	0.027	0.028	0.030	0.031	0.032	0.033	0.034	0.034
2	0.035	0.036	0.037	0.038	0.039	0.040	0.040	0.041	0.042	0.043
3	0.043	0.044	0.045	0.045	0.046	0.047	0.047	0.048	0.049	0.049
4	0.050	0.051	0.051	0.052	0.052	0.053	0.054	0.054	0.055	0.055
5	0.056	0.056	0.057	0.058	0.058	0.059	0.059	0.060	0.060	0.061

Example. To find closure for section of 2.3 miles, run forward and back. Distance leveled, 4.6 miles; closure, 0.054 ft.

Closures in feet; distance leveled in feet

Error, ft	Max. dist., ft	Error, ft	Max. dist., ft	Error, ft	Max. dist., ft	Error, ft	Max. dist., ft
0.008	541	0.023	4,469	0.038	12,199	0.053	23,730
0.009	684	0.024	4,866	0.039	12,849	0.054	24,634
0.010	845	0.025	5,280	0.040	13,517	0.055	25,555
0.011	1022	0.026	5,711	0.041	14,201	0.056	26,493
0.012	1216	0.027	6,159	0.042	14,902	0.057	27,448
0.013	1428	0.028	6,623	0.043	15,620	0.058	28,419
0.014	1656	0.029	7,105	0.044	16,355	0.059	29,407
0.015	1901	0.030	7,603	0.045	17,107	0.060	30,413
0.016	2163	0.031	8,119	0.046	17,876	0.061	31,435
0.017	2441	0.032	8,651	0.047	18,662	0.062	32,474
0.018	2737	0.033	9,200	0.048	19,464	0.063	33,530
0.019	3050	0.034	9,766	0.049	20,284	0.064	34,603
0.020	3379	0.035	10,349	0.050	21,120	0.065	35,693
0.021	3726	0.036	10,949	0.051	21,973	0.066	36,799
0.022	4089	0.037	11,565	0.052	22,843	0.067	37,923

Table II. Curvature and Refraction Corrections for Single-level Sights
$C_c = 0.00000002092d^2$; subtract C_c from rod reading

d, ft	C_c, ft	Stadia 0.6/100	Stadia 0.3/100	d, ft	C_c, ft	Stadia 0.6/100	Stadia 0.3/100
0		0.000	0.000	320		1.922	0.962
	0.0000				**0.0022**		
49		0.293	0.147	328		1.967	0.984
	0.0001				**0.0023**		
85		0.508	0.254	335		2.010	1.005
	0.0002				**0.0024**		
109		0.656	0.328	342		2.053	1.027
	0.0003				**0.0025**		
129		0.776	0.388	349		2.094	1.047
	0.0004				**0.003**		
147		0.880	0.440	409		2.45	1.23
	0.0005				**0.004**		
162		0.973	0.486	464		2.78	1.39
	0.0006				**0.005**		
176		1.057	0.529	513		3.08	1.54
	0.0007				**0.006**		
189		1.136	0.518	557		3.34	1.67
	0.0008				**0.007**		
202		1.209	0.605	559		3.59	1.80
	0.0009				**0.008**		
213		1.278	0.639	637		3.82	1.91
	0.0010				**0.009**		
224		1.343	0.672	674		4.04	2.02
	0.0011				**0.010**		
234		1.406	0.703	708		4.25	2.12
	0.0012				**0.011**		
244		1.466	0.733	741		4.45	2.22
	0.0013				**0.012**		
254		1.525	0.763	773		4.64	2.32
	0.0014				**0.013**		
263		1.580	0.790	803		4.82	2.14
	0.0015				**0.014**		
272		1.633	0.817	833		5.00	2.50
	0.0016				**0.015**		
281		1.684	0.842	861		5.16	2.58
	0.0017				**0.016**		
289		1.735	0.868	888		5.33	2.66
	0.0018				**0.017**		
297		1.783	0.892	915		5.49	2.74
	0.0019				**0.018**		
305		1.832	0.916	940		5.64	2.82
	0.0020				**0.019**		
313		1.878	0.939	965		5.79	2.89
	0.0021				**0.020**		
320		1.922	0.962	990		5.94	2.97
	0.0022				**0.021**		
328		1.967	0.984	1014		6.09	3.04

Table III. Temperature Corrections* Per Foot for Steel Tapes

°F Subtract cor.	Cor.	°F Add cor.	°F Subtract cor.	Cor.	°F Add cor.	°F Subtract cor.	Cor.	°F Add cor.	°F Subtract cor.	Cor.
68	0.00000000	68	48	0.00012900	88	28	0.00025800	108	9	0.00038055
67	0.00000645	69	47	0.00013545	89	27	0.00026445	109	8	0.00038700
66	0.00001290	70	46	0.00014190	90	26	0.00027090	110	7	0.00039345
65	0.00001935	71	45	0.00014835	91	25	0.00027735	111	6	0.00039990
64	0.00002580	72	44	0.00015480	92	24	0.00028380	112	5	0.00040635
63	0.00003225	73	43	0.00016125	93	23	0.00029025	113	4	0.00041280
62	0.00003870	74	42	0.00016770	94	22	0.00029670	114	3	0.00041925
61	0.00004515	75	41	0.00017415	95	21	0.00030315	115	2	0.00042570
60	0.00005160	76	40	0.00018060	96	20	0.00030960	116	1	0.00043215
59	0.00005805	77	39	0.00018705	97	19	0.00031605	117	0	0.00043860
58	0.00006450	78	38	0.00019350	98	18	0.00032250	118	−1	0.00044505
57	0.00007095	79	37	0.00019995	99	17	0.00032895	119	−2	0.00045150
56	0.00007740	80	36	0.00020640	100	16	0.00033540	120	−3	0.00045795
55	0.00008385	81	35	0.00021285	101	15	0.00034185	121	−4	0.00046440
54	0.00009030	82	34	0.00021930	102	14	0.00034830	122	−5	0.00047085
53	0.00009675	83	33	0.00022575	103	13	0.00035475	123	−6	0.00047730
52	0.00010320	84	32	0.00023220	104	12	0.00036120	124	−7	0.00048375
51	0.00010965	85	31	0.00023865	105	11	0.00036765	124	−8	0.00049020
50	0.00011610	86	30	0.00024510	106	10	0.00037410	126	−9	0.00049665
49	0.00012255	87	29	0.00025155	107	9	0.00038055	127	−10	0.00050310

* Based on a coefficient of expansion of 0.00000645 per °F.

Table IV. Right Triangles with Integral Sides

From the theorem (hypotenuse)^2 = (base)^2 + (side)^2, the relationships among the sides of a right triangle may be expressed:

$$\text{Base} = 2ab(a+1)$$
$$\text{Hypotenuse} = \text{base} + b$$
$$\text{Side} = 2ab + b$$

When positive integers are used for a and b, the sides of the right triangle will be expressed in integral numbers.

Dimensions of Right Triangles

Base	Hypotenuse	Side	Base	Hypotenuse	Side
4	5	3	63	65	16
12	13	5	21	29	20
24	25	7	45	53	28
15	17	8	56	65	33
40	41	9	77	85	36
60	61	11	80	89	39
35	37	12	55	73	48
84	85	13	72	97	65

Table V. Conversion of Time to Arc

Hours of Time into Arc

T.	A.	T.	A.	T.	A.	T.	A.	T.	A.	T.	A.
Hr	°	Hr	°	Hr	°	Hr	°	Hr	°	Hr	°
1	15	5	75	9	135	13	195	17	225	21	315
2	30	6	90	10	150	14	210	18	270	22	330
3	45	7	105	11	165	15	225	19	285	23	345
4	60	8	120	12	180	16	240	20	300	24	360

Minutes of Time to Arc
Seconds of Time to Arc

Min Sec	° ′ ′ ″	Min Sec	° ′ ′ ″	Min Sec	° ′ ′ ″
1	0 15	21	5 15	41	10 15
2	0 30	22	5 30	42	10 30
3	0 45	23	5 45	43	10 45
4	1 0	24	6 0	44	11 0
5	1 15	25	6 15	45	11 15
6	1 30	26	6 30	46	11 30
7	1 45	27	6 45	47	11 45
8	2 0	28	7 0	48	12 0
9	2 15	29	7 15	49	12 15
10	2 30	30	7 30	50	12 30
11	2 45	31	7 45	51	12 45
12	3 0	32	8 0	52	13 0
13	3 15	33	8 15	53	13 15
14	3 30	34	8 30	54	13 30
15	3 45	35	8 45	55	13 45
16	4 0	36	9 0	56	14 0
17	4 15	37	9 15	57	14 15
18	4 30	38	9 30	58	14 30
19	4 45	39	9 45	59	14 45
20	5 0	40	10 0	60	15 0

Hundredths of a Second of Time to Arc

Seconds of time	0.00 ″	0.01 ″	0.02 ″	0.03 ″	0.04 ″	0.05 ″	0.06 ″	0.07 ″	0.08 ″	0.09 ″
0.00	0.00	0.15	0.30	0.45	0.60	0.75	0.90	1.05	1.20	1.35
0.10	1.50	1.65	1.80	1.95	2.10	2.25	2.40	2.55	2.70	2.85
0.20	3.00	3.15	3.30	3.45	3.60	3.75	3.90	4.05	4.20	4.35
0.30	4.50	4.65	4.80	4.95	5.10	5.25	5.40	5.55	5.70	5.85
0.40	6.00	6.15	6.30	6.45	6.60	6.75	6.90	7.05	7.20	7.35
0.50	7.50	7.65	7.80	7.95	8.10	8.25	8.40	8.55	8.70	8.85
0.60	9.00	9.15	9.30	9.45	9.60	9.75	9.90	10.05	10.20	10.35
0.70	10.50	10.65	10.80	10.95	11.10	11.25	11.40	11.55	11.70	11.85
0.80	12.00	12.15	12.30	12.45	12.60	12.75	12.90	13.05	13.20	13.35
0.90	13.50	13.65	13.80	13.95	14.10	14.25	14.40	14.55	14.70	14.85

TABLE VI. GRADE CORRECTIONS FOR 100-FT TAPE LENGTHS (SUBTRACT)

For differences of elevation h up to 15 ft, Cor. $= -0.005h^2 - 0.000000125h^4$

h	0.00	0.01	0.02	0.03	0.04	0.05	0.06	0.07	0.08	0.09
Ft	Ft	Ft	Ft	Ft	Ft	Ft	Ft	Ft	Ft	Ft
0.0	0.000	0.000	0.000	0.000	0.000	0.000	0.000	0.000	0.000	0.000
.1	.000	.000	.000	.000	.000	.000	.000	.000	.000	.000
.2	.000	.000	.000	.000	.000	.000	.000	.000	.000	.000
.3	.000	.000	.001	.001	.001	.001	.001	.001	.001	.001
.4	.001	.001	.001	.001	.001	.001	.001	.001	.001	.001
0.5	0.001	0.001	0.001	0.001	0.001	0.002	0.002	0.002	0.002	0.002
.6	.002	.002	.002	.002	.002	.002	.002	.002	.002	.002
.7	.002	.003	.003	.003	.003	.003	.003	.003	.003	.003
.8	.003	.003	.003	.003	.004	.004	.004	.004	.004	.004
.9	.004	.004	.004	.004	.004	.005	.005	.005	.005	.005
1.0	0.005	0.005	0.005	0.005	0.005	0.006	0.006	0.006	0.006	0.006
.1	.006	.006	.006	.006	.006	.007	.007	.007	.007	.007
.2	.007	.007	.007	.008	.008	.008	.008	.008	.008	.008
.3	.008	.009	.009	.009	.009	.009	.009	.009	.010	.010
.4	.010	.010	.010	.010	.010	.011	.011	.011	.011	.011
1.5	0.011	0.011	0.012	0.012	0.012	0.012	0.012	0.012	0.012	0.013
.6	.013	.013	.013	.013	.013	.014	.014	.014	.014	.014
.7	.014	.015	.015	.015	.015	.015	.015	.016	.016	.016
.8	.016	.016	.017	.017	.017	.017	.017	.017	.018	.018
.9	.018	.018	.018	.019	.019	.019	.019	.019	.020	.020
2.0	0.020	0.020	0.020	0.021	0.021	0.021	0.021	0 021	0.022	0.022
.1	.022	.022	.022	.023	.023	.023	.023	.024	.024	.024
.2	.024	.024	.025	.025	.025	.025	.026	.026	.026	.026
.3	.026	.027	.027	.027	.027	.028	.028	.028	.028	.029
.4	.029	.029	.029	.030	.030	.030	.030	.031	.031	.031
2.5	0.031	0.032	0.032	0.032	0.032	0.033	0.033	0.033	0.033	0.034
.6	.034	.034	.034	.035	.035	.035	.035	.036	.036	.036
.7	.036	.037	.037	.037	.038	.038	.038	.038	.039	.039
.8	.039	.039	.040	.040	.040	.041	.041	.041	.041	.042
.9	.042	.042	.043	.043	.043	.044	.044	.044	.044	.045
3.0	0.045	0.045	0.046	0.046	0.046	0.047	0.047	0.047	0.047	0.048
.1	.048	.048	.049	.049	.049	.050	.050	.050	.051	.051
.2	.051	.052	.052	.052	.053	.053	.053	.053	.054	.054
.3	.054	.055	.055	.055	.056	.056	.056	.057	.057	.057
.4	.058	.058	.058	.059	.059	.060	.060	.060	.061	.061
3.5	0.061	0.062	0.062	0.062	0.063	0.063	0.063	0.064	0.064	0.064
.6	.065	.065	.066	.066	.066	.067	.067	.067	.068	.068
.7	.068	.069	.069	.070	.070	.070	.071	.071	.071	.072
.8	.072	.073	.073	.073	.074	.074	.075	.075	.075	.076
.9	.076	.076	.077	.077	.078	.078	.078	.079	.079	.080
4.0	0.080	0.080	0.081	0.081	0.082	0.082	0.082	0.083	0.083	0.084
.1	.084	.084	.085	.085	.086	.086	.087	.087	.087	.088
.2	.088	.089	.089	.090	.090	.090	.091	.091	.092	.092
.3	.092	.093	.093	.094	.094	.095	.095	.096	.096	.096
.4	.097	.097	.098	.098	.099	.099	.100	.100	.100	.101
4.5	0.101	0.102	0.102	0.103	0.103	0.104	0.104	0.104	0.105	0.105
.6	.106	.106	.107	.107	.108	.108	.109	.109	.110	.110
.7	.111	.111	.111	.112	.112	.113	.113	.114	.114	.115
.8	.115	.116	.116	.117	.117	.118	.118	.119	.119	.120
.9	.120	.121	.121	.122	.122	.123	.123	.124	.124	.125
5.0	0.125	0.126	0.126	0.127	0.127	0.128	0.128	0.129	0.129	0.130
.1	.130	.131	.131	.132	.132	.133	.133	.134	.134	.135
.2	.135	.136	.136	.137	.137	.138	.138	.139	.139	.140
.3	.141	.141	.142	.142	.143	.143	.144	.144	.145	.145
.4	.146	.146	.147	.148	.148	.149	.149	.150	.150	.151
5.5	0.151	0.152	0.152	0.153	0.154	0.154	0.155	0.155	0.156	0.156
.6	.157	.157	.158	.159	.159	.160	.160	.161	.161	.162
.7	.163	.163	.164	.164	.165	.165	.166	.167	.167	.168
.8	.168	.169	.170	.170	.171	.171	.172	.172	.173	.174
.9	.174	.175	.175	.176	.177	.177	.178	.178	.179	.180
6.0	0.180	0.181	0.181	0.182	0.183	0.183	0.184	0.184	0.185	0.186
.1	.186	.187	.187	.188	.189	.189	.190	.191	.191	.192
.2	.192	.193	.194	.194	.195	.196	.196	.197	.197	.198
.3	.199	.199	.200	.201	.201	.202	.202	.203	.204	.204
.4	.205	.206	.206	.207	.208	.208	.209	.210	.210	.211
6.5	0.211	0.212	0.213	0.213	0.214	0.215	0.215	0.216	0.217	0.217
.6	.218	.219	.219	.220	.221	.221	.222	.223	.223	.224
.7	.225	.225	.226	.227	.227	.228	.229	.229	.230	.231
.8	.231	.232	.233	.234	.234	.235	.236	.236	.237	.238
.9	.238	.239	.240	.240	.241	.242	.243	.243	.244	.245
7.0	0.245	0.246	0.247	0.247	0.248	0.249	0.250	0.250	0.251	0.252
.1	.252	.253	.254	.255	.255	.256	.257	.257	.258	.259
.2	.260	.260	.261	.262	.262	.263	.264	.265	.265	.266
.3	.267	.268	.268	.269	.270	.270	.271	.272	.273	.273
.4	.274	.275	.276	.276	.277	.278	.279	.279	.280	.281

TABLE VI. GRADE CORRECTIONS FOR 100-FT TAPE LENGTHS (SUBTRACT) (*Continued*)

h	0.00	0.01	0.02	0.03	0.04	0.05	0.06	0.07	0.08	0.09
Ft	Ft	Ft	Ft	Ft	Ft	Ft	Ft	Ft	Ft	Ft
7.5	0.282	0.282	0.283	0.284	0.285	0.285	0.286	0.287	0.288	0.288
.6	.289	.290	.291	.292	.292	.293	.294	.295	.295	.296
.7	.297	.298	.298	.299	.300	.301	.302	.302	.303	.304
.8	.305	.305	.306	.307	.308	.309	.309	.310	.311	.312
.9	.313	.313	.314	.315	.316	.317	.317	.318	.319	.320
8.0	0.321	0.321	0.322	0.323	0.324	0.325	0.325	0.326	0.327	0.328
.1	.329	.329	.330	.331	.332	.333	.333	.334	.335	.336
.2	.337	.338	.338	.339	.340	.341	.342	.343	.343	.344
.3	.345	.346	.347	.348	.348	.349	.350	.351	.352	.353
.4	.353	.354	.355	.356	.357	.358	.359	.359	.360	.361
8.5	0.362	0.363	0.364	0.364	0.365	0.366	0.367	0.368	0.369	0.370
.6	.370	.371	.372	.373	.374	.375	.376	.377	.377	.378
.7	.379	.380	.381	.382	.383	.384	.384	.385	.386	.387
.8	.388	.389	.390	.391	.391	.392	.393	.394	.395	.396
.9	.397	.398	.399	.400	.400	.401	.402	.403	.404	.405
9.0	0.406	0.407	0.408	0.409	0.409	0.410	0.411	0.412	0.413	0.414
.1	.415	.416	.417	.418	.419	.419	.420	.421	.422	.423
.2	.424	.425	.426	.427	.428	.429	.430	.431	.432	.432
.3	.433	.434	.435	.436	.437	.438	.439	.440	.441	.442
.4	.443	.444	.445	.446	.447	.448	.448	.449	.450	.451
9.5	0.452	0.453	0.454	0.455	0.456	0.457	0.458	0.459	0.460	0.461
.6	.462	.463	.464	.465	.466	.467	.468	.469	.470	.471
.7	.472	.473	.474	.474	.475	.476	.477	.478	.479	.480
.8	.481	.482	.483	.484	.485	.486	.487	.488	.489	.490
.9	.491	.492	.493	.494	.495	.496	.497	.498	.499	.500
10.0	0.501	0.502	0.503	0.504	0.505	0.506	0.507	0.508	0.509	0.510
.1	.511	.512	.513	.514	.515	.516	.517	.518	.520	.521
.2	.522	.523	.524	.525	.526	.527	.528	.529	.530	.531
.3	.532	.533	.534	.535	.536	.537	.538	.539	.540	.541
.4	.542	.543	.544	.545	.546	.548	.549	.550	.551	.552
10.5	0.553	0.554	0.555	0.556	0.557	0.558	0.559	0.560	0.561	0.562
.6	.563	.564	.566	.567	.568	.569	.570	.571	.572	.573
.7	.574	.575	.576	.577	.578	.579	.581	.582	.583	.584
.8	.585	.586	.587	.588	.589	.590	.591	.593	.594	.595
.9	.596	.597	.598	.599	.600	.601	.602	.604	.605	.606
11.0	0.607	0.608	0.609	0.610	0.611	0.612	0.613	0.615	0.616	0.617
.1	.618	.619	.620	.621	.622	.624	.625	.626	.627	.628
.2	.629	.630	.631	.633	.634	.635	.636	.637	.638	.639
.3	.641	.642	.643	.644	.645	.646	.647	.648	.650	.651
.4	.652	.653	.654	.655	.657	.658	.659	.660	.661	.662
11.5	0.663	0.665	0.666	0.667	0.668	0.669	0.670	0.672	0.673	0.674
.6	.675	.676	.677	.679	.680	.681	.682	.683	.684	.686
.7	.687	.688	.689	.690	.692	.693	.694	.695	.696	.697
.8	.699	.700	.701	.702	.703	.705	.706	.707	.708	.709
.9	.711	.712	.713	.714	.715	.717	.718	.719	.720	.721
12.0	0.723	0.724	0.725	0.726	0.727	0.729	0.730	0.731	0.732	0.734
.1	.735	.736	.737	.738	.740	.741	.742	.743	.745	.746
.2	.747	.748	.749	.751	.752	.753	.754	.756	.757	.758
.3	.759	.761	.762	.763	.764	.766	.767	.768	.769	.771
.4	.772	.773	.774	.776	.777	.778	.779	.781	.782	.783
12.5	0.784	0.786	0.787	0.788	0.789	0.791	0.792	0.793	0.794	0.796
.6	.707	.798	.800	.801	.802	.803	.805	.806	.807	.808
.7	.810	.811	.812	.814	.815	.816	.817	.819	.820	.821
.8	.823	.824	.825	.826	.828	.829	.830	.832	.833	.834
.9	.836	.837	.838	.839	.841	.842	.843	.845	.846	.847
13.0	0.849	0.850	0.851	0.853	0.854	0.855	0.856	0.858	0.859	0.860
.1	.862	.863	.864	.866	.867	.868	.870	.871	.872	.874
.2	.875	.876	.878	.879	.880	.882	.883	.884	.886	.887
.3	.888	.890	.891	.892	.894	.895	.896	.898	.899	.901
.4	.902	.903	.905	.906	.907	.909	.910	.911	.913	.914
13.5	0.915	0.917	0.918	0.920	0.921	0.922	0.924	0.925	0.926	0.928
.6	.929	.930	.932	.933	.935	.936	.937	.939	.940	.942
.7	.943	.944	.946	.947	.948	.950	.951	.953	.954	.955
.8	.957	.958	.960	.961	.962	.964	.965	.967	.968	.969
.9	.971	.972	.974	.975	.976	.978	.979	.981	.982	.983
14.0	0.985	0.986	0.988	0.989	0.991	0.992	0.993	0.995	0.996	0.998
.1	.999	1.000	1.002	1.003	1.005	1.006	1.008	1.009	1.010	1.012
.2	1.013	1.015	1.016	1.018	1.019	1.021	1.022	1.023	1.025	1.026
.3	1.028	1.029	1.031	1.032	1.034	1.035	1.036	1.038	1.039	1.041
.4	1.042	1.044	1.045	1.047	1.048	1.050	1.051	1.052	1.054	1.055
14.5	1.057	1.058	1.060	1.061	1.063	1.064	1.066	1.067	1.069	1.070
.6	1.071	1.073	1.074	1.076	1.077	1.079	1.080	1.082	1.083	1.085
.7	1.086	1.088	1.089	1.091	1.092	1.094	1.095	1.097	1.098	1.100
.8	1.101	1.103	1.104	1.106	1.107	1.109	1.110	1.112	1.113	1.115
.9	1.116	1.118	1.119	1.121	1.122	1.124	1.125	1.127	1.128	1.130

TABLE VII. CENTESIMAL GRADS TO SEXAGESIMAL

g	°	'	g	'	"	g	"
1	0	54	0.01	00	32.4	0.0001	0.324
2	1	48	0.02	01	04.8	0.0002	0.648
3	2	42	0.03	01	37.2	0.0003	0.972
4	3	36	0.04	02	09.6	0.0004	1.296
5	4	30	0.05	02	42.0	0.0005	1.620
6	5	24	0.06	03	14.4	0.0006	1.944
7	6	18	0.07	03	46.8	0.0007	2.268
8	7	12	0.08	04	19.2	0.0008	2.592
9	8	06	0.09	04	51.6	0.0009	2.916
10	9	00	0.10	05	24.0	0.0010	3.240
11	9	54	0.11	05	56.4	0.0011	3.564
12	10	48	0.12	06	28.8	0.0012	3.888
13	11	42	0.13	07	01.2	0.0013	4.212
14	12	36	0.14	07	33.6	0.0014	4.536
15	13	30	0.15	08	06.0	0.0015	4.860
16	14	24	0.16	08	38.4	0.0016	5.184
17	15	18	0.17	09	10.8	0.0017	5.508
18	16	12	0.18	09	43.2	0.0018	5.832
19	17	06	0.19	10	15.6	0.0019	6.156
20	18	00	0.20	10	48.0	0.0020	6.480
21	18	54	0.21	11	20.4	0.0021	6.804
22	19	48	0.22	11	52.8	0.0022	7.128
23	20	42	0.23	12	25.2	0.0023	7.462
24	21	36	0.24	12	57.6	0.0024	7.776
25	22	30	0.25	13	30.0	0.0025	8.100
26	23	24	0.26	14	02.4	0.0026	8.424
27	24	18	0.27	14	34.8	0.0027	8.748
28	25	12	0.28	15	07.2	0.0028	9.072
29	26	06	0.29	15	39.6	0.0029	9.396
30	27	00	0.30	16	12.0	0.0030	9.720
31	27	54	0.31	16	44.4	0.0031	10.044
32	28	48	0.32	17	16.8	0.0032	10.368
33	29	42	0.33	17	49.2	0.0033	10.692
34	30	36	0.34	18	21.6	0.0034	11.016
35	31	30	0.35	18	54.0	0.0035	11.340
36	32	24	0.36	19	26.4	0.0036	11.664
37	33	18	0.37	19	58.8	0.0037	11.988
38	34	12	0.38	20	31.2	0.0038	12.312
39	35	06	0.39	21	03.6	0.0039	12.636
40	36	00	0.40	21	36.0	0.0040	12.960
41	36	54	0.41	22	08.4	0.0041	13.284
42	37	48	0.42	22	40.8	0.0042	13.608
43	38	42	0.43	23	13.2	0.0043	13.932
44	39	36	0.44	23	45.6	0.0044	14.256
45	40	30	0.45	24	18.0	0.0045	14.580
46	41	24	0.46	24	50.4	0.0046	14.904
47	42	18	0.47	25	22.8	0.0047	15.228
48	43	12	0.48	25	55.2	0.0048	15.552
49	44	06	0.49	26	27.6	0.0049	15.876
50	45	00	0.50	27	00.0	0.0050	16.200

DEGREES (ALL VALUES ARE EXACT)

g	°	′	g	′	″	g	″
51	45	54	0.51	27	32.4	0.0051	16.524
52	46	48	0.52	28	04.8	0.0052	16.848
53	47	42	0.53	28	37.2	0.0053	17.172
54	48	36	0.54	29	09.6	0.0054	17.496
55	49	30	0.55	29	42.0	0.0055	17.820
56	50	24	0.56	30	14.4	0.0056	18.144
57	51	18	0.57	30	46.8	0.0057	18.468
58	52	12	0.58	31	19.2	0.0058	18.792
59	53	06	0.59	31	51.6	0.0059	19.116
60	54	00	0.60	32	24.0	0.0060	19.440
61	54	54	0.61	32	56.4	0.0061	19.764
62	55	48	0.62	33	28.8	0.0062	20.088
63	56	42	0.63	34	01.2	0.0063	20.412
64	57	36	0.64	34	33.6	0.0064	20.736
65	58	30	0.65	35	06.0	0.0065	21.060
66	59	24	0.66	35	38.4	0.0066	21.384
67	60	18	0.67	36	10.8	0.0067	21.708
68	61	12	0.68	36	43.2	0.0068	22.032
69	62	06	0.69	37	15.6	0.0069	22.356
70	63	00	0.70	37	48.0	0.0070	22.680
71	63	54	0.71	38	20.4	0.0071	23.004
72	64	48	0.72	38	52.8	0.0072	23.328
73	65	42	0.73	39	25.2	0.0073	23.652
74	66	36	0.74	39	57.6	0.0074	23.976
75	67	30	0.75	40	30.0	0.0075	24.300
76	68	24	0.76	41	02.4	0.0076	24.624
77	69	18	0.77	41	34.8	0.0077	24.948
78	70	12	0.78	42	07.2	0.0078	25.272
79	71	06	0.79	42	39.6	0.0079	25.596
80	72	00	0.80	43	12.0	0.0080	25.920
81	72	54	0.81	43	44.4	0.0081	26.244
82	73	48	0.82	44	16.8	0.0082	26.568
83	74	42	0.83	44	49.2	0.0083	26.892
84	75	36	0.84	45	21.6	0.0084	27.216
85	76	30	0.85	45	54.0	0.0085	27.540
86	77	24	0.86	46	26.4	0.0086	27.864
87	78	18	0.87	46	58.8	0.0087	28.188
88	79	12	0.88	47	31.2	0.0088	28.512
89	80	06	0.89	48	03.6	0.0089	28.836
90	81	00	0.90	48	36.0	0.0090	29.160
91	81	54	0.91	49	08.4	0.0091	29.484
92	82	48	0.92	49	40.8	0.0092	29.808
93	83	42	0.93	50	13.2	0.0093	30.132
94	84	36	0.94	50	45.6	0.0094	30.456
95	85	30	0.95	51	18.0	0.0095	30.780
96	86	24	0.96	51	50.4	0.0096	31.104
97	87	18	0.97	52	22.8	0.0097	31.428
98	88	12	0.98	52	55.2	0.0098	31.752
99	89	06	0.99	53	27.6	0.0099	32.076
100	90	00	1.00	54	00.0	0.0100	32.400

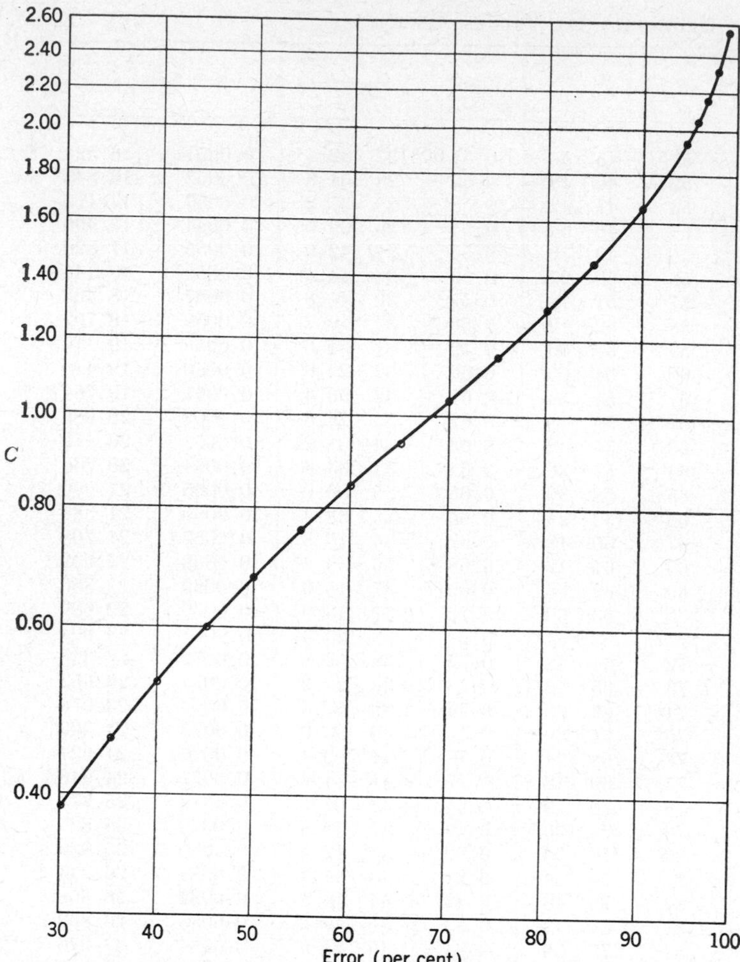

CHART I. Constant C for finding the size of a given per cent error:

$$x\% \text{ error} = c_x \sqrt{\frac{\Sigma v^2}{n - 1}}$$

Example. Assume that, from a test, $\sqrt{\Sigma v^2/(n - 1)} = 3.24$. To find the 83 per cent error: For 83 per cent the chart gives $C = 1.375$. Then the 83 per cent error $= (1.375)\ (3.24) = 4.46$. For the same example, to find how often 4.00 will not be exceeded: $C = 4.00/3.24 = 1.234$. For $C = 1.234$ the chart gives 78 per cent.

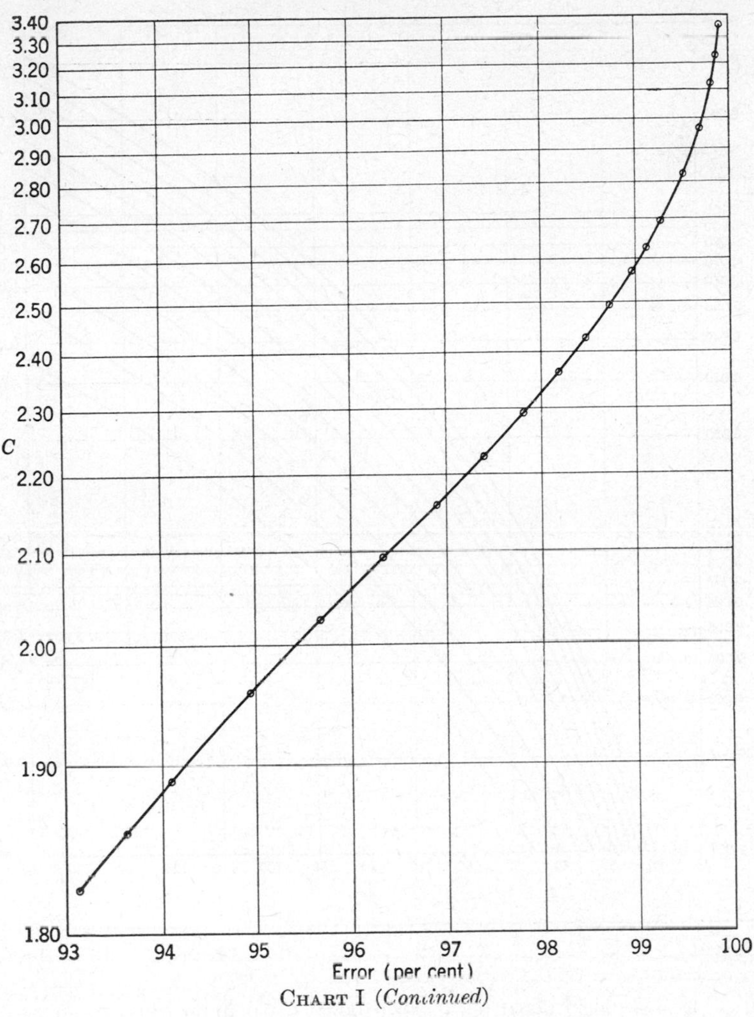

CHART I (*Continued*)

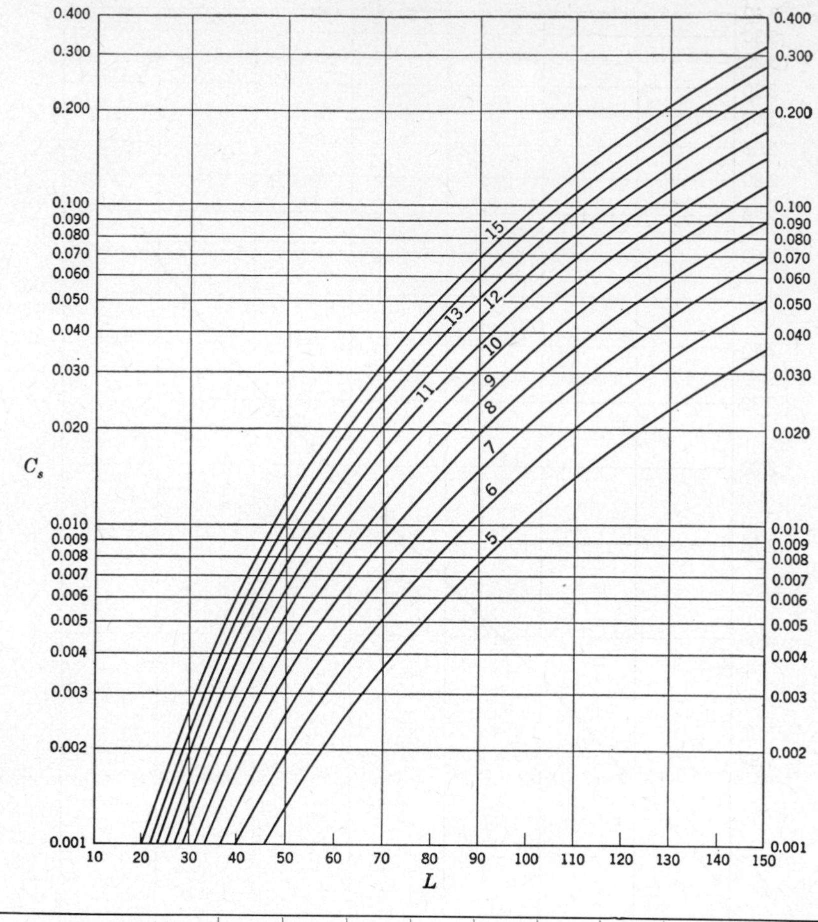

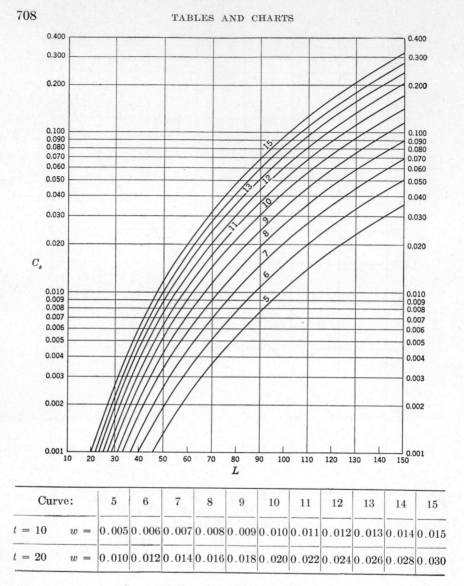

Curve:		5	6	7	8	9	10	11	12	13	14	15
t = 10	w =	0.005	0.006	0.007	0.008	0.009	0.010	0.011	0.012	0.013	0.014	0.015
t = 20	w =	0.010	0.012	0.014	0.016	0.018	0.020	0.022	0.024	0.026	0.028	0.030

CHART II. Sag correction to be subtracted.

C_s = correction for sag, ft
L = length of span, ft
t = tension, lb
w = weight, lb per linear ft

INDEX